QC955 .T67 1997
The Tornado

Alpine Campus
Library
Bristol Hall
P.O. Box 774688
Steamboat Springs
CO 80477

Geophysical Monograph Series

Including

IUGG Volumes

Maurice Ewing Volumes

Mineral Physics Volumes

GEOPHYSICAL MONOGRAPH SERIES

Geophysical Monograph Volumes

1. **Antarctica in the International Geophysical Year** *A. P. Crary, L. M. Gould, E. O. Hulburt, Hugh Odishaw, and Waldo E. Smith (Eds.)*
2. **Geophysics and the IGY** *Hugh Odishaw and Stanley Ruttenberg (Eds.)*
3. **Atmospheric Chemistry of Chlorine and Sulfur Compounds** *James P. Lodge, Jr. (Ed.)*
4. **Contemporary Geodesy** *Charles A. Whitten and Kenneth H. Drummond (Eds.)*
5. **Physics of Precipitation** *Helmut Weickmann (Ed.)*
6. **The Crust of the Pacific Basin** *Gordon A. Macdonald and Hisashi Kuno (Eds.)*
7. **Antarctica Research: The Matthew Fontaine Maury Memorial Symposium** *H. Wexler, M. J. Rubin, and J. E. Caskey, Jr. (Eds.)*
8. **Terrestrial Heat Flow** *William H. K. Lee (Ed.)*
9. **Gravity Anomalies: Unsurveyed Areas** *Hyman Orlin (Ed.)*
10. **The Earth Beneath the Continents: A Volume of Geophysical Studies in Honor of Merle A. Tuve** *John S. Steinhart and T. Jefferson Smith (Eds.)*
11. **Isotope Techniques in the Hydrologic Cycle** *Glenn E. Stout (Ed.)*
12. **The Crust and Upper Mantle of the Pacific Area** *Leon Knopoff, Charles L. Drake, and Pembroke J. Hart (Eds.)*
13. **The Earth's Crust and Upper Mantle** *Pembroke J. Hart (Ed.)*
14. **The Structure and Physical Properties of the Earth's Crust** *John G. Heacock (Ed.)*
15. **The Use of Artificial Satellites for Geodesy** *Soren W. Henricksen, Armando Mancini, and Bernard H. Chovitz (Eds.)*
16. **Flow and Fracture of Rocks** *H. C. Heard, I. Y. Borg, N. L. Carter, and C. B. Raleigh (Eds.)*
17. **Man-Made Lakes: Their Problems and Environmental Effects** *William C. Ackermann, Gilbert F. White, and E. B. Worthington (Eds.)*
18. **The Upper Atmosphere in Motion: A Selection of Papers With Annotation** *C. O. Hines and Colleagues*
19. **The Geophysics of the Pacific Ocean Basin and Its Margin: A Volume in Honor of George P. Woollard** *George H. Sutton, Murli H. Manghnani, and Ralph Moberly (Eds.)*
20. **The Earth's Crust: Its Nature and Physical Properties** *John C. Heacock (Ed.)*
21. **Quantitative Modeling of Magnetospheric Processes** *W. P. Olson (Ed.)*
22. **Derivation, Meaning, and Use of Geomagnetic Indices** *P. N. Mayaud*
23. **The Tectonic and Geologic Evolution of Southeast Asian Seas and Islands** *Dennis E. Hayes (Ed.)*
24. **Mechanical Behavior of Crustal Rocks: The Handin Volume** *N. L. Carter, M. Friedman, J. M. Logan, and D. W. Stearns (Eds.)*
25. **Physics of Auroral Arc Formation** *S.-I. Akasofu and J. R. Kan (Eds.)*
26. **Heterogeneous Atmospheric Chemistry** *David R. Schryer (Ed.)*
27. **The Tectonic and Geologic Evolution of Southeast Asian Seas and Islands: Part 2** *Dennis E. Hayes (Ed.)*
28. **Magnetospheric Currents** *Thomas A. Potemra (Ed.)*
29. **Climate Processes and Climate Sensitivity (Maurice Ewing Volume 5)** *James E. Hansen and Taro Takahashi (Eds.)*
30. **Magnetic Reconnection in Space and Laboratory Plasmas** *Edward W. Hones, Jr. (Ed.)*
31. **Point Defects in Minerals (Mineral Physics Volume 1)** *Robert N. Schock (Ed.)*
32. **The Carbon Cycle and Atmospheric CO_2: Natural Variations Archean to Present** *E. T. Sundquist and W. S. Broecker (Eds.)*
33. **Greenland Ice Core: Geophysics, Geochemistry, and the Environment** *C. C. Langway, Jr., H. Oeschger, and W. Dansgaard (Eds.)*
34. **Collisionless Shocks in the Heliosphere: A Tutorial Review** *Robert G. Stone and Bruce T. Tsurutani (Eds.)*
35. **Collisionless Shocks in the Heliosphere: Reviews of Current Research** *Bruce T. Tsurutani and Robert G. Stone (Eds.)*
36. **Mineral and Rock Deformation: Laboratory Studies—The Paterson Volume** *B. E. Hobbs and H. C. Heard (Eds.)*
37. **Earthquake Source Mechanics (Maurice Ewing Volume 6)** *Shamita Das, John Boatwright, and Christopher H. Scholz (Eds.)*
38. **Ion Acceleration in the Magnetosphere and Ionosphere** *Tom Chang (Ed.)*
39. **High Pressure Research in Mineral Physics (Mineral Physics Volume 2)** *Murli H. Manghnani and Yasuhiko Syono (Eds.)*
40. **Gondwana Six: Structure Tectonics, and Geophysics** *Gary D. McKenzie (Ed.)*

41 **Gondwana Six: Stratigraphy, Sedimentology, and Paleontology** *Garry D. McKenzie (Ed.)*

42 **Flow and Transport Through Unsaturated Fractured Rock** *Daniel D. Evans and Thomas J. Nicholson (Eds.)*

43 **Seamounts, Islands, and Atolls** *Barbara H. Keating, Patricia Fryer, Rodey Batiza, and George W. Boehlert (Eds.)*

44 **Modeling Magnetospheric Plasma** *T. E. Moore and J. H. Waite, Jr. (Eds.)*

45 **Perovskite: A Structure of Great Interest to Geophysics and Materials Science** *Alexandra Navrotsky and Donald J. Weidner (Eds.)*

46 **Structure and Dynamics of Earth's Deep Interior (IUGG Volume 1)** *D. E. Smylie and Raymond Hide (Eds.)*

47 **Hydrological Regimes and Their Subsurface Thermal Effects (IUGG Volume 2)** *Alan E. Beck, Grant Garven, and Lajos Stegena (Eds.)*

48 **Origin and Evolution of Sedimentary Basins and Their Energy and Mineral Resources (IUGG Volume 3)** *Raymond A. Price (Ed.)*

49 **Slow Deformation and Transmission of Stress in the Earth (IUGG Volume 4)** *Steven C. Cohen and Petr Vaníček (Eds.)*

50 **Deep Structure and Past Kinematics of Accreted Terranes (IUGG Volume 5)** *John W. Hillhouse (Ed.)*

51 **Properties and Processes of Earth's Lower Crust (IUGG Volume 6)** *Robert F. Mereu, Stephan Mueller, and David M. Fountain (Eds.)*

52 **Understanding Climate Change (IUGG Volume 7)** *Andre L. Berger, Robert E. Dickinson, and J. Kidson (Eds.)*

53 **Plasma Waves and Istabilities at Comets and in Magnetospheres** *Bruce T. Tsurutani and Hiroshi Oya (Eds.)*

54 **Solar System Plasma Physics** *J. H. Waite, Jr., J. L. Burch, and R. L. Moore (Eds.)*

55 **Aspects of Climate Variability in the Pacific and Western Americas** *David H. Peterson (Ed.)*

56 **The Brittle-Ductile Transition in Rocks** *A. G. Duba, W. B. Durham, J. W. Handin, and H. F. Wang (Eds.)*

57 **Evolution of Mid Ocean Ridges (IUGG Volume 8)** *John M. Sinton (Ed.)*

58 **Physics of Magnetic Flux Ropes** *C. T. Russell, E. R. Priest, and L. C. Lee (Eds.)*

59 **Variations in Earth Rotation (IUGG Volume 9)** *Dennis D. McCarthy and Williams E. Carter (Eds.)*

60 **Quo Vadimus *Geophysics for the Next Generation* (IUGG Volume 10)** *George D. Garland and John R. Apel (Eds.)*

61 **Cometary Plasma Processes** *Alan D. Johnstone (Ed.)*

62 **Modeling Magnetospheric Plasma Processes** *Gordon K. Wilson (Ed.)*

63 **Marine Particles Analysis and Characterization** *David C. Hurd and Derek W. Spencer (Eds.)*

64 **Magnetospheric Substorms** *Joseph R. Kan, Thomas A. Potemra, Susumu Kokubun, and Takesi Iijima (Eds.)*

65 **Explosion Source Phenomenology** *Steven R. Taylor, Howard J. Patton, and Paul G. Richards (Eds.)*

66 **Venus and Mars: Atmospheres, Ionospheres, and Solar Wind Interactions** *Janet G. Luhmann, Mariella Tatrallyay, and Robert O. Pepin (Eds.)*

67 **High-Pressure Research: Application to Earth and Planetary Sciences (Mineral Physics Volume 3)** *Yasuhiko Syono and Murli H. Manghnani (Eds.)*

68 **Microwave Remote Sensing of Sea Ice** *Frank Carsey, Roger Barry, Josefino Comiso, D. Andrew Rothrock, Robert Shuchman, W. Terry Tucker, Wilford Weeks, and Dale Winebrenner*

69 **Sea Level Changes: Determination and Effects (IUGG Volume 11)** *P. L. Woodworth, D. T. Pugh, J. G. DeRonde, R. G. Warrick, and J. Hannah*

70 **Synthesis of Results from Scientific Drilling in the Indian Ocean** *Robert A. Duncan, David K. Rea, Robert B. Kidd, Ulrich von Rad, and Jeffrey K. Weissel (Eds.)*

71 **Mantle Flow and Melt Generation at Mid-Ocean Ridges** *Jason Phipps Morgan, Donna K. Blackman, and John M. Sinton (Eds.)*

72 **Dynamics of Earth's Deep Interior and Earth Rotation (IUGG Volume 12)** *Jean-Louis Le Mouël, D.E. Smylie, and Thomas Herring (Eds.)*

73 **Environmental Effects on Spacecraft Positioning and Trajectories (IUGG Volume 13)** *A. Vallance Jones (Ed.)*

74 **Evolution of the Earth and Planets (IUGG Volume 14)** *E. Takahashi, Raymond Jeanloz, and David Rubie (Eds.)*

75 **Interactions Between Global Climate Subsystems: The Legacy of Hann (IUGG Volume 15)** *G. A. McBean and M. Hantel (Eds.)*

76 **Relating Geophysical Structures and Processes: The Jeffreys Volume (IUGG Volume 16)** *K. Aki and R. Dmowska (Eds.)*

77 **The Mesozoic Pacific: Geology, Tectonics, and Volcanism** *M. S. Pringle, W. W. Sager, W. V. Sliter, and S. Stein (Eds.)*

78 **Climate Change in Continental Isotopic Records** *P. K. Swart, K. C. Lohmann, J. McKenzie, and S. Savin (Eds.)*

Maurice Ewing Volumes

1 **Island Arcs, Deep Sea Trenches, and Back-Arc Basins** *Manik Talwani and Walter C. Pitman III (Eds.)*

2 **Deep Drilling Results in the Atlantic Ocean: Ocean Crust** *Manik Talwani, Christopher G. Harrison, and Dennis E. Hayes (Eds.)*

3 **Deep Drilling Results in the Atlantic Ocean: Continental Margins and Paleoenvironment** *Manik Talwani, William Hay, and William B. F. Ryan (Eds.)*

4 **Earthquake Prediction—An International Review** *David W. Simpson and Paul G. Richards (Eds.)*

5 **Climate Processes and Climate Sensitivity** *James E. Hansen and Taro Takahashi (Eds.)*

6 **Earthquake Source Mechanics** *Shamita Das, John Boatwright, and Christopher H. Scholz (Eds.)*

IUGG Volumes

1 **Structure and Dynamics of Earth's Deep Interior** *D. E. Smylie and Raymond Hide (Eds.)*

2 **Hydrological Regimes and Their Subsurface Thermal Effects** *Alan E. Beck, Grant Garven, and Lajos Stegena (Eds.)*

3 **Origin and Evolution of Sedimentary Basins and Their Energy and Mineral Resources** *Raymond A. Price (Ed.)*

4 **Slow Deformation and Transmission of Stress in the Earth** *Steven C. Cohen and Petr Vaníček (Eds.)*

5 **Deep Structure and Past Kinematics of Accreted Terrances** *John W. Hillhouse (Ed.)*

6 **Properties and Processes of Earth's Lower Crust** *Robert F. Mereu, Stephan Mueller, and David M. Fountain (Eds.)*

7 **Understanding Climate Change** *Andre L. Berger, Robert E. Dickinson, and J. Kidson (Eds.)*

8 **Evolution of Mid Ocean Ridges** *John M. Sinton (Ed.)*

9 **Variations in Earth Rotation** *Dennis D. McCarthy and William E. Carter (Eds.)*

10 **Quo Vadimus Geophysics for the Next Generation** *George D. Garland and John R. Apel (Eds.)*

11 **Sea Level Changes: Determinations and Effects** *Philip L. Woodworth, David T. Pugh, John G. DeRonde, Richard G. Warrick, and John Hannah (Eds.)*

12 **Dynamics of Earth's Deep Interior and Earth Rotation** *Jean-Louis Le Mouël, D.E. Smylie, and Thomas Herring (Eds.)*

13 **Environmental Effects on Spacecraft Positioning and Trajectories** *A. Vallance Jones (Ed.)*

14 **Evolution of the Earth and Planets** *E. Takahashi, Raymond Jeanloz, and David Rubie (Eds.)*

15 **Interactions Between Global Climate Subsystems: The Legacy of Hann** *G. A. McBean and M. Hantel (Eds.)*

16 **Relating Geophysical Structures and Processes: The Jeffreys Volume** *K. Aki and R. Dmowska (Eds.)*

Mineral Physics Volumes

1 **Point Defects in Minerals** *Robert N. Schock (Ed.)*

2 **High Pressure Research in Mineral Physics** *Murli H. Manghnani and Yasuhiko Syona (Eds.)*

3 **High Pressure Research: Application to Earth and Planetary Sciences** *Yasuhiko Syono and Murli H. Manghnani (Eds.)*

Geophysical Monograph 79

The Tornado:
Its Structure, Dynamics, Prediction, and Hazards

C. Church
D. Burgess
C. Doswell
R. Davies-Jones

Editors

American Geophysical Union

Published under the aegis of the AGU Books Board.

Cover: Spearman Texas tornado, May 31st, 1990.
Copyright 1990 Howard B. Bluestein, used by permission.

Library of Congress Cataloging-in-Publication Data

The Tornado : its structure, dynamics, prediction, and hazards / C.
    Church . . . [et al.], editors.
        p.   cm. — (Geophysical monograph ; 79)
    Includes bibliographical references.
    ISBN 0-87590-038-0
    1. Tornadoes—Congresses.   I. Church, Christopher R.   II. Series.
QC955.T67   1993
551.55'3—dc20   93-42302
                                                                             CIP

ISSN: 0065-8448

ISBN 0-87590-038-0

This book is printed on acid-free paper. ∞

Copyright 1993 by the American Geophysical Union, 2000 Florida Avenue, NW, Washington, DC 20009, USA.

Figures, tables, and short excerpts may be reprinted in scientific books and journals if the source is properly cited.

  Authorization to photocopy items for internal or personal use, or the internal or personal use of specific clients, is granted by the American Geophysical Union for libraries and other users registered with the Copyright Clearance Center (CCC) Transactional Reporting Service, provided that the base fee of $1.00 per copy plus $0.10 per page is paid directly to CCC, 21 Congress Street, Salem, MA 10970. 0065-8448/93/$01. + .10.
  This consent does not extend to other kinds of copying, such as copying for creating new collective works or for resale. The reproduction of multiple copies and the use of full articles or the use of extracts, including figures and tables, for commercial purposes requires permission from AGU.

Printed in the United States of America.

CONTENTS

Preface
Christopher Church, Donald Burgess, Charles A. Doswell III, and Robert Davies-Jones xi

Plainfield Tornado of August 28, 1990
T. Theodore Fujita 1

Tornado Vortex Theory and Modeling

Tornado Vortex Theory
W. S. Lewellen 19

Numerical Simulation of Axisymmetric Tornadogenesis in Forced Convection
Brian H. Fiedler 41

Numerical Simulation of Tornadolike Vortices in Asymmetric Flow
R. Jeffrey Trapp and Brian H. Fiedler 49

Discussion 55

Modeling and Theory of Supercell Storms

Supercell Thunderstorm Modeling and Theory
Richard Rotunno 57

Numerical Simulation of Tornadogenesis Within a Supercell Thunderstorm
Louis J. Wicker and Robert B. Wilhelmson 75

Tornado Spin-Up Beneath a Convective Cell: Required Basic Structure of the Near-Field Boundary Layer Winds
Robert L. Walko 89

Environmental Helicity and the Maintenance and Evolution of Low-Level Mesocyclones
Harold E. Brooks, Charles A. Doswell III, and Robert Davies-Jones 97

Mesocyclogenesis From a Theoretical Perspective
Robert Davies-Jones and Harold Brooks 105

Discussion 115

Observations of Tornadic Thunderstorms

Observations and Simulations of Hurricane-Spawned Tornadic Storms
Eugene W. McCaul, Jr. 119

Tornadic Thunderstorm Characteristics Determined With Doppler Radar
Edward A. Brandes 143

Tornadoes and Tornadic Storms: A Review of Conceptual Models
Charles A. Doswell III and Donald W. Burgess 161

Lightning in Tornadic Storms: A Review
Donald R. MacGorman 173

Tornadogenesis via Squall Line and Supercell Interaction: The November 15, 1989, Huntsville, Alabama, Tornado
Steven J. Goodman and Kevin R. Knupp 183

Discussion 201

CONTENTS

Tornado Detection and Warning

Tornado Detection and Warning by Radar
Donald W. Burgess, Ralph J. Donaldson, Jr., and Paul R. Desrochers   203

Single-Doppler Radar Study of a Variety of Tornado Types
Steven V. Vasiloff   223

Radar Signatures and Severe Weather Forecasting
Paul Joe and Mike Leduc   233

The Use of Volumetric Radar Data to Identify Supercells: A Case Study of June 2, 1990
Ron W. Przybylinski, John T. Snow, Ernest M. Agee, and John T. Curran   241

Doppler Radar Identification of Nonsevere Thunderstorms That Have the Potential of Becoming Tornadic
Rodger A. Brown   251

An Examination of a Supercell in Mississippi Using a Tilt Sequence
David A. Imy and Kevin J. Pence   257

Satellite Observations of Tornadic Thunderstorms
James F. W. Purdom   265

Discussion   275

Physical Models and Analogs

Laboratory Models of Tornadoes
Christopher R. Church and John T. Snow   277

Laser Doppler Velocimeter Measurements in Tornadolike Vortices
Donald E. Lund and John T. Snow   297

Vortex Formation From a Helical Inflow Tornado Vortex Simulator
James G. LaDue   307

Discussion   317

Tornado Observations

A Review of Tornado Observations
Howard B. Bluestein and Joseph H. Golden   319

A Comparison of Surface Observations and Visual Tornado Characteristics for the June 15, 1988, Denver Tornado Outbreak
E. J. Szoke and R. Rotunno   353

On the Use of a Portable FM-CW Doppler Radar for Tornado Research
Howard B. Bluestein and Wesley P. Unruh   367

Discussion   377

CONTENTS

Protection of Important or Critical Facilities

Design for Containment of Hazardous Materials
Robert C. Murray and James R. McDonald   379

State-of-the-Art and Current Research Activities in Extreme Winds Relating to Design and Evaluation of Nuclear Power Plants
M. K. Ravindra   389

Wind/Tornado Design Criteria Development to Achieve Required Probabilistic Performance Goals
Dorothy S. Ng   399

Discussion   405

Climatology, Hazards, Risk Assessment

Advances in Tornado Climatology, Hazards, and Risk Assessment Since Tornado Symposium II
Thomas P. Grazulis, Joseph T. Schaefer, and Robert F. Abbey, Jr.   409

Comparative Description of Tornadoes in France and the United States
Jean Dessens and John T. Snow   427

Tornadoes of China
Xu Zixiu, Wang Pengyun, and Lin Xuefang   435

Seasonal Tornado Climatology for the Southeastern United States
Linda Pickett Garinger and Kevin R. Knupp   445

Oregon Tornadoes: More Fact Than Fiction
George R. Miller   453

The Stability of Climatological Tornado Data
Joseph T. Schaefer, Richard L. Livingston, Frederick P. Ostby, and Preston W. Leftwich   459

A 110-Year Perspective of Significant Tornadoes
Thomas P. Grazulis   467

Discussion   475

Damage Surveys

Aerial Survey and Photography of Tornado and Microburst Damage
T. T. Fujita and B. E. Smith   479

Lessons Learned From Analyzing Tornado Damage
Timothy P. Marshall   495

Survey of a Violent Tornado in Far Southwestern Texas: The Bakersfield Valley Storm of June 1, 1990
Gary R. Woodall and George N. Mathews   501

An Observational Study of the Mobara Tornado
H. Niino, O. Suzuki, T. Fujitani, H. Nirasawa, H. Ohno, I. Takayabu, N. Kinoshita, T. Murota, and N. Yamaguchi   511

Discussion   521

CONTENTS

Damage Mitigation and Occupant Safety

Damage Mitigation and Occupant Safety
James R. McDonald   523

Tornado Fatalities in Ohio, 1950–1989
Thomas W. Schmidlin   529

Calculation of Wind Speeds Required to Damage or Destroy Buildings
Henry Liu   535

Risk Factors for Death or Injury in Tornadoes: An Epidemiologic Approach
Sue Anne Brenner and Eric K. Noji   543

Design for Occupant Protection in Schools
Harold W. Harris, Kishor C. Mehta, and James R. McDonald   545

Discussion   555

Tornado Forecasting

Tornado Forecasting: A Review
Charles A. Doswell III, Steven J. Weiss, and Robert H. Johns   557

Some Wind and Instability Parameters Associated With Strong and Violent Tornadoes, 1, Wind Shear and Helicity
Jonathan M. Davies and Robert H. Johns   573

Some Wind and Instability Parameters Associated With Strong and Violent Tornadoes, 2, Variations in the Combinations of Wind and Instability Parameters
Robert H. Johns, Jonathan M. Davies, and Preston W. Leftwich   583

Diurnal Low-Level Wind Oscillation and Storm-Relative Helicity
Robert A. Maddox   591

Tornadoes: A Broadcaster's Perspective
Tom Konvicka   599

The "Short Fuse" Composite: An Operational Analysis Technique for Tornado Forecasting
Jim Johnson   605

The Plainfield, Illinois, Tornado of August 28, 1990: The Evolution of Synoptic and Mesoscale Environments
William Korotky, Ron W. Przybylinski, and John A. Hart   611

Characteristics of East Central Florida Tornado Environments
Bartlett C. Hagemeyer and Gary K. Schmocker   625

Discussion   633

Open Discussion   635

Preface

During the past two decades, remarkable advances have been made in the understanding of the structure and dynamics of tornadoes and tornadic storms. This knowledge has led to improvements in prediction capability, procedures for issue and dissemination of warnings, and the practice of hazard mitigation. This progress can be attributed to the development of Doppler radars, wind profilers, lightning ground-strike location detectors, and automated surface observing systems; to the application of multispectral satellite data; to improvements in numerical simulation of clouds and storms; to the deployment of mobile storm-intercept teams with means to make quantitative observations; and to improved understanding of how structures fail when subjected to tornadoes.

This volume provides a comprehensive account of recent tornado research, documenting the advances made since the symposium on tornadoes held at Texas Technological University in Lubbock, Texas, in 1976 and is based on work presented at the Third Tornado Symposium, held in Norman, Oklahoma, April 2–5, 1991. The 53 papers are organized into 11 topical sections, beginning with the theory and modeling of tornado vortices and tornadic storms. The next four sections cover primarily observational studies and analysis of natural phenomena, and are followed by a section on results of laboratory experiments. The practical aspects of tornadoes as natural hazards, including issues in building design, risk assessment, damage survey and mitigation are addressed in four sections, and the final group of papers is devoted to the techniques of tornado forecasting. Each section is accompanied by an edited transcript of the discussion period that followed each session of the symposium; we hope that the editing has not diminished the spirit of the lively discussions.

The Third Tornado Symposium provided an opportunity to celebrate the appointment of T. Theodore Fujita, a pioneer in tornado research, as Merriam Distinguished Professor at the University of Chicago. As a measure of our esteem for Ted Fujita, we have chosen his presentation on the Plainfield tornado to introduce the volume. Ronald C. Taylor introduced the symposium's honored speaker in the following words:

> Ted's papers, of course, comprise the usual combination of photographs and figures depicting his meteorological analyses. These analyses sustain my interest and inhabit my memory, for they rarely fail to convey, in the words of Napier Shaw, a sense of "the go of things." They could easily be mistaken for a mere summary of empirical experience but to the more attentive eye they offer a conceptual setting, a working hypothesis, or simply invite the reader to consider a plausible speculation. In short, Ted's analyses comprise that elusive combination of critical intellect and active mind, which distinguishes originality from just ready accomplishment.

We are deeply indebted to the many anonymous reviewers for their careful evaluation of the manuscripts and their helpful comments. We also thank Kelly Lynn of the National Severe Storms Laboratory and Rachel Hill of Miami University, Ohio, for their administrative assistance.

We also wish to thank the other members of the program committee for their participation in the organization of the symposium: Howard Bluestein, Joseph Golden, Robert Johns, James MacDonald, Richard Rotunno, John Snow, and Roger Wakimoto. Many thanks to William Beasley, chair of the local organizing committee, for helping the program committee conduct a successful symposium and for facilitating publication of this book. Many organizations and institutions contributed funds and human resources that made the symposium possible and successful. These include the Center for Analysis and Prediction of Storms, the Cooperative Institute for Mesoscale Meteorological Studies, and the School of Meteorology, all at the University of Oklahoma; the NOAA STORM Program Office, the NOAA Office of Chief Scientist, and the NWS/UCAR Cooperative Program for Operational Meteorological Education and Training, the Wind Engineering Council, the National Weather Association, the American Meteorological Society, and the American Geophysical Union. Finally, we acknowledge, with great appreciation, support from the National Science Foundation, through grant ATM-9108873, provided jointly by the Physical Meteorology, Mesoscale Dynamic Meteorology, and Natural and Man-Made Hazard Mitigation programs. This support made possible the participation of many students in the symposium and provided partial support for the publication of this book.

Christopher Church, Donald Burgess
Charles A. Doswell III, Robert Davies-Jones
Editors

Plainfield Tornado of August 28, 1990

T. THEODORE FUJITA

The University of Chicago, Chicago, Illinois 60637

Shortly before 5 p.m. CDT on Tuesday, August 28, 1990, Chicago radio stations began announcing the occurrence of tornado deaths in the Plainfield area, some 60 km (40 miles) southwest of the University of Chicago. By 5:30 p.m. the casualty figures were upgraded to 20 deaths and 200 injuries, giving an impression that it was a major tornado event. The path length of the tornado, sketchy in nature at that time, was given as 13 km (8 miles). I called for a project meeting before ending the workday, reaching the conclusion that Duane Stiegler of my staff (15 years of survey experience) should go directly to Chicago's Midway Airport next morning to fly over the entire path. We assumed that the survey would take 2 hours.

After completing his initial flight, Duane informed me during refueling that the path of the main tornado extended 27 km (17 miles) from Oswego to Joliet. However, an extensive area of storm damage, possibly by a series of downbursts, extended from Oswego toward the northwest. He decided to fly again on the second day to complete the damage survey and photography of both downburst and tornado areas. I am presenting in this talk the results of my storm investigation, making use of Duane's survey data, along with satellite and radar photographs showing the nature of the violent (F5) tornado which left behind 29 deaths, 300 injuries, and $160 million damage in the southwest suburbs of Chicago.

PHOTOGRAPHIC EVIDENCE OF THE STORM

Aerial and ground photographs taken after the Plainfield tornado revealed that the wind effects of the storm system were very complicated. In studying the nature of the storm, over 600 color photographs covering the 600 km^2 areas of DeKalb, Kane, Kendall, and Will counties were examined in detail.

Found in the extensive cornfields in DeKalb and Kane counties are numerous streaks of high and low winds made visible by damaged and undamaged corn crops. The low-wind streaks, consisting of standing crops, are dark when viewed from the direction of shadows. On the other hand, high-wind streaks, consisting of damaged crops, are light colored because they scatter sunlight, especially when photographed from the favorable scattering angle. In general, low-wind streaks are seen in the downwind of isolated and clumps of trees and the high-wind streaks, extending downwind from open areas (Plate 1).

As has been well known, a slanted roof near the center of a microburst deflects the descending airflow, inducing a jet of high winds which extends downwind from the roof (Plate 2). Although I could not confirm the number of the microbursts, a large number of them were involved in producing the large areas of wind damage located to the northwest of Oswego where the major tornado touched down.

Found and photographed in the downburst areas are four vortex marks, 1–7 km long and 10–30 m (30–100 ft) wide. Because these vortex marks were located beneath the path of a well-defined wall cloud, they were rated as F1 and F2 tornadoes (Plate 3). The major tornado, after its touchdown on the Fox River west of Oswego, failed to produce continuous wind damage; instead, it produced a number of strange ground marks in the cornfields. These are identified as comma-shaped (Plate 4), swirl-shaped (Plate 5), and eye-shaped (Plate 6) marks.

Thereafter, a number of isolated suction vortices formed while the parent tornado intensified to F3 intensity before reaching Wheat Plains. As the vortex diameter increased, a series of suction vortices developed, producing numerous vortex marks (Plate 7) which are clearly visible from the air. Apparently, the tornado reached its peak intensity of F5 upon crossing U.S. Highway 30 (Plate 8) northwest of Plainfield. Because near-ground winds were so strong, bean (Plate 9) and wheat (Plate 10) crops were literally flattened, and some even were pulled out of the ground.

A 20-ton trailer (Plate 11) was blown off U.S. 30 and bounced five times before reaching the final position 350 m (1150 ft) from the highway. While traveling over the cornfield, the core of the tornado, evidenced by the debris deposition band (Plate 8) shrank to approximately 10 m (30 ft) in diameter. Found near the path of the tornado center

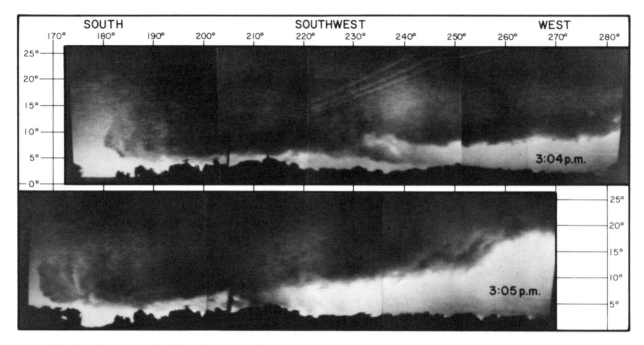

Fig. 1. Two composite video pictures showing the characteristic wall and tail clouds at 3:04 and 3:05 p.m. CDT. The cloud base was as low as 300 m (1000 ft) above ground level, making the visual identification of the large Plainfield tornado very difficult, because it will not be seen as a typical tornado funnel beneath a high cloud base.

was a 25 mm (1 inch) thick plywood board stuck vertically into the ground (Plate 12) where 2 m (6 ft) tall corn crops had existed before the tornado.

F5 Plainfield Tornado

I rated the Plainfield tornado as F5, based on the damage which was comparable to the worst I have ever seen. The damage in the cornfield southeast of U.S. 30 (Plate 8) was entirely different from the damage adjacent to structures affected by the F3 or F4 winds. Some corn crops were stripped of leaves and ears and pushed practically down to the ground. In the worst damage area, corn crops were blown away entirely, leaving behind the remnants of small roots connected to the underground root system.

People often use twisted trees as being the evidence of tornadic winds. What we find in the wake of tornadoes, however, are the results of windshift, rather than windshear. Depending upon the position of the damage site relative to the traveling tornado, a difluence pattern (Plate 13) and a confluence pattern (Plate 14) are commonly seen in the wake of tornadoes. Therefore the patterns shown in these pictures cannot be used as evidence of a tornadic airflow.

St. Mary Immaculate Church in Plainfield was a streamlined quonset structure (Plate 15) affected by F2 to F3 winds. It lost some stained glass without receiving structural damage visible from the air. Crystal Lawns, southeast of Interstate 55, was damaged by F3 to F4 winds. Some houses on Byrum Boulevard were blown off foundations (Plate 16).

However, most frame houses were sitting on top of poorly anchored foundations.

The most tragic event occurred at Crest Hill, where two long apartment buildings were damaged by the tornado (Plate 17). The top floors of the apartment complex were sheared off, leaving behind two wedge-shaped structural remnants, with their heights increasing to three stories at the farthest distance from the path of the tornado center. An-

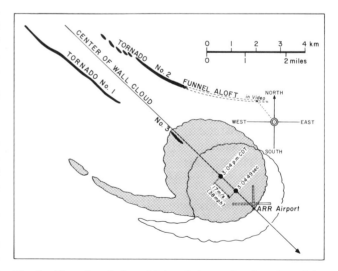

Fig. 2. Tornadoes 1, 2, and 3 in relation to the plan view of the wall cloud shown in Figure 1.

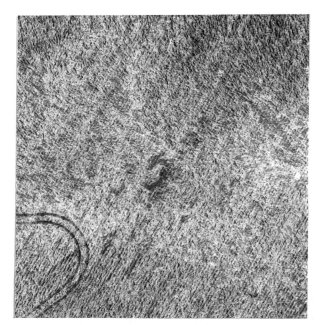

Fig. 3. A close-up aerial photograph of the eye-shaped bare ground inside the core of the Plainfield tornado. (For the location, see Plate 23).

other aerial photograph taken on the following day with identical perspective (Plate 18) shows the removal of corn by the rescue team in search of missing persons.

Mapping the Storm Damage

The area of downbursts on the upwind side of the Plainfield tornado turned out to be much larger than we had estimated. It extended 35 km (22 miles) toward DeKalb, with a maximum width of 19 km (12 miles). The onset time of the downburst winds was 1445 to 1450 CDT. At 1448 CDT, NOAA 11 infrared temperature shows the existence of two cloud tops or "Twin Peaks" over the onset area of the downburst winds (Plate 19).

At 1504 CDT, leaving the downburst storm, Paul Sirvatka, of the College of DuPage, took video photographs, panning his camera through the 160°–270°–350° azimuths. Composite pictures at 1504 and 1505 CDT, 49 s apart (according to the video frame counts), revealed the existence of a well-defined wall and tail clouds (Fig. 1). These clouds are characteristic features of the supercell thunderstorm or mesocyclone clouds. It is estimated that the wall cloud was traveling toward the southeast at 17 m s^{-1}, which is within the computation error of the 20 m s^{-1} movement of the hook echo determined by the Marseilles (MMO) radar imagery.

Aerial photographs revealed the existence of four vortex marks along the path of the wall cloud. In view of the common practice of calling a vortex on the ground beneath a wall cloud a tornado, these vortices were identified as tornados 1–4, all of which occurred inside the downburst area. In fact, a funnel cloud aloft is seen in the video at the extrapolated location of tornado 2 (Fig. 2).

Need for Elevated Scans and Doppler Radar

Marseilles (MMO) radar echoes were depicted by levels 1–6 reflectivity contours (Plate 20) at seven different times, corresponding to the rapid-scan times of the GOES satellite. The result revealed the existence of a hook echo which began forming at 1500 CDT, shortly before the video picture time. Thereafter, the appearance of the echo turned into that of a supercell which could spawn violent tornadoes. The characteristic supercell hook echo was most prominent at 3:11 p.m. CDT (2011 UT) when the weak and narrow tornado 4 was on the ground. When the Plainfield tornado at its F3–F5 intensity was on the ground from 3:25 to 3:40 p.m. CDT (2025–2040 UT), however, the identity of the hook echo was not as clear, in part because of ground clutter.

This evidence and judgment might have been more clear if higher-elevation scans and/or Doppler velocity fields were available for operational use. We expect that the national coverage of the NEXRAD radars and intensive training of storm forecasters will improve the capability for revealing superell-tornado relationships.

Do Tornadoes Grow Upward or Downward?

Like the funnel cloud of a waterspout, a tornado funnel descends from the cloud base, reaching ultimately to the ground. On the other hand, the airflow inside a tornado at the formative stage is predominantly upward. The formation of three isolated rotational winds, comma, swirl, and eye (Plate 21), from F2–F3 intensity presents the following questions: Did they descend from the rotationg winds aloft? Or, did they form on or near the ground first and stretch upward?

I begin to think that the eye (Plate 6) formed near the ground, because it has no sign of the translational characteristics (Plate 3) generated by a fast-moving vortex which descended to the ground. Are comma, swirl, and eye circulations suction vortices? My answer is "yes," because I am hesitant to call them three separate tornadoes. The orbiting suction vortices (Plate 7) probably formed on the ground and stretched upward into the parent tornado, because the aerial photographs suggests that they were slow moving at formation and gained orbital motion thereafter.

While weakening from F5 to F4, the tornado entered Plainfield, passing directly over Plainfield High School (letter S in Plate 22) and St. Mary Immaculate School with its church (letter C). After smashing the community of Lily Cache, the tornado weakened to F2. Then it crossed Interstate 55, injuring six persons in automobiles.

The tornado intensified again while passing through Crystal Lawns, where the Grand Prairie Elementary School (letter S in Plate 22) was damaged by F3 winds. Thereafter, the core diameter shrank while weakening to F2 intensity. In Crest Hill, however, the small-core tornado intensified into F3 while passing across the Cresthill Lake apartments (letter

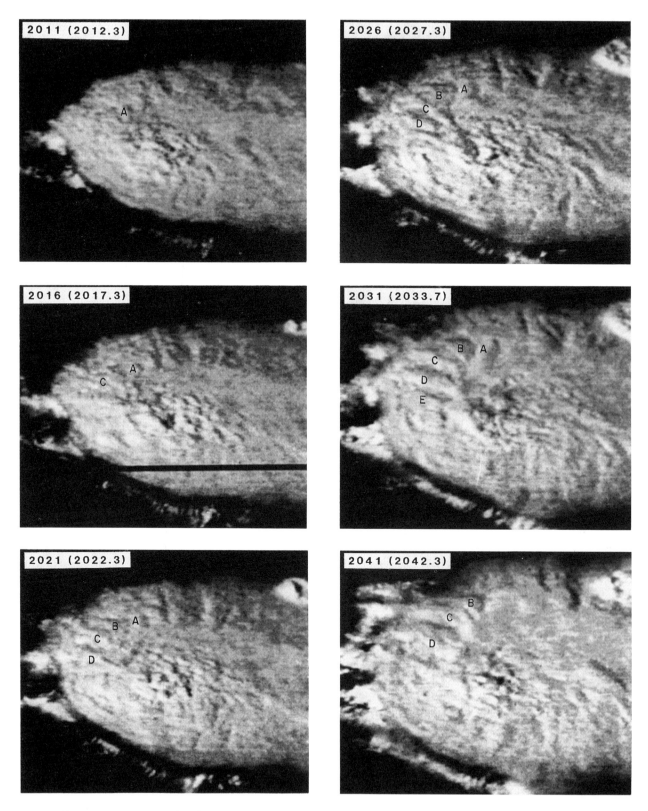

Fig. 4. A sequence of GOES rapid-scan pictures showing fingerlike patterns atop the anvil cloud of the Plainfield cloud. Radar echoes corresponding to these satellite photos are shown in Plate 20.

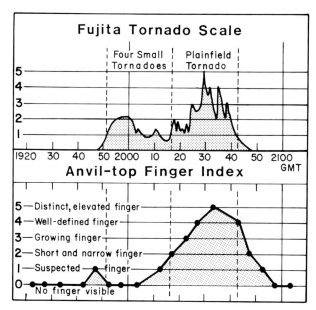

Fig. 5. Variation of anvil-top fingers rated by index 0–5. These fingers are closely related to the F scale variation of the Plainfield storms consisting of tornadoes and downbursts.

A in Plate 22), consisting of two long buildings separated by a 10-m wide pathway. When the tornado crossed the 250-unit apartment complex, several persons were thrown into the cornfield to the southeast (Plate 17). By the morning of August 30 the cornfield was cleared (Plate 18), recovering seven bodies, including a 5-week-old infant. Two persons perished in cars parked by the apartment. Thereafter, the tornado weakened rapidly and disappeared after crossing Larkin Road in Joliet.

What Happened in F5 Winds?

The only location where I estimated F5 winds, based on the corn damage, was to the southeast of U.S. 30 (Plate 23). Prior to the highway crossing of the tornado, its core diameter, evidenced by the width of the debris deposit (Plate 8), shrank from 70 m to 10 m. Meanwhile, the storm intensified up to F5 intensity. Six automobiles were blown into the cornfield. After traveling over the field for about 400 m, the core diameter began increasing again, reaching 100 m.

A series of exposed ground, free from corn crops, was left in the field during the increasing stage of the core diameter (Plate 24). Meanwhile, the corn damage to the southwest of the exposed ground was rated as less than F2, suggesting the existence of an extremely large wind shear, characterized by strong anticyclonic vorticity. Did it induce a series of anticyclonic suction vortices on the right-hand edge of the tornado? Another interesting piece of evidence is that the large-core tornado induced a number of orbiting cyclonic suction vortices shortly thereafter. Aerial photographs also showed an eye-shaped vortex mark consisting of exposed ground (Figure 3) at the location of the eye in Plate 23.

Four persons were killed when automobiles were blown off U.S. 30. A northbound tractor-trailer left two separate impact marks (Plate 25) on the shoulder, indicating that the tractor and trailer had been separated before they were blown off the highway. The tractor flew 100 m before plowing into the debris deposit. Apparently, the tractor left the highway moments after the tornado core crossed the highway. The trailer, after losing most of its scrap metal cargo, bounced in the cornfield five times before resting in the field at position E (Plate 23).

A passenger car, after being blown off the highway, traveled at very low altitude, as seen by the shearing and clipping off of corn crops along its 720-m (2350-ft) flight path. Apparently, the car looped around the core of the tornado before coming to rest at position C (Plate 26) in the right-side-up position with four passenger seats remaining intact. One passenger was reported killed after being thrown out of the car. It is remarkable that a car could fly such a long distance in a violent tornado without either tumbling or bouncing.

Cloud-Top Features by Satellites

Both polar orbiter (NOAA 11) and geostationary GOES satellites were in operation on August 28. The former, with 99° orbital inclination, scanned the storm cloud only once at 2:48.8 p.m. CDT (1948.8 UT) during the northbound orbit from 851 km altitude over southeast Iowa. The latter obtained 5-min rapid-scan data from 1:00 p.m. to 4 p.m. CDT (1300–1600 UT) from above the equator at 97.7°W longitude.

GOES infrared (IR) temperatures at 30-min intervals were contoured and superimposed upon MMO radar echoes and the location of downbursts and tornadoes (Plate 27). Meanwhile, NOAA 11 IR data with 1-km resolution were analyzed with 1°C isotherms below the −50°C cloud-top temperature (Plate 28). Because NOAA 11 isotherms are capable of depicting individual cold tops, they were superimposed upon the isoecho contours of the MMO echo at the same time. Two significant cold tops, identified herein as the "Twin Peaks" (Plate 29), with their coldest temperatures, −67.6°C of the West Peak and −61.5°C of the East Peak, coincided with the high-reflectivity cores of the MMO echo. These peaks were located above the onset area of the downburst winds just to the east of DeKalb, Illinois (Plate 19). The high-temperature ring encircling the Twin Peaks is an interesting feature, but its physical and dynamical meanings are not known at this time.

Fortunately, the 5-cm radar of United Airlines northwest of O'Hare Airport made RHI scans of the Twin Peaks area, obtaining the RHI cross section at the top of Plate 29. It is suspected that the West Peak was in the early sinking stage while the East Peak was in the downburst-inducing stage when its strong echo was already on the ground (Plate 19). It is premature to make a conclusive interpretation of such findings from this case study.

In this particular case, several distinct, elevated fingers extended from above the tornado toward the northwest. On

Plate 1. High- and low-wind streaks seen in the vast area of downbursts.

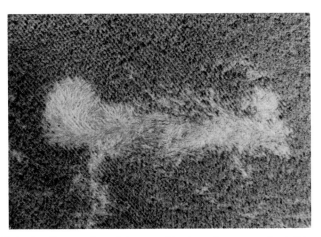

Plate 4. Comma-shaped damage. The tornado was moving left to right.

Plate 2. Corn crops damaged by a jet of high winds deflected by a slanted roof.

Plate 5. A swirling pattern of corn damage. The tornado was moving right to left.

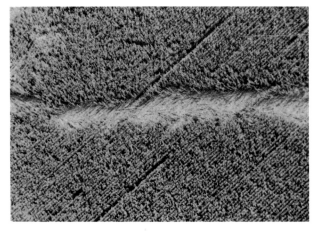

Plate 3. Path of an F2 tornado located inside the area of downbursts.

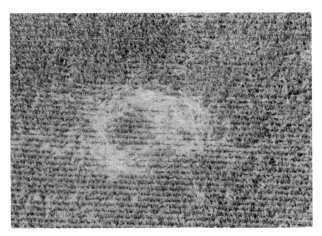

Plate 6. Eye-shaped corn damage. The tornado was moving from top to bottom.

Plate 7. Suction-vortex marks near Whisky Road, looking northeast.

Plate 10. Spring wheat flattened inside the F4–F5 area.

Plate 8. The F5 damage to the southeast of U.S. Highway 30.

Plate 11. A 20-ton trailer blown off U.S. 30. It bounced five times.

Plate 9. A close-up view of bean crops in the F4–F5 area.

Plate 12. A plywood board stuck vertically into the ground in the F4 area.

Plate 13. A difluence pattern produced by the wind shift in the tornado.

Plate 16. Frame houses with weak foundations blown off by F3–F4 winds.

Plate 14. A confluence pattern produced by the wind shift in the tornado.

Plate 17. The Cresthill Lake apartments on August 29 with damaged cornfield.

Plate 15. St. Mary Immaculate church, which lost the cross on the steeple.

Plate 18. The apartment complex on August 30 with cleaned-up cornfield.

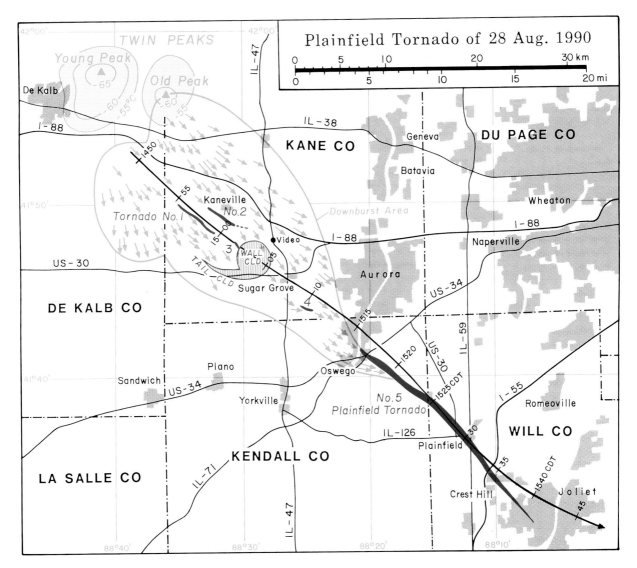

Plate 19. Damage maps of the Plainfield storm. Downburst winds are shown by blue arrows and tornadoes by red areas. The wall cloud depicted in video was located to the northwest of Sugar Grove. The path of the wall cloud center, with 5-min time markers, was estimated by the shape of the MMO radar echoes.

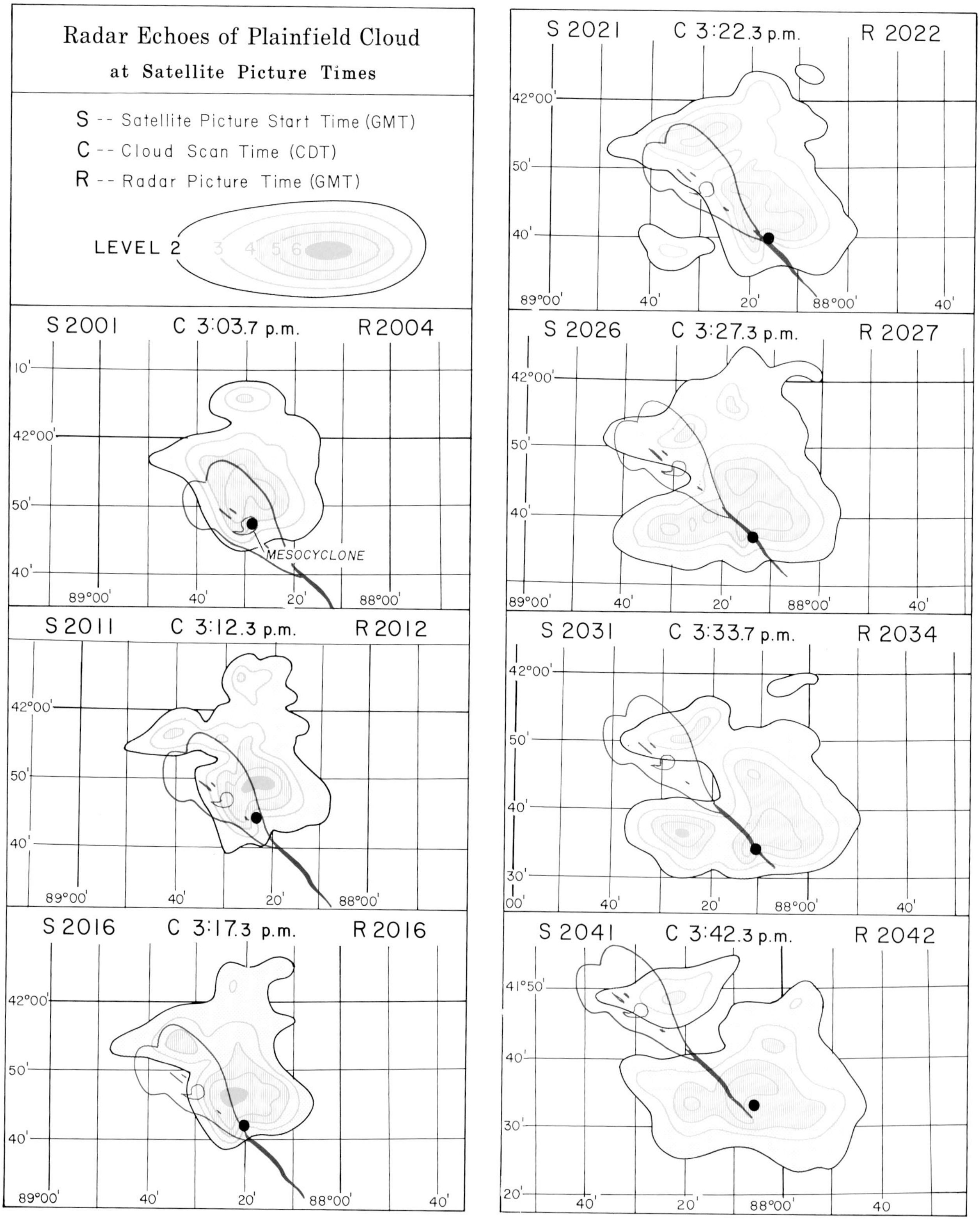

10   PLAINFIELD TORNADO

Plate 20. Reflectivity contours of the Plainfield storm with superimposed positions of the hook-echo centers and damage areas. The hook was most pronounced at 3:12 p.m., shortly after the video pictures were taken.

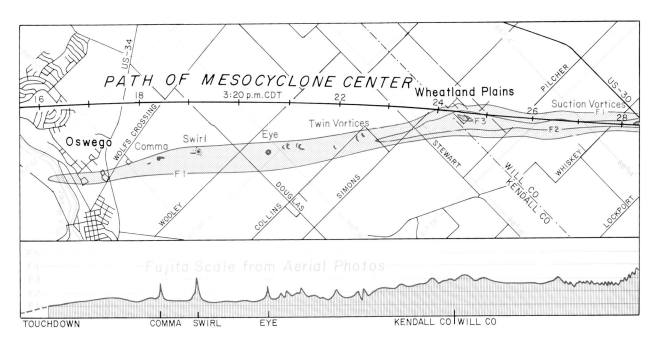

Plate 21. The first half of the damage area of the Plainfield tornado. Variation of the F scale was determined by aerial photographs of cornfields, trees, and structures.

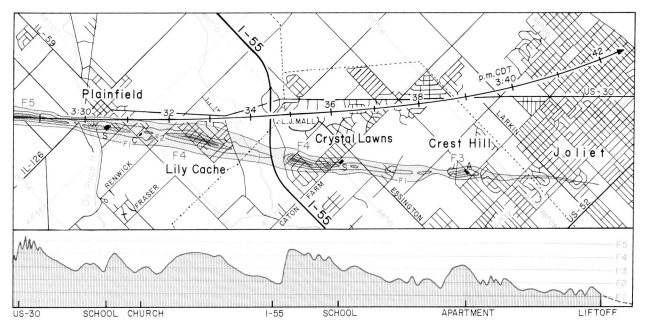

Plate 22. The second half of the damage area of the Plainfield tornado. The tornado at touchdown was 1.7 km to the right of the mesocyclone (hook) center. As the F scale increased to F5, both tornado and mesocyclone axes became close to each other.

12 PLAINFIELD TORNADO

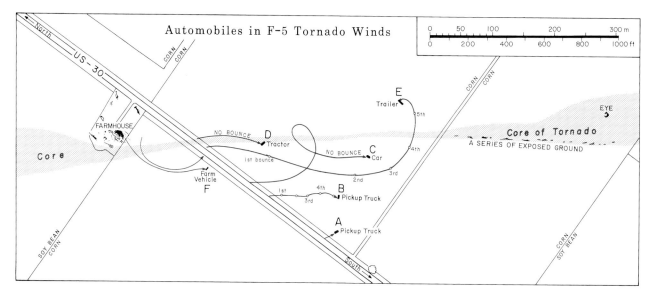

Plate 23. The F5 area of the Plainfield tornado where the core diameter shrank from 70 m to 10 m and increased again to 70 m. When the diameter was increasing, a series of suction vortices formed on the right-hand side of the core, pulling out corn crops and exposing the bare ground.

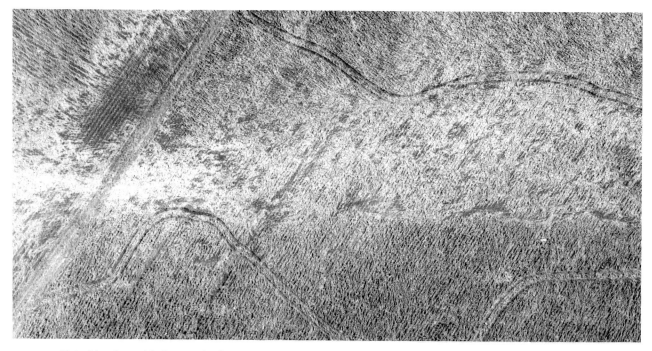

Plate 24. An aerial photograph of a series of exposed ground. Relatively light damage on the southwest side of the core suggests the existence of extremely large anticyclonic wind shear at the core boundary.

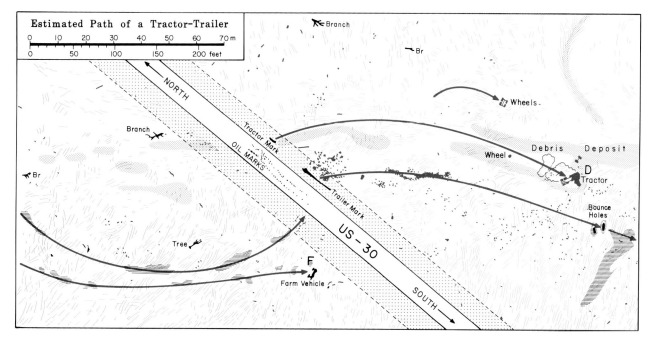

Plate 25. Various ground marks on U.S. 30 and adjacent fields left behind by the Plainfield tornado at its F5 intensity.

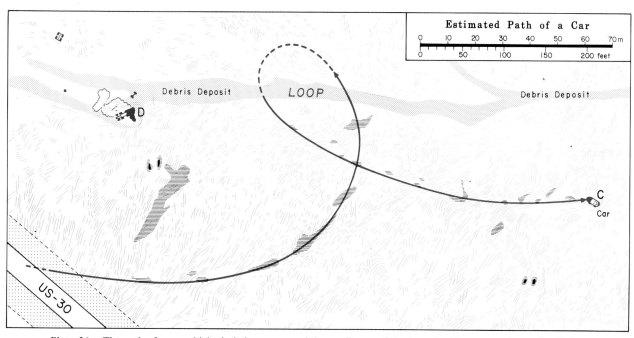

Plate 26. The path of a car which circled once around the small core of the tornado. The car, resting at C with four passenger seats inside, traveled less than 1 m above the ground, probably on the cushion of laminar rising air.

14 PLAINFIELD TORNADO

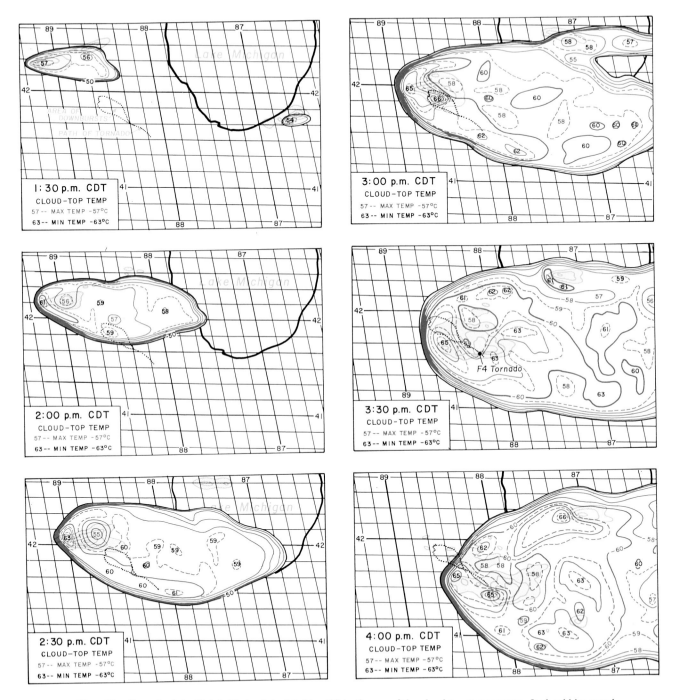

Plate 27. Growth of the Plainfield cloud depicted by 1°C isotherms of the cloud-top temperature. It should be noted that the horseshoe-shaped upwind edge was not characterized by radar echoes of comparable horizontal dimensions.

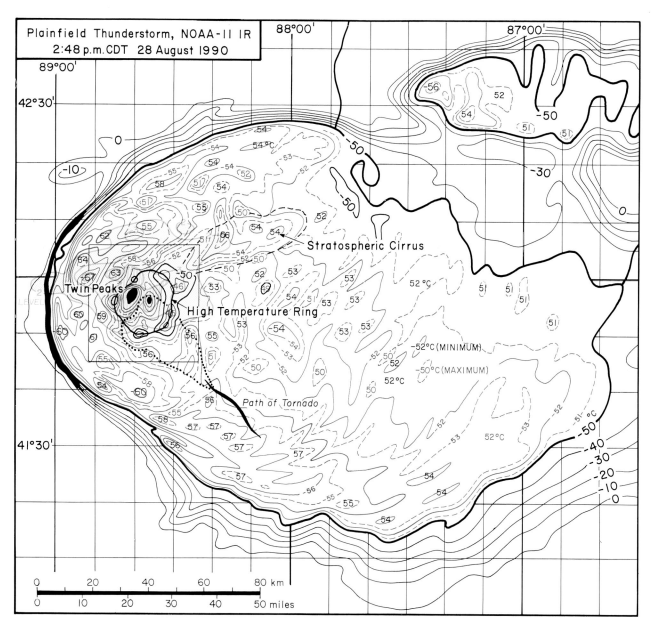

Plate 28. Infrared temperature of the Plainfield thunderstorm obtained by contouring the NOAA 11 data with 1°C resolution. The boxed area near the west edge of the cloud is enlarged in Plate 29.

16 PLAINFIELD TORNADO

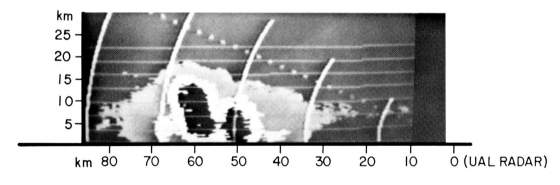

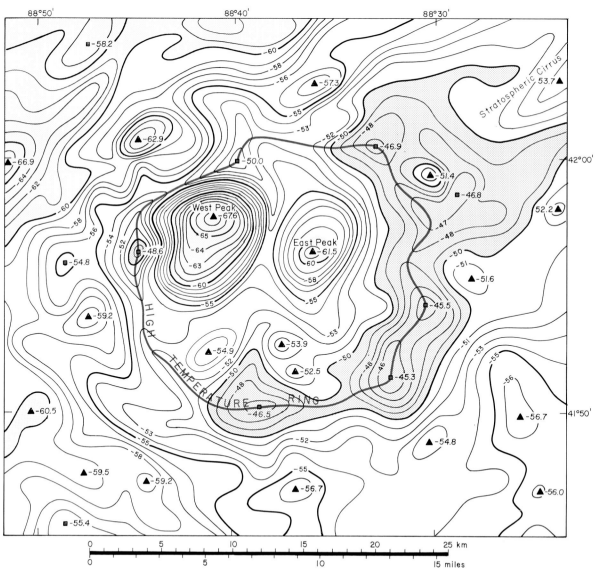

Plate 29. High-resolution isotherms showing two cold-temperature tops identified as the −67.6°C West Peak and the −61.5°C East Peak. An RHI picture of these two peaks by United Airlines radar is shown at the top.

the basis of my Lear Jet experiments in the 1970s (Fujita [1974a, b], to be published in color in 1992), I am assuming that these fingers consist of the anvil material pushed outward from the highly convective region of the cloud. The time-dependent analysis of the cloud top features (Fig. 4) in relation to the temporal F-scale variation resulted in a positive correlation between the F scale and a so-called cloud-top finger index (Fig. 5).

It was an honor to have had an opportunity to present my research on the Plainfield tornado at the banquet of Tornado Symposium III. As in this research case, an investigation of a complicated phenomenon often leads to additional studies, making use of both new data and advanced techniques. This particular tornado was, no doubt, a very complicated one. I hope that this presentation will stimulate a number of questions and studies for the next generation of tornado researchers.

Acknowledgments. The research leading to this presentation before the Tornado Symposium III Banquet has been sponsored since the 1960s by four U.S. government agencies: National Science Foundation, National Aeronautics and Space Administration, National Oceanic and Atmospheric Administration, and Office of Naval Research under their successive grants.

References

Fujita, T. T., Overshooting thunderheads observed from ATS and Lear Jet, *SMRP Res. Pap. 117A*, 29 pp., Univ. of Chicago, Chicago, Ill., 1974a.

Fujita, T. T., Overshooting tops of severe thunderstorms revealed by Lear Jet, satellite, and radar observations, *SMRP Res. Pap. 117B*, 48 pp., Univ. of Chicago, Chicago, Ill., 1974b.

Tornado Vortex Theory

W. S. LEWELLEN

Department of Physics and Atmospheric Science, Drexel University, Philadelphia, Pennsylvania 19104

1. INTRODUCTION

A number of reviews of the dynamics of the tornado vortex are available in the literature. The purpose of the present review is to update my earlier review [*Lewellen*, 1976]. That paper, which will be referred to as L76, attempted to provide a critical assessment of the existing theoretical models to see how well they described the wind and pressure field in a tornado, to see what understanding they provided as to the parameters which govern the flow, and to clarify some of the essential questions which remain to be answered by future research. Although it was possible to piece together a relatively consistent, qualitative model of the flow in different regions of the tornado in 1976, it was not possible to provide a definitive model. Here I will concentrate on the considerable progress which has been made in this direction in the research published in the last 15 years. Again, I will restrict attention to the immediate vicinity of the tornado, within a radius of approximately 1 km. A companion review on thunderstorm modeling by *Rotunno* [this volume] will update the review by *Lilly* [1976] on the larger-scale sources of vorticity and energy for the tornado. Since a number of relatively recent reviews have been published [*Bengtsson and Lighthill*, 1983; *Davies-Jones*, 1986; *Deissler*, 1977; *Lugt*, 1983; *Maxworthy*, 1986; *Rotunno*, 1986; *Smith and Leslie*, 1978; *Snow*, 1982, 1984, 1987], I will feel free to incorporate my own biases in this present review.

The most essential element of tornadolike, vortex flow is the convergence of ambient, axial vorticity. This is present in natural flows ranging from familiar drain vortices [*Lundgren*, 1985] to hurricanes [*Anthes*, 1982]. There are clear demarcations between the tornado and the much larger hurricane. The hurricane is a complete, large storm system, while the tornado is an appendage to a thunderstorm. However, the sparseness of clear distinguishing features between the core dynamics of weak tornadoes and other natural vortices makes it difficult to distinguish weak tornadoes. This common element allows the more benign dust devils [*Idso*, 1974] and waterspouts [*Golden*, 1971] to be investigated as surrogates for some elements of tornado flow [e.g., *Bilbro et al.*, 1977; *Schwiesow*, 1981; *Schwiesow et al.*, 1981]. It also makes it difficult to maintain tornado statistics [*Grazulis and Abbey*, 1983]. This review is primarily concerned with the well-developed, strong tornado.

The essential elements of the tornado vortex are illustrated in the schematic in Figure 1, taken from *Whipple* [1982]. The flow spirals radially inward into a core flow which is basically a swirling rising plume but may include a downward jet along the axis. The radial flow is greatly intensified in the surface layer. The whole flow is driven by the thunderstorm into which the vortex feeds through a rotating wall cloud. The tornado vortex allows a significant fraction of the potential energy of the parent storm to be converted into wind kinetic energy very close to the surface where it can cause great damage. Although there is some evidence that the tornado vortex may intensify the storm, the much more important effect of the tornado vortex is this lowering of maximum wind speeds from high in the storm almost to the surface. Local concentration of the vorticity can lead to wind speeds exceeding 100 m/s within a cylinder of approximately 100 m radius. As will be discussed in section 6, there is even the opportunity for the unique interaction between the vortex and the surface to permit maximum low-level wind speeds to exceed that achievable by a straight exchange of storm potential energy for kinetic energy.

The sharp gradients in particulate concentrations occurring in tornadoes make them very photographic. The schematic may be compared with the photograph, Plate 1, of a western Kansas tornado on August 28, 1979 [*Whipple*, 1982]. It shows a large cloud of dust and debris engulfing the bottom of the funnel. The penetration of the clear region to smaller radius at the surface provides evidence of the strong radial inflow in the surface layer. The top of the dense surface cloud is sharper than might be implied by the upward vertical velocities inferred for these regions from Figure 1. This is influenced both by particle size of the debris and by unsteady interception of appropriate debris along the torna-

The Tornado: Its Structure, Dynamics, Prediction, and Hazards.
Geophysical Monograph 79
Copyright 1993 by the American Geophysical Union.

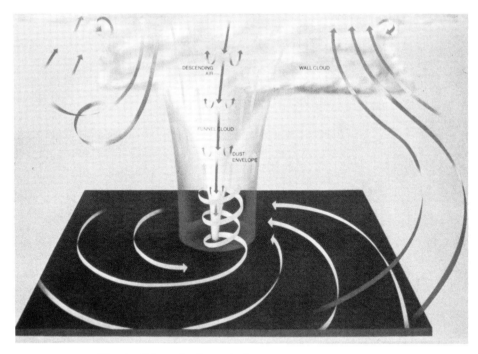

Fig. 1. Schematic of tornado flow taken from *Whipple* [1982].

do's path. Above this dense cloud it is possible to see a sharp water cloud funnel boundary surrounded by a faint dust sheath. The funnel cloud boundary does not denote a streamline, but more nearly corresponds to a constant pressure surface. It can occur in either the upflow or the downflow region of the schematic in Figure 1. The perception of a clear annulus between the cloud funnel and the dust sheath suggests that the funnel is occurring within the inner downflow at this particular time. The sharp boundaries indicate that the level of turbulence in this core flow is low. The bend in the funnel cloud at height is not unusual for tornadoes, particularly in the decaying phase. This is evidence that even when the thunderstorm flow pattern forces a nonvertical core, its interaction with the surface tends to force a perpendicular intersection.

Perhaps the greatest advances in the last 15 years have been in the area of numerical simulations, which were just beginning to make significant contributions in the mid 1970s. Thanks in great measure to the roughly 2 orders of magnitude increase in hardware computational capabilities during this time period, fully three-dimensional, unsteady simulations are now easier than steady state, axisymmetric calculations were at the beginning of the period. I will present some new "large-eddy simulation" results for turbulence in the tornado vortex made during the preparation of this review, which complement the axisymmetric, turbulent transport results of *Lewellen and Sheng* [1980]. I think the technology is now at hand, where it should be possible to take boundary conditions from a numerical storm simulation and obtain valid estimates of the probability of different levels of extreme winds occurring in different storms.

It is convenient to divide the flow into the same four regions as in L76: region I, the core flow; region II, the surface boundary layer flow; region III, the central corner flow; and region IV, the top layer. Three of these four regions are seen in Figure 1 and Plate 1. The top layer is out of the picture in both cases. I will concentrate on the central corner flow region, because this is the region where the maximum velocities occur and where the dynamical mechanisms are most distinctive.

2. Core Flow, Region I

In natural vortices the horizontal convergence tends to be induced by positively buoyant plumes or updrafts. Conservation of angular momentum with respect to the center of these updrafts then amplifies any ambient vertical vorticity. In terms of the dynamics of vorticity [e.g., *Morton*, 1969, 1984], the existing vortex tubes are stretched, so that the cross section becomes smaller and the vorticity amplified to keep the product of the vorticity and the cross-sectional area constant. At the barely discernable level this leads to the small whirls in the steam fog rising from a warm water pond on a cold morning. Under large air/water temperature differences, as may occur after a winter cold front passes over open water [e.g., *Lyons and Pease*, 1972], steam devils, which are quite similar to dust devils, can occur. *Hess et al.* [1988] have looked at the conditions necessary for such boundary layer vortices to occur. They report that the environment is favorable for their development when $-h/L > 100$, where h is the height of the mixed layer and L is the Monin-Obukhov length, consistent with the suggestions of

Deardorff [1972]. *Schmidt and Schumann* [1989] have used a large-eddy simulation of the convective boundary layer to look at these conditions where coherent plumes are initiated. Entrainment into such a developing plume provides the convergence which can amplify any ambient vertical vorticity. The tilting and stretching of horizontal vorticity can also be a major source of vertical vorticity for these boundary layer vortices [*Maxworthy*, 1973].

Once a buoyant plume vortex is formed, the strong stability of the flow allows the flow to persist and even amplify. This feature can allow the swirling plume to be quite distinctive. As a consequence of their high Reynolds number, most atmospheric flows are turbulent, with a cascading of energy from large scales to small scales. Although tornadolike vortices occur in a region of relatively strong, three-dimensional turbulence, the rotational forces occurring within the vortex core are able to impede the normal cascade of turbulent energy and allow the core vortex to dominate. When the vortex is driven by its core buoyancy, this damping of core turbulence also prevents the core buoyancy from diffusing as much as it otherwise would. This provides a natural feedback; the greater the height to which the core buoyancy persists, the stronger the radial pressure drop available to drive the vortex. It also helps explain why fires may be intensified by the introduction of swirl [*Church et al.*, 1980]. In the normal inertial range of three-dimensional cascading turbulence there is an increase of enstrophy (squared vorticity) and decrease in energy with decreasing scale. In a tornadolike vortex there appears to be sufficient stability to allow the normal cascade to terminate and the coherent vortex to intensify. I will return to this stability discussion in section 2.3.

A major problem in modeling the tornado is in determining how the thunderstorm generates the high levels of vertical vorticity necessary to supply the tornado vortex. However, this question is predominantly outside the domain of the current review, since it involves the storm dynamics on a scale larger than 1 km. In order to hint at how this generation and concentration may proceed, Figure 2 shows a conceptional schematic of possible flow in the vicinity of a waterspout from *Simpson et al.* [1986]. The convergence in the immediate vicinity of the waterspout is supplied by an active cumulus cloud. The problem (and probably the reason waterspouts do not occur more frequently) comes in supplying sufficient vertical vorticity, that the stretching associated with the convergence can increase it to order 1 s^{-1} within the 10 min the cumulus is most active. On the basis of GATE data and extensive three-dimensional numerical cumulus simulations (their grid resolution of 0.6 km is unable to resolve the waterspout, but they were looking to obtain sufficiently high vorticity levels on that grid scale), they conclude that conditions which bring together cool descending and warm ascending air side by side are more likely to increase both low-level convergence and low-level vertical vorticity sufficiently to achieve waterspout intensity in the lifetime of the cumulus updraft aloft. A quite similar type of conceptual model is given by *Wakimoto and Wilson*

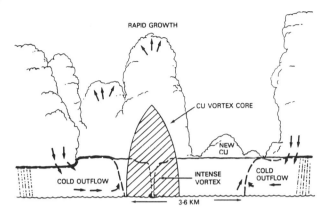

Fig. 2. Schematic illustration (not to scale) of cumulus-outflow interactions in relation to vorticity generation for a waterspout according to *Simpson et al.* [1986].

[1989] for nonsupercell tornadoes. The same type of close proximity between the cold downdraft and the warm updraft appears to be involved in a thunderstorm initiating the strong tornado [*Davies-Jones*, 1986], but I will leave that to *Rotunno* [this volume] to explain.

2.1. Dimensionless Parameters

Most researchers agree that the primary tornado vortex is essentially axisymmetric within a radius of 1 km. Although asymmetric effects are important, this axisymmetric nature of the primary vortex reduces the geometrical variation of the core flow to a single parameter, the geometrical ratio of the domain radius to the domain height, i.e., an aspect ratio, $AR = r_0/h$, with r_0 the radius of the domain and h the height. The aspect ratio has a strong influence on the interaction between the core flow and the top and bottom boundary conditions.

The core flow through a cylindrical domain with fixed aspect ratio is of course largely determined by the inflow/outflow boundary conditions. When this open tornado domain is considered to be drawn from a larger-scale, ambient flow, one of the most important parameters influencing the inflow is some measure of the ratio of the ambient vertical vorticity to the ambient horizontal convergence. Such a swirl parameter, S, is usually taken as

$$S = (\omega r_0)/(2a_h h) = v_0/w_0 \qquad (1)$$

with ω, the ambient vertical vorticity, and a_h, the ambient horizontal convergence. This may also be written as the ratio of the tangential velocity, v_0, of the inflow at r_0, divided by the average vertical velocity, w_0, at the top of the domain. The value of the swirl ratio has a strong effect on the interaction between the vertical and the horizontal components of vorticity. In the work of *Lewellen* [1962] I showed that when S is large, any axisymmetric core flow must have its axial component of vorticity essentially independent of the axial coordinate, z, over large stretches of z. Any substantial z variations in the axial component of vorticity must occur in narrow layers in the limit of large S.

The influence of physical scale size comes through some Reynolds number, such as

$$Re = v_0 r_0 / \nu \qquad (2)$$

Here ν is the kinematic viscosity of the fluid. This is certainly an important parameter, for laboratory and numerical simulations. However, only the fact that it is a large value may be important for actual tornadoes. With a few exceptions [e.g., *Chi and Costopolous*, 1975; *Lewellen and Sheng*, 1979, 1980; *Lewellen and Teske*, 1977; *Lewellen et al.*, 1979], numerical simulations are at relatively low Reynolds numbers since some type of constant eddy viscosity is used to adjust for atmospheric turbulence. I believe the dynamical variations in the effective eddy viscosity in different parts of the flow are important, but this is not universally accepted. More of the influence of Re will be considered when the corner flow is discussed.

A fourth parameter involves some measure of the relative importance of buoyancy forces within the flow. In laboratory flows this is usually given in terms of a Froude number:

$$Fr \equiv (\Delta p / 2g \Delta \rho h)^{1/2} \qquad (3)$$

where Δp is the pressure drop driving the flow, $\Delta \rho$ is the density change within the flow, and g is the gravitational constant. It is not clear how important this parameter is to the tornado vortex. Certainly, buoyancy due to latent heat release by condensation, or absorption by evaporation, plays a dominant role in the thunderstorm which precipitates the tornado. However, much of this driving energy is transferred to the smaller tornado scale by pressure forces. The Froude number depends upon the relative ratio of these pressure forces to the buoyancy within the tornado core. The Froude number was varied in the laboratory simulations of a dust devil by *Mullen and Maxworthy* [1977] and in the numerical simulations of *Lewellen and Sheng* [1980]. It is my guess that Fr is sufficiently large in the low-level corner flow of the tornado that it is of relatively little importance here, but that it becomes quite important at higher levels. Its most important low-level role is likely to be associated with stabilizing the core flow.

A few researchers have considered the influence of compressibility through the Mach number [e.g., *Eagleman et al.*, 1975]. However, since I will argue in the later sections that the maximum low-level velocities are limited to less than 110–130 m/s, the Mach number squared should be less than 0.1, and thus such compressibility effects should be relatively small. Also, we might add a number of additional parameters in attempts to measure the influence of such things as the shape of the radial inflow, the shape of the axial outflow at the top of the domain, and the buoyancy released due to condensation of water vapor within the domain, but such parameters are not easy to formulate in a simple way. Further, I am willing to speculate that they are not as important as the four parameters just defined.

2.2. Simple Models

In the limiting case of incompressible, irrotational flow, both the circulation and the total pressure are conserved as the vortex flow converges radially inward. This implies a minimum radius, r_c, inside which the flow cannot penetrate. This ideal minimum radius, where the maximum velocity occurs, is given by

$$r_c = \Gamma (2 \Delta p / \rho)^{(-1/2)} \qquad (4)$$

where Γ is the product of the tangential velocity times the local radius (circulation/2π), ρ is the density, and Δp is the pressure drop driving the vortex. This provides a simple, approximate relationship between three of the important variables defining the vortex. For flows with roughly equal pressure drop, the core radius is proportional to circulation, while for flows with similar circulation, the stronger the pressure drop, the smaller the core radius. The pressure drop, Δp, is divided between driving the swirl and driving the throughflow, so (4) assures that the swirl parameter, S, will have a strong influence on the flow. In fact, when the pressure drop driving the vortex flow is held fixed, (4) can be manipulated to show that the swirl ratio governs the minimum radius to which the maximum tangential velocity can penetrate. However, I expect turbulent dynamics to play an important role in determining the detailed velocity distributions, particularly close to the surface.

Within the core flow the tangential velocity sufficiently dominates the radial velocity so that there is an essentially cyclostrophic balance between the radial pressure gradient and the centrifugal force. Radial integration of the radial pressure gradient for any assumed tangential velocity distribution then provides a relationship similar to (4), with a numerical factor depending on the assumed distribution. As discussed in L76, a favorite simple model of velocities in the core flow is the Rankine vortex [*Rankine*, 1882] which sets the radial and vertical velocities equal to zero and contains an inner region of solid-body rotation, $v \sim r$, with an outer potential flow region, $v \sim r^{-1}$. This can be readily improved upon by solving for the radial balance between advection and diffusion in the presence of sink flow with a constant viscosity to obtain the Burgers-Rott model [*Burgers*, 1948; *Rott*, 1958] which yields a smooth transition between solid-body rotation and potential flow in the annular region of maximum velocity. This provides for some simple relationships between maximum velocity, core radius, and total pressure drop similar to (4) which are surprisingly good, considering that it ignores turbulence and all of the tight coupling between the top and the bottom boundary conditions which must prevail. The core radius, defined as the radius where maximum velocity occurs, implied by the Burgers-Rott model is approximately 30% larger than that given by (4) when Γ is taken as the value at large radius and Δp is the total pressure drop obtained by integrating the cyclostrophic balance between $r = 0$ and infinity. This is true in spite of the fact that the Burgers-Rott model determines the core radius by the radial balance between advection and diffusion of tangential momentum, with no direct influence of S on the velocity field. However, a strong effect of S is implied when the radial momentum equation is integrated to obtain the pressure drop.

It might be argued by some that these simple models should not even be considered as tornado models because they leave out so much. However, I think that any model which provides useful relationships between important variables within the flow may justifiably be referred to as a tornado model. If only the complete model deserves the tag, then we would still be without any models of a tornado. As discussed in more detail in L76, the radial balance between advection and diffusion of tangential momentum has been solved for a number of different assumptions about the radial velocity [*Bellamy-Knights*, 1970, 1971; *Dergarabedian and Fendell*, 1967; *Einstein and Li*, 1955; *Lewellen*, 1962; *Marchenko*, 1961; *Sullivan*, 1959]. A model which has a stronger coupling between the swirling velocity and the axial velocity is that by *Long* [1956, 1958, 1960, 1961] which assumes conical similarity. *Burggraf and Foster* [1977] show that for some relatively arbitrary initial velocity conditions at $z = 0$, laminar flows tend to approach one of Long's similarity solutions asymptotically, as long as the ratio of the axial momentum of the jet to the angular momentum is above a critical value (~ 3.75). For smaller values they surmised that some type of vortex breakdown occurred. A number of similarity solutions also include buoyancy to generate the updraft [*Franz*, 1969; *Fulks*, 1962; *Gutman*, 1957; *Kuo*, 1966, 1967], but, when a two-cell solution occurs in one of these buoyant models, the downdraft at the center is forced by colder temperatures, rather than being forced by the strong low-level swirl as occurs in the tornado.

Some new similarity solutions have been presented for swirling axisymmetric flows [*Paull and Pillow*, 1985; *Wang*, 1991; *Wu*, 1985; *Yih et al.*, 1982], but in general these one-dimensional flows require more artificial restrictions than appear justified now that higher-dimensional flows may be readily integrated numerically. The most interesting, recent core flow solutions are finite difference solutions, but since these also involve interaction with the top and bottom regions, I will defer discussing these until section 5, which deals with the corner flow where much of the strong interactions occur.

2.3. Stability Considerations

Although the tornado occurs in a region of relatively strong, three-dimensional turbulence, the rotational forces occurring within the tornado core are able to impede the normal cascade of turbulent energy and allow the core vortex to dominate. This phenomenon is apparent in the results of the rotating tank experiments of *Hopfinger et al.* [1982], which demonstrated the blockage of the normal cascade of turbulent energy in favor of the appearance of coherent vortices. They reported a dramatic transition in the turbulent flow field when the local Rossby number was decreased below about 0.2, with a collection of coherent vortices appearing with their axes approximately parallel to the rotation axis.

This transition in the turbulent cascade may be partially explained in terms of helicity, the vector inner product of velocity and vorticity. *Lilly* [1986a, b] has proposed that helicity may be an important descriptive variable in rotating storms. Fundamental research on turbulence [e.g., *Andre and Lesieur*, 1977; *Polifke and Shtilman*, 1989] has shown that high levels of helicity tend to retard the cascading of turbulent energy to smaller scales and thus reduce dissipation. The stretching and tilting terms in the vorticity equation are usually considered to be most responsible for the cascading of energy into the inertial subrange. However, purely helical flow, when the vorticity vector is a constant times the velocity vector (Beltrami flow), allows these two terms in the vorticity equation to exactly cancel with the advection terms. *Davies-Jones et al.* [1984] has suggested that helicity can be used as a predictor of severe storms, with at least some preliminary success [*Davies-Jones and Burgess*, 1990]. I am sure more will be written about this storm scale helicity in the rest of this volume. It may be just as important on the tornado vortex scale. Certainly, there appears to be strong damping of turbulent eddies by local rotation in the core from the smooth appearance of the central funnel in many photographs.

The rotational damping of turbulence may be even stronger in the presence of a positive radial gradient in density which may exist in any buoyantly generated vortex plume. Thus even when the Froude number of the tornado flow is moderately large so that buoyancy effects may be neglected in the mean flow dynamics, it may still be important in damping the turbulence in the tornado core. The centrifugal acceleration of a 70-m/s tangential velocity at a radius of 50 m is approximately 10 times Earth's gravitational acceleration. Thus strong damping of radial motions should be expected when these would carry with them a negative radial buoyant flux. When the Richardson number criterion of stratified turbulence is carried over to axisymmetric swirling flow, it says that turbulence is damped when the radial gradients (denoted by primes) of potential temperature, θ, axial velocity, w, and circulation satisfy the relation

$$\theta'/\theta_0 < 2\Gamma'/\Gamma - w'^2 r^3/(4\Gamma^2) \qquad (5)$$

This oversimplifies the stability problem [*Leibovich*, 1984], but it does illustrate that radial buoyancy gradients do not have to be very large to be as important as incompressible rotational considerations.

More details on the stability of tornadolike vortices may be found in the literature [*Foster and Smith*, 1989; *Gall*, 1983, 1985; *Gall and Staley*, 1981; *Leibovich and Stewartson*, 1983; *Rotunno*, 1978; *Snow*, 1978; *Staley*, 1985; *Staley and Gall*, 1979, 1984; *Steffens*, 1988; *Stewartson*, 1982; *Stewartson and Leibovich*, 1987; *Walko and Gall*, 1984]. When stable conditions exist, inertial waves may travel along the core and provide an important part of any unsteady dynamics [*Maxworthy et al.*, 1985; *Snow and Lund*, 1989].

It is important to note that satisfying a stability criterion such as (5) does not assure a laminar vortex, but rather establishes conditions for which we can expect any existing turbulence to be damped. *Fiedler* [1989] has argued that the central jet eruption from the turbulent boundary layer in a full-scale tornado corner flow, to be discussed in section 5, may be laminar on the basis of a local stability criterion. I think this is unlikely, because of the time scales required to damp the turbulent eddies existing in the radial jet which is converging to form the vertical jet. Although, the details of a laminar corner flow are not appropriate for flow in a full-scale tornado, the core flow, which the corner flow transforms into, may have its turbulence sufficiently damped for the flow to appear laminar, as indicated in the top part of the funnel in the tornado in Plate 1.

3. Surface Boundary Layer, Region II

It has long been recognized that the retardation of a rotating flow at a surface perpendicular to the axis of rotation will induce a radial inflow within the boundary layer [*Ekman*, 1905]. A number of different solutions for the rotating boundary layer were discussed in L76 [*Barcilon*, 1967; *Belcher et al.*, 1972; *Bellamy-Knights*, 1974; *Bödewadt*, 1940; *Burggraf et al.*, 1971; *Carrier*, 1971; *Chaussee*, 1972; *Chi and Glowacki*, 1974; *Chi and Jih*, 1974; *Chi et al.*, 1969; *Goldshtik*, 1960; *Hsu and Tesfamariam*, 1975; *Jischke and Parang*, 1975; *Kidd and Farris*, 1968; *Kuo*, 1971; *Rao and Raymond*, 1975; *Rott*, 1962; *Rott and Lewellen*, 1965, 1966; *Schwiderski*, 1968; *Smith and Smith*, 1965]. The boundary layer beneath a vortex also has continued to be a popular subject since 1976 [*Baker*, 1981; *Carrier et al.*, 1988; *Chi*, 1977; *Hsu and Tesfamariam*, 1976; *Kuo*, 1982; *Leslie*, 1977; *Monji and Yunkuan*, 1989; *Phillips*, 1984, 1985; *Phillips and Khoo*, 1987; *Prahlad*, 1976; *Rostek and Snow*, 1985; *Rotunno*, 1980; *Sychev*, 1989; *Wilson and Rotunno*, 1986]. A majority of these boundary layer analyses have been for laminar flow. These give a valid qualitative view of the boundary interaction with vortex flow, but the variation in turbulence close to the surface is important in determining the detailed structure of the flow. I will not attempt an individual critique of all of these efforts here, but instead will only look at what I consider some key features of flow in this region.

First, let us consider how the surface layer is affected by rotation. The presence of a boundary imposes a strong constraint on the dynamics of the turbulence. In the absence of any body forces on the flow this leads to a logarithmic law of the wall region in which the flux of momentum is constant while the velocity decreases logarithmically as the surface is approached. This logarithmic variation is valid for the tangential velocity in the boundary layer under a tornadolike vortex but is only valid for the radial velocity extremely close to the surface. This may be demonstrated by the following simple analysis from *Lewellen* [1977].

In the surface layer the radial momentum equation may be approximated as

$$\rho \partial \overline{u'w'}/\partial z = \rho v^2/r - \partial p/\partial r \quad (6)$$

This represents the balance between the vertical divergence of the turbulent flux of radial momentum and the difference between the local centrifugal force, which varies with z, and the pressure gradient which is imposed by the centrifugal force above the boundary layer. Close to the surface we also expect a simple mixing-length, turbulent model to be reasonably valid for the shear stress, so that

$$\overline{u'w'} = -c_1 z \partial u/\partial z \quad (7)$$

If $\overline{u'w'}$ were constant, (7) would yield the logarithmic result as it does for the tangential velocity:

$$v = c_2 \ln (z/z_0) \quad (8)$$

Figure 3 gives the results of integrating (6) and (7) with the aid of (8) and compares this result with the laboratory observations of *Savino and Keshock* [1965]. In this comparison, three data points are used to determine the effective surface roughness, z_0, and the two components of the surface shear stress, which are related to the two constants c_1 and c_2. This is the same information required to determine the usual logarithmic layer. The pressure gradient is given by the free stream conditions above the boundary layer.

The comparison in Figure 3 shows that the addition of the pressure gradient influence in the surface layer permits a good representation of the radial velocity for a much greater range of z, distance from the surface, than is the case for the leading logarithmic term. Naturally, this turbulent variation produces velocities with much sharper gradients close to the surface than are produced in a laminar boundary layer (the inherent result produced by a constant eddy viscosity model). The sharp peak in the radial velocity very close to the surface in Figure 3 indicates that if a constant eddy solution is to be used, it is probably better to combine a free-slip boundary condition on the radial velocity, with the no-slip condition only applied to the tangential velocity.

Local conditions at any radius provide an estimate of the radial volume flow induced into the linear Ekman layer, but this is not generally adequate for the full nonlinear rotating boundary layer. This requires information on the history of the boundary layer. A simple estimate of this was reported in L76 as given by *Rott* [1962]:

$$Q_{bl} = -2.5 c_f \Gamma (r_0 - r) \quad (9)$$

where the negative sign indicates that the radial volume flow, Q_{bl} in the boundary layer is radially inward, c_f is the turbulent skin friction coefficient, Γ is the circulation in the outer flow, and r_0 is the outer radius from which the boundary layer develops. Equation (9) provides a direct estimate of the flow which is diverted into the boundary layer at large radii and must be ejected vertically into the corner flow to be discussed in section 5. It shows that there must be negative vertical velocity into the boundary layer outside the radius where the circulation remains essentially constant. When the circulation is not constant above the

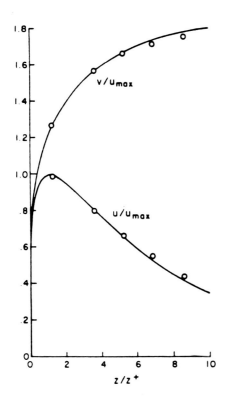

Fig. 3. Surface layer, turbulent profiles of tangential, v, and radial, u, velocities beneath a strong vortex. Comparison of the solution predicted by *Lewellen* [1977] with the profiles measured by *Savino and Keshock* [1965] is shown.

boundary layer, (9) is strongly modified, and some of the radial flow may be ejected from the boundary layer.

The primary additional parameter introduced into the boundary layer flow, not discussed in section 2.1, is some measure of the surface roughness, which may be introduced as a second Reynolds number with r_0 in (2) replaced by z_0.

4. Upper Flow, Region IV

The upper flow region, which must necessarily include the top boundary condition, contains a large degree of uncertainty, because it is buried in the parent thunderstorm. The most popular solution for this in laboratory simulations is to include a baffle, which dissipates the angular momentum, in the exhausting flow at the top of the domain. This, of course, cannot occur in the real storm. The angular momentum brought into the vortex at low levels must be transferred outward, either by diverging flow at the top or by turbulent transport. I argued in L76 that at the top of a swirling updraft where the flow is forced to diverge outward, there should be a secondary downdraft induced along the axis of rotation by the low central pressure at lower levels. If this pressure-induced downdraft from the top penetrates to the surface, then the storm will have an "eye" like a miniature hurricane. There has been and continues to be controversy as to when or if an eye exists in strong tornadoes [*Anderson*, 1985; *Bluestein*, 1985; *Carrier and Fendell*, 1985; *Carrier et al.*,

1979; *Dergarabedian and Fendell*, 1973; *Lilly*, 1976; *Mitsuita et al.*, 1987; *Walko*, 1988; *Wylie and Anderson*, 1983].

It is my opinion that most moderate tornadoes are not associated with an eye extending through the thunderstorm, even when there is a central downdraft at low levels. Rather, they are capped by a buoyant plume, a process that still remains to be accurately simulated. Such a process may be expected to occur when the core radius increases with height sufficiently that (5) is no longer satisfied; then turbulence is allowed to grow and break up the core. The inner eye would be capped at this altitude, with angular momentum transported outward by turbulence. This capping process may have been almost captured in the laboratory results of *Pauley* [1989], where the vertical pressure gradient along the axis was observed up to a region where the strong inner downflow appeared to be originating. I am willing to speculate that the vertical velocity would have been upward along the axis just above the highest position reported.

There appears to be no physical barrier to prohibit exceptionally strong tornadoes from being associated with an eye which penetrates the entire depth of the storm. It is quite possible, even probable, that a few F-5 class tornadoes (the strongest tornadoes in *Fujita*'s [1971a] rating system) have had such well-developed eyes. Indeed, there is some supporting evidence that the tornado core extends far up into the thunderstorm for such storms. *MacGorman* [this volume] reports that lightning underwent a sharp transition from intracloud flashes during a strong tornado to negative cloud-to-ground flashes after the tornado. He also suggests that the initial domination by intracloud flashes may be associated with higher penetration of the warm moist central jet (updraft) from the surface. This suggestion is consistent with the stability discussion in section 2. The stability of the vortex core structure would reduce the entrainment between the rising warm, moist air and the cold downdraft which appears to be wrapping around it in the thunderstorm simulations.

The connection to the thunderstorm makes it difficult to separate out the dynamics in the upper region of the tornado vortex. Rather than attempt to model this, I prefer to consider the tornado vortex model as a small inner nest of a numerical thunderstorm model which cannot yet be modeled as an interactive part of the thunderstorm because of insufficient grid resolution. *Wicker and Wilhelmson* [1990] have been able to obtain some of the coarse features of a tornado vortex in a simulation which nested down to an inner grid resolution of 67 m in the horizontal directions. As will be seen in section 7, a resolution approaching 10 m appears to be needed to incorporate the full dynamics of the tornado vortex.

If the tornado vortex is considered an inner nest of a thunderstorm model, then the upper boundary conditions as well as the outer radial boundary conditions can come directly from the thunderstorm output. Of course, one still needs to look at the sensitivity of the resulting tornado vortex to variations in these boundary conditions which would be subgrid scale to the thunderstorm model. This

26 TORNADO VORTEX THEORY

Plate 1. Photograph of a tornado funnel over the western Kansas plains [*Whipple*, 1982], taken from a distance of $3\frac{1}{2}$ mi (5.6 km).

can lead to a local separation in the presence of the slight adverse pressure gradient resulting from the stagnating radial velocity above the boundary layer. Relatively little swirl is required to keep the flow attached as in Figure 4b. As S is increased, the ratio of the volume of the throughflow passing through the boundary layer to that above the boundary layer increases, and the center jet is intensified. At a larger value of S, which depends on both the inflow and the top outflow conditions, the conditions for Figure 4c are reached. Here the central erupting jet is stronger than can be supported throughout the core flow by the top boundary conditions. The result is a sharp transition in the flow, a type of vortex breakdown, at some distance above the ground. At still higher values of S, the boundary layer eruption occurs in an annulus around the center, and the flow pattern of Figure 4d appears. When instabilities occur in the annular region of

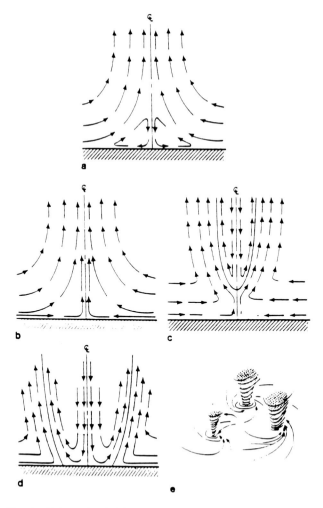

Fig. 4. Schematic of different types of corner flow [*Davies-Jones*, 1986] for increasing swirl: (a) very weak swirl so that the boundary layer separates, (b) one-cell vortex, (c) vortex breakdown above the surface, (d) two-cell vortex with downdraft penetrating to the surface, and (e) multiple vortices rotating about the annulus separating the two cells.

leaves open the question of how much the thunderstorm model depends upon its subgrid parameterization. *Lilly* [1983] has argued that the reluctance of helical turbulence to cascade in an inertial range makes rotating supercell storms much less susceptible to turbulent uncertainties and therefore more predictable by numerical simulations without uncertainties due to subgrid turbulent closure assumptions. The extent to which this is true remains to be determined.

5. Corner Flow, Region III

The most interesting interaction with the surface occurs in the neighborhood of $r = 0$, where the radial flow induced by the boundary layer must turn and produce some type of vertical jet. The structure of this jet depends on the relative fraction of the volume flow which has been diverted to the boundary layer. This is qualitatively determined by the swirl ratio. Figure 4, taken from *Davies-Jones* [1986], shows a schematic plot of the variation of the corner flow with swirl ratio. If $S = 0$, there is no excess flow in the boundary layer. Instead, there is a deficit within the boundary layer which

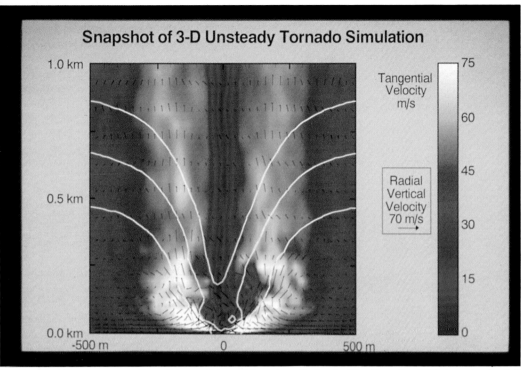

Plate 2. Meridional cross section of tangential, radial, and vertical velocities from the LES tornado simulation described in the text. Pressure isobars are indicated by the light blue lines.

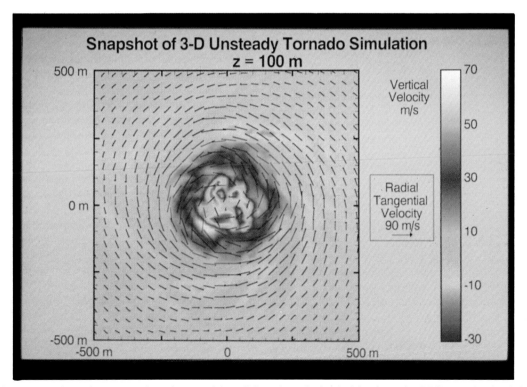

Plate 3. Horizontal cross section of tangential, radial, and vertical velocities from the LES tornado simulation described in the text.

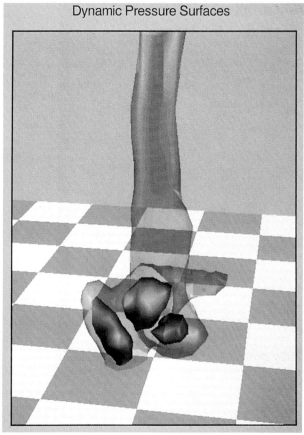

Plate 4. Three-dimensional perspective of surfaces of constant dynamic pressure from the LES tornado simulation described in the text. The squares on the surface are 100 m on a side.

concentrated vorticity, coherent vortices, called suction vortices [Forbes, 1978; Fujita, 1970, 1971b; Fujita et al., 1976], may rotate around the primary vortex as in Figure 4e.

The sharp transition a short distance above the surface at the center of the sketch in Figure 4c is an example of a type of vortex breakdown. Vortex breakdowns in tubes, in wing tip vortices, and in natural vortices have provided a fascinating subject for research [Benjamin, 1962; Brown and Lopez, 1990; Burggraf and Foster, 1977; Escudier, 1986, 1988; Hall, 1972; Leibovich, 1978, 1984; Lopez, 1990; Lugt, 1989; Sarpkaya, 1971]. The breakdown provides a jump between two stable states of swirling flow. Upstream of the jump the axial velocity is supercritical, in the sense that it is faster than the speed at which inertial waves may travel in this vortex environment. The axial velocity is reduced to a subcritical value across the transition, analogous to a hydraulic jump in stratified flow. Even when the classic breakdown does not occur, axial variations in a strongly swirling flow tend to be either quite small or quite abrupt because of the tight coupling between the velocity components forced by axial variation in the tangential velocity.

The swirl ratio is easy to control in both laboratory and numerical simulations, but it becomes somewhat ambiguous in the real tornado vortex because of its dependence on domain geometry. Results of a laboratory investigation of the characteristics of tornadolike vortices as a function of swirl ratio are given by Church et al. [1979]. A plot of minimum pressure versus S is given in Figure 5, as given by Wilkens and Diamond [1987]. They show that at least the laboratory vortex goes through a number of transitions as S is increased, with both the geometry and the throughflow held fixed. This figure also shows that variations with S depend upon the particular geometry. The Purdue chamber, with its smaller aspect ratio, shows a much more distinct minimum in pressure, where there is a transition from the flow type in Figure 4c to that in Figure 4d [Pauley et al., 1982]. Numerical simulations [Wilson and Rotunno, 1986] have been able to essentially duplicate the Purdue results at low Reynolds number.

The flow in the corner flow is sufficiently complex that solutions have required some degree of numerics even when part of the analysis is analytical. When attempting finite difference simulations, the difference between a corner flow simulation and the full tornado vortex simulation is relatively arbitrary since it is largely a matter of how the outer radial and upper vertical boundary conditions are specified. I will not distinguish between the two, nor will I repeat the discussion of the pre-1976 finite difference solutions here. Numerical solutions appearing in the last 15 years include one-dimensional axisymmetric simulations by Carrier et al. [1988], Chen and Watts [1979], and Gall [1982]. These are similar to the inviscid interaction model of Lilly [1969] in that enough assumptions are made about the core flow and the boundary layer flow to allow the interaction problem to be reduced to a set of ordinary differential equations. Gall [1982] is successful in matching the behavior of core radius as a function of swirl obtained in the laboratory. Goldshtik [1990] has reconsidered his conical similarity solution for a line vortex perpendicular to a flat nonslip surface [Goldshtik, 1960] with some apparent modifications in the critical Reynolds number above which no such similarity solution can exist. This nonexistence of such a similarity solution at even moderate Reynolds numbers suggests that something more dramatic such as vortex breakdown may occur.

Laminar, two-dimensional axisymmetric, numerical simulations have been presented by Howells and Smith [1983], Howells et al. [1988], Leslie [1977], Leslie and Smith [1982], McClellan et al. [1990], Proctor [1979], Rotunno [1977, 1979], Rotunno and Lilly [1981], Smith [1987], Smith and Howells [1983], Smith and Leslie [1979], Walko [1988], and Wilson and Rotunno [1982, 1986]. Turbulent, two-dimensional axisymmetric simulations have been presented by Chi [1977], Lewellen and Sheng [1979, 1980], Lewellen and Teske [1977], and Lewellen et al. [1979]. Three-dimensional simulations have been presented by Rotunno [1982, 1984], Walko [1990], and Wicker and Wilhelmson [1990].

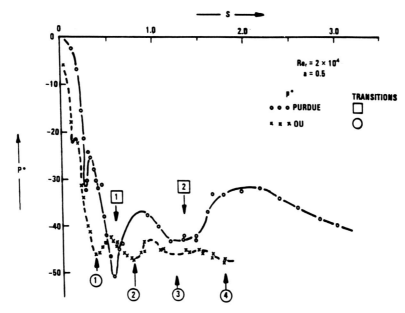

Fig. 5. Comparison of the central pressure drop as a function of swirl ratio for the Oklahoma University and Purdue University laboratory simulations [*Wilkens and Diamond*, 1987].

5.1. *Influence of the No-Slip Condition*

One of the most interesting features of the corner flow is that the radial inflow induced in the boundary layer forces the strongest tangential velocities to occur close to the surface as turning occurs. The no-slip surface boundary condition can actually increase the maximum velocity a short distance above the surface. Velocity distributions obtained when free-slip conditions are applied; for example, the results of *Walko and Gall* [1986] show tangential velocity distributions with a maximum on the surface, with a slow decrease as the vertical coordinate increases and the core slowly expands. This may be contrasted with the no-slip results of *Howells et al.* [1988] exhibited in Figure 6. The no-slip conditions result in an overshoot in the corner flow which can more than compensate for end-wall losses. Howells et al. show that, at least for their strong swirl case and assumed constant viscosity, the no-slip boundary condition produces a maximum velocity approximately 25% greater than that produced by a similar simulation with free slip at the surface. At lower values of swirl the no-slip boundary led to even larger increases in the value of the maximum velocity.

5.2. *Influence of Turbulence*

Most of the numerical simulations which include the corner flow are for constant eddy viscosity, i.e., laminar flow. A comparison between a constant eddy viscosity simulation by *Rotunno* [1979] and a turbulent transport simulation by *Lewellen and Sheng* [1980] for conditions similar to those obtained in the Purdue simulator [*Church et al.*, 1979] showed that the maximum velocity was a little larger and occurred much closer to the wall and at somewhat smaller radius in the turbulent transport model than in the laminar simulation. Not surprisingly, the reduction in effective eddy viscosity predicted by turbulent transport theory close to a surface allows the maximum velocity to occur closer to the surface. The differences occurring at full-scale atmospheric Reynolds numbers should be even larger than the differences evidenced in this simulation of the moderate value Reynolds number appropriate for this laboratory experiment. This comparison was made for swirl conditions similar to those of Figure 4d. Probably an even bigger difference would have been seen at lower swirl conditions approximating Figure 4c. In a more recent, high-resolution, laminar simulation of conditions more appropriate to this case which incorporates a vortex breakdown, *Wilson and Rotunno* [1986] have obtained a maximum vertical velocity which is twice the maximum tangential velocity and occurs between the surface and the breakdown transition. I would not expect a high Reynolds number, turbulent simulation to exhibit such a strong axial jet.

Howells et al. [1988] show that even in their laminar simulation, the details of the corner flow are quite sensitive to the simulation viscosity. They conclude that "due to the extreme sensitivity of the flow to the value of the eddy diffusivity coefficient, any attempt to simulate actual atmospheric vortices using a model which incorporates self-regulating turbulence closure should include a thorough sensitivity test of the parameters assumed within the closure approximation" [*Howells et al.*, 1988, p. 819]. I agree with this comment and add that modelers who use the simplest turbulent closure of constant eddy viscosity should not expect their simulations to provide very valid details of the flow, particularly close to the surface.

NUMERICAL TORNADO-VORTEX MODELS

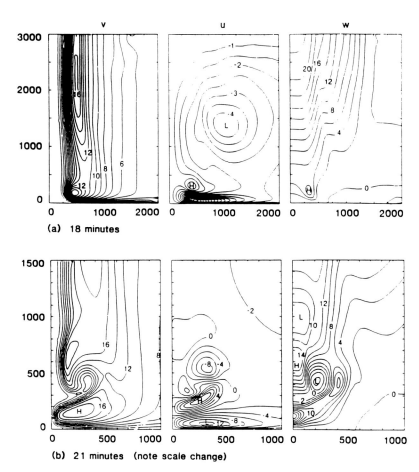

Fig. 6. Transient meridional cross sections of tangential, radial, and vertical velocities from a simulation with no-slip lower boundary condition by *Howells et al.* [1988]. Coordinates are in meters, and velocity contours in meters per second.

The turbulent transport simulations of *Lewellen and Sheng* [1980] for low Reynolds number approximating the laboratory conditions of *Mullen and Maxworthy* [1977] are similar to that shown in Figure 6, as seen in Figure 7. The simulation which led to Figure 7 not only had a relatively low Reynolds number to simulate the laboratory conditions but also was stabilized by relatively high heat flux in this dust devil simulation. The simulated flow was very unsteady, with the height of the vortex breakdown oscillating over a large region. A similar oscillation was reported for the laboratory results. At the simulated time corresponding to Figure 7, the maximum vertical velocity was more than 60% larger than the maximum tangential velocity.

The turbulent transport simulations of *Lewellen and Sheng* [1980] over rough surfaces at high Reynolds numbers were typically more conical at low levels as illustrated in Figure 8. This figure is also normalized by conditions at the position of maximum tangential velocity. The simulation was done with $S = 1.35$, uniform u and v set at 1 km radius, and uniform outflow set at $z = 1$ km. The value of S was chosen to correspond to that deduced from *Brandes'* [1979] observations of the Harrah, Oklahoma, storm of June 8, 1974. The vertical velocity for the same conditions was quite small on the axis with the largest values ($0.3 v_{max}$) occurring along a cone just inside the cone on which the maximum tangential velocity occurs.

The main point I wish to make in the present section is that the corner flow is quite sensitive to turbulence. Rather than show more results from axisymmetric turbulent transport simulations, I will show new results from a more reliable turbulent model in section 7.

6. ESTIMATES OF MAXIMUM VELOCITY

One of the most important features desired from a tornado model is an estimate of the maximum wind speed within a few meters of the surface. In L76 I argued that no existing model was adequate to provide a reliable estimate. *Davies-Jones* [1986] provides a review of the results from several different techniques which have attempted to measure this parameter: photogrammetry, ground marks, direct passage over instruments, remote sensing, direct probing, damage analysis, and funnel cloud analysis. Each of these methods requires some knowledge of the structure of the tornado vortex to properly interpret the measurement. Davies-Jones concludes that most scientists agree that wind speeds in even the strongest tornadoes do not exceed 110–130 m/s^{-1}. Additional careful statistical analysis of the available measurements are required to estimate the probabilities of extreme winds [*Abbey*, 1976; *Abbey and Fujita*, 1979].

The maximum velocity to zeroth order may be determined by the total pressure drop available to the flow as implied by (4). A number of researchers [e.g., *Lilly*, 1976] have discussed the fact that an estimate of this pressure drop can be made by taking the difference between hydrostatically integrating the environmental thermodynamic sounding and integrating the model sounding which could be generated by lifting a surface air parcel to its ideal equilibrium level. The pressure deficit (typically about 50 mbar) provides a rough guess but is subject to considerable uncertainty. If the two-cell core structure persists vertically throughout the storm, i.e., if the storm has an eye, then the resulting subsidence of dry, near-stratospheric air could increase the available pres-

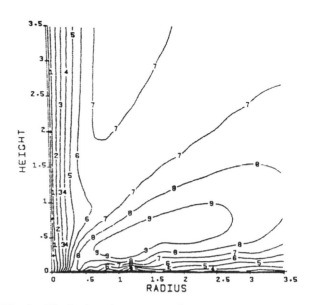

Fig. 8. Distribution of mean tangential velocity from a turbulent transport simulation of *Lewellen and Sheng* [1980] for $S = 1.35$ and $z_0 = 0.04$ m. Contours are labeled in tenths of the maximum tangential velocity, and coordinates are normalized by the radius at which the maximum velocity occurs.

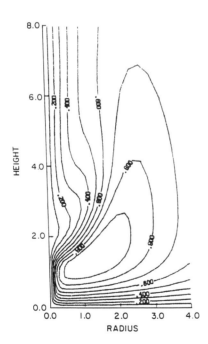

Fig. 7. Transient meridional cross sections of tangential velocity obtained in the turbulent transport simulation of *Lewellen and Sheng* [1980] for a relatively low Reynolds number approximating the laboratory conditions of *Mullen and Maxworthy* [1977]. Velocity is normalized by the maximum value and coordinates by the value of the radius at which the maximum occurs.

sure drop by up to a factor of 3 [*Davies-Jones*, 1986]. Also, when the immediate tornado core region is surrounded by a cold downdraft created by precipitation falling into dry air in the midtroposphere, then the contrast between precipitation-laden downdraft and a central dry subsidence could be even larger. The other difficulty with this simple estimate is that the central corner flow region in the tornado is neither hydrostatic nor cyclostrophic, as seen in section 5.

One of the principal conclusions from my 1976 review was that "the maximum velocities will occur below the top of the ground boundary layer (~100 m above surface) and may be significantly higher than that predicted by a cyclostrophic balance" [*Lewellen*, 1976, p. 136]. There were many doubters, since one is conditioned to expect the boundary layer to reduce wind speeds. In fact, as recently as 1988 *Carrier et al.* [1988] regarded this "as unlikely" to be true, although they did admit to the possibility of a small overshoot in the corner flow region. Nevertheless, numerous numerical simulations [e.g., *Howells et al.*, 1988; *Lewellen and Sheng*, 1980] have confirmed that the overshoot can be quite significant. Laboratory models have also noted that low-level pressures on the centerline may be significantly lower than that obtained by cyclostrophic balance above the boundary layer [*Baker*, 1981]. This led *Snow and Pauley* [1984] also to criticize the hydrostatic, thermodynamic method of estimating maximum tornado velocities.

Fiedler and Rotunno [1986] provide a physical theory for why the most intense laboratory vortex velocities occur close to the surface for the flow type represented in Figure 4c. They argue that the erupting boundary layer flow is supercritical and thus permitted to have an appreciably lower pressure than that occurring along the centerline in the subcritical flow, which is in cyclostrophic balance on the other side of a vortex breakdown transition. Their analysis shows that the maximum azimuthal velocity at low levels of a laminar vortex can be approximately 1.7 times the maximum azimuthal velocity in the subcritical vortex aloft. They point out that a turbulent boundary may result in a somewhat smaller effect, but demonstrate "in any event, that the boundary layer does not simply act as an energy sink—in a much more subtle way it acts to focus the energy of the larger-scale cyclone toward the smaller-scale tornado" [*Fiedler and Rotunno*, 1986, p. 2339].

As stated in section 2.3, I do not accept the subsequent argument by *Fiedler* [1989] that the full-scale tornado corner flow may be laminar. However, I do agree with *Fiedler and Rotunno*'s [1986] basic analysis. Also, I believe the intense velocities occurring in the corner region can be viewed in somewhat simpler and more general terms. The corner flow velocities are allowed to exceed those supported by the driving pressure deficit available aloft, because the expanding, swirling flow directly above the region of maximum velocity behaves as a diffuser, providing a mechanism for pressure recovery. This is true whether there is a very efficient pressure recovery as occurs in a sharp, laminar vortex breakdown or there is the less efficient pressure recovery which occurs in a region of nearly conical, turbulent flow evident in the results of *Lewellen and Sheng* [1980]. If swirling flow is forced through a conventional nozzle, one is not surprised that the maximum velocities occur in the minimum cross section of the nozzle. In this analogy the central corner flow is the region of minimum cross section for the flow. The subsequent maximum velocities which occur at this cross section are determined by the efficiency of the pressure recovery process, as well as by the driving pressure drop. Thus the unique interactions which occur in the corner flow focusing provide a subtle flow arrangement for obtaining a local maximum in kinetic energy which exceeds that generally expected from the available storm energy.

If the tornado vortex domain is considered as a small nested region within the thunderstorm, as suggested in section 4, then the maximum velocity should be determined by the boundary conditions imposed by the thunderstorm on such nests. Two primary inputs required for such a limited domain simulation are the ambient vertical vorticity and horizontal convergence from the desired region of the thunderstorm. Although thunderstorm models were not capable of supplying these parameters in the late 1970s, *Lewellen and Sheng* [1980] did attempt to see how the maximum velocity would depend upon these two variables. The results from their turbulent transport model, which are shown in Figure 9, were intended to help interpret Doppler radar results. For purposes of this figure the inner nest radius and the top of the domain were both set at 0.8 km, about the limits of Doppler radar at the time. The strongest velocities require both significant vorticity and convergence. These results may be expected to vary some with the details of the inflow and outflow which would be subgrid scale to any

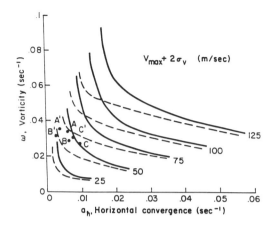

Fig. 9. Maximum tangential velocity to be expected in a tornado as a function of ambient vertical vorticity (ω) and horizontal convergence (a_h) measured in a cylinder with 0.8-km radius for a relatively smooth surface ($z_0 = 0.05$ m, dashed lines) and a rough surface ($z_0 = 1.4$ m, solid lines) according to the simulations of *Lewellen and Sheng* [1980].

thunderstorm model. In the study resulting in Figure 9, increasing the ratio of the vertical velocity along the centerline at the top of the domain to the average across the top by a factor of 2 produced a 15% increase in maximum velocity, while decreasing the same ratio to zero produced a 15% decrease in maximum velocity. On the basis of these results we concluded that the maximum velocity is less sensitive to the details of the outflow at 0.8 km than it is to the ambient vorticity and convergence.

Fujita [1970] has shown that maximum damage is generally associated with unsteady peak velocities where the "suction vortex velocity" adds to the mean tornado velocity. In our axisymmetric model these suction vortices were interpreted as turbulent eddies. Since the turbulent transport model provided a prediction of the root-mean-square (rms) fluctuations about the mean velocity, twice the rms tangential velocity was added to the mean tangential velocity in constructing Figure 9, which was intended to provide maximum wind speeds to be used for damage estimates.

The dashed lines in Figure 9 denote the contours for winds over a relatively smooth surface and the solid lines those for a rough surface. If a thunderstorm model or analysis of a Doppler return indicates an average value of vertical vorticity equal to 0.03 s^{-1} and horizontal convergence equal to 0.01 s^{-1} for a cylinder with a 0.8-km radius and a 0.8-km height above the surface, then Figure 9 would predict a maximum probable wind of 75 m/s, if the tornado is occurring over smooth prairie grassland ($z_0 = 7$ cm). This maximum would be reduced to 60 m/s over rough terrain with $z_0 = 1.6$ m. The labeled points are values supplied by Brandes from the analysis of dual Doppler radar observations of the Hurrah and Del City tornadoes [*Brandes*, 1979, 1981]. It is not clear whether sufficient storms have been captured in this manner in the last 10 years [e.g., *Hennington et al.*, 1982; *Uttal et al.*, 1989; *Zrnic et al.*, 1985] to provide

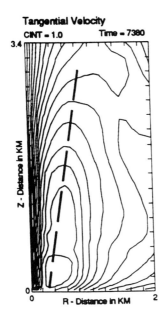 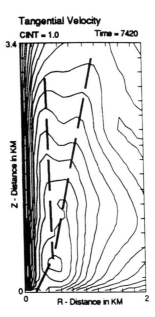

Fig. 10. Tangential velocities exhibited by *Wicker and Wilhelmson* [1990] at different times in their tornado nest within a thunderstorm simulation. Minimum resolution is 67 m.

a confident bound on the combination of these two variables. However, it appears that numerical storm simulations such as those given by *Wicker and Wilhelmson* [1990] should now be adequate to supply their own estimates of these limits.

It may be noted that the constant maximum velocity curves in Figure 9 have the same general shape as the tornado intensity maps given by *Johns et al.* [this volume], where the convergence coordinate is replaced by convective available potential energy (CAPE) and the vorticity coordinate is replaced by the ambient vertical wind shear. Although these plots are for quite different domain scales, their similarity is much more than just coincidental. CAPE provides a direct measure for possible convergence into coherent plumes, and low-level vertical wind shear, horizontal vorticity provides a direct source of vertical vorticity via tilting [*Davies-Jones*, 1986]. On the other hand, I do not wish to imply that Figure 9 provides the definitive answer for the maximum velocity question. Too many uncertainties remain about the sensitivity to domain boundary conditions and the turbulent transport model. If one were to attempt to produce a figure equivalent to Figure 9 appropriate for laboratory vortices, it would have many kinks relating to the transitions in Figure 5 and a dependence on both Reynolds number and geometry.

7. LARGE-EDDY SIMULATION TORNADO MODEL

As mentioned in the introduction, fully three-dimensional, unsteady simulations are now easier, than steady state, axisymmetric calculations were 15 years ago. This presents one way of at least partially circumventing the problems associated with modeling turbulent transport within the tornado, that is, to provide sufficient resolution to simulate the largest eddies responsible for the principle transport within the flow. It is still not possible to do a direct simulation of turbulence at full-scale Reynolds numbers, so some subgrid closure model is still required. However, the more resolution which can be incorporated into the simulation, the less importance the subgrid model assumes.

Rotunno [1984] made a three-dimensional simulation of the Purdue simulation chamber under sufficiently large swirl conditions to exhibit secondary "suction" vortices spinning around the central core. He used a constant eddy diffusivity with a Reynolds number $(u_0 r_0 / \nu) = 150$. More recently *Walko* [1990] and *Wicker and Wilhelmson* [1990] have made three-dimensional tornado simulations, but both of these were emphasizing somewhat larger flow features. Wicker and Wilhelmson were looking at a tornado nested grid within a supercell storm. Walko was looking at ways of introducing the vorticity into the tornado. It is quite impressive that Wicker and Wilhelmson were able to zoom in from the thunderstorm scale to the tornado scale and see the features exhibited in Figure 10. They were able to see that the surface no-slip condition increases the maximum velocity and that the maximum velocity occurs at low levels. In fact, they found that the no-slip surface boundary condition increased the vorticity at low levels by 50% over that obtained in a free-slip simulation. However, to resolve the core dynamics we have been discussing in this review, it is necessary to zoom in to even smaller scale.

While preparing for this review, I asked my colleagues at Aeronautical Research Associates of Princeton to use the large-eddy simulation (LES) code described by *Sykes and Henn* [1989] to attempt an LES model of tornado dynamics. As mentioned earlier, *Schmidt and Schumann* [1989] have used a large-eddy simulation of the convective boundary layer to look at the conditions where coherent plumes are initiated. One might look at the conditions which allow these to become quite distinct dust devils, but our purpose here is quite different. It is to provide sufficient resolution in the corner flow region to allow the dominant turbulent features to be resolved. We set the horizontal boundary conditions to approximate those expected on a 1-km cube with the tornado at the center of this cube. These horizontal boundary conditions were taken as axisymmetric about the center. The tangential velocity was taken to have a vertical variation proportional to ln (z/z_0) as expected in a fully turbulent boundary layer. The radial velocity was given a variation approximating that of Figure 3, with a maximum inward radial velocity at approximately 30 m height. The vertical velocity was set to zero. The top of the domain was set at 2 km where the vertical velocity was assumed uniform within a cylinder with a 500-m radius. Outflow boundary conditions are applied to the remaining variables at this top boundary. The grid included 74 × 74 uniform horizontal points and 50 points stretched in the vertical. Initial conditions were taken as approximating a Burgers-Rott vortex above the surface layer.

For the results shown in Plate 2, the horizontal velocities at the outer boundary were consistent with axisymmetric conditions with both the circulation and the radial volume flow above the surface layer constant at $vr = 1.5 \times 10^4$ m^2/s and $ur = 2 \times 10^3$ m^2/s. During the course of the model simulation spin-up, these horizontal boundary conditions were held fixed except a 20% random variation was permitted to simulate this level of turbulence in the inflow. The surface boundary conditions assumed a surface moving with a constant translation of 15 m/s and with an equivalent surface roughness of 0.2 m corresponding to conditions in between the rough and the smooth surface conditions considered by *Lewellen and Sheng* [1980]. The vortex spun up to essentially a quasi-stationary turbulent flow after about 3 min of simulated time (~2000 time steps). After this quasi-stationary state was achieved, the velocities were saved on a vertical slice through the center and on a horizontal slice at 100 m height at frequent intervals to gather statistics on the turbulence in the flow. Some 26 samples were taken at intervals corresponding to approximately one third of the period of rotation in the region of maximum velocity.

The question of how successful we were at resolving the primary turbulence in the simulation may be answered by comparing the turbulence resolved by the simulation with that represented by the subgrid parameterization of the flow. The maximum rms of the resolved velocity fluctuations was approximately 3 times the maximum subgrid rms velocity. If this ratio held everywhere in the flow, it would indicate that 90% of the turbulent energy had been resolved. We are not claiming to have done this well, but we believe that we have captured enough of the turbulence for the simulation to have reasonable validity.

Plate 2 shows a snapshot of a vertical plane through the center of the vortex. The color coding represents the absolute tangential velocity out of the plane, and the arrows represent the radial/vertical velocities in the plane. The solid lines represent three pressure contours when a hydrostatic vertical pressure gradient is added to the dynamic pressure gradients generated in the simulation. One of these pressure contours might approximate the funnel contour depending upon the condensation level at the time. Even with the translation, the flow looks like an axisymmetric flow with superimposed turbulence. The nature of the turbulence is clearer in the horizontal slice shown in Plate 3. In this plate the color coding represents the vertical velocity, while the arrows represent the horizontal velocity vectors. There are at least three coherent eddies, each with their own up and down vertical velocities rotating about the primary vortex. These three eddies show up even more clearly in the three-dimensional perspective of two contours of the dynamic pressure in Plate 4. The lowest dynamic pressures occur just a little above the surface as illustrated by the dark blue contours. The central dark blue region represents the center of the main vortex, while the third "suction vortex" is not quite strong enough to show this contour, but the lighter color arm is clearly in evidence.

As was evident in the simulations discussed in section 5, the radial flow penetrates to a smaller radius near the surface and results in a higher tangential velocity in this region. The maximum vertical velocities occur slightly off the center and are somewhat lower than the maximum tangential velocity which reaches a local maximum of 90 m/s. The overall flow pattern is quite similar to the results of the axisymmetric, turbulent transport model of *Lewellen and Sheng* [1980] shown in Figure 8. It is interesting to note that although a uniform vertical outflow was imposed within a radius of 500 m at the top of the computational domain (at 2 km), the vertical velocity tends to be concentrated in an annulus about an eyelike central region over most of the lowest kilometer. On the basis of the study of the sensitivity of our axisymmetric simulations to variations in the top boundary condition [*Lewellen and Sheng*, 1980], previously discussed in section 6, I expect these low-level results to depend on details in the prescribed distribution of the outflow at 2 km but not be highly sensitive to them.

A snapshot of the helicity in the vertical slice (not shown) indicated that the helicity is primarily concentrated in braids within the core. Somewhat surprisingly, the low-level region where the maximum value occurs indicated a strong asymmetry. It is not clear how much of this asymmetry was associated with the rotating "suction vortices" evidenced in Plate 3 and how much was associated with the fact the tornado is translating to the right.

This simulation raises as many questions as it answers. Some of these questions are as follows: What is the unsteady behavior during start up and decay? How much influence would buoyancy have on damping the turbulence? How sensitive is the flow to input parameters such as the swirl ratio, surface roughness, details of the horizontal inflow, and details of the upper outflow conditions? The only variation we tried in this preliminary investigation was a reduction in the swirl input ratio. The result was a decrease in the core radius and a modest decrease in the maximum velocities. However, since this variation also was computed with a decrease in horizontal resolution, the combination of decreased core size and decreased resolution forced most of the turbulence into the subgrid unresolved mode.

8. Conclusions and Recommendations

To summarize, I will review how much my conclusions about tornado vortex dynamics have changed in the last 15 years and recommend some directions for future research.

8.1. *Updating the 1976 Conclusions*

There were six conclusions listed in my earlier review, L76. It is interesting to see how each of these should be changed in light of research results over the last 15 years. I will first quote the earlier conclusion and then discuss its present applicability.

1. "A relatively consistent qualitative model of the flow in different regions of the tornado has been pieced together from previous theoretical work" [*Lewellen*, 1976, p. 136]. The qualitative view of the tornado vortex flow has not changed that much in the last 15 years. The most important qualitative changes have been on the slightly larger scale which connects the tornado to the parent storm. There has been considerable progress made toward defining the details of the interactions in the different regions.

2. "The main parameter governing the flow is a swirl ratio (which should be determined by some ratio of the ambient instability to the ambient rotation) with the detailed structure modified by the surface roughness and stability of the ground boundary layer" [*Lewellen*, 1976, p. 136]. There is perhaps a little more recognition now that a single parameter cannot be isolated from the rest of the flow inputs, but the swirl ratio certainly remains one of the most important parameters.

3. "At least three types of wind distributions are likely to occur immediately above the ground in the center of different tornadoes. One with a central updraft, a second with a central downdraft, and a third with a weak central downdraft and secondary vortices rotating around the primary vortex" [*Lewellen*, 1976, p. 136]. This should probably be expanded to four types by counting two types with a low-level central updraft, depending upon whether or not a vortex breakdown occurs a short distance above the surface. The fifth type of central flow illustrated by *Davies-Jones* [1986] and included as in Figure 4a cannot be classified as a tornado.

4. "The maximum velocities will occur below the top of the ground boundary layer (~100 m above surface) and may be significantly higher than that predicted by a cyclostrophic balance" [*Lewellen*, 1976, p. 136]. This conclusion generated some controversy in 1976 but has been confirmed by subsequent numerical and laboratory simulations, along with theoretical analyses. As discussed in section 6, a very unique feature of the corner flow is its tendency to overexpand and achieve speeds that exceed those given by a straight conversion of available potential energy to kinetic energy.

5. "No existing model is adequate to provide a reliable theoretical estimate of the maximum wind velocities which may occur in a tornado" [*Lewellen*, 1976, p. 136]. The key word here is reliable. There have been a number of approaches used to estimate maximum velocities, but each is subject to a number of questionable assumptions. After reviewing the differing approaches to estimating tornado winds, *Davies-Jones* [1986] concludes that most scientists agree that wind speeds in even the strongest tornadoes do not exceed 110–130 m/s. I agree with this statement. Further, I think the technology is now at hand, where it should be possible to apply boundary conditions from a numerical storm simulation to a tornado domain simulation and obtain valid estimates of the probability of different levels of extreme winds occurring in different storms.

6. "The role of turbulence appears to be critical in determining both the maximum velocity and the detailed structure of a tornado" [*Lewellen*, 1976, p. 136]. I still believe this is true. I believe that growth and damping of turbulence is crucial to the details of the overexpansion in the corner flow, which has a major impact on the maximum velocity. I also expect turbulence to be critical to capping the core of the tornado, when an eye does not extend through the storm.

8.2. *Recommendations*

The current numerical capabilities should make the next few years a fertile time for expanding our knowledge of tornado dynamics. I offer the following partial list of research problems which might very usefully be completed in the next few years: (1) turbulent dynamics in the tornado, (2) mapping of wind field sensitivities to various inputs with a tornado model utilized as an inner nest of a numerical storm model, (3) unsteady behavior during start-up and decay, (4) prediction of maximum winds for observable inputs, (5) Design Basis Tornado for wind-engineering studies to replace *Fujita*'s [1977] model, (6) dispersion of debris by tornado wind field, (7) statistical model for winds at 10 m height exceeding 100 m/s, and (8) interaction of tornado winds with typical buildings.

This list is not intended to be all-inclusive, but to indicate the types of problems which might be solved in the near future given a modest level of effort. The first items in this list deal with obtaining a more complete understanding of the tornado wind field and what controls it, while the last items deal with providing useful wind-engineering information.

Acknowledgments. I want to thank R. I. Sykes and S. F. Parker for the LES calculations and graphics. The LES code which made these simulations possible was developed under ONR support with R. F. Abbey as technical monitor.

References

Abbey, R. F., Risk probabilities associated with tornado windspeeds, in *Proceedings, Symposium on Tornadoes: Assessment of*

Knowledge and Implications for Man, edited by R. E. Peterson, pp. 177–236, Texas Tech University, Lubbock, 1976.

Abbey, R. F., and T. T. Fujita, The DAPPLE Method for computing tornado hazard probabilities: Refinement and theoretical considerations, in *Preprints, 11th Conference on Severe Local Storms*, pp. 241–248, American Meteorological Society, Boston, Mass., 1979.

Anderson, C. E., The Barneveld tornado: A new type of tornadic storm in the form of a spiral mesolow, in *Preprints, 14th Conference on Severe Local Storms*, pp. 289–292, American Meteorological Society, Boston, Mass., 1985.

Andre, J. C., and M. Lesieur, Influence of helicity on the evolution of isotropic turbulence at high Reynolds number, *J. Fluid Mech.*, 81, 1987–2007, 1977.

Anthes, R. A., *Tropical Cyclones: Their Evolution, Structure and Effects*, Meteorol. Monogr. Ser., vol. 19, American Meteorological Society, Boston, Mass., 1982.

Baker, G. L., Boundary layers in laminar vortex flows, Ph.D. dissertation, Purdue Univ., West Lafayette, Indiana, 1981.

Barcilon, A. I., Vortex decay above a stationary boundary, *J. Fluid Mech.*, 27, 155–175, 1967.

Belcher, R. J., O. R. Burggraf, and K. Stewartson, On generalized-vortex boundary layers, *J. Fluid Mech.*, 52, 753–780, 1972.

Bellamy-Knights, P. G., An unsteady two-cell vortex solution of the Navier-Stokes equations, *J. Fluid Mech.*, 41, 673–687, 1970.

Bellamy-Knights, P. G., Unsteady multicellular viscous vortices, *J. Fluid Mech.*, 50, 1–16, 1971.

Bellamy-Knights, P. G., An axisymmetric boundary layer solution for an unsteady vortex above a plane, *Tellus*, 26, 318–324, 1974.

Bengtsson, L., and J. Lighthill (Eds.), *Intense Atmospheric Vortices*, Springer-Verlag, New York, 1982.

Benjamin, T. B., Theory of the vortex breakdown phenomenon, *J. Fluid Mech.*, 14, 593–629, 1962.

Bilbro, J. W., H. B. Jeffreys, J. W. Kaufman, and E. A. Weaver, Laser Doppler dust devil measurements, *NASA Tech. Note, D-8429*, 1977.

Bluestein, H. B., Wall clouds with eyes, *Mon. Weather Rev.*, 113, 1081–1085, 1985.

Bödewadt, U. T., Die Drehströmung über festem Grund, *Z. Angew Math. Mech.*, 20, 241–253, 1940.

Brandes, E. A., Tornadic mesocyclone finestructure and implications for tornadogenesis, in *Preprints, 11th Conference on Severe Local Storms*, pp. 549–556, American Meteorological Society, Boston, Mass., 1979.

Brandes, E. A., Fine structure of the Del City–Edmond tornado mesocirculation, *Mon. Weather Rev.*, 109, 635–647, 1981.

Brown, G. L., and J. M. Lopez, Axisymmetric vortex breakdown, 2, Physical mechanisms, *J. Fluid Mech.*, 221, 553–576, 1990.

Burgers, J. M., A mathematical model illustrating the theory of turbulence, *Adv. Appl. Mech.*, 1, 197–199, 1948.

Burggraf, O. R., and M. R. Foster, Continuation or breakdown in tornado-like vortices, *J. Fluid Mech.*, 80, 685–703, 1977.

Burggraf, O. R., K. Stewartson, and R. Belcher, Boundary layer induced by a potential vortex, *Phys. Fluids*, 14, 1821–1833, 1971.

Carrier, G. F., Swirling flow boundary layers, *J. Fluid Mech.*, 49, 133, 1971.

Carrier, G., and F. Fendell, Conditions for two-cell structure in severe atmospheric vortices, in *Preprints, 14th Conference on Severe Local Storms*, pp. 220–223, American Meteorological Society, Boston, Mass., 1985.

Carrier, G. F., P. Dergarabedian, and F. E. Fendell, Analytical studies on satellite detection of severe two-cell tornadoes, *NASA Tech. Note, CR-3127*, 1979.

Carrier, G. F., F. E. Fendell, and P. S. Feldman, Analysis of the near-ground wind field of a tornado vortex, in *Preprints, 15th Conference on Severe Local Storms*, pp. 331–334, American Meteorological Society, Boston, Mass., 1988.

Chaussee, D. S., Numerical solution of axisymmetric vortex formation normal to a solid boundary, Ph.D. dissertation, Iowa State Univ., Ames, 1972.

Chen, J. M., and R. G. Watts, A steady vortex driven by a heat line source, *Int. J. Heat Mass Transfer*, 22, 77–87, 1979.

Chi, J., Numerical analysis of turbulent end-wall boundary layers of intense vortices, *J. Fluid Mech.*, 82, 209–222, 1977.

Chi, S. W., and T. Costopolous, Wind loadings on ground structures in intense atmospheric vortices, 1, annual report, grant GK-41469, Natl. Sci. Found., Washington, D. C., 1975.

Chi, S. W., and W. J. Glowacki, Applicability of mixing length theory to turbulent boundary layers beneath intense vortices, *J. Appl. Mech.*, 41, 15–19, 1974.

Chi, S. W., and J. Jih, Numerical modeling of the three-dimensional flows in the ground boundary layer of a maintained axisymmetric vortex, *Tellus*, 26, 444–445, 1974.

Chi, S. W., S. J. Ying, and C. C. Chang, The ground turbulent boundary layer of a stationary tornado-like vortex, *Tellus*, 21, 693–700, 1969.

Church, C. R., J. T. Snow, G. L. Baker, and E. M. Agee, Characteristics of tornado-like vortices as a function of swirl ratio: A laboratory investigation, *J. Atmos. Sci.*, 36, 1755–1766, 1979.

Church, C. R., J. T. Snow, and J. Dessens, Intense atmospheric vortices associated with a 1000 MW fire, *Bull. Am. Meteorol. Soc.*, 61, 682–694, 1980.

Davies-Jones, R. P., Tornado dynamics, in *Thunderstorm Morphology and Dynamics*, 2nd ed., edited by E. Kessler, pp. 197–236, University of Oklahoma Press, Norman, 1986.

Davies-Jones, R., and D. Burgess, Test of helicity as a tornado forecast parameter, in *Preprints, 16th Conference on Severe Local Storms*, pp. 588–592, American Meteorological Society, Boston, Mass., 1990.

Davies-Jones, R. P., R. Rabin, and K. Brewster, A short-term forecast model for thunderstorm rotation, in *Proceedings, Nowcasting-II Symposium*, pp. 367–371, European Space Agency, Neuilly, France, 1984.

Deardorff, J. W., Numerical investigation of neutral and unstable planetary boundary layers, *J. Atmos. Sci.*, 29, 91–115, 1972.

Deissler, R. G., Models for some aspects of atmospheric vortices, *J. Atmos. Sci.*, 34, 1502–1517, 1977.

Dergarabedian, P., and F. E. Fendell, Parameters governing the generation of free vortices, *Phys. Fluids*, 10, 2293–2299, 1967.

Dergarabedian, P., and F. E. Fendell, One- and two-cell tornado structure and funnel-cloud shape, *J. Astronaut. Sci.*, 21, 26–31, 1973.

Eagleman, J. R., V. U. Muirhead, and N. Williams, Theoretical Compressible Flow Tornado Vortex Model, in *Thunderstorms, Tornadoes, and Building Damage*, pp. 101–131, D. C. Heath, Lexington, Mass., 1975.

Einstein, H. A., and H. Li, Steady vortex flow in a real fluid, *Houille Blanche*, 10, 483–496, 1955.

Ekman, V. W., On the influence of the Earth's rotation on ocean currents, *Ark. Math. Astron. Fys.*, 2, 1–52, 1905.

Escudier, M., Vortex breakdown in technology and nature—Theories, in *Introduction to Vortex Dynamics*, vol. 2, edited by R. Granger and J. F. Wendt, von Kármán Institute, Rhode Saint Genèse, Belgium, 1986.

Escudier, M. P., Vortex breakdown: Observations and explanations, *Prog. Aerosp. Sci.*, 25, 189–229, 1988.

Fiedler, B. H., Conditions for laminar flow in geophysical vortices, *J. Atmos. Sci.*, 46, 252–259, 1989.

Fiedler, B. H., and R. Rotunno, A theory for the maximum windspeed in tornado-like vortices, *J. Atmos. Sci.*, 43, 2328–2340, 1986.

Forbes, G. S., Three scales of motion associated with tornadoes, *Rep. NUREG/CR-0363*, U.S. Nucl. Regul. Comm., Washington, D. C., 1978.

Foster, M. R., and F. T. Smith, Stability of Long's vortex at large flow force, *J. Fluid Mech.*, 206, 405–432, 1989.

Franz, H. W., Die Zellstruktur stationarer Konvektionswirbel, *Beitr. Phys. Atmos.*, 42, 36–66, 1969.

Fujita, T. T., The Lubbock tornadoes: A study of suction spots, *Weatherwise*, 23, 160–173, 1970.

Fujita, T. T., Proposed characterization of tornadoes and hurricanes by area and intensity, SMRP Pap. 91, Dep. of Geophys. Sci., Univ. of Chicago, Chicago, Ill., 1971a.

Fujita, T. T., Proposed mechanisms of suction spots accompanied by tornadoes, SMRP Pap. 102, Dep. of Geophys. Sci., Univ. of Chicago, Chicago, Ill., 1971b.

Fujita, T. T., Tornado structure for engineering applications with design-basis tornado model (DBT-77), report, contract AT(49-24)-0239, Univ. of Chicago, Chicago, Ill., 1977.

Fujita, T. T., G. S. Forbes, T. A. Umenhofer, E. W. Pearl, and T. T. Tecson, Photogrammetric analysis of tornadoes, in *Proceedings, Symposium on Tornadoes: Assessment of Knowledge and Implications for Man*, edited by R. E. Peterson, pp. 43–88, Texas Tech University, Lubbock, 1976.

Fulks, J. R., On the mechanics of the tornado, NOAA Tech. Memo, ERLTM-NSSL 4, 33 pp., 1962.

Gall, R. L., Internal dynamics of tornado-like vortices, *J. Atmos. Sci.*, 39, 2721–2736, 1982.

Gall, R. L., A linear analysis of the multiple vortex phenomenon in simulated tornadoes, *J. Atmos. Sci.*, 40, 2010–2024, 1983.

Gall, R. L., Linear dynamics of the multiple-vortex phenomenon in tornadoes, *J. Atmos. Sci.*, 42, 761–772, 1985.

Gall, R. L., and D. O. Stalcy, Nonlinear barotropic instability in a tornado vortex, in *Abstracts, Third Conference on Atmospheric and Oceanic Waves and Stability*, p. 34, American Meteorological Society, Boston, Mass., 1981.

Golden, J. H., Waterspouts and tornadoes over south Florida, *Mon. Weather Rev.*, 99, 146–154, 1971.

Goldshtik, M. A., A paradoxical solution of the Navier-Stokes equations, *Prikl. Mat. Mekh.*, 24, 610–621, 1960.

Goldshtik, M. A., Viscous-flow paradoxes, *Annu. Rev. Fluid Mech.*, 22, 441–472, 1990.

Grazulis, T. P., and R. F. Abbey, Jr., 103 years of violent tornadoes ... patterns of serendipity, population, and mesoscale topography, in *Preprints, 13th Conference on Severe Local Storms*, pp. 124–127, American Meteorological Society, Boston, Mass., 1983.

Gutman, L. N., Theoretical model of a waterspout, *Bull. Acad. Sci., USSR, Geophys. Ser.*, Engl. Transl., 1, 87–103, 1957.

Hall, M. G., Vortex breakdown, *Annu. Rev. Fluid Mech.*, 4, 195–218, 1972.

Hennington, L., D. Zrnic, and D. Burgess, Doppler spectra of a maxi-tornado, in *Preprints, 12th Conference on Severe Local Storms*, pp. 433–436, American Meteorological Society, Boston, Mass., 1982.

Hess, G. D., K. T. Spillane, and R. S. Lourensz, Atmospheric vortices in shallow convection, *J. Appl. Meteorol.*, 27, 305–317, 1988.

Hopfinger, E. J., F. K. Browand, and Y. Gage, Turbulence and waves in a rotating tank, *J. Fluid Mech.*, 125, 505–534, 1982.

Howells, P., and R. K. Smith, Numerical simulations of tornado-like vortices, I, Vortex evolution, *Geophys. Astrophys. Fluid Dyn.*, 27, 253–284, 1983.

Howells, P. A. C., R. Rotunno, and R. K. Smith, A comparative study of atmospheric and laboratory-analogue numerical tornado-vortex models, *Q. J. R. Meteorol. Soc.*, 114, 801–822, 1988.

Hsu, C. T., and H. Tesfamariam, Turbulent modeling of a tornado boundary layer flow, Preprint ERI-76126, Eng. Res. Inst., Iowa State Univ., Ames, 1975.

Hsu, C. T., and H. Tesfamariam, Computer simulation of a tornado-like vortex boundary layer flow, Preprint ERI-76231, Iowa State Univ., Ames, 1976.

Idso, S. B., Tornado or dust devil: The enigma of desert whirlwinds, *Am. Sci.*, 62, 530–541, 1974.

Jischke, M. C., and M. Parang, Fluid dynamics of a tornado-like vortex flow, final report, NOAA grants N22-200-72(G) and 04-4--22-13, Univ. of Okla., Norman, 1975.

Johns, R. H., P. W. Leftwich, and J. M. Davies, Some wind and instability parameters associated with strong and violent tornadoes, this volume.

Kidd, G. H., Jr., and G. J. Farris, Potential flow adjacent to a stationary surface, *J. Appl. Mech.*, 35, 209–215, 1968.

Kuo, H. L., On the dynamics of convective atmospheric vortices, *J. Atmos. Sci.*, 23, 25–42, 1966.

Kuo, H. L., Note on the similarity solutions of the vortex equations in an unstably stratified atmosphere, *J. Atmos. Sci.*, 24, 95–97, 1967.

Kuo, H. L., Axisymmetric flows in the boundary layer of a maintained vortex, *J. Atmos. Sci.*, 28, 20–41, 1971.

Kuo, H. L., Vortex boundary layer under quadratic surface stress, *Boundary Layer Meteorol.*, 22, 151–169, 1982.

Leibovich, S., The structure of vortex breakdown, *Annu. Rev. Fluid Mech.*, 10, 221–246, 1978.

Leibovich, S., Vortex stability and breakdown: Survey and extension, *AIAA J.*, 22, 1192–1206, 1984.

Leibovich, S., and K. Stewartson, A sufficient condition for instability of columnar vortices, *J. Fluid Mech.*, 126, 335–356, 1983.

Leslie, F. W., Surface roughness effects on suction vortex formations: A laboratory simulation, *J. Atmos. Sci.*, 34, 1022–1027, 1977.

Leslie, L. M., and R. K. Smith, Numerical studies of tornado structure and genesis, in *Intense Atmospheric Vortices*, edited by L. Bengtsson and J. Lighthill, pp. 205–211, Springer-Verlag, New York, 1982.

Lewellen, W. S., A solution for three-dimensional vortex flows with strong circulation, *J. Fluid Mech.*, 14, 420–432, 1962.

Lewellen, W. S., Theoretical models of the tornado vortex, in *Proceedings, Symposium on Tornadoes: Assessment of Knowledge and Implications for Man*, edited by R. E. Peterson, pp. 107–143, Texas Tech University, Lubbock, 1976.

Lewellen, W. S., Influence of body forces on turbulent transport near a surface, *J. Appl. Math. Phys.*, 28, 825–834, 1977.

Lewellen, W. S., and Y. P. Sheng, Influence of surface conditions on tornado wind distributions, in *Preprints, 11th Conference on Severe Local Storms*, pp. 375–381, American Meteorological Society, Boston, Mass., 1979.

Lewellen, W. S., and Y. P. Sheng, Modeling tornado dynamics, Rep. NUREG/CR-2585, U. S. Nucl. Regul. Comm., Washington, D. C., 1980.

Lewellen, W. S., and M. E. Teske, Turbulent transport model of low-level winds in a tornado, in Preprints, *10th Conference on Severe Local Storms*, pp. 291–298, American Meteorological Society, Boston, Mass., 1977.

Lewellen, W. S., M. E. Teske, and Y. P. Sheng, Wind and pressure distribution in a tornado, in *Proceedings, Fifth International Conference on Wind Engineering*, pp. II-3-1–II-3-11, Colorado State University, Fort Collins, 1979.

Lilly, D. K., Tornado dynamics, Manuscript 69-117, Natl. Cent. for Atmos. Res., Boulder, Colo., 1969.

Lilly, D. K., Sources of rotation and energy in a tornado, in *Proceedings, Symposium on Tornadoes: Assessment of Knowledge and Implications for Man*, edited by R. E. Peterson, pp. 145–150, Texas Tech University, Lubbock, 1976.

Lilly, D. K., Dynamics of rotating thunderstorms, in *Mesoscale Meteorology—Theories, Observations, and Models*, edited by D. K. Lilly and T. Gal-Chen, pp. 531–544, D. Reidel, Norwell, Mass., 1983.

Lilly, D. K., The structure, energetics, and propagation of rotating convective storms, 1, Energy exchange with the mean flow, *J. Atmos. Sci.*, 43, 113–125, 1986a.

Lilly, D. K., The structure, energetics, and propagation of rotating convective storms, II, Helicity and storm stabilization, *J. Atmos. Sci.*, 43, 126–140, 1986b.

Long, R. R., Sources and sinks at the axis of a rotating liquid, *Q. J. Mech. Appl. Math.*, *9*, 385–393, 1956.

Long, R. R., Vortex motion in a viscous fluid, *J. Meteorol.*, *15*, 108–112, 1958.

Long, R. R., Tornadoes and dust whirls, *ONR Ser. Tech. Rep. 10*, Johns Hopkins Univ., Baltimore, Md., 1960.

Long, R. R., A vortex in an infinite viscous fluid, *J. Fluid Mech.*, *11*, 611–624, 1961.

Lopez, J. M., Axisymmetric vortex breakdown, 1, Confined swirling flow, *J. Fluid Mech.*, *221*, 533–552, 1990.

Lugt, H. J., *Vortex Flow in Nature and Technology*, 297 pp., Wiley-Interscience, New York, 1983.

Lugt, H. J., Vortex breakdown in atmospheric columnar vortices, *Bull. Am. Meteorol. Soc.*, *70*, 1526–1537, 1989.

Lundgren, T. S., Flow above the drain hole in a rotating vessel, *J. Fluid Mech.*, *155*, 381–412, 1985.

Lyons, W. A., and S. R. Pease, "Steam devils" over Lake Michigan during a January Arctic outbreak, *Mon. Weather Rev.*, *100*, 235–237, 1972.

MacGorman, D. R., Observations of lightning in tornadic storms, this volume.

Marchenko, A. S., Determination of maximum wind velocities in tornadoes and tropical cyclones (in Russian), *Meteorol. Hydrol.*, *5*, 11–16, 1961.

Maxworthy, T., Vorticity source for large scale dust devils and other comments on naturally occurring columnar vortices, *J. Atmos. Sci.*, *30*, 1717–1722, 1973.

Maxworthy, T., Geophysical vortices and vortex rings, in *Introduction to Vortex Dynamics*, vol. 1, edited by R. Granger and J. F. Wendt, von Karman Institute, Rhode Saint Genèse, Belgium, 1986.

Maxworthy, T., E. J. Hopfinger, and L. P. Redekopp, Wave motions on vortex cores, *J. Fluid Mech.*, *151*, 141–165, 1985.

McClellan, T. M., M. E. Akridge, and J. T. Snow, Streamlines and vortex lines in numerically-simulated tornado vortices, in *Preprints, 16th Conference on Severe Local Storms*, pp. 572–577, American Meteorological Society, Boston, Mass., 1990.

Mitsuita, Y., N. Monji, and H. Ishikawa, On the multiple structure of atmospheric vortices, *J. Geophys. Res.*, *92*, 14,827–14,831, 1987.

Monji, N., and W. Yunkuan, A laboratory investigation of characteristics of tornado-like vortices over various rough surfaces, *Acta Meteorol. Sinica*, *47*, 34–42, 1989.

Morton, B. R., The strength of vortex and swirling core flows, *J. Fluid Mech.*, *38*, 315–333, 1969.

Morton, B. R., The generation and decay of vorticity, *Geophys. Astrophys. Fluid Dyn.*, *28*, 277–308, 1984.

Mullen, J. B., and T. Maxworthy, A laboratory model of dust devil vortices, *Dyn. Atmos. Oceans*, *1*, 181–214, 1977.

Pauley, R. L., Laboratory measurements of axial pressures in two-celled tornado-like vortices, *J. Atmos. Sci.*, *46*, 3392–3399, 1989.

Pauley, R. L., C. R. Church, and J. T. Snow, Measurements of maximum surface pressure deficits in modeled atmospheric vortices, *J. Atmos. Sci.*, *39*, 368–377, 1982.

Paull, R., and A. F. Pillow, Conically similar viscous flows, 3, Characterization of axial causes in swirling flow and the one-parameter flow generated by a uniform half-line source of kinematic swirl angular momentum, *J. Fluid Mech.*, *155*, 359–379, 1985.

Phillips, W. R. C., The effusing core at the center of a vortex boundary layer, *Phys. Fluids*, *27*, 2215–2220, 1984.

Phillips, W. R. C., On vortex boundary layers, *Proc. R. Soc. London, Ser. A*, *400*, 253–261, 1985.

Phillips, W. R. C., and B. C. Khoo, The boundary layer beneath a Rankine-like vortex, *Proc. R. Soc. London, Ser. A*, *411*, 177–192, 1987.

Polifke, W., and L. Shtilman, The dynamics of helical decaying turbulence, *Phys. Fluids A*, *1*, 2025–2033, 1989.

Prahlad, T. S., Numerical solutions for boundary layers beneath a potential vortex, *Comput. Fluids*, *4*, 157–169, 1976.

Proctor, F. H., The dynamic and thermal structure of an evolving tornado, in *Preprints, 11th Conference on Severe Local Storms*, pp. 389–396, American Meteorological Society, Boston, Mass., 1979.

Rankine, W. J., *A Manual of Applied Mechanics*, Charles Griffin, London, 1882.

Rao, G. V., and W. H. Raymond, An investigation of the boundary layer of an atmospheric mesoscale vortex, in *Preprints, Ninth Conference on Severe Local Storms*, pp. 234–237, American Meteorological Society, Boston, Mass., 1975.

Rostek, W. F., and J. T. Snow, Surface roughness effects on tornado-like vortices, in *Preprints, 14th Conference on Severe Local Storms*, pp. 252–255, American Meteorological Society, Boston, Mass., 1985.

Rott, N., On the viscous core of a line vortex, *Z. Angew Math. Mech.*, *9*, 543–553, 1958.

Rott, N., Turbulent boundary layer development on the end walls of a vortex chamber, *Rep. ATN-62 (9202)-1*, Aerosp. Corp., El Segundo, Calif., 1962.

Rott, N., and W. S. Lewellen, Examples of boundary layers in rotating flows, in Recent Developments in Boundary Layer Research, *AGARDograph 97*, Advis. Group for Aeronaut. Res. and Dev., NATO, Brussels, 1965.

Rott, N., and W. S. Lewellen, Boundary layers and their interactions in rotating flows, *Prog. Aeronaut. Sci.*, *7*, 111–144, 1966.

Rotunno, R., Numerical simulation of a laboratory vortex, *J. Atmos. Sci.*, *34*, 1942–1956, 1977.

Rotunno, R., A note on the stability of a cylindrical vortex sheet, *J. Fluid Mech.*, *87*, 761–771, 1978.

Rotunno, R., A study in tornado-like vortex dynamics, *J. Atmos. Sci.*, *36*, 140–155, 1979.

Rotunno, R., Vorticity dynamics of a convective swirling boundary layer, *J. Fluid Mech.*, *97*, 623–640, 1980.

Rotunno, R., A numerical simulation of multiple vortices, in *Intense Atmospheric Vortices*, edited by L. Bengtsson and J. Lighthill, pp. 215–228, Springer-Verlag, New York, 1982.

Rotunno, R., An investigation of a three-dimensional asymmetric vortex, *J. Atmos. Sci.*, *41*, 283–298, 1984.

Rotunno, R., Tornadoes and tornadogenesis, in *Mesoscale Meteorology and Forecasting*, edited by P. S. Ray, pp. 414–436, American Meteorological Society, Boston, Mass., 1986.

Rotunno, R., Supercell thunderstorm modeling and theory, this volume.

Rotunno, R., and D. K. Lilly, A numerical model pertaining to the multiple vortex phenomenon, *Final Rep. NUREG/CR-1840*, U.S. Nucl. Regul. Comm., Washington, D. C., 1981.

Sarpkaya, T., Vortex breakdown in swirling conical flows, *AIAA J.*, *9*, 1792–1799, 1971.

Savino, J. M., and E. G. Keshock, Experimental profiles of velocity components and radial pressure distributions in a vortex contained in a short cylindrical chamber, *NASA Tech. Note, TN D-3072*, 1965.

Schmidt, H., and U. Schumann, Coherent structure of the convective boundary layer derived from large-eddy simulations, *J. Fluid Mech.*, *200*, 511–562, 1989.

Schwiderski, E. W., On the axisymmetric vortex flow over a flat surface, *Rep. TR-2210*, Nav. Weapons Lab., Washington, D. C., 1968.

Schwiesow, R. L., Horizontal velocity structure in waterspouts, *J. Appl. Meteorol.*, *20*, 349–360, 1981.

Schwiesow, R. L., R. E. Cupp, P. C. Sinclair, and R. F. Abbey, Waterspout velocity measurements by airborne Doppler lidar, *J. Appl. Meteorol.*, *20*, 341–348, 1981.

Simpson, J., B. R. Morton, M. C. McCumber, and R. S. Penc, Observations and mechanisms of GATE waterspouts, *J. Atmos. Sci.*, *43*, 753–782, 1986.

Smith, D. R., Effect of boundary conditions on numerically simulated tornado-like vortices, *J. Atmos. Sci.*, *44*, 648–656, 1987.

Smith, R. C., and P. Smith, Theoretical flow pattern of a vortex in the neighborhood of a solid boundary, *Tellus*, *17*, 213–219, 1965.

Smith, R. K., and P. Howells, Numerical simulations of tornado-like vortices, II, Two-cell vortices, *Geophys. Astrophys. Fluid Dyn.*, *27*, 285–298, 1983.

Smith, R. K., and L. M. Leslie, Tornadogenesis, *Q. J. R. Meteorol. Soc.*, *104*, 189–199, 1978.

Smith, R. K., and L. M. Leslie, A tornadogenesis in a rotating thunderstorm, *Q. J. R. Meteorol. Soc.*, *105*, 107–127, 1979.

Snow, J. T., On inertial instability as related to the multiple vortex phenomenon, *J. Atmos. Sci.*, *35*, 1660–1671, 1978.

Snow, J. T., A review of recent advances in tornado vortex dynamics, *Rev. Geophys.*, *20*, 953–964, 1982.

Snow, J. T., The tornado, *Sci. Am.*, *250*, 86–97, 1984.

Snow, J. T., Atmospheric columnar vortices, *Rev. Geophys.*, *25*, 371–385, 1987.

Snow, J. T., and D. E. Lund, Inertial motions in analytical vortex models, *J. Atmos. Sci.*, *46*, 3605–3610, 1989.

Snow, J. T., and R. L. Pauley, On the thermodynamic method for estimating maximum tornado windspeeds, *J. Clim. Appl. Meteorol.*, *23*, 1465–1468, 1984.

Staley, D. O., Effect of viscosity on inertial instability in a tornado vortex, *J. Atmos. Sci.*, *42*, 293–297, 1985.

Staley, D. O., and R. L. Gall, Barotropic instability in a tornado vortex, *J. Atmos. Sci.*, *36*, 973–981, 1979.

Staley, D. O., and R. L. Gall, Hydrodynamic instability of small eddies in a tornado vortex, *J. Atmos. Sci.*, *41*, 422–429, 1984.

Steffens, J. L., The effect of vorticity-profile shape on the instability of a two-dimensional vortex, *J. Atmos. Sci.*, *45*, 254–259, 1988.

Stewartson, K., The stability of swirling flows at large Reynolds number when subjected to disturbances with large azimuthal wavenumber, *Phys. Fluids*, *25*, 1953–1957, 1982.

Stewartson, K., and S. Leibovich, On the stability of a columnar vortex to disturbances with large azimuthal wavenumber: The lower neutral points, *J. Fluid Mech.*, *178*, 549, 1987.

Sullivan, R. D., A two-cell vortex solution of the Navier-Stokes equations, *J. Aerosp. Sci.*, *26*, 767–768, 1959.

Sychev, V. V., Viscous interaction of an unsteady vortex with a rigid surface, *Fluid Dyn.*, Engl. Transl., *24*, 548–559, 1989.

Sykes, R. I., and D. S. Henn, Large eddy simulation of turbulent sheared convection, *J. Atmos. Sci.*, *46*, 1106–1118, 1989.

Uttal, T., B. E. Martner, B. W. Orr, and R. M. Wakimoto, High resolution dual-Doppler radar measurements of a tornado, in *Preprints, 24th Conference on Radar Meteorology*, pp. 62–65, American Meteorological Society, Boston, Mass., 1989.

Wakimoto, R. M., and J. W. Wilson, Non-supercell tornadoes, *Mon. Weather Rev.*, *117*, 1113–1140, 1989.

Walko, R. L., Plausibility of substantial dry adiabatic subsidence in a tornado core, *J. Atmos. Sci.*, *45*, 2251–2267, 1988.

Walko, R. L., Generation of tornado-like vortices in nonaxisymmetric environments, in *Preprints, 16th Conference on Severe Local Storms*, pp. 583–587, American Meteorological Society, Boston, Mass., 1990.

Walko, R., and R. Gall, A two-dimensional linear stability analysis of the multiple vortex phenomenon, *J. Atmos. Sci.*, *41*, 3456–3471, 1984.

Walko, R. L., and R. L. Gall, Some effects of momentum diffusion on axisymmetric vortices, *J. Atmos. Sci.*, *43*, 2137–2148, 1986.

Wang, C. J., Exact solutions of the steady-state Navier-Stokes equations, *Annu. Rev. Fluid Mech.*, *23*, 159–178, 1991.

Whipple, A. B. C., *Storm*, vol. 4, *Planet Earth*, Time-Life Books, New York, 1982.

Wicker, L. W., and R. B. Wilhelmson, Numerical simulation of a tornado-like vortex in a high resolution three dimensional cloud model, in *Preprints, 16th Conference on Severe Local Storms*, pp. 263–268, American Meteorological Society, Boston, Mass., 1990.

Wilkens, E. M., and C. J. Diamond, Effects of convection cell geometry on simulated tornadoes, *J. Atmos. Sci.*, *44*, 140–147, 1987.

Wilson, T., and R. Rotunno, Numerical simulation of a laminar vortex flow, in *International Conference on Computer Methods and Experimental Measurements*, pp. 203–215, Springer-Verlag, New York, 1982.

Wilson, T., and R. Rotunno, A numerical simulation of a laminar end-wall vortex and boundary layer, *Phys. Fluids*, *29*, 3993–4005, 1986.

Wu, J., Exact vortex solutions of the Navier-Stokes equations, *Acta Aerodyn. Sinica*, *1*, 80–84, 1985.

Wylie, D. P., and C. E. Anderson, Cloud top anomalies associated with the Binger, OK tornado 22 May 1981, in *Preprints, 13th Conference on Severe Local Storms*, pp. 150–153, American Meteorological Society, Boston, Mass., 1983.

Yih, C.-S., F. Wu, A. K. Garg, and S. Leibovich, Conical vortices: A class of exact solutions of the Navier-Stokes equations, *Phys. Fluids*, *25*, 2147–2158, 1982.

Zrnic, D. S., D. W. Burgess, and L. Hennington, Doppler spectra and estimated windspeed of a violent tornado, *J. Clim. Appl. Meteorol.*, *24*, 1068–1081, 1985.

Numerical Simulation of Axisymmetric Tornadogenesis in Forced Convection

BRIAN H. FIEDLER

School of Meteorology, University of Oklahoma, Norman, Oklahoma 73019

1. THE ISSUE OF TORNADO INTENSITY

Intense tornadoes require a pressure deficit in the core greater than that which can be provided by latent heat release alone [*Snow and Pauley*, 1984]; in other words, tornado wind speeds exceed the thermodynamic speed limit. This paper reports on numerical studies of a simple axisymmetric convection model with a well-defined thermodynamic speed limit. A fixed field of buoyancy localized about the axis drives the convection. The results show that in the strong axial flow of an end-wall vortex, the pressure deficit can be 7 times that provided by the buoyancy, and the thermodynamic speed limit can be exceeded by a factor of 2. This dynamical enhancement of the pressure deficit and wind field could account for the intensity of some natural tornadoes.

Howells et al. [1988] review numerical modeling of tornadoes within an open domain with imposed upstream and downstream conditions. However, neither the open-domain numerical models nor the open-domain laboratory experiments driven by a fan are ideal for addressing questions of tornado intensity. Specifically, in the numerical simulations a certain environment outside the domain is implied by the solution. The question of whether or not the core pressure deficit at the outflow boundary is consistent with a realistic convective environment is left unanswered.

Recently, there have been several attempts to model tornadolike vortices in a closed domain. *Neitzel* [1988] calculated numerically the flow and concomitant vortex breakdown generated by rotation of one of the end walls of a closed circular cylinder. *Walko* [1988] modeled tornadogenesis in axisymmetric storms with prognostic atmospheric thermodynamics and thus was able to address the relationship between the tornado intensity and the environment. Solutions were found with hurricanelike tornadoes containing cloud-free cores in which the subsidence warming was augmenting the buoyancy in the core. The model used here is also a closed system, but, unlike that of *Walko* [1988],

The Tornado: Its Structure, Dynamics, Prediction, and Hazards.
Geophysical Monograph 79
Copyright 1993 by the American Geophysical Union.

the present model avoids the complications of modeling storm evolution because the buoyancy field is held fixed in each computation. The model therefore does not contain provisions for studying the process of subsidence warming in the core of tornadoes. However, the results indicate that subsidence warming is not a necessary process for intense tornadogenesis.

The experimental procedure will be to compare three integrations, one with no rotation, one with rotation and a free-slip lower boundary, and one with rotation and a no-slip lower boundary. Although the effect of the slip condition at the lower boundary has been investigated before, the closed-domain model used here allows for a more credible interpretation of the results.

2. THE MODEL

The incompressible, axisymmetric Navier-Stokes equations are applied in a cylinder rotating at constant rate Ω:

$$\frac{du}{dt} - \frac{v^2}{r} - 2\Omega v = -\frac{\partial p}{\partial r} + \nu\left(\frac{\partial}{\partial r}\frac{1}{r}\frac{\partial ru}{\partial r} + \frac{\partial^2 u}{\partial z^2}\right) \quad (1)$$

$$\frac{dv}{dt} + \frac{uv}{r} + 2\Omega u = \nu\left(\frac{\partial}{\partial r}\frac{1}{r}\frac{\partial rv}{\partial r} + \frac{\partial^2 v}{\partial z^2}\right) \quad (2)$$

$$\frac{dw}{dt} = -\frac{\partial p}{\partial z} + b + \nu\left(\frac{1}{r}\frac{\partial}{\partial r}r\frac{\partial w}{\partial r} + \frac{\partial^2 w}{\partial z^2}\right) \quad (3)$$

$$\frac{1}{r}\frac{\partial ru}{\partial r} + \frac{\partial w}{\partial z} = 0. \quad (4)$$

Here p is the pressure fluctuation from that of the motionless, equilibrium state divided by the constant density, b is a fixed external force (or buoyancy), and ν is the inverse Reynolds number. The equations have been made dimensionless using a convective velocity scale and the depth of the cylinder. The dimensionless depth is 1 and the radius is r_0. In all computations to be discussed here

$$b(r, z) = 1.264 e^{-20[r^2 + (z - 0.5)^2]}, \quad (5)$$

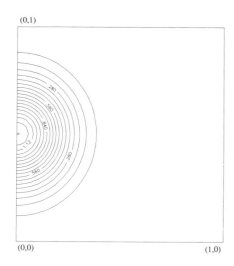

Fig. 1. Contours of the time-independent buoyancy field. View is $0 < r < 1$, $0 < z < 1$, as indicated.

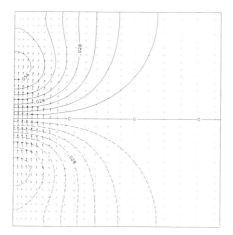

Fig. 2. The impulsive solution. Acceleration vectors (maximum 0.814) and contours of p. View is $0 < r < 1$, $0 < z < 1$.

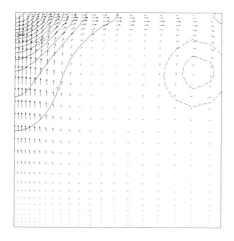

Fig. 3. Steady solution without rotation. Velocity vectors and contours of p; $u_{max} = 0.689$, $w_{max} = 0.849$, and $p_{max} = 0.408$. View is $0 < r < 1$, $0 < z < 1$.

which is plotted in Figure 1. The dimensionless convective available potential energy is

$$\int_0^1 b(0, z)\, dz = 0.5. \qquad (6)$$

The top and radial boundaries are always no-slip. The lower boundary can be either no-slip or free-slip.

In stretched coordinates $R(r)$ and $Z(z)$, the stream function vorticity formulation is

$$\frac{\partial v}{\partial t} + \frac{R'Z'}{r^2}\frac{\partial \psi}{\partial Z}\frac{\partial rv}{\partial R} - \frac{R'Z'}{r^2}\frac{\partial \psi}{\partial R}\frac{\partial rv}{\partial Z} = \nu\left(R'\frac{\partial}{\partial R}R'\frac{\partial v}{\partial R}\right.$$
$$\left. + \frac{R'}{r}\frac{\partial v}{\partial R} - \frac{v}{r^2} + Z'\frac{\partial}{\partial Z}Z'\frac{\partial v}{\partial Z}\right), \quad (7)$$

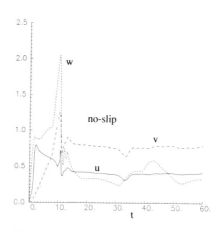

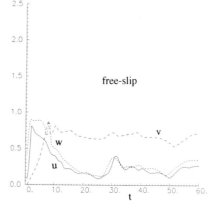

Fig. 4. Time history of u_{max}, v_{max}, and w_{max} in the free-slip and no-slip integrations with $\Omega = 0.2$ and $\nu = 0.0005$.

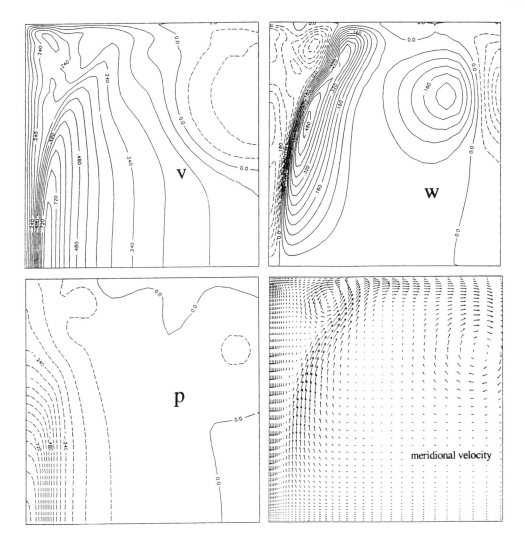

Fig. 5. Free-slip integration with $\Omega = 0.2$ and $\nu = 0.0005$ at $t = 10.5$; $u_{\max} = 0.395$, $v_{\max} = 0.815$, $w_{\max} = 0.495$, and $p_{\max} = -0.975$. View is $0 < r < 1$, $0 < z < 1$.

$$\frac{\partial \eta}{\partial t} + R'Z' \frac{\partial \psi}{\partial Z} \frac{\partial}{\partial R}\left(\frac{\eta}{r}\right) - R'Z' \frac{\partial \psi}{\partial R} \frac{\partial}{\partial Z}\left(\frac{\eta}{r}\right) = \frac{Z'}{r} \frac{\partial}{\partial Z} v^2$$

$$+ \nu \left(R' \frac{\partial}{\partial R} R' \frac{\partial \eta}{\partial R} + \frac{R'}{r} \frac{\partial \eta}{\partial R} - \frac{\eta}{r^2} + Z' \frac{\partial}{\partial Z} Z' \frac{\partial \eta}{\partial Z} \right)$$

$$- R' \frac{\partial b}{\partial R}, \tag{8}$$

and

$$r\eta = R' \frac{\partial}{\partial R} R' \frac{\partial \psi}{\partial R} - \frac{R'}{r} \frac{\partial \psi}{\partial R} + Z' \frac{\partial}{\partial Z} Z' \frac{\partial \psi}{\partial Z} \tag{9}$$

where

$$\eta \equiv (\partial u/\partial z) - (\partial w/\partial r), \tag{10}$$

$$u \equiv (1/r)(\partial \psi/\partial z), \tag{11}$$

$$w \equiv -(1/r)(\partial \psi/\partial r), \tag{12}$$

$$R' \equiv dR/dr, \tag{13}$$

and

$$Z' \equiv dZ/dz. \tag{14}$$

The following coordinate transformations are used in all the computations:

$$r = r_0 \frac{\lambda R + (2 - \lambda) R^3}{1 + R^2} \tag{15}$$

$$z = 0.5 \left[\frac{\tanh(\gamma Z - 0.5\gamma)}{\tanh(0.5\gamma)} + 1 \right] \tag{16}$$

where λ and γ are constants.

The model is integrated using leapfrog time differencing

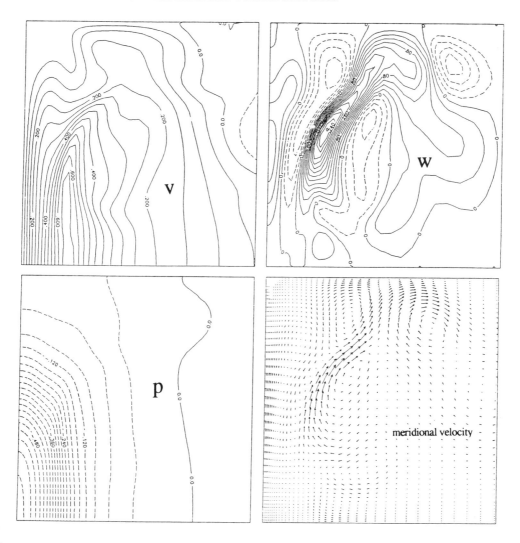

Fig. 6. Free-slip integration with $\Omega = 0.2$ and $\nu = 0.0005$ at $t = 40.0$; $u_{max} = 0.395$, $v_{max} = 0.682$, $w_{max} = 0.268$, and $p_{max} = -0.612$. View is $0 < r < 1$, $0 < z < 1$.

and an Arakawa Jacobian on a grid with constant spacing ΔR and ΔZ. The time step is maintained small enough to insure both stability and sufficient accuracy. Except for the boundary conditions, the additional body force b, and some additional minor exceptions, the model equations and the numerical scheme are those of *Wilson and Rotunno* [1986]. The additional minor differences are as follows. The model uses Dufort-Frankel time differencing for the diffusion terms. When the no-slip condition is applied, at the lower boundary, for example, the following modification of *Pearson* [1965] is used:

$$\eta(R, 0, t + \Delta t) = 0.5 \Big\{ \eta(R, 0, t)$$
$$+ \frac{Z'(0)}{2r(\Delta Z)^2} [-7\psi(R, 0, t) + 8\psi(R, \Delta Z, t)$$
$$- \psi(R, 2\Delta Z, t] \Big\} \quad (17)$$

in which the averaging with the previous time step was sufficient to maintain stability. Also, in the advection terms in (8),

$$\lim_{R \to 0} \frac{\eta(R, Z, t)}{r} = \frac{R'(0)}{2\Delta R} [4\eta(\Delta R, Z, t) - \eta(2\Delta R, Z, t)]. \quad (18)$$

3. Model Solutions

3.1. *No Rotation*

A linear, impulsive, solution with $\Omega = 0$ and $\nu = 0$ (Figure 2) is obtained by first solving

$$\frac{1}{r} \frac{\partial}{\partial r} r \frac{\partial p}{\partial r} + \frac{\partial^2 p}{\partial z^2} = \frac{\partial b}{\partial z} \quad (19)$$

for p and then solving for the accelerations in (1) and (3). The solution represents the response of the fluid at the instant b

is "turned on" in a motionless state. The vertical symmetry and the weak pressure fluctuation of the impulsive solution will be in contrast with the nonlinear solutions shown below.

A nearly steady state, nonrotating solution with $\nu = 0.0005$, $r_0 = 2$, and a no-slip lower boundary is achieved by $t = 10.0$ (Figure 3). The pressure in the nonlinear integrations is diagnosed by solving a Poisson equation derived from the primitive equations, with an additional constraint that $p(r_0, 0.5) = 0$. In steady, inviscid flow,

$$\frac{w^2(0, z)}{2} = p(0, 0) - p(0, z) + \int_0^z b(0, z')\, dz' \quad (20)$$

and the pressure difference between the two axial stagnation points would be

$$p(0, 1) - p(0, 0) = \int_0^1 b(0, z)\, dz = 0.5. \quad (21)$$

In the numerical integration the difference is 0.433, being slightly less as a result of the viscous terms. The maximum value of w in the domain is $w_{max} = 0.849$. The high pressure of the upper stagnation point extends significantly down into the buoyancy field and prevents w from achieving the upper limit of 1.0 allowed by the buoyancy term alone. This value of 1.0 will be referred to as the thermodynamic speed limit of the model. Note that in the formal inviscid limit, the kinetic energy in the domain could increase without bound and the thermodynamic speed limit would not be a bound on the wind speed. It turns out that the value of viscosity used in these calculations is enough to dissipate most of the kinetic energy of the fluid before it recycles into the storm so that the thermodynamic speed limit is especially relevant. Furthermore, we shall see in the rotating solutions that the kinetic energy production term wb is diminished because w is retarded or even negative near the axis.

Perhaps the most appropriate analog to isolated thunderstorms would be the case with $r_0 \to \infty$. When the above experiment is repeated with $r_0 = 1$, the result is $w_{max} = 0.851$. With both $r_0 = 1.5$ and $r_0 = 2.0$, the result is $w_{max} = 0.849$. We will use $r_0 = 2$ in the remaining integrations as an approximation for an infinite domain.

3.2. Rotation and a Free-Slip Lower Boundary

In the above steady, nonrotating solution the largest magnitudes in velocity and pressure are skewed toward the top of the domain, without much "severe weather" near the surface. The opposite is true for the rotating solutions. Time histories of the maximum wind speeds in the domain in integrations with $\Omega = 0.2$, $\nu = 0.0005$, and free-slip and no-slip lower boundary conditions are shown in Figure 4. A peak in v_{max} occurs in both integrations at $t = 10.5$.

The free-slip solution is shown at $t = 10.5$ in Figure 5 and at $t = 40$ in Figure 6. At $t = 10.5$ the vortex is narrower, permitting a large vertical velocity to occur outside the core where the buoyancy is relatively unimpeded by a vertical pressure gradient. The solutions at later times, of which $t = 40$ is an example, are more similar to the steady, hydrostatic, inviscid solutions (or balanced solutions) obtained from

$$\partial p/\partial r = (v^2/r) + 2\Omega v \quad (22)$$

and

$$\partial p/\partial z = b. \quad (23)$$

Figure 7 shows a balanced solution for $\Omega = 0.2$ with boundary conditions $v = 0$ and $p = 0$ at $z = 1$. Note that the pressure difference along the axis is 0.500. In contrast, in Figure 5 the pressure difference between the stagnation points is 0.82, with both inertial and viscous forces substantially contributing to the pressure drop. In Figure 6 the pressure difference between the stagnation points is 0.57, and the weak flow in the core confirms that it is nearly hydrostatic. Further integration shows that the free-slip case

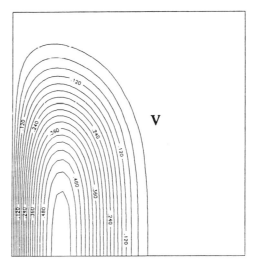

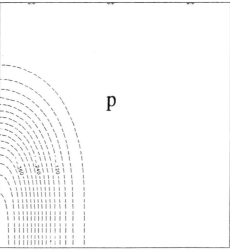

Fig. 7. Balanced solution with $\Omega = 0.2$; $v_{max} = 0.561$ and $p_{max} = -0.500$. View is $0 < r < 1$, $0 < z < 1$.

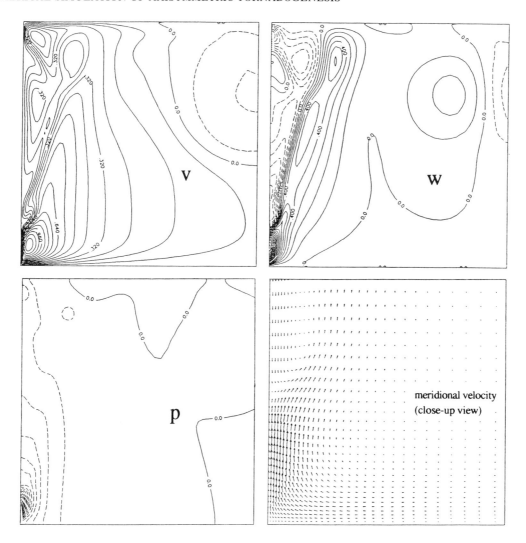

Fig. 8. No-slip integration with $\Omega = 0.2$ and $\nu = 0.0005$ at $t = 10.5$; $u_{max} = -0.568$, $v_{max} = 1.261$, $w_{max} = 2.042$, and $p_{max} = -3.617$. Close-up, corner view of meridional velocity is $0 < r < 0.3$, $0 < z < 0.3$; otherwise, view is $0 < r < 1$, $0 < z < 1$.

does not come to steady state, but rather goes through cycles with a period of about 140 in which wind speeds remain less than or equal to those shown here after $t = 30$. Although the core pressure deficit, particularly in the initial transient response, can be greater than that provided by buoyancy, the wind speed remains less than the thermodynamic speed limit.

3.3. Rotation and a No-Slip Lower Boundary

When the no-slip boundary condition replaces the free-slip boundary condition, significantly larger peak maximum wind speeds result. The peak maximum wind speeds at $t = 10.5$ are associated with an end-wall vortex (Figure 8), the dynamics of which are discussed by *Wilson and Rotunno* [1986] and *Fiedler and Rotunno* [1986]. The vortex at $t = 40$ (Figure 9) is similar to many previous published solutions for the no-slip boundary case [*Rotunno*, 1986; *Howells et al.*, 1988]. Although it has some qualitative similarity with the solution at $t = 10.5$, the corner region of the solution at $t = 40$ would be better described as a case of "overshoot" rather than as a genuine supercritical end-wall vortex. As such, the overshoot solutions provide only a modest amplification of the free-slip intensity, a result that was also found by *Howells et al.* [1988]. The dynamics in the solution at $t = 10.5$ is something fundamentally different from overshoot. The collision of the end-wall boundary layer results in an axial jet that allows for a core pressure deficit that is more than 7 times greater than that in the balanced vortex and more than 5 times greater than that in the overshoot solution. Although this transient supercritical vortex exists near its maximum intensity for only a few dimensionless time units, this corresponds to about 5 min when scaled to a thunderstorm, and air cycles through the core about

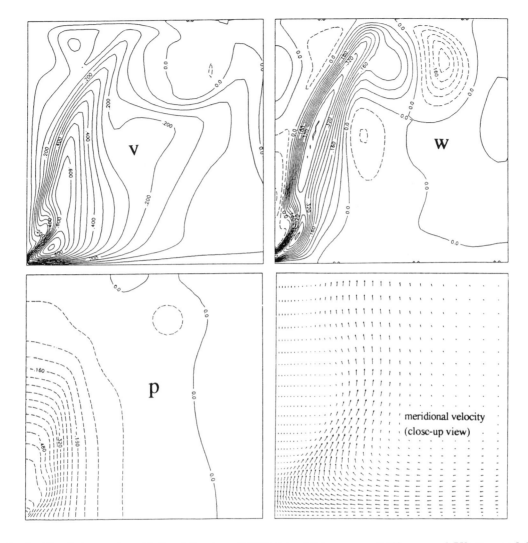

Fig. 9. No-slip integration with $\Omega = 0.2$ and $\nu = 0.0005$ at $t = 40.0$; $u_{max} = -0.402$, $v_{max} = 0.759$, $w_{max} = 0.488$, and $p_{max} = -0.696$. Views as in Figure 8.

20 times during this phase. Perhaps with other parameters the vortex breakdown could remain in steady state suspension away from the lower boundary, which could allow for a more impressive, long-lasting supercritical vortex. Nevertheless, the dynamics proposed by *Fiedler and Rotunno* [1986] to be a possible cause of intense tornadoes has been captured, albeit only momentarily.

4. Implications for Tornadoes

Table 1 shows the results for the peak wind speeds in the free-slip and no-slip experiments and in the three control experiments. The results are in accord with those of *Fiedler and Rotunno* [1986]. They deduced from laboratory experiments and theoretical analysis that v_{max} in supercritical

TABLE 1. The Peak Maximum Wind Speeds and Pressure Fluctuations

Experiment	t	u_{max}	v_{max}	w_{max}	p_{max}
Impulse ($\nu = 0$, $\Omega = 0$, free-slip)	0	0	0	0	0.118
Balanced ($\nu = 0$, $\Omega = 0.2$, free-slip)	steady	0	0.561	0	−0.500
No rotation ($\nu = 0.0005$, no-slip)	steady	0.689	0	0.849	0.408
Free-slip ($\nu = 0.0005$, $\Omega = 0.2$)	10.5	0.229	0.815	0.495	−0.975
No-slip ($\nu = 0.0005$, $\Omega = 0.2$)	10.5	−0.568	1.261	2.042	−3.62

tornadoes could be 1.7 times that in the downstream subcritical vortex. Although an exact height of transition from supercritical to subcritical flow is difficult to identify in Figure 8, this ratio can be estimated to be $1.261/0.7 \cong 1.7$. They also imply that the dynamics past the breakdown point would be essentially that of the free-slip solution, which is nearly the case here. The ratio of v_{max} in the peak no-slip solution to v_{max} in the peak free-slip solution is 1.55. Another prediction was that $w_{max} \cong 2v_{max}$ in the most intense tornadoes; in the most intense vortex found in these simulations, $w_{max} = 1.62 v_{max}$.

These results add credibility to the hypotheses put forward by *Fiedler and Rotunno* [1986] that some instances of the most intense tornadoes could be supercritical end-wall vortices with axial jets in the core. *Fiedler and Rotunno* [1986] used 3800 m^2 s^{-2} as an approximate upper limit to convective available potential energy in the atmosphere, which implies a thermodynamic speed limit of 87.2 m s^{-1}. Using this speed limit to dimensionalize the no-slip result implies that the most intense tornado could have $u_{max} = 50$ m s^{-1}, $v_{max} = 110$ m s^{-1}, $w_{max} = 178$ m s^{-1}, and a core pressure deficit of 330 mbar.

Acknowledgments. This work was supported by the National Severe Storms Laboratory and the University of Oklahoma.

REFERENCES

Fiedler, B. H., and R. Rotunno, A theory for the maximum windspeeds in tornado-like vortices, *J. Atmos. Sci., 43*, 2328–2340, 1986.

Howells, P. A. C., R. Rotunno, and R. K. Smith, A comparative study of atmospheric and laboratory-analogue numerical tornado-vortex models, *Q. J. R. Meteorol. Soc., 114*, 801–822, 1988.

Neitzel, G. P., Streak-line motion during steady and unsteady axisymmetric vortex breakdown, *Phys. Fluids, 31*, 958–960, 1988.

Pearson, C. E., A computational method for viscous flow problems, *J. Fluid Mech., 21*, 611–622, 1965.

Rotunno, R., Tornadoes and tornadogenesis, in *Mesoscale Meteorology and Forecasting*, edited by P. S. Ray, pp. 414–436, American Meteorological Society, Boston, Mass., 1986.

Snow, J. T., and R. L. Pauley, On the thermodynamic method for estimating maximum tornado windspeeds, *J. Clim. Appl. Meteorol., 23*, 1465–1468, 1984.

Walko, R. L., Plausibility of substantial dry adiabatic subsidence in a tornado core, *J. Atmos. Sci., 45*, 2251–2267, 1988.

Wilson, T., and R. Rotunno, Numerical simulation of a laminar end-wall vortex and boundary layer, *Phys. Fluids, 29*, 3993–4005, 1986.

Numerical Simulation of Tornadolike Vortices in Asymmetric Flow

R. JEFFREY TRAPP AND BRIAN H. FIEDLER

School of Meteorology, University of Oklahoma, Norman, Oklahoma 73019

1. INTRODUCTION

Most current knowledge of the dynamics of tornadolike vortices has been obtained from axisymmetric models with initial vertical vorticity. One question which continues to elude researchers is how tornadoes, which appear at least locally axisymmetric, are born out of nonaxisymmetric ambient flow with initial horizontal vorticity only.

A three-dimensional numerical model will be used to investigate this question. The work of *De Siervi et al.* [1982] (hereinafter referred to as DS) on the existence of the inlet (or ground) vortices associated with jet engines serves as the template for our numerical experiments. DS showed that an inlet vortex may form as a loop of vorticity transverse to the advecting flow, is drawn into a horizontally oriented inlet facing the flow, and then is stretched and deformed. This stretching and deformation process could result in a pair of counterrotating vortices, but only a single vortex is usually visualized. Indeed, as illustrated by DS, the upper vortex legs become spread out and diffuse in comparison with the lower legs which are concentrated about a line and which extend downward toward a point on the ground beneath; the latter is the inlet vortex. The "spin-up" of this vortex might be explained by the secondary flow attendant with rotation just above a viscous boundary. Because only one vortex appears even though the net circulation around the inlet is zero, and since the vortex is produced through the tilting and stretching of vortex lines, we believe that the formation of inlet vortices may be similar to many, but not all, instances of tornadogenesis.

Here we present results of the first stage of this tornado modeling approach. Our discussion will focus on the evolution and structure of a vertical vortex and its image, which are generated in laminar, nonaxisymmetric flow with no initial vertical vorticity. The model employed represents an idealized version of the DS experiments.

The Tornado: Its Structure, Dynamics, Prediction, and Hazards.
Geophysical Monograph 79
Copyright 1993 by the American Geophysical Union.

2. EXPERIMENTAL PROCEDURE

2.1. Physical Model

The physical model is meant to be representative of both the region beneath a thunderstorm and the DS experiments (Figure 1). A fixed shear flow (Figure 2) enters the cubic domain of unit dimension through one wall. The vorticity associated with this shear is horizontal and transverse to the inflow velocity.

Fluid is extracted from the cube through an exhaust region along the top boundary. To establish the viability of this procedure and for reasons of computational efficiency, we model only half of a symmetrical domain and place a semicircular exhaust region at the edge of the top boundary, adjacent to a vertical symmetry plane. The lifting of loops of horizontal vorticity yields two symmetrical vortices, only one of which appears within the modeled domain.

2.2. Numerical Model

The Navier-Stokes equations for compressible flow of constant viscosity are numerically integrated, using an explicit scheme with fourth-order accurate advection terms. At the initial time the dependent variables consisting of perturbation pressure (p) and the three momentum components (ρu, ρv, ρw) all have zero value.

Boundary conditions are shown in Figure 1. The lower boundary ($z = 0$) is either free-slip ($w = 0$; $\partial u/\partial z = \partial v/\partial z = 0$) or no-slip ($u = v = w = 0$). At the inflow boundary ($x = 1$), $v = w = 0$, and $u = -U(z)$, where

$$U(z) = z/0.5 \quad 0 \le z < 0.5$$

$$U(z) = 1 \quad \text{elsewhere}$$

Free-slip conditions for u and v are applied at the top boundary, while the exhaust condition on w is specified by

$$w = pg(x, y)$$

where

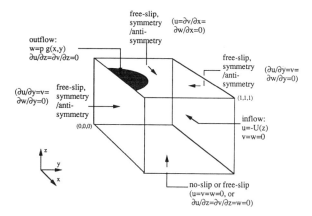

Fig. 1. Boundary conditions and schematic of physical domain.

$$g(x, y) = \exp\left\{-\left[\left(\frac{x - x_0}{r}\right)^2 + \left(\frac{y - y_0}{r}\right)^2\right]\right\}$$

and $x_0 = 0.5$, $y_0 = 0$, and $r = (0.05)^{1/2}$. On the remaining boundaries ($x = 0$; $y = 0$ and $y = 1$), free-slip, symmetry/antisymmetry conditions for u, v, and w are applied. Hence at the symmetry plane ($y = 0$), for example, $v = \partial u/\partial y = \partial w/\partial y = 0$, implying that an image solution lies across this boundary (as well as across the other free-slip boundaries).

A few aspects of the top boundary condition deserve further explanation. First, the exhaust region is positioned midway between the $x = 0$ and the $x = 1$ boundaries to minimize any interference of the boundaries which could be deleterious to the development of the vortex. Also, the e-folding radius of $r = (0.05)^{1/2}$ is the smallest possible value that affords numerical stability yet still provides for a considerable exhaust velocity. The exhaust is driven through this condition, then, as follows. The entering flow causes

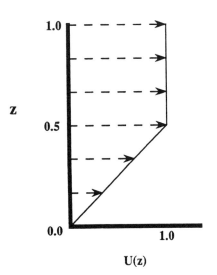

Fig. 2. Inflow velocity profile.

convergence and a pressure rise in the domain which, when felt at the top boundary, requires fluid to be evacuated. As the fluid is then released via the exhaust region, convergence and pressure are reduced, and the volume flow rates of the inflow and exhaust tend toward equality.

A finite difference grid consisting of $30 \times 30 \times 30$ points is used in all the cases. For experiments involving higher Reynolds number flow ($Re \geq 500$), the resolution is locally enhanced near the symmetry wall. Such enhancement is accomplished by the following coordinate transformation in the y direction:

$$Y = y \exp(sy - s),$$

where Y replaces y as the new nondimensional coordinate in that direction, and s is the stretching constant, assigned a value of 0.75.

The modeled flow has a Mach number of approximately 0.3. The Mach number is defined as

$$M = u_{\max}/c,$$

where c is the sound speed, and $u_{\max}$ is the magnitude of the maximum velocity found in the domain. Also, the Reynolds number here is

$$Re = VL/\nu,$$

where V is the reference velocity of unit value (the maximum inflow velocity), L is the unit length scale, and ν is the viscosity of air. For scaling purposes a dimensional velocity of 10 m s^{-1}, a length of 1 km, and a density of 1 kg m^{-3} have been assumed.

3. Results

In experiment N500 we assign $Re = 500$ and choose a no-slip lower boundary. A time series of maximum values for this integration is given in Figure 3. Here "low-level" is used to denote those levels below $z = 0.5$. The evolution of the flow changes significantly at $t = 8$. Notice the increase in the maximum low-level convergence out to this time. Initially, fluid moves in the negative x direction toward the $x = 0$ boundary. Fluid which impacts this boundary, or "stagnation wall," is forced to turn parallel to the wall. A downward moving branch of the flow spreads horizontally as it meets the lower surface and then converges with the inflow. The zone of maximum convergence propagates in the positive x direction out to $x \sim 0.75$, then stops. It should be noted that such a convergence zone is similar to a thunderstorm outflow boundary, a feature believed by some to be important, if not essential, for tornadogenesis in the atmosphere.

Referring back to Figure 3, now consider the rise in maximum low-level vertical vorticity, ζ_l, prior to $t = 8$. During this time, horizontal vorticity present in the inflow and generated by pressure gradients at the no-slip boundary is tilted into the vertical. Apparently, the horizontal gradients in the vertical velocity, w, responsible for such tilting

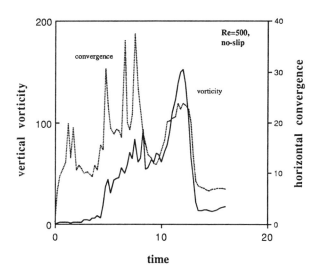

Fig. 3. Experiment N500: time series of peak values of low-level vertical vorticity, ζ_l, and low-level horizontal convergence. The values are dimensionless.

are due both to the rising motion in the convergence zone and to the exhaust. Despite the increasing values of ζ_l, a well-defined circulatory pattern in the horizontal velocity field is not evident prior to $t = 8$.

Rapid changes occur in the flow after $t = 8$. Low-level cyclonic circulation appears at $t = 9$ in an area previously occupied by large horizontal shear. Coincident with this developing vortex is an updraft region of nearly the same size and shape, which extends to the exhaust above. The vortex becomes well-defined by $t \sim 10$, as can be seen in Figure 4. Local axisymmetry is clearly shown, but also note an intriguing vertical velocity field. Outside of a vertical velocity minima are two maxima. Comparison of the velocity field in Figure 4 with the three-dimensional vorticity field at the same time and level (Figure 5) shows that there is a high degree of upward vertical tilting of horizontal vorticity where the horizontal gradient of w is large and positive. Moreover, vertical stretching of tilted vorticity is occurring in positive vertical gradients of w. Note further, in Figure 5, the convergence of horizontal vortex lines into a region of

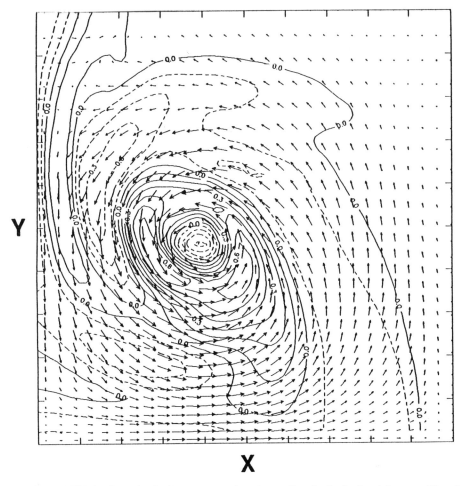

Fig. 4. Experiment N500: horizontal velocity vectors and contours of vertical velocity at time $t = 10$ and level $z = 0.03$. The magnitude of the maximum vector is 1.8, and the vertical velocity contour interval is 0.1.

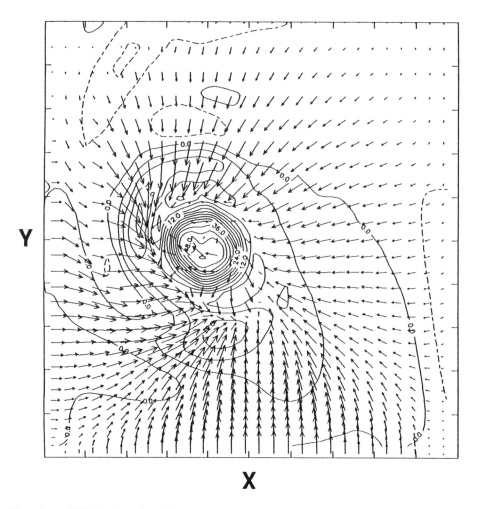

Fig. 5. Experiment N500: horizontal vorticity vectors and contours of vertical vorticity at time $t = 10$ and level $z = 0.03$. The magnitude of the maximum vector is 30.1, and the vertical vorticity contour interval is 4.0.

maximum vertical vorticity. This reflects well the three-dimensional nondivergence of vorticity just above a no-slip boundary.

Horizontal sections of the velocity field taken at $t = 10$ (Figure 6) show the three-dimensional structure. A slanted columnar vortex is quite evident. Over time the vortex straightens as its lower and mid sections move toward the symmetry plane. The low-level vortex reaches its peak intensity ($\zeta_l = 152$ units at $t = 12$) while very near the wall (vortex center at $x \sim 0.75$, $y \sim 0.1$). Within the next 1–2 time units, however, ζ_l decreases dramatically (see Figure 3), and by $t = 13$ the vortex can no longer be visualized. One reason for such decay could be cross diffusion with the image vortex across the symmetry plane.

In N500 the maximum low-level convergence rises after the vortex forms because of radial inflow, beneath $z = 0.1$, toward the vortex core. An imbalance between the pressure gradient and centrifugal forces within this viscous boundary layer results in a net inward force which drives the inflow. In some tornadolike vortices that form in axisymmetric numerical models for certain values of swirl ratio [*Davies-Jones*, 1973], radially converging flow in the viscous boundary layer erupts to produce an axial jet in the core of the vortex [*Howells et al.*, 1988; *Wilson and Rotunno*, 1986]. Here $w_{max} \approx v_{max}/3$, where w_{max} is the maximum vertical velocity within the core of the low-level vortex, and v_{max} is the effective maximum azimuthal velocity. In intense single-celled tornadoes, $w_{max} \approx 2v_{max}$ is anticipated by *Fiedler and Rotunno* [1986]. For such an axial jet to occur in our model, we believe that the Reynolds number of the flow must be increased by 1–2 orders of magnitude (which will in turn decrease the depth of the boundary layer), that the exhaust radius must be decreased, and that the vortex and boundary layer must be resolved better.

With the preceding discussion in mind, it is now useful to introduce results of experiments involving lower Reynolds number flow. Though not physically significant, preliminary cases with $Re < 100$ were performed. A vortex did not form

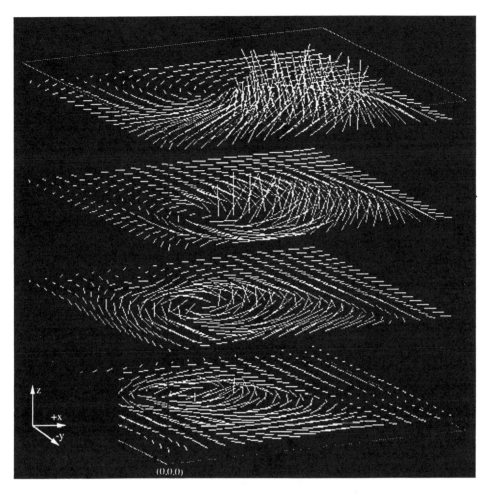

Fig. 6. Experiment N500: three-dimensional velocity vectors at $t = 10$. The levels are $z = 0.05$, 0.35, 0.65, and 0.95.

in this flow. However, small amounts of vertical vorticity were located near the top boundary. An experiment with $Re = 200$ and a no-slip lower boundary (N200) did yield a vortex which in fact was similar to that of N500 in terms of its local axisymmetry and vertical structure. The evolution of N200 progressed very rapidly, with the peak ζ_l of 25 units occurring at $t \sim 7$. Also, the maximum wind speed of about 1 unit associated with that peak was comparatively weak: in N500 the horizontal velocity at $z = 0.03$ was in excess of 3 units at the time of the peak ζ_l (of 152 units).

4. Summary

Our model results show that a vertical columnar vortex and its image can arise readily from the low-level horizontal vorticity present in the fixed shear inflow and generated by pressure gradients at the no-slip boundary. The horizontal vortex lines were tilted into the vertical by sharp gradients within the updraft at low levels, organized into a locally axisymmetric state, and stretched. The resultant vortex was initially inclined toward the symmetry plane, with the lowest position of the vortex near the center of the domain. The vortex did straighten, however, as its lower and mid sections moved toward the symmetry wall. In this position and location the vortex suffered a rapid drop in intensity, apparently through cross diffusion with the image vortex.

During the formation of the vortex the low-level convergence increased because of radial inflow into the core. A net inward force owing to an imbalance between the pressure gradient and the centrifugal forces within the viscous boundary layer was responsible for this inflow. The converging flow was not strong enough, however, to produce the axial jet seen in some end-wall vortices. The trend shown in our experimentation with the Reynolds number of the flow was that of an increase in such converging flow with increasing Re and hence with decreasing boundary layer depth.

The continuation of this research will first involve the replacement of the stagnation wall, even though it aids the formation of the gust front feature, with an open boundary.

In hindsight it has been recognized that the wall lacks a physical analog within the region beneath a thunderstorm (as well as in the DS model). Future experiments will decide whether a gust front needs to be explicitly modeled. Next, the symmetry plane will be removed so that, instead of studying a vertical vortex pair, we may investigate why one sign of vertical vorticity is favored and how it results from horizontal vorticity. Many studies concerning the origin of storm rotation [e.g., *Davies-Jones*, 1984] have discussed the importance of storm-relative environmental winds, which veer with height, and of the associated streamwise component of vorticity in the low-level inflow. This perhaps implies the need here for such a variation in the inflow winds. Note, however, that the vorticity associated with the upstream flow in the DS experiments was entirely crosswise, but the symmetry breaking mechanism leading to the formation of a single intense vortex is unclear. Finally, subsequent work will involve higher Reynolds number flow on a finer grid. This will be accomplished efficiently through a technique known as dynamic grid adaption in which grid points continually flow from areas of little interest to, in this case, the area surrounding the vortex.

Acknowledgments. The authors wish to thank Robert Davies-Jones and the two anonymous reviewers for many helpful comments and suggestions. This research was funding in part by the Center for the Analysis and Prediction of Storms and by the National Science Foundation, grant ATM-9002391.

REFERENCES

Davies-Jones, R. P., The dependence of core radius on swirl ratio in a tornado simulator, *J. Atmos. Sci.*, *30*, 1427–1430, 1973.

Davies-Jones, R. P., Streamwise vorticity: The origin of updraft rotation in supercell storms, *J. Atmos. Sci.*, *41*, 2991–3006, 1984.

De Siervi, F., H. C. Viguier, E. M. Greitzer, and C. S. Tan, Mechanisms of inlet-vortex formation, *J. Fluid Mech.*, *124*, 173–207, 1982.

Fiedler, B. H., and R. Rotunno, A theory for the maximum windspeeds in tornado-like vortices, *J. Atmos. Sci.*, *43*, 2328–2340, 1986.

Howells, P. A. C., R. Rotunno, and R. K. Smith, A comparative study of atmospheric and laboratory-analogue numerical tornado-vortex models, *Q. J. R. Meteorol. Soc.*, *114*, 801–822, 1988.

Wilson, T., and R. Rotunno, Numerical simulation of a laminar end-wall vortex and boundary layer, *Phys. Fluids*, *29*, 3993–4005, 1986.

Discussion

LOU WICKER, SESSION CHAIR

University of Illinois

PAPER A1

Presenter, Steve Lewellen, Drexel University [*Lewellen*, this volume, Tornado vortex theory]

(Joe Golden, National Oceanic and Atmospheric Administration, Office of the Chief Scientist) Did you give a number of 125 m s^{-1} for what you believe the models are presently predicting for probable wind speeds in tornadoes?

(Lewellen) That is the number I gave. I can't say that I got it from all of the modelers.

(Golden) I will say something about that in my paper [paper F1]. Second question: That intriguing three-dimensional conceptual model of the flow in a tornado indicated subsidence in the core all the way to the surface. Could you comment on the possibility that (a) the axial flow may depend upon the life cycle of the tornado and (b) there may be one or more axial stagnation points.

(Lewellen) The core flow is very unsteady. There are large variations from minute to minute and stagnation points at different points along the axis. That was why I showed the detailed results from the large-eddy simulation. The features don't remain fixed; they change on a very fast time scale.

(Brian Fiedler, University of Oklahoma) Your 3-D experiments, although very impressive, seem to be overly preoccupied with the radial equation of motion, not the vertical equation of motion. The design of the experiment seems to imply that large-scale swirl, radial convergence, and turbulent diffusion in the vicinity of the core are the only factors that limit the wind speeds in tornadoes. Could you comment on how your model is coupled to the storm that should be above it and how the buoyancy in a storm also is related to the maximum wind speeds that could be achieved?

(Lewellen) I'm not quite sure of the question. I agree that all of the factors you mention are in fact involved. The vertical acceleration certainly is as important as the radial acceleration. Turbulence is one of the uncertainties that still remain to be resolved as a goal of research. So I'm not quite sure where your differences with the model really come in.

(Fiedler) I'll answer my question during my talk [paper A4].

PAPER A4

Presenter, Brian Fiedler, University of Oklahoma [*Fiedler*, this volume, Numerical simulation of axisymmetric tornadogenesis in forced convection]

(Bob Davies-Jones, National Severe Storms Laboratory) How appropriate is the use of a no-slip boundary condition in a numerical model with a turbulent boundary layer? It seems that there should be partial slip at the ground.

(Fiedler) The real atmosphere does not have partial slip; it has no-slip and varying viscosity.

(Davies-Jones) But it has no-slip only in the lowest millimeters.

(Fiedler) You are throwing out the baby with the bath water if you use a surface layer parameterization and think that the atmosphere works in this way. There is an important mass flux in the boundary layer that is not represented by a surface layer parameterization. It might be better to use a sophisticated turbulence model with the no-slip boundary condition.

(Davies-Jones) If you did that, would it lower the extreme wind speeds that you are getting?

(Fiedler) I find these results to be independent of Reynolds number above 2000 or so. The curves of wind speed versus Re flatten out, and wind speeds do not exceed twice the thermodynamic speed limit. So I find that the results are pretty much independent of diffusion.

(Bob Walko, Colorado State University) I would like to disagree with your evaluation of the way that the surface layer is related to intensification. I found, in some results that I presented at the 15th Conference on Severe Local

Storms, that the intensification factor is very highly dependent on the type of surface layer parameterization that is used. With constant diffusion with height, which is similar to the low-Re simulations that you are doing and also laboratory tornado models, the intensification is indeed very strong. However, in more realistic models in which the diffusion coefficient increases linearly with height in the friction layer, there is much weaker intensification than a factor of 2.

(Fiedler) Right, but again you are using a parameterization that is more appropriate for the fair-wind boundary layer and does not include the effects of helicity on the turbulence. So I think that both of our turbulence parameterizations are equally as bad. More turbulent laboratory experiments are needed.

(Walko) I would agree that both models are bad, but my model showed a sensitivity to the parameterization. I think that you have to take the factor of 2 intensification with a grain of salt because it is so intense and sensitive to the surface layer parameterization.

(Fiedler) Yes, we can take it with a grain of salt, but if I can pull a rabbit out of the hat here, Joe, would you like to comment on waterspouts? I don't know what the CAPE [convective available potential energy] was in the Florida Keys at the time of this 90 m s^{-1} waterspout. I don't know what the thermodynamic speed limit was on that day, but $\cdots$

(Joe Golden, National Oceanic and Atmospheric Administration, Office of the Chief Scientist) The [operational] soundings do not give much information concerning the environment of Florida Keys waterspouts. Irrespective of whether there are many small, weak waterspouts or one very intense one, as on this day, the Key West soundings are all very similar. They all show very light, east to northeast flow and conditionally and convectively unstable stratification. We were able to define the waterspout environment better on the few occasions that we were able to get aircraft data in the subcloud layer. I found that the subcloud layer was nearly adiabatic and, in some cases, was superadiabatic in the immediate vicinity of waterspouts. An environmental sounding needs to be within a few hundred meters of the portion of the cloud line that produces the vortex.

Paper A5

Presenter, Jeff Trapp, University of Oklahoma [*Trapp and Fiedler*, this volume, Numerical simulation of tornadolike vortices in asymmetric flow]

(Joe Golden, National Oceanic and Atmospheric Administration, Office of the Chief Scientist) It appears to me that the nature of the underlying surface is almost incidental to the modeling results. Does it make any difference whether the lower boundary is solid or water, or whether it is a heat source?

(Trapp) I'm sure that it does. We decided to keep our experiment simple, i.e., incompressible, neutral stratification, constant viscosity, no-slip or free-slip boundaries. As time goes on and we understand the cause and effect in the simple formulation, we can start becoming more elaborate with more complicated parameterizations.

Supercell Thunderstorm Modeling and Theory

RICHARD ROTUNNO

National Center for Atmospheric Research, Boulder, Colorado 80307

1. INTRODUCTION

Tornadoes occur in thunderstorms. *Ferrel* [1889] theorized that tornadoes form when the thunderstorm updraft encounters a preexisting "gyratory" wind field. Only lately has it been found that tornadoes/waterspouts can be produced by nonrotating thunderstorms forming in environments with a preexisting low-level gyratory wind field [*Wakimoto and Wilson*, 1989]. However, the most intense, and long-lived, tornadoes occur in a special type of thunderstorm known as the "supercell," which generates its own gyratory wind field. That it does so is interesting, but perhaps the most fascinating aspect of rotation in the supercell, which became clear in the past decade or so, is the rotating wind field's vital role in producing the supercell's extraordinary properties of long life and deviate motion. Thus the present review will focus on what was learned from modeling and theory about the rotation and propagation of, and the relation of tornadoes to, supercell thunderstorms.

2. A BRIEF CHRONOLOGY

In previous tornado symposia there was no discussion of thunderstorm rotation, although the literature extends back to the last century. To understand better the significance of the discoveries made in the last decade, one must appreciate the context within which those discoveries were made. Hence I think it worthwhile to give a brief chronology here of the development of ideas of thunderstorm rotation.

2.1. Early Ideas

In the earliest writings, no real distinction was made among the various forms of convective phenomena (hurricanes, tornadoes, waterspouts, etc.) Figure 1 contains an early sketch from *Espy* [1841] which shows that rotation was not even recognized as an important property of these waterspouts. The recognition that rotation is important was made by *Ferrel* [1889]. As was mentioned, he made the connection between the tornado and a parent thunderstorm

The Tornado: Its Structure, Dynamics, Prediction, and Hazards.
Geophysical Monograph 79
Copyright 1993 by the American Geophysical Union.

updraft. He attributed rotation in the funnel to the rotation of the Earth.

Association of rotation with the variation of the horizontal wind with height (wind shear) was made by a number of European authors in the 1920s, most notably *Wegener* [1928] (Figure 2). Noticing the intense shear, as evidenced by the anvil outflow, he conjectured that there is a "mutterwirbl" (mothervortex) that gets bent down at the edges of the thunderstorm. Figure 2 is from a very thoughtful essay by *Fulks* [1962] on this topic.

A good deal of the intellectual tension in our subject started with this idea since by the symmetry of the picture (cyclonic vortex on the south, anticyclonic on the north), one could legitimately ask, "Why isn't there an anticyclonic tornado associated with the anticyclone on the other side?"

During and after World War II the advent of radar and rawinsondes confirmed the idea that a special type of thunderstorm occurs when the wind shear is large. The radar echoes showed that intense tornadoes were associated with large coherent echoes that traveled to the right of the winds in the cloud-bearing layer. Rotation was inferred from surface observations, and radar signatures such as the "hook" echo. Figure 2 contains a diagram from a lecture given by *Stout* [1957] showing the rotational structure of the tornado-bearing thunderstorm as known at the time. Note the cyclone-anticyclone at midlevels, reminiscent of Wegener's model.

2.2. Discovery of the Supercell

Continuing studies by radar meteorologists concerned with hail formation focused attention on the thunderstorm's internal flow structure. *Browning* [1964] proposed a special airflow pattern for certain especially severe, long-lived thunderstorms he called "supercells" [*Browning*, 1968]. His idea was that longevity is achieved by a flow pattern (Figure 3) that allows the updraft to unload its rain so it will not disrupt the storm's inflow; the airflow is also consistent with rightward propagation since, from a frame of reference moving with the storm, there is a component of the airflow from the south at all levels shown. Also note the placement of the

 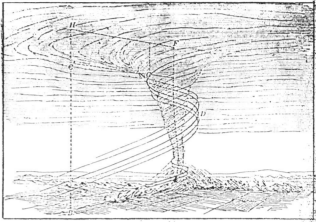

Espy, 1841 **Ferrel, 1889**

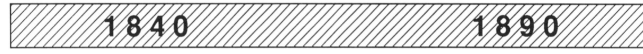

Fig. 1. Figures 1–6 show a "time line" showing significant events in the development of ideas on thunderstorm rotation. Other significant contributions are listed below the time line. The lithographs in Figure 1 illustrate *Espy*'s [1841] view that lowered pressure within intensely rising air accounts for the condensation funnel and *Ferrel*'s [1889] recognition that the funnels mark rotating columns of air.

tornado at the microfront based on the visual observations of *Ward* [1961].

In the same period, *Browning and Landry* [1963] and *Barnes* [1968] developed the idea that the inflow branch of the supercell model carried horizontal vorticity that could be tilted upward and hence account for the rotation in the updraft (Figure 3). The idea was taken to be incomplete in some quarters since no accounting for the anticyclonic vorticity (implicit in any vortex-tilting theory) was made.

Also, in this same period, a study which I think was viewed as a curiosity at the time would later provide a vital clue in unraveling the mystery of the supercell. The paper by *Fujita and Grandoso* [1968] (Figure 4) showed that a growing thunderstorm may split into a rightward propagating, cyclonic thunderstorm and a leftward propagating, anticyclonic thunderstorm. *Browning* [1968] also pointed out this possibility of a leftward moving counterpart to the "supercell" shown in Figure 3.

2.3. *Early Doppler Radar Investigations and Three-Dimensional Cloud Models*

Moving into the 1970s, the technological advance of Doppler radar allowed for a view of the actual three-dimensional flow within the supercell. Figure 5 contains a figure adapted from *Ray* [1976] showing the cyclone-anticyclone structure of the midlevel flow, with the dominance of the cyclonic vortex at low levels, much like the *Stout* [1957] picture. In addition, the indications were that the cyclonic rotation was mostly in updraft and the anticyclonic rotation was mostly in downdraft. At around the same time, computer power had advanced to the point where three-dimensional cloud models could be contemplated. Figure 5 shows results from the early simulation by *Schlesinger* [1975] showing a vortex pair centered on an updraft at midlevels but little else that resembled the supercell.

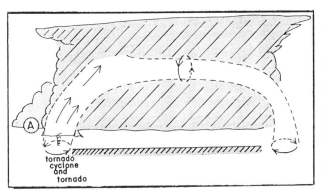
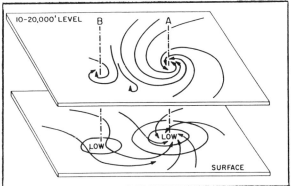

Wegener, 1928   **Stout, 1957**

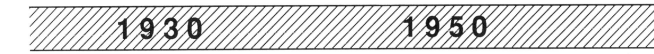

Markgraf, 1928

van Everdingen, 1925

Brooks, 1949

Stout and Huff, 1953

Huff, Hiser, and Bigler, 1954

Fujita, 1950s

Fig. 2. According to *Wegener* [1928] the effects of wind shear produce a horizontally oriented "mutterwirbl" that gets bent down at the sides [*Markgraf*, 1928] implying rotation about a vertical axis (cyclonic to the south/anticyclonic to the north) [from *Fulks*, 1962]. Weather radar, radiosondes, and surface observations allowed *Stout* [1957] to deduce the circulation within a tornado-bearing thunderstorm. Note the cyclonic (anticyclonic) flow on the south (north) side (as in the Wegener picture). (The Fujita papers are discussed by *Fujita* [1963].)

2.4. *Discovery of the Numerical-Model-Equivalent Supercell*

It was not until the three-dimensional modeling papers by *Klemp and Wilhelmson* [1978a, b], showing the connections among thunderstorm splitting, supercell-flow structure, and the rotational properties of these storms did the various pieces of the puzzle begin to fall into place. These later simulations (e.g., Figure 6) were carried out for much longer periods than were the *Schlesinger* [1975] simulations. One of the critical insights of Klemp and Wilhelmson was that the supercellular structure shown by *Browning* [1964] is acquired during the act of cell splitting. From the southward traveling storm's point of view (Figure 6) there is inflow from the south at all levels. Since the environmental winds are all east-west, this means there is deviate motion, and judging from the figure, the cell lives a long time. Also, there is a

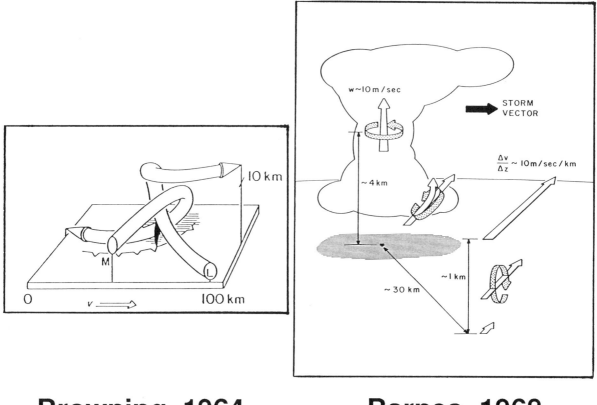

Browning, 1964 **Barnes, 1968**

1960

Browning and Ludlam, 1962

Browning and Landry, 1963

Browning and Donaldson, 1963

Fig. 3. *Browning* [1964] proposed that a special three-dimensional airflow structure allows for a storm cell to be long-lived. The flow diagram is drawn from the frame of reference moving with the storm. The flow arrows indicate that the storm motion has a velocity component across the shear (in addition to the usual component with the shear); in the parlance of severe-storm forecasters, it exhibits deviate motion. *Barnes* [1968] carried out a systematic data analysis and showed that a storm moving across the shear is thus able to tilt up the horizontal vorticity associated with the shear and so account for the cyclonic rotation in these severe rightward deviating thunderstorms.

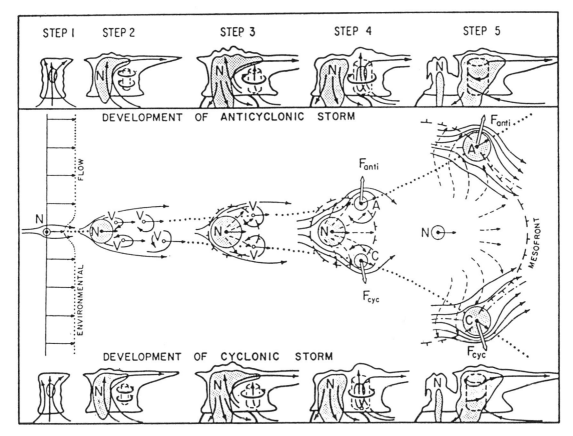

Fujita and Grandoso, 1968

Splitting:

Hitschfeld, 1960

Achtemeier, 1969

Charba and Sasaki, 1971

Fankhauser, 1971

Left moving:

Hammond, 1967

Browning, 1968

Fig. 4. Observations of *Fujita and Grandoso* [1968] showing that a storm cell may split into two cells, one rotating cyclonically and propagating to the right, the other anticyclonically, propagating to the left, facing in the general direction of storm movement.

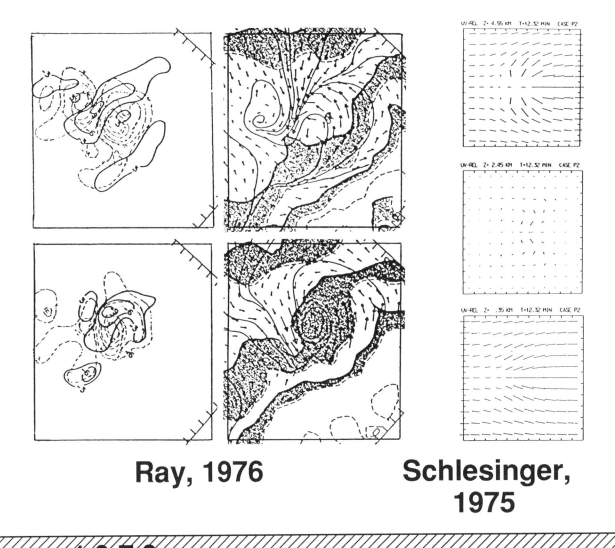

Fig. 5. In the 1970s the technological advances of Doppler radar and more powerful computers provided fresh impetus for severe-storm research. *Ray*'s [1976] deduction of the winds in the supercell showed a vortex pair at midlevels (upper panels; vorticity in dashed lines, divergence in solid lines) and a flow dominated by cyclonic rotation a low levels (lower panels). The early simulation by *Schlesinger* [1975] shows the development of a vortex pair at midlevels (middle panel) in a three-dimensional numerical cloud model.

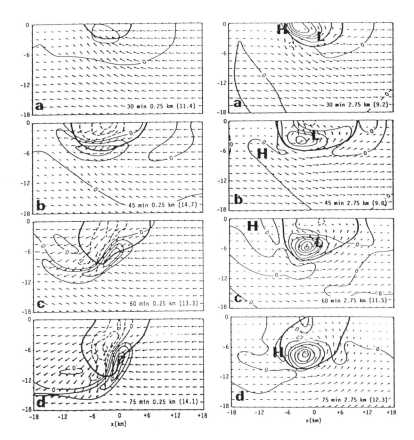

Wilhelmson and Klemp, 1978

Thorpe and Miller, 1978

Klemp and Wilhelmson, 1978

Fig. 6. In a series of papers, *Klemp and Wilhelmson* [1978a, b] showed that although the vortex pair is a signature of the nascent thermal, mature supercell-like structure is not obtained until after rainy downdrafts develop and the initial cell splits into two cells propagating to the left and right, respectively, of the mean shear. (Only the rightward propagating member is shown here.)

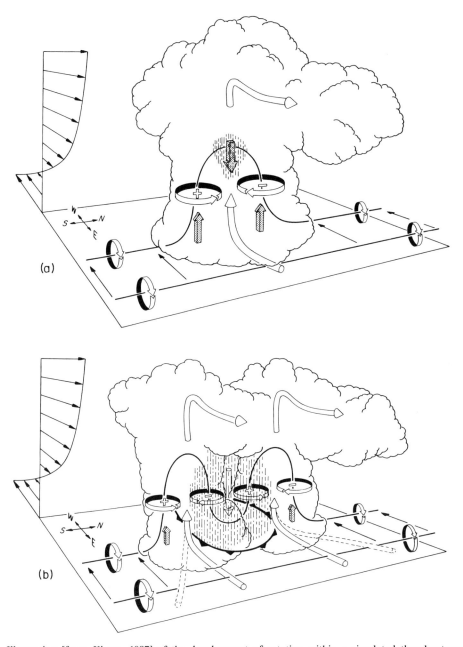

Fig. 7. Illustration [from *Klemp*, 1987] of the development of rotation within a simulated thunderstorm through vortex-line tilting. (*a*) In the early stage a vortex pair forms through tilting of the horizontal vorticity associated with the mean shear. (*b*) As rainy downdrafts form and the cell splits, vortex lines are tilted downward, and the original updraft-centered vortex pair is transformed into two vortex pairs. The rightward (facing with the shear) moving member propagates towards the positive vorticity on the right flank, and thus a correlation between updraft and positive vorticity develops.

mirror image anticyclonic storm (not shown) as observed by *Fujita and Grandoso* [1968].

3. Modeling and Theory of Thunderstorm Rotation

3.1. *Midlevel Rotation*

Although these simulations showed across-shear propagation, the reasons for this were not clear at the time. It occurred to a few of us in the early 1980s that even if we did not understand the across-shear propagation, we could ask how the rotation of the storm develops given the storm motion. Later, it was discovered that rotation in turn induces propagation, about which more appears later.

Figure 7 is an artistic rendering of the thunderstorm evolution shown in Figure 4 taken from the review by *Klemp*

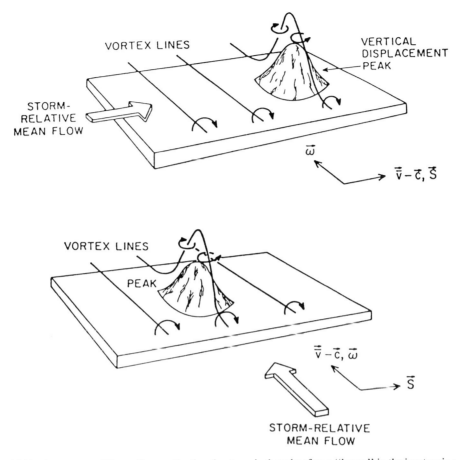

Fig. 8. Within the context of linear theory, the thunderstorm is thought of as a "bump" in the isentropic surface (or, more precisely, moist isentropic surface) [from *Davies-Jones*, 1984]. If one gives the storm motion vector $\vec{c}$, then one can deduce the phase relation between vertical vorticity and vertical velocity, since the flow lines and vortex lines follow the isentropic surface. The vortex lines over the bump always imply a vertically oriented vortex pair. In the upper panel the storm motion is purely in the shear direction, and so the updraft is colocated with the bump and so straddles (is out of phase with) the vortex pair. In the lower panel the storm motion is across the shear direction, the vertical velocity is up on the right (facing downshear) and down on the left; hence the vertical vorticity is in phase with the vertical velocity.

[1987]. As shown in Figure 7a, during the early stage of updraft development there is a tilting of the vorticity associated with the mean shear flow which gives rise to a vortex pair straddling the updraft, much like the Wegener/Markgraf picture. As shown in Figure 7b, subsequent storm splitting implies decreased vertical velocity, and eventually downdraft, developing at the location of the original updraft. This behavior of the vertical velocity implies a downward turning of the vortex lines that in turn implies that the original vortex pair, too, splits into two vortex pairs. Next, the two new updraft-downdraft couplets begin to propagate across the shear toward the vorticity centers of the original cell. This across-shear propagation of the updraft/downdraft pair towards the vorticity-production sectors tends to correlate vertical velocity and vertical vorticity. Hence the rightward propagating cell has a cyclonically rotating updraft, as in the Barnes diagram, but also an anticyclonically rotating downdraft as suggested in Ray's dual-Doppler analysis.

These ideas were developed on the basis of a linearization of the vertical vorticity equation about the mean shear flow [*Rotunno*, 1981; *Lilly*, 1982]. The most complete development of this line of reasoning was given by *Davies-Jones* [1984]. Figure 8 clarifies this distinction between storms that move with the shear from storms that move across the shear. Here the thunderstorm is the "bump" is the isentropic surface; the top panel shows that with along-shear propagation the updraft is maximum on the centerline upshear; hence the vortex pair straddles the updraft. The bottom panel shows that with across-shear propagation the updraft is maximum across the shear (on the right); hence it tends to be colocated with the vertical vorticity. Viewed from the storm's reference frame the inflow from the right brings the

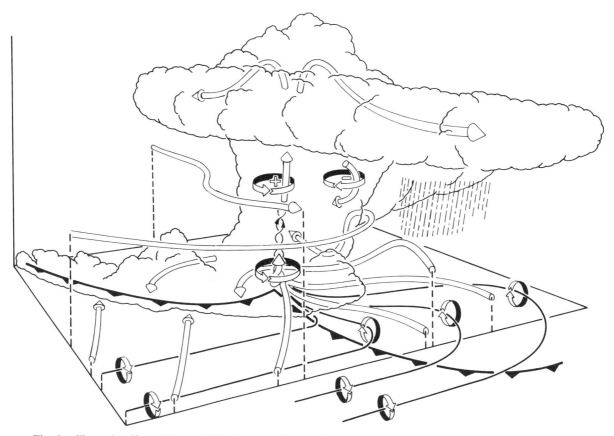

Fig. 9. Illustration [from *Klemp*, 1987] of how the low-level horizontal vorticity is modified by the temperature gradient beneath the thunderstorm. The ambient mean shear vorticity points across from south to north; however, cold air on the north side means that vorticity directed from east to west is created along the cold-air boundary. This vorticity can then be tilted upward as shown and is responsible for the strong low-level thunderstorm rotation.

mean shear vorticity in with it; hence the term "streamwise vorticity" (this quantity multiplied by the storm-relative wind speed is the now-popular "storm-relative helicity"). It is important to keep in mind that we have taken storm motion as a "given" and without that given one does not know which way "streamwise" is.

Davies-Jones [1984] and *Rotunno and Klemp* [1985] noticed that the conservation of equivalent potential vorticity accounts for the fact that these predictions from the linear models are much better they should have been since the latter are based upon a small-amplitude assumption.

3.2. *Low-Level Rotation*

The three-dimensional numerical simulations also revealed a previously unsuspected source of low-level rotation. The relatively simple picture of the tilting of mean-shear vorticity by the updraft worked pretty well for midlevel rotation but not so for low levels. As noted in the early three-dimensional simulations, a feature of the long-lived supercell is that the rain-cooled surface outflow of the thunderstorm stays positioned beneath the main updraft [*Thorpe and Miller*, 1978]. As shown in the schematic in Figure 9, the relative inflow from the east has its horizontal vorticity changed by baroclinic production. This baroclinically produced vorticity can then be tilted upward to produce strong positive vertical vorticity. Hence the dominance of the cyclone at low levels as in the Stout picture and in Ray's Doppler radar study.

At this point we are still stuck with the symmetry of the picture, since there is an implied mirror image storm (not shown) to the left of the shear vector. One could still reasonably ask, "Why isn't there always an anticyclonic supercell (and tornado) when there is a cyclonic supercell?"

4. CYCLONIC BIAS AND PROPAGATION

The second critical insight of the Klemp-Wilhelmson papers is the distinction they drew between veering wind and veering shear. A straight-line hodograph may represent a veering wind, but there is complete mirror symmetry of a disturbance about the shear vector. They found through numerical experiment (assuming a horizontally homogeneous, nonrotating environment) that the right-moving member of the splitting pair is enhanced when the shear vector veers (when the hodograph curves clockwise, which is the

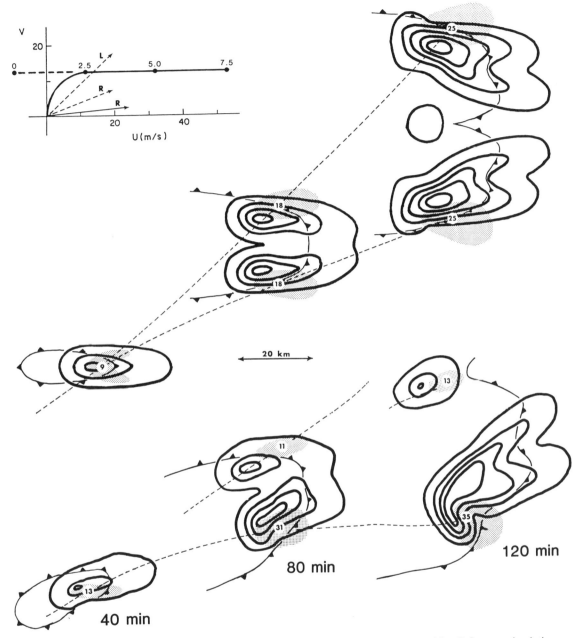

Fig. 10. Plan views of cloud-model-produced rainwater fields (at 1.8 km above ground level) for two simulations using, respectively, a straight-line hodograph and one with low-level clockwise curvature. The straight-line hodograph produces storms with complete mirror symmetry, while the curved-line hodograph enhances the rightward moving storm [from *Klemp*, 1987].

typical case). This is illustrated in Figure 10 [from *Klemp*, 1987].

Rotunno and Klemp [1982] noticed that the bias in these simulations occurs from the earliest time (when the updraft's amplitude is small); this suggested that a linear analysis might shed light on the matter. Since the updraft is arguably the defining feature of a thunderstorm, we looked at the vertical momentum equation and deduced that the asymmetry in updraft production is related to the vertical pressure gradient. In brief, we found that according to linear theory the perturbation pressure goes like the dot product of the shear and the gradient of vertical velocity w. As Figure 11a shows, this indicates high-pressure upshear and low-pressure downshear of the updraft; in this example of nonveering shear there is complete mirror symmetry about the shear vector.

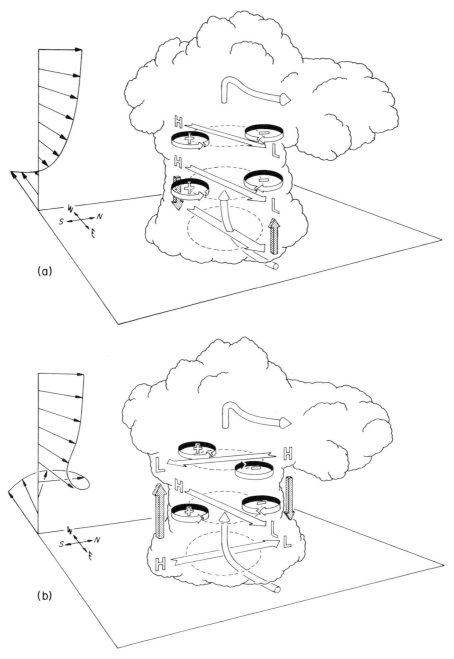

Fig. 11. Illustration [from *Klemp*, 1987] of the pressure perturbations arising as an updraft interacts with an environmental wind shear that (*a*) does not change direction with height and (*b*) turns clockwise with height. The high (H) to low (L) horizontal pressure gradients parallel to the shear vectors (flat arrows) are labeled. The shaded arrows indicate the implied vertical pressure gradients.

However, if the shear veers with height, Figure 11*b* shows that the high-low couplet also veers creating an enhanced upward pressure gradient force on the right and a negative such force on the left. Hence the bias. (The basic idea of looking for preferential vertical pressure gradients to explain deviate motion goes back to *Newton and Newton* [1959]).

Rotunno and Klemp [1982] also found, consistent with *Schlesinger* [1980], that there are lifting pressure gradients on the storm flanks at work even in the nonveering shear case. They identified the lowered midlevel pressure on the flanks with the strong rotation there. *Weisman and Klemp* [1982, 1984] carried out detailed sensitivity studies and

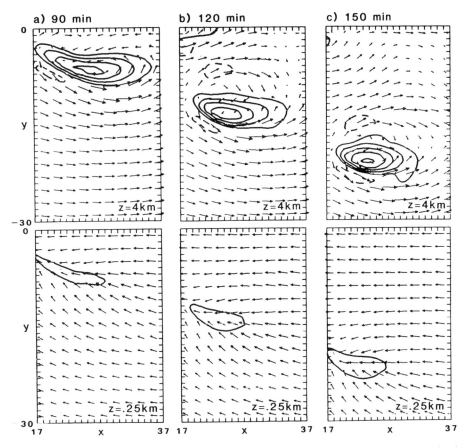

Fig. 12. Plan view of vertical velocity contours and horizontal flow arrows from a special no-rain cloud simulation [from *Rotunno and Klemp*, 1985]. The time sequence shows across-shear propagation of the updraft, even though there is no rain loading or evaporatively cooled cold air at the surface.

analysis showing that strong shear is a critical factor that allows a supercell to form and that these storms are dynamically different than their nonsupercellular cousins by virtue of the importance of the rotationally induced lifting pressure gradients. In fact, *Rotunno and Klemp* [1985] (Figure 12), carried out a special simulation with the rain-making process set to zero and showed that splitting and across-shear propagation can occur without the rain loading and rain-cooled outflow. (Also, there was little low-level rotation to speak of, confirming the importance of the low-level baroclinic generation mechanism for producing low-level rotation.)

This mechanism of dynamically induced uplift is finding application in such thermodynamically and geographically diverse thunderstorms as the low-precipitation (LP) thunderstorms of the high plains discussed by *Weisman and Bluestein* [1985] and the tornado-producing thunderstorms associated with land-falling hurricanes studied by *McCaul* [1991].

These simulations show that the essence of across-shear propagation is that new updraft be continuously forced on the updraft's flank by the enhanced rotation there (Figure 13). On the other hand, the linear models of storm rotation discussed above predict that a steadily propagating updraft will be in phase with the vorticity. It is clear that the linear theory can tell one only that there is a tendency for the updraft to "catch up with the vorticity" center, since if the updraft actually did catch the vorticity, the mechanism for propagation is cancelled (like what happens if the dog actually catches the car).

5. ENERGETICS

I think most observers would judge the supercell (this solitary, rotating and propagating thunderstorm) an extraordinary, possibly singular, meteorological phenomenon. Although the work just described gives, I believe, a good picture of how the machine works, one suspects that there might be some kind of "super" principle at work. Since 1986, D. K. Lilly and collaborators have looked for an energy principal.

Through the investigation of several simple models of buoyant thermals in shear, *Lilly* [1986a] (Figure 14) inferred the very different energetics occurring in three-dimensional

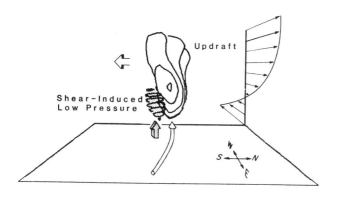

Fig. 13. Analysis of the no-rain simulation shown in Figure 12. This figure shows that rotationally induced low pressure on the south (across-shear side) of the updraft induces low pressure there that forces conditionally unstable air upward (shaded arrow) that in turn regenerates the updraft there [from *Rotunno and Klemp*, 1985].

versus two-dimensional flow. A two-dimensional disturbance in shear gives up its energy to the mean flow, while a three-dimensional disturbance is able to derive energy from the mean flow, this effect being tantamount to the production of rotation about the vertical axis derived through the tilting of mean flow horizontal vorticity. On the basis of a simple picture of a rotating updraft, Lilly further conjectured that part of the rotational kinetic energy could be transferred to the divergent kinetic energy, the latter of which, as mentioned, goes back to the mean flow; hence a positive-feedback loop is described.

These ideas were evaluated by *Wu* [1990] by an analysis of two simulations (straight hodograph versus curved hodograph, but also different thermodynamics) of supercells using the Klemp-Wilhelmson model. In both cases, rotational kinetic energy definitely derives from the mean flow, and divergent kinetic energy definitely derives from the pressure transfer term. However, the direction of the exchange between the divergent and rotational energy and

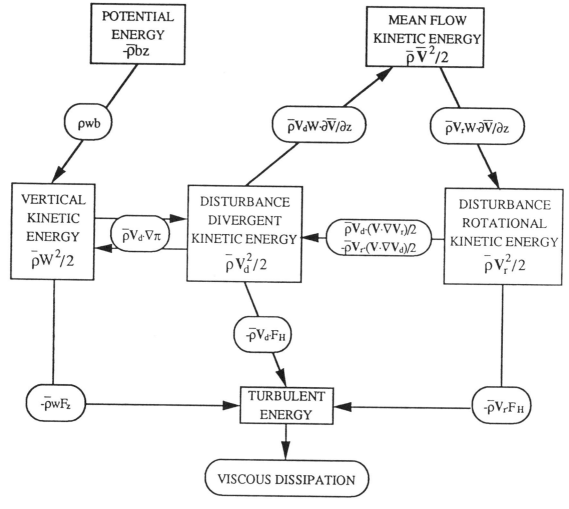

Fig. 14. *Lilly*'s [1986a] proposal for the energetics of the supercell. See text for discussion.

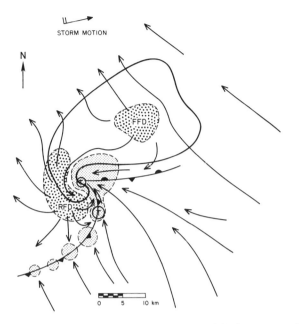

Fig. 15. Structure of low-level flow and precipitation pattern of a supercell thunderstorm as inferred from surface observers and radar analysis (from *Lemon and Doswell* [1979], as adapted by *Davies-Jones* [1986]). FFD is forward flank downdraft, RFD is rear flank downdraft, and T is tornado.

between the divergent energy and the mean flow is different in both cases. Also in both cases the divergent kinetic energy did not recycle energy back to the vertical kinetic energy. Hence the feedback aspects of the Lilly picture are perhaps not essential for the supercell. I think it would be useful to perform the same energetics analysis on the *Rotunno and Klemp* [1985] no-rain case which had an erect-to-downshear leaning updraft.

Lilly [1986b] proposed that since supercells are, by their nature, helical (strong alignment between velocity and vorticity vectors), and since helicity suppresses turbulent energy cascade, supercells are therefore less prone to dissipation than ordinary cells. *Wu* [1990] made a crude estimate of the inertial transfer term, which shows that the grid-resolved transfer is reduced as the simulated storm cell acquires helicity. These results are very interesting, and a more complete spectral energy budget is certainly worth doing.

6. WHERE THE TORNADO FITS IN ...

When the supercell produces a tornado, there are a number of attendant features of the finer-scale structure of the supercell that change. These were reviewed and emphasized in the important essay by *Lemon and Doswell* [1979] (Figure 15), which was based on a decade of their and others' personal visual observations of tornadic thunderstorms in conjunction with concurrent Doppler radar observations. In particular, they affirmed *Ward*'s [1961] observation that the tornado forms between the inflow and outflow, and they

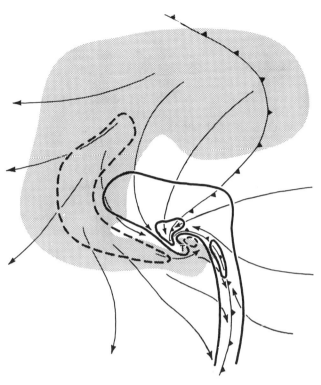

Fig. 16. Results from a fine-resolution simulation of the tornado-bearing region of a supercell [from *Klemp and Rotunno*, 1983]. Note the similarities with the observed structure shown in Figure 15. T is tornado.

emphasized that the entire gust front undergoes a contortion reminiscent of the occlusion of a large-scale cold and warm front with the tornado appearing at the tip of the occlusion.

Klemp et al. [1981] used the Klemp-Wilhelmson three-dimensional cloud model to simulate the May 20, 1977, tornadic thunderstorm observed with dual-Doppler data. Although many of the important features of the thunderstorm, such as the general flow structure and rotational character, were obtained, the fine-scale features associated with tornadogenesis could not be captured with 1 km horizontal grid resolution. *Klemp and Rotunno* [1983] took the model solution near the time of storm maturity, interpolated the model data from the vicinity of the mesocyclone to a nested finer grid (250 m), and then continued the integration. What was found (Figure 16) was a sequence of developments that were remarkably in parallel with those described by *Lemon and Doswell* [1979]. A rear flank downdraft developed in association with an occluded structure to the low-level temperature field and with very high values of vertical vorticity at the tip of the occlusion.

Although this was encouraging, and allowed *Klemp and Rotunno* [1983] to come up with a plausible physical model of the process, the results were less than totally satisfying because the nesting is turned on at a time when the storm is fully developed. One could wonder when, or even whether, the same flow scenario would occur if high resolution were used throughout the simulation. Here the specter of predictability looms since the tornado forms near where the cool air and the warm air collide (a region of large wind shear that is almost certainly unstable by any linear stability analysis). *Wicker and Wilhelmson* [1991] have carried out simulations of a supercell using interactive, nested grids with a horizontal grid interval of 120 m in the innermost grid system. The simulations show vividly that the precise prediction of the time and location of the tornado circulation is difficult, but that, once started, a very intense vortex (44 m s^{-1}) can be obtained.

One final observation: *Rotunno* [1986] pointed out that even with very high resolution, a cloud model using free-slip lower boundary probably could not simulate the important structural features of the tornado. Although the free-slip simulations by *Wicker and Wilhelmson* [1991] described above show that tornadic velocities can be achieved, I am still impressed by the early photogrammetry of tornadoes showing vertical velocities of 50 m s^{-1} 50 m off the ground (see *Rotunno* [1986] for references). Moreover, there is good evidence that horizontal winds in a strong tornado may exceed 100 m s^{-1}. Such structure certainly cannot be simulated by a model with 120 m vertical resolution, and on the basis of a large body of theoretical and laboratory studies I believe a no-slip condition is necessary to produce such a structure.

Future research will continue on the smaller scale to reach a better understanding of how the tornado fits into the thunderstorm. (I still do not know where the tornado vortex signature (MVS) fits into the picture [see *Rotunno*, 1986]). I believe we are at the point where the modeler can start looking upscale, to understand how these thunderstorms behave in horizontally inhomogeneous, time-dependent environments.

Acknowledgments. Discussions with M. L. Weisman are gratefully acknowledged. The National Center for Atmospheric Research is sponsored by the National Science Foundation.

REFERENCES

Achtemeier, G. L., Some observations of splitting thunderstorms over Iowa on August 25-26, 1965, in *Preprints, 6th Conference on Severe Local Storms*, pp. 89-94, American Meteorological Society, Boston, Mass., 1969.

Barnes, S. L., On the source of thunderstorm rotation, *ESSA Tech. Memo. ERLTM-NSSL 38*, 28 pp., Natl. Severe Storms Lab., Norman, Okla., 1968.

Brandes, E. A., Mesocyclone evolution and tornadogenesis: Some observations, *Mon. Weather Rev.*, 106, 995-1011, 1978.

Brooks, E. M., The tornado cyclone, *Weatherwise*, 2, 32-33, 1949.

Browning, K. A., Airflow and precipitation trajectories within severe local storms which travel to the right of the mean wind, *J. Atmos. Sci.*, 21, 634-639, 1964.

Browning, K. A., The organization of severe local storms, *Weather*, 23, 429-434, 1968.

Browning, K. A., and R. J. Donaldson, Jr., Airflow and structure of a tornadic storm, *J. Atmos. Sci.*, 20, 533-545, 1963.

Browning, K. A., and C. R. Landry, Airflow within a tornadic storm, in *Preprints, 10th Weather Radar Conference*, pp. 116-122, American Meteorological Society, Boston, Mass., 1963.

Browning, K. A., and F. H. Ludlam, Air flow in convective storms, *Q. J. R. Meteorol. Soc.*, 88, 117-135, 1962.

Charba, J., and Y. Sasaki, Structure and movement of the severe thunderstorms of 3 April 1964 as revealed from radar and surface mesonetwork analysis, *J. Meteorol. Soc. Jpn.*, 49, 191-213, 1971.

Davies-Jones, R. P., Streamwise vorticity: The origin of updraft rotation in supercell storms, *J. Atmos. Sci.*, 41, 2991-3006, 1984.

Davies-Jones, R. P., Tornado dynamics, in *Thunderstorm Morphology and Dynamics*, edited by E. Kessler, chap. 10, pp. 197-236, University of Oklahoma Press, Norman, 1986.

Espy, J. P., *The Philosophy of Storms*, 552 pp., Little, Brown, Boston, Mass., 1841.

Fankhauser, J. C., Thunderstorm-environment interactions determined from aircraft and radar observations, *Mon. Weather Rev.*, 99, 171-192, 1971.

Ferrel, W., *A Popular Treatise on the Winds*, 505 pp., John Wiley, New York, 1989.

Fujita, T., Analytical mesometeorology: A review, *Meteorol. Monogr.*, 5, 77-128, 1963.

Fujita, T., and H. Grandoso, Split of a thunderstorm into anticyclonic and cyclonic storms and their motion as determined from numerical model experiments, *J. Atmos. Sci.*, 25, 416-439, 1968.

Fulks, J. R., On the mechanics of the tornado, *Rep. 4*, 33 pp., Natl. Severe Storms Proj., Dep. of Commer., Washington, D. C., 1962.

Hammond, G. R., Study of a left moving thunderstorm of 23 April 1964, *ESSA Tech. Memo. ERLTM-NSSL 31*, 75 pp., Natl. Severe Storms Lab., Norman, Okla., 1967.

Hitschfeld, W., The motion and erosion of convective storms in severe vertical wind shear, *J. Meteorol.*, 17, 270-282, 1960.

Huff, F. A., H. W. Hiser, and S. G. Bigler, Study of an Illinois tornado using radar, synoptic weather and field data, 73 pp., State Water Surv., Urbana, Ill., 1954.

Klemp, J. B., Dynamics of tornadic thunderstorms, *Annu. Rev. Fluid Mech.*, 19, 369-402, 1987.

Klemp, J. B., and R. Rotunno, A study of the tornadic region within a supercell thunderstorm, *J. Atmos. Sci.*, 40, 359-377, 1983.

Klemp, J. B., and R. B. Wilhelmson, The simulation of three-dimensional convective storm dynamics, *J. Atmos. Sci.*, 35, 1070-1096, 1978a.

Klemp, J. B., and R. B. Wilhelmson, Simulations of right- and left-moving storms produced through storm splitting, *J. Atmos. Sci.*, 35, 1097-1110, 1978b.

Klemp, J. B., R. B. Wilhelmson, and P. S. Ray, Observed and numerically simulated structure of a mature supercell thunderstorm, *J. Atmos. Sci.*, 38, 1558-1580, 1981.

Kropfli, R. A., and L. J. Miller, Kinematic structure and flux quantities in a convective storm from dual-Doppler radar observations, *J. Atmos. Sci.*, 33, 520-529, 1976.

Lemon, L. R., and C. A. Doswell III, Severe thunderstorm evolution and mesocyclone structure as related to tornadogenesis, *Mon. Weather Rev.*, 107, 1184-1197, 1979.

Lilly, D. K., The development and maintenance of rotation in convective storms, in *Topics in Atmospheric and Oceanographic Sciences: Intense Atmospheric Vortices*, edited by L. Bengtsson and M. J. Lighthill, pp. 149-160, Springer-Verlag, New York, 1982.

Lilly, D. K., The structure, energetics and propagation of rotating convective storms, I, Energy exchange with the mean flow, *J. Atmos. Sci.*, 43, 113-125, 1986a.

Lilly, D. K., The structure, energetics and propagation of rotating convective storms, II, Helicity and storm stabilization, *J. Atmos. Sci.*, 43, 126-140, 1986b.

Markgraf, H., Ein Beitrag zu Wegeners mechanischer Trombentheorie, *Meteorol. Z.*, 45, 385-388, 1928.

McCaul, E. W., Buoyancy and shear characteristics of hurricane-tornado environments, *Mon. Weather Rev.*, 119, 1954-1978, 1991.

Miller, M. J., and R. Pearce, A three-dimensional primitive equation

model of cumulus convection, *Q. J. R. Meteorol. Soc.*, *100*, 133–154, 1974.
Newton, C. W., and H. R. Newton, Dynamical interactions between large convective clouds and environment with vertical shear, *J. Meteorol.*, *16*, 483–496, 1959.
Patushkov, R. S., The effects of vertical wind shear on the evolution of convective clouds, *Q. J. R. Meteorol. Soc.*, *101*, 281–291, 1975.
Ray, P. S., Vorticity and divergence within tornadic storms from dual Doppler radar, *J. Appl. Meteorol.*, *15*, 879–890, 1976.
Rotunno, R., On the evolution of thunderstorm rotation, *Mon. Weather Rev.*, *109*, 577–586, 1981.
Rotunno, R., Tornadoes and tornadogenesis, in *Mesoscale Meteorology and Forecasting*, edited by P. S. Ray, pp. 414–436, American Meteorological Society, Boston, Mass., 1986.
Rotunno, R., and J. B. Klemp, The influence of the shear-induced pressure gradient on thunderstorm motion, *Mon. Weather Rev.*, *110*, 136–151, 1982.
Rotunno, R., and J. B. Klemp, On the rotation and propagation of simulated supercell thunderstorms, *J. Atmos. Sci.*, *42*, 271–292, 1985.
Schlesinger, R. E., A three-dimensional numerical model of an isolated deep convective cloud: Preliminary results, *J. Atmos. Sci.*, *32*, 934–957, 1975.
Schlesinger, R. E., A three-dimensional numerical model of an isolated deep thunderstorm, II, Dynamics of updraft splitting and mesovortex couplet evolution, *J. Atmos. Sci.*, *37*, 395–420, 1980.
Stout, G. E., Mesometeorological systems from dense network stations, paper presented at IUGG meeting, Int. Union of Geod. and Geophys., Toronto, Ont., Canada, 1957.
Stout, G. E., and F. A. Huff, Radar records Illinois tornado genesis, *Bull. Am. Meteorol. Soc.*, *34*, 281–284, 1953.
Thorpe, A. J., and M. J. Miller, Numerical simulations showing the role of the downdraught in cumulonimbus motion and splitting, *Q. J. R. Meteorol. Soc.*, *104*, 873–893, 1978.
van Everdingen, E., The cyclone-like whirlwinds of August 10, 1925, *Proc. R. Acad. Amsterdam Sect. Sci.*, *25*, 871–889, 1925.
Wakimoto, R. M., and J. W. Wilson, Non-supercell tornadoes, *Mon. Weather Rev.*, *117*, 1113–1140, 1989.
Ward, N. B., Radar and surface observations of tornadoes of May 4, 1961, in *Preprints, Ninth Weather Radar Conference*, pp. 175–180, American Meteorological Society, Boston, Mass., 1961.
Wegener, A., Beitrage zur Mechanik der Tromben und Tornados, *Meteorol. Z.*, *45*, 201–214, 1928.
Weisman, M. L., and H. B. Bluestein, Dynamics of numerically simulated LP storms, in *Preprints, 14th Conference on Severe Local Storms*, pp. 167–170, American Meteorological Society, Boston, Mass., 1985.
Weisman, M. L., and J. B. Klemp, The dependence of numerically simulated convective storms on vertical wind shear and buoyancy, *Mon. Weather Rev.*, *110*, 504–520, 1982.
Weisman, M. L., and J. B. Klemp, The structure and classification of numerically simulated convective storms in directionally varying wind shears, *Mon. Weather Rev.*, *112*, 2479–2498, 1984.
Wicker, L. J., and R. B. Wilhelmson, Numerical simulation of tornadogenesis within a supercell thunderstorm, this volume.
Wilhelmson, R. B., The life cycle of a thunderstorm in three dimensions, *J. Atmos. Sci.*, *31*, 1629–1651, 1974.
Wilhelmson, R. B., and J. B. Klemp, A three-dimensional numerical simulation of splitting that leads to long-lived storms, *J. Atmos. Sci.*, *35*, 1037–1063, 1978.
Wu, W.-S., Helical buoyant convection, Ph.D. thesis, 161 pp., Univ. of Okla., Norman, 1990.

Numerical Simulation of Tornadogenesis Within a Supercell Thunderstorm

LOUIS J. WICKER AND ROBERT B. WILHELMSON

Department of Atmospheric Sciences and the National Center for Supercomputing Applications, University of Illinois, Urbana-Champaign, Illinois 61801

1. INTRODUCTION

During the past 15 years, numerical simulations have captured many features observed in supercell storms [*Wilhelmson and Klemp*, 1978; *Weisman and Klemp*, 1982, 1984]. Encouraged by the success in simulating large-scale supercell features, *Klemp and Rotunno* [1983] (hereafter referred to as KR) used a limited region fine-mesh simulation with 250 m horizontal resolution to investigate small features within the Del City supercell [*Klemp et al.*, 1981]. ("Resolution" refers to the spacing between grid points. The smallest scales adequately resolved by the grid are roughly 4 times the grid increment.) The model successfully reproduced observed features such as the divided mesocyclone (half updraft, half downdraft), the rear flank downdraft, and the low-level occlusion process. Further, strong low-level convergence concentrated preexisting vertical vorticity into a ringlike region with maximum values of 0.06 s^{-1}. The results demonstrated that small features often observed with a tornadic storm can be generated with a numerical storm model if sufficient resolution is employed. However, the vertical resolution used was rather coarse ($\Delta z = 500$ m), and no effort was made to study the impact of surface friction on storm structure, particularly at low levels.

Wicker [1990], in a similar manner to Klemp and Rotunno but with a higher resolution (70 m horizontal fine-grid resolution), examined the fine-scale features of a supercell storm on April 3, 1964, that was originally simulated by *Wilhelmson and Klemp* [1981]. The resolution near the ground was increased to 50 m by using a vertically stretched grid. The numerical simulation having surface friction at the lower boundary produces the strongest low-level rotation. A single concentrated vortex with a maximum vorticity value of 0.35 s^{-1} forms within the mesocyclone. This vortex lasts for several minutes before dissipating. A comparison between a simulation having a free-slip lower boundary and the simulation having surface friction shows that the vortex which forms in the surface friction run is stronger and has slightly larger updrafts near the ground. These differences were due to surface friction, which caused the low-level wind to be subcyclostrophic, allowing the development of strong low-level inflow into the vortex and enhancing the updraft within the vortex.

Snow [1982] divides the study of tornado dynamics into two categories on the basis of scale. The larger-scale vortex, called the tornado cyclone, is an intense vortex present within the mesocyclone circulation of the supercell storm. (Note that other researchers define a tornado cyclone as a mesocyclone that produces one or more tornadoes. We use Snow's definition here.) The smaller-scale vortex, the tornado itself, is the intense columnar vortex which forms within the tornado cyclone. Results from *Klemp and Rotunno* [1983] and *Wicker* [1990] demonstrate that high-resolution numerical simulations of supercell storms can produce intense vortices having horizontal scales about a kilometer in diameter within the mesocyclone, even in the absence of surface friction. We believe that these vortices represent tornado cyclones and conclude that surface friction is not crucial in their creation. However, it is well known that friction at the lower boundary plays a crucial role in the dynamics of tornadoes [*Rotunno*, 1979; *Snow*, 1982; *Fiedler and Rotunno*, 1986]. Therefore it is necessary to include the effects of surface friction to study the complete process of tornadogenesis.

The simulations reported in this paper were made with a new numerical cloud model having a stretched vertical grid and a nested horizontal grid in order to resolve storm features on scales of several hundred meters. It provides higher resolution than used by KR and extends the approach of *Wicker* [1990] by incorporating a fully interactive nested grid. To shed light about the evolution of the tornado cyclone within the storm, a supercell simulation was done using a frictionless lower boundary. A second simulation was then made with a surface-friction parameterization to investigate the role of friction in creating the tornado. These two numerical simulations will be used to discuss the generation of strong low-level rotation during tornadogenesis.

The Tornado: Its Structure, Dynamics, Prediction, and Hazards.
Geophysical Monograph 79
Copyright 1993 by the American Geophysical Union.

In order to clarify the discussion, we adopt definitions of tornadic-storm features which are consistent with those used in observational studies. *Lemon and Doswell* [1979] give an overview of some of these features such as the divided mesocyclone and the rear flank downdraft. The mesocyclone is associated with a strong rotating updraft with the vertical vorticity exceeding 0.01 s^{-1}, based on Doppler radar analysis [*Brandes*, 1984; *Brandes et al.*, 1988]. A few cases exist where the velocity field of a large tornado or its parent circulation was directly measured by the Doppler radar. Maximum shear values from these tornado cyclones are typically between 0.1 and 0.2 s^{-1} [*Brandes*, 1981; *Ray et al.*, 1981]. By using these measurements as a basis, tornado cyclones in the simulations are denoted as shear regions having vertical vorticity greater than 0.125 s^{-1}.

2. Computational Methodology

2.1. Numerical Model

The new cloud model is similar in design and construction to the Klemp-Wilhelmson cloud model [*Klemp and Wilhelmson*, 1978a] (hereafter referred to as KW78). A new feature in the cloud model is the capability to integrate horizontally nested grids using the adaptive grid techniques described by *Skamarock and Klemp* [1989]. Vertical grid stretching enables increased vertical resolution near the lower surface. Other differences from the KW78 model are listed briefly below.

1. The scalar variables are now integrated using a forward time scheme. Scalar advection is computed using a second-order upwind monotonic advection scheme [*van Leer*, 1977; *Wicker*, 1990].

2. The turbulent mixing coefficient is computed using a diagnostic formulation [*Clark*, 1979; *Tripoli and Cotton*, 1982].

3. A geometric mapping is used to stretch the grid vertically [*Walko*, 1988].

The upwind monotonic advection scheme is more accurate than the centered time-space scheme used by KW78 [*Wicker*, 1990]. The upwind scheme has smaller phase errors than the centered scheme, and the monotonicity constraint prevents the generation of spurious negative values of nonnegative scalar quantities in regions containing sharp gradients. The turbulent mixing scheme was changed to the diagnostic formulation because of the difficulty in implementing the prognostic approach used by KW78 in a model containing nested grids. Geometric vertical grid stretching method was preferred to the scheme used by *Wilhelmson and Chen* [1982] because it permits higher vertical resolution near the ground at a lower computational cost.

A simple bulk parameterization for surface friction was also added to the model physics. This parameterization is identical to the formulation used by *Wilhelmson and Chen* [1982] and *Droegemeier* [1985], i.e.,

$$\left(\frac{\partial u_1}{\partial t}, \frac{\partial u_2}{\partial t}\right)_{fric} = -\frac{C_d \bar{V}}{\Delta z}(u_1, u_2); \qquad (1)$$

where $\bar{V}$ is the ground-relative wind speed averaged to the appropriate staggered positions of u_1 (x component) and u_2 (y component) winds and Δz is the vertical grid spacing at the lowest model level. The friction term is applied only to the horizontal velocity variables at the lowest grid level. For the free-slip simulation, $C_d = 0.0$. The surface-friction simulation uses a value of $C_d = 0.04$. This value is similar to that used by *Wilhelmson and Chen* [1982] and *Droegemeier* [1985]. A sensitivity test showed that using a value of $C_d = 0.004$ changed the solution very little from the free-slip solution.

2.2. Adaptive Grid Methodology

In order to capture fine features in their supercell simulation, KR used interpolated coarse grid model data (1000 m in the horizontal) at the time of peak low-level rotation to initialize a small domain with fine resolution (250 m in the horizontal). The model was then integrated for a period of 6 min using the KW78 open lateral boundary conditions. Interpreting the initial evolution on their fine grid is difficult due to flow adjustments in response to decreases in turbulent mixing coefficients (which are based on the grid size). With the two-way interactive grid system used here, a fine grid mesh can be introduced into the calculation 10–15 min before the maximum low-level rotation occurs in the simulation, and integration on both the coarse-mesh and fine-mesh domains can be carried out indefinitely. Thus adjustments due to a reduction in grid size occur well before the maximum low-level rotation develops. *Klemp and Rotunno* [1983] and *Wicker* [1990] could not do this because the boundary conditions for the high-resolution grid were determined solely by flow inside the small domain.

Experiences with nested-grid simulations show that the solution collapses to the smallest scales which can be represented on the fine grid. This phenomenon is caused by the use of a grid-dependent mixing length in the calculation of the turbulent mixing coefficient [*Skamarock and Klemp*, 1989]. The philosophy employed here is that fine features should be well resolved. Thus a method is needed to prevent the solution from collapsing to the smallest resolved scales represented on the grid. We accomplish this by imposing a constant (not dependent on grid size) mixing coefficient on all grids. Experiences with modeling other types of small-scale convective flows indicate that a constant mixing coefficient of the order of 1 m s^{-1} Δ_f, where Δ_f is the horizontal grid spacing of the finest grid, is sufficient to force the solution to be well resolved on the finest grid. The simulations presented here use a constant mixing coefficient of 100 m^2 s^{-1} only on perturbation of variables, so that the initial, horizontally homogeneous base state (winds, temperature, and moisture) is preserved. We did several tests which showed that supercell evolution is insensitive to modest changes in this parameter. The constant mixing coefficient is used in addition to the grid-dependent eddy viscosity provided by the turbulence parameterization scheme.

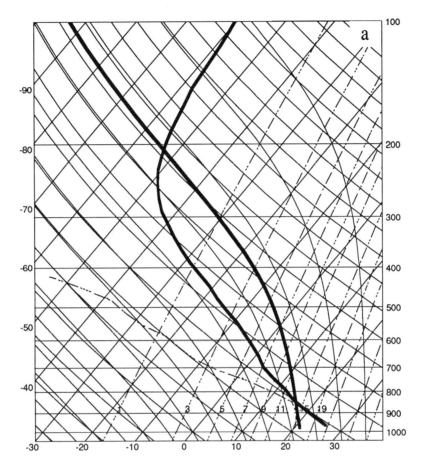

Fig. 1. (a) Skew-temperature Log-pressure plot of the initial vertical profile of temperature and moisture used to initialize the model's domain for the supercell simulation. Heavy line indicates the moist adiabat for surface parcel lifted to free-convection level. Medium line is the temperature profile. Dashed line is the moisture profile. (b) Hodograph of winds used to initialize the model's domain. Axis labels are in units of meters per second. Height of profile is indicated in decameters next to profile. Arrow indicates storm motion.

3. SIMULATION RESULTS

The model is initialized with a thermodynamic profile (Figure 1a) similar to the one used by *Weisman and Klemp* [1982, 1984]. The vertical profile of the horizontal wind (Figure 1b) is a composite obtained from examining hodographs from the Binger, Oklahoma, tornadic storm [*Wicker et al.*, 1984], the Raleigh, North Carolina, tornadic storm [*Davies-Jones et al.*, 1990], and the Davis, Oklahoma, tornadic storm on April 19, 1972 [*Brown et al.*, 1973]. The environment has a convective available potential energy of approximately 2900 J kg^{-1} and a bulk Richardson number of 23 [*Weisman and Klemp*, 1982; *Moncrieff and Green*, 1972]. Calculations using the storm motion observed in the simulation yield a storm relative total helicity value of 580 m^2 s^{-2} [*Davies-Jones et al.*, 1990]. Similar values of helicity have been observed in environments of supercell storms that spawned significant tornadoes [*Davies-Jones et al.*, 1990].

3.1. *Coarse-Resolution Simulation (Supercell Simulation)*

The coarse-grid domain is a 75 × 75 × 16 km box. Horizontal resolution is 600 m. Vertical resolution in the lowest grid volume is 120 m and increases to 200 m at 1 km and then to 700 m at 7.5 km. Above this level it remains constant. There are seven grid levels below 1 km in the simulation. The storm is initiated by placing a "standard" thermal bubble, 4°K in amplitude, at $z = 1.5$ km in the center of the domain. The horizontal "radius" of the bubble is 10 km, and the vertical "radius" is 1.5 km. The Coriolis force is set to zero for the simulations, since it has little impact on the solution [*Klemp and Wilhelmson*, 1978b]. The lower boundary condition is free slip. The model is integrated on the coarse grid for 120 min. During the first 30 min, the initial impulse grows and splits into left- and right-moving cells. After the split, the right-moving cell dominates, always containing the largest vertical velocity,

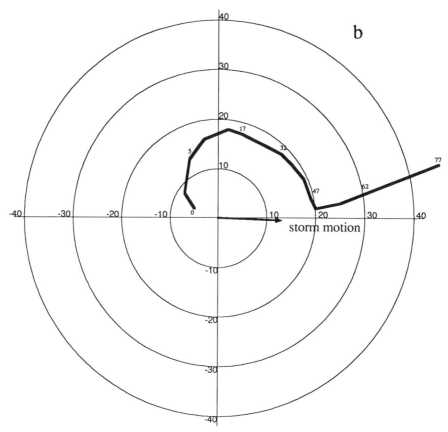

Fig. 1. (continued)

rain echo, and cyclonic rotation. Maximum updrafts during the first 90 min exceed 55 m s^{-1}. After 110 min the right-moving storm weakens somewhat, and its low-level rotation decreases.

Horizontal slices of vertical velocity at $z = 1.0$ km and the surface rainwater mixing ratio are shown for 70 (Figure 2a) and 90 (Figure 2b) min (for convenience, $z = 60$ m, the first vertical grid level for all variables but vertical velocity, is referred to as the surface). As a reference, the $-1.0°$K potential temperature isotherm, representative of the surface gust-front boundary, is shown as the cold-front symbol. Between 70 and 90 min, updrafts and downdrafts at $z = 1.0$ km increase. The updraft at $z = 1.0$ km develops into a horseshoe shape, and the mesocyclone divides into regions of updraft and downdraft. The surface rainfall field is somewhat disorganized at 70 min and then evolves into a classic "hook"-echo shape at 90 min. Aloft, a persistent bounded weak-echo region, 5 km tall, develops, while the surface vertical vorticity increases from 0.038 s^{-1} to more than 0.057 s^{-1}. These vorticity values are indicative of a strong low-level mesocyclone. The development of a divided mesocyclone, hook echo, rear flank downdraft, and the increases in low-level rotation are consistent with observational indicators of tornadogenesis [*Lemon and Doswell*, 1979]. To capture this evolution using higher spatial resolution, a horizontally nested fine-mesh grid is introduced at 70 min. The model is then integrated forward in time on both the coarse-mesh and the fine-mesh grids in order to study smaller-scale features that are unresolved on the coarse grid.

3.2. *Fine-Resolution Simulation (Tornado Cyclone Simulation)*

The fine mesh is 15 km square in horizontal section and is centered on the low-level mesocyclone. This domain is nearly large enough to contain the convectively active part of the storm. Horizontal resolution of the fine domain is 120 m, while the vertical resolution remains the same as in the coarse grid (e.g., Δz at the surface is 120 m). The integration on both the fine-mesh and coarse-mesh domains continues from 70 min to 110 min. During this period, two intense vortices develop within the mesocyclone. Each vortex lasts for approximately 10 min, is approximately 1 km in diameter, and has vertical vorticity greater than 0.125 s^{-1} from the surface through a depth of several kilometers. The size and intensity of these vortices are similar to tornado cyclones and are referred to as such.

During the first 10 min of integration on the fine grid, several small-scale horseshoe-shaped updrafts develop along the western edge of the storm's low-level updraft (Figure

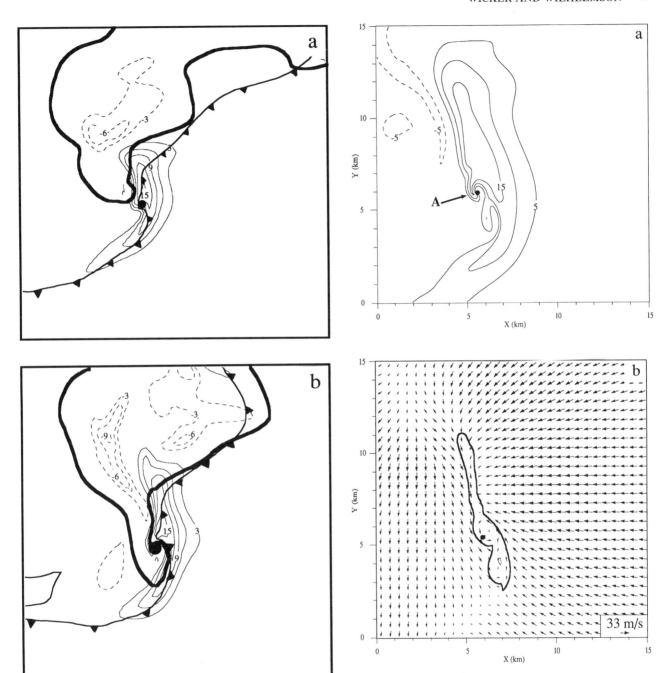

Fig. 2. (a) Horizontal section of coarse-grid solution at 70 min. Contour plot of vertical velocity at $z = 1$ km with a contour interval of 3 m s^{-1}. Heavy solid line represents the 0.25 g kg^{-1} surface-rainwater mixing ratio contour at $z = 60$ m. Cold-frontal boundary locates the position of the $-1°$K potential temperature isotherm at the surface. Region where vertical vorticity greater than 0.03 s^{-1} at surface is indicated by the solid dot. Peak vertical vorticity is 0.038 s^{-1}. (b) Same as Figure 2a except at 90 min. Peak vertical vorticity is now 0.06 s^{-1}.

Fig. 3. (a) An 80-min horizontal section of vertical velocity at $z = 1$ km within fine-grid domain. Contour interval is 5 m s^{-1}. The zero contour is not plotted. Solid region indicates area where vertical vorticity at $z = 1$ km is greater than 0.125 s^{-1}. Peak vertical vorticity at this level is 0.125 s^{-1}. "A" is the northern horseshoe-shaped updraft described in text. (b) An 80-min horizontal wind vectors at surface. Solid region indicates area where vertical vorticity at surface is greater than 0.125 s^{-1}. Peak vertical vorticity at this level is 0.125 s^{-1}. Heavy line is the 0.01 s^{-1} vertical-vorticity contour that delineates the mesocyclone.

3a). The mesocyclone, located along the shear zone created by the forward and rear flank gust fronts, is elliptical (Figure 3b). The region of vorticity greater than 0.125 s^{-1} is a column which tilts to the north between the surface and $z = 1$ km. A more detailed examination of the vorticity field (see Figure 7c) shows that the region of enhanced vertical vorticity extends upward to only 1 km in height. The shallow vertical extent of enhanced vorticity suggests that this feature may be analogous to a "gustnado" [Wilson, 1986].

From 80 to 88 min the areal coverage of vorticity greater than 0.125 s^{-1} doubles in size, and an intense circulation develops through a depth of several kilometers. At 88 min the horseshoe-shaped updraft at $z = 1$ km expands to 4 km across and extends from the tip of the occlusion updraft eastward to the gust front updraft (Figure 4a). Just east of the tornado cyclone center (hereafter denoted as TC), an intense downdraft greater than -25 m s^{-1} is present. Updrafts are 5 m s^{-1} just to the south of the vortex center. At the surface an intense circulation is present (Figure 4b). The TC is at the southern end of the mesocyclone. It is defined by the region of vorticity greater than 0.125 s^{-1} and is 720 m in diameter. The maximum vorticity exceeds 0.25 s^{-1}. Horizontal winds are greater than 40 m s^{-1} (ground-relative flow) to the west of the TC and exceed 53 m s^{-1} along the southern edge of the TC. The central pressure of the TC is -16 mbar, and horizontal pressure gradients at the surface are approximately 14 mbar km^{-1}.

The first TC decays between 90 and 95 min in the simulation. The rapid decay occurs when the first TC moves southeastward away from the main storm updraft into a region of strong downdraft. The mesocyclone becomes somewhat disorganized for a few minutes and then redevelops in an elliptical shape (Figure 5b). The development of the second TC begins when a downdraft develops on the west side of the circulation center and cool air spreads toward the east. The outflow increases the low-level convergence along the western side of the mesocyclone. Horizontal winds greater than 55 m s^{-1} (ground relative) are present there. The increased convergence stretches the vertical vorticity along the western side of the mesocyclone, producing the elongated regions of vorticity greater than 0.125 s^{-1} seen at the surface and at $z = 1$ km (Figure 5). The -5 m s^{-1} downdraft behind the rear flank gust front in Figure 5a is associated with the first TC. The developing TC is located 3–4 km northwest of this position where the storm's main updraft is reintensifying. The next 7 min show continued intensification of the circulation at the surface and at higher levels.

The most intense stage of the second TC (at 103 min) occurs when surface vorticity is greater than 0.3 s^{-1} (Figure 6). At $z = 1$ km (Figure 6a), the TC has an average diameter of 840 m and is located in the maximum vertical-velocity gradient between the updraft and downdraft. Updrafts along the western side of the TC are 5–15 m s^{-1}, and downdrafts along the eastern edge are about -5 m s^{-1}. The magnitude of the downdraft east of the cyclone center is much smaller for the second TC than for the first one (see Figure 4a). The

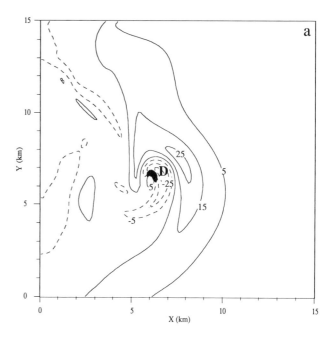

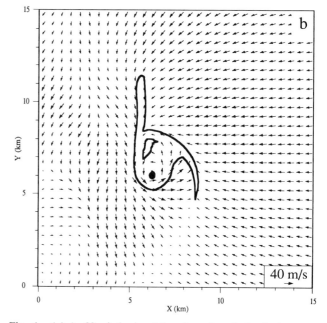

Fig. 4. (a) An 88-min horizontal section of vertical velocity at $z = 1$ km within fine-grid domain. Contour interval is 5 m s^{-1}. The zero contour is not plotted. Solid region indicates area where vertical vorticity at $z = 1$ km is greater than 0.125 s^{-1}. Peak vertical vorticity at this level is 0.20 s^{-1}. Strong-downdraft region east of tornado cyclone is indicated by "D". (b) An 88-min horizontal wind vectors at surface. Solid region indicates area where vertical vorticity at surface is greater than 0.125 s^{-1}. Peak vertical vorticity at this level is 0.25 s^{-1}. Heavy line is the 0.01 s^{-1} vertical-vorticity contour that delineates the mesocyclone.

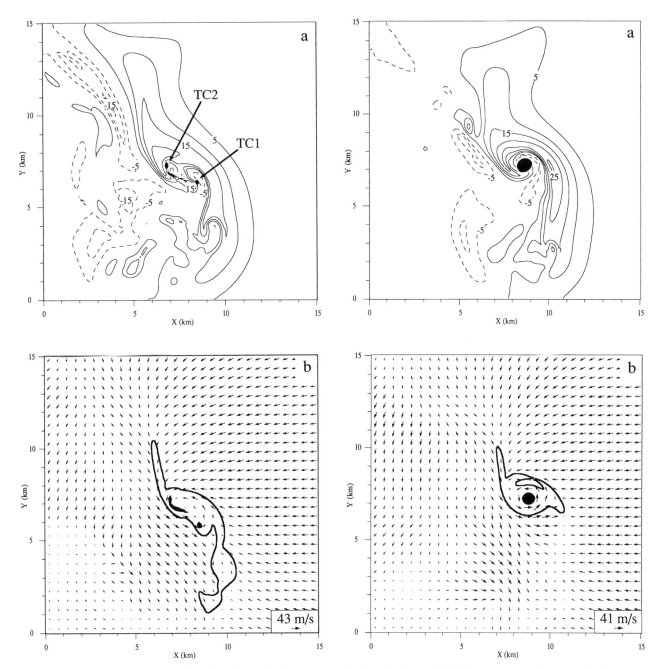

Fig. 5. (a) A 96-min horizontal section of vertical velocity at $z = 1$ km within fine-grid domain. Contour interval is 5 m s^{-1}. The zero contour is not plotted. Solid region indicates area where vertical vorticity at $z = 1$ km is greater than 0.125 s^{-1}. Peak vertical vorticity at this level is 0.17 s^{-1}. TC1 and TC2 indicate the positions of the decaying and developing tornado cyclones, respectively. (b) A 96-min horizontal wind vector at surface. Solid region indicates area where vertical vorticity at surface is greater than 0.125 s^{-1}. Peak vertical vorticity at this level is 0.20 s^{-1}. Heavy line is the 0.01 s^{-1} vertical-vorticity contour that delineates the mesocyclone.

Fig. 6. (a) A 103-min horizontal section of vertical velocity at $z = 1$ km within fine-grid domain. Contour interval is 5 m s^{-1}. The zero contour is not plotted. Solid region indicates area where vertical vorticity at $z = 1$ km is greater than 0.125 s^{-1}. Peak vertical vorticity at this level is 0.24 s^{-1}. (b) A 103-min horizontal wind vectors at surface. Solid region indicates area where vertical vorticity at surface is greater than 0.125 s^{-1}. Peak vertical vorticity at this level is 0.3 s^{-1}. Heavy line is the 0.01 s^{-1} vertical-vorticity contour that delineates the mesocyclone.

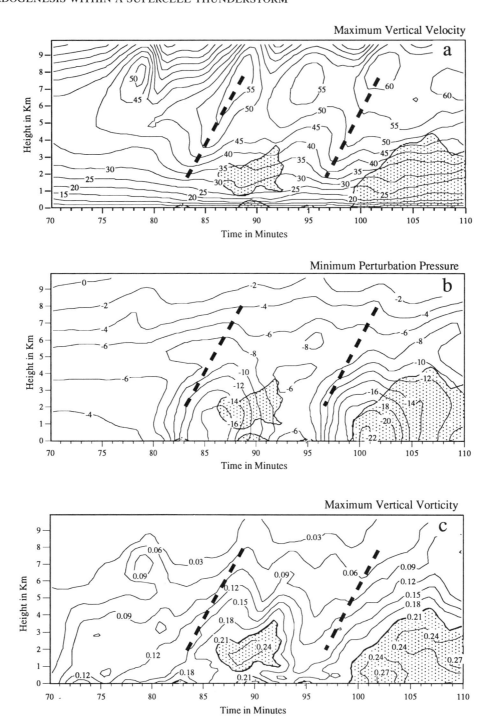

Fig. 7. Time-height cross section of variables from fine-grid simulation using free-slip lower boundary conditions. (a) Maximum updraft. Contour interval is 5 m s^{-1}. Heavy dashed line indicates updraft pulses that occur near the mesocyclone. (b) Minimum pressure. Contour interval is 2 mbar. (c) Maximum vertical vorticity. Contour interval is 0.03 s^{-1}. Regions where shear is greater than 0.21 s^{-1} are shaded.

genesis of the second TC is similar to the first one; the surface mesocyclone contracts in diameter and changes from an elliptical to a circular shape. The TC has a vorticity core with nearly vertical orientation between the surface and 1 km. Peak ground-relative wind speeds now exceed 55 m s^{-1}.

In order to examine more completely the changes in the storm's structure in time and space, a time-height cross section is created of the model data from the fine grid during the 40-min integration period (Figure 7). Maximum or minimum values of the vertical velocity, pressure, and vertical vorticity at each horizontal level were used to create the cross sections. The values were checked to make certain that their locations are in or very near the mesocyclone circulation. A 1-2-1 filter was applied in both directions on each variable to remove noise from the fields. During the model integration the domain average of the perturbation pressure field is forced to be zero at each time step because the perturbation pressure field is determined only to within an arbitrary constant in the cloud model. Examination of the actual pressure fields show that falls (rises) are associated with increases (decreases) in the horizontal gradient of the pressure field. For example, at 100 min the lowest surface pressure is −22 mbar, and the TC has a horizontal gradient of 22 mbar km^{-1}; at 93 min the corresponding values are −7 mbar and 6 mbar km^{-1}.

Several prominent features are apparent in the time-height cross section of the maximum vertical velocity within the storm (Figure 7a). There are four distinct updraft pulses between 70 and 105 min. These pulses are of two distinct types. The first and third updraft pulses develop above 5 km in height near the level of maximum buoyancy. The increases in updraft intensity aloft are not reflected at low levels. During the development of the second and fourth pulses, large upper level vertical-velocity increases are preceded by increases in low-level vertical velocities. For example, during the period between 96 and 100 min, updrafts greater than 35 m s^{-1} exist within the storm at the 2-km level, and updrafts greater than 45 m s^{-1} are present at the 3-km level. Further examination of data shows that the second and fourth updraft pulses are located in the northwestern portion of the mesocyclone near the intersection of the forward and rear flank gust fronts. The first and third pulse are updrafts located southeast of the mesocyclone along the southern flank of the storm. At any given time, the northwest and southeast updrafts are connected by a region of weaker updrafts. At midlevels ($z = 4$–5 km) they form the characteristic horseshoe pattern described by others (e.g., Del City storm [*Ray et al.*, 1981] and Harrah storm [*Brandes*, 1977]).

The second and fourth pulses develop at the onset of pressure falls within the mesocyclone (Figure 7b). Minimum pressures at low levels occur during the period of maximum updraft intensity. After the occurrence of minimum pressures, peak surface vorticity values are attained (Figure 7c). The development of the first and second TCs are very different. Vertical vorticity greater than 0.2 s^{-1} develops first at $z = 2$ km at 87 min and during the next several minutes a 2-km-deep region of enhanced vorticity develops aloft. One minute after maximum vorticity aloft, the peak surface vorticity of 0.26 s^{-1} occurs. Vorticity greater than 0.2 s^{-1} persists near the surface from 88 to 91 min. The development of strong rotation aloft with a subsequent extension downward to the surface is similar to Doppler radar observations of the Union City tornado cyclone evolution [*Brown et al.*, 1978]. In contrast, the second TC develops strong rotation nearly simultaneously from the surface to 2 km at 99 min. During the period from 99 to 107 min the depth, over which vertical vorticity exceeds 0.2 s^{-1}, increases from 2 km to 4 km. Peak vorticity at the surface is 0.31 s^{-1} at 103 min, and vertical vorticity greater than 0.2 s^{-1} persists near the surface from 99 to 110 min.

The second TC's evolution, with strong rotation developing at low levels and then extending rapidly upward into the storm, is very similar to the Del City mesocyclone and tornado signature observed on May 20, 1977. Figure 8, from *Johnson et al.* [1987], is a time-height cross section obtained from Doppler observations of the Del City storm. The observed maximum vertical-vorticity value at each level within the storm is shown in Figure 8a. At 1830 CST, vertical vorticity rapidly increases from the surface through 8 km. From 1830 CST to 1845 CST the low-level vorticity continues to increase between the surface and $z = 4$ km. Tornado touchdown is at 1845 CST. Figure 8b shows the observed maximum updraft at each level within the Del City storm. Beginning at 1830 CST, updrafts greater than 40 m s^{-1} develop rapidly between 3 and 4 km prior to touchdown of the Del City tornado. The development of an updraft pulse prior to TC development in the Del City storm is qualitatively similar to the rapid increase in 2–4 km vertical-velocity maxima around 97 min in the model simulation.

3.3. *Surface-Friction Simulation (Tornado Run)*

Results from *Wicker* [1990] indicate that the inclusion of surface friction increases the low-level convergence and vertical vorticity within the tornado cyclone, compared to a free-slip simulation. By using initial conditions from the tornado cyclone simulation on the fine grid, surface friction is added to the model physics. The surface drag increases low-level convergence in the tornado cyclone, resulting in the formation of a concentrated vortex, which we call a tornado. To conserve computational resources, the friction simulation is started at 94 min, well after the start of the nested fine-mesh simulation. At 94 min the first TC has nearly dissipated, and the intensification of the stronger, second TC is 5 min in the future. The new simulation is integrated from 94 to 114 min. After 112 min the tornado weakens considerably from its maximum intensity.

A time-height cross section is constructed for the friction simulation for the period from 94 to 114 min (Figure 9). Similar structure in the updraft fields is present between the friction and free-slip simulations (Figures 7a and 9a). Minimum surface pressure during the most intense rotation (102 min) is −16 mbar, with associated horizontal pressure gra-

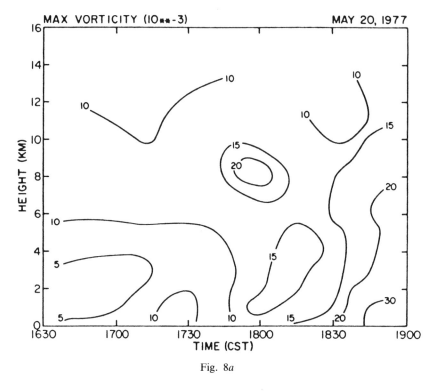

Fig. 8a

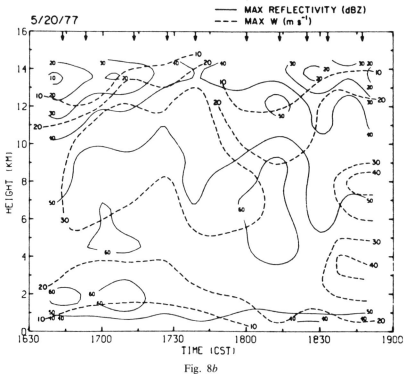

Fig. 8b

Fig. 8. (a) Time-height cross section of maximum vertical vorticity for the Del City storm. Contour values are in units of 10^{-3} s^{-1}. (b) Time-height section of maximum reflectivity and maximum updraft speeds for the Del City storm. Dashed line is the maximum updraft. Solid line is the maximum reflectivity. Arrows at top of the figure indicate Doppler analysis times. Figures 8a and 8b were taken from *Johnson et al.* [1987].

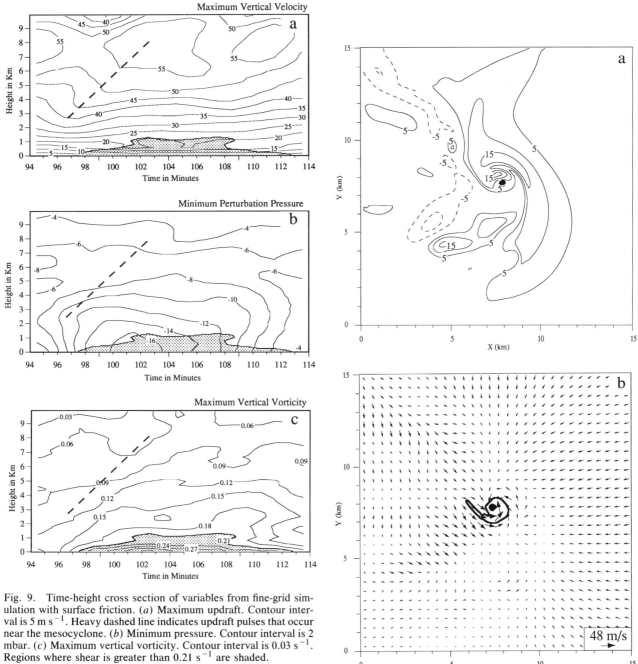

Fig. 9. Time-height cross section of variables from fine-grid simulation with surface friction. (a) Maximum updraft. Contour interval is 5 m s^{-1}. Heavy dashed line indicates updraft pulses that occur near the mesocyclone. (b) Minimum pressure. Contour interval is 2 mbar. (c) Maximum vertical vorticity. Contour interval is 0.03 s^{-1}. Regions where shear is greater than 0.21 s^{-1} are shaded.

Fig. 10. (a) A 102-min horizontal section of vertical velocity at z = 1 km within fine-grid simulation with surface friction. Contour interval is 5 m s^{-1}. The zero contour is not plotted. Solid region indicates area where vertical vorticity at z = 1 km is greater than 0.125 s^{-1}. Peak vertical vorticity at this level is 0.25 s^{-1}. (b) A 102-min horizontal wind vector at surface. Solid region indicates area where vertical vorticity at surface is greater than 0.125 s^{-1}. Peak vertical vorticity at this level is 0.54 s^{-1}. Heavy line is the 0.01 s^{-1} vertical-vorticity contour that delineates the mesocyclone.

dients of about 16 mbar km^{-1}. During most of the tornado event the pressure at z = 250 m is 1–2 mbar lower than the surface pressure. During the period of peak rotation between 101 to 102 min, the pressure is lowest at the surface. This result is different from the free-slip simulation where the pressure is always lowest at the surface during the TC events. Higher pressures at the surface for the friction simulation are the result of significantly larger values of low-level convergence at the base of the vortex than in the

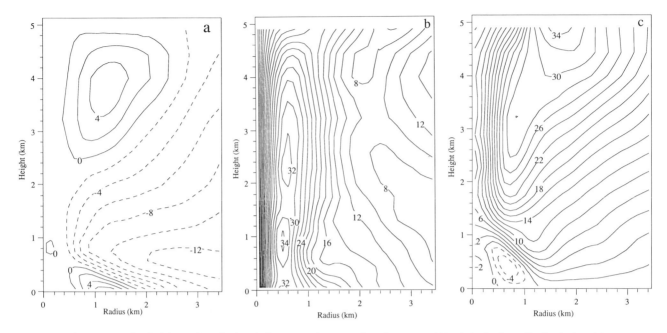

Fig. 11. Radius-height section of azimuthally averaged mesocyclone features at 103 min in the free-slip simulation. (a) Radial velocity. Contour interval is 2 m s^{-1}. (b) Tangential velocity with contour interval of 2 m s^{-1}. (c) Vertical velocity with contour interval of 2 m s^{-1}.

free-slip vortex. The vertical-vorticity evolution (Figure 9c) is very different from that in the free-slip simulation (Figure 7c). Maximum surface vorticity values are 70% larger than in the free-slip simulation. However, the vertical extent of strong vorticity is much shallower. Surface vorticity values in the friction run exceed 0.36 s^{-1} in the cross section, and unsmoothed vorticity values exceed 0.5 s^{-1} between 101 and 102 min. Vertical vorticity greater than 0.2 s^{-1} during the period from 98 to 112 min extends from the surface to 1.5 km in height, compared to a depth of 4 km in the free-slip case.

Figure 10 depicts the vertical-velocity field at $z = 1$ km and the surface wind vector field at 102 min. Peak vertical vorticity is 0.54 s^{-1} and maximum ground-relative wind speeds are 62 m s^{-1} along the southern edge of the tornado. At the surface the area of the mesocyclone is 2–3 times smaller than in the free-slip case (see Figure 6b), and the wind field has a significantly stronger inward component. The vertical velocity at $z = 1$ km (Figure 10a) has upward velocities of 15 m s^{-1} north of the tornado center. Weaker updrafts of 5 m s^{-1} exist to the south and east of the tornado, so the tornado is positioned in the vertical-velocity gradient. Rather than being located between the mesocyclone's updraft and downdraft as the tornado cyclone is in the free-slip simulation at 103 min (Figure 6a), the tornado is completely embedded in updraft. At 1 km, no significant downdrafts are present within the mesocyclone. At $z = 180$ m, vertical velocities along the northern side of the tornado exceed 19 m s^{-1}.

To compare the differences in the structure of the mesocyclone, tornado cyclone, and tornado between simulations an azimuthal average of model output is computed for the free-slip and surface-friction simulations. The center point for the averaging at each level is taken to be the position of the maximum vorticity value at that level. Pressure, vertical-velocity, tangential-velocity, and radial-velocity values at each level are averaged into rings that are 120 m wide and extend outward 3 km from the maximum vorticity value. This averaging procedure is similar to one used by Brandes [1977] and Wicker [1990].

The radial velocity for the free-slip simulation at 103 min shows inflow below $z = 2.5$ km and outflow aloft (Figure 11a). Near the surface and between $r = 1$ km and $r = 1.8$ km the flow is outward from the axis. This is a manifestation of the downdraft east of the tornado cyclone center (see Figure 11c). The tangential-velocity field (Figure 11b) has a maximum velocity of 34 m s^{-1} aloft at $z = 1$ km and has a value of 32 m s^{-1} near the surface at $r = 0.5$ km. The core radius of the tornado cyclone is nearly constant with height and extends upward to 4.5 km. The storm's main updraft is the dominant feature in the vertical-velocity field (Figure 11c). Vertical motions are very weak in the center of the tornado cyclone below $z = 1$ km. The azimuthal average of the pressure field (not shown) shows a downward pressure gradient along the central axis. This causes the maximum vertical velocity to be displaced from the axis.

The radial-velocity field for the surface-friction simulation at 102 min indicates stronger low-level inflow than in the free-slip simulation (Figure 12a). Between the surface and $z = 200$ m, radial inflow exceeds 14 m s^{-1}. Above the inflow layer and close to the axis, weak outflow exists from $z = 1.5$ km to $z = 5$ km. The tangential-velocity plot (Figure 12b) shows that the tornado cyclone has a larger diameter above

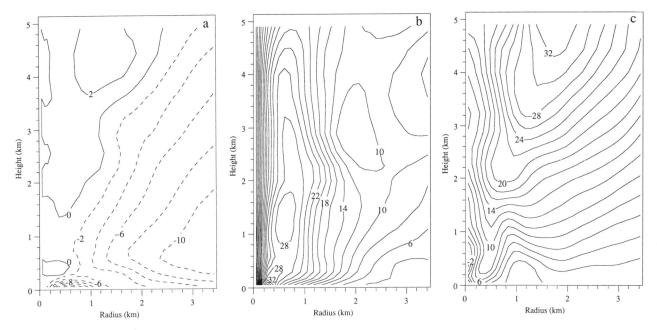

Fig. 12. Radius-height section of azimuthally averaged mesocyclone features at 102 min in the surface-friction simulation. (a) Radial velocity. Contour interval is 2 m s^{-1}. (b) Tangential velocity with contour interval of 2 m s^{-1}. (c) Vertical velocity with contour interval of 2 m s^{-1}.

$z = 1$ km than in the free-slip simulation, while near the surface the radius of maximum winds has contracted to a smaller core radius of 250 m. Maximum tangential velocities are 36 m s^{-1} and are confined to a shallow layer near the surface. The vertical-velocity field (Figure 12c) has updrafts of 10 m s^{-1} near the axis below $z = 1$ km, with a slight downdraft (-2 m s^{-1}) on the central axis at $z = 600$ m.

The results described in Figures 11 and 12 are consistent with the results from *Howells et al.*'s [1988] axisymmetric vortex simulations. The inclusion of surface friction creates significant radial inflow into the base of the tornado cyclone. At low levels the radial inflow transports angular momentum closer to the axis than in the free slip simulation, resulting in higher tangential velocities within the tornado at the surface. An increase in low-level convergence is associated with stronger upward motion in the tornado. At higher levels, radial outflow is present closer to the axis in the friction simulation than in the free-slip simulation. This flow divergence causes the core radius for the tornado cyclone to increase with height and explains the reduction in vertical vorticity above $z = 1$ km in the friction simulation.

4. SUMMARY

Numerical simulations of tornadogenesis within a supercell thunderstorm are carried out using a three-dimensional nested-grid numerical model with very high horizontal and vertical resolution in the low-level mesocyclone region. Simulations using free-slip and surface-friction lower boundary conditions are compared to assess the impact of surface friction on the fine-scale structure of the storm.

The high-resolution storm simulation using free-slip boundary conditions develops two separate tornado cyclones over a 40-min period. The first tornado cyclone develops aloft and then extends to the surface in a similar manner to the one in the Union City storm [*Brown et al.*, 1978]. Maximum surface wind speeds exceed F2 intensity during both events. The second tornado cyclone is stronger than the first and maintains its intensity over a 10-min period. Unlike the first tornado cyclone, the intensification of the second circulation occurs simultaneously from the surface to 2 km in height. This evolution is similar to the Del City tornado cyclone [*Johnson et al.*, 1987].

Time series analyses of the model output show that the development of the two tornado cyclones within the modeled storm is preceded 5–7 min by intensification of updrafts to magnitudes of 30–40 m s^{-1} at an altitude of 3–4 km. A similar phenomenon is noted in the updraft evolution of the Del City storm [*Johnson et al.*, 1987]. Simultaneously with the updraft intensification, rapid pressure falls develop within the mesocyclone through a depth of 4 km. After the development of these low-pressure cores, maximum vorticity values are observed within the storm, with maximum values at the surface and somewhat smaller values aloft. The development of these low-pressure regions and the intensification of updrafts at 3–4 km appear to be linked together. Further analyses are being done to understand this relationship.

To assess the impact of surface friction on the tornado cyclone, the fine-grid simulation is restarted from 94 min with surface friction added. An intense vortex core with many characteristics of tornadoes such as strong low-level radial inflow, large vorticity, and large low-level updrafts develops. At low levels the diameter of the vortex is approximately half the size of the vortex in the free-slip simulations. Surface wind speeds are 10–15% larger, and low-level updrafts around the tornado are 5 times greater at low levels than in the free-slip simulation. These changes in the low-level vortex structure due to the inclusion of surface friction are similar to previous studies of axisymmetric tornadolike vortices [Howells et al., 1988]. Further analysis of the reported simulations is underway, and new higher-resolution simulations are planned to investigate in detail the generation and maintenance of tornadoes.

Acknowledgments. The authors wish to thank Rich Rotunno, Jerry Straka, Bill Skamarock, Brian Fiedler, Harold Brooks, Don Burgess, Brian Jewett, and Robert Davies-Jones for many useful discussions over the past several years about this work. The authors also wish to thank the anonymous reviewers and the editors who were very helpful with their constructive criticisms of the paper. This work was supported by the National Science Foundation under the NSF grant ATM-87-00778 and by the National Center for Supercomputing Applications at the University of Illinois.

References

Brandes, E. A., Mesocyclone evolution and tornado generation within the Harrah, Oklahoma storm, *NOAA Tech. Memo. ERL NSSL-81*, 28 pp., Natl. Severe Storms Lab., Norman, Okla., 1977.

Brandes, E. A., Finestructure of the Del City-Edmond tornadic mesocirculation, *Mon. Weather Rev., 109*, 635–647, 1981.

Brandes, E. A., Relationships between radar derived thermodynamic variables and tornadogenesis, *Mon. Weather Rev., 112*, 1033–1052, 1984.

Brandes, E. A., R. P. Davies-Jones, and B. C. Johnson, Streamwise vorticity effects on supercell morphology and persistence, *J. Atmos. Sci., 45*, 947–963, 1988.

Brown, R. A., D. W. Burgess, and K. C. Crawford, Twin tornado cyclones within a severe thunderstorm, *Weatherwise, 26*, 63–71, 1973.

Brown, R. A., L. R. Lemon, and D. W. Burgess, Tornado detection by pulsed Doppler radar, *Mon. Weather Rev., 106*, 29–38, 1978.

Clark, T. L., Numerical simulations with a three-dimensional cloud model: Lateral boundary condition experiments and multicellular severe storm simulations, *J. Atmos. Sci., 36*, 2191–2215, 1979.

Davies-Jones, R. P., D. W. Burgess, and M. P. Foster, Test of helicity as a tornado forecast parameter, in *Preprints, 16th Conference on Severe Local Storms*, pp. 588–592, American Meteorological Society, Boston, Mass., 1990.

Droegemeier, K. K., The numerical simulation of thunderstorm outflow dynamics, Ph.D. thesis, 695 pp., Univ. of Ill., Urbana-Champaign, 1985.

Fiedler, B. H., and R. Rotunno, A theory for maximum windspeeds in tornado-like vortices, *J. Atmos. Sci., 43*, 2328–2340, 1986.

Howells, P. A. C., R. Rotunno, and R. K. Smith, A comparative study of atmospheric and laboratory-analogue numerical tornado-vortex models, *Q. J. R. Meteorol. Soc., 114*, 801–822, 1988.

Johnson, K. W., P. S. Ray, B. C. Johnson, and R. P. Davies-Jones, Observations related to the rotational dynamics of the 20 May 1977 tornadic storms, *Mon. Weather Rev., 115*, 2463–2478, 1987.

Klemp, J. B., and R. Rotunno, A study of the tornadic region within a supercell thunderstorm, *J. Atmos. Sci., 40*, 359–377, 1983.

Klemp, J. B., and R. B. Wilhelmson, The simulation of three-dimensional convective storm dynamics, *J. Atmos. Sci., 35*, 1070–1096, 1978a.

Klemp, J. B., and R. B. Wilhelmson, Simulations of right- and left-moving storms produced through storm splitting, *J. Atmos. Sci., 35*, 1097–1110, 1978b.

Klemp, J. B., R. B. Wilhelmson, and P. S. Ray, Observed and numerically simulated structure of a mature supercell thunderstorm, *J. Atmos. Sci., 38*, 1558–1580, 1981.

Lemon, L. R., and C. A. Doswell III, Severe thunderstorm evolution and mesocyclone structure as related to tornadogenesis, *Mon. Weather Rev., 107*, 1184–1197, 1979.

Moncrieff, M. W., and J. S. A. Green, The propagation and transfer properties of steady convective overturning in shear, *Q. J. R. Meteorol. Soc., 98*, 336–352, 1972.

Ray, P. S., B. C. Johnson, K. W. Johnson, J. S. Bradberry, J. J. Stephens, K. K. Wagner, R. B. Wilhelmson, and J. B. Klemp, The morphology of several tornadic storms on 20 May 1977, *J. Atmos. Sci., 38*, 1644–1663, 1981.

Rotunno, R., A study in tornado-like vortex dynamics, *J. Atmos. Sci., 36*, 140–155, 1979.

Skamarock, W. C., and J. B. Klemp, Adaptive models for 2-D and 3-D nonhydrostatic atmospheric flow, in *6th International Conference on Numerical Methods in Laminar and Turbulent Flow*, pp. 1413–1424, Pineridge, Swansea, Wales, 1989.

Snow, J. T., A review of recent advances in tornado dynamics, *Rev. Geophys. Space Phys., 20*, 953–964, 1982.

Tripoli, G. J., and W. R. Cotton, The Colorado State University three-dimensional cloud/mesoscale model—1982, I, The general theoretical framework and sensitivity experiments, *J. Rech. Atmos., 16*, 185–220, 1982.

van Leer, B., Towards the ultimate conservative difference scheme, IV, A new approach to numerical convection, *J. Comput. Phys., 23*, 276–299, 1977.

Walko, R. L., Plausibility of substantial dry adiabatic subsidence in a tornado core, *J. Atmos. Sci., 45*, 2251–2267, 1988.

Weisman, M. L., and J. B. Klemp, The dependence of numerically simulated convective storms on vertical wind shear and buoyancy, *Mon. Weather Rev., 110*, 504–520, 1982.

Weisman, M. L., and J. B. Klemp, The structure and classification of numerically simulated convective storms in directionally varying wind shears, *Mon. Weather Rev., 112*, 2479–2498, 1984.

Wicker, L. J., A numerical simulation of a tornado-scale vortex in a three-dimensional cloud model, Ph.D. thesis, 264 pp., Univ. of Ill., Urbana-Champaign, 1990.

Wicker, L. J., D. W. Burgess, and H. Bluestein, The pre-storm environment of the severe weather outbreak in western Oklahoma on May 22, 1981, in *Preprints, 13th Conference on Severe Local Storms*, pp. 281–282, American Meteorological Society, Boston, Mass., 1984.

Wilhelmson, R. B., and C. S. Chen, A simulation of the development of successive cells along a cold outflow boundary, *J. Atmos. Sci., 39*, 1446–1483, 1982.

Wilhelmson, R. B., and J. B. Klemp, A numerical study of storm splitting that leads to long-lived storms, *J. Atmos. Sci., 35*, 1974–1986, 1978.

Wilhelmson, R. B., and J. B. Klemp, A three-dimensional analysis of splitting severe storms on 3 April 1964, *J. Atmos. Sci., 38*, 1581–1600, 1981.

Wilson, J. W., Tornadogenesis by nonprecipitation induced wind shear lines, *Mon. Weather Rev., 114*, 270–284, 1986.

Tornado Spin-Up Beneath a Convective Cell: Required Basic Structure of the Near-Field Boundary Layer Winds

ROBERT L. WALKO

Department of Atmospheric Science, Colorado State University, Fort Collins, Colorado 80523

1. INTRODUCTION

Tornadoes occur in many sizes, shapes, and configurations, but all have in common a central core of highly organized, intense vorticity. The various fundamental fluid dynamical processes capable of producing such a vorticity field are well known. How those processes are actually brought about by the tornadic storm and its environment remain topics of study. The present paper addresses this problem with the aid of three-dimensional (3-D) numerical simulations.

The related topic of vertical vorticity in a convective cell updraft has been discussed by *Lilly* [1982], *Davies-Jones* [1984], *Rotunno and Klemp* [1985] and others and has apparently been adequately explained. Such vorticity is theorized to arise from ambient horizontal vorticity present in vertically sheared boundary layer winds, when horizontal gradients of updraft strength tilt some vorticity into the vertical. Although it has been speculated that the vertical vorticity in a tornado core may arise from the same mechanism, there is a basic weakness to this explanation, as pointed out by *Davies-Jones* [1982]. He argues that if only updraft and no downdraft is causing the tilting of vorticity, vertical vorticity remains zero close to the ground and is acquired by a parcel only as it rises away from the surface. Thus vorticity tilting by an updraft alone can account for rotation of a thunderstorm convective updraft above the planetary boundary layer but appears to fall short of explaining the vertical vorticity in a tornado very close to the ground.

The more traditional and simpler explanation for the intense vorticity of a tornado core is that it becomes concentrated from a larger horizontal area of weaker vertical vorticity, in some cases identifiable as a tornado cyclone [*Church et al.*, 1979], by the field of horizontal convergence beneath the thunderstorm updraft. However, this explanation is incomplete unless it also accounts for the origin of that weaker vertical vorticity. *Wakimoto and Wilson* [1989] documented a case where vertical vorticity along a quasi-stationary convergence line preceded convective development, and subsequent convection locally intensified the convergence to the point of spinning up a tornado from the background vorticity field. In other cases it has been argued that the thunderstorm and its immediate environment interact to generate a local field of vertical vorticity. For example, *Davies-Jones* [1982] argues theoretically and cites observational evidence that a rain-cooled downdraft may introduce vertical vorticity by tilting environmental horizontal vorticity, and *Rotunno and Klemp* [1985] describe numerical simulations in which solenoids at the edge of a rain-cooled thunderstorm outflow generate horizontal vorticity, which is subsequently tilted into the vertical.

From the preceding discussion, we may identify several possible general mechanisms for tornadogenesis.

1. Weak to moderate vertical vorticity is present in the prestorm environment and is concentrated into a tornado by horizontal convergence beneath the convective updraft.

2. Horizontal vorticity is present in the boundary layer winds of the prestorm environment. It gets tilted into the vertical by a vertical motion field which contains no subsidence.

3. Horizontal vorticity is present in the boundary layer winds of the prestorm environment. It gets tilted into the vertical by a vertical motion field which contains some subsidence.

4. The vorticity responsible for the tornado is generated solenoidally by the effects of the thunderstorm and is subsequently tilted into the vertical.

The above list, as specifically worded, does not describe all possible mechanisms of generating an intense tornadolike vortex, but it addresses those mechanisms proposed in the foregoing discussion. Mechanisms 2–4, which involve vorticity tilting, can each be divided into two subcategories.

1. Vorticity tilting results in only weak-to-moderate ver-

The Tornado: Its Structure, Dynamics, Prediction, and Hazards.
Geophysical Monograph 79
Copyright 1993 by the American Geophysical Union.

tical vorticity, which must be further concentrated by horizontal convergence to form a tornado.

2. Concentration of vertical vorticity by horizontal convergence is not required for tornadogenesis. A tornado core can be produced by tilting horizontal vorticity of equal intensity.

The present paper explores possible mechanisms of tornadogenesis through a series of 3-D numerical simulations of an idealized storm and its environment. Each simulation is designed to focus on one of the above mechanisms in order to examine whether it shows any tendency toward tornadic development. Although tornadoes occur under a wide variety of circumstances, including apparently some instances where strong buoyant convection aloft is absent, we confine attention here to tornadoes which form beneath a strong updraft, as in a supercell storm. In all simulations the convective updraft is generated and maintained in the center of the computational domain, located at $(x, y) = (0, 0)$, by imposing a constant-in-time heat source of roughly cylindrical shape to model latent heat release. The heating rate and the dimensions of the heated volume approximate those of a supercell storm updraft, and quasi-steady updraft speeds around 50 m s^{-1} are obtained over a 50-km^2 area. The environmental lapse rate is 1 K km^{-1} in the lowest 10 km of the domain and 10 K km^{-1} above. The individual simulations differ from each other in the initial low-level ambient wind and temperature fields and in the presence or absence of a low to middle level heat sink to approximate evaporative cooling of precipitation and a consequent cool downdraft. The results of these simulations indicate which categories of ambient conditions will interact with the convection to produce a tornadolike vortex and which will not. For each category, two or more distinct simulations were run. Except for cases which included a heat sink, simulations within the same category yielded the same qualitative result.

Computations are performed with the Colorado State University Regional Atmospheric Modeling System (RAMS) using a series of three telescoping nested grids having horizontal grid cell dimensions of 1600 m, 400 m, and 100 m. Vertical spacing on all grids begins at 20 m near the ground, stretches gradually to 500 m, and remains at 500 m up to the model top 20 km above the ground. All grids contain 50 × 50 × 54 mesh points. A slightly coarser mesh with fewer grid points is used for some of the experiments.

2. RESULTS

The first numerical experiments to be described belong to the category in which vertical vorticity is contained in the initial low-level wind field. For comparison with previous vortex simulations using axisymmetric models [e.g., *Howells and Smith*, 1983; *Walko*, 1988], I first choose as an initial condition a field of solid rotation (uniform vertical vorticity of the order of 100 times that of the Earth) centered on a vertical axis in the center of the domain. Horizontal convergence at low levels induced by the convection spins up a tornadolike vortex within 15 min. The flow retains a high

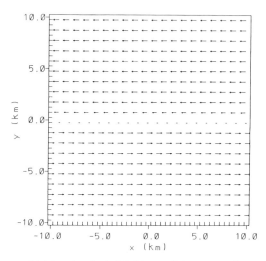

Fig. 1. Initial low-level wind field containing two regions of uniform velocity with a shear zone in between. Winds decrease linearly with height, reaching and remaining at zero above 2 km.

degree of axial symmetry and qualitatively resembles the previous axisymmetric results.

Next, an experiment is initialized with a second type of wind field which contains vertical vorticity: straight-line flow with a region of horizontal shear as shown in Figure 1. (Except where noted, this and all other figures are plotted from fields on the intermediate nested grid. All "low-level" and "middle level" cross sections were taken from the 90-m and 4-km levels, respectively, but are each representative of fairly thick layers.) The shear zone contains equal amounts of vertical vorticity and deformation, but the vorticity alone is the crucial ingredient for vortex spin-up. This category of flow is intended to approximate the vertical-vorticity distribution of the observed wind field prior to the July 2 Convective Initiative and Downburst Experiment (CINDE) tornado [*Uttal et al.*, 1989]. As in the previous axisymmetric experiment, the convectively driven low-level convergence concentrates ambient vertical vorticity in the center of the domain, producing a tornadolike vortex. Figure 2 illustrates the low-level horizontal wind vectors at 36 min into the simulation when the strongest wind is obtained. A noticeable feature is that the winds 5–10 km away from the vortex are asymmetric, consisting of two main inflow jets feeding the vortex: an easterly jet to the north and northeast of the center and a westerly jet to the south and southwest. These grew out of the initial two uniform-wind regions. Due east and west of the vortex at the same distance, the winds are light. Closer to the vortex center, the winds become much more axisymmetric, a feature clearly seen in the winds of the highest-resolution nested grid (Figure 3). Additional experiments were performed with similar initial wind fields, differing in actual wind strength and in the width of the shear zone; all resulted in the spin-up of a concentrated vortex as described above. I note here that the presence of small closed circulations, apparently due to Kelvin-Helmholtz

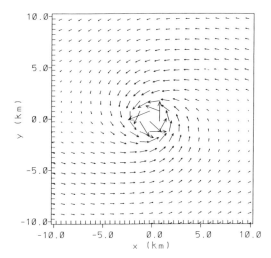

Fig. 2. Low-level wind field 36 min after the initial condition shown in Figure 1.

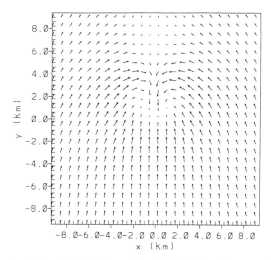

Fig. 4. Horizontal velocity vectors at middle levels in experiment with low-level baroclinic region.

instability, which were reported along the convergence line prior to the July 2 CINDE tornado [*Wakimoto and Wilson*, 1989], is incidental to tornadogenesis. The tornado originates from a much larger circulation (see section 4) than the Kelvin-Helmholtz vortices.

The next experiment is initialized with no vertical or horizontal vorticity but with a pool of air 1 km deep, which has been cooled by 10 K, in the north half of the domain. Because such a temperature field would induce a northerly density current, a southerly wind is initialized at all levels to keep the baroclinic zone situated beneath the main updraft in the center of the domain. This experiment is designed to examine the possible role in tornadogenesis of horizontal vorticity production by solenoids along the baroclinic zone. The main results of this experiment are summarized by

Figures 4 and 5. The low-level baroclinically generated horizontal vorticity gets tilted into the vertical along the eastern and western edges of the updraft, resulting in a pair of counterrotating vortices at middle levels. Near the ground, however, no significant vertical vorticity is generated. The tilting in this case is done by a vertical-motion field which contains no significant downdraft. Other similar experiments were performed with different initial temperature differences and southerly wind speeds, and with the cold air pool occupying a quarter or three-quarter pie-shaped section of the domain. In every case there was no tendency for vertical vorticity develop close to the surface.

To investigate the possible role of ambient vertical shear (horizontal vorticity) in tornadogenesis, a series of simulations was performed having no horizontal gradients of either initial wind or temperature but containing various amounts

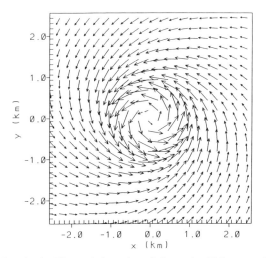

Fig. 3. As in Figure 2 but plotted from the highest-resolution nested grid.

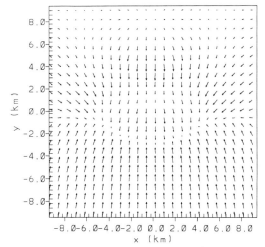

Fig. 5. As in Figure 4 but at low levels.

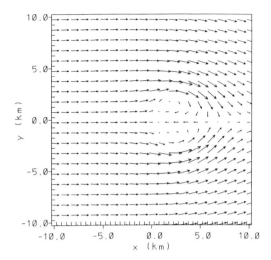

Fig. 6. Horizontal velocity vectors at middle levels in experiment with low-level westerly shear.

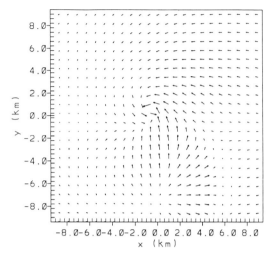

Fig. 8. Horizontal velocity vectors at low levels in experiment with low-level westerly shear and a heat sink southwest of the convective updraft.

of (westerly) vertical wind shear in the lowest 1–3 km. Winds at the lowest levels were easterly and graduated to westerlies higher up, so that the vertically averaged wind velocity was similar to the (zero) horizontal velocity of the zone of convective forcing.

Figures 6 and 7 characterize the results of these experiments. A pair of counterrotating vortices forms at middle levels (Figure 6) within the main updraft. The easterly flow between the vortices can be explained as resulting from the upward transport of the ambient easterly momentum in the lower boundary layer. The surface winds (Figure 7) are strongly convergent but show no tendency to develop significant vertical vorticity. Variations of the experiment shown were performed with different strengths and vertical limits of wind shear. Some experiments contained an ambient southerly wind, as if the main updraft were propagating to the right of the shear vector, thus encountering vertically veering winds. This caused the southern (cyclonic) vortex of the pair to be stronger, as explained by *Davies-Jones* [1984] and *Rotunno and Klemp* [1985]. None of these experiments, however, resulted in vertical vorticity production near the ground. As in the previous set of experiments, tilting was caused by a field of vertical motion in which subsidence was insignificant.

A final set of experiments was run with low-level westerly shear as in the previous set and, in addition, with a heat sink located to the southwest of the main updraft between 2 and 4 km in height. The heat sink represents diabatic cooling due to evaporation of precipitation and induces a downdraft approximately 5 km in diameter.

Figure 8 shows the low-level horizontal velocity field in one of these experiments 16 min after initialization. The winds at this level were easterly initially, and the winds remain generally easterly at the time shown. However, a region south of the domain center contains strong westerly, divergent flow. The temperature field at the same level and time is shown in Figure 9. A pool of cool air is present south of the domain center and coincides with the westerly current. The cool air is the outflow from the heat-sink-induced downdraft, and the westerly momentum, which has carried the cool air eastward from the heat sink location, results from that downdraft transporting the ambient westerly momentum from aloft downward to the surface. Vertical momentum transport in a vertically sheared flow was discussed by *Wiin-Nielsen* [1973, pp. 96–97] and is associated with tilting of ambient horizontal vorticity into the vertical. The vertical vorticity thus produced resides along the boundary between the distinct air masses and is responsible in the present experiment for the formation of a tornadolike vortex near the center of the domain.

Other experiments performed in this set differed in the

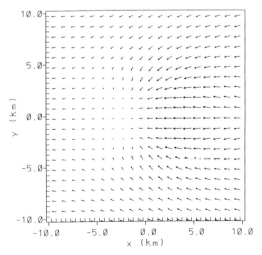

Fig. 7. As in Figure 6 but at low levels.

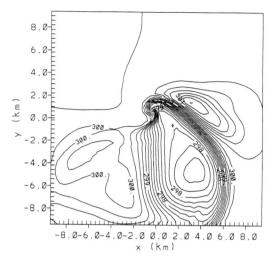

Fig. 9. Temperature field (in degrees Kelvin) at low levels corresponding to Figure 8.

strength, size, and location of the heat sink and in the strength of the westerly shear. Not all of these resulted in a well-defined tornadolike vortex, but the majority did. A critical factor in preventing vortex development in some simulations was a rapidly propagating gust front. This prevented the newly generated vertical vorticity from spending sufficient time beneath the convective updraft where horizontal convergence would concentrate it.

3. Vortex-Line Analysis

In this section and the next I investigate in greater detail the mechanism by which the vortex shown in Figure 8 was produced. The vertical vorticity 90 m above the ground (outside the friction layer, which is less than 20 m deep) is plotted in Figure 10. As can be seen, values in the vortex

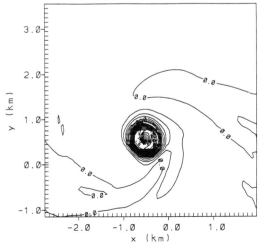

Fig. 10. Vertical vorticity corresponding to Figures 8 and 9.

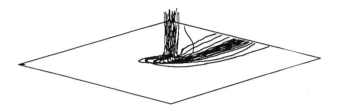

Fig. 11. Three-dimensional perspective of vortex lines present in the flow from which Figures 8–10 are extracted. The view is from the southeast, and the vertical scale is exaggerated by a factor of 2.

core, centered at $(x, y) = (-500 \text{ m}, 500 \text{ m})$, are far greater than elsewhere.

To further illustrate the kinematics of tornadogenesis, I construct a field of vortex lines, each of which represents a vortex tube, say, of strength S. Beginning at the horizontal surface in Figure 10, a starting point for each vortex line is selected on the basis of the strength and distribution of vertical vorticity across the surface. A total of N starting points are determined, where NS is equal to the area-integrated vertical vorticity across the horizontal slice through the core. Then, a curve is constructed through each of the starting points, such that the curves are everywhere parallel to the local 3-D vorticity vector. Although physically a vortex line either is a closed curve or terminates at a rotating solid surface (which is not applicable to the present work since Earth rotation is neglected), I have, for clarity, constructed only segments of the curves, terminating them when they reach a height of 2 km as they rise through the tornado core, and when they reach a lateral boundary of the nested grid on which they are computed. These vortex line segments are illustrated in 3-D perspective in Figures 11 and 12. Note in particular that as the vortex lines are followed downward through the vortex core, they turn into the horizontal within a few meters of the ground and continue laterally rather than terminating at the ground. This follows from the facts that velocity and vertical vorticity are zero at the ground surface (which was observed in the construction of the vortex lines) and that the friction layer adjacent to the ground is a region of strong horizontal vorticity. In the present simulation it can be seen (in Figure 11) that the vortex lines all eventually lead northward from the vortex

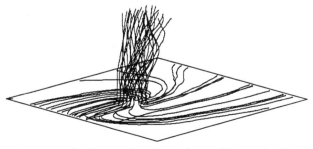

Fig. 12. As in Figure 11 but plotted for vorticity on the highest-resolution nested grid and with no vertical exaggeration.

core, in keeping with the fact that the friction layer shear is easterly over most of the domain. However, in the immediate vicinity of the core, the vortex lines spiral outward almost uniformly over all azimuths for a distance of several core radii, before they finally turn toward the north. This is consistent with the general feature of tornadoes that just above the friction layer and out to several core radii, their azimuthal wind component is strong and nearly axisymmetric. Only if a tornado were to translate horizontally at a speed comparable to its peak azimuthal wind speed would vortex lines spread outward mostly on one side of the core. For example, consider a tornado which is axisymmetric just above the friction layer and has a rotational speed at the edge of the core (relative to the axis) equal to the translation speed of the axis relative to the ground. For counterclockwise rotation the ground-relative wind at the left edge of the core would be zero, and the mean radial vorticity component within the friction layer on that side must likewise be zero. Vortex lines descending through the core to the friction layer would turn outward mostly on the right side of this vortex.

Consider a flow like that shown in Figure 8 with the same slowly evolving, larger-scale structure but at a short time prior to the occurrence of a concentrated tornadolike vortex. Follow the kilometers-wide collection of vortex lines in Figure 11 southward toward the future location of the tornado. These lines do not yet converge radially toward a concentrated bundle because the tornado does not yet exist. Thus they either continue away from the region while remaining in the friction layer or rise into the free atmosphere without being concentrated. Eventually, they must all rise into the free atmosphere if not already there because they will soon extend upward through the tornado core. In the free atmosphere where diffusion is negligible, and in cases where solenoidal effects can also be ignored, vortex lines advect with the flow [*Batchelor*, 1967]. Thus radial inflow must advect the vortex lines from their initial broad spread to their location in the tornado core. This argument, although admittedly not a proof, implies that the computed configuration of vortex lines is evidence that flow convergence toward the vortex center from a distance of several core radii operates to concentrate vorticity into the core as the final step in tornadogenesis. Hence, while tilting of vortex lines is probably responsible for initiating vertical vorticity beneath the storm, that tilting must take place on a horizontal scale of at least several tornado core radii, and horizontal convergence (radial inflow) concentrates that vorticity as the final stage of tornadogenesis.

4. Circulation Analysis

According to Stokes' theorem the circulation along a closed curve circling the perimeter of the vortex core is a measure of the mean axial vorticity in the core. The circulation theorem states that the circulation along any closed material curve in a fluid changes only as a result of solenoidal or viscous terms. By selecting a horizontal circular path of radius 500 m centered at the point $(x, y, z) = (-500$ m, 500

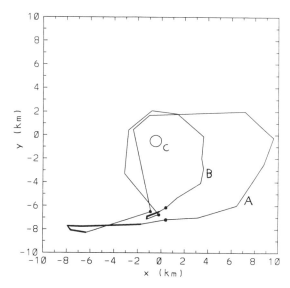

Fig. 13. Succession of three approximate paths traversed by a material curve. Curves A, B, and C denote the paths at simulation times of 5, 11, and 16 min, respectively, the final curve closely surrounding a tornadolike vortex just above the friction layer. Heavy segment of curve A denotes section which is approximately 3 km above the ground, and the two dots indicate where the curve becomes close to the ground. Between the heavy segment and each dot the curve is sloping, while the remainder of the curve is near the ground. Curve B is similarly notated, but the heavy segment denotes a height near 2 km.

m, 90 m), tracing that path as a material curve backward in time, and evaluating the terms in the equation for the circulation around the curve throughout its history, the origin of the concentrated vorticity in Figure 10 can be determined. This technique was applied by *Rotunno and Klemp* [1985] in a less idealized numerical simulation of a tornadic storm. Here, the material curve is approximated by a sequence of 3000 Lagrangian tracer particles. Each particle is advected backward in time using a time series of complete 3-D velocity fields obtained from the intermediate grid at 15-s intervals. Advection is performed in 1-s steps, and the particle velocity is recomputed each step by trilinear spatial interpolation from the gridded velocity components. The total number of tracer particles is sufficient for adjacent ones along the material curve to remain within one grid interval throughout their history, with only minor exceptions.

Upon tracking the material curve backward in time, it is found to exceed the lateral boundaries of the intermediate grid for the first 5 min of the numerical simulation. The curve at $t = 5$ min is depicted as curve A in Figure 13. Because of the initial westerly shear, which is still essentially intact this early in the simulation, it is obvious that such a path has positive circulation. The actual numerical value of the circulation is 4×10^5 m^2 s^{-1}, which is comparable to probably the largest documented circulation around a tornado core [*Walko*, 1988]. As the numerical simulation advances for the first several minutes, and the material curve evolves toward its final circular shape immediately outside the vortex core, the circulation around the material curve decreases steadily due almost entirely to a solenoidal influence. The cold air in the southwest and later southern part of the domain is primarily responsible for this. The material curve at $t = 11$ min (depicted by curve B of Figure 13) has nearly all descended to the surface, such that solenoids can no longer strongly influence the circulation. The circulation at this time is 1.5×10^5 m^2 s^{-1}. The curve contracts steadily toward its final position under the strongly convergent wind field beneath the updraft. The circulation decreases slowly at first, primarily due to surface drag, until the final minute when velocities along the curve become stronger, and the circulation decreases very rapidly. At $t = 16$ min, the material curve reaches its designated "final" position (depicted as curve C in Figure 13) and has a circulation of 5×10^4 m^2 s^{-1}. Hence in this simulation, virtually all circulation in the vortex was derived from the initial low to middle level shear, with solenoidal effects acting to weaken the circulation. This result is opposite to the results of *Rotunno and Klemp* [1985] in which solenoids actually generated the circulation within 15 min prior to the tornado. Although the solenoidal term in the present case was a circulation sink, the downdraft, which is instrumental in producing the solenoids, played the crucial role in tilting horizontal vorticity into the vertical or, equivalently, tilting the circulating material curve into the horizontal, so that vertical vorticity at the surface then could be concentrated through horizontal convergence.

The circulation analysis may appear, at first glance, to be partly in contradiction with the vortex-line analysis of the previous section. In Figure 11 the vortex lines shown that ascend through the vortex core approach the general region from large distances to the north, while the vortex lines that pass through the area enclosed by curve A in Figure 13 approach the general region from large distances to the south. It seems likely that solenoidal recombination of vortex lines in baroclinic regions and diffusive recombination near the ground may account for this difference. For example, the initial flow consists of a shallow friction layer with densely packed vortex lines pointing toward the south and a much deeper shear layer above with sparsely packed vortex lines pointing toward the north. Both solenoidal and viscous effects are capable of combining segments of oppositely oriented vortex lines such that a line in the friction layer approaching from the north turns upward and returns to the north in the deeper overlying shear layer rather than continuing to the south in the friction layer as it did initially. A similar phenomenon occurs simultaneously with vortex lines approaching from the south. A detailed analysis of the vortex-line evolution would reveal the processes involved but has not been performed here.

5. Conclusion

The experimental results presented here, and several others in the same categories but with differing parameters, support the argument of *Davies-Jones* [1982] that tilting of horizontal vorticity by an updraft alone is not capable of producing a tornado vortex extending to the ground. Either vertical vorticity must be present in the ambient winds, or it must be tilted from horizontal vorticity by motions which include a downdraft. I have argued here that such tilting must occur at a horizontal scale of at least several tornado core radii and that the tornado is then formed by radial inflow concentrating vorticity from this scale to the core scale.

Acknowledgments. This research was supported by Army Research Office contract DAAL03-86-K-0175. Computations were performed on the Stardent 3040 at the Department of Atmospheric Science at Colorado State University. The Stardent 3040 was purchased under support of ARO. The model used in this research, the RAMS, was developed at CSU under support of the ARO and NSF.

References

Batchelor, G. K., *An Introduction to Fluid Dynamics*, 615 pp., Cambridge University Press, New York, 1967.

Church, C. R., J. T. Snow, G. L. Baker, and E. M. Agee, Characteristics of tornado-like vortices as a function of swirl ratio: A laboratory investigation, *J. Atmos. Sci.*, 36, 1755–1776, 1979.

Davies-Jones, R. P., A new look at the vorticity equation with application to tornadogenesis, in *Preprints, 12th Conference on Severe Local Storms*, pp. 249–252, American Meteorological Society, Boston, Mass., 1982.

Davies-Jones, R. P., Streamwise vorticity: The origin of rotation in supercell storms, *J. Atmos. Sci.*, 41, 2991–3006, 1984.

Howells, P., and R. K. Smith, Numerical simulations of tornado-like vortices, 1, Vortex evolution, *Geophys. Astrophys. Fluid Dyn.*, 27, 253–284, 1983.

Lilly, D. K., The development and maintenance of rotation in convective storms, in *Topics in Atmospheric and Oceanographic Sciences: Intense Atmospheric Vortices*, edited by L. Bengtsson and J. Lighthill, pp. 149–160, Springer-Verlag, New York, 1982.

Rotunno, R., and J. Klemp, On the rotation and propagation of simulated supercell thunderstorms, *J. Atmos. Sci.*, 42, 271–292, 1985.

Uttal, T., B. E. Martner, B. W. Orr, and R. M. Wakimoto, High resolution dual-Doppler radar measurements of a tornado, in *Preprints, 24th Conference on Radar Meteorology*, pp. 62–65, American Meteorological Society, Boston, Mass., 1989.

Wakimoto, R. M., and J. W. Wilson, Non-supercell tornadoes, *Mon. Weather Rev.*, 117, 1113–1140, 1989.

Walko, R. L., Plausibility of substantial dry adiabatic subsidence in a tornado core, *J. Atmos. Sci.*, 45, 2251–2267, 1988.

Wiin-Nielsen, A., *Compendium of Meteorology for Class I and II Personnel. Vol I, Part 1: Dynamic Meteorology*, WMO Publ. 364, 334 pp., World Meteorol. Organ., Geneva, 1973.

Environmental Helicity and the Maintenance and Evolution of Low-Level Mesocyclones

HAROLD E. BROOKS, CHARLES A. DOSWELL III, AND ROBERT DAVIES-JONES

National Severe Storms Laboratory, NOAA, Norman, Oklahoma 73069

1. INTRODUCTION

Significant tornadoes are almost always associated with supercell thunderstorms. Supercells are characterized by their long life and the presence of a mesocyclone, a persistent region of rotation several kilometers across through a deep layer of the storm. Approximately 90% of mesocyclones observed during the Joint Doppler Operational Project (JDOP) were associated with severe weather, and 50% produced tornadoes [*Burgess and Lemon*, 1990]. As a result, Doppler radar identification of a mesocyclone is considered a useful tool for issuing severe weather warnings. However, an important question is why radar-observed mesocyclones do not all produce tornadoes. As the WSR-88D Doppler radars are installed across the United States, our understanding of the connection between radar-observed mesocyclones and surface severe weather will become even more important. In particular, establishing why mesocyclones fail to produce significant tornadoes can reduce the possibility of high false alarm rates based on radar signatures of mesocyclones.

The detection problem of distinguishing between tornadic and nontornadic thunderstorms is paralleled by the forecast problem of distinguishing between environments that have the potential to support tornadic and nontornadic severe storms. Improving our ability to solve this forecast problem has obvious benefits in the public warning process. Most of the research community's focus has been on tornadic rather than nontornadic environments [e.g., *Darkow and Fowler*, 1971; *Darkow and McCann*, 1977]. *Turcotte and Vigneux* [1987] examined all of the severe weather days for 1984–1986 in the Quebec Weather Centre forecast area and a large number of nonsevere thunderstorm days (mostly from 1986) and considered the amount of convective available potential energy (CAPE) and the mean shear over the lowest 4 km in the morning soundings. While they agreed with the result of *Rasmussen and Wilhelmson* [1983] that tornadic environments are characterized by moderate to high CAPE and shear, they found that they could not discriminate between tornadic and nontornadic severe thunderstorms (Figure 1). Using a shear/CAPE diagram, they could discriminate between severe and nonsevere thunderstorms with a high level of accuracy. On one hand, this result is extremely encouraging, implying that forecasters can have some confidence in a forecast of severe thunderstorms just by looking at these two parameters. On the other hand, it implies that something additional must be done to discriminate tornadic from nontornadic environments.

Patrick and Keck [1987] found that parameters describing low-level curvature of the hodograph could be useful in making a tornado/no-tornado discrimination. On the basis of recent theoretical work [*Davies-Jones*, 1984; *Lilly*, 1986] and preliminary operational testing at the National Weather Service Forecast Office in Norman, Oklahoma, *Davies-Jones et al.* [1990] proposed that the most appropriate measure of hodograph curvature is storm-relative helicity. The environmental storm-relative helicity, $\mathcal{H}$, is defined by

$$\mathcal{H}(\mathbf{c}) = -\int_0^h \mathbf{k} \cdot (\mathbf{V} - \mathbf{c}) \times \frac{\partial \mathbf{V}}{\partial z} \, dz \tag{1}$$

where h is an assumed inflow depth, $\mathbf{c}$ is the storm motion vector, $\mathbf{V}(z)$ is the environmental wind profile (from a sounding, profiler, Doppler radar, or other sensor), and $\mathbf{k}$ is the unit vector in the vertical. Helicity has a simple geometric representation as minus twice the area swept out by the storm-relative velocity vector between 0 and h on a hodograph. This property allows the operational forecaster to estimate values of helicity from a hodograph and use whatever information is available to him or her, such as numerical model guidance or asynoptic observations, to update helicity rapidly. *Davies-Jones et al.* [1990] found that the storm-relative helicity through 3 km altitude shows promise as a tornado forecasting tool. *Leftwich* [1990] and *Woodall* [1990] also reported encouraging results in their use of helicity with forecast winds. Because of the success of the

The Tornado: Its Structure, Dynamics, Prediction, and Hazards.
Geophysical Monograph 79
Copyright 1993 by the American Geophysical Union.

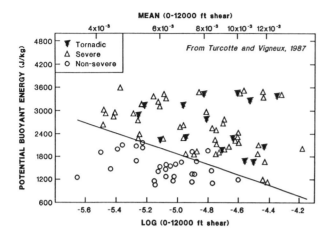

Fig. 1. Scatter diagram of potential buoyancy energy and low-level shear magnitude for thunderstorm days and associated weather events in Quebec Weather Centre's forecast area for 1985–1987. Sloping line shows approximate division between severe and non-severe storms.

operational tests we have continued a series of numerical model studies [*Brooks*, 1990] that look at how thunderstorm behavior varies with helicity. *Brooks* [1990] focused on possible physical mechanisms linking the environmental helicity to the development of significant low-level rotation in thunderstorms. In this paper we extend his work to look at cases where significant, long-lasting, low-level rotation does not develop.

We consider a three-stage qualitative model of the production of long-lived, significant tornadoes by supercells. The first is the development of strong midlevel ($\sim$3–7 km) rotation in the mesocyclone (vorticity $\sim$0.01 s^{-1}). This problem has been treated by *Lilly* [1982] and *Davies-Jones* [1984]. The second stage is the development and maintenance of surface mesocyclonic vorticity. The development aspect in particular is treated theoretically by *Rotunno and Klemp* [1985] and *Davies-Jones and Brooks* [this volume]. The final stage is the intensification of the surface vorticity to tornadic values ($\sim$1 s^{-1}). Given the constraints of radar coverage, the result of the first stage is the only one that can be observed routinely by Doppler radar. Since a critical warning issue is the nature of the surface weather threat, implicit assumptions must be made about the connection between the first and later stages. Lack of development or persistence at either of the first two stages would result in no significant, long-lived tornado occurring. We refer to such cases as "failure modes." Our focus here is on the second stage, examining whether a low-level mesocyclone develops and is maintained. We describe two failure modes and show how they differ from the development of a "classic" supercell tornado case, using a three-dimensional numerical cloud model. The failure modes are related to the location of the gust front relative to the main storm updraft, and to the sources of low-level rotation.

2. EXPERIMENTAL DESCRIPTION

The simulations that form the basis of this work are part of a set of 21 supercell cases described by *Brooks* [1990]. All 21 have the same initial thermodynamic profile, similar to, but slightly drier than, that used by *Weisman and Klemp* [1982, 1984]. The surface moisture is 15 g kg^{-1}, resulting in a CAPE of about 2100 J kg^{-1}. The vertical wind profile has been varied to provide a range of helicities while maintaining a bulk Richardson index [*Weisman and Klemp*, 1982, 1984] in a range associated with numerically simulated supercells (values of 8–40). Since the helicity is a function of the storm motion, its value cannot be known a priori. After the completion of the simulation the helicity as a function of time can be computed.

The numerical model used is that of *Klemp and Wilhelmson* [1978], as modified by *Wilhelmson and Chen* [1982] and *Brooks* [1990]. It is three-dimensional, fully compressible, and uses a Kessler microphysical parameterization. The horizontal domain is 70 × 70 × 16 km, with 1 km horizontal resolution and a vertical resolution that varies as a hyperbolic tangent function from 200 m near the ground to 600 m near the top. The horizontal grid spacing is too large to resolve the tornadic circulation itself but is sufficient for resolving large mesocyclonic circulations. The Coriolis effect and surface drag are excluded for simplicity. Each simulation was carried out until at least 9000 s, with a large time step of 5 s. Other details of the model parameters are given by *Brooks* [1990].

The three hodographs used here and associated storm motions at maturity are shown in Figure 2. All three of the simulations produce supercells, with long-lasting (greater than 2 1/2 hours), intense (greater than 30 m s^{-1}), rotating updrafts that extend from cloud base to the tropopause. Two hours into the simulations, case A has the largest helicity (705 m^2 s^{-2}), with B and C substantially less (550 and 120 m^2 s^{-2}, respectively). The helicities of A and B are well within the values associated with tornadoes by *Davies-Jones et al.* [1990], but only A develops a significant, long-lasting, low-level circulation. Simulations A and B had identical strongly curved hodographs below 3 km. From 3 to 7 km, case A has a constant shear of 10 × 10^{-3} s^{-1}, while the shear for B is half that, 5 × 10^{-3} s^{-1}. Case C has an environment characterized by a straight hodograph with shear of 10 × 10^{-3} s^{-1} over the lowest 7 km, which is nearly equal to the mean shear over the lowest 7 km in case A, 9.7 × 10^{-3} s^{-1}. We focus on the differences in low-level structure in these simulations.

3. RESULTS

The early stages of development of the storms in A and B are similar, but after 1 hour of simulation time, differences in the surface pressure field and rain water distribution have dramatic impacts on the further evolution of the two storms. Case A provides a classic example of an intensifying and occluding low-level mesocyclone. The origins of initial vertical vorticity development at the surface in this mesocy-

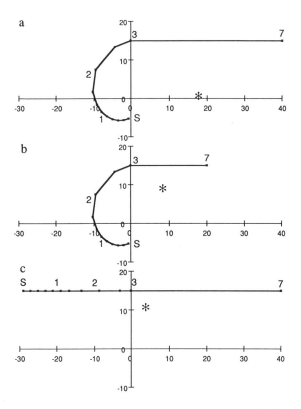

Fig. 2. Hodographs and storm motion for three storms (A, B, and C) discussed in paper. S indicates surface wind, and numbers are winds at various heights in kilometers. Individual model levels below 3 km are shown by squares. A and B differ only in A having 10×10^{-3} s^{-1} shear from 3 to 7 km and B having 5×10^{-3} s^{-1} over that layer. Storm motion during mature phase of storm for each case is indicated by stars.

clone are discussed by *Davies-Jones and Brooks* [this volume]. Here we look at the morphological characteristics of the storm during the later, mature stage of the storm. A time series of horizontal cross sections of the velocity and pressure fields at the second model level (300 m) shows the development of long-lived, low-level rotation (Figure 3). Two separate updraft maxima are apparent at this level. One is at the "nose" of the gust front, and the other is at the foot of the main updraft. As the mesocyclone becomes better developed, an extension of the rear flank downdraft wraps around the main updraft as it intensifies from 4200 to 6600 s (Figures 3a–3c). Dramatic changes occur in the horizontal velocity field simultaneously. Initially, the flow west of the gust front is northerly. In the region between the two updraft maxima, westerly winds soon develop in the outflow from the rainy region associated with the rear flank downdraft. North of the gust front updraft maximum, the flow has a southerly component, leading to a closed circulation centered on the updraft/downdraft interface [*Lemon and Doswell*, 1979]. By 6600 s this region of closed circulation is large (5–7 km diameter) and intense, with storm-relative winds of the order of 15–20 m s^{-1}. As the gust front

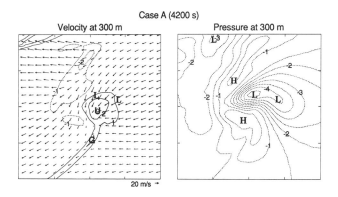

Fig. 3a

Fig. 3. Velocity (left) and perturbation pressure (right) fields at 300 m in 40 × 40 km section of domain in case A for times (a) 4200 s, (b) 5400 s, (c) 6600 s, and (d) 7800 s in life cycle of low-level mesocyclone. Contours in left panel show vertical velocity (interval 1 m s^{-1}) and vectors show storm-relative horizontal velocity at every other gridpoint. U indicates main storm updraft, and G indicates gust front updraft. Right panel shows pressure with a contour interval of 0.5 hPa. M indicates the location of the mesocyclone low pressure region, L other relative lows, and H relative high pressure. Features from the pressure field are indicated on the velocity figure for reference. Selected contours are labeled. Dashed contours are negative and zero contour is suppressed.

continues to move eastward relative to the main updraft, the circulation becomes more elongated, and the southerly wind component west of the gust front weakens. By 9000 s the vorticity maximum in the region is now associated with a long region of high shear instead of the more nearly circular flow seen at 6600 s (not shown).

An important feature in the surface pressure field is the inflow low. It is co-located with high wind speeds in the region of inflow of warm, moist environmental air (high θ_e) to the east of the updraft. The inflow low is a common feature in observations of tornadic thunderstorms [e.g., *Charba and Sasaki*, 1971]. Theoretical implications of the inflow low have been discussed by *Davies-Jones* [1985].

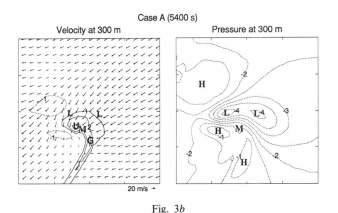

Fig. 3b

Fig. 3. (continued)

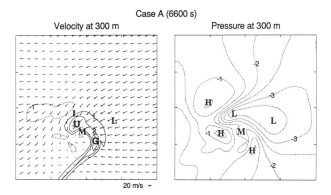

Fig. 3c

Fig. 3. (continued)

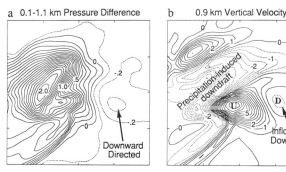

Fig. 4. (a) Vertical perturbation pressure difference between 0.1 and 1.1 km in 40 × 40 km section of domain in case A at 3600 s. Contour interval is 0.1 hPa, and dashed contours are negative. Sign convention such that positive values indicate higher perturbation pressure at 0.1 km. Selected contours are labeled. (b) Vertical velocity at 0.9 km in same region. Contour interval is 0.5 m s^{-1}.

From consideration of the Bernoulli relationship [*Milne-Thomson*, 1968] it can be shown that the magnitude of the low is proportional to the inner product of the storm-relative environmental wind and the perturbation velocity. The perturbation pressure gradient is downward over the lowest kilometer in the region just upwind of the surface low at 3600 s. Downward motion in the inflow of the storm is associated with this pressure gradient (Figure 4).

In contrast to the pronounced inflow low of storm A, the low in the inflow of storm B remains weak (Figure 5). This is a result of weaker storm-relative environmental winds, due to slower storm motion, which in turn is a consequence of weaker midlevel winds. With the weaker inflow low, the perturbation pressure gradient is essentially zero over the lowest kilometer, and there is no region of downward motion ahead of the storm. Soon after this time, the gust front begins to move rapidly ahead of the main updraft, cutting off its inflow and destroying any well-organized circulation at low levels. Only a region of shear along the gust front is apparent later in the storm's lifetime (Figure 6). The absence of a large, long-lived region of mesocyclonic rotation in low levels in case B argues that it would not support a significant,

long-lived tornado. It is interesting to note that experienced storm chasers frequently have observed developing cumulus in the inflow region of storms a short time before the rapid movement of the gust front away from the main updraft. Such developments are not seen in storms with long-lived mesocyclones. The downward motion ahead of storm A would be consistent with the suppression of cloud development in that region. On the other hand, there is no downdraft in the inflow of B to suppress growing cumulus. Note also that mass continuity implies that there will be an enhancement of the component of flow in storm A toward the main updraft and northern end of the gust front beyond the already strong environmental inflow.

There are also significant differences in rain water distribution between A and B. With weaker storm-relative winds aloft, considerable precipitation is present in B west and southwest of the main updraft (Figure 7). The rain water maximum at the tip of the hook at 3.3 km in case B has cut into the updraft at this level and can also be seen in a precipitation maximum and a larger area of downdraft southwest of the updraft at the lowest model level (Figure 8). The associated stronger, rain-cooled outflow (low θ_e) helps to

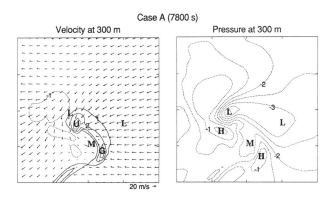

Fig. 3d

Fig. 3. (continued)

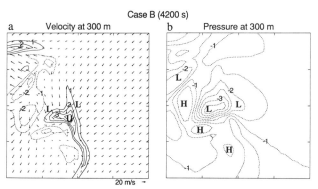

Fig. 5. Same as Figure 3 for case B at 4200 s.

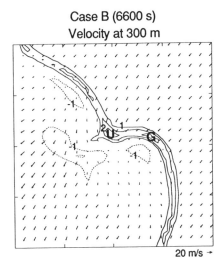

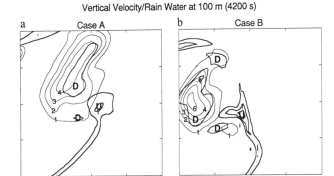

Fig. 8. Same as Figure 7 except at 100 m and with contour interval for rain water 1 g kg^{-1} and for velocity 1 m s^{-1}.

Fig. 6. Velocity field at 300 m in 40 × 40 km section of domain in case B at 6600 s. Contours show vertical velocity (interval 1 m s^{-1}), and vectors show storm-relative horizontal velocity at every other gridpoint. U indicates main storm updraft, and G indicates gust front updraft.

advance the gust front. While the difference in intensity of the outflow behind the gust front between A and B probably is the largest factor contributing to the faster gust front movement in B, we observe that the difference between the inflows, discussed above, also contributes to faster gust front movement in B. The precipitation pattern in B, with a large amount of rain in the vicinity of the mesocyclone, on the south side of the updraft, is similar to some high-precipitation (HP) supercells [*Moller et al.*, 1990]. We speculate that this represents one environmentally controlled mechanism in the development of HP supercells, namely, the generation of a strong mesocyclone in an environment with moderately weak mid to upper level storm-relative winds. In relatively weak environmental winds aloft, little precipitation is thrown far downstream of the updraft, and much of the precipitation is carried around the updraft by the mesocyclone, leading to the characteristic rain-embedded circulation of an HP supercell.

The results of *Turcotte and Vigneux* [1987] suggest that some storms in high-shear environments may not produce significant low-level rotation. An example of such a storm is case C. Storm C is slower to develop than A but has updrafts greater than 20 m s^{-1} from 1 hour until the end of the simulation (2 hours 50 min). The midlevel vorticity reaches mesocyclonic values (0.01 s^{-1}) after only 20 min into the simulation and is 0.02 s^{-1}, similar in magnitude to A and B, from 70 min onward. The low-level vorticity, on the other hand, never exceeds 0.005 s^{-1}, while A and B reach 0.02 s^{-1}. We stress that despite the similarity in the mean shear in A and C, the surface features associated with the storms are dramatically different. The case C low-level velocity field at 9600 s shows a nearly east-west aligned, weak gust front (Figure 9) with the cold air never advancing to the west and south of the updraft. Because of the strong environmental storm-relative winds, parcels of air traveling into the updraft from the north side spend only a short time in the strong temperature gradient generated by the evaporatively cooled air north of the updraft. As a result, there is little baroclinic

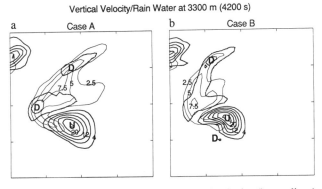

Fig. 7. Rain water (light lines) and vertical velocity (heavy lines) at 3.3 km at 4200 s in 40 × 40 km section of domain for (*a*) case A and (*b*) case B. Contour interval for rain water is 2.5 g kg^{-1} and for velocity is 4 m s^{-1}. Dashed contours indicate downdraft. Selected contours are labeled. Updrafts and downdrafts are indicated by U and D.

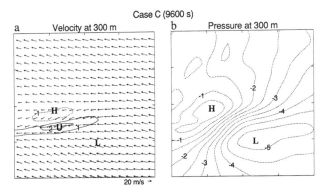

Fig. 9. Same as Figure 3 except for case C at 9600 s.

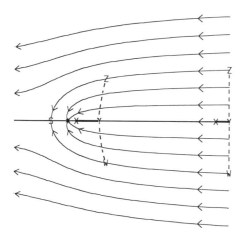

Fig. 10. Streamlines for a sink (marked by an asterisk) in a uniform easterly flow in a horizontal cross section. The stagnation point is labeled S, and X, Y, Z, and W mark the position of four material points at two different times. The material line XY, aligned with the flow, is stretched, and the material line ZYW, transverse to the flow, is shrunk. Thus, when the environmental winds veer height (streamwise vorticity case), the perturbation vorticity is westward, implying upward (downward) motion on the left (right) side of the inflow. In a westerly sheared environment (crosswise vorticity toward north), the perturbation vorticity is southward, implying downward motion on the east side of the sink.

generation of horizontal vorticity. Further, the environmental vorticity vector (perpendicular to the hodograph shown in Figure 2c), is nearly perpendicular to the storm-relative winds at this level, so that flow has little streamwise vorticity in this region. Because of these two effects, air rising into the updraft has little positive vorticity, and a mesocyclone fails to develop at low levels.

The low-pressure area in the inflow of C is even stronger than in A, and associated with the low-pressure area is a strong downdraft of the order of 1 m s^{-1}. The downward motion in the inflow of storms A and C can be explained using a simple inviscid flow model consisting of the environmental winds with a sink flow (representing the effects of the updraft on the inflow) superimposed, as illustrated by the schematic shown in Figure 10. A similar model was used by *Rothfusz and Lilly* [1989] to interpret the flow in a helical-vortex chamber. It can be shown that the basic flow is not a steady state solution of the Euler equations and that vertical circulations, which increase in strength with proximity to the sink, must develop owing to unbalanced pressure forces.

When the environmental vorticity is nearly streamwise, as in case A, vortex line stretching in the accelerating inflow amplifies the streamwise vorticity, giving rise to helical flow with upflow and downflow on the left and right sides of the inflow, respectively. The surface inflow low, being associated with the maximum surface wind speed, is located on the center line of the surface inflow. Because the environmental winds and hence the inflow veer with height, the surface inflow low is overlain by downdraft, as observed.

In case C the ambient vorticity is nearly crosswise in storm-relative coordinates. Confluence of the streamlines shrinks the vortex lines, i.e., generates perturbation vorticity in the opposite sense to the ambient vorticity. Because the strength of the circulation increases inward toward the sink, the perturbation vorticity must be associated with downward motion on the inflow side.

4. SUMMARY AND IMPLICATIONS

We have shown two different evolutions that result in storms with midlevel mesocyclones not producing significant, long-lived low-level rotation, which we refer to as "failure modes." The intensity of the storm-relative winds appears to play a critical role. A comparison of various features of the storms isolates the nature of the failure modes (Table 1). Storm C fails because of the absence of mechanisms to generate low-level rotation [*Davies-Jones and Brooks*, this volume]. Storm B fails because the outflow overwhelms the storm inflow, and the gust front surges out, undercutting the low-level mesocyclone.

The movement of the gust front, which affects the length of time that the low-level mesocyclone exists, is a complex problem. We have focused on two aspects, the strength of the storm-generated outflow and the strength of the environmental inflow. While the outflow intensity probably is the major factor, the inflow may make a significant contribution

TABLE 1. Basic Desciption of Features of Interest in the Three Simulations

Case	Storm-Relative Inflow Winds	Precipitation	Inflow Low Character	Flow Behind Gust Front	Environmental Low-Level Vorticity	Baroclinic Streamwise Vorticity Generation
A	strong	less than in B	strong low with downdraft	parallel	mostly streamwise but less so than in B	large
B	weaker than in A	heavy to west and south of updraft	weak low with very weak updraft	perpendicular and strong	mostly streamwise	less than in case A
C	very strong	moderate (no hook)	very strong with strong downdraft	parallel and strong	nearly crosswise	small

Case A produces a large, strong, long-lived, low-level mesocyclone. In case B the gust front surges out ahead of the low-level mesocyclone; no low-level mesocyclone forms in case C.

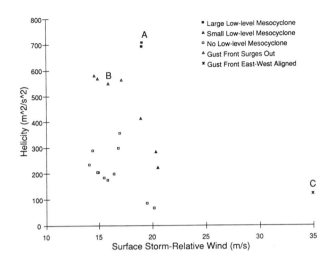

Fig. 11. Classification of low-level flow in storms as a function of the environmental helicity and surface storm-relative wind. Letters indicate storms discussed in detail in paper.

to the movement of the gust front. In the case of A and B the two effects are in the same sense, namely, to accelerate the gust front in B more than in A.

While we have discussed only three simulations, we have characterized the behavior of all of the storms in the larger data set of *Brooks* [1990] as a function of helicity and storm inflow speed (Figure 11). There is a tendency within the group that produces mesocyclones to have larger mesocyclones as helicity increases. Storm C is unique in the data set, while the failure mode exhibited by storm B is limited to high-helicity, low-inflow speed storms in the present study. This latter point may have relevance to the observation of *Moller et al.* [1990] that significant tornadoes in HP supercells are less common than in "classic" supercells. The results of case B indicate that one reason may be that strong outflow associated with HP storms inhibits the maintenance of a long-lived, low-level mesocyclone. More study of observed and modeled HP supercells is necessary to test this hypothesis.

Our examination of the spectrum of helicity and storm inflow speed is limited at this time. Of particular interest is the high-inflow-speed, high-helicity region in Figure 11 for which we have no data. Rapidly moving storms, frequently seen in the early spring, such as the one which produced the long-track Selmer, Tennessee, tornado in March 1991, are characterized by inflow speeds of the order of 30 m s^{-1} and high helicity. To date, we have not modeled any of these storms and only can speculate as to the relevance of the concepts discussed here. Clearly, some storms with high inflow speeds produce significant long-lived, low-level rotation. The ability to identify environmental conditions associated with their production would be of benefit to operational meteorologists.

The observation that shear and buoyancy, combined into a single parameter, cannot distinguish between tornadic and nontornadic environments [*Turcotte and Vigneux*, 1987] may be explained in part by these results. Our efforts are just beginning to reveal the relationship between the environment and the storm, and the conditions under which rotating storms produce or do not produce tornadoes. As Doppler radar observations become an important part of the severe weather warning system, these issues must be addressed in more detail. The results here reinforce the notion that the question of whether an environment will support tornadic thunderstorms is not the same as the question of whether it will support supercells.

Acknowledgments. This work was done while one of the authors (H.E.B.) held a National Research Council-NOAA Research Associateship. The computer simulations were carried out on the CRAY-2 at the National Center for Supercomputer Applications (NCSA) and software developed at NCSA was used in the analysis of the data. The simulations were supported, in part, by NSF grant ATM-87-00778. We thank Doug Lilly, Richard Rotunno, and Michael Foster for their comments, and Robert Wilhelmson for his assistance with the simulations. Viateur Turcotte, Denis Vigneux, and Andre Sevigny of the Quebec Weather Centre gave permission to use Figure 1.

References

Brooks, H. E., Low-level curvature shear and supercell thunderstorm behavior, Ph.D. thesis, 233 pp., Univ. of Ill., Urbana-Champaign, 1990.

Burgess, D. W., and L. R. Lemon, Severe thunderstorm detection by radar, in *Radar in Meteorology*, edited by D. Atlas, pp. 619–647, American Meteorological Society, Boston, Mass., 1990.

Charba, J., and Y. Sasaki, Structure and movement of the severe thunderstorms of 3 April 1964 as revealed from radar and surface mesonetwork data analysis, *J. Meteorol. Soc. Jpn.*, 49, 191–214, 1971.

Darkow, G. L., and M. G. Fowler, Tornado proximity wind analysis, in *Preprints, 7th Conference on Severe Local Storms*, pp. 148–151, American Meteorological Society, Boston, Mass., 1971.

Darkow, G. L., and D. W. McCann, Relative environmental winds for 121 tornado bearing storms, *Preprints, 8th Conference on Severe Local Storms*, pp. 413–417, American Meteorological Society, Boston, Mass., 1977.

Davies-Jones, R. P., Streamwise vorticity: The origin of updraft rotation in supercell storms, *J. Atmos. Sci.*, 41, 2991–3006, 1984.

Davies-Jones, R. P., Dynamical interaction between an isolated convective cell and a veering environmental wind, in *Preprints, 14th Conference on Severe Local Storms*, pp. 216–219, American Meteorological Society, Boston, Mass., 1985.

Davies-Jones, R. P., and H. Brooks, Mesocyclogenesis from a theoretical perspective, *Proceedings, Tornado Symposium III*, edited by C. Church, this volume.

Davies-Jones, R. P., D. W. Burgess, and M. P. Foster, Test of helicity as a tornado forecast parameter, in *Preprints, 16th Conference on Severe Local Storms*, pp. 588–592, American Meteorological Society, Boston, Mass., 1990.

Klemp, J. B., and R. B. Wilhelmson, The simulation of three-dimensional convective storm dynamics, *J. Atmos. Sci.*, 35, 1070–1096, 1978.

Leftwich, P. W., Jr., On the use of helicity in operational assessment of severe local storm potential, in *Preprints, 16th Conference on Severe Local Storms*, pp. 306–310, American Meteorological Society, Boston, Mass., 1990.

Lemon, L. R., and C. A. Doswell III, Severe thunderstorm evolution and mesocyclone structure as related to tornadogenesis, *Mon. Weather Rev.*, *107*, 1184–1197, 1979.

Lilly, D. K., The development and maintenance of rotation in convective storms, in *Topics in Atmospheric and Oceanographic Sciences: Intense Atmospheric Vortices*, edited by L. Bengtsson and J. Lighthill, pp. 149–160, Springer-Verlag, New York, 1982.

Lilly, D. K., The structure, energetics and propagation of rotating convective storms, II, Helicity and storm stabilization, *J. Atmos. Sci.*, *43*, 126–140, 1986.

Milne-Thomson, L. M., *Theoretical Hydrodynamics*, 743 pp., MacMillan, New York, 1968.

Moller, A. R., C. A. Doswell III, and R. Pryzybylinski, High-precipitation supercells: A conceptual model and documentation, in *Preprints, 16th Conference on Severe Local Storms*, pp. 52–57, American Meteorological Society, Boston, Mass., 1990.

Patrick, D., and A. J. Keck, The importance of the lower level windshear profile in tornado/nontornado discrimination, in Proceedings, Symposium on Mesoscale Analysis and Forecasting, *Eur. Space Agency Spec. Publ. ESA SP-282*, 393–397, 1987.

Rasmussen, E. N., and R. B. Wilhelmson, Relationships between storm characteristics and 1200 GMT hodographs, low level shear, and stability, in *Preprints, 13th Conference on Severe Local Storms*, pp. J5–J8, American Meteorological Society, Boston, Mass., 1983.

Rothfusz, L., and D. K. Lilly, Quantitative and theoretical analyses of an experimental helical vortex, *J. Atmos. Sci.*, *49*, 2265–2279, 1989.

Rotunno, R., and J. B. Klemp, On the rotation and propagation of simulated supercell thunderstorms, *J. Atmos. Sci.*, *42*, 271–292, 1985.

Turcotte, V., and D. Vigneux, Severe thunderstorms and hail forecasting using derived parameters from standard RAOBS data, in *Preprints, 2nd Workshop on Operational Meteorology*, pp. 142–153, Atmospheric Environment Service and Canadian Meteorological and Oceanographic Society, Downsview, Ont., 1987.

Weisman, M. L., and J. B. Klemp, The dependence of numerically simulated convective storms on vertical wind shear and buoyancy, *Mon. Weather Rev.*, *110*, 504–520, 1982.

Weisman, M. L., and J. B. Klemp, The structure and classification of numerically simulated convective storms in directionally varying wind shears, *Mon. Weather Rev.*, *112*, 2479–2498, 1984.

Wilhelmson, R. B., and C. S. Chen, A simulation of the development of successive cells along a cold outflow boundary, *J. Atmos. Sci.*, *39*, 1466–1483, 1982.

Woodall, G. R., Qualitative forecasting of tornadic activity using storm-relative environmental helicity, in *Preprints, 16th Conference on Severe Local Storms*, pp. 311–315, American Meteorological Society, Boston, Mass., 1990.

Mesocyclogenesis From a Theoretical Perspective

ROBERT DAVIES-JONES AND HAROLD BROOKS

National Severe Storms Laboratory, NOAA, Norman, Oklahoma 73069

1. INTRODUCTION

Since the last tornado symposium in 1976, considerable progress has been made toward understanding tornadogenesis in supercells. The genesis of mesocyclones and associated tornadoes is now known through numerous observations to be a three-stage process. The mesocyclone develops first at midlevels through tilting by an updraft of streamwise vorticity present in low levels of the environment [*Browning and Landry*, 1963; *Barnes*, 1970; *Lilly*, 1982, 1986; *Davies-Jones*, 1984]. The mesocyclone does not simply build down to low levels; instead, low-level rotation appears to be caused by a separate, more complicated mechanism [*Rotunno and Klemp*, 1985; *Johnson et al.*, 1987] requiring downward as well as upward motions and also strong horizontal temperature gradients. Finally, in perhaps the least known process, a tornado develops within the mesocyclone.

The theoretician has a number of tools that he can use in attempting to uncover the origins of rotation. These include vector equations for vorticity and entropy gradient [*Davies-Jones*, 1982a], the streamwise vorticity equation [*Scorer*, 1978], vorticity and circulation theorems [*Dutton*, 1976, pp. 364–374], Beltrami solutions of the Euler equations [*Lilly*, 1982, 1986; *Davies-Jones*, 1985], integrals of the vorticity equation for isentropic flow [*Dutton*, 1976, pp. 385–390] and conservation laws for entropy, potential vorticity [*Dutton*, 1976, p. 383], and helicity (the volume integral of the scalar product of velocity and vorticity) [*Lilly*, 1986]. Even though rotation is a kinematic property ostensibly unrelated to entropy, the above list implies, rather surprisingly, that specific entropy S is as important a variable as vorticity ω. In fact, vorticity and entropy are closely entwined in geophysical flows [*Dutton*, 1976, pp. 338, 383]. Unfortunately, several of the theorems hold only for dry, inviscid convection (e.g., there is no strict equivalent potential vorticity conservation law for moist convection) and may not be useful when moisture effects are essential to a particular process.

In this paper we first review the midlevel mesocyclogenesis mechanism. We then apply vorticity and circulation theorems to determine how near-ground rotation develops in a three-dimensional numerical simulation of a supercell. We conclude with a brief survey of current tornadogenesis theories.

2. THE INITIAL MIDLEVEL MESOCYCLONE

Linear theory of dry convection in shear virtually explains initial net rotation of the storm's updraft at midlevels as a consequence of tilting toward the vertical of low-level environmental streamwise (horizontal) vorticity by the updraft [*Lilly*, 1982, 1986; *Davies-Jones*, 1984]. Only the flow relative to the updraft is physically relevant because the updraft is the tilting agent. Therefore streamwise refers here to the direction parallel to the local storm-relative wind. In a horizontally uniform wind that veers (backs) with height without a change in speed, an air parcel experiences a differential velocity that causes it to spin clockwise (anticlockwise) about its direction of motion like a propeller or a spiraling American football thrown right- (left-)handedly, indicating the vorticity is purely streamwise (antistreamwise). When the wind increases with height without turning, the differential velocity is parallel to the wind and the vorticity is purely crosswise (like a wheel or a ball with top spin). These solid body analogies are limited because air parcels change shape as well as spin; strictly speaking, vorticity is twice the angular velocity of a spherical parcel that is instantaneously solidified while its spin angular momentum is conserved. In a horizontally homogeneous, strongly sheared environment with the Earth's rotation "switched off," an incipient, isolated updraft of roughly circular cross section acquires cyclonic (anticyclonic) rotation in the streamwise (antistreamwise) case and no net rotation in the crosswise case. *Davies-Jones* [1984] was the first to prove this result for general wind profiles; further, he did it from a full set of (linearized) equations instead of from the vertical vorticity equation alone. A key step in the proof

The Tornado: Its Structure, Dynamics, Prediction, and Hazards.
Geophysical Monograph 79
Copyright 1993 by the American Geophysical Union.

was the introduction of vertical displacement as a dependent variable. *Davies-Jones* [1984, 1986] explained the process physically in terms of the updraft uplifting the initially horizontal isentropic surfaces and vortex lines, thereby establishing the link between the thermal and vorticity fields. Lifting up loops of horizontal vorticity naturally produces a cyclonic-anticyclonic vortex pair. In the streamwise (antistreamwise) case the anticyclonic (cyclonic) vortex is downstream in lesser updraft or in downdraft, and in the crosswise case the vortices are on the left and right downstream sides of the updraft [see *Rotunno*, this volume, Figure 8]. Potential vorticity $\alpha \omega \cdot \nabla S$ (α is specific volume) is conserved even without linearization and is zero because of the initial conditions [*Davies-Jones*, 1982a, 1984; *Rotunno and Klemp*, 1985]. Thus the vortex lines remain in their original isentropic surfaces. This result has important implications for tornadogenesis (see below). It also permits generalization of the above theory to a semilinear one by allowing finite vertical displacements. Even simpler explanations for updraft rotation, without recourse to the concepts of isentropic surfaces and vortex lines, are given by *Rabin and Davies-Jones* [1986] and *Davies-Jones* [1992a, b].

The above assumes that the disturbance is roughly circular and semilinear. These assumptions are partly justifiable because (1) supercells are quite circular in their initial stages and, furthermore, the shape dependence disappears in the limit where the vorticity is purely streamwise, and (2) the environmental winds are as large as the storm-generated ones for mature severe storms, so the linear terms are always important; also, the excluded nonlinear terms have little qualitative influence on updraft rotation.

According to the linear theory the spatial relationships between important variables at each level are as depicted in Figure 1. In the storm's reference frame and relative to the location of maximum vertical displacement (or peak buoyancy if the lapse rate is unstable at that level) the maximum vertical velocity and vertical perturbation pressure gradient force (VPPGF) are upwind, while the maximum vertical vorticity is on the right side with respect to the shear vector. When streamwise environmental vorticity is present, (horizontal) environmental helicity is converted into vertical helicity (the volume integral of vertical velocity times vertical vorticity) through a term, which is the integral of vertical vorticity times VPPGF. This is analogous to the transformation of horizontal into vertical kinetic energy through a similar pressure term. The buoyant production of helicity (essentially the covariance of buoyancy and vertical vorticity) is zero in linear theory and in a nonlinear extension thereof [*Davies-Jones*, 1985]. *Lilly* [1986] suggested that vertical helicity receives a contribution from only the buoyancy term and gives away part of that to horizontal disturbance helicity through the pressure term, because buoyancy and the vertical perturbation pressure gradient (VPPG; $\propto -$ VPPGF) are generally positive in peak updraft regions. However, as shown in Figure 1, vertical vorticity and VPPG can be negatively correlated, even though both these variables are positively correlated with vertical velocity. While

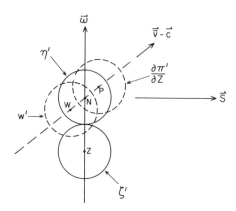

Fig. 1. Schematic diagram illustrating the linear theory relationships between vertical displacement η', vertical velocity w', vertical vorticity ζ', and vertical perturbation pressure gradient (VPPG) $\partial \pi'/\partial z$ for a level where the storm-relative winds, $\mathbf{v} - \mathbf{c}$, increase and veer with height. Environmental shear and vorticity vectors are denoted by $\mathbf{S}$ and ω. Centers, N, W, Z, and P, of circles indicate the relative positions of the η', w', ζ', $\partial \pi'/\partial z$ maxima, respectively. Overlapping, touching, or disjointed circles signify positive, zero, or negative correlation between the two variables involved. Note, however, that the correlation between w' and $\partial \pi'/\partial z$ becomes negative for increasing storm-relative wind speed [from *Davies-Jones*, 1984].

buoyant production of helicity is observed to be positive (instead of zero) in numerical simulations of storms with mesocyclones, the other correlations shown in Figure 1 appear to be qualitatively correct. For instance, new updrafts (due to upward VPPGF) and cyclonic vorticity both are observed typically on the storms' right sides.

Storm motion, which affects the "streamwiseness" of the vorticity, thus far has been treated as given, even though in the linear theory it is determined by a complex eigenvalue. *Rotunno and Klemp* [1982] physically and *Davies-Jones* [1984] mathematically deduced that in linear theory, storm motion should lie to the concave side of a curved hodograph and on the hodograph for a straight one. Storm splitting and movement off a straight hodograph are nonlinear effects. Furthermore, boundary layer convergence zones associated with fronts, outflow boundaries of other storms, etc., may affect storm motions considerably, sometimes altering them in the right way for significant updraft rotation to develop. Currently, no satisfactory method for quantitatively estimating storm motion has been devised. Existing methods either are not Galilean-invariant or fail to predict storm splitting.

Given observed or crudely estimated storm motions, the linear theory may be used to nowcast mesocyclones from proximity upper air data. *Davies-Jones et al.* [1984] adapted a formula for the correlation coefficient between vertical velocity and vertical vorticity derived from linear theory. The correlation coefficient was estimated from soundings as the relative helicity (cosine of the angle between the environmental vorticity and storm-relative wind vectors) times a factor between zero and one that penalizes for light storm-relative environmental winds compared to storm-generated horizontal winds (estimated from the convective available potential energy). An average value over the inflow layer of the storm (taken to be the lowest 3 km) was used. *Davies-Jones et al.* [1990] used the storm-relative helicity of the 0- to 3-km layer instead, because it may be computed as a simple integral without any differentiation and hence is a far more stable parameter. In the linear theory the environmental helicity density (streamwise vorticity times wind speed in storm-relative coordinates, i.e., the integrand in the helicity definition) is proportional to the covariance of vertical velocity and vertical vorticity. Once again, weak storm-relative winds result in low values of the forecast parameter. In addition to computational stability, helicity has another advantage over other parameters. It is minus twice the signed area swept out by the storm-relative winds between 0 and 3 km (the assumed inflow depth) on a hodograph diagram. Area is well defined even for fractal hodographs and is easily visualized by forecasters. Also, helicity is a linear function of storm motion. Thus helicity contours as a function of storm motion (for a fixed wind profile) are parallel straight lines on a hodograph diagram, and the effects of deviant storm motion in a given environment are easily assessed in advance.

In contrast to linear theory, which describes small growing disturbances, exact steady state Beltrami solutions of the Euler equations [*Lilly*, 1982, 1986; *Davies-Jones*, 1985] (R. P. Davies-Jones and H. E. Brooks, manuscript in preparation, 1993) are useful for modeling mature, midlevel mesocyclones. A Beltrami flow is one where the vorticity and velocity vectors are aligned with each other at every point in the fluid. The advantage of having an exact analytical solution is partly offset by omission of buoyancy forces. Furthermore, practical Beltrami solutions are intrinsically steady, so mesocyclone development is not described, storm propagation is not modeled, and the coordinate system must be storm relative. The pressure and wind fields are linked by a universal Bernoulli relationship. Beltrami flows are useful for modeling the dynamic pressure distribution around a rotating updraft and for explaining observed characteristics of mesocyclones such as diameter, circulation value, low surface pressure in the inflow region [*Davies-Jones*, 1985] (R. P. Davies-Jones and H. E. Brooks, manuscript in preparation, 1993), and anticyclonic trajectories in cyclonic updrafts (owing to the effects of environmental wind veering with height overcoming those of cyclonic rotation [*Klemp et al.*, 1981; *Lilly*, 1986]).

The streamwise vorticity equation shows that a parcel's streamwise vorticity changes in three ways in steady inviscid flow. Streamwise vorticity is amplified as the parcel accelerates and is stretched along the streamline. It is generated baroclinically by buoyancy torques along its direction of motion. Lastly, exchanges between crosswise and streamwise vorticity occur in curved flow. Horizontal flow with cyclonic curvature, e.g., inflow to a mesocyclone, is similar to the flow around a cyclonic river bend [*Scorer*, 1978]. Crosswise vorticity (speed shear) is converted into streamwise vorticity (directional shear). Thus the streamwise vorticity of inflowing air may increase as it accelerates and curves into a mesocyclone. This might be a significant effect after the development of intense low-level rotation and low pressure. At this stage, the streamlines in the warm air may bend up more abruptly as they enter the updraft because the low has descended from midlevels, and tilting of streamwise vorticity in the warm air might contribute now to rotation at fairly low levels.

3. BAROTROPIC AND BAROCLINIC VORTICITY

Before discussing the development of low-level rotation, it is convenient to introduce the concepts of barotropic and baroclinic vorticity. *Dutton* [1976] shows from an integral of the vector vorticity equation for dry, inviscid, isentropic flow that the vorticity is the sum of barotropic and baroclinic components. This decomposition is dependent on the choice of an arbitrary initial reference time t_0. The barotropic vorticity of a parcel at a later time t is the vorticity that the parcel would have from the amplification and reorientation of initial vorticity, if the baroclinic term were ignored. Barotropic vortex lines are "frozen" into the fluid. The barotropic vorticity field at time t is given by the Cauchy formula in terms of the initial (i.e., reference) vorticity field and the mapping from Lagrangian to Eulerian coordinates. It is independent of parcel positions at intermediate times $\tau (t_0 < \tau < t)$. On the other hand, baroclinic vorticity consists of vorticity that has been generated as horizontal vorticity since the initial time by buoyancy torques and subsequently affected by vortex tube stretching and tilting. The baroclinic vorticity of a parcel depends on its integrated thermal history from t_0 to t. The baroclinic vortex lines lie in isentropic surfaces; hence the baroclinic component does not contribute to potential vorticity. The generation of vertical vorticity through the tilting of baroclinic vorticity is a nonlinear effect and so has little effect on the initiation of midlevel rotation discussed in section 2.

Incorporation of latent heat release and evaporative cooling complicates the physical interpretation, but it does not invalidate the vorticity decomposition because barotropic vorticity is still well defined. Baroclinic vorticity may be redefined simply as the total vorticity minus its barotropic component so as to include these effects. For nondiffusive flow with horizontally homogeneous initial conditions, equivalent potential vorticity (defined in terms of the entropy of moist air) is conserved almost exactly in saturated air and in air in dry adiabatic ascent or descent but only approximately in evaporatively cooled air that has descended in unsaturated downdrafts [*Rotunno and Klemp*, 1985]. This implies that the baroclinic vortex lines no longer lie exactly in isentropic surfaces.

4. THE DEVELOPMENT OF ROTATION NEXT TO THE GROUND

Davies-Jones [1982*a*, *b*] realized that in a sheared environment with negligible background vertical vorticity, an "in, up, and out" type circulation driven by forces primarily aloft would fail to produce vertical vorticity close to the ground. This conclusion, which depends on eddies being too weak to transport vertical vorticity downward against the flow, has been verified in numerical models by *Rotunno and Klemp* [1985] and *Walko* [this volume]. The Beltrami model illustrates the point well. Since the vortex lines are coincident with the streamlines, parcels flowing into the updraft at very low levels do not have significant vertical vorticity until they have ascended a few kilometers. Otherwise, there would have to be abrupt upward turning of the streamlines, strong pressure gradients, and large vertical velocities next to the ground. These features actually do develop in axisymmetric tornado models, in which ambient vertical vorticity is concentrated by an updraft [*Smith and Leslie*, 1979]. In this case the vortex core acts as a pipe with cyclostrophic balance preventing inflow through the sides. The "pipe" draws air into it from below, establishing cyclostrophic balance at successively lower levels. In this manner, the vortex builds downward to the ground. This mechanism does not operate in a supercell updraft that derives its rotation from ambient streamwise vorticity because the updraft is quite porous to the ambient flow and cyclostrophic balance is never established at any level. This is also evident from Bjerknes' circulation theorem applied to a horizontal fluid circuit, say, a few kilometers across and 100 m above the ground at a given time. In an "in, up, and out" type flow without background vertical vorticity, any such circuit must be quasi-horizontal at all earlier times and should have little circulation. Therefore a downdraft appears essential for the development of rotation in the lowest few hundred meters.

Davies-Jones [1982a, b] neglected baroclinic vorticity. He suggested that the downdraft had the following roles in low-level mesocyclogenesis. First, tilting of horizontal vorticity by the downdraft produces vertical vorticity; second, subsidence transports air with vertical vorticity closer to the surface; third, this air flows out from the downdraft and enters the updraft where it is stretched vertically; fourth, convergence beneath the updraft is enhanced by the outflow. He also showed kinematically that the flow responsible for tilting and concentrating the vortex lines also tilts and packs the isentropic surfaces, thus explaining observations of strong entropy gradients across mesocyclones at the ground and low levels (also see *Davies-Jones* [1985]). Strong tilting of the isentropic surfaces also follows from the zero potential vorticity constraint. A favorable condition for such tilting is midlevel, cool, dry air overlying warm, moist, surface-based air, long known to be a propitious stratification for tornadoes.

Klemp and Rotunno [1983] and *Rotunno and Klemp* [1985] showed that baroclinic vorticity could not be ignored; in fact, it is the main source for low-level rotation in the case they simulated. Spiraling rain curtains and downdraft form in the midlevel mesocyclone. In the lowest kilometer, air that has been moderately cooled by rain acquires a great amount of streamwise baroclinic vorticity as it subsides and travels southward parallel to closely packed isotherms just behind the outflow boundary. As this relatively cool air is ingested into the updraft, its vorticity is tilted upward and amplified by stretching (see also *Weisman and Klemp* [1984], *Rotunno* [1986], and *Klemp* [1987]). In some fortuitous storm encounters, the outflow may come from a neighboring storm instead of the storm's own downdraft. Note the contrast in the origins of midlevel and low-level rotation. The midlevel rotation stems from barotropic vorticity present on the warm inflow side of the storm, while the near-ground rotation originates from baroclinic vorticity on the cool side.

Once it develops, the low-level rotation is often more intense than the midlevel circulation. According to the cyclostrophic approximation the vertical perturbation pressure gradient force between low levels and midlevels should reverse sign, resulting in weakening updraft and formation of an "occlusion" downdraft within the mesocyclone.

Mesocyclones propagate discretely to the right by a new core periodically forming along the bulging gust front to the south (roughly every 40 min) as cold air wraps around and occludes the old core [*Burgess et al.*, 1982; *Johnson et al.*, 1987]. The first core takes much longer to develop (~2 hours) than do subsequent ones because of the time that elapses before the downdraft and cool outflow develop. Some new cores form aloft first, like the first one. Others form almost simultaneously over their whole depth, consistent with new updrafts seizing some of the cyclonically rotating air flowing out of the downdraft, spinning it up, and advecting it upward.

We decided to investigate the Rotunno and Klemp mechanism further because we still did not understand how rotation developed next to the ground. It seemed to us that even though the horizontal vorticity is greatly enhanced by baroclinic generation, tilting in an updraft still would not produce substantial vertical vorticity in the lowest kilometer. Therefore we undertook a diagnostic study of *Brooks et al.*'s [this volume] strong-shear simulation (their case A) to determine how vertical vorticity first reaches the ground.

5. Circulation Analysis

In Brooks et al.'s simulation, the horizontal grid spacing is 1 km, 100 m is the lowest level at which vertical vorticity is defined, and 50 min is the first time significant vertical vorticity (5×10^{-3} s^{-1}) appears at this level (Figure 2). Adopting *Rotunno and Klemp*'s [1985] approach, we identify at 50 min a square material circuit with 4-km sides around the "surface" vorticity maximum and a horizontal material surface enclosed by this circuit. We traced the circuit backward in time to 35 min. This was done by computing backward trajectories from model data saved every 2 min and by using linear spatial and temporal interpolation. The maximum surface vorticity at 35 min is 5×10^{-4} s^{-1}, an order of magnitude smaller than at 50 min. We then used Bjerknes' circulation theorem, which states that the rate of increase of the circulation around the circuit is equal to the work done by buoyancy forces on a hypothetical parcel that travels once around the circuit instantaneously [*Dutton*, 1976]. Note that as time decreases, this circuit cannot stay horizontal because this is inconsistent with its nonzero circulation (a permanently horizontal circuit encloses no barotropic vortex lines and has a zero baroclinic generation term). Thus the parcels that make up the circuit must

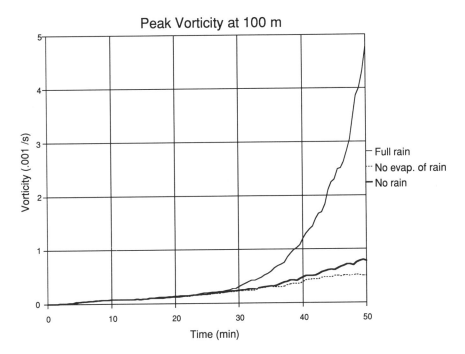

Fig. 2. Peak vertical vorticity at 100-m level versus time for simulations with full Kessler microphysics, without evaporation of rain, and without rain.

originate from different levels and hence have different equivalent potential temperatures (θ_e). Once again, we see that the vorticity maximum must lie in a strong θ_e gradient [*Davies-Jones*, 1982a, b, 1985; *Rotunno and Klemp*, 1985].

The hodograph for the simulation is shown in Figure 2 (case A) of *Brooks et al.* [this volume]. The storm motion vector in meters per second is (6, 4) at 35 min and (12, 4) at 50 min. Storm-relative winds are northeasterly at the surface and veer to southeasterly at 3 km. The environmental vorticity vector (same magnitude as and 90° to left of the shear vector) is southward in the lowest few hundred meters.

Comparison runs were made with the full Kessler microphysics, without evaporation of rain, and without formation of raindrops from cloud droplets (Figure 2). The peak vorticity at 100 m is 6 or more times greater at 50 min in the full microphysics case than it is in the other runs. *Rotunno and Klemp* [1985] found a similar result. Thus strong low-level rotation requires evaporative cooling.

At 50 min the vorticity maximum at 100 m is in strong gradients of updraft and temperature with the maximum updraft to the northeast, downdrafts to the northwest and southwest, and the coldest air to the west (Figures 3 and 4). Notice that at this level there is positive vorticity even within the downdrafts. Storm-relative flow converging into the vorticity maximum is northeasterly to northwesterly. Most of the positive contribution to the circulation around the fluid circuit is from winds on the west (cool) side, indicating that the source of rotation at this level is in the cool air. The vorticity maximum is located on the gust front and leading edge of the θ_e gradient (Figure 5). The 300 m horizontal vorticity field (Figure 5) contains high values in the rain-cooled air behind the gust front. The same field at 100 m (not shown) is very similar but weaker. The horizontal vorticity in the cool air has the same general direction as the environmental vorticity (unlike Rotunno and Klemp's straight line hodograph, splitting storm case), but is much larger (roughly 5 times) due partly to baroclinic generation. The vorticity vectors cross the θ_e contours at large angles near the θ_e minimum, indicating that equivalent potential vorticity is not conserved well in air that has been cooled considerably by evaporation of rain (see section 3).

Figure 6 shows the fluid circuit (ABCD) at 35 and 50 min. Air parcels on the east side (BC) and most of the north and south sides (CD and AB) at 50 min originate from lower levels (less than 30 m), while those on the west side (DA) descend in the 15-min interval to 100 m from heights of 400–650 m. The northwest part of the material surface at 35 min is tilted up the most, so the environmental vorticity flux is greatest through this part of the surface. The corresponding circulation is positive so that cyclonic vorticity is generated as the surface levels out at 50 min. Note that since the surface is not folded at 50 min, the elevated northwest part of the surface must slide down "feet first" rather than "head first" (i.e., the higher points must move southward more slowly than the lower points). The area of the material surface shrinks by a factor of 10 in the 15-min period, implying a tenfold increase in the average vorticity normal to the surface if only barotropic effects are considered.

The circulation around the fluid circuit increases by 70% from 35 to 40 min as the parcels on the side DA sink, then

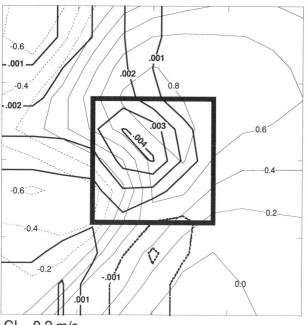

Fig. 3. Vertical velocity and vertical vorticity fields at 100-m level after 50 min. Contour intervals are given at lower left. Zero-vertical-vorticity contour is suppressed. Also shown is outline of 4 km × 4 km box around vorticity maximum.

of these descending parcels become more anticyclonic at first, then less anticyclonic, and actually cyclonic in the lowest 50–125 m of their descent.

How is cyclonic vorticity generated in a downdraft? If the downdraft and outflow are axisymmetric, the horizontal vorticity (both baroclinic and barotropic) is crosswise and the vertical vorticity zero because the tilting term vanishes by symmetry. If purely crosswise horizontal vorticity enters a downdraft from one direction, tilting produces a vortex pair with the cyclonic vortex on the left side (looking downwind), as shown in Figure 9. This case does not apply here because the vorticity and storm-relative velocity vectors are generally parallel in the cool air (cf. Figures 4 and 5). If, on the other hand, purely streamwise vorticity enters a downdraft and baroclinity is negligible, the vortex lines (which are frozen into the fluid) and trajectories coincide, so the downdraft is anticyclonic (Figure 9). We are able to explain how the vertical spin of parcels reverses during descent only by including baroclinic generation of streamwise vorticity. Consider, for example, a mass of evaporatively cooled air that descends through the mesocyclone on its north to northwest sides and spreads out "feet first" (like a gravity current) toward the south nearly parallel to the isotherms with the warm air to the east (Figure 9). Such a flow occurs on the left side of cold-air outflow that advances cyclonically around a mesocyclone. Normally, a gravity

remains more or less constant during the next 10 min (Figure 7). The time integral of Bjerknes' theorem is

$$\Gamma(t) = \Gamma(t_0) + \int_{t_0}^{t} \left[\oint b\mathbf{k} \cdot d\mathbf{l} \right] dt \quad (1)$$

where $\Gamma(t)$ is the circulation, b is buoyancy, and t_0 is an initial time (35 min here). We refer to the last term as the buoyantly generated circulation. Direct computation of this term shows that it accounts quite well for the increase in circulation (Figure 7). Thus we are confident that our trajectory computations are fairly accurate.

6. Vorticity Changes in Individual Parcels

The circulation analysis, although enlightening, does not portray the acquisition of vertical vorticity by individual parcels. Therefore, to gain further insights into the development of low-level rotation, we examined how the vertical vorticities of the labeled parcels in Figure 6 change with time from 35 to 50 min. As expected, the rising parcels on the east side acquire little vertical vorticity, while the subsiding ones on the west side show appreciable gains. Graphs of vertical vorticity versus height for three parcels on the west side DHA of the fluid circuit (Figure 8) show that the vorticities

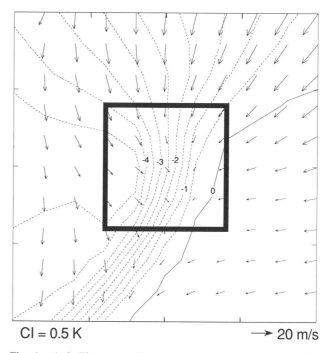

Fig. 4. As in Figure 3 but for potential temperature and horizontal wind fields. Scale for vectors is given at lower right.

current moves normal to the isotherms; however, the cold air in this case has considerable momentum along the isotherms, as it must in a quasi-steady storm. It descends in a cyclonic spiral around the rear of the mesocyclone because, as it falls, it experiences both environmental winds that back (with decreasing height) and the cyclonic rotation of the mesocyclone (cf. anticyclonic trajectories in the cyclonic updraft, section 2). Strong, storm-relative inflow winds also hold the outflow back under the updraft [*Brooks et al.*, this volume]. Baroclinic generation produces a circulation normal to the isotherms, so the flow actually is not confined to two dimensions as depicted in the simplified schematic. As the air subsides, the vortex lines turn downward due to the barotropic "frozen vortex lines" effect, but with less inclination than the trajectories. This is because horizontal southward vorticity is being generated continually by the baroclinic effect (which introduces slippage between the fluid and vortex lines). Because the gravity current subsides feet first and the vortex lines now cross the streamlines from lower to higher ones, the barotropic effect acts to turn the vortex lines upward even during descent. The baroclinic effect acts to increase horizontal vorticity further but does not control the sign of vertical vorticity. Thus air with cyclonic vorticity appears close to the ground. As this

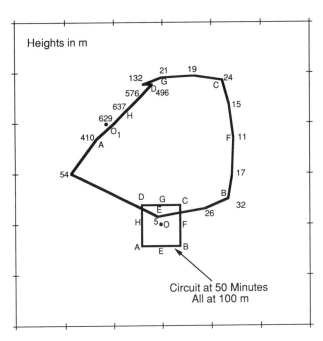

Fig. 6. Fluid circuit AEBFCGDHA at 35 and 50 min. Numbers along circuit denote heights in meters. At 50 min, O is center of square, A, B, C, and D are corners, and E, F, G, and H are midpoints of sides.

air passes from the downdraft into the updraft, its cyclonic spin is greatly amplified by vertical stretching [*Davies-Jones*, 1982a]. This mechanism explains why the horizontal convergence term in the vertical vorticity equation is large next to the ground in the low-level mesocyclone. If the tilting of vorticity were confined to the updraft, then there would be no ground-level vertical vorticity to stretch!

7. Tornadogenesis

Tornadogenesis is the last of the three stages in the "spin-up" process (section 1). With the development of ground-level rotation in an updraft, it is tempting to explain mesocyclone tornadoes simply as the result of frictional interaction of the mesocyclone with the ground, i.e., as end-wall vortices [*Rotunno*, 1986]. However, this explains neither the development of small-scale regions of large vorticity ("tornado cyclones") within the mesocyclone in high-resolution simulations with a free-slip lower boundary [*Wicker and Wilhelmson*, this volume] nor the foreshadowing of some violent tornadoes by tornadic vortex signatures (TVSs) that first appear aloft in Doppler radar observations [*Rotunno*, this volume] and are, perhaps, indications of tornado cyclones rather than tornadoes themselves. However, surface friction does play an important role in the "spin-up" of tornado cyclones into tornadoes [*Wicker and Wilhelmson*, this volume] and in the flow structure of torna-

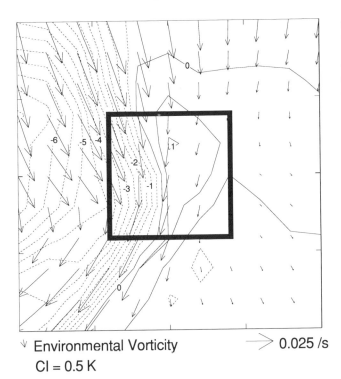

Fig. 5. Same as Figure 4 but for equivalent potential temperature at 100 m and horizontal vorticity at 300 m.

112 MESOCYCLOGENESIS FROM A THEORETICAL PERSPECTIVE

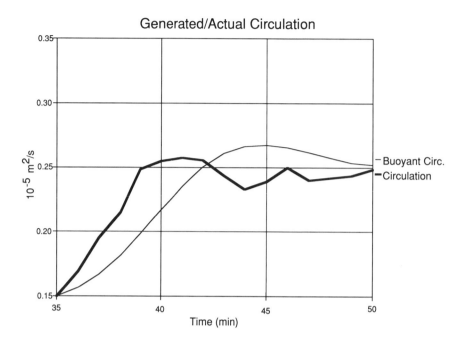

Fig. 7. Circulation (left scale) and buoyantly generated circulation since 35 min (left scale, 0.15×10^{-5} m^2 s^{-1}) of fluid circuit versus time.

does. *Davies-Jones* [1982a, b] and *Walko* [this volume] conclude that tornadogenesis is caused by concentration by a convergent wind field of vertical vorticity advected at very low levels into the updraft from the downdraft. This would explain why tornadoes are located on the side of the updraft nearest to the downdraft and why anticyclonic tornadoes, when present, are generally in a flanking line updraft further south along the gust front on the right side of the rear flank downdraft (where

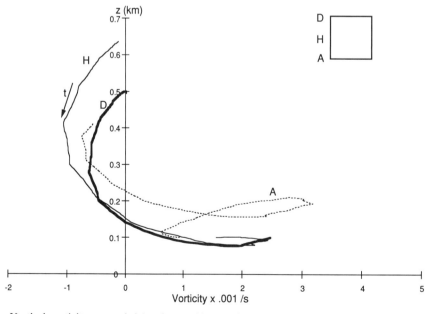

Fig. 8. Vertical vorticity versus height of west side parcels A, H, and D over 35- to 50-min time interval.

CYCLONIC VORTICITY GENERATION IN DOWNDRAFT

CROSSWISE VORTICITY

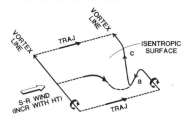

STREAMWISE BAROTROPIC VORTICITY

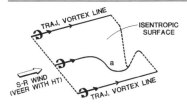

STREAMWISE VORTICITY WITH BAROCLINITY

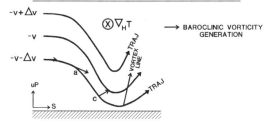

Fig. 9. Schematic diagram showing how cyclonic vorticity may be generated from tilting of barotropic and baroclinic horizontal vorticity in a downdraft. For crosswise barotropic vorticity (top), tilting produces cyclonic (anticyclonic) vorticity on the left (right) side, looking downwind in storm-relative reference frame. In the streamwise barotropic vorticity with steady flow (middle), tilting produces a purely anticyclonic downdraft. In the streamwise vorticity case with flow to the right of the horizontal buoyancy gradient and a southerly shear component as shown (bottom), a combination of tilting and baroclinic generation causes the vorticity of parcels to change from anticyclonic to cyclonic while still descending.

tilting of horizontal baroclinic and barotropic vorticity produces anticyclonic vorticity in our simulation).

Another possible mechanism is horizontal shearing instability causing the vertical sheet of vorticity at a wind-shift line to roll up into individual vortices, which are stretched vertically by convective updrafts if overhead [*Davies-Jones and Kessler*, 1974; *Brandes*, 1977]. This process explains the formation of multiple vortices in tornado simulators (analogous to multivortex tornadoes or multiple tornadoes around the rim of a mesocyclone core), gust front tornadoes, and, at least partly, landspouts and waterspouts [*Wakimoto and Wilson*, 1989]. Wakimoto and Wilson suggested that mesocyclone tornadoes might originate as vortices that are generated under flanking-line clouds by shearing instability, and move with their parent clouds as they merge into the main updraft. However, this is inconsistent with the tornado being embedded in a low-level mesocyclone, which derives its vorticity from a cool, rainy downdraft to the north. *Walko* [this volume] makes another objection, based on the initial small scale and weak circulations of the flanking-line vortices. Note that none of the above processes explains the cases where the TVS descends from midlevels to the ground in 20–30 min, instead of developing near the surface first or at the same time as at low to mid levels.

At first sight, landspouts and most waterspouts appear to have completely different origins from mesocyclone tornadoes. They typically occur in low-shear environments, along almost stationary, preexisting, frontlike boundaries early in a developing storm's life cycle, and only during afternoons with steep low-level lapse rates. However, like tornadoes, neighboring heavy rain showers appear critical to their formation. *Wilczak et al.* [1992] suggest that the dissimilarities with misocyclone tornadoes are not as great as they seem. The low-level, misoscale (<4 km diameter) vortex resulting from horizontal shearing instability in a deep convective boundary layer and the preexisting pseudofront act as surrogates for the midlevel mesocyclone and gust front of supercells, respectively. At least some of the spout's vorticity appears to be derived from tilting of baroclinically generated vorticity on the cool-air side of the misocyclone.

Clearer understanding of tornadogenesis and cyclic mesocyclogenesis awaits better numerical cloud models. The waters may be muddied by there being several distinct mechanisms and combinations thereof with the mixture of processes varying from case to case.

8. CONCLUSIONS

The initial midlevel mesocyclone originates from tilting by the updraft of environmental streamwise vorticity. Near-ground rotation awaits development of a cool downdraft. In the present case it derives from both baroclinic and barotropic vorticity on the cool-air side of the storm. In this region the baroclinic vorticity appears stronger than, but in the same direction as, the barotropic vorticity because of strong low-level curvature of the hodograph. In the straight hodograph cases of *Rotunno and Klemp* [1985] and *Walko* [this volume] the two vorticity components were in opposition in the cold air, with low-level rotation stemming from baroclinic vorticity in the former and barotropic vorticity in the latter.

Our mechanism for low-level rotation is quite similar to that described by *Rotunno and Klemp* [1985], *Weisman and Klemp* [1984, p. 2489], *Rotunno* [1986, p. 425], and *Klemp* [1987, p. 396]. There are, however, subtle but important differences. Rotunno and Klemp's [1985, p. 281] description, restated in the other articles, is as follows: "Thus, as the air approaches from the northeast, it first acquires horizontal vorticity directed towards the southwest from baroclinic generation along the cold air boundary. This horizontal vorticity is then tilted upward to produce cyclonic vertical vorticity as the air encounters the updraft. Finally, the air is subjected to intense stretching . . ." In contrast, we believe that tilting of horizontal vorticity (either barotropic or baroclinic) toward the vertical by an updraft does not produce rotation very near the ground because vertical vorticity is generated in rising air. We find in our simulation that the first cyclonic vorticity at 100 m is generated by tilting of quasi-streamwise vorticity in descending, evaporatively cooled air. This air originates from ~500 m (probably higher at later times) and is on the east edge of the cool air outflow. Its vorticity changes sign from anticyclonic to cyclonic below 250 m, while it is still in the downdraft, owing to simultaneous baroclinic generation and tilting of vorticity. The air is then entrained into the southwest side of the main updraft below 100 m. Its cyclonic vorticity is amplified by the convergence term, with a minor contribution from further tilting. Note that nothing in this scenario is inconsistent with the analysis of numerical results presented by Rotunno and Klemp.

The production and downward transport of cyclonic vorticity in the downdraft is important for tornadogenesis because of the supply of cyclonically spinning air practically to ground level. Tornadoes may result as this air flows into the foot of the updraft and undergoes considerable vertical stretching.

Acknowledgments. We thank Charles A. Doswell III for several helpful discussions during the course of this work, which was done while one of the authors (H.E.B.) held a National Research Council-NOAA Research Associateship. We gratefully acknowledge Bob Walko and an anonymous reviewer for pointing out places where we needed to bolster our arguments. Our thanks to Joan Kimpel for drafting Figure 9. The computer simulations were made on the Cray-2 at the National Center for Supercomputing Applications (NCSA), and some were supported by NSF grant ATM-87-00778.

REFERENCES

Barnes, S. L., Some aspects of a severe, right-moving thunderstorm deduced from mesonetwork rawinsonde observations, *J. Atmos. Sci.*, 27, 634–648, 1970.

Brandes, E. A., Gust front evolution and tornado genesis as viewed by Doppler radar, *J. Appl. Meteorol.*, 16, 333–338, 1977.

Brooks, H. E., C. A. Doswell III, and R. P. Davies-Jones, Environmental helicity and the maintenance and evolution of low-level mesocyclones, this volume.

Browning, K. A., and C. R. Landry, Airflow within a tornadic storm, in *Preprints, 10th Weather Radar Conference*, pp. 116–122, American Meteorological Society, Boston, Mass., 1963.

Burgess, D. W., V. T. Wood, and R. A. Brown, Mesocyclone evolution statistics, in *Preprints, 12th Conference on Severe Local Storms*, pp. 422–424, American Meteorological Society, Boston, Mass., 1982.

Davies-Jones, R. P., Observational and theoretical aspects of tornadogenesis, in *Topics in Atmospheric and Oceanographic Sciences: Intense Atmospheric Vortices*, edited by L. Bengtsson and J. Lighthill, pp. 175–189, Springer-Verlag, New York, 1982a.

Davies-Jones, R. P., A new look at the vorticity equation with application to tornadogenesis, in *Preprints, 12th Conference on Severe Local Storms*, pp. 249–252, American Meteorological Society, Boston, Mass., 1982b.

Davies-Jones, R. P., Streamwise vorticity: The origin of updraft rotation in supercell storms, *J. Atmos. Sci.*, 41, 2991–3006, 1984.

Davies-Jones, R. P., Dynamical interaction between an isolated convective cell and a veering environmental wind, in *Preprints, 14th Conference on Severe Local Storms*, pp. 216–219, American Meteorological Society, Boston, Mass., 1985.

Davies-Jones, R. P., Tornado dynamics, in *Thunderstorm Morphology and Dynamics*, 2nd ed., edited by E. Kessler, chap. 10, pp. 197–236, University of Oklahoma Press, Norman, 1986.

Davies-Jones, R. P., Thunderstorm, in *Encyclopedia of Science and Technology*, 7th ed., McGraw-Hill, New York, 1992a.

Davies-Jones, R. P., Tornado, in *Encyclopedia of Science and Technology*, 7th ed., McGraw-Hill, New York, 1992b.

Davies-Jones, R. P., and E. Kessler, Tornadoes, in *Weather and Climate Modification*, edited by W. N. Hess, chap. 16, pp. 552–595, John Wiley, New York, 1974.

Davies-Jones, R. P., D. W. Burgess, and M. P. Foster, Test of helicity as a tornado forecast parameter, in *Preprints, 16th Conference on Severe Local Storms*, pp. 588–592, American Meteorological Society, Boston, Mass., 1990.

Davies-Jones, R., R. Rabin, and K. Brewster, A short-term forecast model for thunderstorm rotation, in Proceedings, Nowcasting II Symposium, Norrköping, Sweden, *Eur. Space Agency Spec. Publ. ESA SP-208*, 367–371, 1984.

Dutton, J. A., *The Ceaseless Wind*, 579 pp., McGraw-Hill, New York, 1976.

Johnson, K. W., P. S. Ray, B. C. Johnson, and R. P. Davies-Jones, Observations related to the rotational dynamics of the 20 May 1977 tornadic storms, *Mon. Weather Rev.*, 115, 2463–2478, 1987.

Klemp, J. B., Dynamics of tornadic thunderstorms, *Annu. Rev. Fluid Mech.*, 19, 369–402, 1987.

Klemp, J. B., and R. Rotunno, A study of the tornadic region within a supercell thunderstorm, *J. Atmos. Sci.*, 40, 359–377, 1983.

Klemp, J. B., R. B. Wilhelmson, and P. S. Ray, Observed and numerically simulated structure of a mature supercell thunderstorm, *J. Atmos. Sci.*, 38, 1558–1580, 1981.

Lilly, D. K., The development and maintenance of rotation in convective storms, in *Topics in Atmospheric and Oceanographic Sciences: Intense Atmospheric Vortices*, edited by L. Bengtsson and M. J. Lighthill, pp. 149–160, Springer-Verlag, New York, 1982.

Lilly, D. K., The structure, energetics and propagation of rotating convective storms, II, Helicity and storm stabilization, *J. Atmos. Sci.*, 43, 126–140, 1986.

Rabin, R. M., and R. P. Davies-Jones, Atmospheric structure ahead of thunderstorms, *Endeavour*, 10(1), 20–27, 1986.

Rotunno, R., Tornadoes and tornadogenesis, in *Mesoscale Meteorology and Forecasting*, edited by P. S. Ray, chap. 18, pp. 414–436, American Meteorological Society, Boston, Mass., 1986.

Rotunno, R., Supercell thunderstorm modeling and theory, this volume.

Rotunno, R., and J. B. Klemp, The influence of the shear-induced pressure gradient on thunderstorm motion, *Mon. Weather Rev.*, 110, 136–151, 1982.

Rotunno, R., and J. B. Klemp, On the rotation and propagation of simulated supercell thunderstorms, *J. Atmos. Sci.*, 42, 271–292, 1985.

Scorer, R. S., *Environmental Aerodynamics*, 488 pp., Ellis-Horwood, Chichester, England, 1978.

Smith, R. K., and L. M. Leslie, A numerical study of tornadogenesis in a rotating thunderstorm, *Q. J. R. Meteorol. Soc.*, 105, 107–127, 1979.

Wakimoto, R. M., and J. W. Wilson, Non-supercell tornadoes, *Mon. Weather Rev.*, 117, 1113–1140, 1989.

Walko, R. L., Tornado spin-up beneath a convective cell: Required basic structure of the near-field boundary layer winds, this volume.

Weisman, M. L., and J. B. Klemp, The structure and classification of numerically simulated convective storms in directionally varying wind shears, *Mon. Weather Rev.*, 112, 2479–2498, 1984.

Wicker, L. J., and R. B. Wilhelmson, Numerical simulation of tornadogenesis within a supercell thunderstorm, this volume.

Wilczak, J. M., T. W. Christian, D. E. Wolfe, and R. J. Zamora, Observations of a Colorado tornado, I, Mesoscale environment and tornadogenesis, *Mon. Weather Rev.*, 120, 497–520, 1992.

Discussion

MORRIS WEISMAN, SESSION CHAIR

National Center for Atmospheric Research

PAPER B1

Presenter, Rich Rotunno, National Center for Atmospheric Research [*Rotunno*, this volume, Supercell thunderstorm modeling and theory]

(Doug Lilly, University of Oklahoma) You still believe in [storm] splitting and vortex development without rain, yet all laboratory experiments in unidirectional shear produce line convection. To my knowledge, splitting never occurs in either laboratory experiments or numerical simulations of them.

(Rotunno) The critical factor missing in the laboratory experiments, which are essentially Rayleigh-Benard convection with shear, and the distinct feature of the numerical cloud simulations are conditional instability. Air has to be lifted to its level of free convection before updrafts form. In Figure 13, there is lower pressure on the south flank of the updraft. All the air outside the updraft is not going to rise on its own without being lifted. It needs a lift [provided by upward pressure gradient forces] to overcome the inversion that is there, and once the air penetrates the inversion, it becomes buoyant. It rises and regenerates the updraft on the south side. Thus the updraft propagates to the right of the mean shear. You need conditional instability; with absolute instability, convection would go up everywhere.

(Joe Golden, National Oceanic and Atmospheric Administration, Office of the Chief Scientist) The model results of the last 10 years are incredibly impressive. In the future, can we use a model to predict the probability that a given supercell will produce a tornado in the next 1 to 3 hours, assuming that we have data from profilers and NEXRAD radars to initialize the model?

(Rotunno) I can't predict that [general laughter]. I think that you have to try and see. It's not out of the question.

PAPER B2

Presenter, Lou Wicker, University of Illinois [*Wicker and Wilhelmson*, this volume, Numerical simulation of tornadogenesis within a supercell thunderstorm]

(Don Burgess, National Severe Storms Laboratory) This question is for both Rich [Rotunno] and Lou. I have always been impressed by the agreement between observations and numerical modeling for the Del City storm, but there are some observational data sets that are somewhat different from Del City. I want to make sure that the numerical modeling community does not quit with the Del City case. I'd like to see some of these other storms that have different features analyzed. It's important that we look at a spectrum of storms because I definitely think that there is a spectrum of storms out there.

(Wicker) The honest reason why I am showing Del City is because other cases didn't work as well. For example, in the April 3, 1964, case which I did for my thesis, there is similar evolution, but the mesocyclone is half the diameter of the one in the Del City case. So I didn't run the April 3 case because I needed twice the grid resolution in order to resolve the same structures and I didn't have the computer time to do this. We are definitely going to look at different cases. Some of the other cases, which we expected to produce intense low-level mesocyclones because storm-relative helicity is large, don't work as well as Del City.

(Brian Fiedler, University of Oklahoma) I noticed that the vortex that you are quick to call a tornado vortex looks like . . .

(Wicker) In this particular case, I wouldn't call it a tornado vortex.

(Fiedler) Okay, but it looked like the safest place to be in a storm. It looks more like a stagnation point or an F minus 1 tornado. How do you distinguish between a tornado vortex and an inflection point instability? Maybe the vortex has nothing to do with the tornado that actually occurred. It could be a barotropic instability along a gust front.

(Wicker) It could be. Certainly such a mechanism could be important at low levels. However, this vortex has time and height continuity. It extends to levels well above the gust front. So the low-level vortex may develop from an instability along the boundary. But it must connect with features aloft that are not connected with the gust front. This particular vorticity maximum extended upward 4–5 km into the storm. In the April 3 simulation, vorticity greater than $0.1\ s^{-1}$ was present up to 3.5 km. So there is definitely some vertical continuity associated with these vortices.

(Chair) I believe that your vortex is within about half a kilometer or so of the observed location of many tornadoes, so there is some consistency there.

(Rich Rotunno, National Center for Atmospheric Research) The thrust of the question is whether or not the simulated vorticity maximum equals tornado. I don't think that you think that.

(Wicker) No, I don't think so. Determining whether a vorticity maximum in the model is really a tornado is difficult.

(Rotunno) Can I answer Don Burgess' question?

(Chair) No! [Out of time.]

Paper B3

Presenter, Bob Walko, Colorado State University [*Walko*, this volume, Tornado spin-up beneath a convective cell: Required basic structure of the near-field boundary layer winds]

(Brian Fiedler, University of Oklahoma) Based on the thermodynamic speed limit and the depth of the layer, what is the Reynolds number of your simulations?

(Walko) I haven't computed it, but based on the number of grid points, it is probably around 1000.

(Fiedler) We didn't have any luck making tornadoes in 3-D until we used a Re above about 2000. Below that, we had to insert vertical vorticity, but if Re is high enough, we could get from horizontal vorticity to shear along a gust front to a tornado.

(Walko) That's very interesting. It seems to me that these simulations were generally robust, that I could get a vortex provided that I put the downdraft in the correct sector, etc. It took some tuning, but then I could get a tornadolike vortex with very low Re, even a few hundred.

(Rich Rotunno, National Center for Atmospheric Research) The circulation in the Rotunno and Klemp simulation was baroclinically generated. Analysis showed that the curve around the vortex at the time of maximum vorticity had circulation produced by the baroclinic generation term. This is the mechanism in the full cloud model. The relationship between that and your result is subject to further study.

Paper B4

Presenter, Harold Brooks, National Severe Storms Laboratory [*Brooks et al.*, this volume, Environmental helicity and the maintenance and evolution of low-level mesocyclones]

(Howie Bluestein, University of Oklahoma) According to your argument, if there is a very strong inflow low, then the gust front won't surge out. But it seems to me that if the gust front has a strong high behind it and if the inflow is very intense, there will be a strong pressure gradient force which will accelerate the gust front rapidly.

(Brooks) It depends on the balance between the two effects. If the high behind the gust front is weak, the gust front won't advance. If this high is strong, the gust front will move rapidly.

(Bluestein) But if the inflow low is very strong and the high behind the gust front is very weak, won't the gust front also surge out?

(Bob Davies-Jones, National Severe Storms Laboratory) Not if the air parcels ascend at the gust front. The pressure gradient force acts on parcels, which ascend at the gust front rather than passing through it.

(Chair) I have a bias against helicity because of its storm-relative nature. If one takes the divergence of the momentum equations, one can derive a diagnostic pressure equation that describes the pressure field. The forcing function in this equation for pressure is dependent on buoyancy and velocity gradients but is totally independent of any terms that are dependent on reference frame. Your Bernoulli relationship [for pressure in terms of velocity] is only correct in a steady state reference frame.

(Brooks) We are assuming that it is a steady state storm. In a ground-relative frame, there is an extra term, $\partial \phi/\partial t$ [ϕ is the velocity potential], in the Bernoulli equation. In the storm-relative frame, $\partial \phi/\partial t$ drops out automatically [because $\partial/\partial t \equiv 0$].

(Chair) But when you go back to the original diagnostic pressure equation, there is no storm-relative term. I see a contradiction there.

(Davies-Jones) The diagnostic pressure equation is a Poisson's equation involving second derivatives of the pressure field, not pressure (p) itself. It can't be integrated simply by assuming that $\nabla^2 p \propto -p$, as is frequently done, because some time-dependent terms reappear after integration.

(Chair) I think that if you make the same steady state assumptions, though, there is still a contradiction there.

(Rich Rotunno, National Center for Atmospheric Research) The virtue of the diagnostic pressure equation is that the time dependency, $d(\nabla \cdot \mathbf{v})/dt$, drops out. Numerical models use this equation [to recover pressure at each time step]. The time dependency doesn't reappear.

(Brooks) We'll let the two theoreticians fight this out!

(Davies-Jones) You are really considering $\nabla^2 p$, not p itself; p is found by integrating the forcing function over the whole volume. Admittedly, it is weighted by [inverse] distance, but this integration completely changes the picture. It gives an inflow low at the ground which is forced entirely from above like Dudhia and Moncrieff [*J. Atmos. Sci.*, *46*, 3363–3391, 1989] said.

(Doug Lilly, University of Oklahoma) You said that storm motion cannot be predicted. I guess you mean by that from environmental conditions only or that the evolution as such is maybe too complicated to predict. But, surely, if you have observations of the storm at a given time, the equations of motion are capable of predicting the motion.

(Brooks) I'm saying that you cannot predict storm motion from the environmental conditions alone.

PAPER B5

Presenter, Robert Davies-Jones, National Severe Storms Laboratory [*Davies-Jones and Brooks*, this volume, Mesocyclogenesis from a theoretical perspective]

(V. G. Blanchette, retired professional engineer) Your analyses [of the numerical model results] were for 100-m elevation. Do they also apply at much higher elevations, say, 2–4 km?

(Davies-Jones) At 2–4 km, there is a lot of vorticity that has been generated on the warm side of the storm. The analyses presented here apply just to very low levels, which are the important ones for investigating how rotation gets down to the ground.

(Rich Rotunno, National Center for Atmospheric Research) Where does the TVS fit into all of this?

(Davies-Jones) I don't know. The TVS mystifies me because, according to Doppler radar observations, it begins aloft, and I don't understand why.

Observations and Simulations of Hurricane-Spawned Tornadic Storms

EUGENE W. MCCAUL, JR.

Universities Space Research Association, Huntsville, Alabama 35806

1. INTRODUCTION

The occurrence of tornadoes within landfalling tropical cyclones has been recognized for a long time, with some reports extant from as early as 1811 [*Tannehill*, 1950]. However, the widespread rain, low clouds, and generally poor visibility attending most landfalling tropical cyclones has hindered direct observations of these tornadoes and impeded progress in understanding the character of their parent convective storms. Still, as more and more reports have become available, it has become possible to characterize the tornadoes and tornadic storms with growing accuracy.

In the early studies, limited data forced investigators to consider only tornadoes spawned by tropical cyclones of full-fledged hurricane intensity. *Malkin and Galway* [1953] noted mounting evidence that most hurricane-spawned tornadoes were weaker than their counterparts from the United States Great Plains and Midwest. *Smith* [1965] identified the right-front (RF) quadrant of the hurricane as being the favored area for tornado development, while *Pearson and Sadowski* [1965] found that most of the tornadoes occurred outside the area of hurricane-force surface winds. *Hill et al.* [1966] suggested that dry intrusions entering the circulations of landfalling hurricanes might contribute to the development of tornadic storms by augmenting the convective instability. The same authors also noted the tendency for greater numbers of tornadoes to occur in the more intense hurricanes.

Novlan and Gray [1974] employed compositing techniques on a set of data comprising a 25-year period to arrive at the first representative look at upper air conditions accompanying hurricane-spawned tornadoes. They documented the presence of very strong low level vertical shear and the relative absence of thermodynamic instability in the composite hurricane tornado environment. The strong vertical shears were attributed to thermal wind adjustment to the adiabatic core cooling experienced by air parcels entering the landfalling hurricane center subsequent to its loss of contact with the warm ocean surface. *Gentry* [1983] argued that a more important factor in the generation of the strong shears was the increase in surface drag experienced by the low level flow as it encounters a coastline.

While most of the early research on hurricane-spawned tornadoes focused on their temporal and spatial patterns of occurrence within the parent hurricane, little progress was made regarding the specific characteristics of the tornadoes until recently. *Stiegler and Fujita* [1982] examined the damage from some of Hurricane Allen's (1980) tornadoes in Texas and observed strong evidence of multiple-vortex structure reminiscent of intense Midwestern tornadoes. The first direct photographic confirmation of this finding was given by *McCaul* [1987], who documented two of the tornadoes spawned by 1985's Hurricane Danny in northern Alabama (Figure 1). *McCaul* [1987] also discovered that some of Danny's tornadic storms exhibited persistent distinctive radar echoes (Figure 2) resembling those of small supercell thunderstorms. Wind profiles obtained from nearby soundings indicated the presence of large amounts of helicity [*Lilly*, 1986] and streamwise vorticity [*Davies-Jones*, 1984], parameters which by that time had been found to be related to the development of rotation in the updrafts of storms that occur in the Midwest and Great Plains of the United States.

The accumulation of new observations and severe storm conceptual models that became available by the late 1980s prompted a comprehensive new study of hurricane-spawned tornado environments, along with the first numerical simulations of hurricane-spawned tornadic convection. This paper reviews the results of these recent research efforts.

With respect to observations there remains a lack of radar data suitable for use in investigating individual hurricane-spawned tornadic storms. As a consequence, recent efforts have concentrated on extracting more information about storm environments from already existing standard data sources. *McCaul* [1991], hereinafter abbreviated as M91, has examined 39 years of United States sounding data taken in proximity to hurricane-spawned tornadoes and has constructed composite temperature, moisture, and wind profiles

Fig. 1. Tornado spawned by remnants of Hurricane Danny is photographed at approximately 2039 UTC August 16, 1985, as it moves northward across Redstone Arsenal, near Huntsville, Alabama. Note that tornado is pendent from a well-defined wall cloud formation. The tornado was one in a series of tornadoes produced by this supercell-like storm, which was one of several spawned by Danny. For more details, see *McCaul* [1987]. Photo is courtesy of A. Junkins, Huntsville, Alabama.

for a variety of stratifications of the data. A major goal of that work was to find out how well various environmental parameters such as convective available potential energy (CAPE [*Moncrieff and Miller*, 1976; *Weisman and Klemp*, 1982]), vertical shear and Bulk Richardson Number (BRN [*Weisman and Klemp*, 1982, 1984]), and helicity [*Lilly*, 1986; *Davies-Jones et al.*, 1990] and streamwise vorticity [*Davies-Jones*, 1984] describe the tornadic potential of the landfalling hurricane environment. Detailed comparisons were also made with the Great Plains tornado proximity environment.

McCaul [1990] also conducted numerical simulations of hurricane-spawned severe storms in order to study their structure and dynamics. The simulations were initiated with a selection of soundings from his 39-year data sample, along with some of his composite soundings. Comparison of the results with simulations of Great Plains severe storms showed important similarities but also many differences. These similarities and differences may prove useful in identifying the essential kinematic and thermodynamic properties of pretornadic environments in general.

The remainder of this paper is organized into four major parts. Section 2 summarizes the major findings of the M91 observational work. Section 3 contains results of previous [*McCaul*, 1990] and more recent numerical simulation studies of hurricane-spawned convection. Section 4 consists of discussion and interpretation of the results of the previous two sections. Section 5 gives a summary and outlook for future work.

2. OBSERVATIONS

2.1. *Background*

McCaul's [1991] recent study of the characteristics of hurricane-spawned tornado environments employed all available United States rawinsounding data taken near those tornadoes that were reported during the 39-year period from 1948 through 1986. Even though only tornadoes associated with Atlantic and Gulf Coast tropical cyclones were included in that study, the data set is still the largest ever assembled for the study of hurricane-spawned tornadoes. Only the highlights of this study will be described here; for additional details the reader is referred to the original work. The set of tornado-producing tropical cyclones included not only those

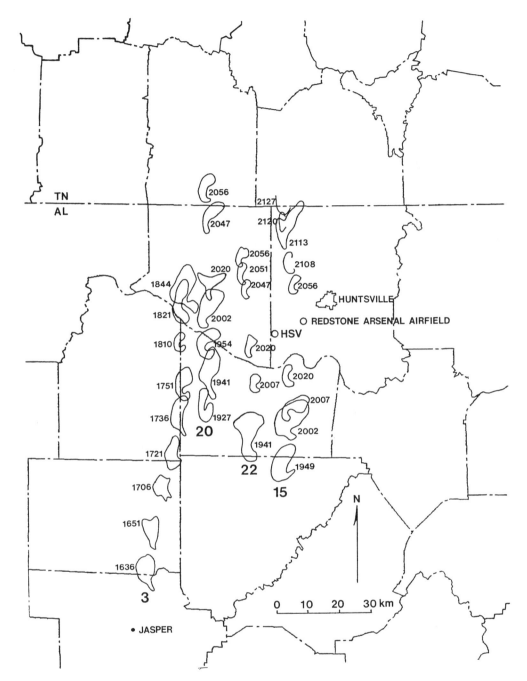

Fig. 2. Radar echo history of four distinctive-echo storms spawned by Hurricane Danny on August 16, 1985, as seen by National Weather Service radar at Huntsville, Alabama. Storm labeled "22" produced tornado shown in Figure 1. See *McCaul* [1987, Figure 3] for further discussion of this event.

cyclones of full-fledged hurricane intensity but also some tropical storms and subtropical cyclones. Nevertheless, throughout the remainder of this paper all these cyclones will be referred to using the generic term "hurricanes" for convenience.

The 39-year data sample yielded some 199 rawinsonde observations ("raobs") that were made within 185 km and 3 hours of reported hurricane-spawned tornadoes. These criteria match those established by *Novlan and Gray* [1974] in their earlier composite study. An additional 1097 raobs taken within one-day buffer periods before and after those of the actual proximity soundings were included in the data base to help delineate the larger-scale structure of the tornadic hurricanes. All raobs were inspected and erroneous data

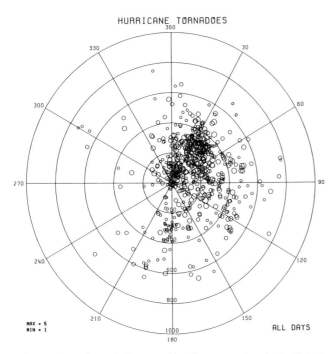

Fig. 3. Locations of all reported hurricane tornadoes in the United States, 1948–1986. Tornadoes are plotted at azimuths taken relative to headings of hurricane centers at tornado times. These headings have been rotated to coincide with the 360° azimuth. Areas of circles are proportional to unity plus the F scale rating of the tornado. Note clustering of data points in right-front quadrant. Range rings, labeled along the 180° azimuth, are 200 km apart.

intensity associated with each raob. In M91 the term "tornado outbreak" refers simply to the collection of all tornadoes reported with a landfalling hurricane and does not imply the occurrence of more than some minimum number of tornadoes. The subjectively assigned F scale values undoubtedly contain numerous errors [*Doswell and Burgess*, 1988], but they are found to correlate significantly with various meteorological parameters and thus may be useful indicators of the intensity of tornadic activity.

The three measures of ambient helicity examined were the convective cell-relative total and relative helicity and streamwise vorticity. Mathematically, the cell-relative total helicity was defined [*Lilly*, 1986] as $H_t = (\mathbf{V} - \mathbf{V_c}) \cdot (\mathbf{k} \times \partial \mathbf{V}/\partial z)$, the relative helicity as $H_r = H_t/[|\mathbf{V} - \mathbf{V_c}||\partial \mathbf{V}/\partial z|]$, and streamwise vorticity as $\omega_s = H_t/|\mathbf{V} - \mathbf{V_c}|$, where $\mathbf{V}$ is the ground-relative flow, $\mathbf{V_c}$ is cell motion, and $\mathbf{k}$ is the unit vector along the vertical (z) coordinate. Because of the similarities between these parameters they will be referred to in the remainder of this paper collectively as "helicity parameters." M91 presented estimates of H_t, H_r, and ω_s that were computed by averaging the profiles of point estimates of those quantities over the lowest 3, 6, or 12 km. In computations of quantities requiring knowledge of convective cell motions $\mathbf{V_c}$, the 0- to 6-km mass-weighted mean winds were used as estimates of the cell motions. Because cell motions were believed to be slower than those mean winds (and consequently to be represented by vectors that lay farther off the hodograph than the mean wind vector [see

deleted. The 199 tornado proximity raobs were close to some 366 reported tornadoes. This is more than half of the total of 626 tornadoes reported in all the Atlantic and Gulf hurricanes affecting the United States during the 39 years of data coverage. The hurricane-relative spatial distribution of those tornadoes, given in Figure 3, confirms the findings of *Smith* [1965] and others that the right-front quadrant is the preferred sector of a hurricane for tornado formation.

During the analysis of the data all raob winds were converted from zonal and meridional components to radial and tangential components relative to the hurricane. Then, for each sounding, bulk parameters such as CAPE, surface to midtropospheric vertical shear, various measures of ambient helicity, and BRN were computed. The raob location relative to the hurricane's position and heading were also recorded. The hurricane-relative spatial distributions of these quantities were then mapped and cross correlations computed against the intensity of tornadic activity reported in proximity to each raob. Although the per-raob frequency distribution of tornadic activity was skewed, M91 found that the logarithm of the sum of the tornadoes' Fujita scale (F scale) numbers [*Fujita*, 1973, 1981], each augmented by unity to avoid nullifying the F0 events, was sufficiently close to Gaussian to permit application of parametric statistical analysis techniques. Thus, in McCaul's study, the log (F-sum) per raob is taken as the measure of tornado outbreak

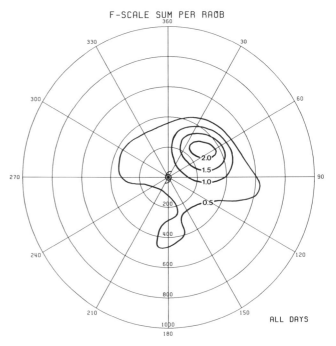

Fig. 4. Spatial distribution, relative to hurricane motion, of objectively analyzed values of F scale sum per raob for 1296 hurricane raobs. Objective analysis grid mesh had spacing of 100 km. As in Figure 3, hurricanes are centered at origin and are moving toward 360°.

McCaul, 1987]), the computed helicity values tended to underestimate the true values. M91 argued that for strongly curved hodographs the H_r estimates should be increased by 20–50% to obtain values corresponding to realistic storm motions. Furthermore, the estimates of 0–3-km total helicity could be converted to the units used by *Davies-Jones et al.* [1990] by multiplying by the depth of the averaging layer, 3000 m.

Once the raobs were analyzed, they were stratified in a variety of ways, then composited. In constructing the composite profiles the raw temperature, dew point, and wind component data were transformed into a 101-level sigma coordinate system, then averaged, then transformed back into pressure coordinates using the mean of the surface pressures as a base value. The root-mean-square variability of each quantity was also computed during the compositing process to facilitate assessment of statistical reliability.

2.2. Hurricane-Relative Spatial Distributions of Raob Parameters

All 1296 raobs in the M91 data base were used in preparing objectively analyzed maps of the spatial distribution of the various raob parameters. The individual raob bulk values were interpolated to a 100-km Cartesian mesh using a distance weighting function with cutoff at 200 km. The results were then contoured throughout those regions where data density was sufficient to insure representativeness.

Results of the objective analyses of F-sum, CAPE, BRN shear, BRN, 0- to 3-km H_t are shown in Figures 4–8. BRN

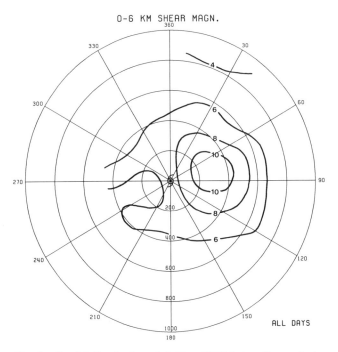

Fig. 6. Spatial distribution of 0–6-km "BRN shear" magnitude (see text) for the 1296 hurricane raobs. Format is as in Figure 3. Units of contours are meters per second.

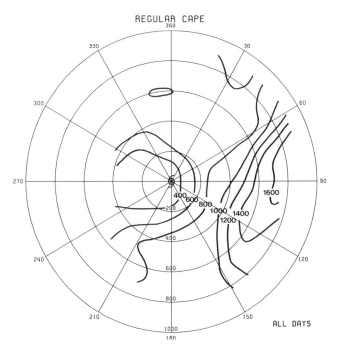

Fig. 5. Spatial distribution of convective available potential energy for the 1296 hurricane raobs. Format is as in Figure 3. Units of contours are joules per kilogram.

shear is defined as the magnitude of the vector difference of the density-weighted mean winds between 0 and 6 km and between 0 and 500 m (M91). Plots of H_r and ω_s (see M91) generally show patterns resembling that of H_t and are not shown here. The pattern of contours seen in Figure 4, representing the distribution of F-sum, was taken as a measure of the spatial distribution of tornado outbreak intensity. Parameters showing maxima and minima arranged similarly to those found in Figure 4 may be considered possibly useful as predictors of hurricane tornadoes.

In general, the patterns of BRN shear and the helicity parameters more closely resemble that of F-sum than do those of CAPE or BRN. Analyzed CAPEs reach a minimum of less than 400 J kg^{-1} near and just to the left of the hurricane centers, relative to their motion. CAPE reaches a maximum well to the right of the hurricane track and is generally larger at the rear of the hurricane than on the forward side. The BRN shear and helicity parameters, on the other hand, reach distinct maxima in the right front or right quadrants. The occurrence of maximum shear in these areas is roughly consistent with the findings of *Novlan and Gray* [1974, Figures 10 and 12]. Because of the behavior of BRN shear and CAPE the BRN contour pattern exhibits its maximum in the distal portions of the right and rear quadrants, with its minimum in the right-rear (RR) quadrant of the hurricane, near the center. The BRN minimum is located close to that of the maximum in F-sum and has a value well within the supercell range [*Weisman and Klemp*, 1982], suggesting that supercell storms could indeed be responsible for some hurri-

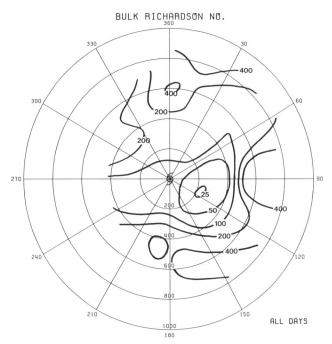

Fig. 7. Spatial distribution of Bulk Richardson Number for the 1296 hurricane raobs. Format is as in Figure 3.

cane tornadoes. However, the location of the BRN minimum and its overall pattern do not correspond as well with F-sum as do the BRN shear and helicity parameter patterns.

2.3. Correlations of Raob Parameters With Tornado Outbreak Intensity

While the results of Figures 4–8 depict the spatial relationships between F-sum and the other raob parameters, linear cross-correlation coefficients give a concise quantitative measure of how parameter variations intercompare individually. In order to examine how the various raob parameters changed with the reported intensity of hurricane tornado outbreaks, M91 computed the cross correlations between his selection of parameters from the 199 proximity raobs and their respective log (F-sum) data. Data from the other nonproximity raobs were excluded in order to avoid biasing the results with an excessive number of nonproximity raobs.

Table 1 contains a listing of the computed correlations, along with the t values from a Student t test of the significance of the correlation's deviation from an assumed zero hypothetical value. As the table shows, many of the listed variables exhibit correlations with log (F-sum) which are significantly different from zero. The results of the t tests and some additional correlation simulations show that correlations of approximately 0.14 magnitude are significant at the 5% level (i.e., have only a 5% chance of occurring as a result of random effects).

As expected from examination of Figures 4–8, the raob parameters that correlate best with tornado outbreak severity are the helicity parameters and BRN shear. The simple parameter, 700-hPa wind speed, also displayed a high correlation with log (F-sum), 0.31. Interestingly, the helicity parameters averaged through 12 km show the highest correlations, ranging from 0.31 to 0.34, even though the mean values of the helicity parameters attain their highest values in the 0- 3-km layer. This suggests that favorable shears through deeper layers make for an environment especially conducive to development of more numerous and more intense hurricane-spawned tornadoes. It should be noted, however, that only for H_r are the 0- to 12-km layer mean values larger than their 0–3-km counterparts in a statistically significant sense.

Consistent with Figure 5, CAPE shows a weak negative correlation of -0.07 with log (F-sum). Such a correlation is significant at approximately the 23% level. Because of this negative correlation between CAPE and log (F-sum), M91 did not automatically exclude from his data base soundings having zero CAPE. However, exclusion of the zero-CAPE soundings from the correlation computations was found to have no major impact on the results.

Although all the highest ranking correlations are statistically significant at better than the 1% level, their absolute magnitudes are not very large. M91 suggested four possible reasons for the smallness of the correlations: (1) mesoscale variability within the hurricanes, with potentially important variations having scales not resolved by the raob network or amplitudes not much larger than the accuracy of the raob measurements, (2) errors in F scale assignments and probable underreporting of the tornadoes, (3) for the helicity parameters, errors in the estimates of storm cell motions,

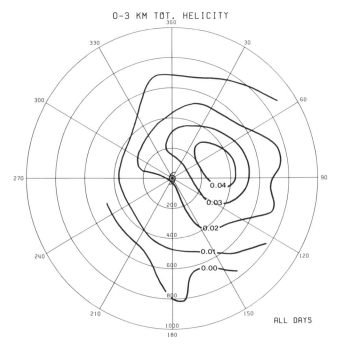

Fig. 8. Spatial distribution of 0–3-km total helicity H_t for the 1296 hurricane raobs. Format is as in Figure 3. Units of contours are meters per square second.

TABLE 1. Cross-Correlations Between Log (F-Sum) and Raob Parameters Using Only Raobs in General Proximity to Tornadoes

Variable	Correlation With Log (F-Sum)	t Value
CAPE (J kg^{-1})	−0.07	−0.91
BRN shear (m s^{-1})	0.32	4.47
BRN	−0.18	−2.50
0- 3-km H_r	0.16	2.18
0- 6-km H_r	0.22	3.04
0- 12-km H_r	0.31	4.34
0- 3-km H_t (m s^{-2})	0.26	3.58
0- 6-km H_t (m s^{-2})	0.28	3.99
0- 12-km H_t (m s^{-2})	0.32	4.51
0- 3-km ω_s (s^{-1})	0.26	3.54
0- 6-km ω_s (s^{-1})	0.30	4.18
0- 12-km ω_s (s^{-1})	0.34	4.77
850-hPa windspeed (m s^{-1})	0.22	3.08
700-hPa windspeed (m s^{-1})	0.31	4.40
500-hPa windspeed (m s^{-1})	0.25	3.46
Sfc-850-hPa shear (m s^{-1})	0.19	2.54
Sfc-700-hPa shear (m s^{-1})	0.28	3.92
Sfc-500-hPa shear (m s^{-1})	0.20	2.69

CAPE is convective available potential energy; BRN is Bulk Richardson Number; Sfc is surface.

and (4) inherently limited predictability of details of tornado intensity using larger-scale ambient single-parameter data derived only from raob profiles. Many of the parameters examined are designed to assess only probability of mesocyclone formation, rather than actual tornado formation, and thus should not be expected to correlate straightforwardly with tornado activity.

2.4. Proximity Composite Soundings

The 199 proximity raobs were used to construct a "general proximity" hurricane tornado composite sounding. These soundings include those made under a variety of conditions, with no stratifications of the data applied except for proximity to tornadoes. A secondary set of criteria, 40 km and 2 hours, applied to the general proximity set yielded a smaller but apparently representative set of 10 "close proximity" soundings. These latter soundings were also screened to make sure they were made at greater distances from the hurricane centers than the tornado occurrences. This restriction was designed to maximize the probability that the soundings were representative of "inflow" conditions feeding into the tornado-bearing convective storms. A listing of the specific soundings in the "close proximity" group is given in Table 2.

The general and close proximity composites are shown in Figure 9 in skew-T log-p and polar hodograph format. Both profiles feature a conditionally unstable layer below about 650 hPa, surmounted by an absolutely stable layer. Above 500 hPa the lapse rates are close to moist adiabatic. Relative humidities are high throughout much of the troposphere, especially in the close proximity profile, where dew point depressions are often less than 6°C. Maximum buoyancies of boundary layer parcels, approximately 3°C, are achieved around 650 hPa. The tropopause is found on average near 150 hPa. All these patterns resemble those found by *Sheets* [1969] in his study of hurricane soundings.

In both proximity composites the winds are characterized by inflow toward the hurricane center below approximately 2-km altitude, with outflow above. The hodograph has a distinct "loop" or "horseshoe" shape, with veering winds and shears from the surface through at least 10-km altitude. The strongest winds occur near 2- to 3-km altitude, and very strong vertical shears exist in the lowest 1 km. These results are consistent with those of *Novlan and Gray* [1974] but have the advantage of showing explicitly how the winds are organized relative to the hurricane center.

In both composites the altitude of maximum tangential winds, near 2–3 km, is seen to be almost twice that observed in mature maritime hurricanes [*Frank*, 1977, Figure 9]. This suggests an increase in the depth of the layer containing positive shear of the tangential wind during and after landfall. Such a redistribution of tangential momentum is consistent with either the postlandfall adiabatic core cooling of the hurricane described by *Novlan and Gray* [1974] or the increasing influence of surface friction [*Gentry*, 1983].

Comparison of the two composite profiles in Figure 9 also shows many important differences. Of special note is the significantly larger wind speed and shear in the close prox-

TABLE 2. Soundings Taken in Close Proximity to Hurricane Tornadoes

Raob Site	Hurricane	Time/Date	Azran	Maxwnd
Jacksonville, Florida	unnamed	0000 UTC June 9, 1957	33 at 256	18.0
Charleston, South Carolina	Cleo	0600 UTC Aug 29, 1964	40 at 205	43.8
Cape Canaveral, Florida	Isbell	0200 UTC Oct. 15, 1964	301 at 186	56.6
Miami, Florida	Alma	1200 UTC June 8, 1966	23 at 412	41.2
Miami, Florida	Alma	1800 UTC June 8, 1966	52 at 279	41.2
Victoria, Texas	Beulah	0000 UTC Sept. 21, 1967	59 at 207	72.1
Chantilly, Virginia	David	0000 UTC Sept. 6, 1979	26 at 234	43.8
Victoria, Texas	Allen	1200 UTC Aug. 10, 1980	79 at 265	56.6
Nashville, Tennessee	Danny	0000 UTC Aug. 17, 1985	344 at 305	41.2
Athens, Georgia	Danny	1200 UTC Aug. 17, 1985	37 at 298	41.2

Azran is raob azimuth, range (in kilometers) relative to hurricane heading. Maxwnd is peak surface winds at landfall (in meters per second).

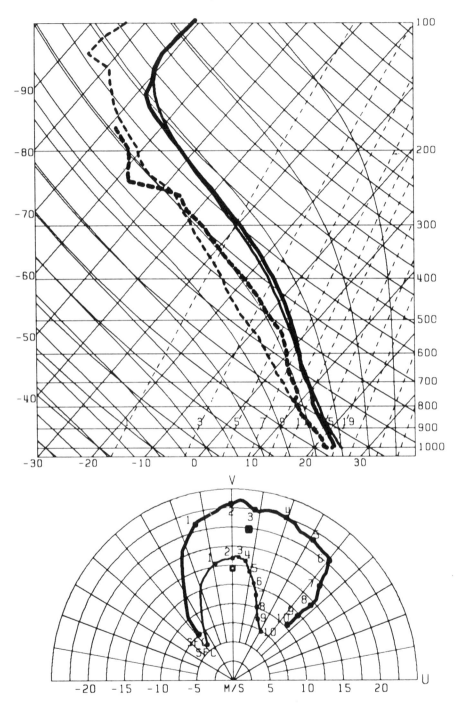

Fig. 9. Skew-T log-p and hodograph diagrams for the composite soundings taken in close proximity (heavier lines) and general proximity (lighter lines) to hurricane tornado events. U and V components represent the radial and tangential winds relative to hurricane centers at raob time. Small squares on hodograph diagrams mark 0–6-km mean winds used as estimates of convective cell motion. Solid (open) square applies to thick- (light-) line hodograph.

TABLE 3. Mean Parameters of Tornado Proximity Soundings
Data Stratified by Proximity

Variable	General Proximity	Close Proximity
Number of temperature profiles	180	10
Number of wind profiles	169	10
Range from center	329.3	264.8
CAPE (J kg^{-1})	760.4	253.4
BRN shear (m s^{-1})	9.8	14.0
BRN	66.3	3.3
0- 3-km H_r	0.28	0.52
0- 6-km H_r	0.22	0.42
0-12-km H_r	0.15	0.39
0- 3-km H_t (m s^{-2})	0.035	0.078
0- 6-km H_t (m s^{-2})	0.023	0.055
0-12-km H_t (m s^{-2})	0.016	0.046
0- 3-km ω_s (s^{-1})	0.0037	0.0068
0- 6-km ω_s (s^{-1})	0.0025	0.0051
0-12-km ω_s (s^{-1})	0.0017	0.0040

CAPE is convective available potential energy. BRN is Bulk Richardson Number.

imity case. In addition, temperatures in the close proximity composite are slightly warmer in the middle and upper troposphere and slightly cooler below 700 hPa. The relatively small difference in the mean radial distances of the two Figure 9 composites from the hurricane centers (264.8 km versus 329.3 km) cannot explain all the differences. The close proximity winds are evidently stronger because they are derived from data taken mostly within the RF quadrants of relatively intense hurricanes (mean landfall sustained winds of 45.6 m s^{-1}), while the general proximity winds include data collected in a variety of quadrants in both weak and intense hurricanes (mean winds of 35.4 m s^{-1}). The differences in the temperature profiles can also be attributed at least in part to differences in hurricane intensity.

Table 3 gives listings of the values of mean CAPE, BRN shear, BRN, and layer-averaged H_r, H_t, and ω_s for 0- to 3-km, 0- to 6-km, and 0- to 12-km layers of the soundings used in generating each composite. CAPE and BRN were found to be smaller and the shear and helicity parameters much larger for the close proximity data. Mean total helicity converted to the form used by *Davies-Jones et al.* [1990] was approximately 300 J kg^{-1} for the close proximity composite. This value is near the Davies-Jones proposed threshold for "strong" (at least F2 intensity) tornadoes.

M91 pointed out that in spite of the simplicity of the composite profiles the large variability of the data made it inevitable that some individual hurricane tornado soundings deviate considerably from the composite profiles. CAPEs were found to range from zero to more than 3000 J kg^{-1}, while maximum winds were under 10 m s^{-1} in strength in some of the soundings. However, it was noted that the soundings with weak winds were never associated with more than one or two weak (F0-F1) tornadoes; on the other hand, all the large tornado outbreaks (more than eight tornadoes apiece) were found to occur in environments having wind maxima in excess of 15 m s^{-1}.

The mean hurricane tornado proximity sounding and hodograph were found to differ in many important respects from their tornado proximity counterparts on the Great Plains. These differences are highlighted in Figure 10, which shows the close proximity hurricane tornado profiles along with the Oklahoma supercell mean profiles of *Bluestein and Jain* [1985]. The latter profiles closely approximate the findings of other composite tornado proximity studies [*Maddox*, 1976; *Darkow and McCann*, 1979; *Schaefer and Livingston*, 1988] and were taken to be representative of tornadic environments on the Great Plains. Hodograph comparisons required some caution because, unlike the hurricane composites, the winds in the Oklahoma profile were composited using true zonal and meridional components.

With regard to the thermodynamic data, M91 found that the hurricane tornado events displayed much less CAPE than the Oklahoma events. The mean of the close proximity hurricane tornado CAPEs (Table 3) was 253 J kg^{-1}, or only about 10% of the 2542 J kg^{-1} value found in the Oklahoma supercell composite of *Bluestein and Jain* [1985] (their Table 3, which gives 2490 J kg^{-1}, using other software than that used in M91's calculations). The hurricane CAPEs were, however, comparable in magnitude to those found by *Barnes and Stossmeister* [1986] near the rainbands of a maritime hurricane. Surface wet-bulb potential temperatures θ_w were 22.7°C in the Oklahoma composite and 22.3°C in the hurricane composite, but the temperature and dew point profiles showed significant differences in many other respects. Most prominent among the differences were the extensive dry layer aloft, layer of enhanced static stability below 700 hPa, and the steep lapse rates in the Oklahoma composite.

Much of the CAPE in the hurricane tornado cases is realized in the conditionally unstable layer that exists between the surface and about 650 hPa. Boundary layer parcels ascending through this layer can achieve buoyancies of 2°-3°C, thanks to a minimum in ambient saturation wet-bulb potential temperature θ_w^* that occurs near 650 hPa. This relatively "cool" layer is apparently created by the hurricane's own circulation, which draws in unperturbed lower tropospheric tropical air while warming the mid and upper troposphere with the anvils from intense hurricane-core convection. Thus it is apparently not the lower troposphere that is being cooled, but rather the mid and upper troposphere that are being warmed. This warming and attendant reduction of upper tropospheric contribution to CAPE is most pronounced in intense hurricanes, which are just the ones most likely to produce large tornado outbreaks (with large F sums) during and after landfall.

Comparison of the hurricane profile with that of *Jordan*'s [1958] mean Atlantic hurricane season sounding (Figure 11) reveals how hurricanes modify the mid and upper level temperature and moisture profiles. The hurricane temperature profile is slightly cooler than the Jordan profile below 600 hPa but becomes 2°-3°C warmer aloft. The dew point profiles are virtually identical below about 850 hPa. Citing the absence of any significant evidence of midlevel drying, both in the composites and in most individual soundings,

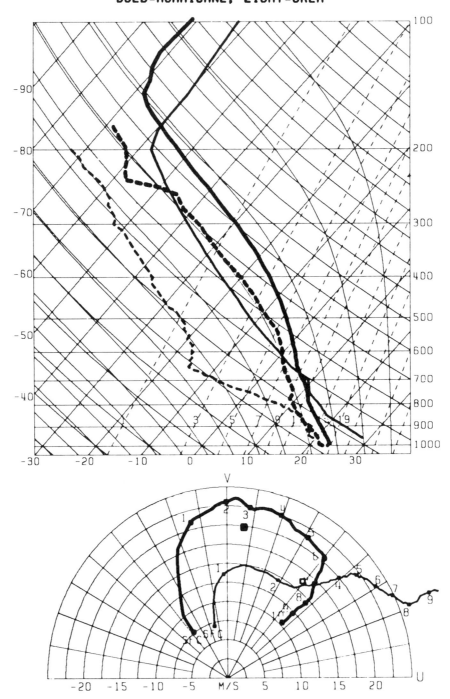

Fig. 10. Skew-T log-p and hodograph diagrams for the hurricane tornado close proximity composite (heavy line) and Oklahoma supercell composite of Bluestein and Jain (light line). U and V components of the Oklahoma composite are relative to true zonal and meridional directions. Squares mark 0–6-km mean winds as in Figure 9.

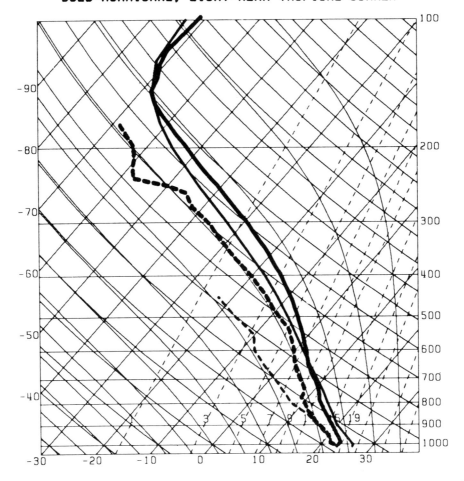

Fig. 11. Skew-T log-p diagram showing the hurricane tornado close proximity composite sounding (heavy line) and Jordan mean hurricane season sounding (light line).

M91 suggested that subsidence does not play a major role in producing this warming aloft. Despite the increased overall moisture in the hurricane sounding, the CAPE decreases from 1345 J kg^{-1} in the Jordan profile to the hurricane value of 253 J kg^{-1}. The residual lower tropospheric minimum in θ_w^* and warm layer aloft have also been observed in reconnaissance flight soundings taken in maritime hurricanes [e.g., *Hawkins and Imbembo*, 1976, Figures 5, 6, and 10], other mean hurricane soundings [*Sheets*, 1969], and in the hurricane tornado study of *Novlan and Gray* [1974].

With regard to winds the hurricane tornado close proximity composite displays greater shear magnitude, at least at low levels, than the Oklahoma supercell composite. Wind vector magnitude differences between the surface and 1 km are 14.8 m s^{-1} for the hurricane profile and 9.0 m s^{-1} for the Oklahoma profile. The Oklahoma profile has 12.5 m s^{-1} of BRN shear and the hurricane profile 14.0 m s^{-1}. Because of the small CAPE and large shears found in the close proximity hurricane tornado cases, the BRN of the hurricane cases (3.3) is much smaller than that of the Great Plains tornado cases (32.8). For both data sets, however, the BRN falls in the "possible supercell" range [*Weisman and Klemp*, 1982, 1984].

2.5. Variations With Hurricane Intensity and Size

Novlan and Gray [1974], *Gentry* [1983], and *Weiss* [1985] all pointed out that the likelihood of hurricane tornado activity increases with the "intensity," that is, peak sustained surface winds, of the landfalling hurricane. M91's data support this notion but also suggest that hurricane "size" [*Merrill*, 1983], measured, for example, in terms of the radius of outer closed isobar (ROCI), is important. Hurricanes notable for both their size and intensity have been among the most prolific tornado producers in the data base: Audrey (1957), Carla (1961), Beulah (1967), and Allen (1980).

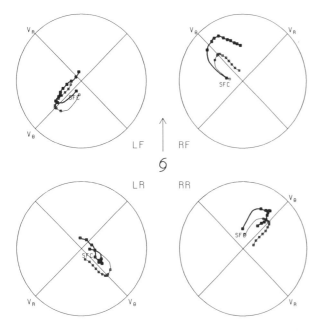

Fig. 12. Quadrant composite hodographs (heavy lines) arranged in their respective quadrants around the composite hurricane, which is assumed to be moving to top of diagram as shown. Mean hodograph of the four quadrants (light lines) is shown in each quadrant for reference. Boxes mark winds at each 1-km increment of altitude from the surface through 10 km. The labels V_r and V_θ indicate the radial and tangential components of wind, elsewhere referred to as U and V. The label "SFC" is plotted just below the surface wind box for the individual quadrant hodographs. Radii of hodograph circles correspond to a wind speed of 25 m s^{-1}.

M91 surveyed all hurricanes that made landfall in the United States between 1964 and 1986 and that produced zero, one, or more than eight tornadoes, and found the major tornado outbreak producers (those with more than eight tornadoes) were consistently larger and more intense at landfall than those producing one or no tornadoes. In fact, McCaul's data indicate that the hurricanes that produced major tornado outbreaks had average peak sustained winds of 47.1 m s^{-1} at landfall, compared to only 21.9 m s^{-1} for the nontornadic "hurricanes." The major tornado outbreak producers also had average ROCIs of 2.93° latitude, more than twice that for nontornadic hurricanes (1.40°). These ROCIs are smaller than those described by *Merrill* [1983], probably because of differences in the cyclones sampled and, especially, in the resolution of pressure contour analyses used in measuring ROCI. The size and intensity of those cyclones that produced only one tornado resembled those of the nontornadic cyclones more than they did the major outbreak producers.

While some of the hurricanes responsible for major tornado outbreaks were not particularly intense, they all had ROCIs larger than the mean of the nontornadic class of hurricanes. M91 thus concluded that in addition to intensity the area exposed to favorable winds and vertical shears was an important factor influencing overall tornado productivity.

2.6. Azimuthal Variations Around the Hurricanes

It has already been shown that most hurricane-spawned tornadoes occur in the RF quadrants of the hurricanes in Figure 3. The spatial distribution of raobs used in constructing Figure 4 shows a similar pattern (not shown [see M91, Figure 11]). Insofar as the mean landfall heading of tornadic hurricanes in this study is 357.5°, this pattern also indicates that the northeast quadrant of hurricanes is the favored geographic quadrant for tornadoes. The preference for tornado activity in the northeast quadrants of landfalling hurricanes has been documented by *Hill et al.* [1966], *Orton* [1970], *Novlan and Gray* [1974], and *Gentry* [1983].

To investigate the strong preference for hurricane tornado formation in the RF quadrants of hurricanes, M91 constructed composite profiles in each of the quadrants relative to calculated hurricane motion. The composites in the four quadrants were also combined, using unit weighting for each quadrant's composite, to produce an azimuthally averaged overall profile for landfalling tornadic hurricanes. In the latter case the effects of hurricane translation were effectively removed by the averaging process, so that the resulting profile was felt to approximate one that might be obtained around a stationary, symmetric hurricane. The average of the quadrants formed a standard against which each of the quadrant profiles could be compared.

The temperature profiles (not shown) differed in the lower troposphere, where the rear quadrants were significantly warmer than the front quadrants. In addition, the right quadrants were slightly warmer than the left quadrants in the mid and upper troposphere, with the most significant differences at upper levels. These temperature differences, along with enhanced moisture in the front quadrants, suggest considerable quadrant-to-quadrant variation in thermodynamic instability. These variations in low level temperature and moisture are consistent with the patterns of CAPE variation seen earlier in Figure 5.

The differences between the hodographs were even more striking than those between the thermal properties of the quadrants. To highlight these differences, Figure 12 shows the four hodographs drawn at the centers of their respective quadrants relative to the mean hurricane center, with each hodograph rotated appropriately to reflect the systematic changes in orientation of the local (U, V) coordinates. The mean of the four quadrants, which serves as a reference, is also shown.

It is evident from Figure 12 that the hodograph in the left-rear (LR) quadrant tends to fold back on itself as winds first veer, then back, with height. This is quite different from what occurs in the RF quadrant, where winds veer continuously with height. In general, the RF hodograph shows particularly large amplitudes of wind speed and shear and a more "open" shape than the other quadrants. The comparison between quadrants suggests that the four hodographs would become nearly equal to the mean if (1) a sheared mean flow from rear to front of the hurricane and varying in magnitude from 1.4 m s^{-1} at the surface to 7.6 m s^{-1} at

TABLE 4. Mean Parameters of Tornado Proximity Soundings Data Stratified by Quadrant

Variable	LF	RF	RR	LR
Number of temperature profiles	36	86	44	21
Number of wind profiles	32	81	42	20
Range from center	255.9	330.6	421.1	314.6
CAPE (J kg^{-1})	299.8	684.2	1031.4	1212.3
BRN shear (m s^{-1})	8.4	11.3	8.9	6.5
BRN	32.6	33.5	119.0	190.7
0- 3-km H_r	0.30	0.34	0.23	0.07
0- 6-km H_r	0.19	0.29	0.18	0.06
0-12-km H_r	0.10	0.21	0.14	0.02
0- 3-km H_t (m s^{-2})	0.031	0.049	0.022	0.012
0- 6-km H_t (m s^{-2})	0.019	0.033	0.013	0.009
0-12-km H_t (m s^{-2})	0.011	0.023	0.010	0.002
0- 3-km ω_s (s^{-1})	0.0036	0.0048	0.0027	0.0008
0- 6-km ω_s (s^{-1})	0.0023	0.0034	0.0018	0.0006
0-12-km ω_s (s^{-1})	0.0014	0.0023	0.0013	0.0004

Quadrants are LF, left front; RF, right front; RR, right rear; and LR, left rear. CAPE is convective available potential energy. BRN is Bulk Richardson Number.

10-km altitude were removed from all quadrants; the shear in this component of flow is not uniform with height, being larger at lower altitudes; and (2) an additional across-hurricane mean flow from right to left, ranging from 1.4 m s^{-1} at the surface to roughly 2 m s^{-1} at 1-km altitude, were removed from the low-level winds; this component of flow becomes negligible above 4 km.

Thus M91 found that most of the systematic azimuthal variation in the hodographs could apparently be explained in terms of the superimposition of an almost unidirectionally sheared steering current on the circularly symmetric internal circulation of the hurricane itself.

To compare the wind profiles shown in Figure 12, a t test was applied at each sigma level, under the assumption that the variances from each profile were not necessarily the same [e.g., *Dixon and Massey*, 1969, p. 119]. Application of this technique to all possible combinations of hodographs from the four quadrants confirms that the differences between the wind profiles are highly significant statistically, often at better than the 1% level, especially in the mid and upper troposphere.

A summary of key sounding parameters for the four quadrants is given in Table 4. The table shows that the mean distance from raob site to hurricane center ranges from 256 to 421 km in the various quadrants. However, most of the raobs were made sufficiently far from the hurricane centers that the patterns in the composites and their characteristic parameters would not change greatly if the data used in the compositing were more carefully stratified to insure that the mean distances to hurricane centers became equal.

As Table 4 indicates, the mean CAPEs vary considerably in the four quadrants, with a minimum of 300 J kg^{-1} in the left front (LF) quadrant, and a maximum of 1212 J kg^{-1} in the LR. Mean CAPE in the right rear (RR) quadrant is 1031 J kg^{-1}, while that in the RF quadrant is 684 J kg^{-1}. All these values are considerably smaller than the mean CAPE of the Great Plains tornado environment. The mean LR CAPE is the most uncertain, owing to the relatively small sample size (21 raobs) in that quadrant and the presence of several very unstable profiles there. Examination of mean values of θ_w in the lowest 500 m (not listed in Table 4) shows that the RR quadrant has the highest value, 23.2°C, while the LF quadrant has the lowest, 21.5°C. Surface and low level relative humidities are distinctly higher in the RF quadrant, and the lifting condensation level is at its lowest altitude (highest pressure) there.

The reduction of CAPE in the LF quadrant is consistent with the lower surface temperatures analyzed by *Novlan and Gray* [1974] in tornado-producing hurricanes. Mechanisms that might be responsible for the cooler boundary layer in the LF quadrant include the adiabatic core cooling experienced by inflowing air parcels that have been separated from the warm sea surface for relatively long times [*Novlan and Gray*, 1974], rain cooling of relatively dry air entrained into the hurricane circulation from the fringes of the storm [*McCaul*, 1987], and the weak background baroclinicity that would be expected if sheared westerlies were influencing the hurricane.

As Table 4 shows, the mean BRN shears varied from quadrant to quadrant in a pattern much different than that seen with CAPE. The mean shear is greatest, 11.3 m s^{-1}, in the RF quadrant, and least, 6.5 m s^{-1}, in the LR. Mean values of BRN range from 32.6 and 33.5 in the LF and RF quadrants to 119.0 and 190.7 in the RR and LR quadrants. The BRNs in the front quadrants are within the "possible supercell" range, according to *Weisman and Klemp* [1982, 1984]. They are also somewhat smaller than the corresponding values inferred for the RF quadrant from Figure 7, primarily because the data used in constructing that figure included many nontornadic raobs and BRN tends to be larger for such raobs. The means of each of the helicity parameters exhibit large quadrant-to-quadrant variations, with values in the RF quadrant being typically about 4 times larger than those in the LR quadrant. Mean 0- to 3-km H_r and H_t in the RF quadrant are 0.34 and 0.049 m s^{-2}, respectively.

2.7. Composites Based on Other Data Stratifications

M91 also generated composite profiles for various other stratifications of the data. Because raobs were plentiful in the RF quadrants of the hurricanes but spotty elsewhere, these other composites were constrained to use RF quadrant data only. Only a general review of the results is possible here; for further details the reader is referred to M91.

In order to see how the hurricane thermodynamic and kinematic profiles evolved following landfall, M91 stratified and composited the RF quadrant raob data with respect to time after landfall. The results showed little difference in temperature and moisture but significant changes in hodograph structure. In particular, there was a distinct tendency for mid and upper level winds to veer with time, with only a slight loss of strength relative to the low level winds, which weakened substantially. These changes have the effect of maintaining, sometimes for three days or more, the large values of helicity which prevail in those portions of the circulations of landfalling hurricanes which lie downshear of the centers relative to the steering flow. This pattern of evolving hodograph structure was evident to some degree in all the specific cases of individual tornadic hurricanes which could be traced through several days following landfall.

Because previous investigators [*Fujita et al.*, 1972; *Weiss*, 1987] noted diurnal variations in hurricane tornado occurrences, M91 also composited the RF quadrant raob data with respect to time of day. Only data from the standard times 0000 and 1200 UTC were available, so that only two composites could be constructed. The temperatures in the 0000 UTC composite were found to be slightly warmer at virtually all levels. Thus there was no major difference in CAPE between the two composites. Winds likewise did not show major differences. These results suggested a possible diurnal observational bias with respect to reporting of hurricane tornadoes. However, M91 also pointed out that the observed frequencies of occurrence of the tornadoes are nearly identical at 0000 and 1200 UTC, so that the results required cautious interpretation. Raobs would be needed at approximately 0800 and 2000 UTC to be able to draw more definitive conclusions about the effects of diurnal temperature changes on occurrences of tornadic convection.

Weiss [1987] noted a positive correlation between hurricane forward translational speed and tornado occurrences. To study how the environments of fast- and slow-moving hurricanes differed, M91 composited raob data taken during the period from the day of landfall through the following two days, within the RF quadrants of hurricanes that were observed to be moving at either less than 4 m s^{-1} or greater than 8 m s^{-1} at raob time. The mid and upper level winds in the fast hurricane composite were distinctly stronger than those of the slow group, consistent with the notion of interactions with sheared steering currents of greater strength. The middle and upper portions of the troposphere were also significantly drier in the fast composite. M91 found a slight tendency for faster hurricanes to produce more tornadoes, but the relationship broke down for hurricanes moving faster than approximately 15 m s^{-1}. The reasons for this complex relationship require additional research.

Gentry [1983] and *Weiss* [1985] noted differences in tornado frequency for hurricanes making landfall in different areas, with more tornadoes associated with southern landfalls. M91 concluded that the most significant differences were the result of whether the landfall occurred along the Gulf of Mexico or Atlantic coasts, with Gulf of Mexico landfalls showing a pronounced tendency to generate more tornadoes. RF quadrant composites for both regions showed no major differences, prompting M91 to ascribe the differences in tornado frequency to other factors. Among the leading candidates were systematic differences in hurricane intensity, intensity tendency, and ability to expose the favorable RF quadrant winds to land. In each of these areas, Gulf of Mexico hurricanes were found to be more favorable for tornado formation than their Atlantic counterparts.

3. NUMERICAL SIMULATIONS

3.1. Simulation Specifications

McCaul [1990] also performed numerical simulations of hurricane-spawned severe convection. He used the three-dimensional cloud model developed by *Klemp and Wilhelmson* [1978], with modifications reported by *Wilhelmson and Chen* [1982]. The model solves nine prognostic equations for the basic variables, potential temperature θ, dimensionless pressure π, the Cartesian velocity components u, v, and w, water substance mixing ratios for vapor q_v, cloud water q_c, and rainwater q_r, and subgrid turbulent mixing coefficient K_m. Ice phases processes are not included. Open lateral boundary conditions are used at the sides of the integration domain, while rigid lid conditions are specified at the top and bottom boundaries.

To give satisfactory resolution, the model runs were conducted on a mesh having 500-m horizontal spacing and smoothly varying vertical spacing ranging from 250 m at the bottom to 750 m at the top. To capture the high tropopauses often observed in hurricane environments, a vertical domain depth of 20 km was specified. Horizontally, the domain was a 50-km square. These were approximately the largest domain dimensions feasible with the available computing and data analysis resources. Large domain size was found to be desirable as a means of insuring that the developing simulated storms remained within the simulation area while growing through atmospheric layers with considerably differing winds. For all simulations of hurricane convection the y axes of the model grids were aligned parallel to the tangential component of wind flow relative to the parent hurricane. For the control Great Plains simulation the y axis was oriented toward true north.

In each simulation, convection was triggered by release of ellipsoidal thermal bubbles of horizontal radius of either 7.07 or 10.0 km, vertical semiaxis length of 1.4 km, and maximum thermal excess of 2.0°C. A few additional test runs were made using either 1.0° or 4.0°C bubbles. Unless otherwise noted, results described here represent runs using 2.0°C

TABLE 5. Characteristics of Environmental Soundings Used in Initializing Simulations

Variable	FSI	CKL	VCT	AVG
CAPE (J kg^{-1})	2472.8	1833.9	522.0	610.8
BRN shear (m s^{-1})	12.3	12.9	17.1	12.3
BRN	32.6	22.1	3.6	8.0
0- 3-km H_r	0.48	0.50	0.53	0.86
0- 3-km H_t (m s^{-2})	0.015	0.077	0.063	0.071
0- 3-km ω_s (s^{-1})	0.0024	0.0073	0.0070	0.0076

FSI is Fort Sill, Oklahoma, composite, 2100 UTC May 20, 1977. CKL is Centreville, Alabama, 0000 UTC August 17, 1985. VCT is Victoria, Texas, partially interpolated (above 500 hPa), 1200 UTC September 20, 1967. AVG is close proximity composite, modified to reduce low level dew point depressions. CAPE is convective available potential energy. BRN is Bulk Richardson Number.

bubbles. A Coriolis parameter of 0.0001 s^{-1} was used in all simulations except for two tests where zero and 0.0002 s^{-1} were used. All simulations were allowed to run for 7200 s (2 hours) of simulated time, with fields stored each 600 s (10 min) for postanalysis.

The model storms were initialized in environments selected from the observational study of M91. Results reviewed here will consist of simulations initialized with (1) a tornado proximity sounding at Centreville, Alabama (CKL) from 1985's Hurricane Danny, where supercells were observed, (2) the close proximity composite (denoted "AVG") described in section 2, and (3) a proximity sounding at Victoria, Texas (VCT) from Hurricane Beulah in 1967. These three soundings may be said to represent much of the range of parcel buoyancy conditions seen in the spectrum of hurricane-spawned tornado environments. Even the high-buoyancy hurricane soundings, however, generally contain less CAPE than typical Great Plains tornado proximity soundings.

For purposes of comparison, a control simulation of a Great Plains severe storm (the "Del City Storm [see *Klemp et al.*, 1981]) was made also, using a composite sounding from Fort Sill, Oklahoma (FSI) and the same model mesh used for the hurricane convection simulations. Results were essentially similar to those obtained by *Klemp et al.* [1981] and are discussed here only in terms of how they contrast with the results of the hurricane-spawned storm simulations. A summary of the characteristics of the environments used to initialize all the simulations is given in Table 5. Note that H and ω_s are computed using the 0- to 6-km mean wind (as in section 2) and not the actual storm motion.

In all the hurricane convection simulations made by *McCaul* [1990], persistent convective activity occurred. The abundance of low-level moisture in the initiating environments encouraged the development of numerous convective cells, most of which were relatively transient but some of which displayed supercell characteristics. McCaul's results emphasized only the characteristics of the most intense and persistent convection noted in each model run. In this review, only the most general aspects of the simulated storms are discussed; further details are forthcoming in papers to be published elsewhere.

3.2. *Simulation of Hurricane Danny's Storms*

The environment which supported the Hurricane Danny tornado outbreak of August 16, 1985, in northern Alabama apparently was well represented by the CKL raob from 0000 UTC August 17 (see Figure 13). This sounding featured CAPE of 1834 J kg^{-1} and a veering hodograph containing 0–3-km total helicity of 0.076 m s^{-2}. The CAPE reading is large compared to most other hurricane soundings but still smaller than the mean of Oklahoma supercell or Great Plains tornado environments, which are about 2500 J kg^{-1}. It is also larger than the value reported by *McCaul* [1987], who used a preliminary sounding that underestimated the depth of the very moist boundary layer. The H_t reading is large enough to suggest the possibility of tornadoes, according to the criteria devised by *Davies-Jones et al.* [1990] for Great Plains storms.

In the simulation a weak supercell develops during the first hour, then strengthens in a pulsating manner during the second. By 7200 s the storm still shows a trend of slow intensification. Peak updraft strength of 27 m s^{-1} is noted in a burst of growth just prior to the end of the simulation period. Classic supercell characteristics are evident in the q_r field at 7200 s, with a "weak echo" area embraced by a "hook" at $z = 0.5$ km (Figure 14). A mesocyclonic region of enhanced vertical vorticity ζ is also apparent, with peak ζ exceeding 0.016 s^{-1}. Comparison of the CKL storm with a control simulation of the FSI storm [*Klemp et al.*, 1981] reveals similarities in structure but differences in horizontal scale (Figure 15). The FSI storm is noticeably larger in most respects than the CKL storm.

Differences in the vertical extent and intensity of the updrafts of the CKL and FSI storms are also apparent. The vertical profiles of updraft intensity along columns representative of flow in the main updraft cores of the two storms are shown in Figure 16. The CKL storm achieves peak updraft amplitude of approximately 22 m s^{-1} near $z = 3.0$ km, while the FSI storm reaches 44 m s^{-1} near 9.0 km. Because the FSI storm has little buoyancy at low levels, the CKL storm updraft is actually more intense everywhere below 4-km altitude. The levels of condensation and free convection are near the lowest model level in the CKL environment, while the level of free convection is above 2 km in the FSI storm. The CKL storm updraft does not extend above an altitude of 10 km, whereas the FSI storm reaches above 14 km.

The perturbation θ field at $z = 0.5$ km in the CKL storm (Figure 17) reveals only a rather weak pool of rain-cooled outflow. Minimum θ' values are only $-2.4°C$, only a third of the magnitudes seen in the corresponding field (not shown) of the FSI storm. The high relative humidities and relative lack of dry, potentially cold air at midlevels in the hurricane environments makes only weak rain-cooled outflow a common feature of the hurricane convection simulations.

The CKL mesocyclone has an even stronger, more clearly defined, closed circulation at 2.0 km (Figure 18) than at 0.5 km. Also seen in Figure 18 is the crescent-shaped updraft

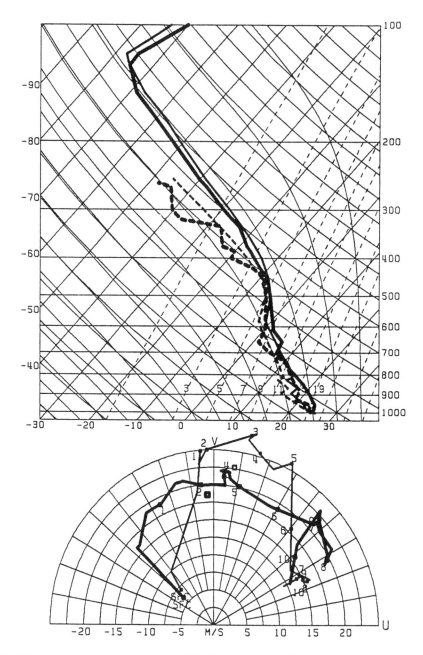

Fig. 13. Skew-T log-p and hodograph diagrams for 0000 UTC August 17, 1985, at Centreville, Alabama (heavy line), during passage of Hurricane Danny's remnants, and for 1200 UTC September 20, 1967 at Victoria, Texas (light line), during passage of Hurricane Beulah. Data above 500 hPa in Beulah are interpolated from adjacent 6-hourly raobs, which showed little change in the upper troposphere. U and V components of the winds are radial and tangential to the hurricane centers at raob time. Squares mark 0- to 6-km mean winds as in Figure 9.

core characteristic of mesocyclonic storms. Peak updraft intensity at this altitude is a respectable 18.0 m s^{-1}.

A vertical cross section of the flow vectors and perturbation pressure field of the CKL storm, taken looking toward negative x (Figure 19), reveals clearly defined S-shaped streamlines wrapping around and through low level and upper level pressure minima. The pressure minima are located on opposite sides of the main updraft core, but because the vertical shear in the environment reverses direction with height, both minima are located on the downshear side of the updraft at their respective levels. The low level downshear mesolow, centered near $z = 2.0$ km, features a pressure deficit of 2.8 hPa. The flow around these mesolows, especially the low level one, actually contains many streamlines which appear to form well-defined closed loops. The centers of these loops are not stagnant but rather contain strong flow normal to the plane of the cross section. The three-dimensional character of the storm flow is thus highly helical.

The principal features of the updraft dynamics are well described by the inviscid momentum equation:

$$\frac{d\mathbf{V}}{dt} = -c_p \theta_0 \nabla \pi + B\mathbf{k}, \qquad (1)$$

where $\mathbf{V}$ is the velocity vector, π is the dimensionless perturbation pressure, c_p the isobaric heat capacity of air, θ_0 the ambient potential temperature, B the buoyancy, and $\mathbf{k}$ a unit vector along the z axis. Following the general approach used by *Rotunno and Klemp* [1982] and *Weisman and Klemp* [1984], a diagnostic expression for the perturbation pressure may be obtained by taking the divergence of (1). The diagnostic pressure equation thus obtained is

$$c_p \nabla \cdot (\theta_0 \nabla \pi) = -\nabla \cdot (\mathbf{V} \cdot \nabla \mathbf{V}) - \frac{\partial}{\partial t}(\nabla \cdot \mathbf{V}) + \frac{\partial B}{\partial z}, \qquad (2)$$

which is a Poisson equation forced by buoyancy divergence (last term on the right) and dynamic effects (other terms on the right). The total perturbation pressure π may be parti-

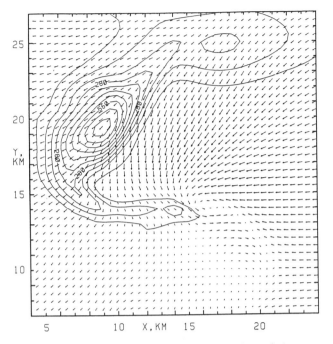

Fig. 15. Rainwater mixing ratio (contour interval 0.7 g/kg) at $z = 0.5$ km and $t = 7200$ s in control simulation of Oklahoma supercell storm. Wind vectors analogous to those in Figure 14 are also shown. Simulation was initialized with data from Fort Sill, Oklahoma (FSI) raob, as described by *Klemp et al.* [1981]. Note increased horizontal size of hook echo configuration and its greater displacement from apparent center of mesocyclone, as compared to Figure 14.

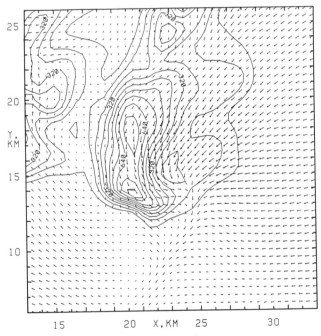

Fig. 14. Rainwater mixing ratio (contour interval 0.8 g/kg) at $z = 0.5$ km and $t = 7200$ s in simulation of supercell-like storms spawned by Hurricane Danny on August 16, 1985. Wind vectors relative to simulation domain translation velocity are also shown. Simulation was initialized with sounding from 00 UTC August 17, 1985, at Centreville, Alabama (CKL). Note warping of contours into a distinctive "hook" configuration at lower end of rain area, adjacent to region of high cyclonic vorticity.

tioned into buoyancy and dynamic contributions π_b and π_d by solving the Poisson equation (2) individually for the respective forcing terms on its right-hand side. The pressure contributions π_b and π_d will be called buoyancy pressure and dynamic pressure [*Schlesinger*, 1984]. This dynamic pressure is not to be confused with the similarly named pressure associated with parcel energy changes found in flows satisfying a Bernoulli equation.

Most of the vertical accelerations responsible for the strong updraft in the CKL storm are caused by gradients of the dynamic pressure. This is strongly suggested by comparison of Figures 20 and 21, which display contoured analyses of the vertical accelerations due to dynamic pressure and total buoyancy effects respectively. The data are displayed on the same cross section as in Figure 19. The total buoyancy effects include vertical pressure gradients of π_b as well as buoyancy itself [see *Weisman and Klemp*, 1984]. Detailed analyses of updraft forcing along parcel trajectories confirm that dynamic pressure forcing in the mature CKL storm is approximately 3 times as strong a contributor to net updraft intensity at the level of maximum updraft as is total buoyancy forcing.

In Figure 21, strong wavelike patterns are evident in the buoyancy forcing field. In particular, a lobe of strong negative forcing extends downward through the low level downshear mesolow gyre, where vertical velocities are becoming

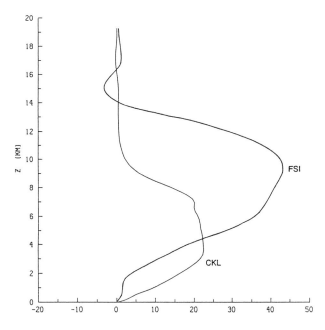

Fig. 16. Comparative vertical profiles of maximum updraft velocity (in meters per second) from CKL and FSI simulated storms. Note that CKL storm is actually more intense below $z = 4$ km than is the FSI storm. Depth of CKL storm updraft is much smaller than that of FSI storm (see also Figures 22 and 23).

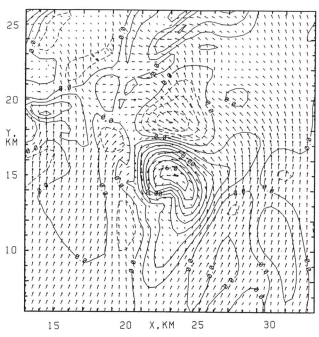

Fig. 18. Vertical velocity of CKL storm contoured every 2 m s^{-1} (negative values dashed) at 7200 s and an altitude $z = 2.0$ km. Storm-relative wind vectors are also shown for reference. Note existence of a strong mesocyclone with closed circulation on concave side of crescent-shaped updraft. Peak updraft exceeds 18 m s^{-1} at this altitude.

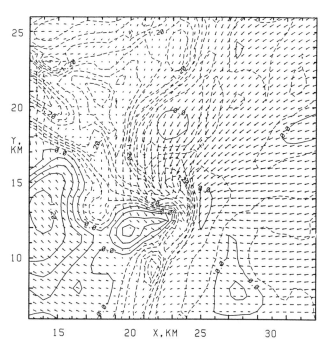

Fig. 17. Perturbation potential temperature field for CKL storm at 7200 s and an altitude of 0.5 km (contour interval is 0.3°C; negative values are dashed). Wind field of Figure 14 is also shown for reference. Note cold pool at this level has minimum temperature deficit of only 2.4°C, which is only about one third that in the FSI storm simulation.

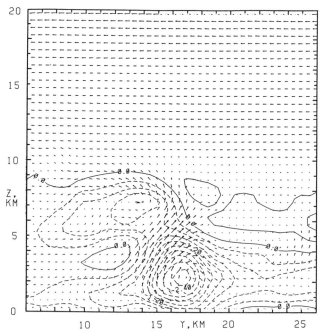

Fig. 19. Vertical cross section of wind flow and perturbation pressure contours (contour interval is 0.3 hPa; negative values are dashed) of CKL storm at 7200 s. Section viewed toward negative x. Note S-shaped streamline pattern in updraft, with flow wrapping around two closed gyre circulations, one at low levels, the other aloft. Both gyres occur on downshear sides of updraft at their respective altitudes, consistent with reversal of shear with height. Pressure minimum at $z = 2.0$ km has amplitude -2.8 hPa.

negative. Farther to the right, buoyant forcing again becomes positive, and streamlines begin to recover an upward tendency. The reason for this pattern is revealed in Figure 22, which displays the q_r field for the CKL storm on the same cross-section plane. The negatively buoyant lobe is associated with a concentrated fallout of heavy rain in a narrow sheath just downshear of the main updraft. The downdraft accompanying this rain helps close the loop of the gyre circulation, then, after being purged of rain, becomes positive buoyancy as it continues to subside. At the top of the updraft the negative buoyancy lobe extends through the region where rain is accumulating due to matching of terminal velocity with updraft speed. Updraft air that manages to rise through this region after shedding its rain once again becomes positively buoyant owing to latent heating and absence of rain loading.

A small vault is also evident in the q_r field of Figure 22. Comparison of this feature with the corresponding one from the FSI storm, Figure 23, again illustrates the differences in scale of the hurricane-spawned and Great Plains storms. The FSI storm also generates a sheath of heavy rain, but because the sheath is broader (storm is larger) and smaller in amplitude (less ambient moisture), and because the updraft is less tilted, it is less effective at generating the concentrated downdrafts at the proper downstream position needed to produce a closed flow gyre.

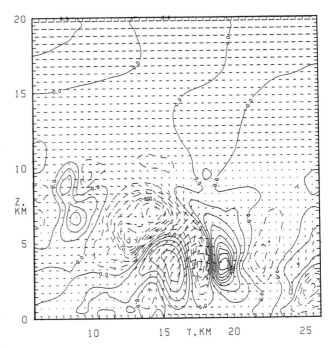

Fig. 21. Contours of vertical accelerations due to buoyancy effects (contour interval is 0.01 m s^{-2}; negative values are dashed), on same cross section as in Figure 18, for CKL storm. Buoyancy effects include those of perturbation pressure gradients produced by vertical gradient of actual buoyancy. Storm-relative wind vectors are also shown for reference. Note smallness of upward accelerations near base of main updraft and wavelike pattern of accelerations at midlevels of storm.

3.3. Simulation of "Close Proximity" Hurricane-Spawned Storms

Convection was also simulated using the "close proximity" ("AVG") hurricane tornado sounding (see Figure 9), moistened at low levels to facilitate storm development. The simulated storms were weaker than the CKL convection but still exhibited many features suggestive of incipient supercell structure. The q_r field at $z = 1.0$ km at 7200 s (Figure 24) has a distinctive curved shape, much like that seen in the CKL simulation. Enhanced vorticity is also present in the area of the mesocyclone. The storm is again quite small compared to the FSI supercell and appears not to be fully mature. Like the CKL storm, it was continuing to increase in intensity at the end of the simulation.

The absence of low level rain-cooled outflow is even more pronounced in the AVG storm than it was in the CKL storm. The 0.5 km θ' field (Figure 25) shows no significant negative temperature perturbations at all. In fact, the rear flank of the storm contains only a pool of air warmed as much as 2.0°C by subsidence. However, there is some low level cool air beneath the storm, with a -1.8°C temperature perturbation noted at the lowest model level, 0.125 km.

The updraft dynamics of the AVG storm show many similarities to those of the CKL storm. Dynamic pressure forcing again appears to predominate over buoyancy forcing

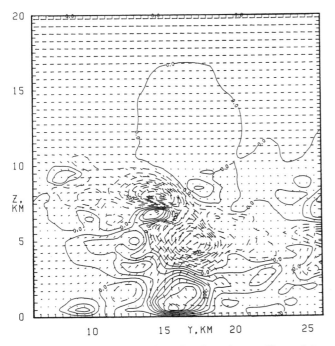

Fig. 20. Contours of vertical accelerations due to effects of dynamically forced pressure (contour interval is 0.01 m s^{-2}; negative values are dashed) on same cross section as in Figure 18 for CKL storm. Storm-relative wind vectors are also shown for reference. Note large upward accelerations near base of main updraft.

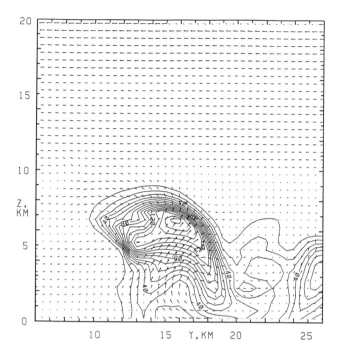

Fig. 22. Vertical cross section of wind flow and rainwater contours (contour interval is 1.0 g/kg) of CKL storm at 7200 s. Section is same as in Figures 18–20. Note shieldlike canopy of heavy rain forming in updraft, then falling out on downshear side, producing a shallow vault under the main updraft and downward motions in low level gyre to its north (right). Farther to the north, subsiding air that has shed its rain load acquires positive buoyancy (see Figure 21).

in establishing total updraft velocity. A cross section through the updraft, looking towards negative x as before, again shows the double gyre flow structure (Figure 26) witnessed before in the CKL results. This time, however, the storm is even smaller than the CKL storm. Updrafts in the AVG storm barely reach 7-km altitude and form a stark contrast with the massive updraft of the FSI storm (Figure 23). The q_r field shows a small vaulted overhang in the lowest 2 km and a sheath of rain falling into the downward branch of the low level downshear gyre. The overall morphology and dynamics of the AVG storm appear to be quite similar to those of the CKL storm.

3.4. *Simulation of Hurricane Beulah's Storms*

Simulations were also conducted using sounding data from 1200 UTC, September 20, 1967, at VCT during Hurricane Beulah. The sounding contained, even for hurricane tornado environments, only a "moderate" amount of CAPE (see Table 5), while the hodograph again contained considerable helicity. The environmental conditions are shown in Figure 13.

Storms simulated in this environment (not shown) are numerous but generally multicellular in character. Considerable vorticity is generated along the elongated flanks, but clear-cut supercell features are not apparent. However, a sensitivity test conducted using a value of Coriolis parameter f doubled to 0.0002 s^{-1} from the value used in the other runs began to display some supercell features after $t = 6000$ s. The test was intended to serve as a simplified attempt to mimic conditions associated with the strong background rotation found in hurricane circulations.

The "high-f" VCT supercell has many features in common with the CKL and AVG storms described earlier. The q_r and wind flow fields at $z = 0.25$ km and $t = 7200$ s are given in Figure 27. Noteworthy are the well-defined "hook" in the rainwater and the highly symmetric mesocyclone flow adjacent to it. Maximum vorticity in the mesocyclone was 0.044 s^{-1}, somewhat larger than that achieved in the CKL or AVG storms. As before, there is little evidence of a gust front in association with the rear flank of this mesocyclone. Many other storms coexist with this incipient supercell in the model domain, but none of the others acquired as many supercell traits.

4. Discussion

The concepts of helicity and streamwise vorticity appear to offer helpful insights into mechanisms underlying the genesis of hurricane-spawned tornadoes as well as other types of tornadoes. Values of ambient helicity are often quite high in landfalling hurricanes, suggestive of considerable

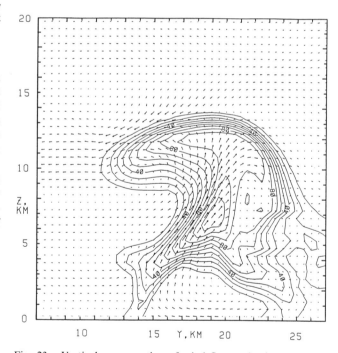

Fig. 23. Vertical cross section of wind flow and rainwater contours (contour interval is 1.0 g/kg) of FSI storm at 7200 s. Comparison with Figure 22 shows differences in vertical scale between this Great Plains supercell and the simulated hurricane-spawned CKL storm. Note also the diffuse character of the precipitation shaft to the right of the main updraft; this shaft is less effective at producing a concentrated downdraft than the one in the corresponding portion of the CKL storm in Figure 22.

potential for updraft rotation. Correlations between hurricane tornado outbreak intensity and helicity parameters are positive and rank among the leading indicators of tornado activity. Hurricane tornado activity is also positively correlated with both the intensity and size of the parent hurricane circulation, but these two factors are not independent of raob-derived wind and shear parameters.

The good spatial agreement between the patterns of hurricane tornado occurrence and enhanced ambient helicity represents additional evidence in favor of the connection between helicity and tornadoes. The fact that the maximum ambient helicity occurs in the right-front quadrant of the landfalling hurricane offers a new explanation for why that quadrant bears the brunt of most of the tornado activity. Examination of the azimuthal variation of hodograph structure around the hurricane centers suggests that it is the superimposition of a sheared large-scale steering flow on the otherwise circularly symmetric hurricane circulation that produces the helicity maximum in the right-front quadrant. This explanation is fundamentally different from any proposed by previous investigators of hurricane tornadogenesis. However, the findings of M91 do not disprove the hypotheses of shear enhancement by core cooling [*Novlan and Gray*, 1974] or increased surface drag [*Gentry*, 1983] as the hurricane encounters land.

The negative correlation between hurricane tornado activ-

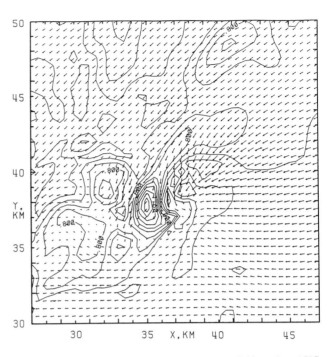

Fig. 25. Perturbation potential temperature field under AVG storm at 7200 s and an altitude of 0.5 km. Wind field of Figure 23 is also shown for reference. Note virtual absence of cold pool usually seen at this level beneath precipitating storms; only a small shallow region of cool air was found at the lowest model level ($z = 0.125$ km; not shown). Downdrafts have produced anomalously warm temperatures in area to left of mesocyclone.

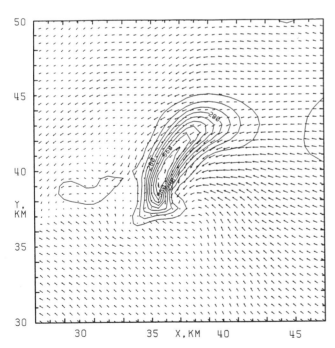

Fig. 24. Rainwater mixing ratio (contour interval is 0.7 g/kg) at $z = 1.0$ km and $t = 7200$ s for storm simulated in "AVG" environment, similar to that of "close proximity" hurricane tornado sounding. Wind vectors relative to simulation domain translation speed are also shown. As in CKL simulation, there is warping of contours into a distinctive "hook" configuration at lower end of rain area, adjacent to region of enhanced cyclonic vorticity.

ity and CAPE is apparently the result of the meager upper tropospheric contributions to CAPE in the large, warm anvils of the more intense hurricanes. The smallness of CAPE in the hurricane environments does not, however, prevent severe convection, because the level of free convection is low and abundant mesoscale convergence in the fully developed hurricane is always available to trigger storms. Furthermore, the simulations indicate that convective-scale perturbation-pressure forcing in the hurricane-spawned storms is capable of compensating for the reduced buoyancy. In fact, the perturbation-pressure minima in the hurricane-spawned storms may achieve amplitudes comparable to those found in Great Plains storms, owing primarily to the favorable combination of vertical profiles of buoyancy and shear which prevail in landfalling hurricanes. The fact that buoyancy is not the only factor governing severe storm intensity is also consistent with the recent findings of other investigators [*Johns et al.*, 1990; *Lazarus and Droegemeier*, 1990].

Other factors also influence how buoyancy operates in the hurricane cases. Because buoyancy is small and environmental moisture abundant, the effects of rainwater loading have a more pronounced and direct impact on storm dynamics than is the case with Great Plains storms. There is less production of negative buoyancy through evaporative cooling of air by rain than in the Great Plains cases, with attendant differences in the spatial distribution of negative

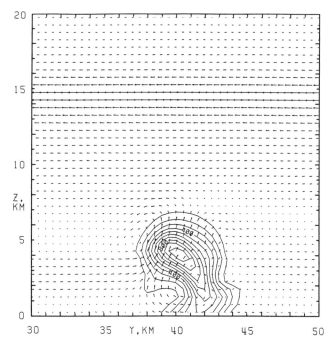

Fig. 26. Vertical cross section of wind flow and rainwater contours (contour interval is 1.0 g/kg) of AVG storm at 7200 s. Section is same as in Figure 22. Note shieldlike canopy of heavy rain forming in updraft, then falling out on downshear side, producing a shallow overhang under the main updraft and downward motions in low level gyre to its north.

buoyancy regions in the storms. Simulations of the hurricane-spawned storms commonly show the tilted helical updrafts quickly shedding heavy rain into the downshear environment. Negative buoyancy associated with this rainwater loading produces a downdraft that creates a helical gyre circulation in the low-level storm inflow. Simulation results indicate that this gyre contains at least as much horizontal vorticity as the mesocyclone does vertical vorticity. The horizontal vorticity arises from baroclinic effects associated not only with latent heat release in the updraft but also with the extra forcing produced by the concentration of rainwater loading just downshear of the updraft. These effects are also important in Great Plains severe storms, although in those storms the horizontal buoyancy gradients are usually associated with large temperature gradients as well.

An additional factor influencing hurricane-spawned storm dynamics is the existence of a stable layer that lies typically just above 650 hPa. This stable layer acts to confine the intense convection in the hurricane environments to the lower troposphere. It also has a strong impact on the vertical changes of buoyancy experienced by storm updraft parcels and thereby restricts the vertical extent of the low level downshear mesolow, whose shape, intensity, and position play such a pivotal role in the dynamics of the hurricane-spawned storms. By way of contrast, for Great Plains cases, convection is usually constrained only by the tropopause, and the storms there occupy the entire depth of the troposphere, with updrafts reaching maximum intensity in middle and upper layers. Thus, the morphology of convection in both the Plains and hurricane-type environments may be seen to be strongly controlled by not only the bulk or average values of characteristic parameters but also by their actual distributions in the vertical. This conclusion is reminiscent of some of the findings reported by *Simpson et al.* [1985] regarding tropical waterspouts and their parent cumulus clouds. It is also reminiscent of the findings of *Szoke et al.* [1986] for general maritime tropical convection, although here the key finding is that most of the buoyancy in low-buoyancy tropical environments is concentrated in the lower troposphere, where the vertical shear is strongest. The presence of small buoyancy in hurricane environments, in combination with relatively large perturbation-pressure fields, also encourages the development of convective storm circulations that resemble in some respects the idealized Beltrami model of *Davies-Jones* [1985]. In the Beltrami model, buoyancy forces are zero, yet a swirling updraft supported entirely by pressure forces can exist. A summary listing of typical values of what appear to be the most important buoyancy and shear parameters, gradients, and scale lengths governing convective storm dynamics in Great Plains and hurricane environments is given in Table 6.

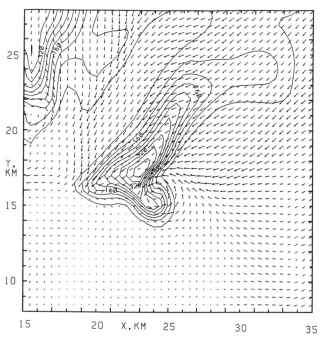

Fig. 27. Rainwater mixing ratio (contour interval is 0.4 g/kg) at $z = 0.25$ km and $t = 7200$ s in simulation of tornadic storms spawned by Hurricane Beulah on September 20, 1967. Wind vectors taken relative to simulation domain translation speed are also shown. Simulation was initialized with sounding from 1200 UTC September 20, 1967, at Victoria, Texas (VCT), using a Coriolis parameter double that used in the other simulations. Note warping of contours into a distinctive "hook" configuration at lower end of rain area, adjacent to highly symmetric mesocyclone circulation.

TABLE 6. Characteristic Values of Atmospheric Parameters in Hurricane and Great Plains Tornado Environments

Variable	Hurricane	Great Plains
CAPE (J kg^{-1})	500	2500
BRN shear (m s^{-1})	12.0	12.0
BRN	7.0	30.0
0- 3-km H_r	0.35	0.35
0- 3-km H_t (m s^{-2})	0.05	0.05
0- 3-km ω_s (s^{-1})	0.005	0.005
LFC (km)	0.5	2.0
Maximum θ'	3.0	9.0
Maximum buoyancy (m s^{-2})	0.10	0.30
Height of maximum buoyancy (km)	3.0	9.0
$\partial B/\partial z$ (10^{-4} s^{-2})	0.33	0.33
Maximum wind speed (m s^{-1})	25.0	35.0
Height of maximum wind speed (km)	3.0	10.0
$\partial V/\partial z$ (s^{-1})	0.010	0.004

CAPE is convective available potential energy. BRN is Bulk Richardson Number. LFC is level of free convection.

The simulated storms agree well with those few hurricane-spawned tornadic storms that have been documented in the literature. In both the simulations and the observations it is seen that their vertical and horizontal scale is reduced compared with typical Great Plains supercells. The hurricane-spawned storms may nevertheless acquire many of the characteristics of classic supercell storms. The simulations indicate that it is in the lowest 2–3 km where the hurricane-spawned supercells give their strongest and clearest radar signatures. Therefore detection of these storms using the existing network of weather radars will probably be more difficult than detection of Great Plains-type supercells because of problems of incomplete surveillance at low levels in much of the region between radars. The finer beam widths of the new National Weather Service Doppler radars will make it somewhat easier to resolve the circulations of these storms, but only a denser network of radars can close the gaps in the surveillance. In the absence of uniform surveillance it may still be necessary to attempt identification of potentially tornadic hurricane-spawned storms through analysis of gross features such as echo intensity and longevity.

5. Summary and Recommendations

McCaul's results have offered a new explanation for why hurricanes spawn tornadoes preferentially in their right-front quadrants: helicity enhancement caused by the interaction of the hurricane's swirling flow with a steering current containing shear roughly parallel to the hurricane heading. Questions remain, however, regarding how and when relative to landfall this helicity enhancement occurs. Further observational and simulation work is required before answers to these questions can be obtained. Such investigations could significantly improve our ability to predict which landfalling hurricanes will become major tornado producers and which will not.

The simulation findings suggest that relationships between the vertical distributions of ambient vertical shear, buoyancy, and moisture play an important role in governing the morphology of convective storms. Although there is less buoyancy in the hurricane environments, its concentration in the lower troposphere, along with most of the vertical shear, encouraged the development of a form of miniature supercell storm not previously seen in numerical simulations. Apparently, large buoyancy is not a prerequisite for supercell storms, but what buoyancy there is must be distributed in a propitious way in the vertical.

Additional research into the details of the mesoscale environments containing hurricane-spawned severe storms is also desirable. Although the numerical simulations indicate that potentially tornadic convection can develop from warm bubbles alone, it is possible that other dynamical effects not represented in the initial conditions are playing significant roles. Furthermore, because of limitations to the horizontal domain size used, the simulations may not capture realistically the mesoscale organization of the groups of tornadic storms often observed simultaneously in outbreaks of hurricane-spawned tornadoes.

Acknowledgments. The observational analyses were done in conjunction with the author's dissertation at the University of Oklahoma, under the auspices of National Science Foundation grant ATM-850130 and a teaching assistantship. The results were extended and refined during the author's tenure as an Advanced Study Program postdoctoral fellow at the National Center for Atmospheric Research's Mesoscale and Microscale Meteorology Division. The numerical simulations shown here were conducted on Cray X-MP supercomputers at both NCAR and the National Center for Supercomputing Applications at the University of Illinois. Analysis of the simulation results was done using VAX 11/780, VAX 8550, and VAX 6000-410 computers at OU's Geosciences Computing Network, MMM, and the Engineering and Analysis Data System at the National Aeronautics and Space Administration's Marshall Space Flight Center, respectively, the latter under support from Universities Space Research Association through grant NAS8-38769. The author gratefully acknowledges guidance, support, and constructive criticism from Doug Lilly, Howie Bluestein, Claude Duchon, Kelvin Droegemeier, John Firor, Joe Klemp, Rich Rotunno, Morris Weisman, Gary Barnes, Ed Zipser, Bob Wilhelmson, Bob Davies-Jones, Mike Kalb, and Jim Arnold. Many useful discussions were had with others such as Bill Gray, Joe Golden, Chuck Doswell, Don Burgess, Steve Weiss, Bob Johns, Mark Powell, Frank Marks, and John Flueck. Many, especially Wes Roberts, Dwight Moore, Pat Waukau, Ken Hansen, Bill Boyd, Dennis Joseph, Will Spangler, Ramesh Jayaraman, Jayanthi Srikishen, Chris Pizzano, and Bob Wilson, provided assistance with computing needs. Thanks also are due to anonymous reviewers, who made thorough and insightful comments on earlier versions of this manuscript.

References

Barnes, G. M., and G. J. Stossmeister, The structure and decay of a rainband in Hurricane Irene (1981), *Mon. Weather Rev.*, 114, 2590–2601, 1986.

Bluestein, H. B., and M. H. Jain, Formation of mesoscale lines of precipitation: Severe squall lines in Oklahoma during the spring, *J. Atmos. Sci.*, 42, 1711–1732, 1985.

Darkow, G. L., and D. W. McCann, Relative environmental winds for 121 tornado bearing storms, in *Preprints, 11th Conference on*

Severe Local Storms, pp. 413–417, American Meteorological Society, Boston, Mass., 1979.

Davies-Jones, R. P., Streamwise vorticity: The origin of updraft rotation in supercell storms, J. Atmos. Sci., 41, 2991–3006, 1984.

Davies-Jones, R. P., Dynamical interaction between an isolated convective cell and a veering environmental wind, in Preprints, 14th Conference on Severe Local Storms, pp. 216–219, American Meteorological Society, Boston, Mass., 1985.

Davies-Jones, R., D. Burgess, and M. Foster, Test of helicity as a tornado forecast parameter, in Preprints, 16th Conference on Severe Local Storms, pp. 588–592, American Meteorological Society, Boston, Mass., 1990.

Dixon, W. J., and F. J. Massey, Jr., Introduction to Statistical Analysis, 638 pp., McGraw-Hill, New York, 1969.

Doswell, C. A., and D. W. Burgess, On some issues of United States tornado climatology, Mon. Weather Rev., 116, 495–501, 1988.

Frank, W. M., The structure and energetics of the tropical cyclone, I, Storm structure, Mon. Weather Rev., 105, 1119–1135, 1977.

Fujita, T. T., Tornadoes around the world, Weatherwise, 26, 56–62, 78–83, 1973.

Fujita, T. T., Tornadoes and downbursts in the context of generalized planetary scales, J. Atmos. Sci., 38, 1511–1534, 1981.

Fujita, T. T., K. Watanabe, K. Tsuchiya, and M. Shimada, Typhoon-associated tornadoes in Japan and new evidence of suction vortices in a tornado near Tokyo, J. Meteorol. Soc. Jpn., 50, 431–453, 1972.

Gentry, R. C., Genesis of tornadoes associated with hurricanes, Mon. Weather Rev., 111, 1793–1805, 1983.

Hawkins, H. F., and S. M. Imbembo, The structure of a small, intense hurricane—Inez 1966, Mon. Weather Rev., 104, 418–442, 1976.

Hill, E. L., W. Malkin, and W. A. Schulz, Jr., Tornadoes associated with cyclones of tropical origin—Practical features, J. Appl. Meteorol., 5, 745–763, 1966.

Johns, R. H., J. M. Davies, and P. W. Leftwich, An examination of the relationship of 0–2 km AGL "positive" wind shear to potential buoyant energy in strong and violent tornado situations, in Preprints, 16th Conference on Severe Local Storms, pp. 269–274, American Meteorological Society, Boston, Mass., 1990.

Jordan, C. L., Mean soundings for the West Indies area, J. Meteorol., 15, 91–97, 1958.

Klemp, J. B., and R. B. Wilhelmson, The simulation of three-dimensional convective storm dynamics, J. Atmos. Sci., 35, 1070–1096, 1978.

Klemp, J. B., R. B. Wilhelmson, and P. S. Ray, Observed and numerically simulated structure of a mature supercell thunderstorm, J. Atmos. Sci., 38, 1558–1580, 1981.

Lazarus, S. M., and K. K. Droegemeier, The influence of helicity on the stability and morphology of numerically simulated storms, in Preprints, 16th Conference on Severe Local Storms, pp. 269–274, American Meteorological Society, Boston, Mass., 1990.

Lilly, D. K., The structure, energetics and propagation of rotating convective storms, II, Helicity and storm stabilization, J. Atmos. Sci., 43, 126–140, 1986.

Maddox, R. A., An evaluation of tornado proximity wind and stability data, Mon. Weather Rev., 104, 133–142, 1976.

Malkin, W., and J. G. Galway, Tornadoes associated with hurricanes, Mon. Weather Rev., 81, 299–303, 1953.

McCaul, E. W., Jr., Observations of the Hurricane "Danny" tornado outbreak of 16 August 1985, Mon. Weather Rev., 115, 1206–1223, 1987.

McCaul, E. W., Jr., Simulations of convective storms in hurricane environments, in Preprints, 16th Conference on Severe Local Storms, pp. 334–339, American Meteorological Society, Boston, Mass., 1990.

McCaul, E. W., Jr., Buoyancy and shear characteristics of hurricane-tornado environments, Mon. Weather Rev., 119, 1954–1978, 1991.

Merrill, R. T., A comparison of large and small tropical cyclones, Mon. Weather Rev., 111, 1408–1418, 1983.

Moncrieff, M. W., and M. J. Miller, The dynamics and simulation of tropical cumulonimbus and squall lines, Q. J. R. Meteorol. Soc., 102, 373–394, 1976.

Novlan, D. J., and W. M. Gray, Hurricane-spawned tornadoes, Mon. Weather Rev., 102, 476–488, 1974.

Orton, R., Tornadoes associated with Hurricane Beulah on September 19–23, 1967, Mon. Weather Rev., 98, 541–547, 1970.

Pearson, A. D., and A. F. Sadowski, Hurricane-induced tornadoes and their distribution, Mon. Weather Rev., 93, 461–464, 1965.

Rotunno, R., and J. B. Klemp, The influence of the shear-induced pressure gradient on thunderstorm motion, Mon. Weather Rev., 110, 136–151, 1982.

Schaefer, J. T., and R. L. Livingston, The typical structure of tornado proximity soundings, J. Geophys. Res., 93, 5351–5364, 1988.

Schlesinger, R. E., Effects of the perturbation pressure field in numerical models of unidirectionally sheared thunderstorm convection: Two versus three dimensions, J. Atmos. Sci., 41, 1571–1587, 1984.

Sheets, R. C., Some mean hurricane soundings, J. Appl. Meteorol., 8, 134–146, 1969.

Simpson, J., M. C. McCumber, and J. H. Golden, Tropical waterspouts and tornadoes, in Preprints, 14th Conference on Severe Local Storms, pp. 284–288, American Meteorological Society, Boston, Mass., 1985.

Smith, J. S., The hurricane-tornado, Mon. Weather Rev., 93, 453–459, 1965.

Stiegler, D. J., and T. T. Fujita, A detailed analysis of the San Marcos, Texas, tornado induced by Hurricane "Allen" on 10 August 1980, in Preprints, 12th Conference on Severe Local Storms, pp. 371–374, American Meteorological Society, Boston, Mass., 1982.

Szoke, E. J., E. J. Zipser, and D. P. Jorgensen, A radar study of convective cells in mesoscale systems in GATE, I, Vertical profile statistics and comparison with hurricanes, J. Atmos. Sci., 43, 182–197, 1986.

Tannehill, I. R., Hurricanes, 303 pp., Princeton University Press, Princeton, N. J., 1950.

Weisman, M. L., and J. B. Klemp, The dependence of numerically simulated convective storms on vertical wind shear and buoyancy, Mon. Weather Rev., 110, 504–520, 1982.

Weisman, M. L., and J. B. Klemp, The structure and classification of numerically simulated convective storms in directionally varying shears, Mon. Weather Rev., 112, 2479–2498, 1984.

Weiss, S. J., On the operational forecasting of tornadoes associated with tropical cyclones, in Preprints, 14th Conference on Severe Local Storms, pp. 293–296, American Meteorological Society, Boston, Mass., 1985.

Weiss, S. J., Some climatological aspects of forecasting tornadoes associated with tropical cyclones, in Preprints, 17th Conference on Hurricanes and Tropical Meteorology, pp. 160–163, American Meteorological Society, Boston, Mass., 1987.

Wilhelmson, R. B., and C. S. Chen, A simulation of the development of successive cells along a cold outflow boundary, J. Atmos. Sci., 39, 1466–1483, 1982.

Tornadic Thunderstorm Characteristics Determined With Doppler Radar

EDWARD A. BRANDES[1]

National Severe Storms Laboratory, Norman, Oklahoma 73069

1. INTRODUCTION

The absence of papers concerned with multiple-Doppler radar observations at the 1976 Tornado Symposium points out that much of what is known about tornadic thunderstorms from this data source is fairly recent. A review is in order to summarize these observations, to interpret the observations in view of recent developments in numerically simulated thunderstorms, and to determine which of the current tornadogenesis theories are consistent with the observations. Here the contributions of dual-Doppler radar observations are emphasized. A discussion of thunderstorm dynamics is given by *Klemp* [1987]. Tornadogenesis theories are discussed by *Rotunno* [1986] and *Davies-Jones* [1986]. Field observations of tornadic storms are described by *Davies-Jones* [1988].

The entire archive of dual-Doppler radar observations from well-sampled supercell storms, the most violent of the tornadic storms, probably consists of less than 10 storms. There are a number of reasons for this, but basically it has proven difficult to capture a significant portion of a long-lived tornadic storm within a relatively small, ground-based, dual-Doppler radar network.

Also, many of the observed supercell storms in the archive occurred years ago when Doppler radar signal processing was in its infancy. Yet, these storms constitute some of the best available data sets. One of these storms, the Harrah storm of June 8, 1974, remains an outstanding example of a supercell thunderstorm and is used extensively in this review. Other accounts of this storm are given by *Ray* [1976] and *Heymsfield* [1978]. Other documented tornadic storms include the Oklahoma City storm of April 20, 1974 [*Ray*, 1975], the Spencer-Luther storm of June 8, 1974 [*Brandes*, 1978], the Del City storm of May 20, 1977 [*Brandes*, 1981; *Ray et al.*, 1981; *Johnson et al.*, 1987], the Fort Cobb storm of May 20, 1977 [*Johnson et al.*, 1987], and the Lahoma and Orienta storms of May 2, 1979 [*Brandes et al.*, 1988].

The evolution of supercell thunderstorms may not apply to all tornadic storms. Indeed, many storms, for example, those that produce Denver, Colorado, area tornadoes, form in environments that are much different than those of supercells. Storm formation and evolution often are tied to orographic features, and storm lifetimes are relatively short compared to supercells. Denver area storms offer research opportunities for studying tornadogenesis not possible in other regions or with other storm types.

It is important to note that tornado flow patterns are not well resolved in Doppler radar measurements. Hence kinematic and thermodynamic properties fostering tornado genesis and dissipation necessarily are deduced from observations of the larger-scale storm flow.

2. LIFE CYCLE OF SUPERCELL STORMS

For the sake of discussion we define several nonoverlapping stages of supercell evolution. Development will refer to the entire period from storm initiation up to the first time that tornadoes become likely. Many of the characteristics associated with supercells (e.g., weak echo regions, hook echoes, and overhangs) develop during this period. In general, early development has received little attention in either observational or modeling studies. It is impossible when radar "first echoes" are observed to determine whether or not a storm will become tornadic. Also, numerically simulated thunderstorms are most influenced by the initialization process during early development. The mature stage will refer to that critical period in storm evolution at which the basic updraft and vertical vorticity patterns associated with supercells have evolved and the storm is primed for tornadogenesis. Elevated tornadolike vortices may be observed at this stage. The onset of damaging rotary surface winds heralds the tornadic stage. We will then review storm properties associated with the posttornadic stage. This stage represents dissipation in many storms but merely a quiescent phase in storms with cyclic tornado development.

---

[1] Now at the National Center for Atmospheric Research, Boulder, Colorado.

The Tornado: Its Structure, Dynamics, Prediction, and Hazards.
Geophysical Monograph 79
This paper is not subject to U.S. copyright. Published in 1993 by the American Geophysical Union.

2.1. Development

Historically, there has been a tendency to categorize severe thunderstorms as either multicell or supercell. Multicell storms are composed of several convective cells (the elemental convection unit [see *Byers and Braham*, 1949]) at various stages of development [e.g., *Browning and Ludlam*, 1962; *Dennis et al.*, 1970; *Chisholm and Renick*, 1972]. New cells form on a preferred flank, pass through the storm, and eventually dissipate on the opposite storm flank as they complete their life cycle. The addition of new cells gives a propagative component to the storm's motion and causes a motion that deviates from the mean wind to the side on which the storm is growing.

Supercell storms are often conceptualized as being composed of a steady, single, supercell with right-deviate motion [*Browning*, 1964; *Marwitz*, 1972]. However, many supercells on close inspection display multicellular structure, particularly in their upper levels [e.g., *Lemon*, 1976; *Burgess et al.*, 1977; *Weaver and Nelson*, 1982; *Foote and Wade*, 1982]. *Foote and Frank* [1983] presented Doppler radar observations of a multicellular Colorado hailstorm that had a number of features normally ascribed to supercells. They propose to distinguish multicell and supercell storms by the ratio of the distance between successive updraft perturbations (L) and the diameter of the updraft perturbations (D). In multicell storms, $L/D > 1$, while in supercells the spacing between successive updraft perturbations is much less than their diameters, that is, $L/D \ll 1$. As the ratio L/D decreases, persistent background regions of updraft and radar reflectivity (henceforth reflectivity) develop and impart a certain steadiness to the storm (Figure 1). In this interpretation, supercells form by the overlap of neighboring perturbations rather than the evolution of a single large, and enduring cell.

Vasiloff et al. [1986] studied an Oklahoma supercell storm that began as a cluster of seemingly unrelated ordinary cells. Although individual cells and updraft centers always could be identified (e.g., Figure 2a), the increase in size of these features resulted in large persistent regions of updraft and reflectivity whose motion differed from the individual centers by 65° (Figure 2b). The inflow air to the persistent updraft region had a significantly larger storm-relative streamwise component of vorticity than the inflow air in the cell-relative reference frame. Strong vertical vorticity concentrated at the leading edge of the background updraft region (Figure 3). A large supercell with middle level updraft rotation, a hook echo, and radar reflectivity in excess of 60 dBZ was produced; however, no tornadoes were reported.

The role of strong vertical vorticity in determining the characteristics of radar-observed supercells was also studied by *Brandes et al.* [1988]. Individual reflectivity and updraft centers were ill-defined and did not propagate across the storm. Instead, they were swept by the rotational flow to stagnation points upwind of the updraft.

2.2. Mature Stage

The discussion of the mature evolutionary stage uses the exceptional radar observations collected from the Harrah

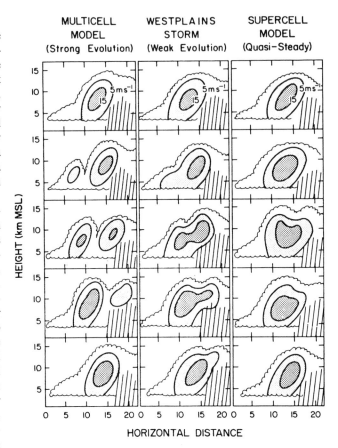

Fig. 1. Conceptualized updraft models of multicell, hybrid (west plains), and supercell thunderstorms [*Foote and Frank*, 1983]. Areas of parallel lines represent precipitation. Note that individual updraft cells lose their identify in the continuum from multicell storm to supercell.

storm of June 8, 1974. Figure 4a shows the storm-relative horizontal wind flow near cloud base (1.3 km) at 1530. (All heights are above ground level, and unless otherwise noted all times are Central Standard.) The Harrah storm moved from 230° at 18 m s^{-1}. The 40-dBZ reflectivity contour has been accentuated to help orient the reader when other wind field parameters are discussed. The heavy dashed line (near $x = 11$, $y = 22$ km) marks the location of the mesocyclone within which the tornado formed at ~1546. For our purposes, a mesocyclone is defined as the region where the vertical vorticity is $\geq 10^{-2}$ s^{-1}. The storm's first echo was detected at 1406; thus the storm was ~1½ hours old when the data in Figure 4 were obtained.

At this pretornadic stage, the coverage of reflectivity >50 dBZ is small but growing. An arc-shaped gust front between southerly inflow and rain-cooled air with a westerly component begins to the west of the northern end of the mesocyclone, passes through it, and then extends to the southwest. A rudimentary hook echo on the west side of the circulation ($x = 10$, $y = 23$ km) no doubt responds to the circulation.

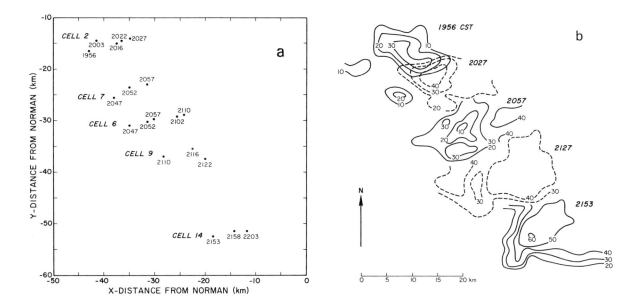

Fig. 2. Motion of (a) individual reflectivity cells and (b) storm motion for the Lindsay, Oklahoma, storm of June 19, 1980 [*Vasiloff et al.*, 1986].

The distribution of vertical velocity at 1.3 km, with the 40-dBZ reflectivity contour and the mesocyclone position superimposed, is shown in Figure 4b. The long arclike region of updrafts, which extends from the mesocyclone southwestward along the gust front and northwestward to the small region of updrafts near $x = 7$, $y = 25$ km, is typical of tornadic storms. (Vertical velocity contours are broken in the vicinity of the mesocyclone because of low signal strength near ground (upward integration of the continuity equation is used here). Horizontal wind vectors to the east of the mesocyclone (Figure 4a) have been computed by assuming that the vertical wind is zero. The error in the horizontal wind introduced by this assumption, approximately equal to the sine of the radar antenna elevation angle times the vertical velocity, is estimated to be <1 m s^{-1}.) The intruding area of weak reflectivity ($x = 12$, $y = 23$ km), dubbed the weak-echo region or WER by *Marwitz* [1972] and vault by *Browning* [1964], marks the primary inflow and updraft region of the storm. Downdrafts within the reflectivity core are relatively weak (>-3 m s^{-1}) compared to those that subsequently develop.

The distribution of vertical vorticity at 1.3 km is displayed in Figure 5. The arresting structural feature is the large arc of positive vertical vorticity that roughly coincides with the updraft arc.

The several vorticity centers (near $x = 8$, $y = 24$ km; $x = 11.5$, $y = 22$ km; and perhaps at $x = 10$, $y = 17$ km) are apparently related to development of elementary cells (updrafts) within the Harrah storm. New centers are thought to develop in right-hand portions of the arc (when viewed in the direction of storm motion), while old centers dissipate on the left. Clearly, the evolution of centers is tied to storm propagation. Individual centers do not appear to move along the arc; rather, the characteristic pattern periodically reestablishes itself as new centers intensify, deform the gust front between outflow and inflow, and, ultimately, choke the flow to older centers. This evolution, described conceptually by *Burgess et al.* [1982] and illustrated with observations by *Johnson et al.* [1987], may cause tornado families [*Agee et al.*, 1976].

An anomalous shear zone in the radial velocity measurements from both radars at 1530 indicates the presence of an elevated tornadolike circulation [*Brown et al.*, 1978]. The vortex signature, between 1.5 and 3 km, was located within the northernmost vorticity maximum. Note that the circulation is located on the storm's left rear flank. (A similarly behaved and placed circulation appeared above 3 km in 1826 measurements from the Del City storm [*Brandes*, 1981]). As is often observed with tornadoes [e.g., *Brandes*, 1978], the elevated vortex lies in the vertical velocity gradient between the weak downdraft behind the updraft arc and the small updraft at the left end of the arc. Photographs, taken at ~1500, when the storm passed to the west of Norman, Oklahoma, show a small funnellike appendage that protruded from the storm's cloud base. No surface damage was reported with the elevated circulations at either 1500 or 1530. Curiously, intense elevated vortices have not been reported for simulated supercell storms. Perhaps this is due to poor spatial resolution.

In the Harrah storm, negative vertical vorticity concentrates on the left flank of the storm, for example, from $x = 5.5$, $y = 30$ to $x = 5$, $y = 23$ km and beneath a small adjoining cell to the southwest ($x = 4$, $y = 19$ km). Anticyclonic vorticity is also found behind the positive

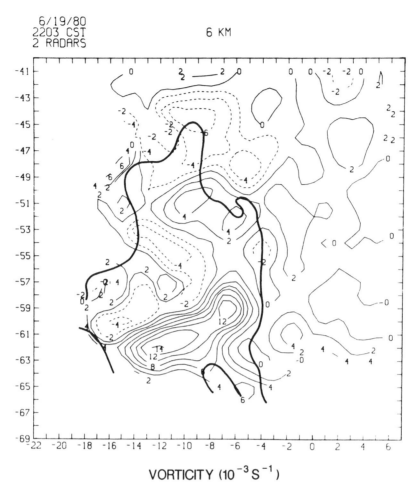

Fig. 3. Vertical vorticity at 6-km elevation in the Lindsay, Oklahoma, storm [*Vasiloff et al.*, 1986]. The vorticity contour interval is 2×10^{-3} s^{-1}. The region of updrafts greater than 5 m s^{-1} is shown by the heavy line.

vorticity arc. The basic configuration of positive vorticity on the upwind side of updrafts and negative vorticity downwind agrees with the simple twisting of ambient horizontal vorticity by the storm updraft [see *Davies-Jones*, 1984, Figure 8].

The vertical-vorticity maximum that spawns the Harrah tornado resides at the noses of the vorticity and updraft arcs ($x = 11.5$, $y = 22$ km). Maximum vorticity increases slightly with height at 1530 (Figure 6). At each radar analysis level, the distribution of divergence was determined and the mean value for all grid points within the arbitrarily defined mesocyclone was computed. This analysis (Figure 7) reveals that the low-level mesocyclone flow is wholly convergent below 2 km and convergent in the mean up to 3.8 km elevation.

Figure 8 presents the mean vertical-vorticity tendencies within the mesocyclone due to convergence, twisting, and turbulent mixing. Results for the twisting and turbulent vorticity terms are suspect due to the absence of scatterers at low levels and the subsequent loss of the vertical wind component in regions adjacent to the mesocyclone. Nonetheless, the profiles, except for the twisting term, are readily reproduced for other storms. In the absence of strong vertical gradients of vorticity, the turbulent diffusion of vorticity is usually small. This mechanism will be ignored in the discussion which follows.

Especially at low levels, the convergence term dominates the twisting term. This is primarily due to the small horizontal gradients of vertical velocity at low levels. The maximum vorticity production by the twisting of horizontal vorticity occurs at the leading edge and just upwind of the mesocyclone. Most often the contribution of twisting to the vorticity budget increases toward middle storm levels.

At 1543, approximately 3 min before tornado touchdown, the Harrah storm was sampled again. Prominent changes in storm structure included an increase in the areal coverage of radar reflectivity >50 dBZ and the development of 3 m s^{-1} rainy downdrafts within the enhanced reflectivity core. Vertical vorticity increased slightly at all levels (Figure 6). Two small-scale shear anomalies were present in the raw radial-velocity measurements. The stronger of the two sig-

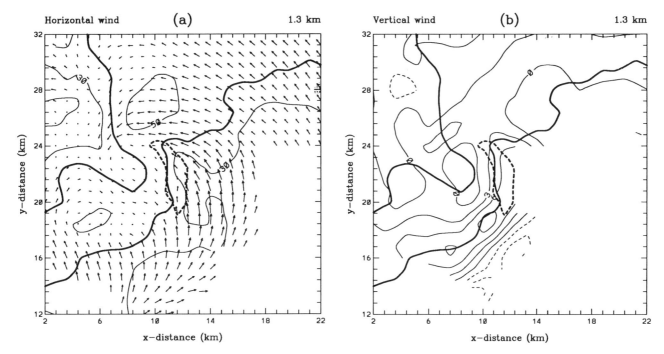

Fig. 4. The wind field at 1.3 km in the Harrah prior to tornadogenesis (1530 CST). (a) Horizontal wind vectors relative to the storm. The longest vector represents 24 m s^{-1}. Radar reflectivity contours (in dBZ) are superimposed, and the 40-dBZ contour is accentuated. (b) Vertical velocity (in meters per second) with the 40-dBZ contour superimposed. The mesocyclone that produced the Harrah tornado, defined as the region with vertical vorticity $\geq 10^{-2}$ s^{-1}, is dashed in Figures 4a and 4b. The tornado damage path (beginning near $x = 22$, $y = 28$ km) is shown by stippling in Figure 4a.

natures persisted at the extreme left end of the updraft and vertical-vorticity arcs. The second circulation, thought to be the incipient Harrah tornado, was located within the mesocyclone at the nose of the vorticity arc.

2.3. Tornadic Stage

Figure 9 depicts the radar reflectivity and wind patterns in the Harrah storm 7 min after damaging winds began. Although not readily apparent at 1.3 km, the tornado has formed along a rolled-up or occluded section of the gust front. Details of the surface wind field at this stage are discussed by *Brandes* [1978]; storm-scale flow patterns are described by *Ray* [1976] and *Heymsfield* [1978]. High-resolution radar radial-velocity measurements during the tornadic stage in the Del City storm [*Brandes*, 1981] suggest that the mesocyclone flow consists of wind bands that spiral inward about the tornado.

A prominent structural change in the Harrah storm is the development of a strong rainy downdraft within the reflectivity core. Outflow from the downdraft caused the mesocyclone to elongate in the north-to-south direction. The more easterly tornado path after 1553 may also respond to the intense outflow. Although usually not quite this strong, rainy downdrafts are salient features in all supercells at this stage.

Maximum updrafts, located within the mesocyclone, exceed 12 m s^{-1} (Figure 9b). The tornado, displaced from the strongest updraft, resides in the vertical-velocity gradient between the updraft and a downdraft that is forming behind the occluded gust front. This configuration of vertical drafts seems typical of tornadic storms, for example, the Spencer-Luther storm [see *Brandes*, 1978, Figure 13] and the Del City storm [*Brandes*, 1984b, Figure 4]. Desiccation of cloud material within the downdraft causes a "clear slot" often seen to the right of tornadoes [e.g., *Lemon and Doswell*, 1979]. Such downdrafts are thought to be induced by downward pressure gradients which respond to the sudden low-level buildup of vertical vorticity and associated pressure fall [*Klemp and Rotunno*, 1983; *Brandes*, 1984a].

At 1.3 km a ringlike region of strong vertical vorticity with several maxima $>150 \times 10^{-4}$ s^{-1} has developed (Figure 10a). This shape is also seen in a fine-scale simulation of the Del City storm [see *Klemp and Rotunno*, 1983, Figure 7] and, at least in the Harrah storm, seems linked to the intensification of the rainy outflow and its influence on the distribution of updrafts and vorticity. The vertical profile of maximum vertical vorticity (Figure 6) reveals that a rapid increase in vorticity has occurred below 4 km. Little change is evident above that level. The observations do not show the descent of a strong mesocyclone from the upper middle troposphere [e.g., *Burgess et al.*, 1977], but rather they show an intensification of the mesocyclone in the lower troposphere. Although the transition to the tornadic stage in the

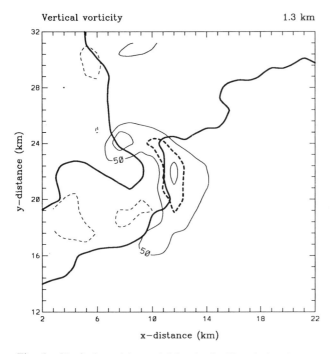

Fig. 5. Vertical vorticity at 1.3 km in the Harrah thunderstorm prior to tornadogenesis (1530 CST). Contour intervals are 50×10^{-4} s^{-1}. The 40-dBZ contour and mesocyclone are superimposed as in Figure 4b.

Del City storm was marked by increased vertical vorticity throughout a deep layer, vorticity amplification on the mesocyclone scale was most pronounced below 3 km. A factor of 4 increase occurred at 0.8 km between 1826 and 1847 [*Brandes*, 1984a, Figure 3]. The radar vortex signature of the incipient Del City tornado, a maximum at 0.6 km elevation (1838), could not be distinguished from mesocyclone flow above 3 km [*Brandes*, 1981].

Numerical simulations have shown that the vertical wind shear (horizontal vorticity) of the ambient air upon which the storm feeds is a primary source of rotation in supercells [*Schlesinger*, 1975, 1978; *Klemp and Wilhelmson*, 1978a, b; *Wilhelmson and Klemp*, 1978]. Vertical vorticity is generated when the horizontal vorticity is tipped into the vertical by the storm updraft. This rotation source was first suggested in the observational studies of *Browning and Landry* [1963], *Browning* [1968], and *Barnes* [1970]. However, if twisting of horizontal vorticity by updrafts were the only mechanism operating, only elevated vertical vorticity maxima should result [*Davies-Jones*, 1982]. Obviously, other mechanisms are required to produce the buildup of vertical vorticity necessary for tornadoes to develop at ground. *Rotunno and Klemp* [1985] argue that the origin of low-level vorticity lies with baroclinically generated horizontal vorticity that is produced locally as environmental air from the storm's right flank is cooled differentially by the evaporation of precipitation falling from the storm. Temperature gradients are produced that create strong horizontal streamwise vorticity near ground in air that flows parallel to the isotherms. Again, the generated vorticity must be twisted into

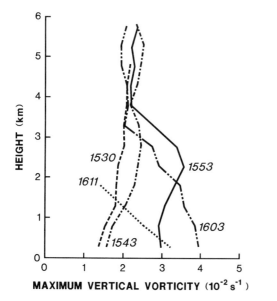

Fig. 6. Vertical distribution of maximum vertical vorticity in the tornado-spawning mesocyclone of the Harrah storm. The 1530 and 1543 CST profiles represent the pretornadic (mature) stage, the profile at 1553 CST is nearly midway during the life of the tornado, and 1603 and 1611 CST are early and late posttornadic stages, respectively.

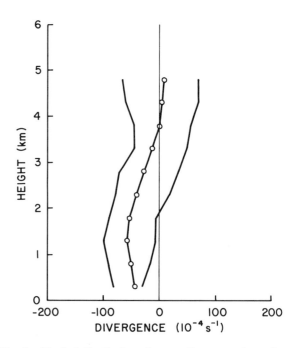

Fig. 7. Vertical distribution of mean divergence (central curve) and range in grid-point values for the pretornadic stage of mesocyclone development (1530 CST).

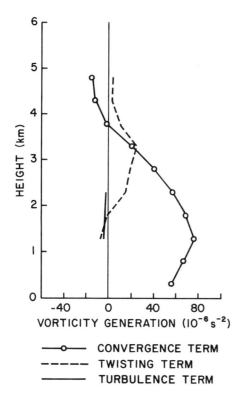

Fig. 8. Vertical distribution of mean vertical-vorticity production by convergence, twisting, and turbulent diffusion in the tornado-spawning mesocyclone prior to tornadogenesis (1530 CST).

the vertical by updrafts to produce tornadoes; however, in this case the vertical vorticity maximum is close to the ground. The twisted vorticity and any preexisting vertical vorticity are then amplified by convergence between outflow and inflow air masses. The final intensification of the wind to tornado strength occurs as the circulation interacts with the ground [*Rotunno*, 1986].

Figure 10b shows the strong horizontal vorticity, associated with a veering wind, that exists in the storm's environment. Some air parcels that enter the updraft and mesocyclone at low levels (below 0.5 km) have, at times, horizontal vorticity in excess of 200×10^{-4} s^{-1} (e.g., Figure 11). Vorticity maxima at $x = 25$, $y = 34$ and $x = 29$, $y = 36$ km lie near strong radar-reflectivity gradients at the edge of the storm, between the axis of the weak-echo intrusion and heavier precipitation on the right (when facing in the direction of the storm-relative wind). Baroclinic generation of horizontal vorticity is likely in this region as precipitation falls into the inflow air and evaporative cooling takes place. A component of vorticity that aligns with the flow results. With increasing altitude the source of the inflow air shifts to the storm's right rear [e.g., *Brandes*, 1984b, Figure 14]. The implication is that the baroclinic generation of horizontal vorticity was confined only to a shallow layer near ground.

In the Harrah storm, rapid generation of positive vertical vorticity takes place when the inflow air encounters the intense updraft in northern sections of the mesocyclone and the strong horizontal vorticity is tipped into the vertical (Figure 10c). Twisted vorticity and preexisting vertical

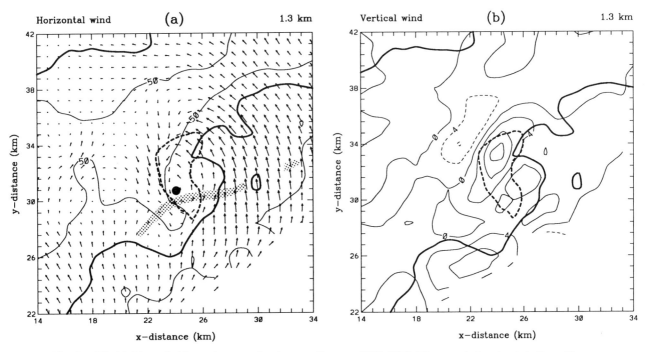

Fig. 9. Wind fields, as in Figure 4, except for the tornadic stage (1553 CST). The tornado location is indicated by a dot. The longest vector represents 36 m s^{-1}.

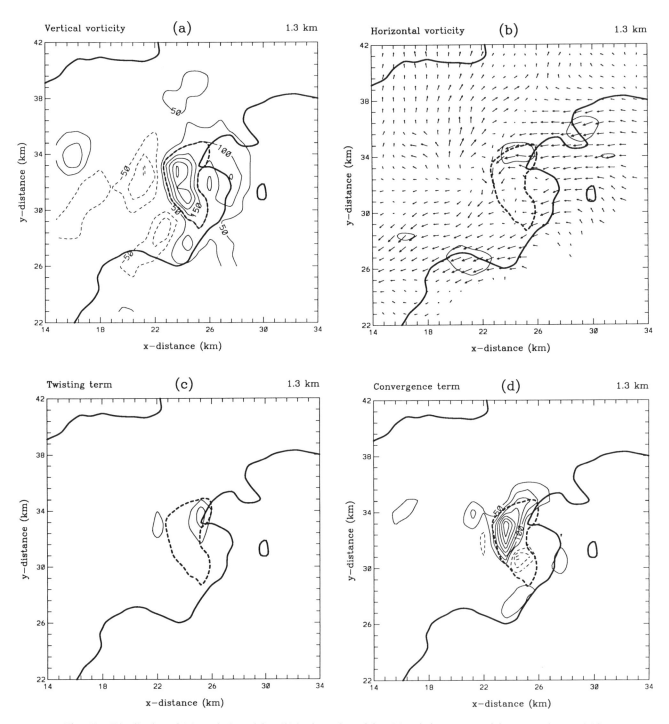

Fig. 10. Distribution of (*a*) vertical vorticity, (*b*) horizontal vorticity, (*c*) twisting-term vorticity generation, and (*d*) convergence-term vorticity amplification during the tornadic stage of the Harrah storm (1553 CST). The contour interval in Figure 10*a* is 50×10^{-4} s^{-1}. The 200×10^{-4} s^{-1} horizontal-vorticity contour has been added to Figure 10*b*. The contour interval in Figures 10*c* and 10*d* is 50×10^{-6} s^{-2}. The 40-dBZ contour and mesocyclone location are superimposed as in Figure 4*b*.

vorticity are then amplified by convergence (Figure 10d). Mesocyclone-mean convergence shows little change from the pretornadic stage (Figures 7 and 12). The larger range in individual values within the mesocyclone stems from the intrusion of the induced downdraft and from strong local convergence as the rainy outflow collides with inflow air. The maximum vertical-vorticity production by the convergence term occurs where the rainy outflow interacts with the low-level inflow. Mesocyclone flow is convergent in the mean throughout the layer of marked vorticity increase. Mean vorticity amplification by convergence below 2 km shows a corresponding increase (Figure 13). At this stage, the mean twisting-term contribution to the vorticity tendency shows the customary increase with height.

2.4. Posttornadic Stage

The Harrah storm was next sampled at 1603. The areas with echo coverage >50 dBZ, and regions with rainy downdrafts <-4 m s^{-1} had shrunken considerably. An intense occlusion downdraft had developed to the southeast of the principal updraft. The downdraft principally was fed by air parcels that overtook the storm from the rear [*Brandes*, 1984b, Figure 18].

Although tornado-intensity damage ended at ~1559, mesocyclone vertical vorticity near the ground continued to increase (Figure 6). Further, several shear anomalies persisted along the elongated (horizontal) axis of the mesocy-

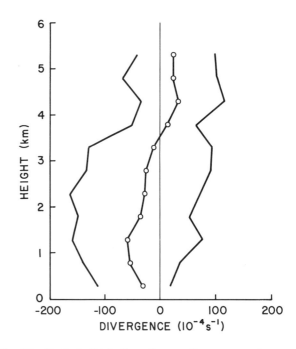

Fig. 12. Vertical distribution of mean divergence and range in grid-point values, as in Figure 7, except for the tornadic stage of mesocyclone development (1553 CST).

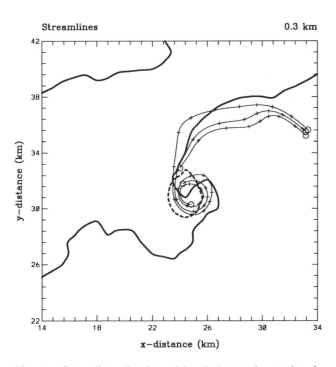

Fig. 11. Streamlines showing origin of air parcels entering the mesocyclone at 0.3-km elevation. Tick marks given parcel locations at 2-min intervals. The 40-dBZ contour and mesocyclone location are superimposed as in Figure 4b.

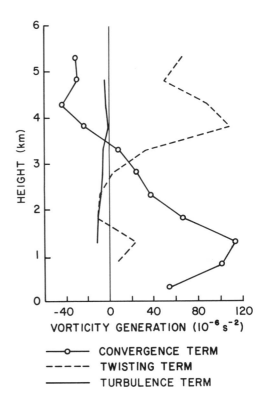

Fig. 13. Vertical distribution of mean vertical vorticity tendency due to convergence, twisting, and turbulent diffusion, as in Figure 8, except for the tornadic stage (1553 CST).

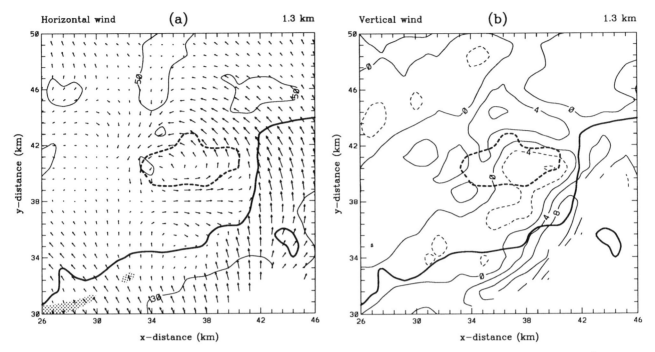

Fig. 14. Wind fields, as in Figure 4, except for the posttornadic stage (1611 CST). The longest vector represents 38 m s^{-1}.

clone. Low-level mesocyclone intensification causes a vertical lapse of vorticity between ground and 3 km. The growth of vorticity toward ground is associated with a downward pressure-gradient force that is thought to have induced the occlusion downdraft [*Brandes*, 1984a]. Radar observations from the Del City storm [*Brandes*, 1984a; *Hane and Ray*, 1985] also support this conclusion. The range of divergence values, the mean divergence, and the vorticity tendencies due to convergence and twisting were not much changed from the tornadic stage.

When last sampled (1611), the gust front had moved well to the east of the mesocyclone (Figure 14a), and the rotational flow had begun to decrease. The principal updraft had slowed to 8 m s^{-1} (Figure 14b), and the rear downdraft had declined to 8 m s^{-1}. Updrafts were confined to the western third of the mesocyclone, while downdrafts in excess of 4 m s^{-1} filled the eastern half. All air parcels passing through the mesocyclone at 1.3 km originated at higher levels on the storm's rear [*Brandes*, 1984a, Figure 21].

Mesocyclone vertical vorticity was still strong near ground but rapidly diminished with height. The vortex core could no longer be detected above 2 km (Figure 6). The range of divergence values and the mean divergence with the mesocyclone are shown in Figure 15. The flow, except for a neutral layer near ground, was divergent in the mean; the convergence term of the vorticity tendency equation had become negative. No new mesocyclonic vorticity centers appeared at the gust front nose or along its southern extension, and no additional severe weather was reported with the storm.

2.5. *Summary*

Low-level features observed during the tornadic stage of supercell development are summarized in Figure 16. The primary region of tornadogenesis (the uncircled T) is along an occluded or rolled-up section of the gust front that separates rain-cooled outflow from ambient air. Low-level inflow to the mesocyclonic circulation, in which the tornado is embedded, travels parallel to the isotherms and through a region where the temperature gradient, created by the evaporation of precipitation, enhances the ambient horizontal streamwise vorticity (in the vicinity of the stationary frontal boundary). The horizontal vorticity is then twisted into the vertical when the flow encounters the primary storm udpraft. The tipped vorticity and preexisting vertical vorticity are amplified further as the flow continues to converge in the updraft. The tornado lies on the updraft side of the vertical velocity gradient between the principal storm updraft and the rotation-induced downdraft. Many tornadoes are confined to the lowest 3 or 4 km of the troposphere, but tornadolike vortices have been observed to form at great heights [*Burgess et al.*, 1977; *Lemon et al.*, 1982]. The tornadic stage often is preceded (sometimes by tens of minutes) by public reports of funnels aloft and/or radar signatures of tornadolike vortices.

Specific events probably trigger tornadoes in storms primed for development. Tornadogenesis may be tied to the life cycle of individual elemental cells that compose the parent thunderstorm. Spreading outflow from elemental cells in their dissipating stages would alter the low-level flow properties in several ways. Where the outflow most interacts with storm inflow, local temperature gradients would be enhanced and the baroclinic generation of horizontal vorticity increased. Low-level convergence and updrafts would also strengthen. The twisting of horizontal vorticity and the convergent amplification of vertical vorticity would be promoted. The size of the interaction zone where tornadoes form would be determined by the scale of the individual convective elements, that is, would have dimensions of several kilometers. Tornadoes would simply develop from the enhanced vorticity. Or, possibly, tornadoes begin as shearing instabilities along locally perturbed sections of the gust front that separates the inflow and outflow air [*Brandes*, 1978]. Development would be much like vortices along disturbed vortex sheets. Some tornadoes are probably triggered by outflow boundaries from neighboring storms which interfere with the low-level circulation of the parental storm.

The very processes which lead to the buildup of low-level vorticity and tornadogenesis promote storm dissipation. The rotation-induced downdraft increasingly infiltrates the mesocyclone with time. Radar observations [*Brandes*, 1984a] show that the air that enters the downdraft overtakes the storm from the rear. This air is potentially cold and has little streamwise vorticity. Eventually, this air mixes with and reduces the intensity of the updraft and the mesocyclone.

Not all supercell tornadoes fit the pattern described in previous sections. *Brandes* [1978] shows dual-Doppler wind fields for a multiple tornado producing thunderstorm while a "gust front" tornado was in progress. The tornado formed near the nose of the gust front, well outside the mesocy-

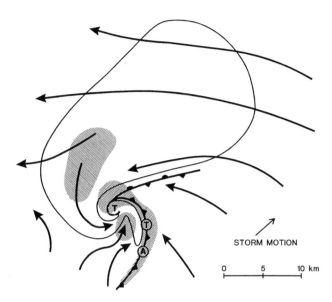

Fig. 16. Schematic diagram of prominent low-level features observed in supercell thunderstorms during tornadogenesis. Arrows show storm-relative flow. Rainy downdrafts and rotation-induced downdraft (behind the gust front) are shown by parallel lines that run from the upper left to the lower right. Updrafts are shown by parallel lines that run from the lower left to the upper right. The primary tornadogenesis is shown by an uncircled T. The locations of less frequently observed gust-front cyclonic tornadoes (circled T) and anticyclonic tornadoes (circled A) are also indicated.

clone. Its relative position is shown by the encircled T in Figure 16. Rarer still are anticyclonic tornadoes which appear to favor gust front locations to the right of the strong winds that gird the mesocyclone (the encircled A; see *Fujita and Wakimoto* [1982] for an example).

Statistics show [*Burgess et al.*, 1977] that more than 50% of the storms with middle level mesocyclones, including many with prominent hook echoes, do not produce tornadoes. Clearly, mesocyclones are neither a sufficient nor necessary condition for tornadoes. Why then are they important and so often associated with tornadoes? Perhaps their primary role is in the sculpturing of the low-level wind flow and temperature fields so that air parcel trajectories and the isotherms become parallel and the horizontal vorticity generation process described by *Rotunno and Klemp* [1985] can take place.

An important motivation for deploying a national network of Doppler radars is the prospect of improved tornado warnings. The purported lead time of tens of minutes is based primarily on the detection of the middle level mesocyclonic circulation and the average lag for the onset of damaging winds at ground. If tornadoes are largely low-level phenomena, warning schemes based on the detection of middle level rotation are going to have a high failure rate.

3. Nonsupercell Tornadoes

Some tornadolike vortices are not associated with supercells. Waterspouts are a common phenotype. Tornadoes

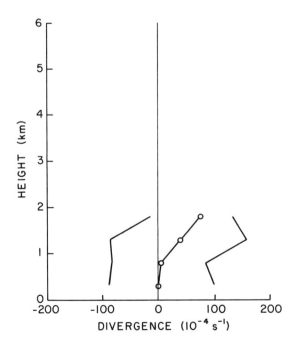

Fig. 15. Vertical distribution of mean divergence and range in grid-point values, as in Figure 7, except for the posttornadic stage (1611 CST).

observed with thunderstorms along cold fronts [*Carbone*, 1983] and with cold upper level synoptic circulations [*Cooley*, 1978] are other examples. Some tornadoes that occur in the Denver area appear to be more closely related to waterspouts than to supercell tornadoes. Environmental forcing is weak. The initial circulation often forms near ground along convergence boundaries, at the intersection of two boundaries, or when boundaries collide. (There are exceptions, of course; some vortices form aloft and later descend to ground [e.g., *Wakimoto and Wilson*, 1989, Table 1].) A possible sequence of events, taken from the study of *Brady and Szoke* [1989], is shown in Figure 17. The initial circulation precedes cloud formation and develops upward from ground as convection builds.

Figure 18, from the study of *Wilson and Roberts* [1990], shows a radar-derived horizontal wind field shortly after two outflow boundaries had collided on July 15, 1988. The wind field has been reconstructed from essentially "clear air" radar returns since precipitation had not yet developed. Strong convergence and positive vertical vorticity marks the resultant boundary. Labeled vorticity centers T1, T2, and T3 subsequently produced tornadoes. The evolution of center T1, which produced a tornado between 2204 and 2210 GMT, is presented in Figure 19. Low-level rotation was detected ~20 min before tornadogenesis and developed upward with time. As in supercells, the vorticity tendency within the mesocyclone, that is, the region of vertical vorticity $>10^{-2}$ s^{-1}, was dominated by convergence term amplification.

Note that if the pretornadic (1530) profile of the Harrah supercell storm in Figure 6 were subtracted from the remaining profiles, the resulting positive perturbations could be plotted to yield a vorticity distribution similar to that in Figure 19. The 1530 profile then approximates the influence of the supercell environment, particularly the low-level veering of the shear vector, while the perturbations represent changes associated with tornadogenesis.

Details of low-level storm structure during tornadogenesis in the Colorado storm of July 2, 1987, have been described in a preliminary study by *Wilczak and Christian* [1990]. The storm formed when a preexisting stationary convergence boundary (pseudofront), represented by a radar "fine line" of weak reflectivity, was overtaken by an outflow boundary from a nearby storm (Figure 20).

Rapid convective growth ensued along a bend that developed in the stationary boundary so that when the tornado was just becoming visible, maximum reflectivity exceeded 50 dBZ (Figure 21a). Although the intruding reflectivity minimum near $x = 11$ and $y = 15.5$ km and the reflectivity maximum centered at $x = 11.5$ and $y = 13.5$ km resemble the WER and hook echo in supercells, in this case they apparently result in part from the juxtaposition of two cells. (Often, pronounced hook echoes are observed [e.g., *Wakimoto and Wilson*, 1989, Figure 2d].) The vertical-velocity field (Figure 21b) shows several updraft centers >3 m s^{-1} that are distributed about a mesocyclone and in a trailing line of updrafts. A velocity minimum resides within the mesocyclone ($x = 8.2$, $y = 13.6$ km). A downdraft existed at this location above 2 km. A corresponding arclike region of strong vertical vorticity is not present (Figure 21c); rather, there are several maxima, the largest of which coincides with the tornado. The distribution of strong and weak updrafts about the tornado is similar to the updraft/downdraft pattern in supercells.

Scale and magnitude in the Colorado storm are small, but the arclike region of vorticity amplification by stretching (convergence) is like that in supercells (compare Figures 21d and 10d). Curiously, at this time, when the tornado is just becoming visible, mesocyclone vertical vorticity is being decreased by both the stretching and tilting (twisting) mechanisms (see also Figure 21e). Radar reflectivity values near the mesocyclone are fairly high (>10 dBZ), suggesting that deduced flow characteristics are based on strong meteorological signals. At earlier times, prior to the development of the updraft minimum within the mesocyclone, vorticity tendencies due to stretching and tilting were both positive [*Wilczak and Christian*, 1990, Figure 2].

The origin of the cyclonic rotation in Denver tornadoes is the subject of speculation. *Wakimoto and Wilson* [1989] and *Brady and Szoke* [1989] attribute the ultimate source of the rotation to horizontal wind shear across preexisting convergent boundaries. Shearing instabilities are thought to produce small vortices which amplify to tornadic intensity when they encounter updrafts. Because tornado proximity soundings may show only weak winds and little wind shear [e.g., *Brady and Szoke*, 1989, Figure 2], possible influences of vertical wind shear are discounted.

An important mesoscale feature on many Colorado tornado days is the orographically induced Denver Cyclone which forms under southeasterly wind conditions in the lee of an east-west ridge to the south of Denver. *Szoke and Augustine* [1990] note that tornadoes are 3 times more likely on cyclone days.

Numerical simulations of the Denver Cyclone [*Crook et al.*, 1990] indicate that the cyclone forms only after nocturnal inversions are removed by daytime heating. Rotation originates with baroclinically produced horizontal vorticity that develops when isentropic surfaces are deformed by the flow about the obstructing ridge. In a subsequent study, *Crook et al.* [1991] show that continued heating leads to the development of boundary layer rolls that may intersect the convergence zone. Concentrated vortices (mesocyclones), thought capable of producing tornadoes, develop at the intersection points.

Wilczak et al. [1992] found a temperature gradient of 2.5°C/10 km and strong horizontal vorticity of 100×10^{-4} s^{-1} with the boundary that ultimately spawned the July 2 tornado. They suppose that the horizontal vorticity was baroclinically generated and that this vorticity played a significant role in tornado formation. They assert that the surface convergence boundary, baroclinicity, related horizontal vorticity, and the parent vortex in which the tornado formed are surrogates for similar features in supercell storms.

Thus the observational studies (see also *Carbone* [1983])

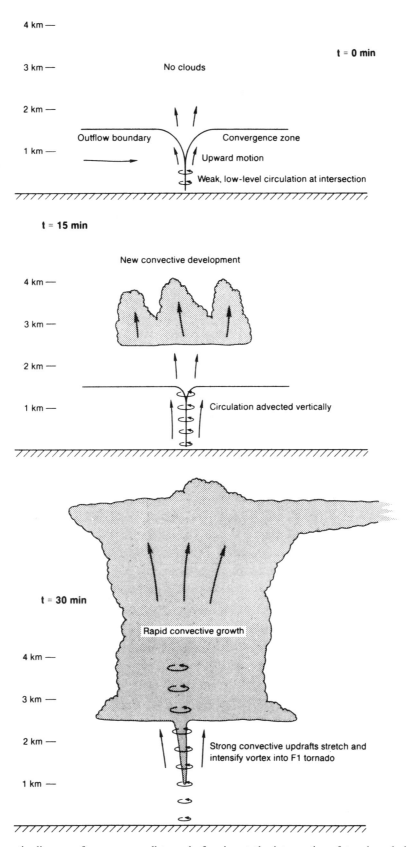

Fig. 17. Schematic diagram of a nonsupercell tornado forming at the intersection of two boundaries [*Brady and Szoke*, 1989].

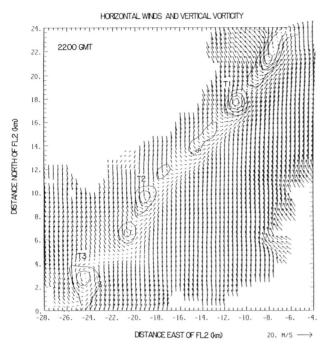

Fig. 18. Radar-derived wind field at 0.2-km elevation shortly after the collision of two boundaries (2200 GMT). The contours depict vertical-vorticity maxima. The contour interval is 5×10^{-3} s^{-1}. Labeled centers spawned tornadoes [*Wilson and Roberts*, 1990].

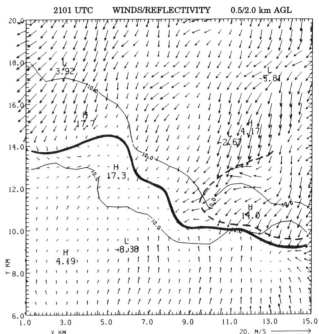

Fig. 20. Horizontal wind field at 0.5-km elevation and reflectivity (in dBZ) at 2 km prior to tornadogenesis in a Colorado thunderstorm observed on July 2, 1987, at 2101 UTC [*Wilczak and Christian*, 1990].

suggest that local concentrations of baroclinically generated vorticity, likely in all cases involving outflow boundaries and boundary layer rolls, also may be the primary source of rotation for nonsupercell tornadoes. Supercell tornadoes, waterspouts, and nonsupercell tornadoes may not be separate genre. The differences in intensity among these vortices may lie with the strength of the updrafts that tilt and amplify the horizontal vorticity.

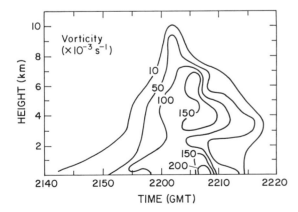

Fig. 19. Time versus height plot of maximum vertical vorticity, estimated from radial wind measurements, in circulation T1 of Figure 18. The tornado occurred between 2204 and 2210 GMT [*Wilson and Roberts*, 1990].

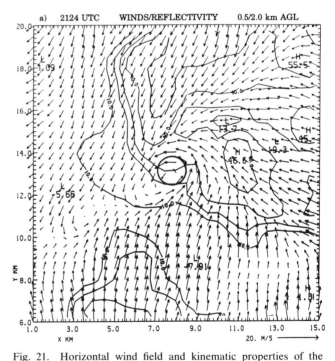

Fig. 21. Horizontal wind field and kinematic properties of the Colorado tornadic thunderstorm of July 2, 1987, at 2124 UTC [*Wilczak and Christian*, 1990]. The horizontal wind field and reflectivity (Figure 21a) are as in Figure 20. Figures 21b–21e are for 1-km elevation. Contour intervals are 1 m s^{-1} in Figure 21b, 2.5×10^{-3} s^{-1} in Figure 21c, and 5×10^{-6} s^{-6} in Figures 21d and 21e. The mesocyclone, defined by vorticity $\geq 10 \times 10^{-3}$, is indicated by an open circle.

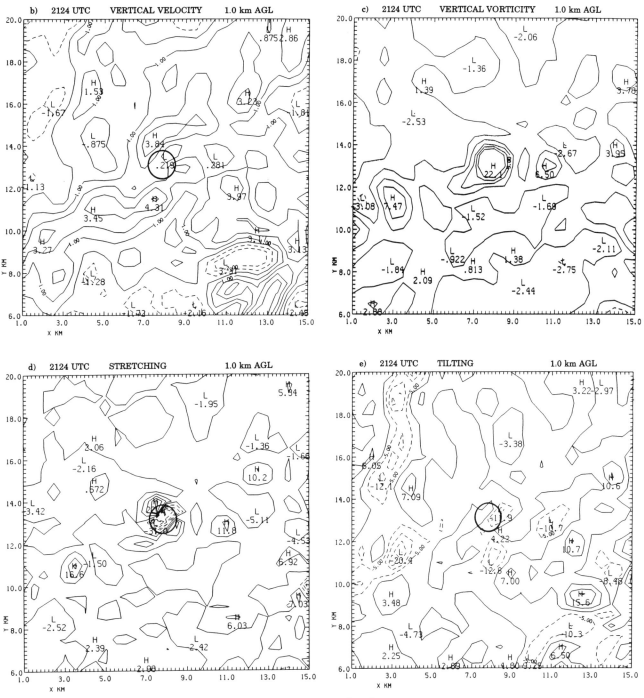

Fig. 21. (continued)

4. Summary and Recommendations

Doppler radar has become the primary sensor for observing tornadic storms. The observations are an important complement to numerical simulations as we seek to complete our knowledge of tornadogenesis. Moreover, Doppler radar observations will remain the principal basis for issuing severe weather warnings in the foreseeable future.

Although tornadolike vortices can develop at any height within thunderstorms, most supercell and nonsupercell tornadoes develop in the lower troposphere. These circulations apparently tap the rich supply of baroclinically generated horizontal vorticity that exists near ground in regions of strong temperature gradients. This vorticity is tipped and amplified within the storm updraft. Paradoxically, surface friction further enhances the intensity of the surface winds by allowing low-level air to approach closer to the axis of rotation.

Why some tornadolike vortices form and stay aloft is not clear. They develop along vertical extensions of the same boundaries that produce tornadoes, suggesting that at least initially they respond to the same mechanisms that act near ground. Because they do not make contact with the ground, these circulations are probably weaker than tornadoes.

While much has been learned about supercell and nonsupercell tornadic storms, some important questions remain. We still cannot predict when tornadoes will form within storms, precisely where they will form, and what their size and strength will be.

To complete our understanding of tornadic storms, we need new high-resolution data sets from state-of-the-art radar systems that permit determination of the fine-scale structure of storms. Of particular interest is the sequence of events that lead to tornadogenesis. Coordinated photogrammetric observations from the field are necessary to relate the visual characteristics and kinematic properties of tornadic storms. Finally, more studies comparing radar-observed and numerically simulated storms are needed. Understanding situations where models fail to reproduce observed storms are just as important as the successes.

Acknowledgment. This study benefitted from indispensable suggestions and comments provided by Robert P. Davies-Jones. Bob's in-depth and thought-provoking reviews are always a source of inspiration.

REFERENCES

Agee, E. M., J. T. Snow, and P. R. Clare, Multiple vortex features in the tornado cyclone and the occurrence of tornado families, *Mon. Weather Rev.*, *104*, 552–563, 1976.

Barnes, S. L., Some aspects of a severe, right-moving thunderstorm deduced from mesonetwork rawinsonde observations, *J. Atmos. Sci.*, *27*, 634–648, 1970.

Brady, R. H., and E. J. Szoke, A case study of nonmesocyclone tornado development in northeast Colorado: Similarities to waterspout formation, *Mon. Weather Rev.*, *117*, 843–856, 1989.

Brandes, E. A., Mesocyclone evolution and tornadogenesis: Some observations, *Mon. Weather Rev.*, *106*, 995–1011, 1978.

Brandes, E. A., Finestructure of the Del City-Edmond tornadic mesocirculation, *Mon. Weather Rev.*, *109*, 635–647, 1981.

Brandes, E. A., Relationships between radar-derived thermodynamic variables and tornadogenesis, *Mon. Weather Rev.*, *112*, 1033–1052, 1984*a*.

Brandes, E. A., Vertical vorticity generation and mesocyclone sustenance in tornadic thunderstorms: The observational evidence, *Mon. Weather Rev.*, *112*, 2253–2269, 1984*b*.

Brandes, E. A., R. P. Davies-Jones, and B. C. Johnson, Streamwise vorticity effects on supercell morphology and persistence, *J. Atmos. Sci.*, *45*, 947–953, 1988.

Brown, R. A., L. R. Lemon, and D. W. Burgess, Tornado detection by pulsed Doppler radar, *Mon. Weather Rev.*, *106*, 29–38, 1978.

Browning, K. A., Airflow and precipitation trajectories within severe local storms which travel to the right of the winds, *J. Atmos. Sci.*, *21*, 634–639, 1964.

Browning, K. A., The organization of severe local storms, *Weather*, *23*, 429–434, 1968.

Browning, K. A., and C. R. Landry, Airflow in convective storms, *Q. J. R. Meteorol. Soc.*, *88*, 117–135, 1963.

Browning, K. A., and F. H. Ludlam, Airflow in convective storms, *Q. J. R. Meteorol. Soc.*, *88*, 117–135, 1962.

Burgess, D. W., R. A. Brown, L. R. Lemon, and C. R. Safford, Evolution of a tornadic thunderstorm, in *Preprints*, *10th Conference on Severe Local Storms*, pp. 84–89, American Meteorological Society, Boston, Mass., 1977.

Burgess, D. W., V. T. Wood, and R. A. Brown, Mesocyclone evolution statistics, in *Preprints*, *12th Conference on Severe Local Storms*, pp. 422–424, American Meteorological Society, Boston, Mass., 1982.

Byers, H. R., and R. R. Braham, *The Thunderstorm*, 287 pp., U.S. Department of Commerce, Washington, D. C., 1949.

Carbone, R. E., A severe frontal rainband, II, Tornado parent vortex circulation, *J. Atmos. Sci.*, *40*, 2639–2654, 1983.

Chisholm, A. J., and J. H. Renick, The kinematics of multicell and supercell Alberta hailstorms *Alberta Hail Stud. Rep. 72-2*, pp. 24–31, Res. Counc. of Alberta, Edmonton, 1972.

Cooley, J. R., Cold air funnel clouds, *Mon. Weather Rev.*, *106*, 1368–1372, 1978.

Crook, N. A., T. L. Clark, and M. W. Moncrieff, The Denver cyclone, I, Generation in low Froude number flow, *J. Atmos. Sci.*, *47*, 2725–2742, 1990.

Crook, N. A., T. L. Clark, and M. W. Moncrieff, The Denver cyclone, II, Interaction with the convective boundary layer, *J. Atmos. Sci.*, *48*, 2109–2126, 1991.

Davies-Jones, R. P., Observational and theoretical aspects of tornadogenesis, in *Intense Atmospheric Vortices*, edited by L. Bengtsson and J. Lighthill, pp. 175–189, Springer-Verlag, New York, 1982.

Davies-Jones, R. P., Streamwise vorticity: The origin of updraft rotation in supercell storms, *J. Atmos. Sci.*, *41*, 2991–3006, 1984.

Davies-Jones, R. P., Tornado dynamics, in *Thunderstorms: A Social, Scientific, and Technological Documentary*, vol. 2, *Thunderstorm Morphology and Dynamics*, 2nd ed., edited by E. Kessler, pp. 197–236, University of Oklahoma Press, Norman, 1986.

Davies-Jones, R. P., Tornado interception with mobile teams, in *Instruments and Techniques for Thunderstorm Observation and Analysis*, 2nd ed., pp. 23–32, University of Oklahoma Press, Norman, 1988.

Dennis, A. S., C. A. Schock, and A. Kosceilski, Characteristics of hailstorms of western South Dakota, *J. Appl. Meteorol.*, *9*, 127–135, 1970.

Foote, G. B., and H. W. Frank, Case study of a hailstorm in Colorado, III, Airflow from triple-Doppler measurements, *J. Atmos. Sci.*, *40*, 686–707, 1983.

Foote, G. B., and C. G. Wade, Case study of a hailstorm in Colorado, I, Radar echo structure and evolution, *J. Atmos. Sci.*, *39*, 2828–2846, 1982.

Fujita, T. T., and R. M. Wakimoto, Anticyclonic tornadoes in 1980 and 1981, in *Preprints*, *12th Conference on Severe Local Storms*, pp. 401–404, American Meteorological Society, Boston, Mass., 1982.

Hane, C. E., and P. S. Ray, Pressure and buoyancy fields derived from Doppler radar data in a tornadic thunderstorm, *J. Atmos. Sci.*, *42*, 18–35, 1985.

Heymsfield, G. M., Kinematic and dynamic aspects of the Harrah tornadic storm analyzed from dual-Doppler radar data, *Mon. Weather Rev.*, *106*, 233–254, 1978.

Johnson, K. W., P. S. Ray, B. C. Johnson, and R. P. Davies-Jones, Observations related to the rotational dynamics of the 20 May 1977 tornadic storms, *Mon. Weather Rev.*, *115*, 2463–2478, 1987.

Klemp, J. B., Dynamics of tornadic thunderstorms, *Annu. Rev. Fluid Mech.*, *19*, 369–402, 1987.

Klemp, J. B., and R. Rotunno, A study of the tornadic region within a supercell thunderstorm, *J. Atmos. Sci.*, *40*, 359–377, 1983.

Klemp, J. B., and R. B. Wilhelmson, The simulation of three-dimensional convective storm dynamics, *J. Atmos. Sci.*, *35*, 1070–1096, 1978*a*.

Klemp, J. B., and R. B. Wilhelmson, Simulations of right- and left-moving storms produced through storm splitting, *J. Atmos. Sci.*, *35*, 1097–1110, 1978*b*.

Lemon, L. R., The flanking line, a severe thunderstorm intensification source, *J. Atmos. Sci.*, *33*, 686–694, 1976.

Lemon, L. R., and C. A. Doswell, III, Severe thunderstorm evolution and mesocyclone structure as related to tornadogenesis, *Mon. Weather Rev.*, *107*, 1184–1197, 1979.

Lemon, L. R., D. W. Burgess, and L. D. Hennington, A tornado extending to extreme heights as revealed by Doppler radar, in *Preprints, 12th Conference on Severe Local Storms*, pp. 430–432, American Meteorological Society, Boston, Mass., 1982.

Marwitz, J. D., The structure and motion of severe hailstorms, I, Supercell storms, *J. Atmos. Sci.*, *11*, 166–179, 1972.

Ray, P. S., Dual-Doppler observation of a tornadic storm, *J. Appl. Meteorol.*, *14*, 1521–1530, 1975.

Ray, P. S., Vorticity and divergence fields within tornadic storms from dual-Doppler observations, *J. Appl. Meteorol.*, *15*, 879–890, 1976.

Ray, P. S., B. C. Johnson, K. W. Johnson, J. S. Bradberry, J. J. Stephens, K. K. Wagner, R. B. Wilhelmson, and J. B. Klemp, The morphology of several tornadic storms on 20 May 1977, *J. Atmos. Sci.*, *38*, 1643–1663, 1981.

Rotunno, R., Tornadoes and tornadogenesis, in *Mesoscale Meteorology and Forecasting*, edited by P. S. Ray, pp. 414–436, American Meteorological Society, Boston, Mass., 1986.

Rotunno, R., and J. B. Klemp, On the rotation and propagation of simulated supercell thunderstorms, *J. Atmos. Sci.*, *42*, 271–292, 1985.

Schlesinger, R. E., A three-dimensional numerical model of an isolated deep convective cloud: Preliminary results, *J. Atmos. Sci.*, *32*, 934–957, 1975.

Schlesinger, R. E., A three-dimensional numerical model of an isolated thunderstorm, I, Comparative experiments for variable ambient wind shear, *J. Atmos. Sci.*, *35*, 690–713, 1978.

Szoke, E. J., and J. A. Augustine, A decade of tornado occurrence associated with a surface mesoscale flow feature—The Denver Cyclone, in *Preprints, 16th Conference on Severe Storms*, pp. 554–559, American Meteorological Society, Boston, Mass., 1990.

Vasiloff, S. V., E. A. Brandes, R. P. Davies-Jones, and P. S. Ray, An investigation of the transition from multicell to supercell storms, *J. Clim. Appl. Meteorol.*, 1022–1036, 1986.

Wakimoto, R. M., and J. W. Wilson, Non-supercell tornadoes, *Mon. Weather Rev.*, *117*, 1113–1140, 1989.

Weaver, J. F., and S. P. Nelson, Multiscale aspects of thunderstorm gust fronts and their effects on subsequent storm development, *Mon. Weather Rev.*, *110*, 707–718, 1982.

Wilczak, J., and T. Christian, A vorticity analysis of the non-supercell, 2 July 1987 tornado, in *Preprints, 16th Conference on Severe Local Storms*, pp. 560–565, American Meteorological Society, Boston, Mass., 1990.

Wilczak, J. M., T. W. Christian, D. E. Wolfe, R. J. Zamora, and B. Stankov, Observations of a Colorado tornado, I, Mesoscale environment and tornadogenesis, *Mon. Weather Rev.*, *120*, 497–520, 1992.

Wilhelmson, R. B., and J. B. Klemp, A numerical study of storm splitting that leads to long-lived storms, *J. Atmos. Sci.*, *35*, 1974–1986, 1978.

Wilson, J. W., and R. D. Roberts, Vorticity evolution of a non-super cell tornado on 15 June 1988 near Denver, in *Preprints, 16th Conference on Severe Local Storms*, pp. 479–484, American Meteorological Society, Boston, Mass., 1990.

Tornadoes and Tornadic Storms: A Review of Conceptual Models

CHARLES A. DOSWELL III AND DONALD W. BURGESS

National Severe Storms Laboratory, Norman, Oklahoma 73069

1. INTRODUCTION

The definition of a tornado in the Glossary of Meteorology [*Huschke*, 1959] begins with the following: "A violently rotating column of air, pendant from a **cumulonimbus** cloud, and nearly always observable as a 'funnel cloud' or **tuba**." This definition seems relatively straightforward, but in reality distinctions implied by the words we use tend to blur.

Our ability to make distinctions among tornadic storm types and among tornadoes has been affected by three important developments since the last tornado symposium: (1) high-resolution Doppler radar observations, (2) extensive visual observations by storm intercept teams, and (3) detailed three-dimensional numerical cloud models. The definition and classification of convective vortices and the storms which produce them have become more complex and, in many ways, more troublesome than ever before. On the other hand, the complexity we are encountering is really a positive sign that our understanding is growing.

While it was well known long before the early 1960s that tornadoes were associated with deep, moist convection, the relationship between these intense vortices and the convective storms with which they are associated was not well understood. Beginning in the 1950s, with the deployment of weather radar, it became clear that at least some tornadic storms exhibited special characteristics [see *Stout and Huff*, 1953; *Garrat and Rockney*, 1962]. Careful, systematic examination of such storms began in the early 1960s with the work of Browning [*Browning and Donaldson*, 1963; *Browning*, 1964] using reflectivity radar as the primary observational tool. Research Doppler radars began to make detailed, systematic observations of airflow in tornadic storms in the early 1970s [e.g., *Burgess and Brown*, 1973], just prior to the last symposium.

In 1972, building on early pioneering work like that of *Ward* [1961], an organized program to intercept tornadic storms began at the National Severe Storms Laboratory, with the involvement of the Department of Meteorology at the University of Oklahoma [see *Golden and Morgan*, 1972; *Moller et al.*, 1974]. Storm chasing (which has become a part of many scientific programs outside of Oklahoma) has provided an opportunity for meteorologists to observe the visually recognizable characteristics of both tornadoes and the storms associated with them.

Finally, the development of computer models capable of realistic simulations of tornadic storms [*Schlesinger*, 1975; *Klemp and Wilhelmson*, 1978] and tornadoes [*Rotunno*, 1979] provided a means for careful quantitative evaluation of the physical processes which yield tornadoes. With time, the sophistication of these models has continued to grow, and it is now possible to produce "tornadolike vortices" within a numerical model of the entire storm [*Wicker*, 1990].

One thing has become quite clear during this evolution: the storm type Browning called a "supercell" produces by far the most intense convective vortices and certainly is the type of storm most likely to produce them. Indeed, most of the radar reflectivity structures that were associated empirically with tornadoes have come to be recognized as characteristic of supercells.

However, the quantitative evaluation of physical processes made possible by numerical models has made it clear that the morphology of radar reflectivity alone is inadequate for categorizing convective storms. Instead, a persistent correlation (positive or negative) between vertical velocity and vertical vorticity has come to be recognized as the most useful definition of a supercell [*Weisman and Klemp*, 1984]. That is, the airflow pattern most characteristic of a supercell is a deep and persistent cyclone, called a "mesocyclone" [see *Burgess and Lemon*, 1990].

Many convective vortices are associated with nonsupercellular convection, however. The ways in which these nonsupercell-related events arise remain quite inadequately understood, with some efforts to deal systematically with one type having only just begun [*Brady and Szoke*, 1988; *Wakimoto and Wilson*, 1989].

As we study the tornadic vortices themselves, it is becoming evident that not all tornadoes are the same, and a proper classification of them is not simple. Understanding that a

The Tornado: Its Structure, Dynamics, Prediction, and Hazards.
Geophysical Monograph 79
This paper is not subject to U.S. copyright. Published in 1993 by the American Geophysical Union.

vortex is a kinematically defined process rather than an object, with different air parcels participating in the flow from moment to moment, turns out to be an important notion in trying to classify observed events. This, in turn, affects our climatological record of events, as we shall discuss below.

The dynamics of vortices, such as those simulated numerically [e.g., *Rotunno*, 1979; *Gall*, 1983] and in the laboratory [e.g., *Ward*, 1972; *Church et al.*, 1979], certainly are pertinent to tornadoes. Such studies have improved our understanding of phenomena like multiple vortices and vortex breakdown, but they concern properties of vortices in general, not necessarily those of tornadoes. From a purely dynamical viewpoint, tornadoes arise from amplification of either existing or locally created vorticity [*Rotunno*, 1986; *Davies-Jones*, 1986]. However, this is a somewhat abstract framework for understanding tornadoes. This paper attempts to review tornadoes in the context of the convective events giving rise to them. In particular, we shall distinguish between events associated with supercell storms and those produced in association with nonsupercellular convection. Not everything we present will have been thoroughly investigated in the scientific literature, especially nonsupercellular events; in fact, we wish to mention some of these lesser known phenomena in hopes of stimulating their systematic study.

2. Supercell Storms

2.1. *The Supercell Spectrum*

We have chosen the presence of a deep, persistent mesocyclone to be the single distinguishing characteristic of supercells. (By "deep," we mean a significant fraction of the depth of the cumulonimbus cloud in which the circulation is embedded (several kilometers). By "persistent," we mean in comparison to a convective time scale, defined by the time it takes for air parcels to rise from within the inflow layer of the updraft to the anvil outflow (a few tens of minutes).) Even within the category of supercell storms, however, it turns out that distinctions that appear to have significance can be made [see *Doswell et al.*, 1990]. While we concur with the kinematic-dynamic approach for defining a supercell, first advocated by *Browning* [1977] and recently reemphasized by *Weisman and Klemp* [1984], it appears that the amount and spatial distribution of precipitation with the convection are important indicators of the weather phenomena associated with a particular storm.

Some supercell storms produce relatively little precipitation and yet show clear visual signs of rotation (Figure 1). Such storms have come to be called low-precipitation (LP) supercells [*Bluestein and Parks*, 1983]. LP supercells occur most often near the surface dryline and, owing to the sparse precipitation and relatively dry environments with little or no intervening cloudiness, cloud structures showing rotation are visible readily to a suitably positioned observer.

On the other hand, precisely because of the sparse precipitation, radar reflectivity may not reveal the circulation adequately, if at all. LP storms frequently are nontornadic, and many are nonsevere despite exhibiting persistent rotation.

At the other end of the supercell spectrum are the so-called high- (or heavy-) precipitation (HP) supercells (Figure 2). Whereas LP storms have little or no precipitation (and hence low reflectivity) within their mesocyclones, HP storms are characterized by substantial precipitation within their mesocyclonic circulations. When HP storms have a recognizable hook echo (many do not), reflectivities in the hook will be comparable to those in the precipitation core.

HP supercells are probably the most common form of supercell, occurring not only in the humid half of the United States east of the Mississippi but also westward into the high plains. They produce severe weather of all types (including tornadoes) and, unlike other types of supercells, also may produce torrential, flash flood-producing rainfalls [*Moller et al.*, 1990]. Some of the distinctive radar echoes [*Forbes*, 1981] traditionally associated with tornadic storms, like the so-called bow echoes, comma echoes, and line-echo wave patterns (or LEWPs [*Nolen*, 1959]) can be associated with HP supercell storms. The rationale for including these forms in the HP supercell class is that they result from persistent mesocyclones embedded within precipitation-filled regions of the storm.

Because HP supercells often occur in humid, cloud-filled environments, visible signs of rotation may difficult to detect. In contrast, since the circulation in HP storms is embedded within precipitation, radar reflectivity usually depicts the HP storm circulations readily, sometimes even as curved bands apparently aligned with the flow.

Finally, between these two extremes is the classic supercell, which exhibits moderate precipitation production (Figure 3). Such storms typically match the traditional supercell conceptual models [e.g., *Browning*, 1964; *Lemon and Doswell*, 1979] and are most common in the transitional environments of the Great Plains. Many of the tornadic storms in major tornado outbreaks east of the Mississippi River are of the classic variety, however.

Although there may be some precipitation within a classic supercell's mesocyclone, it typically is not heavy. Note that because a radar (even at its lowest elevations) scans a storm above the surface, a region with little or no surface precipitation may still be within radar-detectable precipitation aloft. If such a storm has a hook echo (and many do), the hook reflectivities will be less than those of the precipitation core. Late in a classic supercell's life cycle, during collapse of its updraft [see *Lemon*, 1977], the mesocyclone may fill with precipitation, but this should not be considered a transition to an HP supercell unless the mesocyclone persists well after the collapse phase.

Classic supercells are readily detectable both visually and via radar reflectivity and produce a full range of severe weather, but only rarely are they associated with flash flooding. Classic supercells probably account for the majority of violent (F4-F5) tornadoes.

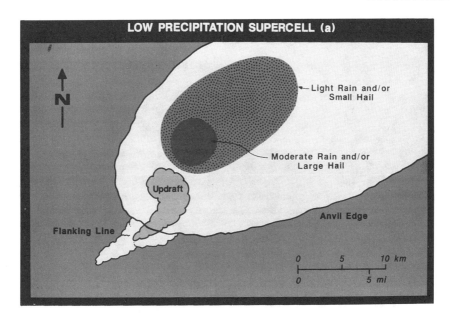

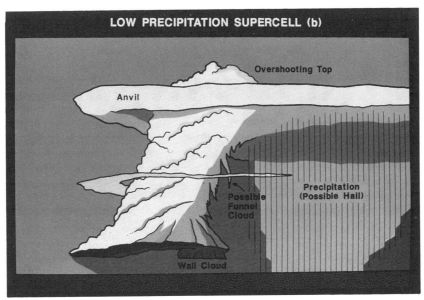

Fig. 1. Low-precipitation supercell schematics for (a) low-level radar structure and cloud features looking down from above and (b) visual structures from the viewpoint of a nearby observer on the ground.

2.2. Hybrid Events

Since class distinctions are much less obvious in the real atmosphere than they are in the abstract, it is quite common to see events that do not fit the preceding prototypes precisely. For example, it is likely that most LP storms do not become tornadic unless they evolve along the supercell spectrum toward the classical structure. The June 5, 1982, Borger, Texas, tornadic storm (Figure 4) had a visual appearance that might suggest it to be an LP storm, but its appearance on radar was more like a classic supercell, exhibiting a substantial hook echo. Exceptions are inevitable, naturally. The tornadic storms reported on by *Burgess and Davies-Jones* [1979] and *Burgess and Donaldson* [1979] produced intense tornadoes, and yet, as LP storms, they had little or no distinctive radar structure.

Supercell storms seem capable of evolving from LP to classical, from classical to HP, and so on. As noted by *Doswell et al.* [1990], the variety of radar reflectivity morphologies, especially within the HP group, can be quite confusing (see also *Imy and Burgess* [1991]). Nonsupercell convection can evolve into supercells [*Burgess and Curran*, 1985] and vice versa. The common factor in all supercells is

164 CONCEPTUAL MODELS OF TORNADOES AND TORNADIC STORMS

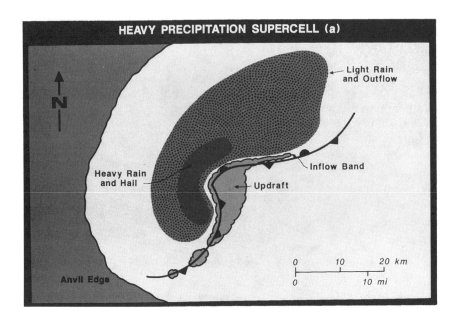

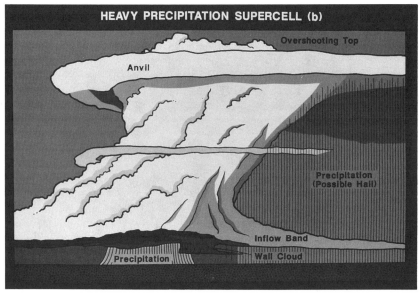

Fig. 2. High-precipitation supercell schematics as in Figure 1.

the deep, persistent mesocyclone, regardless of the storm's precipitation characteristics. However, the observed variations in precipitation amount and distribution make supercell recognition a challenge, especially when that recognition depends mostly on non-Doppler radar observations; this situation that will be remedied in time with the deployment of the operational Doppler radar (the WSR-88D network).

2.3. *Supercell Identification Criteria*

Having focused on the mesocyclone as the criterion for identifying supercells, we wish to review some of the traditionally accepted supercell characteristics. The presence of a single, persistent "cell" is arguably the most commonly accepted radar characteristic associated with supercells. The difficulty with this as a defining characteristic is that when observing convective storms visually or with especially high resolution radar, it turns out that a multicellular structure can be observed to be superimposed on most convective storms, including supercells. Although the Byers-Braham prototype convective "cell" typically is depicted as "plume"-like [e.g., *Weisman and Klemp*, 1986, Figure 15.1], such cells really are more "bubble"-like, even in supercells (compare Figure 12b with Figure 12a of *Newton* [1963] and see *Hane and Ray* [1985], especially their Figure 13).

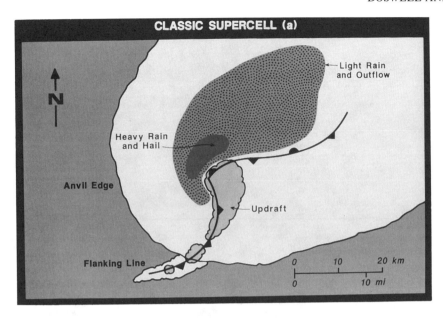

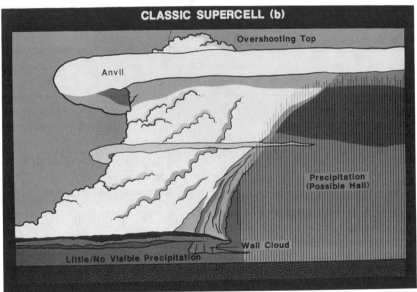

Fig. 3. Classic supercell schematics as in Figure 1.

Associated with the single-cell notion is another commonly employed yardstick to identify supercells: their tendency for "steady state" character. As with the single-cell criterion, this simply does not hold up to detailed observations. Supercells producing "tornado families" (cyclic tornado-producing storms) undergo an evolution over time scales of the order of several tens of minutes (as described by *Lemon and Doswell* [1979] and *Burgess et al.* [1982]). There also is the constantly evolving cellular structure superimposed on the overall storm evolution that has a time scale of several minutes. These subprocesses make even an approximately steady state unlikely. Nevertheless, it can be argued that supercells do exhibit a long-lasting "background" process that evolves only slowly over periods of a few hours, the characteristic lifetime of the constantly regenerating supercell structure [*Foote and Frank*, 1983]. In extreme cases, supercells evolve very slowly indeed and may have tornadoes on the ground for periods approaching (and occasionally exceeding) an hour. Events of this character are quite rare, and it is not yet understood how such steadiness arises. If such steadiness is a defining characteristic, then the supercell class is quite sparsely populated. In our opinion, there has been so much published emphasis on the steadiness and unicellularity criteria for supercells that operational identification of supercells using these characteristics often misses less prototypical (but still clearly supercellular) events.

Fig. 4. A tornadic supercell in the vicinity of Borger, Texas, on June 5, 1982. Photograph copyrighted 1982 by A. Moller, used by permission.

On occasion, supercells arise in environments with relatively modest instability, as in tropical cyclones [*McCaul*, 1991] and in strongly baroclinic systems [e.g., *Gonski et al.*, 1989]. Although the instability may be weak, there can be little doubt of the supercellular character of storms in such events. Although there is always a question about the existence of a small-scale, unobserved region of strong instability, it seems unnecessary to postulate some mechanism not supported by the existing data. The evidence is substantial that supercells do not require an environment with strong instability.

Finally, there often has been considerable emphasis on the deviate motion (from the direction of the mean flow in which the storm is embedded) of the supercell, even to the point of suggesting it as a defining characteristic. Not all supercells deviate significantly from the mean wind direction [see *Davies and Johns*, this volume]; therefore deviate motion is not required for development of a deep, persistent mesocyclone. This is especially so when hodographs are curved [see *Doswell*, 1991b].

Hook echoes and other "distinctive" structures (as discussed by *Forbes* [1981]) are the direct result of the mesocyclone circulations of a supercell. The distribution of precipitation quite clearly depends on airflow within the storm, which we have agreed is the most characteristic feature of a supercell, so such structures certainly are potentially useful in identifying supercells. Sometimes, though, such features as hook-shaped reflectivity structure can arise as a result of "configuration" instead of circulation. Thus *Lemon*'s [1977] emphasis on the three- and four-dimensional echo structure of storms: it is with such structural knowledge that a radar observer can separate bona fide supercell structures from "imposters" created by particular echo juxtapositions. Since LP storms typically exhibit few, if any, distinctive reflectivity features, and since detection of the classic echo features is so range- and resolution-dependent, these distinctive structures do not seem appropriate criteria for supercell identification.

Thus on the basis of the above arguments, we advocate a deemphasis for many of the traditional supercell identification criteria. With Doppler radar data, a time- and space-continuous mesocyclone is the best way for identifying such storms. With reflectivity alone, the three- and four-dimensional echo structure can be used to infer the presence of mesocyclones in many cases. The optimum situation for interpretation is when combining velocity and reflectivity information with a knowledge of characteristic storm structures. LP storms present a problem to any purely radar-based identification process, so visual recognition (spotters) still has an important role to play, even in the era of operational Doppler radars.

2.4. Tornadoes Within Supercells

The common association between mesocyclones and tornadoes in supercells hints that conservation of angular momentum may explain tornadoes associated with mesocyclones. However, even supercell tornadogenesis may be more complicated than that. Tornado development in the vicinity of the so-called wall cloud [*Fujita*, 1960] suggests

Fig. 5. An example of a nonsupercell tornado event (sometimes called a "landspout") near Sublette, Kansas, on May 15, 1991. Note that this is a relatively high cloudbase, estimated at about 5000 feet (~1525 m). Photograph copyrighted 1991 by C. Doswell.

that nearby downdrafts play an important role in getting tornadic/mesocyclonic vorticity to low levels in the storm [*Lemon and Doswell*, 1979; *Davies-Jones*, 1982; *Davies-Jones and Brooks*, this volume].

Intense vortices associated with supercells do not necessarily all develop via identical processes. (Recall the discussion in the introduction, distinguishing between the limited number of abstract mechanisms for creating intense vortices and the processes operating at storm scales to allow the vortex dynamics to operate.) There may well be more than one mechanism operating for any given vortex associated with a supercell, or within different regions of the same supercell. Moreover, those events leading to tornado initiation may not be the same as those maintaining the large vorticity. *Moller et al.* [1974] have described funnel clouds on the northwest side of the Union City tornadic storm, with cool outflow at the surface, even as the primary tornado was developing on the southwestern updraft flank, near the inflow/outflow interface of the same storm. It is hard to imagine the same storm-scale processes operating in these areas, although both were intense vortices in the abstract sense.

3. NONSUPERCELL STORMS

A variety of intense atmospheric vortices can develop in association with nonsupercell storms. Terminology can be a controversial topic, but we do not wish to get involved in terminology debates. (See the discussion among *Fankhauser et al.* [1983a, b], *Doswell* [1983], and *Moller* [1983] for some sense of the terminology issues; although that debate concerns the names for cloud features, its flavor is characteristic of terminology debates in general.) While we use certain terms that have been common in the vernacular and/or the literature, we do not necessarily endorse those terms. These events comprise several categories, and we will attempt to give a brief description of each.

3.1. Landspouts

In an analogy with the common waterspout [*Bluestein*, 1985], most of which develop from nonsupercell storms, many nonsupercell tornadic events (e.g., Figure 5) arise via intensification of preexisting, shallow vertical vortices near the surface, through simple vortex stretching when a developing convective updraft moves over them [see *Brady and Szoke*, 1988]. (Of course, some waterspouts do arise from supercells. They have been called tornadic waterspouts by *Golden* [1971] and appear to be virtually identical to tornadoes associated with supercells over land. The distinction between a tornado and a waterspout is basically of little or no scientific value.) Doppler radar evidence shows the pretornadic existence of these vortices on convergence boundaries [*Wilczak et al.*, 1991]. The details of the origin of these "misoscale" [*Fujita*, 1981] vortices are as yet unclear, but such preexisting vortices may explain the "dark spots" seen on the sea surface prior to the development of common waterspouts [*Golden*, 1974], as noted by *Wakimoto and Wilson* [1989].

Perhaps a related phenomenon is the weakly unstable, linearly convective tornadic event first documented by *Carbone* [1983]. As with landspouts, a frontal boundary may develop locally enhanced circulation centers, which subsequently can attain tornadic proportions. What makes these events distinctive is the weak buoyancy in their environment; the updrafts are forced along the frontal zone [see *Carbone*, 1982], and the tornadic circulations are comparable in depth to the updraft (which was shallow to begin with, only a few kilometers). Again, such events have not been observed often enough to have been subjected to systematic study.

Preexisting vortices at low levels also may be associated with tornadoes arising as convergence boundaries collide [e.g., *Holle and Maier*, 1980]. Such events are associated with multicellular lines and clusters, and the resulting flows can be quite complex. Although multicell storms have been the subject of many observational studies [e.g., *Marwitz*, 1972], they have not yet been given the attention they deserve in three-dimensional numerical modeling. Therefore the dynamics of interacting convective cells are as yet poorly understood. Tornadogenesis under such circumstances is therefore correspondingly poorly understood.

3.2. Cold Pool Vortices

To our knowledge, the sole reference to these is that by *Cooley* [1978]. They seem to be associated with cold pools aloft, which frequently pass overhead with clear skies in the wake of cold fronts. Such cold pools aloft (not necessarily coincident with the upper circulation center) may be associated with high lapse rates if skies are clear and surface heating couples the boundary layer with the cold air aloft. In such cases there is enough residual moisture in the postfrontal environment that deep convection ensues. In most cases the cloud base is high owing to lack of abundant moisture, while cloud tops are low because of a cool troposphere, giving rise to a low tropopause.

The mechanism by which these cold pool vortices form is quite unclear, because of a lack of quantitative observational studies. Since the midtropospheric environment in which they occur may be rich in vertical vorticity, they might result from simple vertical vortex tube stretching. The rarity of tornado touchdowns from these cold pool vortices may be associated with the relative weakness of the initial vorticity at low levels (as discussed by *Smith and Leslie* [1978]).

These events are distinct from those along and ahead of cold fronts. If a cold pool aloft is situated over a front rather than behind it, low-topped storms can develop in such an environment. Whereas wind shears usually are weak beneath cold pools in the postfrontal region, storms along and near fronts often arise in relatively highly sheared environments; funnel clouds and tornadoes developing in these conditions simply are shallow versions of supercells. (At the risk of being repetitious, it is the presence of a deep, persistent mesocyclone which defines a supercell, not the depth of convection. When the mesocyclonic circulation exists through a substantial fraction of the depth of the storm, it does not matter if the storm is relatively shallow; it is a supercell. Storms poleward of, say, 45° latitude often have low tops because the environment is relatively cold, with a correspondingly low tropopause.) The specific class of cold pool vortices to which we refer only arise more or less directly under the upper circulation center (where the vertical shear usually is weak) and well poleward of the surface cold front. Systematic investigation of such events has not been done, to our knowledge.

3.3. Gustnadoes

Very small scale, shallow vortices (Figure 6) may develop near the surface along outflow boundaries and/or cold fronts, with or without deep convection overhead [see *Idso*, 1974, 1975; *Meaden*, 1981; *Doswell*, 1985]. The boundary develops "lobes" and bulges, with cyclonic circulations at the cusps created by those lobes. Sometimes, for reasons that essentially are not known, those circulations become quite intense; at least as intense as weak tornadoes. If they are associated with a damaging outflow, they may create short, narrow zones of even more intense damage than is common along the rest of the outflow. They also can produce damage swaths along an otherwise nondamaging outflow.

Although we have no documentation for making this distinction, we propose that they are distinguishable from "landspouts" by remaining quite shallow. Virtually no circulations can be seen at cloud base, visually or on Doppler radar. Such events seem not to depend on the superpositioning of a developing updraft above them. If such a vortex is, indeed, deepened and intensified by an overriding updraft, we believe it will undergo a transition to a landspout. Obviously, considerably greater documentation and study of these events are needed.

3.4. Fair Weather Vortices

There is a substantial variety of distinct fair weather convective vortices, even ignoring the "dust devil" phenomenon [see *Idso*, 1974]. Dust devils arise in association with dry rather than moist convection, of course. (Interestingly, some citizens observing the deadly Cheyenne, Wyoming, tornado of July 16, 1979, thought they were seeing a dust devil; this confusion may have arisen because of the relative rarity of tornadoes in Wyoming, along with the absence of a visible condensation funnel for the early part of the tornado's life.)

Meteorologists operating on storm intercept teams have observed relatively long-lived funnel clouds in association with quite ordinary cumulus clouds (Figure 7). A rather different phenomenon has been observed on fair weather days, the so-called "horseshoe vortex" (Figure 8). These may arise in much the same way as "mountainadoes" [*Bergen*, 1976]: tilting and the associated stretching of an enhanced region of horizontal vorticity over some upward protruding object, or perhaps by an isolated updraft (a small cumulus-scale version of the process depicted by *Klemp* [1987, Figure 3a].

With most of these fair weather vortices, it seems unlikely they ever would reach damaging proportions at the surface, and so it is improbable that they would (or should) be

Fig. 6. An example of a circulation along a gust front (sometimes called "gustnadoes") near Welch, Texas, on May 23, 1982. In contrast to Figure 5, this cloud base is quite low, around 500 feet (150 m) or less. Photograph copyrighted 1982 by C. Doswell.

classified as tornadoes. Knowledge that they exist may be important in responding appropriately to citizen reports of such events, however.

4. CLASSIFICATION OF VORTICES

At present, our perspective has come to paraphrase Richardson's famous limerick about vortices: the extratropical cyclone contains mesolows, the convection within the vicinity of mesolows develops mesocyclones, the tornado cyclone develops within the mesocyclone, the tornado within the tornado cyclone, the subvortex (or suction vortex [see *Fujita*, 1971] within the tornado, and so on (presumably, to viscosity). Thus the processes associated with tornadoes (at least those developing from supercells) can be seen in a context of a larger vortex and contains smaller subprocesses within. In such a hierarchy of processes the boundaries between events can become blurry when observed in the natural world.

Forbes and Wakimoto [1983] have presented a quite insightful discussion on classification of tornadoes. We are basically in agreement with their conclusions, which advocate a more pragmatic approach to defining a tornado than implied by the glossary definition, namely, any damaging vortex associated with a convective storm, including its accompanying wind field, should be called a tornado. *Forbes and Wakimoto* [1983, p. 233] also suggested, and we agree, that "Damaging vortices not associated with thunderstorms [ought to be] considered tornadic vortices of a particular type." However, we believe the issues can be even more difficult to resolve than they have described. These difficulties arise from improvements in observations and understanding, so the problems are really the sign of progress.

Many supercells produce tornadoes from and/or near the so-called wall cloud (see section 2.1). It has been shown observationally and numerically [see *Rotunno and Klemp*, 1985] that the wall cloud arises from the admixture of outflow and inflow within the mesocyclone. Now, suppose the mesocyclonic circulation becomes so intense that it reaches damaging proportions, with a wall cloud reaching near, or perhaps down to, the surface. Is such a damaging circulation a tornado? It certainly meets the definition given in the introduction, as well as that advocated by Forbes and Wakimoto. Both storm intercepts and eyewitness accounts suggest that mesocyclonic vortices can be damaging whether or not they ever produce a visible cloud to the ground. Are such damaging events "straight line" winds? How large does the radius of curvature have to be to call an event "straight winds" as opposed to a tornado?

Moreover, what about a large wall cloud that spins out visible funnels that develop damaging ground circulations every few minutes (either one at a time or several at once) over a period of a few tens of minutes? Are we seeing one tornado with many subvortices, or are we seeing several different tornadoes? Again, such events have been observed and recorded, but how one classifies such an event seems unclear to us.

As visual observations of tornadoes accumulate, it is clear that tornadoes virtually never "skip" in the sense of the circulation "lifting and descending"; instead, the circulation at the surface may strengthen and weaken on time scales of

Fig. 7. An example of a fair weather vortex with an ordinary cumulus cloud near Sayre, Oklahoma, on May 17, 1983. Photograph copyrighted 1983 by C. Doswell.

Fig. 8. An example of a vortex associated with a dissipating cumulus cloud (sometimes referred to as a "horseshoe vortex") near Shamrock, Texas, on April 13, 1976. Photograph copyrighted 1976 by C. Doswell.

a few seconds or more, but a significant circulation typically remains on the ground for the lifetime of the event. The funnel cloud aloft associated with the event may be continuous during such a weakening and strengthening cycle, or it, too, may dissipate and redevelop. If the winds cease to be damaging as a result of a weakening circulation and then redevelop, is this redevelopment a new tornado or should we say that the gap is a "skip" in the path of a continuous tornado? If the answer depends on the distance and/or time between damage, is there a nonarbitrary way to establish criteria for making such classifications?

There are numerous movies and videos showing quite clearly the dissipation of one damaging funnel cloud/tornado with the nearly simultaneous development of another within close proximity. A ground survey of the track would probably reveal a continuous damage swath, perhaps with a small offset. Are these subvortices within a larger, more or less continuous tornado, or are we seeing two different tornadoes? This issue is complicated by the existence of multiple-vortex phases interspersed with single-vortex modes.

5. Discussion

A tornado, no matter how one chooses to define it, is a kinematic structure that renews itself from instant to instant via one or more dynamic processes. It is not a "thing" in the sense that a table or a book (neglecting atomic or molecular fluctuations) is the same from one moment to the next. Much confusion about tornadoes comes from thinking of tornadoes as objects rather than as the kinematic manifestation of dynamic processes. The actual physical processes are not heedful of our somewhat arbitrary classification schemes and, as scientists, we need constantly to remind ourselves that our understanding of tornadoes and tornadic storms can be clouded by an inappropriate classification scheme (see the discussion by *Doswell* [1991a]).

The only scientific justification for a classification scheme is if that scheme proves to be useful in developing our understanding and/or in application of that understanding. While we probably have muddied the waters by mentioning additional difficulties with event classification, we believe that an appreciation for classification problems is needed in any proper use of the data derived from classification.

The more we learn about tornadoes and tornadic storms, the more they seem to be terribly complicated processes. It is possible that some insight we have yet to find will simplify our understanding of tornadoes and tornadic storms. On the other hand, new observations may not result in some simple reconciliation but may raise new and even more confusing issues with which to deal. There is nothing that guarantees simplicity in nature.

Despite the confusion it has caused, however, our new understanding developed since the last symposium (as a result of radar, storm chasing, and numerical and laboratory modeling) has been applicable in both a research and an operational sense. The recognition of a range of processes at the scale of the convective storm and at the tornado scale has been valuable to our science and to society as a whole. It is likely that numerical cloud models soon will be able to resolve tornadic flows, offering the chance for new insights into tornadoes. As new operational and research observing systems are implemented, it is virtually certain that we shall come to know much more about nonsupercell events than at present. We close by noting that a considerable challenge confronts us in applying any new understanding of tornadoes and tornadic storms to benefit society; efforts to do so have been painfully slow, up to the present. We hope that operational deployment of new technologies will be associated with concomitant accelerations in the application of scientific understanding to serve society.

References

Bergen, W. R., Mountainadoes: A significant contribution to mountain windstorm damage?, *Weatherwise*, 29, 64–69, 1976.

Bluestein, H. B., The formation of a "landspout" in a "broken-line" squall line in Oklahoma, in *Preprints, 14th Conference on Severe Local Storms*, pp. 267–270, American Meteorological Society, Boston, Mass., 1985.

Bluestein, H. B., and C. R. Parks, A synoptic and photographic climatology of low-precipitation severe thunderstorms in the southern plains, *Mon. Weather Rev.*, 111, 2034–2046, 1983.

Brady, R. H., and E. J. Szoke, The landspout—A common type of northeast Colorado tornado, in *Preprints, 15th Conference on Severe Local Storms*, pp. 312–315, American Meteorological Society, Boston, Mass., 1988.

Browning, K. A., Airflow and precipitation trajectories within severe local storms which travel to the right of the winds, *J. Atmos. Sci.*, 21, 634–639, 1964.

Browning, K. A., The structure and mechanism of hailstorms, *Meteorol. Monogr.*, 38, 1–39, 1977.

Browning, K. A., and R. J. Donaldson, Jr., Airflow and structure of a tornadic storm, *J. Atmos. Sci.*, 20, 533–545, 1963.

Burgess, D. W., and R. A. Brown, The structure of a severe-right-moving thunderstorm: New single Doppler radar evidence, in *Preprints, 8th Conference on Severe Local Storms*, pp. 40–43, American Meteorological Society, Boston, Mass., 1973.

Burgess, D. W., and E. B. Curran, The relationship of storm type to environment in Oklahoma on 26 April 1984, in *Preprints, 14th Conference on Severe Local Storms*, pp. 208–211, American Meteorological Society, Boston, Mass., 1985.

Burgess, D. W., and R. P. Davies-Jones, Unusual tornadic storms in eastern Oklahoma on 5 December 1975, *Mon. Weather Rev.*, 107, 451–457, 1979.

Burgess, D. W., and R. J. Donaldson, Jr., Contrasting tornadic storm types, in *Preprints, 11th Conference on Severe Local Storms*, pp. 189–192, American Meteorological Society, Boston, Mass., 1979.

Burgess, D. W., and L. R. Lemon, Severe thunderstorm detection by radar, in *Radar in Meteorology*, edited by D. Atlas, pp. 619–647, American Meteorological Society, Boston, Mass., 1990.

Burgess, D. W., V. T. Wood, and R. A. Brown, Mesocyclone evolution statistics, in *Preprints, 12th Conference on Severe Local Storms*, pp. 422–424, American Meteorological Society, Boston, Mass., 1982.

Carbone, R. E., A severe frontal rainband, I, Stormwide hydrodynamic structure, *J. Atmos. Sci.*, 39, 258–279, 1982.

Carbone, R. E., A severe frontal rainband, II, Tornado parent vortex circulation, *J. Atmos. Sci.*, 40, 2639–2654, 1983.

Church, C. R., J. T. Snow, G. L. Baker, and E. M. Agee, Characteristics of tornado-like vortices as a function of swirl ratio: A laboratory investigation, *J. Atmos. Sci.*, 36, 1755–1766, 1979.

Cooley, J. R., Cold air funnel clouds, *Mon. Weather Rev.*, 106, 1368–1372, 1978.

Davies, J. M., and R. H. Johns, Some wind and instability parameters associated with strong and violent tornadoes, 1, wind shear and helicity, this volume.

Davies-Jones, R. P., A new look at the vorticity equation with application to tornadogenesis, in *Preprints, 12th Conference on Severe Local Storms*, pp. 249–252, American Meteorological Society, Boston, Mass., 1982.

Davies-Jones, R. P., Tornado dynamics, in *Thunderstorms: A Social, Scientific, and Technological Documentary*, vol. 2, *Thunderstorm Morphology and Dynamics*, 2nd ed., edited by E. Kessler, pp. 197–236, University of Oklahoma Press, Norman, 1986.

Davies-Jones, R. P., and H. E. Brooks, Mesocyclogenesis from a theoretical perspective, this volume.

Doswell, C. A., III, Comments on "Photographic documentation of some distinctive cloud forms observed beneath a large cumulonimbus", *Bull. Am. Meteorol. Soc.*, 64, 1389–1390, 1983.

Doswell, C. A., III, The operational meteorology of convective weather, vol. II, Storm-scale analysis, *NOAA Tech. Memo. ERL ESG-15*. (Available from Natl. Severe Storms Lab., Norman, Okla.)

Doswell, C. A., III, Comments on "Mesoscale convective patterns of the southern High Plains", *Bull. Am. Meteorol. Soc.*, 72, 389–390, 1991a.

Doswell, C. A., III, A review for forecasters on the application of hodographs to forecasting severe thunderstorms, *Natl. Weather Dig.*, 16, 2–16, 1991b.

Doswell, C. A., III, A. R. Moller, and R. Przybylinski, A unified set of conceptual models for variations on the supercell theme, in *Preprints, 16th Conference on Severe Local Storms*, pp. 40–45, American Meteorological Society, Boston, Mass., 1990.

Fankhauser, J. C., G. M. Barnes, L. J. Miller, and P. M. Rostkowski, Photographic documentation of some distinctive cloud forms observed beneath a large cumulonimbus, *Bull. Am. Meteorol. Soc.*, 64, 450–462, 1983a.

Fankhauser, J. C., G. M. Barnes, and L. J. Miller, Response to comments on "Photographic documentation of some distinctive cloud forms observed beneath a large cumulonimbus", *Bull. Am. Meteorol. Soc.*, 64, 1391–1392, 1983b.

Foote, G. B., and H. W. Frank, Case study of a hailstorm in

Colorado, III, Airflow from triple-Doppler measurements, *J. Atmos. Sci.*, *40*, 686–707, 1983.

Forbes, G. S., On the reliability of hook echoes as tornado indicators, *Mon. Weather Rev.*, *109*, 1457–1466, 1981.

Forbes, G. S., and R. M. Wakimoto: A concentrated outbreak of tornadoes, downbursts and microbursts, and implications regarding vortex classification, *Mon. Weather Rev.*, *111*, 220–235, 1983.

Fujita, T., Detailed analysis of the Fargo tornadoes of June 20, 1957, *Weather Bur. Res. Pap. 42*, 67 pp., U.S. Dept. of Commerce, U.S. Govt. Printing Office, Washington, D. C., 1960.

Fujita, T. T., Proposed mechanism of suction spots accompanied by tornadoes, in *Preprints, 7th Conference on Severe Local Storms*, pp. 208–213, American Meteorological Society, Boston, Mass., 1971.

Fujita, T. T., Tornadoes and downbursts in the context of generalized planetary scales, *J. Atmos. Sci.*, *38*, 1511–1534, 1981.

Gall, R. L., A linear analysis of the multiple vortex phenomenon in simulated tornadoes, *J. Atmos. Sci.*, *40*, 2010–2024, 1983.

Garrett, R. A., and V. D. Rockney, Tornadoes in northeastern Kansas, May 19, 1960, *Mon. Weather Rev.*, *110*, 118–135, 1962.

Golden, J. H., Waterspouts and tornadoes over south Florida, *Mon. Weather Rev.*, *99*, 146–154, 1971.

Golden, J. H., The life cycle of Florida Keys' waterspouts, *J. Appl. Meteorol.*, *13*, 676–692, 1974.

Golden, J. H., and B. J. Morgan, The NSSL-Notre Dame tornado intercept program, spring 1972, *Bull. Am. Meteorol. Soc.*, *53*, 1178–1180, 1972.

Gonski, R. F., B. P. Woods, and W. D. Korotky, The Raleigh tornado—28 November 1988: An operational perspective, in *Preprints, 12th Conference on Weather Analysis and Forecasting*, pp. 173–178, American Meteorological Society, Boston, Mass., 1989.

Hane, C. E., and P. S. Ray, Pressure and buoyancy fields derived from Doppler radar data in a tornadic thunderstorm, *J. Atmos. Sci.*, *42*, 18–35, 1985.

Holle, R. L., and M. W. Maier, Tornado formation from downdraft interaction in the FACE network, *Mon. Weather Rev.*, *108*, 1010–1028, 1980.

Huschke, R. E., *Glossary of Meteorology*, p. 505, American Meteorological Society, Boston, Mass., 1959.

Idso, S. B., Tornado or dust devil: The enigma of desert whirlwinds, *Am. Sci.*, *62*, 530–541, 1974.

Idso, S. B., Whirlwinds, density currents, and topographic disturbances: A meteorological melange of intriguing interactions, *Weatherwise*, *28*, 61–65, 1975.

Imy, D. A., and D. W. Burgess, The structural evolution of a tornadic supercell with a persistent mesocyclone, in *Preprints, 25th Conference on Radar Meteorology*, pp. 408–411, American Meteorological Society, Boston, Mass., 1991.

Klemp, J. B., Dynamics of tornadic thunderstorms, *Annu. Rev. Fluid Mech.*, *19*, 369–402, 1987.

Klemp, J. B., and R. B. Wilhelmson, The simulation of three-dimensional convective storm dynamics, *J. Atmos. Sci.*, *35*, 1070–1096, 1978.

Lemon, L. R., New severe thunderstorm radar identification techniques and warning criteria: A preliminary report, *NOAA Tech. Memo. NWS NSSFC-1*, 60 pp. (Available as *PB-273049* from Natl. Tech. Inf. Serv., Springfield, Va.)

Lemon, L. R., and C. A. Doswell III, Severe thunderstorm evolution and mesocyclone structure as related to tornadogenesis, *Mon. Weather Rev.*, *107*, 1184–1197, 1979.

Marwitz, J. D., The structure and motion of severe hailstorms, II, Multicell storms, *J. Appl. Meteorol.*, *11*, 180–188, 1972.

McCaul, E. W., Jr., Buoyancy and shear characteristics of hurricane-tornado environments, *Mon. Weather Rev.*, *119*, 1954–1978, 1991.

Meaden, G. T., Whirlwind formation at a sea breeze front, *Weather*, *36*, 47–48, 1981.

Moller, A. R., Comments on "Photographic documentation of some distinctive cloud forms observed beneath a large cumulonimbus", *Bull. Am. Meteorol. Soc.*, *64*, 1390–1391, 1983.

Moller, A., C. Doswell, J. McGinley, S. Tegtmeier, and R. Zipser, Field observations of the Union City tornado in Oklahoma, *Weatherwise*, *27*, 68–77, 1974.

Moller, A. R., C. A. Doswell III, and R. Przybylinski, High-precipitation supercells: A conceptual model and documentation, in *Preprints, 16th Conference on Severe Local Storms*, pp. 52–57, American Meteorological Society, Boston, Mass., 1990.

Newton, C. W., Dynamics of severe convective storms, *Meteorol. Monogr.*, *5*, 33–58, 1963.

Nolen, R. H., A radar pattern associated with tornadoes, *Bull. Am. Meteorol. Soc.*, *40*, 277–279, 1959.

Rotunno, R., A study of tornado-like vortex dynamics, *J. Atmos. Sci.*, *36*, 140–155, 1979.

Rotunno, R., Tornadoes and tornadogenesis, in *Mesoscale Meteorology and Forecasting*, edited by P. S. Ray, pp. 414–436, American Meteorological Society, Boston, Mass., 1986.

Rotunno, R., and J. B. Klemp, On the rotation and propagation of simulated supercell thunderstorms, *J. Atmos. Sci.*, *42*, 271–292, 1985.

Schlesinger, R. E., A three-dimensional numerical model of an isolated thunderstorm: Preliminary results, *J. Atmos. Sci.*, *32*, 835–850, 1975.

Smith, R. K., and L. M. Leslie, Tornadogenesis, *Q. J. R. Meteorol. Soc.*, *104*, 189–198, 1978.

Stout, G. E., and F. A. Huff, Radar records an Illinois tornado, *Bull. Am. Meteorol. Soc.*, *34*, 281–284, 1953.

Wakimoto, R. M., and J. W. Wilson, Non-supercell tornadoes, *Mon. Weather Rev.*, *117*, 1113–1140, 1989.

Ward, N. B., Radar and surface observations of tornadoes on May 4, 1961, in *Proceedings, 9th Weather Radar Conference*, pp. 175–180, American Meteorological Society, Boston, Mass., 1961.

Ward, N. B., The exploration of certain features of tornado dynamics using a laboratory model, *J. Atmos. Sci.*, *29*, 1194–1204, 1972.

Weisman, M. L., and J. B. Klemp, The structure and classification of numerically simulated convective storms in directionally varying wind shears, *Mon. Weather Rev.*, *112*, 2479–2498, 1984.

Weisman, M. L., and J. B. Klemp, Characteristics of isolated convective storms, in *Mesoscale Meteorology and Forecasting*, edited by P. S. Ray, pp. 331–358, American Meteorological Society, Boston, Mass., 1986.

Wicker, L. J., A numerical simulation of a tornado-scale vortex in a three-dimensional cloud model, Ph.D. thesis, 264 pp., Univ. of Ill., Urbana, 1990.

Wilczak, J. M., T. W. Christian, D. E. Wolfe, R. J. Zamora, and B. Stankov, Observations of a Colorado tornado, Part I, Mesoscale environment and tornadogenesis, *Mon. Weather Rev.*, *120*, 497–520, 1992.

Lightning in Tornadic Storms: A Review

DONALD R. MACGORMAN

National Severe Storms Laboratory, NOAA, Norman, Oklahoma 73069

1. INTRODUCTION

There have been many reports of unusual lightning characteristics in tornadic storms. For example, eyewitnesses have reported scorching beneath tornado funnels, a steady or rapidly oscillating glow inside funnels, rapidly recurring small patches of light on the side of the thunderstorm, or unusually high or low flash rates [e.g., *Church and Barnhart*, 1979; *Vaughan and Vonnegut*, 1976; *Vonnegut and Weyer*, 1966]. It is difficult, however, to quantify relationships between lightning and tornadic storms from these eyewitness reports.

Prior to 1975 most quantitative measurements of lightning in tornadic storms examined the rates and characteristics of sferics, the electromagnetic noise radiated by lightning. These measurements were difficult to interpret, because results depended partly on the characteristics of the sferics receiver, and because sferics instrumentation provided at best only the bearing to the lightning flash generating the sferics and often provided no location information at all. Furthermore, many of the sferics studies lacked radar data, and only one had any Doppler radar data. Equally important, there was not an adequate appreciation of the variety of storms that produce tornadoes.

Some of these limitations were overcome when systems were developed to map the location of lightning flashes. For example, beginning in the late 1970s, various organizations began deploying newly developed systems for automatically mapping where lightning channels strike ground over ranges of a few hundred kilometers [e.g., *Krider et al.*, 1980; *Orville et al.*, 1983; *Bent and Lyons*, 1984; *MacGorman and Taylor*, 1989]. Now coverage by these systems is continual across the contiguous United States, with errors in strike locations typically less than 10 km [e.g., *Mach et al.*, 1986; *MacGorman and Rust*, 1989].

In this paper we first review what was learned about tornadic storms from sferics studies and then consider what has been learned from analysis of data from lightning mapping systems.

2. LIGHTNING TERMINOLOGY

This section explains lightning terminology and concepts that will be used in the rest of the paper. Lightning can be classified into two types: (1) Cloud-to-ground lightning has at least one channel spanning from the cloud to the ground. (2) Intracloud lightning does not have a channel to ground. Because the locations of lightning channels relative to visible cloud boundaries are difficult to detect remotely, scientists who study lightning and storm electrification normally group in-cloud, cloud-to-air, and cloud-to-cloud lightning together in the intracloud lightning classification. Cloud-to-ground lightning can be further classified by the polarity of charge that it effectively lowers to ground: Positive cloud-to-ground lightning lowers positive charge; negative cloud-to-ground lightning lowers negative charge. Most cloud-to-ground flashes are negative.

A cloud-to-ground lightning flash usually begins inside the cloud and is first apparent when a faint channel, called the stepped leader, moves from the cloud to the ground in jumps roughly 50–100 m long. When the stepped leader connects with the ground, a bright pulse moves back up the lightning channel in a process called a return stroke. After a pause of roughly 20–150 ms, another leader can travel back down the already established lightning channel, followed by another return stroke. The combination of a leader and return stroke is called a stroke. All strokes going through essentially the same channel to ground make up a single cloud-to-ground flash. There can be anywhere from one to a few tens of strokes in a flash.

Sferics are generated by changes in the vector electric current, the frequency of the radiation depending on the time scale of the changes. Since currents in a lightning channel increase faster than they decay, a lightning channel will generally radiate higher frequencies during an initial current surge than during its decay. The time required for a current pulse to propagate through a channel segment depends on the length of the segment, so longer channels radiate more energy at lower frequencies, if all else is equal.

The Tornado: Its Structure, Dynamics, Prediction, and Hazards.
Geophysical Monograph 79
This paper is not subject to U.S. copyright. Published in 1993 by the American Geophysical Union.

The spectrum of sferics from a typical return stroke peaks at roughly 5 kHz. Most strong signals at frequencies below about 30 kHz are from cloud-to-ground return strokes and so occur in discrete bursts, separated by tens of milliseconds. At frequencies near or above 1 MHz, the amplitudes of sferics from return strokes, from other processes of a cloud-to-ground flash, and from intracloud flashes are all comparable. Furthermore, at these higher frequencies, sferics are more numerous and more continual throughout a lightning flash, and relatively few are associated with return strokes.

The vertical component of sferics measured at the surface of the conducting Earth is generally much larger than the horizontal, even if the two components are comparable at the source. Any horizontal component that exists is reduced by currents that are induced in the ground by the sferics. Therefore a sferics receiver at the ground receives little of the energy radiated by horizontal components of current change. Airborne receivers are less affected by currents induced in the ground and so are better able to detect the horizontal components of sferics.

3. Sferics Studies

Initial studies of sferics from tornadic storms examined radio bands from 10 to 500 kHz. (Sferics normally are classified by the frequency of the receiver used to detect them.) *Dickson and McConahy* [1956] found that sferics rates detected by a 10-kHz receiver often increased as storms grew rapidly taller, but rates peaked earlier in storms that had more violent severe weather. Rates peaked about 1.5 hour before tornadoes and decreased during tornadoes to about 40% of peak value. At frequencies of 150 kHz and higher, sferics rates were found to increase until tornadoes were produced, and rates became exceptionally high during periods when tornadoes occurred [*Jones*, 1951, 1958; *Jones and Hess*, 1952; *Kohl*, 1962; *Kohl and Miller*, 1963]. *Kohl and Miller* [1963] observed that sferics rates at 150 kHz usually peaked during severe weather and began decreasing prior to the end of severe weather. As might be expected from the above observations, *Jones and Hess* [1952] found that the ratio of the numbers of higher-frequency to lower-frequency sferics increased for sferics of large amplitude when storms were more severe; the ratio was 1:20 before and after tornadoes and in nonsevere thunderstorms versus 1:1 during tornadoes.

Later studies at frequency bands up to 150 MHz [*Silberg*, 1965; *Taylor*, 1973; *Stanford et al.*, 1971; *Johnson et al.*, 1977] found that the increase in sferics rates during severe weather was greatest at frequencies above 1 MHz. This is shown in Figure 1, which presents a summary of Taylor's data concerning the dependence of sferics rates on receiver frequency. *Silberg* [1965] and *Taylor* [1973] also found that the energy radiated as sferics (not shown) increases at higher frequencies as storm severity increases; radiated energy peaks at roughly 5 kHz for nonsevere storms [*Taylor*, 1972], but the energy at higher frequencies gradually increases with

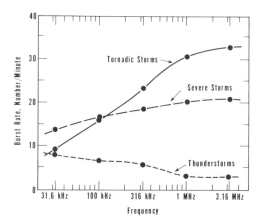

Fig. 1. Average sferics burst rate as a function of receiver frequency for various storm classifications [from *MacGorman et al.*, 1989]. Data are from four tornadic storms, three nontornadic severe storms, and seven nonsevere thunderstorms. Sferics burst rates were measured at 10 receiver frequencies between 10 kHz and 137 MHz, indicated by dots. There was little change above 3 MHz, so higher frequencies are not shown.

increasing severity, until for tornadic storms, the energy is independent of frequency.

There were exceptions to all of these findings. In several cases, high sferics rates did not occur in tornadic storms [*Ward et al.*, 1965; *Stanford et al.*, 1971; *Lind et al.*, 1972; *Taylor*, 1973; *Greneker et al.*, 1976; *Johnson et al.*, 1977; *MacGorman et al.*, 1989]. Taylor, who examined the largest number of storms, found that roughly 20% of tornadic storms did not have high sferics rates [*MacGorman et al.*, 1989]. It also was possible for high sferics rates to occur in storms without tornadoes [*Ward et al.*, 1965; *Stanford et al.*, 1971; *MacGorman et al.*, 1989]. Taylor found that although only 23% of nontornadic severe storms and 1% of nonsevere storms had sferics rates comparable to those of tornadic storms, 70% of the storms with high sferics rates were not tornadic, because the number of nonsevere storms was much larger than the number of tornadic storms [*MacGorman et al.*, 1989]. Some tornadic storms without high sferics rates had only weak tornadoes [e.g., *Johnson et al.*, 1977], and the combined effect of several simultaneous storms explains some nonsevere storms with high sferics rates [e.g., *MacGorman et al.*, 1989]. However, these situations did not account for all exceptions to the rule that 3-MHz sferics rates are much higher in tornadic storms than in nonsevere storms.

The most obvious inference from the increasing number of sferics at high frequencies and the decreasing or constant number at low frequencies was that intracloud flash rates usually increase while cloud-to-ground lightning return strokes decrease or remain fairly constant during the tornadic stage of storms. This interpretation was further supported by the theoretical work of *Stanford* [1971] and the airborne observations of *Shanmugam and Pybus* [1971]. Both reported that horizontal polarization of sferics increased and vertical polarization decreased, indicating a greater preponderance of horizontal lightning channels, as the sferics rate and storm severity increased.

Other observations also supported this interpretation. *Jones* [1958] reported watching a storm when it was producing high sferics rates about 1 hour before a tornado; no ground flashes could be seen, but relatively dim circular patches, about 0.5 km in diameter, were illuminated in rapid succession on the side of the storm. He suggested that these were an unusual form of intracloud lightning. Electric field records for this same storm indicated that the lightning flash rate was 12 min^{-1} during the tornado [*Gunn*, 1956]. *Vonnegut and Moore* [1957] inferred from recordings of the closest electric field sensor to the tornado that there were no cloud-to-ground flashes near the tornado. Unusually high lightning flash rates were also observed in two other tornadic storm systems by *Orville and Vonnegut* [1977] and *Turman and Tettelbach* [1980], who used satellite-borne optical detectors. Films of 17 tornadoes and one funnel cloud documented low ground flash rates near tornadoes; *Davies-Jones and Golden* [1975a] (see also *Davies-Jones and Golden* [1975b, c], *Colgate* [1975], and *Vonnegut* [1975] for a discussion) found cloud-to-ground lightning flashes near only two of the 18 filmed vortices. One of these two was a tornado with 12 ground flashes; the other was a funnel cloud with two ground flashes.

From studies of sferics, however, it appears that the tornado itself is not likely to be the cause of unusual sferics rates. For example, *Jones* [1965] noted that sferics rates can be unusually high 60–90 min before tornadoes, although they may increase further when tornadoes occur. *Scouten et al.* [1972] further analyzed the data of *Jones* [1958] at 150 kHz and concluded that touchdown of the tornado did not affect the high sferics rate. Although larger-amplitude sferics clustered more closely about a central core near the azimuth of the tornado, sferics as a whole were spread among azimuths spanning most of the storm. A similar lack of correlation with the time and location of tornado touchdown was reported for sferics in three bands between 300 kHz and 3 MHz [*Taylor*, 1973] and for sferics at 1–50 kHz [*Brown and Hughes*, 1978]. *Brown and Hughes* [1978] also noted a lack of obvious relationships with the tornadic vortex signature [*Brown et al.*, 1978]. In a study of sferics burst rates at high frequencies (30–300 MHz) in two tornadic storms, *Johnson* [1980] found that the tornadoes began near the time of peaks in the sferics burst rate but were not coincident with any peaks.

Several studies considered changes in tornadic storms that were correlated with the high sferics rates. For example, several investigators found periodicities of 10–60 min in the sferics data and suggested that these resulted from the cyclical intensification and decay of thunderstorm cells [*Kohl and Miller*, 1963; *Stanford et al.*, 1971; *Shanmugam and Pybus*, 1971; *Lind et al.*, 1972; *Trost and Nomikos*, 1975]. *Taylor* [1973] reported that sferics at high frequencies originated along azimuths to low-level reflectivity cores and that sferics rates were larger from cores with higher reflectivity. Furthermore, *Brown and Hughes* [1978] found that the pattern in the number of VLF sferics (which are produced predominantly by return strokes) from the Union City tornadic storm was similar both in time and space to patterns in reflectivity at low levels of the storm.

Although the sferics studies clearly demonstrated that lightning tended to evolve in characteristic patterns in tornadic storms, there were a number of uncertainties. (1) There were uncertainties in extracting lightning flash type and rates from sferics data. (2) No sferics study examined the evolution of mesocyclones or considered differences in the types of storms that produced tornadoes in order to examine why sferics rates evolved differently. (3) No sferics study examined the three-dimensional vector wind field for tornadic storms.

4. LIGHTNING MAPPING STUDIES

Studies using lightning mapping systems have begun to address some of the shortcomings of the sferics studies. However, only two of the mapping studies thus far have included intracloud lightning; all others have examined only cloud-to-ground lightning. Furthermore, mapping studies have found considerable variability in the relationship of tornadoes to cloud-to-ground flash rates. In two tornadic storms, ground flash rates were small before and during tornadoes [*Orville et al.*, 1982; *MacGorman et al.*, 1989]. In three, ground flash rates peaked during tornadoes [*Kane*, 1991; *MacGorman and Nielsen*, 1991; *Keighton et al.*, 1991]. In two, tornadoes occurred after ground flash rates had increased but when ground flash rates were not at their maximum [*Kane*, 1991; *MacGorman et al.*, 1985].

Results from two of the studies suggest some reasons for this variability. In the Binger tornadic storm studied by *MacGorman et al.* [1989], negative ground flash rates within 10 km of the mesocyclone center were less than 1 min^{-1} until the tornado began dissipating. Negative cloud-to-ground flash rates reached a relative maximum after the last tornado, as its mesocyclone core dissipated, and reached an absolute maximum (about 4 min^{-1}) as the last mesocyclone core in the storm dissipated (see Figure 2). Intracloud flash rates reached an absolute maximum of approximately 14 min^{-1} during the most violent tornado and were well correlated with low-level cyclonic shear when the shear was greater than 1×10^{-2} s^{-1} during the tornado. (Cyclonic shear is one half vertical vorticity for ideal solid-body rotation. In calculating this shear, tornadic winds were ignored; shear was measured across the diameter of maximum tangential wind speed for the mesocyclone, typically a distance of a few kilometers.) After the last tornado, when there was a large decrease in low-level cyclonic shear but the mesocyclone was still strong at middle levels, intracloud flash rates were only 0–2 min^{-1}.

In the Edmond storm studied by *MacGorman and Nielsen* [1991] there were no intracloud data, but negative ground

176 REVIEW OF LIGHTNING IN TORNADIC STORMS

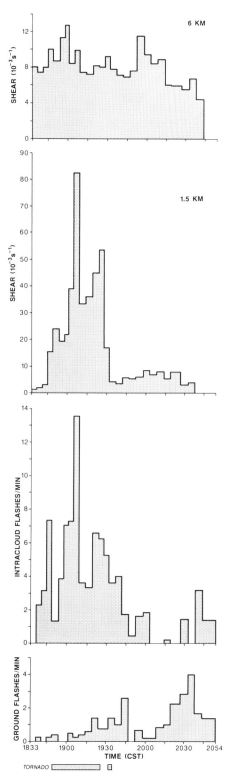

Fig. 2. Time series plots of cyclonic shear at the 1.5- and 6-km levels and of ground and intracloud flash rates within 10 km of the mesocyclone core in the Binger storm of May 22, 1981 [from *MacGorman et al.*, 1989]. The bars on the bottom indicate when tornadoes occurred.

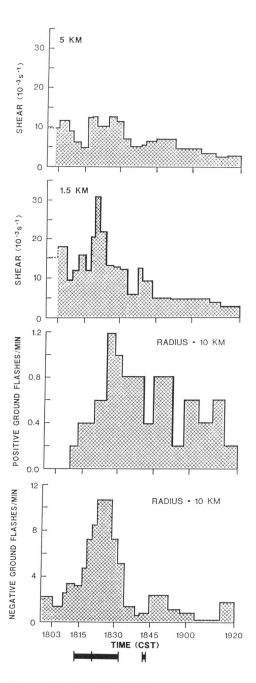

Fig. 3. Time series plots of cyclonic shear at the 1.5- and 5-km levels and of ground flash rates within 10 km of the mesocyclone center in the Edmond storm of May 8, 1986 [from *MacGorman and Nielsen*, 1991]. The bars on the bottom indicate when tornadoes occurred.

flash rates within 10 km of the mesocyclone center evolved differently than in the Binger storm. Rates increased to a peak of 11 min^{-1} during tornadic activity and appeared to be correlated with cyclonic shear at the 5 km level (Figure 3). Ground strike locations tended to cluster in the vicinity of a reflectivity core north of the mesocyclone during tornadic activity and to be more scattered before and after the tornadic stage of the storm (Figure 4). *Keighton et al.* [1991] observed a similar increase in scattering of lightning strikes after tornadic activity on another day.

MacGorman and Nielsen [1991] suggested that observed differences between the Binger and Edmond storms might account for the differences in the evolution of their lightning flash rates. The Binger storm was a classic supercell storm, with a prominent weak-echo region, a strong, deep updraft core, and a long-lasting mesocyclone having a family of mesocyclone cores [*Burgess et al.*, 1982]. The Edmond storm also was a supercell storm but was weaker than the Binger storm by almost any measure: the duration of its supercell stage was only 30% that of the Binger storm; its mesocyclone was weaker, shallower, and shorter lived; and features such as the weak-echo vault and tornadic vortex signature were not as pronounced. In fact, the Edmond storm did not form a tornado until the outflow boundary of another storm began to overtake it from the west. *MacGorman and Nielsen* [1991] suggested that the increase in flash rates in both storms when the mesocyclone was strongest probably was a result of the strong updraft that existed then. They also suggested, however, that the deeper and stronger updraft core and weak echo region of the Binger storm probably delayed ground flash activity and caused intracloud activity to dominate.

To understand the hypothesis that *MacGorman and Nielsen* [1991] offered to explain why ground flash rates behaved differently in these two storms, it is necessary first to understand a little about thunderstorm electrification. One important class of thunderstorm electrification mechanisms relies on microphysical particle interactions to place charge of one sign on graupel and charge of the opposite sign on cloud ice. Such mechanisms are affected (1) by factors that control the sizes and numbers of graupel and ice particles that collide, and (2) by factors that control subsequent transport of the charged particles into different regions of a thunderstorm. Factors affecting the number of relevant collisions include the strength and depth of the updraft and the microphysics of the storm. Subsequent transport is dominated initially by the sedimentation of graupel but may become dominated by the relative velocity of the winds in different regions as the graupel and ice crystals move farther apart. Thunderstorm measurements [e.g., *Krehbiel*, 1981; *Byrnne et al.*, 1983; *Chauzy et al.*, 1985; *Dye et al.*, 1986; *Koshak and Krider*, 1989] suggest that most of the main negative charge at middle levels of the storm is on graupel and precipitation, while the main positive charge is generally on ice crystals at colder temperatures (the next section will discuss one possible exception).

Several investigators have observed that a strengthening updraft at altitudes colder than $-20°C$ increases flash rates (see discussion by *MacGorman et al.* [1989]). This probably explains the observed increases in flash rates during the tornadic stage of the Binger and Edmond storms. *MacGorman et al.* [1989] also observed increasing areas of large reflectivity at heights between 6 km and 8 km shortly before

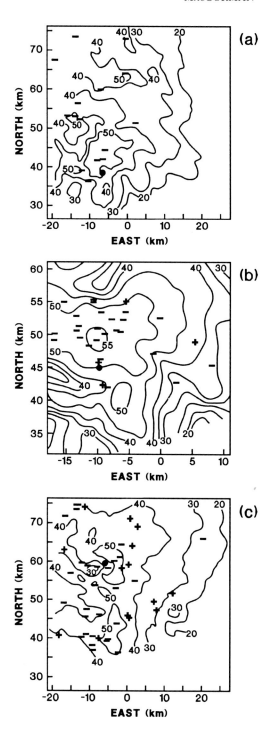

Fig. 4. Lightning ground strikes superimposed on radar reflectivity at the 3-km level of the May 8, 1986, Edmond storm during three periods: (*a*) 1802–1806 CST, (*b*) 1814–1817 CST, and (*c*) 1834–1838 CST [from *MacGorman and Nielsen*, 1991]. There was a tornado only during the period in Figure 4*b*. Minuses indicate the strike point of flashes that effectively lowered negative charge to ground; pluses, flashes that effectively lowered positive charge. The large dot marks the center of the mesocyclone core. Radar reflectivity is labeled in dBZ. Note that Figure 4*b* has a different distance scale.

and during periods of high flash rates. They suggested that the increased flash rates probably were caused by increased particle interactions leading to reflectivity growth at 7–9 km in and near the strong updraft. Similar correlations of lightning rates with the horizontal area of large reflectivity at midlevels of storms have been observed, for example, by Larson and Stansbury [1974], Lhermitte and Krehbiel [1979], and Keighton et al. [1991].

One of the primary problems posed by the Binger and Edmond storms is explaining why ground flash rates were so small near and during the time when total and intracloud flash rates were large in the Binger storm and why this apparently was not true for the Edmond storm. MacGorman et al. [1989] suggested that the strong, deep updraft and the resulting pronounced weak-echo region of the Binger storm enhanced production of intracloud flashes and delayed or suppressed production of ground flashes because of effects on the thunderstorm charge distribution.

For example, one primary effect on the charge distribution is to keep negative charge higher than in most storms. This would increase the energy required for lightning to span the distance to ground and would decrease the electric field at most heights below the negative charge. Negative charge is higher for three reasons. (1) The strong updraft core rapidly lifts all but the largest particles to upper levels of the storm. (2) Large updraft speeds do not allow most hydrometeors to remain long in a given layer. The short residence time prevents hydrometeors from acquiring much charge and prevents positive and negative particles from moving far apart after becoming charged. Therefore there is little net negative or positive charge in a given layer of a strong updraft. (3) Temperatures in stronger updraft cores are generally warmer, and this causes the ice processes thought to be responsible for charging (see discussion of Illingworth [1985]) to occur somewhat higher.

A second important effect of a strong, deep updraft is that regions of net positive and negative charge would be relatively close together near the top of the storm updraft. This occurs for two reasons. (1) As noted above, sedimentation and differing relative velocities of positively and negatively charged particles would not have time to separate the charges very far in a fast updraft. (2) The vertical shear of horizontal winds in a classic supercell storm causes graupel to fall to the side of the updraft core, so much of the storm-scale separation of charge occurs outside the updraft.

When two charged regions are closer together, less net charge is needed in each region to create electric field magnitudes sufficient to initiate lightning between them. This enhances intracloud flash rates. It also reduces the negative charge available for cloud-to-ground lightning near the updraft, because the threshold for initiating lightning regulates the net negative (and positive) charge that can be separated. This, combined with the unusually large height of negative charge discussed above, makes cloud-to-ground lightning much less likely to occur near a very strong, deep updraft core. Ground flash rates increase later, as the negative charge moves closer to ground and as the positive and

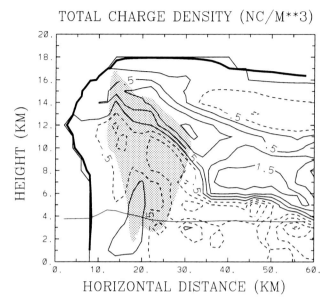

Fig. 5. Total space charge density in a southwest-northeast vertical cross section through the main updraft, from a numerical simulation of the Binger storm of May 22, 1981. Solid contours indicate zero or positive net space charge density; dashed contours, negative net charge density. The contour interval is 0.5 nC m^{-3}. Stippling delineates the region of updraft exceeding 5 m s^{-1}. The thin horizontal curve near a height of 4 km indicates the melting level. The bold curve outlines the cloud and precipitation boundary.

negative charges move farther apart (thereby increasing the amount of net negative charge that can be separated before causing intracloud lightning).

Since MacGorman et al. [1989] first suggested this hypothesis, there has been support from two sources. First, the lowest charge detected by the only published balloon-borne electric field sounding through a mesocyclone updraft was a region of negative charge starting at 9.5 km above mean sea level, where the temperature was approximately −37°C in the storm's environment and −31°C in the updraft [Rust et al., 1990]. This lower boundary is 3–5 km higher, and at environmental temperatures 20°–30°C colder, than the lower boundary normally reported for the main negative charge in continental thunderstorms. Second, Ziegler and MacGorman [1990] used a kinematic retrieval that included electrification processes [Ziegler et al., 1991] to model the electric field and charge structure of the Binger storm. In agreement with the balloon-borne measurements, the lower boundary of negative charge inside the updraft core in the retrieved charge distribution was elevated to a higher altitude than normal (Figure 5).

MacGorman and Nielsen [1991] suggested that the charge distribution near the updraft of the Edmond storm was more like that of an ordinary thunderstorm and that cloud-to-ground flashes therefore were able to occur frequently when the mesocyclone was strong. They concluded that negative charge was lower, a larger fraction of oppositely charged particles had time to separate, and positive and negative charge were farther apart because (1) the updraft core was weaker, shallower, and shorter lived, (2) the resulting weak-echo region was shallower and contained larger reflectivities, and (3) the 55-dBZ reflectivity cores extended to a height of only 5–6 km in the Edmond storm versus over 10 km in the Binger storm. Similarly, *Keighton et al.* [1991] observed another storm in which cloud-to-ground flash rates did not decrease during the tornadic stage and noted that it also had a weaker mesocyclone than the Binger storm.

5. Positive Cloud-to-Ground Lightning in Tornadic Storms

Data on the less common flashes that effectively lower positive charge to ground (+CG flashes) have been studied in detail for only one tornadic storm, the Edmond storm investigated by *MacGorman and Nielsen* [1991]. (Reliable positive cloud-to-ground data were not available for the Binger storm or for other tornadic storms before 1983.) In the Edmond storm, positive cloud-to-ground flashes began to occur a few minutes before the tornado touched down, and positive ground flash rates increased to their maximum value during the tornado, near the time when negative ground flash rates also peaked (see Figure 3). *MacGorman and Nielsen* [1991] noted from watching ground flash displays for several tornadic storms in real time that positive ground flashes often occurred during tornadoes. Sometimes a few positive ground flashes were clustered on the southern flank of storms in which negative cloud-to-ground lightning still dominated ground flash activity (as in the Edmond storm), and sometimes most of the ground flashes in the storm were positive ground flashes.

Although many severe or tornadic storms do not have large numbers of positive ground flashes, preliminary evidence suggests that when relatively high flash rates and densities of positive cloud-to-ground lightning occur, they are associated with severe weather. *Reap and MacGorman* [1989] reported in a climatological study that there was a correlation between a high density of positive ground flashes and severe weather. Furthermore, in August 1990 and spring 1991, *MacGorman and Burgess* [1991] observed several storms that were unusual because (1) most cloud-to-ground flashes were positive ground flashes and (2) ground strike points of positive cloud-to-ground lightning occurred in dense clusters, much like the clusters of negative ground flashes observed in most electrically active storms. (Positive ground flashes are generally infrequent and diffuse.) All of the storms with frequent positive ground flashes produced large hail, and many produced tornadoes. Similar observations were reported by *Rust et al.* [1985], *Curran and Rust* [1992], and *Branick and Doswell* [1992].

These storms with unusual positive lightning activity appear to occur in at least two modes. In one, positive ground flashes dominate cloud-to-ground activity throughout the lifetime of the storm, including all tornadic activity. An example of a storm in this mode is shown in Figure 6, which presents ground flash rates for the storm that produced an F4

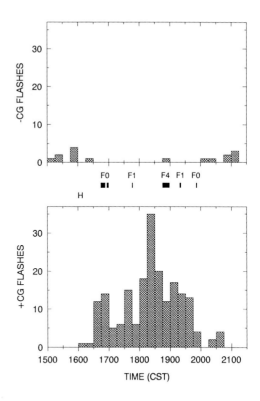

Fig. 6. Time series plots of the number of positive and negative cloud-to-ground flashes in each 15-min interval during a storm that produced a violent tornado near Hutchinson, Kansas, on March 26, 1991. Bars indicate the time of tornadoes, labeled with F scale ratings. The approximate time of large hail reports are indicated by the leading edge of Hs.

tornado near Hutchinson, Kansas, on March 26, 1991. This storm produced severe weather throughout much of its life.

In the second mode the dominant polarity of ground flashes in the storm switches from positive to negative, with the peak negative ground flash rates comparable to or larger than the earlier peak positive ground flash rates. An example of the second mode is given in Figure 7, which shows ground flash rates for the storm that produced an F5 tornado in Plainfield, Illinois, on August 28, 1990. Prior to the F5 tornado most ground flashes in the storm were positive. During this period, large hail and four tornadoes rated up to F2 were produced. Three of these tornadoes occurred during the period of maximum positive ground flash rates.

The dominant polarity of cloud-to-ground lightning switched as the Plainfield tornado began. Most subsequent ground flashes were negative, as shown clearly in the time series plots in Figure 7. Also note that there was a decrease in overall ground flash activity shortly before and during the F5 tornado, much as described earlier for negative ground flashes in some tornadic storms. Negative ground flash rates increased to their largest values at the end of the time series plot, when the storm was merging with other storms to form a squall line (similar to lightning behavior observed by

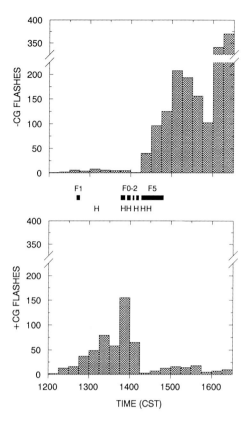

Fig. 7. Same as Figure 6 but for the Plainfield storm on August 28, 1990. Large hail reports whose H symbols would overlap earlier Hs are omitted.

Goodman and MacGorman [1986] when storms merged to form a mesoscale convective complex).

Visual observations indicate that the character of the Plainfield storm changed at approximately the time when the last and most violent tornado began and ground flash polarity switched. Initially, the storm grew rapidly into a supercell, and it remained a supercell for much of its life. During this period the storm had a prominent, deep weak-echo region and produced large hail. At about the time that the Plainfield tornado began, heavy precipitation began to wrap around the mesocyclone, filling in the weak-echo region. Observers of the Plainfield storm noted that the tornado was obscured by heavy precipitation.

Curran and Rust [1992] and *Branick and Doswell* [1992] have observed that low-precipitation supercell storms often produce high densities of positive ground flashes and that nearby supercell storms with heavy precipitation produce the normal preponderance of negative ground flashes. At least some severe storms (including the earlier stages of the Plainfield storm) that have high numbers and densities of positive ground flashes are not low-precipitation storms, but there are no reports of this positive cloud-to-ground lightning signature in high-precipitation supercell storms. Therefore it appears that the transition of the Plainfield storm to a high-precipitation supercell was related to the transition in ground flash polarity. A similar transition was observed by *Curran and Rust* [1992].

MacGorman and Nielsen [1991] suggested two reasons why positive ground flashes might occur in severe storms. First, the structure of supercell storms, in which precipitation at middle to low levels is displaced horizontally from the updraft core, might lead to a tilted bipolar charge distribution, with a lower negative charge occurring in the vicinity of precipitation and positive charge occurring on ice crystals at the top of the updraft core, above the weak echo region. The tilt might be sufficient for the upper positive charge to be effectively exposed to the ground, as for winter storms observed by *Brook et al.* [1982]. *Branick and Doswell* [1992] have suggested a variation of this: not only does the horizontal displacement of the updraft and precipitation tilt the dipole, but the lack of low-to-middle level precipitation in low-precipitation storms inhibits formation of a large region of negative charge.

The second mechanism discussed by *MacGorman and Nielsen* [1991] involved a possible region of significant positive charge beneath the main negative charge of the thunderstorm. Such a region could be formed in different ways. For example, several investigators have suggested that a lower positive charge could be formed in thunderstorms by a noninductive mechanism studied in the laboratory by *Takahashi* [1978], *Jayaratne et al.* [1983], and others. In most regions in which this mechanism for microphysical charge separation might occur in thunderstorms, collisions between graupel and ice crystals place negative charge on graupel and positive charge on ice crystals. These collisions result in the normal tendency for the main positive charge to be generally above the main negative charge. At temperatures warmer than some threshold between $-10°$ and $-20°C$, however, this is reversed: positive charge is placed on graupel; negative charge, on ice crystals. If enough collisions occur between $0°C$ and this reversal temperature, they could create a region of positive charge on graupel beneath the main negative charge of thunderstorms.

6. Conclusion

Obviously, the studies reviewed in this paper fall far short of fully characterizing the diversity of lightning evolution in tornadic storms. Most studies examined lightning behavior relative just to the existence of tornadoes. Only a few studies examined lightning evolution relative to the evolution of the reflectivity structure, mesocyclone, or wind field of tornadic storms. To test adequately the hypothesized relationship between updraft characteristics and suppression of ground flashes requires a more systematic study of large data sets of lightning and storm characteristics, possibly supplemented by numerical simulations. Such studies also may be able to determine better the underlying relationship that caused the observed close correlation of low-level cyclonic shear and intracloud flash rates in the Binger storm.

If the hypotheses discussed in this paper are correct, we

expect that more comprehensive studies of tornadic storms will find the following. Almost all storms that have mesocyclones will have an increase in intracloud and total flash rates as the updraft increases in depth, horizontal extent, and speed at altitudes above the $-20°C$ level, thereby increasing the number of hydrometeor collisions that generate charge. Intracloud flash rates will be enhanced even more in classic supercell storms, which tend to have very large and strong updrafts that are quasi-steady, in contrast to updrafts in lesser thunderstorms. In some tornadic storms, cloud-to-ground flashes will begin and peak shortly after intracloud lightning (roughly 10–15 min later), as in most nonsevere storms. However, in storms that have updrafts large and strong enough to create prominent, deep weak-echo regions (with high lower boundaries of negative charge) we expect cloud-to-ground lightning near the updraft and mesocyclone to be suppressed, so that few ground flashes occur when the updraft is strong and ground flash rates peak when the updraft finally weakens.

There are far fewer data on positive cloud-to-ground lightning in severe storms, so observed relationships and suggested hypotheses should be regarded as more tentative. If a tilted charge distribution causes positive cloud-to-ground lightning, then we would expect to see positive ground flashes in storms in which the top of the main updraft is displaced sufficiently far horizontally from the reflectivity core. If the described mechanism for the lower positive charge causes positive cloud-to-ground (+CG) flashes, then we would expect to see +CG lightning in storms in which the number of collisions between graupel and ice crystals is enhanced at temperatures between freezing and roughly $-15°C$. Such a situation could occur, for example, if there is significant recirculation of graupel, as has been reported for some hail storms.

It may be difficult to determine which of these or other mechanisms are responsible for positive cloud-to-ground lightning. For example, the large separation reported between the mesocyclone and precipitation for some storms in which positive ground flashes dominate [*Branick and Doswell*, 1992] could result from either scenario. There probably would be significant tilt in the charge distribution, and a significant number of recirculating graupel particles might pass through the layer between $0°$ and $-15°C$. Distinguishing which mechanism dominates will require at least some of the following: simultaneous electric field measurements in upper regions of the updraft and in heavy precipitation, improved Doppler radar data, simulations of severe storms, and three-dimensional lightning mapping systems.

References

Bent, R. B., and W. A. Lyons, Theoretical evaluations and initial operational experiences of LPATS (Lightning Position and Tracking System) to monitor lightning ground strikes using a time-of-arrival (TOA) technique, in *Preprints, 7th International Conference on Atmospheric Electricity*, pp. 317–324, American Meteorological Society, Boston, Mass., 1984.

Branick, M. L., and C. A. Doswell III, An observation of the relationship between supercell structure and lightning ground strike polarity, *Weather Forecasting, 7*, 143–149, 1992.

Brook, M., M. Nakano, P. Krehbiel, and T. Takeuti, The electrical structure of the Hokuriko winter thunderstorms, *J. Geophys. Res., 87*, 1207–1215, 1982.

Brown, R. A., and H. G. Hughes, Directional VLF sferics from the Union City, Oklahoma, tornadic storm, *J. Geophys. Res., 83*, 3571–3574, 1978.

Brown, R. A., L. R. Lemon, and D. W. Burgess, Tornado detection by pulsed Doppler radar, *Mon. Weather Rev., 106*, 29–38, 1978.

Burgess, D. W., V. T. Wood, and R. A. Brown, Mesocyclone evolution statistics, in *Preprints, 12th Conference on Severe Local Storms*, pp. 422–424, American Meteorological Society, Boston, Mass., 1982.

Byrnne, G. J., A. A. Few, and M. E. Weber, Altitude, thickness, and charge concentrations of charged regions of four thunderstorms during TRIP 1981 based upon in situ balloon electric field measurements, *Geophys. Res. Lett., 10*, 39–42, 1983.

Chauzy, S., M. Chong, A. Delannoy, and S. Despiau, The 22 June tropical squall line observed during COPT 81 experiment: Electrical signature associated with dynamical structure and precipitation, *J. Geophys. Res., 90*, 6091–6098, 1985.

Church, C. R., and B. J. Barnhart, A review of electrical phenomena associated with tornadoes, in *Preprints, 11th Conference on Severe Local Storms*, pp. 337–342, American Meteorological Society, Boston, Mass., 1979.

Colgate, S. A., Comment on "On the relation of electrical activity to tornadoes," by R. P. Davies-Jones and J. H. Golden, *J. Geophys. Res., 80*, 4556, 1975.

Curran, E. B., and W. D. Rust, Positive ground flashes produced by low-precipitation thunderstorms in Oklahoma on 26 April 1984, *Mon. Weather Rev., 120*, 544–553, 1992.

Davies-Jones, R. P., and J. H. Golden, On the relation of electrical activity to tornadoes, *J. Geophys. Res., 80*, 1614–1616, 1975a.

Davies-Jones, R. P., and J. H. Golden, Reply, *J. Geophys. Res., 80*, 4557–4558, 1975b.

Davies-Jones, R. P., and J. H. Golden, Reply, *J. Geophys. Res., 80*, 4561–4562, 1975c.

Dickson, E. B., and R. J. McConahy, Sferics readings on windstorms and tornadoes, *Bull. Am. Meteorol. Soc., 37*, 410–412, 1956.

Dye, J. E., J. J. Jones, W. P. Winn, T. A. Cerni, B. Gardiner, D. Lamb, R. L. Pitter, J. Hallet, and C. P. R. Saunders, Early electrification and precipitation development in a small isolated Montana cumulonimbus, *J. Geophys. Res., 91*, 1231–1247, 1986.

Goodman, S. J., and D. R. MacGorman, Cloud-to-ground lightning activity in mesoscale convective complexes, *Mon. Weather Rev., 114*, 2320–2328, 1986.

Greneker, E. F., C. S. Wilson, and J. I. Metcalf, The Atlanta tornado of 1975, *Mon. Weather Rev., 104*, 1052–1057, 1976.

Gunn, R., Electric field intensity at the ground under active thunderstorms and tornadoes, *J. Meteorol., 13*, 269–273, 1956.

Illingworth, A. J., Charge separation in thunderstorms: Small-scale processes, *J. Geophys. Res., 90*, 6026–6032, 1985.

Jayaratne, E. R., C. P. R. Saunders, and J. Hallett, Laboratory studies of the charging of soft hail during ice crystal interactions, *Q. J. R. Meteorol. Soc., 109*, 609–630, 1983.

Johnson, H. L., R. D. Hart, M. A. Lind, R. E. Powell, and J. L. Stanford, Measurements of radio frequency noise from severe and nonsevere thunderstorms, *Mon. Weather Rev., 105*, 734–747, 1977.

Johnson, R. L., Bimodal distribution of atmospherics associated with tornadic events, *J. Geophys. Res., 85*, 5519–5522, 1980.

Jones, H. L., A sferic method of tornado identification and tracking, *Bull. Am. Meteorol. Soc., 32*, 380–385, 1951.

Jones, H. L., The identification of lightning discharges by sferic characteristics, in *Recent Advances in Atmospheric Electricity*, edited by L. G. Smith, pp. 543–556, Pergamon, New York, 1958.

Jones, H. L., The tornado pulse generator, *Weatherwise, 18*, 78–79, 85, 1965.

Jones, H. L., and P. N. Hess, Identification of tornadoes by observation of waveform atmospherics, *Proc. IRE, 40*, 1049–1052, 1952.

Kane, R. J., Correlating lightning to severe local storms in the northeastern United States, *Weather Forecasting, 6*, 3–12, 1991.

Keighton, S. J., H. B. Bluestein, and D. R. MacGorman, The evolution of a severe mesoscale convective system: Cloud-to-ground lightning location and storm structure, *Mon. Weather Rev., 119*, 1533–1556, 1991.

Kohl, D. A., Sferics amplitude distribution jump identification of a tornado event, *Mon. Weather Rev., 90*, 451–456, 1962.

Kohl, D. A., and J. E. Miller, 500 kc/sec sferics analysis of severe weather events, *Mon. Weather Rev., 91*, 207–214, 1963.

Koshak, W. J., and E. P. Krider, Analysis of lightning field changes during active Florida thunderstorms, *J. Geophys. Res., 94*, 1165–1186, 1989.

Krehbiel, P. R., An analysis of the electric field change produced by lightning, Ph.D. dissertation, 400 pp., Inst. of Sci. and Technol., Univ. of Manchester, Manchester, England, 1981. (Available as *Rep. T-11*, Geophys. Res. Center, N. Mex. Inst. of Mining and Technol., Socorro.)

Krider, E. P., A. E. Pifer, and D. L. Vance, Lightning direction-finding systems for forest fire detection, *Bull. Am. Meteorol. Soc., 61*, 980–986, 1980.

Larson, H. R., and E. J. Stansbury, Association of lightning flashes with precipitation cores extending to height 7 km, *J. Atmos. Terr. Phys., 36*, 1547–1553, 1974.

Lhermitte, R., and P. R. Krehbiel, Doppler radar and radio observations of thunderstorms, *IEEE Trans. Geosci. Electron., GE-17*, 162–171, 1979.

Lind, M. A., J. S. Hartman, E. S. Takle, and J. L. Stanford, Radio noise studies of several severe weather events in Iowa in 1971, *J. Atmos. Sci., 29*, 1220–1223, 1972.

MacGorman, D. R., and D. W. Burgess, Positive cloud-to-ground flashes during strong convection in severe storms, *Eos Trans. AGU, 72*, Fall Meeting Suppl., 93, 1991.

MacGorman, D. R., and K. E. Nielsen, Cloud-to-ground lightning in a tornadic storm on 8 May 1986, *Mon. Weather Rev., 119*, 1557–1574, 1991.

MacGorman, D. R., and W. D. Rust, An evaluation of the LLP and LPATS lightning ground strike mapping systems, in *Preprints, 5th International Conference on Interactive Information and Processing Systems*, pp. 249–254, American Meteorological Society, Boston, Mass., 1989.

MacGorman, D. R., and W. L. Taylor, Positive cloud-to-ground lightning detection by a direction-finder network, *J. Geophys. Res., 94*, 13,313–13,318, 1989.

MacGorman, D. R., W. D. Rust, and V. Mazur, Lightning activity and mesocyclone evolution, 17 May 1981, in *Preprints, 14th Conference on Severe Local Storms*, pp. 355–358, American Meteorological Society, Boston, Mass., 1985.

MacGorman, D. R., D. W. Burgess, V. Mazur, W. D. Rust, W. L. Taylor, and B. C. Johnson, Lightning rates relative to tornadic storm evolution on 22 May 1981, *J. Atmos. Sci., 46*, 221–250, 1989.

Mach, D. M., D. R. MacGorman, W. D. Rust, and R. T. Arnold, Site errors and detection efficiency in a magnetic direction-finder network for locating lightning strikes to ground, *J. Atmos. Oceanic Technol., 3*, 67–74, 1986.

Orville, R. E., and B. Vonnegut, Lightning detection from satellites, in *Electrical Processes in Atmospheres*, edited by H. Dolezalek and R. Reiter, pp. 750–753, Dietrich Steinkopff Verlag, Darmstadt, Germany, 1977.

Orville, R. E., M. W. Maier, F. R. Mosher, D. P. Wylie, and W. D. Rust, The simultaneous display in a severe storm of lightning ground strike locations onto satellite images and radar reflectivity patterns, in *Preprints, 12th Conference on Severe Local Storms*, pp. 448–451, American Meteorological Society, Boston, Mass., 1982.

Orville, R. E., R. W. Henderson, and L. F. Bosart, An east coast lightning detection network, *Bull. Am. Meteorol. Soc., 64*, 1029–1037, 1983.

Reap, R. M., and D. R. MacGorman, Cloud-to-ground lightning: Climatological characteristics and relationships to model fields, radar observations, and severe local storms, *Mon. Weather Rev., 117*, 518–535, 1989.

Rust, W. D., D. R. MacGorman, and S. J. Goodman, Unusual positive cloud-to-ground lightning in Oklahoma storms on 13 May 1983, in *Preprints, 14th Conference on Severe Local Storms*, pp. 372–375, American Meteorological Society, Boston, Mass., 1985.

Rust, W. D., R. Davies-Jones, D. W. Burgess, R. A. Maddox, L. C. Showell, T. C. Marshall, and D. K. Lauritsen, Testing a mobile version of a cross-chain Loran atmospheric sounding system (MCLASS), *Bull. Am. Meteorol. Soc., 71*, 173–180, 627, 1990.

Scouten, D. C., D. T. Stephenson, and W. G. Biggs, A sferic rate azimuth-profile of the 1955 Blackwell, Oklahoma, tornado, *J. Atmos. Sci., 29*, 929–936, 1972.

Shanmugam, K., and E. J. Pybus, A note on the electrical characteristics of locally severe storms, in *Preprints, 7th Conference on Severe Local Storms*, pp. 86–90, American Meteorological Society, Boston, Mass., 1971.

Silberg, P. A., Passive electrical measurements from three Oklahoma tornadoes, *Proc. IEEE, 53*, 1197–1204, 1965.

Stanford, J. L., Polarization of 500 kHz electromagnetic noise from thunderstorms: A new interpretation of existing data, *J. Atmos. Sci., 28*, 116–119, 1971.

Stanford, J. L., M. A. Lind, and G. S. Takle, Electromagnetic noise studies of severe convective storms in Iowa: The 1970 storm season, *J. Atmos. Sci., 28*, 436–448, 1971.

Takahashi, T., Riming electrification as a charge generation mechanism in thunderstorms, *J. Atmos. Sci., 35*, 1536–1548, 1978.

Taylor, W. L., Atmospherics and severe storms, in *Remote Sensing of the Troposphere*, edited by V. E. Derr, pp. 17-1–17-17, U.S. Government Printing Office, Washington, D. C., 1972.

Taylor, W. L., Electromagnetic radiation from severe storms in Oklahoma during April 29–30, 1970, *J. Geophys. Res., 78*, 8761–8777, 1973.

Trost, T. F., and C. E. Nomikos, VHF radio emissions associated with tornadoes, *J. Geophys. Res., 80*, 4117–4118, 1975.

Turman, B. N., and R. J. Tettelbach, Synoptic-scale satellite lightning observations in conjunction with tornadoes, *Mon. Weather Rev., 108*, 1878–1882, 1980.

Vaughan, O. H., Jr., and B. Vonnegut, Luminous electrical phenomena associated with nocturnal tornadoes in Huntsville, Ala., 3 April 1974, *Bull. Am. Meteorol. Soc., 57*, 1220–1224, 1976.

Vonnegut, B., Comment on "On the relation of electrical activity to tornadoes," by R. P. Davies-Jones and J. H. Golden, *J. Geophys. Res., 80*, 4559–4560, 1975.

Vonnegut, B., and C. B. Moore, Electrical activity associated with the Blackwell-Udall tornado, *J. Meteorol., 14*, 284–285, 1957.

Vonnegut, B., and J. R. Weyer, Luminous phenomena in nocturnal tornadoes, *Science, 153*, 1213–1220, 1966.

Ward, N. B., C. H. Meeks, and E. Kessler, Sferics reception at 500 kc/sec, radar echoes, and severe weather, in Papers on Weather Radar, Atmospheric Turbulence, Sferics, and Data Processing, *Tech. Note 3-NSSL-24*, pp. 39–71, Natl. Severe Storms Lab., Norman, Okla., 1965.

Ziegler, C. L., and D. R. MacGorman, Observed lightning flash rates relative to modeled space charge and electric field distributions in a tornadic storm (abstract), *Eos Trans. AGU, 71*, 1238, 1990.

Ziegler, C. L., D. R. MacGorman, P. S. Ray, and J. E. Dye, A model evaluation of noninductive graupel-ice charging in the early electrification of a mountain thunderstorm, *J. Geophys. Res., 96*, 12,833–12,855, 1991.

Tornadogenesis via Squall Line and Supercell Interaction: The November 15, 1989, Huntsville, Alabama, Tornado

STEVEN J. GOODMAN

NASA Marshall Space Flight Center, Huntsville, Alabama 35812

KEVIN R. KNUPP

Atmospheric Science Program, University of Alabama, Huntsville 35899

1. INTRODUCTION

Even though the motion of intense long-lived thunderstorms (e.g., supercells) can be extrapolated reasonably well, nowcasting the occurrence of tornadogenesis is difficult and generally beyond present capabilities. The tornadic storm that struck the city of Huntsville (HSV) in Madison County, Alabama, at 2230 UTC on November 15, 1989, can be described as the end result of a scenario where an isolated supercell storm developed ahead of and, subsequently, interacted and merged with a squall line. The supercell was identified first by National Weather Service radars in central Mississippi 5 hours before reaching HSV, where it produced an F4 intensity tornado that killed 22 people, injured nearly 500, and caused an estimated $100 million in damage.

Considerable progress in our understanding of tornadic storms has been made over the last decade. Doppler radar observations [*Brandes*, 1984], laboratory simulations [*Ward*, 1962], and numerical modeling [*Klemp and Rotunno*, 1983] have indicated the importance of low-level downdraft and outflow in generation and/or amplification of low-level vorticity within supercell storms. Satellite observations have shown that thunderstorms usually tend to intensify when interacting with an outflow boundary produced by adjacent deep convection [*Purdom*, 1976]. Low-level boundaries also play important roles in nonsupercell tornadic storms [*Wakimoto and Wilson*, 1989]. Thus vorticity generation and amplification within the lowest 1 km of the atmosphere are crucial to tornadogenesis.

The ability to infer the generation of low-level vorticity prior to tornadogenesis may have practical significance for the use of WSR88-D radars by operational meteorologists. Unfortunately, this ability degrades with range since, for example, the height of the radar beam above the surface is nearly 1 km at a distance of 100 km. In the present case we attempt to assess the important low-level processes, especially the interaction between a vigorous squall-line gust front and a preexisting mesocyclone, that appear to have intensified low-level vorticity, thereby strengthening an existing tornado or providing an impetus for tornadogenesis.

This case study analysis serves to reemphasize the existence of a high conditional probability of tornado occurrence, given the merger of a gust front (or storm outflow) with an adjacent moderate to strong thunderstorm (first suggested by *Cook* [1961]). Mesoscale observations of the merger process are used herein to support this hypothesis. Observational data sets include National Weather Service (NWS) composite radar products, local mesonet data, Marshall Space Flight Center (MSFC) cloud-to-ground lightning network data [*Goodman et al.*, 1988a], and visual observations.

Section 2 documents the large-scale environment. Section 3 provides an overview morphology of the squall line and supercell storm. Section 4 presents a detailed mesoscale analysis centered around the time of tornadogenesis. Section 5 discusses other cases where similar squall line and supercell interactions have been observed, and section 6 summarizes the findings.

2. ENVIRONMENTAL SETTING

The observed synoptic environment was of the type commonly observed for widespread severe weather outbreaks. A deepening trough, located over western Oklahoma at 1200 UTC November 15, moved eastward to western Arkansas by 0000 UTC. The 1200 UTC lifted indices at Nashville, Tennessee (BNA) and Centreville, Alabama

The Tornado: Its Structure, Dynamics, Prediction, and Hazards.
Geophysical Monograph 79
Copyright 1993 by the American Geophysical Union.

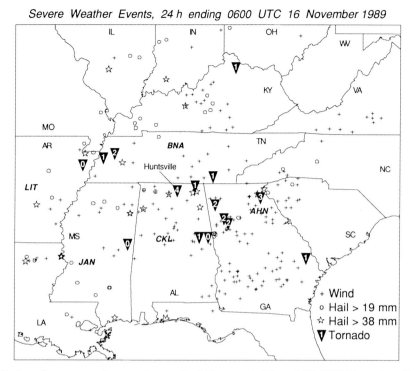

Fig. 1. Distribution of severe weather events over the 24-hour period from 0600 UTC November 15 to 0600 UTC November 16. Tornado locations are indicated by inverted triangles in which the embedded digits indicate the F scale. Straight line (SL) wind damage is indicated by plus signs. Sounding sites are indicated by three-character identifiers defined in the text. The source is the National Severe Storm Forecast Center severe-weather data base tape.

(CKL) were between $-5°$ and $-6°$C. Warm advection and positive vorticity advection occurred east of the trough within a moist, unstable air mass throughout the day. The unstable air mass was widespread over the southeastern United States, as indicated by relatively large values (1000–2000 J kg^{-1}) of convective available potential energy (CAPE) in the 1200 UTC soundings at Little Rock, Arkansas (LIT), and Jackson, Mississippi (JAN). Combining the LIT upper air data with the observed surface conditions over northern Mississippi at 1800 UTC yielded a CAPE of 2800 J kg^{-1} and a corresponding lifted index of $-7°$C. A large number of severe weather events including 16 tornadoes, 199 damaging wind, and 63 large hail events were reported over a 24-hour period. A map of these events is shown in Figure 1.

Figure 2 shows an estimated sounding for northern Alabama at 2200 UTC. Input to this composite was derived from four soundings, two 1200 UTC upstream NWS soundings (LIT and JAN) and two 0000 UTC soundings at Centreville, Alabama (CKL) and Nashville, Tennessee (BNA), which were launched within 1 hour of the tornado occurrence but were both greater than 180 km distant (see Figure 1 for locations). Although the composite sounding is relatively unstable for the time of year (LI of $-4.9°$C and CAPE of 1829 J kg^{-1}), the instability is less than that of the 1200 UTC LIT sounding modified with observed surface conditions over northern Mississippi. The composite sounding also contains significantly more shear than the 1200 UTC LIT sounding, consistent with the fact that a jet streak had entered the synoptic-scale trough during the day. The upper level jet and low-level wind shear are also evident in the 0000 UTC (1900 EST) sounding at Athens, Georgia (AHN) and thus corroborates some salient features of the composite sounding. The AHN sounding displayed a 64 m s^{-1} southwesterly wind maximum at 20 kPa and a change in low-level wind direction from 170° to 235° with a corresponding speed increase from 4 to 22 m s^{-1} in the 0- 3-km AGL (above ground level) layer. The NMC analysis of the geopotential height field over northern Alabama at 0000 UTC displayed even stronger gradients at all levels than that over AHN. The Richardson number ($Ri = 12$) of the composite sounding is quite low by virtue of the appreciable wind shear over the lower troposphere. Such a low number is indicative of high potential for steady (supercell) storms [*Weisman and Klemp*, 1982].

3. Storm System Morphology

The primary focus of this paper to identify the mesoscale events associated with the development of the Huntsville tornado. The most prominent event involved the merger of an active squall-line system with a supercell storm near the time of tornadogenesis. This merger occurred as a result of the different motions of the two systems, as seen in the

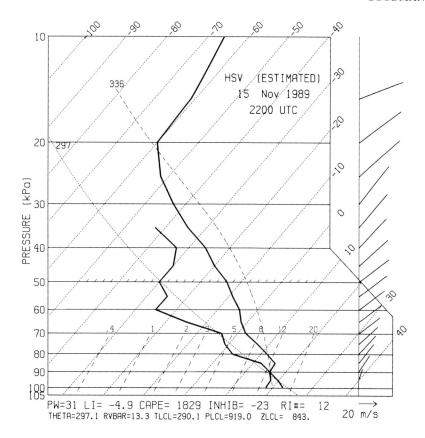

Fig. 2. Composite sounding estimated for the Huntsville region at 2200 UTC November 15, 1989, plotted on a skew-T, ln p diagram. Refer to section 2 of the text for details. The reference dry adiabat ($\theta = 297$ K) represents average subcloud values, and a saturated adiabat ($\theta_e = 336$ K) is drawn through the lifting condensation level (LCL). Parameters plotted at the bottom include precipitable water (PW in millimeters), lifted index (LI, in degrees Celsius), convective available potential energy (CAPE in joules per kilogram), lower negative area (INHIB in joules per kilogram), bulk convective Richardson number (RI), mean values of boundary potential temperature and mixing ratio (theta in degrees Kelvin, and RVBAR in grams per kilogram), and values of temperature, pressure, and height of the LCL (TLCL in degrees Kelvin, PLCL in millibars, and ZLCL in meters).

sequence of composite radar plots in Plate 1. Two isolated VIP-5 intensity supercell storms (SC1 and SC2) located ahead (east-southeast) of the line at 2000 UTC were overtaken as the line moved eastward more rapidly than the isolated cells. The motion vector of cells embedded in the squall line averaged 25 m s^{-1} from 235°, while the line moved due east at 22 m s^{-1}. In contrast, the average motion vector of the supercell was 243° at 18.3 m s^{-1}, 8° to the right of the mean lower tropospheric wind based on the 1200 UTC soundings from BNA and CKL. Thus the squall line eventually overtook the slower supercell storms (Plate 1). The severe weather reported over northern Alabama from each system is shown in Figure 3. Both the squall line and the supercell storm SC1 exhibited a history of severe weather prior to their intersection, some details of which are provided in the following sections.

3.1. Early Development of the Squall Line

The squall line formed around 1500 UTC. It became quite extensive by 1800 UTC, when it consisted of two very long segments, one extending from northeast Louisiana to the Texas gulf coast, and the other from southeast Arkansas to the Indiana-Ohio border. After 1800 UTC, isolated cells and small mesoscale patches of precipitation formed in advance of the squall line. Such features are visible in the sequence of composite radar images from 2000 to 2300 UTC (Plate 1). During this time, individual cells within the squall line produced strong winds and large hail over Mississippi, Tennessee, and Alabama (Figure 1).

3.2. Mature Phase of the Squall Line

At 2000 UTC (Plate 1a) the squall line consisted of a broken configuration of intense cells. From 2100 to 2200 UTC the squall system evolved to a more uniform and linear shape (Plates 1b–1d). Although located both in advance of and behind the convective cores, stratiform precipitation always was more extensive to the rear of the squall line. During the latter stages of the line's mature phase (Plates 3e and 3f), isolated supercell storms formed well in advance of the line, over extreme eastern Alabama and western and

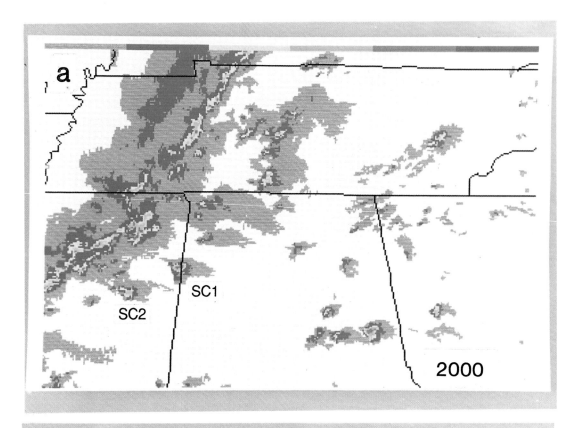

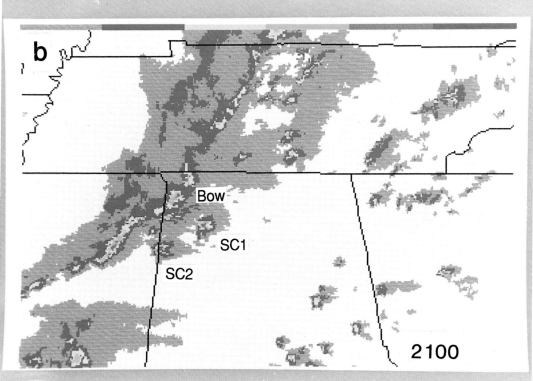

Plate 1. Composite base-level radar reflectivity maps (NOWrad Composite Radar produced by WSI Corporation, Billerica, Massachusetts) obtained from regional NWS radars at (a) 2000 UTC, (b) 2100 UTC, (c) 2130 UTC, (d) 2200 UTC, (e) 2230 UTC, and (f) 2300 UTC. The radar reflectivity is contoured at 18, 30, 40, 45, 50, and 55 dBZ (VIP levels 1–6).

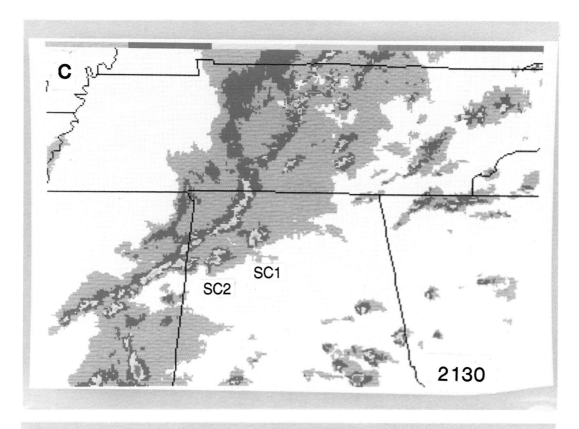

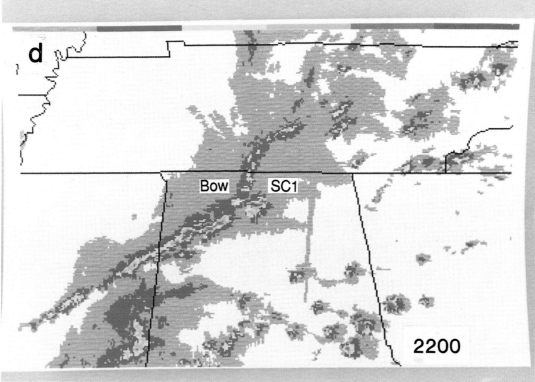

Plate 1. (continued)

188 TORNADOGENESIS VIA SQUALL LINE AND SUPERCELL INTERACTION

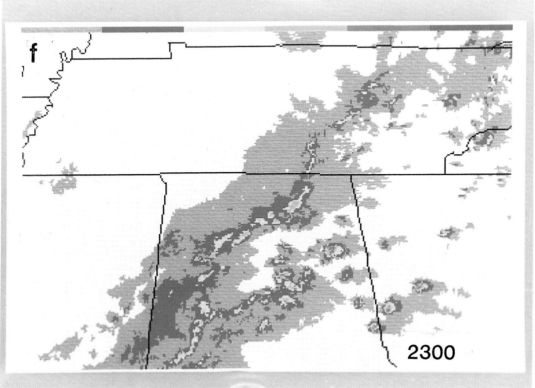

Plate 1. (continued)

northern Georgia, and produced strong F2–F3 tornadoes over northern Georgia (Figure 1) without interacting with the squall line.

Figure 4 shows the 9-hour (1600–0100 UTC) time series of cloud-to-ground lightning produced by the portion of the mesoscale convective system (the squall line plus the supercell) within 200 km radius of HSV. For reference, the domain of this analysis is approximately that shown in Plate 1. The most active hour of cloud-to-ground lightning activity occurred between 2000 and 2100 UTC. The secondary peaks in flash rates during the decline in electrical activity (from 2100 to 2200 UTC) were apparently caused by cell mergers and cell interactions with thunderstorm outflows, both of which helped reinvigorate the thunderstorms. The lightning frequency decreased rapidly after 2200 UTC.

A particularly noteworthy feature of the squall line at 2100 UTC is the bulge (bow echo) appearing over northwest Alabama (Plate 1b). Very strong surface winds producing F1 damage (at 2122 and 2130 UTC at Florence and Killen, Alabama, Figure 3, located north of Muscle Shoals (MSL)) were associated with this bow echo, consistent with the conceptual model of a collapsing storm with surging outflow winds (downburst) proposed by *Fujita* [1981]. Strong winds appear to have been associated with this bow echo for the next hour, up to the time of intersection with the supercell storm over Huntsville, as indicated in Figure 3.

The spatial distribution of cloud-to-ground lightning activity during a 1.5-hour period prior to the tornado is shown in Figure 5. The storm tracks from southwest to northeast are clearly delineated. Isolated lightning discharges of both positive (plus sign) and negative (dot) polarity occurred ahead of the line from thunderstorm anvils, and from the trailing stratiform rain region behind the main line of storms (see Plate 1). Numerous, tightly clustered, positive polarity,

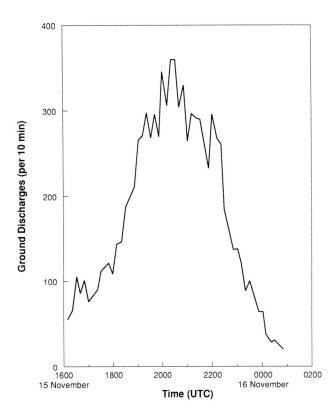

Fig. 4. Cloud-to-ground lightning time series (discharges per 10 min) for the mesoscale convective system observed on November 15, 1989, in the Tennessee Valley.

cloud-to-ground discharges occurred with the cell in southern Tennessee throughout the period shown by Figure 5. Extensive straight-line wind damage was reported from this cell (see Figure 1). *Buechler and Goodman* [1988] have previously documented an increase in the frequency and density of positive polarity cloud-to-ground discharges in association with the arrival of downdrafts at the surface emanating from long-lived microburst-producing storms.

In summary, the squall line was associated with initially intense convection that decreased in strength (but was still severe) after 2200 UTC, prior to intersection with the supercell storm. Strong surface winds and an associated bow echo apparently existed continuously over a 120-km path between northwest Alabama and Huntsville.

3.3. The Tornadic Supercell Storm

The tornadic supercell storm, identified as SC1 in Plate 1, initially formed prior to 1800 UTC, approximately 150 km in advance of the squall line in central Mississippi, and maintained a quasi-steady state after 2000 UTC. A trailing supercell (SC2), which formed at the same time, followed a path nearly identical of that of the leading supercell and dissipated just ahead of the squall line over western Alabama between 2130 and 2200 UTC. The only known severe weather produced by SC2 was large hail over eastern Mississippi.

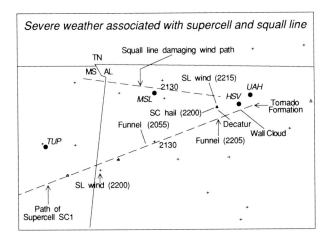

Fig. 3. Severe weather events, as in Figure 1, but plotted only for the northern Alabama region. Additional observations are included in the Huntsville region. The two dashed lines connect severe weather points produced by each system and represent an average linear motion of each system.

190 TORNADOGENESIS VIA SQUALL LINE AND SUPERCELL INTERACTION

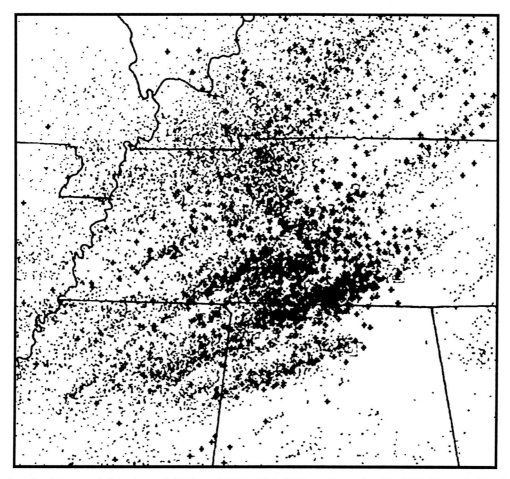

Fig. 5. Cloud-to-ground lightning activity from 2105 to 2236 UTC on November 15, 1989. The polarity of the discharges is denoted by a plus sign for positive and a dot for negative.

Several reports of large hail (up to golf ball size) and strong winds and two reports of funnel clouds near 2100 and 2200 UTC were associated with the leading supercell (SC1), which eventually produced the F4 tornado at Huntsville. Thus supercell SC1 was relatively steady and long-lived but apparently was unable to generate tornadoes during the first 5 hours of its lifetime.

During the 1-hour period prior to tornadogenesis, the NWS local warning radar at HSV revealed that supercell SC1 had high echo tops (~17 km) but no well-defined hook echo signatures. At 2200 UTC the supercell achieved a peak in vertically integrated liquid water (VIL), as measured by the Nashville Radar Data Processor (RADAP) II system located 190 km to the north. Reports of golf ball sized hail were common about 30 km west-southwest of HSV around this time. VIL decreased significantly after 2200 UTC, indicating a weakening in storm vigor. A funnel cloud was observed just south of Decatur, Alabama (see Figure 3 for location) during this decline. The impending merger between the squall line and supercell is clearly indicated by the 5-min totals of cloud-to-ground lightning shown in Figure 6.

Observations around Huntsville between 2210 and 2230 UTC indicate that the largest hail was only pea size, significantly smaller than that observed in the previous 30 minutes. A wall cloud was observed by NWS personnel and NASA scientists between 2215 and 2230 UTC (Plate 2). No rotation was evident as the wall cloud passed NWS observers at ~2215 UTC. A photograph of the wall cloud 5 minutes before tornado touchdown (Plate 2) also reveals the dense precipitation curtain associated with the approaching squall line. For the next 5 min, ropelike funnel-shaped clouds were observed beneath the wall cloud and were associated with sporadic surface damage. Cloud-to-ground lightning was infrequent in the vicinity of the wall cloud, but intracloud lightning flashes, calculated from two separate video tape recordings between 2230 and 2235 UTC, were produced at a rate in excess of 40 per minute. Large ratios of intracloud to ground discharges have been previously documented in Oklahoma supercells [MacGorman et al., 1989], Alabama air-mass storms producing microbursts [Goodman et al., 1988b], and in deep tropical storms during the premonsoon season near Darwin, Australia [Williams et al., 1992].

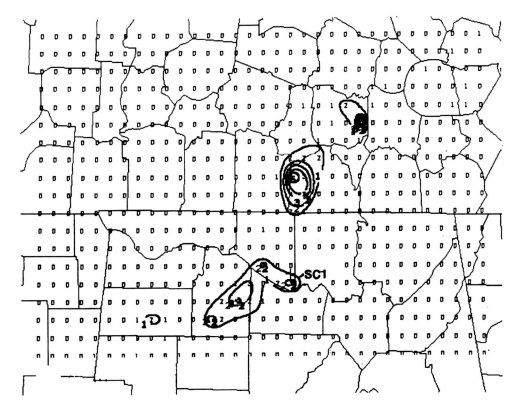

Fig. 6. Contoured lightning density map during the period 2215–2220 UTC. Contour interval is 0.01 discharges per square kilometer, beginning with the value 0.02 discharges per square kilometer.

Observations from UAH (see Figure 7a for location) were made by one of the authors (K.R.K.) as the storm passed to the south. At this location there was an absence of outflow air with low values of θ_e. Winds were generally light and variable as the supercell passed from SW to SSE. The estimated rainfall rate was moderately low (10–15 mm h^{-1}), showers of pea-sized hail were observed, and visibility through the precipitation core was not severely restricted. Cloud-to-ground lightning and low cloud formations were visible in the supercell inflow. As the supercell storm core passed SSE of UAH near 2230 UTC, the squall line reached UAH. Estimated wind gusts immediately behind the gust front were ~30 m s^{-1}, visibility was reduced to ~300 m in heavy rain, and the temperature dropped nearly 6°C.

Other eyewitness observations and a video recording of the squall-line shelf cloud (but not the tornado) suggest that the gust front exhibited appreciable distortion and curvature by ~2237 UTC (Plate 3). When viewed from the southeast, the leading edge of the gust-front segment (labeled "squall line shelf cloud" in Plate 3) south of the tornado is well to the right (east) of the tornado. The easternmost location of the gust-frontal surge is estimated to be midway between the observer and the tornado. This panoramic view also indicates a secondary wall-cloud formation (in the foreground of Plate 3) located along the leading edge of the gust front. Thus the supercell does not appear to be characteristic of the high-precipitation (HP) supercell, examined by *Moller et al.* [1990] and thought to be common in the moist environment typical of the Southeastern United States. Rather, this supercell may be better classified as a cyclic supercell, a term coined to describe the periodic genesis of new mesocyclone cores, wall clouds, funnel clouds, and tornadoes [*Burgess et al.*, 1982; *Jensen et al.*, 1983; *Johnson et al.*, 1987].

The radar summaries in Plates 1d and 1e show merger of the supercell and squall line near the time of tornado formation. As indicated in the previous discussion, both the squall line and supercell were declining in intensity near the time of merger. The squall-line passage was a highly impressive event throughout much of the Huntsville region. Consequently, many eyewitnesses were able to relate the locations of the tornado to the squall line. The next section provides finer-scale details of the interaction between the squall-line gust front and supercell storm.

4. Mesoscale Aspects of the Merger Process

4.1. *Characteristics of the Tornado*

Tornado characteristics and associated cloud patterns were determined using information from a detailed ground survey, aerial survey photographs, and eyewitness reports and videos. The tornado path length was ~30 km, the

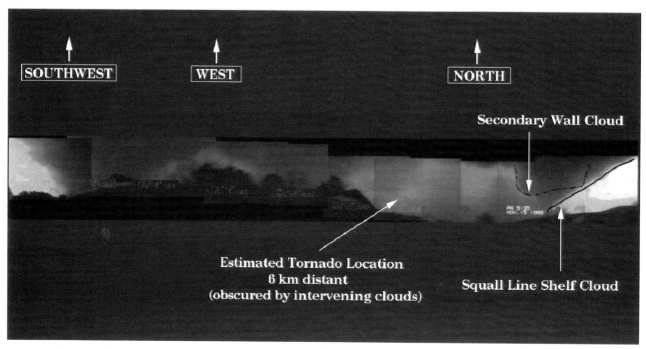

Plate 3. Panorama composite image constructed from 11 video frames taken around 2237 UTC, showing cloud formations associated with the squall line. The view is north through southwest (counter-clockwise) from point GT in Figure 7a. The tornado is located ~6 km north at this time. Video was taken by Greg Talley.

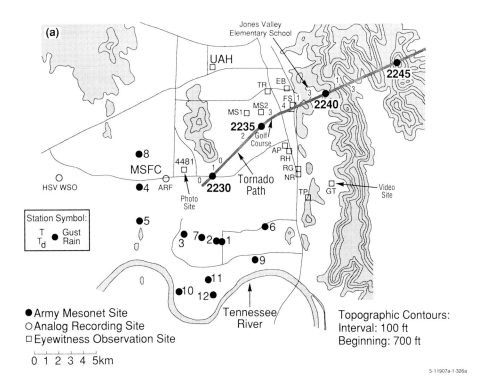

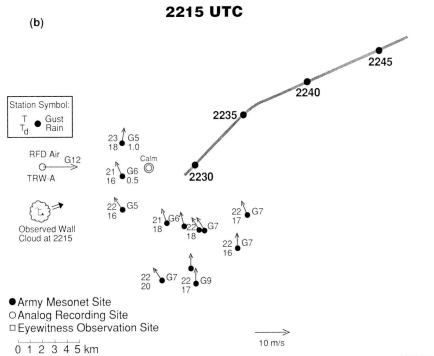

Fig. 7. (a) Station locations and tornado F scale intensity at points along the path of the tornado and analysis of mesonet data for (b) 2215 UTC, (c) 2230 UTC, and (d) 2245 UTC. The thin vectors in Figures 7b–7d represent 15-min wind averages from the Army mesonet sites, while the thick vectors denote measured or estimated (E) instantaneous winds. The dashed line intersecting with the tornado path in Figure 7c marks the range of the time-to-space conversion of 1-min winds presented in Figure 10. Vector magnitudes are such that 1 km represents 5 m s^{-1} for the 15-min average and 10 m s^{-1} for the instantaneous value. Plotted station data at the Army sites include temperature T and dew point T_d (both in degrees Celsius), peak wind gust (in meters per second) over the 15-min period, and 15-min rainfall (in millimeters). In Figures 7c and 7d the estimated location of the squall-line gust front is shown. The tornado track is indicated by the stippling and extends 20 km further to the east-northeast.

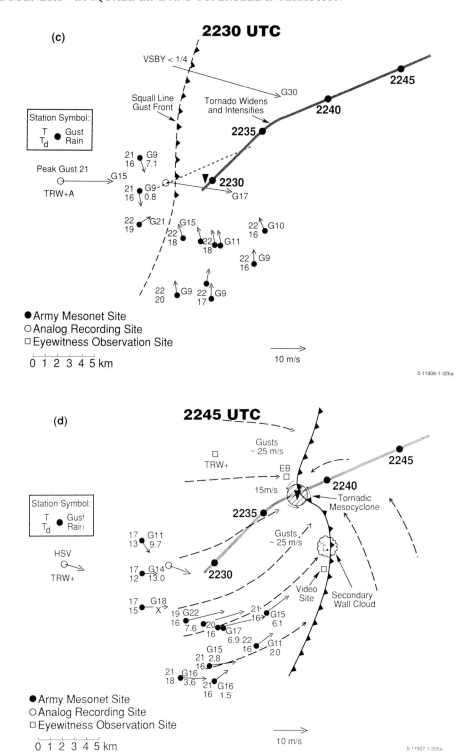

Fig. 7 (continued)

maximum path width was ~800 m, and the peak intensity was marginally F4 over a small portion of the path. A detailed damage survey analysis by the University of Chicago can be found in Storm Data [*National Oceanic and Atmospheric Administration*, 1989]. An exhaustive search failed to uncover photography of the tornado itself. Information on the tornado appearance is based solely on eyewitness observations. The initial 60% of the smoothed center-

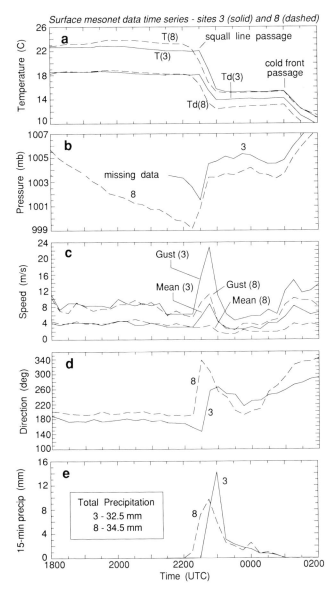

Fig. 8. Time series of surface parameters from mesonet sites 3 (solid lines) and 8 (dashed lines) (mesonet locations are shown in Figure 7): (a) temperature and dew point, (b) sea level pressure, (c) 15-min averaged winds and peak gust within the 15-min interval, (d) wind direction, and (e) precipitation over each 15-min interval.

observed throughout the entire path length. There is little evidence in the surface damage patterns that the tornado evolved to a thin ropelike vortex before dissipating.

4.2. Mesoscale Analysis: General Patterns

Because the supercell passed over a mesonet during the tornadogenesis phase, some surface characteristics of the supercell storm, and of the interaction between the evolving supercell thunderstorm (mesocyclone/tornado) and the squall line, can be defined. Mesonet data consisting of 15-min averages of all basic meteorological parameters, including peak wind gust within the 15-min period, were recorded at 12 sites operated by the U.S. Army on Redstone Arsenal. These data are supplemented with two additional surface stations (from which analog recordings were available), one at the HSV WSO and the other at the MSFC Atmospheric Research Facility (ARF), as well as with eyewitness observations at sites shown in Figure 7a. Because the Army data are 15-min averages, spatial scales less than ~10 km are unresolvable. Finer-scale structures were gleaned from 1-min wind data available from the HSV and MSFC ARF sites.

Time series of 15-min data from two key mesonet stations (3 and 8) are presented in Figure 8. These data clearly show the passage of the squall line near 2230 UTC and the synoptic-scale cold front near 0100 UTC. Important differences in wind characteristics, discussed in further detail below, are apparent at each site.

Merger of the 30-dBZ echoes started at ~2145 UTC, and by 2230 UTC the supercell had become absorbed by the squall line (Plate 1e). At the surface the mesonet station plots displayed in Figures 7b–7d for 2215, 2230, and 2245 UTC depict the passage of the system over the mesonet. Prior to the arrival of the system, the surface analysis at 2215 UTC (Figure 7b) reveals uniform conditions. Averaged winds were light southerly and temperature/dew point values were near 22° and 17°C, respectively. The 2230 UTC analysis (Figure 7c) is more complicated and ambiguous because the 15-min averages (2215–2230 UTC) have smoothed the gradients associated with the mesocyclone and approaching squall-line gust front. The flow is, however, generally cyclonic and convergent. The location of the squall-line gust front, depicted in Figure 7c, was determined primarily from 1-min time series data from HSV and MSFC ARF. Relatively warm and moist air remains over the eastern part of the mesonet, but average winds have increased considerably from 2215 UTC, and wind gusts within the southeasterly flow are relatively uniform in the range 9–11 m s^{-1}. These gusts most likely occurred in response to the increasing pressure gradient around the mesocyclone. One notable exception is the 15 m s^{-1} gust recorded at site 3 at 2230 UTC (Figures 7c and 8). This gust may have been associated with the recent arrival of the squall-line gust front, which displayed an eastward bulge south of the mesocyclone center, as shown in Figure 7c. To the north of the mesocyclone track, gusts ahead of the squall line were produced by the

line of the tornado path, along with estimates of F scale intensity, are shown in Figure 7a.

Prior to tornadogenesis, a well-defined wall cloud was observed from the photo site location at 4481, shown in Figure 7a. Intensification to F3 severity occurred near 2235 UTC, at about the time the squall-line gust front intersected the tornado/mesocyclone circulation (Figure 7d). At this point, the tornado widened, and a rightward turn of ~22° occurred in the direction of movement. A strong southerly inflow of F0 intensity, inferred from sporadic tree falls, was

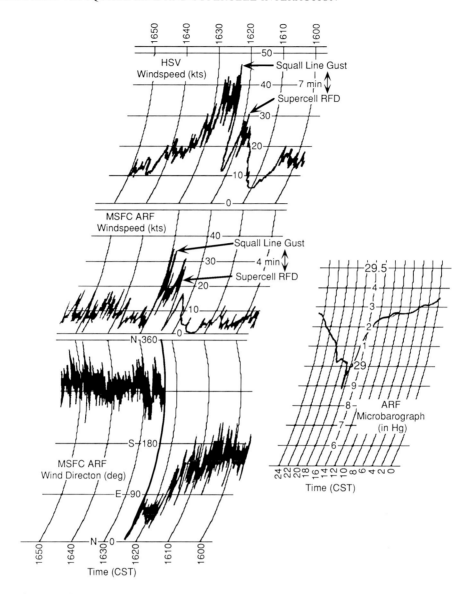

Fig. 9. Analog recordings from 2200 to 2250 UTC (1600–1650 CST) of wind speed (in knots) at the HSV WSO (top), wind speed (in knots) at the ARF (middle), and wind direction (in degrees) at the ARF (bottom). The plot to the right is a microbarograph trace (in Hg) at the ARF from 0000 to 2400 CST. The time lag between the squall line's gust front and rear flank downdraft (RFD) is indicated in the top and middle panels. (10 knots = 5.1 m s^{-1}, 1 in Hg = 33.9 mb.)

supercell's gust front associated with its rear flank downdraft, as happened at HSV and MSFC ARF.

By 2245 UTC (Figure 7d), winds over the region show a westerly component, and the flow is again broadly cyclonic. The reconstructed wind flow (consistent with the video scene in Plate 3) indicates the squall-line gust front had passed over the entire network. The maximum gusts (indicated in both Figures 7 and 8) were most likely associated with the squall line's gust front. Of particular interest is the variation in mean wind direction and speed of the maximum gust over the network. From the time series data in Figure 8, we noted previously that site 8 recorded weaker wind gusts (11 m s^{-1}) with the squall-line passage than sites located south and north of site 8 (Figures 7c and 7d). For example, a wind estimate of ~25 m s^{-1} occurred at UAH, and 22 m s^{-1} was measured at site 3. This observation suggests that the pressure gradients were weaker within the supercell's core than at other locations behind the squall line's gust front.

In summary, there is strong evidence that the supercell's mesohigh blocked the eastward advance of the squall line gust front. The squall-line gust front surged eastward south of the mesocyclone center (Figures 7c and 7d), thus forming a significant perturbation in the gust front. These data

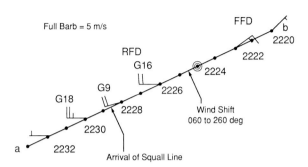

Fig. 10. Time-to-space conversion of the analog wind data from location MSFC, along the dashed line intersecting the tornado path (a and b denote the endpoints of the line) in Figure 7c. An average storm speed of 20 m s^{-1} has been assumed. Wind gusts are in meters per second. The locations of the data relative to storm features at 2230 UTC are shown in Figure 7c.

suggest that tornado formation and/or intensification were associated with the interaction of the squall line with the supercell's mesocyclone. However, the 15-min averages are insufficient to uncover the details of this interaction. Analysis of the 1-min data from the HSV and MSFC ARF sites, augmented by key eyewitness observations, provide some clarification of the merger.

4.3. Surface Features and Timing of the Merger

The time series wind data from HSV and MSFC ARF both show two wind maxima whose time separation decreases from ~7 min at HSV to ~4 min at MSFC ARF (top and middle panels in Figure 9). The initial wind maximum was apparently associated with the rear flank downdraft of the supercell storm, while the second peak was produced by the squall-line gust front. The 1-min data at MSFC, shown in Figure 10 have been converted to a distance scale using the observed storm-motion vector. These data reveal that the supercell exhibits the classical structural details at the surface, including a forward flank downdraft (FFD that gives rise to northeasterly flow) and the rear flank downdraft (RFD, the westerly flow). As noted previously, the squall-line gust front appears to have exhibited appreciable variability over the analysis region. At HSV the supercell and squall-line maximum wind gusts were 14 and 23 m s^{-1}, respectively, while at MSFC they were nearly the same at 16 and 18 m s^{-1} (the wind direction associated with these gusts was ~270°). Over the mesonet the peak squall-line wind gust varied appreciably from 11 m s^{-1} at site 8 to 21 m s^{-1} at site 3 (Figure 8). The strongest wind gust was observed 10 km to the north at UAH (Figure 7b), where a value of 25–30 m s^{-1} (275°) was estimated. No outflow winds were observed in association with the supercell at this location as the storm passed to the south.

The gust front associated with the RFD of the supercell was in close proximity (<1.5 km) to the mesocyclone at location 4481 (shown in Figure 7a) at 2227 UTC. The sequence of events observed there by one of the authors (S.J.G.) was as follows: a wall cloud was observed moving from southwest to the east, and when it was to the southeast, there was a gust front passage associated with the RFD. This was followed ~2 min later by the more vigorous squall gust and associated heavy rain. Thus there were two gust fronts in close proximity to the mesocyclone.

To quantify the relative locations of the mesocyclone and gust fronts (both from the supercell and from the squall line) as a function of time, time series data from the HSV and MSFC locations (depicted in Figure 9) were further analyzed and compared, noting the relative time separation of wind patterns associated with the supercell and squall line. This analysis was supplemented with key eyewitness observations (locations MS1, MS2, and 4481 in Figure 7a) where the time of the squall-line gust frontal passage could be determined from the relative locations of the gust front and tornado. All time difference observations were converted to distances assuming a translational speed of 20 m s^{-1}. The resulting analysis presented in Figure 11 includes plots of (Figure 11a) the distance between the supercell's RFD gust front and the squall-line gust front, (Figure 11b) the distance between the mesocyclone/tornado and the squall line's gust front, and (Figure 11c) one point showing the distance between the supercell's RFD gust front and the mesocyclone. All four points on curve b form a nearly straight line and suggest that the faster squall-line gust front apparently intersected the tornado at the golf course (see Figure 7a for location) near 2235 UTC, just prior to the maximum intensity of the tornado. This intersection was viewed from the NW side of the tornado by a knowledgeable observer (MS2), who watched the tornado pass just to the south, and then observed strong winds and very heavy rain associated with the squall line. By the time the tornado reached F4 intensity (Figure 7a) a common eyewitness observation was that weak winds preceded the damaging tornadic winds and that strong winds and heavy rain occurred for several minutes after the tornado passed. These observations suggest that the tornado, while at maximum F4 intensity, was indeed located at the leading edge of the (distorted) gust front and apparently remained there for the remaining 14 min of its lifetime.

Details of the squall line-supercell merger south of the tornado are less certain because of the lack of good eyewitness reports. The surface data suggest that, to the south, the squall-line gust front infiltrated the mesocyclone at an earlier time. There are observations (e.g., Plate 3 and Figures 7c and 7d) that the squall-line gust front to the south of the tornado was moving faster. Several eyewitnesses positioned a few kilometers south of the tornado path noted that the strongest winds associated with the gust frontal passage were from the south to southwest. During the period of tornado intensification from F1 to F4, the eyewitness referenced in the previous paragraph (MS2 in Figure 7a) observed that the tornado width increased rapidly from ~100 m at 2233 to >500 m by 2236 UTC. Although such a rapid increase is consistent with a pressure reduction associated with an intensifying vortex, the condensation funnel expansion may have been accentuated by the arrival of outflow air (either from the RFD or squall line) that the surface data in Figure 8a show was closer to saturation than that of the air mass preceding the squall line.

5. OTHER CASES

What is the significance of the interaction of the type documented here and how frequently do such interactions occur? This case has served to reemphasize an important scenario, defined previously by *Ward* [1962, 1968] and *Hamilton* [1969], involving the merger of a squall line with an isolated vigorous storm ahead of the line. Hamilton summarized work conducted by *Cook* [1961], regarding observed cases in which cells that developed ahead of a squall line were subsequently overtaken by the squall line. When the cells were intense at the time of the merger, a tornado was reported in eight of 11 cases. In two of the remaining three cases, funnel clouds were observed. A similar interaction has been noted in association with the F4 Edmonton, Alberta, tornado that caused 27 deaths and more than 300 injuries on July 31, 1987 [*Bullas and Wallace*, 1988]. The Huntsville case can be classified similarly and provides a quantification of the process involved in such a merger.

For the November 15, 1989, event it was noted that the active squall line also merged at an earlier time (~2145 UTC, Plate 1) with supercell SC2. In this merger there were no reports of tornadoes, and the supercell appeared to weaken prior to becoming engulfed by the squall line. One may speculate that this supercell, which trailed the Huntsville supercell, may have ingested outflow air left in the wake of the former. The echo structures associated with the 15 other tornadoes reported on this day were also examined to determine potential interactions between the parent cell and the squall line. Interestingly, eight tornadoes occurred near the squall line, either from new cells forming immediately in advance of the primary line or from existing cells (not of prolonged supercell stature, as was the case for SC1) within the line. Tornadoes near the squall line were generally weak (two F0, four F1, and one F2 intensity; see Figure 1), with the exception of the Huntsville storm. Generally stronger tornadoes (F2–F3) were produced by isolated supercell storms over eastern Alabama and Georgia.

6. DISCUSSION AND SUMMARY

This paper has described some of the features associated with the merger of a vigorous squall line with a strong supercell storm, which formed 5 hours earlier in advance of the squall line. At the time of the merger, both systems were declining in strength. To our knowledge, the supercell did not generate tornadoes prior to this merger, although two funnel clouds were observed earlier. We have shown, using

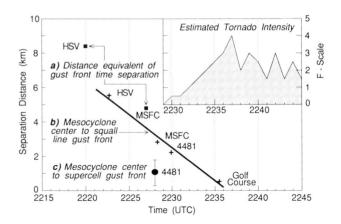

Fig. 11. The interaction between the squall line and the supercell mesocyclone (tornado) is depicted in terms of the time dependence of (*a*) the distance separation of peak wind gusts associated with the rear flank downdraft outflow of the supercell and the squall-line gust front, as measured at HSV and MSFC (the time differences have been converted to space, and data points are marked by solid squares), (*b*) the distance between the squall-line gust front at a point and the estimated location of the mesocyclone or tornado at that time (the line represents the best linear fit to the four data points, denoted by the plus signs), and (*c*) the distance between the supercell's rear flank downdraft (RFD) gust front and the mesocyclone (shown by the solid dot with error bar). These data show that the time of the squall-line gust front merger with the tornado coincided with the time of tornado intensification around 2235 UTC, while the supercell RFD gust front was close to the tornado near the time of tornado formation. The location of the golf course is shown in Figure 7a.

analysis of surface mesonet data, that the radar-echo merger was accompanied by the interaction of a vigorous squall-line gust front with the mesocyclone circulation. As the squall line's gust front intersected the supercell mesocyclone, the gust front became distorted and appeared to wrap around the south side of the mesocyclone. In addition, this interaction was associated with a rapid intensification of the existing tornado. The potential importance of this interaction on the initiation of the tornado is less clear. A less vigorous gust front associated with the rear flank downdraft of the supercell storm was in the vicinity of the tornado at the time of tornado formation. Thus one unresolved aspect of the tornadogenesis is whether it was triggered by the squall gust front or by the supercell's rear flank downdraft.

The importance of the rear flank downdraft in intensifying low-level vorticity within the mesocyclone, prior to the tornadogenesis stage, was suggested by *Lemon and Doswell* [1979] and has since been further quantified with Doppler radar data [*Brandes*, 1984] and with cloud model results [*Klemp and Rotunno*, 1983]. These studies have demonstrated the potential importance of the gust front in increasing vertical vorticity via additional tilting of low-level vorticity, generation via the solenoidal term of the vorticity equation, and additional amplification via convergence along the gust front. The important role of vorticity amplification by convergence also has been suggested in the Doppler radar analysis of *Johnson et al.* [1987]. Our observations are consistent with these studies but bring an additional complicating factor into the model, that is, the introduction of a more vigorous squall-line gust front generated externally, its interaction with the RFD gust front, and its apparent importance in intensifying an existing tornado.

It is possible that this problem may be investigated further with three-dimensional numerical models in order to define the detailed physics of the merger and the precursor conditions needed for tornadogenesis.

Acknowledgments. Irv Watson of NOAA kindly provided NWS composite radar information which made a significant contribution to this study. Mike Botts expertly completed the image analysis of Plate 3. Discussions with Don Burgess and R. Davies-Jones, and a thorough review by R. Davies-Jones, improved the content of this paper. Steve Williams assisted in reduction and interpretation of the mesonet and sounding data, which were furnished by the meteorological team on Redstone Arsenal. Tim Rushing assisted in analysis and reduction of the mesonet data. We also thank the many individuals who kindly provided videos and descriptive information of particular events. Partial funding support was provided by NASA under grant NAG8-654. Reference herein to any specific commercial products, process, or service by trade name, trade mark, manufacturer, or otherwise, does not constitute or imply its endorsement, recommendation, or favoring by the United States Government or the University of Alabama in Huntsville.

References

Brandes, E., Vertical vorticity generation and mesocyclone sustenance in tornadic thunderstorms: The observational evidence, *Mon. Weather Rev.*, *112*, 2253–2269, 1984.

Buechler, D. E., and S. J. Goodman, Initial observations of cloud-to-ground lightning activity in microburst-producing storms, in *Proceedings, 8th International Conference on Atmospheric Electricity*, pp. 842–848, Swedish Natural Science Research Council and Uppsala University, Uppsala, Sweden. (Available as S-755 92 from Univ. Inst. of High Voltage Res., Husbyborg, Uppsala, Sweden.)

Bullas, J. M., and A. F. Wallace, The Edmonton tornado, July 31, 1987, in *Preprints, 15th Conference on Severe Local Storms*, pp. 438–443, American Meteorological Society, Boston, Mass., 1988.

Burgess, D. W., V. T. Wood, and R. A. Brown, Mesocyclone evolution statistics, in *Preprints, 12th Conference on Severe Local Storms*, pp. 422–424, American Meteorological Society, Boston, Mass., 1982.

Cook, B. J., Some radar LEWP observations and associated severe weather, in *Proceedings of the 9th Radar Conference*, pp. 181–185, American Meteorological Society, Boston, Mass., 1961.

Fujita, T. T., Tornadoes and downbursts in the context of generalized planetary scales, *J. Atmos. Sci.*, *38*, 1511–1534, 1981.

Goodman, S. J., D. E. Buechler, and P. J. Meyer, Convective tendency images derived from a combination of lightning and satellite data, *Weather Forecasting*, *3*, 173–188, 1988a.

Goodman, S. J., D. E. Buechler, P. D. Wright, and W. D. Rust, Lightning and precipitation history of a microburst-producing storm, *Geophys. Res. Lett.*, *15*, 1185–1188, 1988b.

Hamilton, R. E., A review of use of radar in detection of tornadoes and hail, *Tech. Memo. WBTM-ER-34*, 64 pp., Weather Bur. East. Reg. Headquarters, Garden City, N. Y., 1969.

Jensen, B., E. Rasmussen, T. P. Marshall, and M. A. Mabey, Storm scale structure of the Pampa storm, in *Preprints, 13th Conference on Severe Local Storms*, pp. 85–88, American Meteorological Society, Boston, Mass., 1983.

Johnson, K. W., P. S. Ray, B. C. Johnson, and R. P. Davies-Jones, Observations related to the rotational dynamics of the 20 May 1977 tornadic storms, *Mon. Weather Rev.*, *115*, 2463–2478, 1987.

Klemp, J. B., and R. Rotunno, A study of the tornadic region in a supercell thunderstorm, *J. Atmos. Sci.*, *40*, 359–377, 1983.

Lemon, L. R., and C. A. Doswell III, Severe thunderstorm evolution and mesocyclone structure as related to tornadogenesis, *Mon. Weather Rev.*, *107*, 1184–1197, 1979.

MacGorman, D. R., W. D. Rust, D. W. Burgess, V. Mazur, W. L. Taylor, and B. C. Johnson, Lightning rates relative to mesocyclone evolution in tornadic storms on 22 May 1981, *J. Atmos. Sci.*, *46*, 221–250, 1989.

Moller, A. R., C. A. Doswell III, and R. Przybylinski, High-precipitation supercells: A conceptual model and documentation, in *Preprints, 16th Conference on Severe Local Storms*, pp. 52–57, American Meteorological Society, Boston, Mass., 1990.

National Oceanic and Atmospheric Administration, *Storm Data*, vol. 31, no. 11, 6–17. (Available from the Natl. Clim. Data Center, Natl. Oceanic and Atmos. Admin., Asheville, N. C.)

Purdom, J. F. W., Some uses of high resolution GOES imagery in the mesoscale forecasting of convection and its behavior, *Mon. Weather Rev.*, *104*, 1474–1483, 1976.

Wakimoto, R. M., and J. W. Wilson, Non-supercell tornadoes, *Mon. Weather Rev.*, *117*, 1113–1140, 1989.

Ward, N. B., The effect of low level wind shear on the formation of atmospheric vortices, in *Proceedings, 2nd Conference on Severe Local Storms*, pp. 1–4, American Meteorological Society, Boston, Mass., 1962.

Ward, N. B., Rotational characteristics of a tornado cyclone, in *Proceedings, 13th Weather Radar Conference*, pp. 183–186, American Meteorological Society, Boston, Mass., 1968.

Weisman, M. A., and J. B. Klemp, The dependence of numerically simulated convective storms on vertical wind shear and buoyancy, *Mon. Weather Rev.*, *110*, 504–520, 1982.

Williams, E. R., S. A. Rutledge, S. G. Geotis, N. Renno, E. Rasmussen, and T. Rickenback, A radar and electrical study of tropical "hot towers," *J. Atmos. Sci.*, *49*, 1386–1395, 1992.

Discussion

ALAN MOLLER, SESSION CHAIR

National Weather Service

PAPER C1

Presenter, Bill McCaul, Universities Space Research Association [*McCaul*, this volume, Observations and simulations of hurricane-spawned tornadic storms]

(Bob Davies-Jones, National Severe Storms Laboratory) You showed that the updrafts are weak in these storms. Are the downdrafts also weak? There is not much evaporative cooling in hurricanes.

(McCaul) No, there is not much. In fact, in a lot of the simulations, there is little evidence of a cold pool.

(Davies-Jones) So how do you explain the wall cloud [in Hurricane Danny]?

(McCaul) That case may have been a little atypical. It had more CAPE and somewhat more potential for producing a downdraft. It would be nice to get more observations of such cases.

(Howie Bluestein, University of Oklahoma) Do you think that there are other instances in nature where there are weak CAPE and hodographs like those in tropical cyclones? Very weak CAPE is quite common, but where else would similarly shaped hodographs occur?

(McCaul) I personally haven't seen any other hodographs with the same shape as the tropical cyclone ones. These have a low-level or 700-mbar jet, with weak winds aloft. In mid-latitude storm environments, there is generally a jet aloft, as in the Bluestein and Jain [*J. Atmos. Sci.*, 42, 1711–1732, 1985] composite sounding. Frequently, there are thermodynamic profiles that are quite similar to the ones in hurricane environments, almost moist adiabatic with a small positive area at low levels. This is seen in a lot of mid-latitude and tropical convective systems. But I don't know of other environments with semicircular or nearly circular hodographs.

The Tornado: Its Structure, Dynamics, Prediction, and Hazards.
Geophysical Monograph 79
This paper is not subject to U.S. copyright. Published in 1993 by the American Geophysical Union.

(Joe Golden, National Oceanic and Atmospheric Administration, Office of the Chief Scientist) Could you say a little more about the detection problem? It is a vexing operational problem. What percentage of the time do you expect a mesocyclone that is detectable by NEXRAD radars, and is the supercell the dominant [tornado-producing] mechanism in general, as it was in Hurricane Danny?

(McCaul) I have looked at several other landfalling tropical cyclones, and Danny is not the only one that produced supercells. I have seen evidence in Hurricanes Beulah and Allen of long-lived storms that produced multiple tornadoes. Concerning detectability, I think that it will be a problem if the storms are not close enough to the radars for the lowest 3 km of the storms to be sampled. If the lowest 3 km cannot be sampled, then the only hope for detection is to examine gross features such as storm longevity. I find that supercells in hurricanes last at least 2 to 3 hours, sometimes even longer.

(Chair) I guess that the storm-relative mid-level winds determine the usual location of the tornado in the right rear quadrant of the individual storm.

(McCaul) There can be some variability from hurricane to hurricane, depending on the direction of the shear of the ambient steering current. Generally, it is almost from rear to front of the hurricane, because it is guiding the hurricane's motion. So the right front quadrant is favored, but this is not always the case.

(Chair) I am talking about an individual storm cell, not an individual tropical cyclone. I want information for tornado spotters. I saw some footage from near Austin, Texas, of some tornadoes that appeared to be on the northeast side of a storm complex in a westward moving hurricane that had come inland. Would you expect the tornadoes to be in the right rear quadrant of their parent cell?

(McCaul) I think that's correct. It's the same as Hurricane Danny, except rotated 90° because of the difference in the motions of the hurricanes.

PAPER C2

Presenter, Ed Brandes, National Severe Storms Laboratory [*Brandes*, this volume, Tornadic thunderstorm characteristics determined with Doppler radar]

(Chair) If we in the National Weather Service detect a mesocyclone, should we issue a tornado warning based on that alone?

(Brandes) I think that I would, but Don Burgess is a better expert on that than I am. Given a mesocyclone, there is a 50% probability of a tornado. That number may be biased more to mid-level circulations than low-level circulations. If there were strong rotation at low levels, I would certainly issue a warning.

(Chair) But we should wait for the low-level circulation?

(Brandes) Yes.

(V. G. Blanchette, retired professional engineer) The data you show are at some distance from the tornado. Have you considered what is going on inside the tornado?

(Brandes) The tornado is not well resolved in the radar data. We need really dense measurements to determine exactly what is going on. The observations are spaced several hundred meters apart. In this particular case [the Harrah storm], I think that the [low-level] mesocyclone flow was excited by rainy downdraft development. The trigger for the tornado itself may have been some type of shearing instability at the side of strong mesocyclone winds.

(Lou Wicker, University of Illinois) From your observations, does the tornado-scale vortex develop nearly simultaneously through a fairly deep layer, say, the lowest 2–3 km, or does it develop first at a particular height?

(Brandes) I suspect that it is the latter. The data that were used here were roughly at 10-minute intervals, and it is far too coarse to really see what is going on. We need rapidly spinning radars that get 2-minute data, and even they may not provide the answer. In the Del City storm where the sampling was roughly every 5 minutes, it appears that the initial formation was at about 2-km elevation. The initial radar signature of the vortex was more intense at 2 km than it was above or below this height. In the Harrah storm, the vortex did not build up to 5 km until later.

PAPER C3

Presenter, Chuck Doswell, National Severe Storms Laboratory [*Doswell and Burgess*, this volume, Tornadoes and tornadic storms: A review of conceptual models]

(Keith Brewster, University of Oklahoma) You discussed the different classifications of storms. Do you think that tornadoes should be classified according to whether they are produced by mesocyclones or whether they are landspouts?

(Doswell) From a practical viewpoint, it is difficult to tell the type of tornado at present. When there is a network of Doppler radars across the country, I would advocate classifying tornadoes into mesocyclone and nonmesocyclone types.

PAPER C4

Presenter, Don MacGorman, National Severe Storms Laboratory [*MacGorman*, this volume, Lightning in tornadic storms: A review]

(Tom Grazulis, Environmental Films) In the [August 28, 1990] Plainfield [Illinois] tornadic storm, was there also a ratio of 50 cloud-to-cloud flashes to every 1 cloud-to-ground flash?

(MacGorman) We don't have data on cloud-to-cloud flashes in the Plainfield case. [The cloud-to-cloud data for the Binger storm were acquired from a special L band radar.] The [cloud-to-ground] data that I presented are from the State University of New York system that operates continuously. So we were able to get the [cloud-to-ground] data, much like we are hoping to be able to do with Doppler radar data when NEXRAD radars are installed.

PAPER C5

Presenter, Steve Goodman, NASA Marshall Space Flight Center [*Goodman and Knupp*, this volume, Tornadogenesis via squall line and supercell interaction: The November 15, 1989, Huntsville, Alabama, Tornado]

No time for questions because session ran overtime.

Tornado Detection and Warning by Radar

DONALD W. BURGESS[1]

National Severe Storms Laboratory, NOAA, Norman, Oklahoma 73069

RALPH J. DONALDSON, JR.

Hughes STX Corporation, Lexington, Massachusetts 02173

PAUL R. DESROCHERS

Geophysics Directorate, Phillips Laboratory, Hanscom Air Force Base, Massachusetts 01776

1. INTRODUCTION

Traditionally, tornadoes have been difficult to detect by radar. Conventional radar signatures, although available for over 30 years, have been only qualitatively associated with actual tornado existence. As recently as the 1980s, it has been suggested that most successful tornado warnings are still based on visual sightings, even when the warning message mentions radar signatures [see *Kelly and Schaefer*, 1982].

The advent of meteorological Doppler radar with velocity information has helped define tornado location considerably, but the small size of the tornado (with respect to radar pulse volume) makes unambiguous identification difficult. However, evidence of tornado potential is readily available from Doppler with the detection of the larger, parent circulation surrounding the tornado (i.e., the mesocyclone or misocyclone). The impending upgrade of the United States operational weather radar network to Doppler (the NEXRAD Program) warrants focus of attention on Doppler signatures instead of those associated with conventional (non-Doppler) radar.

This review will begin with discussion of problems associated with radar detection of tornadoes (section 2), continue with a look at the progress made in tornado detection and warning since the last symposium (section 3), discuss automated tornado detection algorithms (section 4), and conclude with thoughts on future radar-warning improvements (section 5).

2. LIMITATIONS OF DOPPLER WEATHER RADAR

A Doppler radar measures the radial component of target motion. Radar targets, or scatterers, are typically precipitation particles (rain and hail), although sensitive radars may detect return from particulate matter (insects and airborne debris) and refractive index gradients. Cloud droplets, such as the condensation funnel of a tornado, are too small to be detected by all except very specialized radars. Combinations of two or more radars (multiple Doppler networks) can be used to extract full, three-dimensional, air motion from individual radar velocity components. The high cost of maintaining several operational radars in the same area precludes use of real-time, multiple Doppler networks for warning purposes.

Much useful information for tornado detection and warning can be obtained from single-Doppler, radial-component information. An area of rotation stands out in a single-Doppler field of mean velocity as a couplet of strong, localized flow, toward and away from the radar (Figure 1). If the tornado is extremely large and close to the radar, the rotation may be attributable to the tornado itself. Often, however, aspect ratio problems (see below) preclude tornado observation, at least for a well-defined couplet, and the detected signature is the larger area of swirl that surrounds the tornado, defined as the mesocyclone or misocyclone, depending on the size of the swirl.

The first suggested Doppler signature for a tornado [*Atlas*,

---

[1] Now at WSR-88D Operational Support Facility, Norman, Oklahoma.

The Tornado: Its Structure, Dynamics, Prediction, and Hazards.
Geophysical Monograph 79
Copyright 1993 by the American Geophysical Union.

204 TORNADO DETECTION AND WARNING BY RADAR

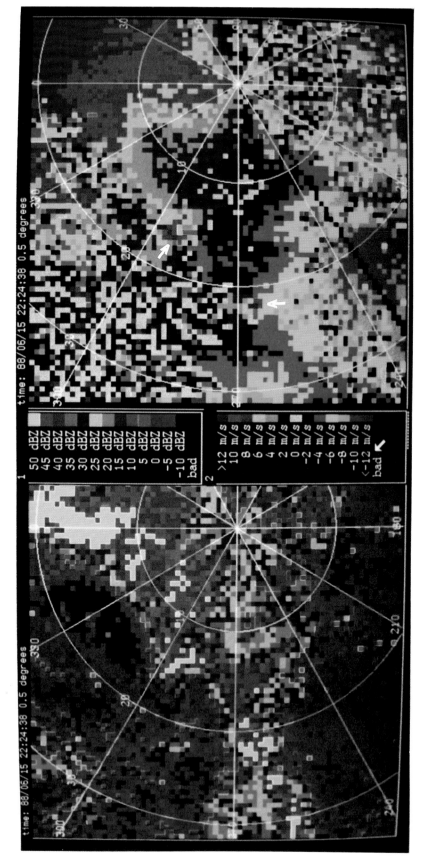

Plate 1. Reflectivity (left) and velocity (right) for misocyclone signatures of June 15, 1988 [from *Wakimoto and Wilson*, 1989]. Color codes are in the middle; misocyclone and tornado locations are marked by arrows on velocity image. Range marks are in kilometers. Data are courtesy of Lincoln Laboratories and the National Severe Storms Laboratory, NSSL.

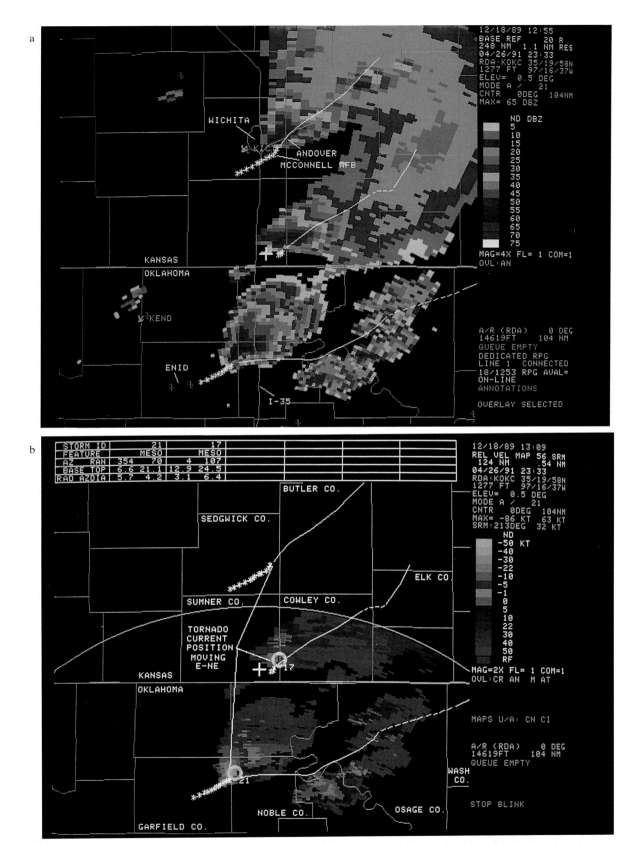

Plate 2. (a) Relectivity and (b) velocity displays from the Oklahoma City, Oklahoma, WSR-88D for the tornado outbreak of April 26, 1991. Mesocyclone algorithm output has been overlaid in Plate 2b; past (future) tornado tracks are asterisks (solid lines). The Wichita tornado is beyond quantitative velocity range (230 km, 125 nm). Violent (F4-F5) tornadoes are underway with each of the three large storms. Future WSR-88D sites (KEND and KICT) are indicated in Plate 2a. Displays are courtesy of WSR-88D Operational Support Facility.

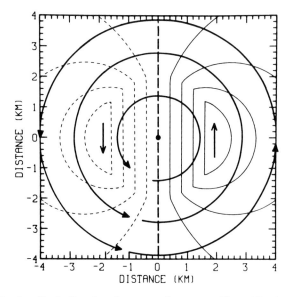

Fig. 1. Single-Doppler signature of a vortex [from *Wood and Brown*, 1983]. Heavy lines are true airflow; light lines are radial velocity contours with dashes for inbound radial component; short arrows mark radial velocity maxima.

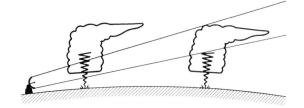

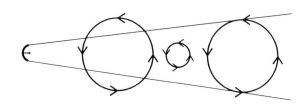

Fig. 2. Schematic depicting limitations of radar.

1963] was not the horizontal pattern of the couplet but a very large spectrum width (the spread of velocities about the mean value). This is reasonable because return from a sample volume containing a tornado would feature both inbound and outbound radial velocities and thus high spectrum width. Radar observations have shown that spectrum width signatures, by themselves, are not good tornado indicators. This occurs because there are several causes of broad spectrum widths, both meteorological and hardware associated (see *Doviak and Zrnić* [1984] for a more thorough discussion). Some smaller Doppler weather radars currently available commercially provide only spectrum-width-related outputs. These radars are thought to have only low skill in tornado detection. During operational Doppler tests, spectrum width, by itself, has never been demonstrated as a useful tornado signature.

All radars, including Doppler, suffer from two main limitations in observing weather echoes. These limitations affect rotation signatures and make them less useful. A short discussion of each follows.

2.1. *Radar Horizon Problems*

Refraction of microwave energy by the atmosphere produces radar beams that do not bend at the same rate as the Earth curves. For this reason, radar beams, even at 0° elevation angle, will increase in height as they travel away from the radar. This process is illustrated by the schematic in the top part of Figure 2. Since tornadoes and mesocyclones are best defined in the lower and middle portions of storms, a radar will detect a rotation signature from a nearby storm but will overshoot a signature from a storm at longer range.

Also, lowest-elevation-angle measurements for ranges greater than 80 km (40 nm) are frequently above cloud base. Because a tornado and funnel are defined only below cloud base, interpretation of longer range observations must be limited to finding rotation in the cloud, without knowing for sure that the rotation extends to the ground. This means that radar observations must be supplemented by other information (i.e., storm spotter reports) to be completely definitive.

Similarly, boundary layer convergence lines, possible tornado formation areas, cannot be detected at longer ranges unless they extend well above cloud base. Typically, these boundaries are hard to detect beyond 100 km (50 nm) range.

2.2. *Aspect Ratio Problems*

The radar recognition of atmospheric vortices depends on the aspect ratio between the radar illumination volume (beam width) and the vortex size (core radius). For more details, see *Burgess and Lemon*, [1990]. Aspect ratios for a Rankine vortex model are graphed in Figure 3. When the radar beam width exceeds the core radius by a factor of 3, less than one half of the true maximum velocity is detected, and the vortex will probably not be recognized.

This is illustrated by the lower portion of Figure 2. The vortex on the left (aspect ratio <1) is at a range where it is definitely large enough to be detected. The vortex on the right (aspect ratio of 2) is at a range where it will be detected only for a favorable combination of radar viewing angles. The center vortex of the lower part of Figure 2 is at a range where it is much smaller than the beam width (aspect ratio >3) and therefore will not be detected by the radar. The center vortex approximates the typical sampling condition for an operational radar scanning an average-sized tornado at moderate range.

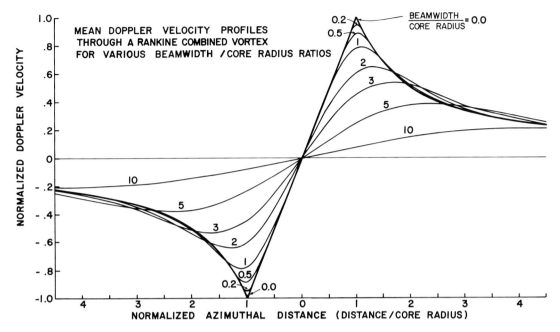

Fig. 3. Continuous velocity profile across a Rankine combined vortex for a number of beam width to core size ratios [from *Burgess and Lemon*, 1990].

In real radar observations the recognition problem is further complicated by discrete azimuthal sampling. A peak in velocity will not be optimally resolved unless one of the discrete sample volumes along an azimuth is centered on the velocity peak.

3. RADAR SIGNATURES FOR TORNADO DETECTION AND WARNING

This section will document the development of Doppler signatures during the 1970s and 1980s, the time since the Lubbock, Texas, tornado symposium (1976). It will begin by reviewing research that was in progress at symposium time and continue through the rapid developments that began soon thereafter. Real-time Doppler applications were, in part, made possible by significant improvements in computer processors, velocity estimators, and display technology. Rapid data processing and real-time, color display made it possible to learn about tornadic storm evolution and to apply the results in public warnings.

3.1. Tornadic Vortex Signatures

An analog display, the plan shear indicator (PSI) developed by *Armstrong and Donaldson* [1969], was employed to record the Doppler velocity field in the storm that spawned the devastating Union City, Oklahoma, tornado of May 24, 1973. Analysis of these observations by *Donaldson* [1978] revealed a disturbed pattern of velocities at midlevels (5–9 km) in the storm 45 min before the earliest tornado damage. Subsequent observations showed intensification of the disturbance, and descent of its base, as tornado touchdown time approached.

The Doppler observations of the Union City tornado alternated between PSI mode and digital recordings of the velocity data. Analysis of the digital data by *Burgess et al.* [1975] revealed a unique and characteristic signature of extreme azimuthal shear, in excess of 0.05 s^{-1}, that first appeared aloft 33 min before the earliest tornado damage. The base of this pattern, now called a tornadic vortex signature (TVS) because of the large magnitude and vertical structure of the shear, descended to the ground coincident in both time and space to tornado touchdown.

This discovery stimulated a search for more TVS examples by *Brown and Lemon* [1976] and was formally discussed by *Brown et al.* [1978]. These authors found 10 TVS appearances in the years subsequent to and including the 1973 Union City tornado. In all but two cases a tornado or funnel cloud was observed to accompany the TVS. The two unverified cases occurred in sparsely populated areas with no report of whether or not a tornado occurred, so they cannot fairly be considered as false alarms. Although TVS detectability is severely limited by resolution, when a TVS does appear, it is prudent to assume that a tornado is on the ground or soon will be.

An example of a very strong TVS is shown in Figures 4a and 4b. The displayed velocities for adjacent radar-viewing azimuths (Figure 4a) indicate a maximum velocity difference of 114 m s^{-1} across the azimuth change increment (0.5°), leading to a shear value of 1.9×10^{-1} s^{-1}. At observation time, the tornado was 1 km wide and was producing damage

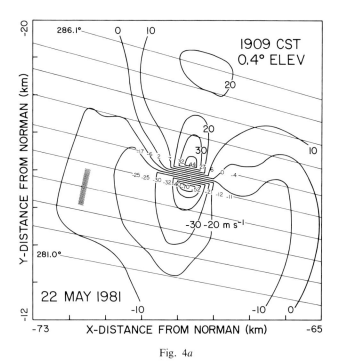

Fig. 4a

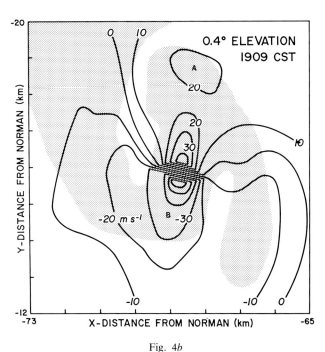

Fig. 4b

Fig. 4. Plot of (a) single-Doppler velocities and (b) overlaid reflectivities for Binger, Oklahoma, tornado of May 22, 1981 [from *Burgess and Lemon*, 1990]. Heavy lines are single Doppler contours; light lines in Figure 4a are radial centers; reflectivity factor >30 dBZ is shaded in Figure 4b. Selected velocities along the vortex centerline are shown for every other gate location in Figure 4a. An example of radar sample volume size (150 m by 0.8°) is shaded in Figure 4a. Peak mesocyclone velocities are marked with "A" and "B" in Figure 4b.

rated at F4 on the Fujita scale (see *Vasiloff* [this volume] for a picture of the tornado and a plot of the TVS signature from another NSSL radar). Doppler velocity data collected in a specially adapted mode, wherein very large wind speed could be more easily resolved, indicated maximum tornado wind speeds of at least 100 m s^{-1} [*Zrnić et al.*, 1985a]. The TVS and tornado location within the mesocyclone may be estimated from Figure 4b, although the mesocyclone signature is somewhat obscured by the strong TVS. The implied mesocyclone diameter is 4 km (see *Wood and Brown* [1983] for a simulation of the two-signature combination). The hook-shaped echo of the reflectivity field (stippled in Figure 4b) is related to the mesocyclone, not the TVS.

It is important to point out that not all tornadoes, even at close range, will have TVSs. Assuming a 1° beam with samples every 1° (adjacent but not overlapping), the maximum detection range for TVSs is estimated at not much more than 100 km (50 nm), even for large tornadoes (1-2 km diameter). Narrow tornadoes (10–50 m diameter) may escape detection at only 20 km (10 nm) range. Thus whole classes of tornadoes may be excluded from observation for most of the radar coverage interval. For example, many nonsupercell tornadoes such as those along gust fronts (weak, narrow, and confined to the boundary layer) do not have detectable TVSs (see *Vasiloff* [this volume] for a plot of one that was detected). Other nonsupercell tornadoes such as landspouts and waterspouts are undetectable beyond close range, and supercell tornadoes cannot be detected at moderate to long range. However, the presence of supercell and nonsupercell tornadoes may be inferred by the strength of the mesocyclones or misocyclones surrounding them. TVS observation can lead to warning lead times of 10 min or more, particularly for supercell tornadoes.

Although there have been a considerable number of TVS observations at National Severe Storms Laboratory (NSSL) [*Vasiloff*, this volume] and at other places (Air Force Geophysics Laboratory [*Kraus*, 1973], National Center for Atmospheric Research [*Wilson et al.*, 1980], the National Weather Service in Alabama [*Petit*, 1990], Colorado [*Dunn*, 1990], and Oklahoma [*Burgess et al.*, 1991]), no complete set of climatological statistics has been compiled. Until that is accomplished, conclusions related to the TVS must be termed tentative and subject to change. The development and interpretation of a large data set will probably await the deployment of the NEXRAD network.

3.2. Mesocyclone Signatures

Mesocyclones serve as temporal bridges between supercell tornadoes and effective forewarning of their destructive power. The initial observation by Doppler radar of a mesocyclone was presented by *Donaldson et al.* [1969]. They observed a pattern suggestive of a mesoscale vortex in a heavily damaging storm with a suspected tornado. The mesocyclone was confirmed by visual observations of cloud base motions offered by an eyewitness.

A single Doppler radar provides only those components of

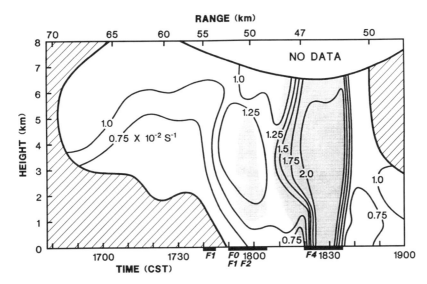

Fig. 5. Time-height plot of shear for the mesocyclone of April 30, 1978 (adapted from *Burgess and Donaldson*, [1979]). Shading indicates the tornadic vortex signature present; solid bars at bottom mark tornado times. The last tornado is at Piedmont, Oklahoma.

the velocity field directed along the radar beam. A pattern suggestive of rotation therefore is subject to ambiguity. Consequently, *Donaldson* [1970] proposed criteria of shear, persistence, and vertical extent to establish confidence in the validity of a mesocyclone signature, by showing that alternative interpretations of this distinctive Doppler velocity pattern are unlikely when the criteria are met.

Shortly thereafter, *Burgess* [1976] initiated a search for mesocyclones in Oklahoma, and during a 5-year period he identified 37 of them using Donaldson's criteria. Tornadoes were associated with 23 of these mesocyclones, with a mean lead time of 36 min from mesocyclone detection to tornado touchdown. In the same study, Burgess found that no verified tornado occurred during radar observations without a preceding mesocyclone signature. Furthermore, all but two of the 37 mesocyclones produced some form of severe weather events at the surface. Recent observations with more sensitive radars and different scanning techniques indicate a somewhat less but still significant association of tornadoes with mesocyclones.

The close association between mesocyclones and tornadoes revealed by the *Burgess* [1976] study, and the promise of warnings offered by the precedence of mesocyclones, provided crucial encouragement for initiation of the Joint Doppler Operational Project (JDOP). The JDOP experiment was designed to determine the improvements, if any, that could be achieved in warning of tornadoes and other severe storm hazards by the introduction of Doppler radar data into an operational scenario. The results reported by the JDOP staff [*Burgess et al.*, 1979] were very impressive, showing improved detection for tornadoes and a remarkably small number of false alarms when compared with using conventional forecasting techniques. Most significantly, the Doppler-aided tornado warnings achieved an average lead time before tornado touchdown of 21 min, compared with the average 2-min lead time for warnings prepared without the benefit of Doppler radar information. This superior JDOP detection and warning performance validated the incorporation of Doppler capability into NEXRAD.

A time-height plot of one of the mesocyclones detected during JDOP (Figure 5) indicates the relationships between the mesocyclone, the TVS, and the accompanying tornadoes. The mesocyclone forms well before the first tornado but is defined only at storm midlevels. The first tornado (F1) occurs as the mesocyclone becomes defined in the boundary layer. The second and third tornadoes form as the mesocyclone shear intensifies (at low levels and midlevels), followed soon by the fourth tornado (F2), accompanied by a brief, midlevel TVS. The fifth and strongest tornado (F4) forms about 15 min after the rapid intensification of a new mesocyclone core and the development of a new TVS. Note that the intense phase of the mesocyclone core and TVS occurrence are almost coincident and that the mesocyclone weakens rapidly as the violent tornado and TVS dissipate. This suggests that vortex stretching and at least partial conservation of angular momentum are associated with the formation of the violent tornado. After the last tornado, the mesocyclone weakens, and its top lowers rapidly.

The case illustrated in Figure 5 is a classic supercell storm. For these storms it can be seen that the mesocyclone signature precedes tornado occurrence by several tens of minutes and that the time of maximum mesocyclone strength is the most likely time for strong and violent tornadoes. Also, the mesocyclone signature extends through a deep layer and therefore is visible at extended ranges from the radar. These characteristics make mesocyclone signatures

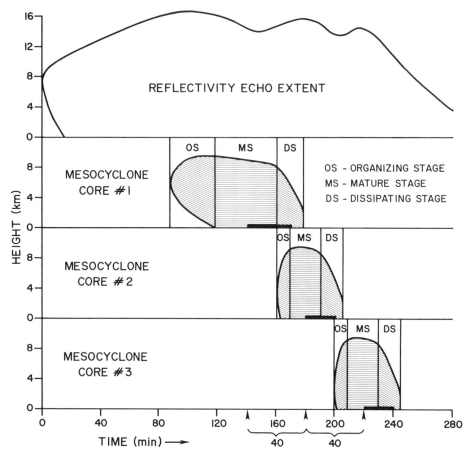

Fig. 6. Time-height evolution of (top) radar echo and (bottom) multiple mesocyclone cores [from *Burgess et al.*, 1982]. Solid horizontal bars are tornado occurrence; tornado formation interval is indicated along bottom.

from supercell storms very useful in the radar contribution to the tornado warning process.

With single-Doppler radar, characteristics of the mesocyclone inner core are readily measured (see Figure 1). The characteristics and evolution of supercell mesocyclone cores have been studied at NSSL [*Burgess et al.*, 1982] and at other research facilities. Core life is composed of three stages [*Burgess and Lemon*, 1990]:

Organizing stage. This is a period of growth, both upward and downward, from midlevel beginnings (5 km height). It is common for convergence to exist below the mesocyclone base. The organizing stage ends when the mesocyclone base extends below cloud base.

Mature stage. This is the period of maximum strength when velocity parameters have their highest values and tornado formation potential is highest. The mesocyclone signature extends through a deep layer, perhaps two thirds of the storm height.

Dissipating stage. This stage begins with a rapid decrease in mesocyclone height and is generally characterized by weakening velocities. At the end of the dissipating stage, the vortex core exists only over a small, shallow depth and is associated with divergence.

Some mesocyclones produce just one core, but others produce a series of cores. These cores occur in a predictable way, acting as the mesocyclone propagation mechanism. The relationship between multiple mesocyclone cores is conceptualized in Figure 6. The mesocyclone forms after the storm echo is mature and near its peak height. The first (and sometimes only) mesocyclone core has a relatively long organizing and mature stage. The second core organizes as the first begins dissipating. Second and succeeding cores have extremely short organizing stages as they quickly form over a large depth and have relatively short mature stages as evolution proceeds rapidly. Some supercell storms have a long succession of cores, and the mesocyclone persists for several hours. These storms can produce "tornado families" with a tornado recurrence interval of about 40 min.

A conceptual model of the evolution of a supercell mesocyclone in horizontal section (Figure 7) reveals the existence of a gust front that wraps cyclonically about the mesocyclone core. The core evolution closely resembles synoptic

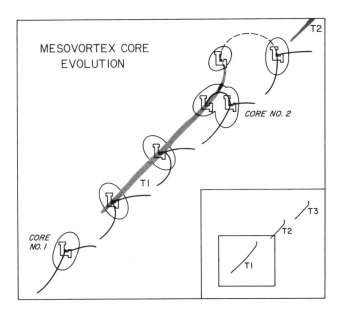

Fig. 7. Conceptual model of mesocyclone core evolution [from *Burgess et al.*, 1982]. Heavy lines are low-level wind discontinuities; tornado tracks (exaggerated for size indication) are shaded. Insert shows tornado family tracks, and the square is the region expanded in the figure.

cyclone development: during the first core mature stage, the gust front accelerates around the right flank; occlusion occurs, and as the mesocyclone warm sector separates from the first core, its dissipation begins; strong convergence localizes at the point of the occlusion, and a second vortex core organizes rapidly because of the vorticity-rich environment in which it forms.

Not all mesocyclones produce tornadoes. In fact, the converse is true; most mesocyclones do not produce tornadoes. Statistics available from 20 years of Doppler radar operation in Oklahoma indicate that only in years with large numbers of significant supercells do 50% of the mesocyclones produce tornadoes. In other years, tornadoes are produced by as few as 30% of the mesocyclones. The long-term average is somewhere between 30% and 50%.

The job of trying to categorize mesocyclones by their tornado potential has not been easy. Data from 45 well-observed signatures (Figure 8) indicate few well-defined boundaries between mesocyclones that produced no tornadoes, weak or strong tornadoes, and violent tornadoes. There seems to be a tendency for tornadic mesocyclones to have higher rotational velocity and stronger shear.

Donaldson and Desrochers [1985], in a search for an indicator of mesocyclone intensity that might correlate with the severity of subsequent tornadoes, proposed calculation of excess rotational kinetic energy (ERKE) of the mesocyclone core. ERKE is the rotational kinetic energy remaining after subtraction of the product of core radius and a selected value of shear from the observed rotational velocity. The resultant diminished velocity can be regarded as excess to the minimum requirement for maintenance of mesocyclonic shear.

Continuation of ERKE study by *Desrochers et al.* [1986] showed that a notable quality of tornadic mesocyclones is

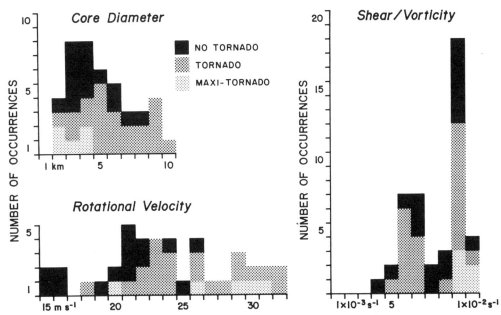

Fig. 8. Characteristic parameters of mature supercell mesocyclones for tornadic and nontornadic storms [from *Burgess et al.*, 1982]. Tornado implies F0–F3 and maxi-tornado implies F4-F5.

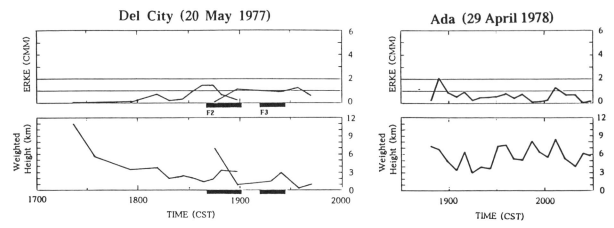

Fig. 9. Examples of mesocyclone excess rotational kinetic energy (ERKE) evolution for a tornadic storm (Del City) and a nontornadic storm (Ada) [from *Desrochers et al.*, 1986]. ERKE is normalized by the value of a climatological mature mesocyclone (CMM). Tornado durations are indicated by the solid bars.

their intensification at low levels prior to tornado formation. The evolution disparity between tornadic and nontornadic storms is borne out well by the Del City (May 20, 1977) and Ada (April 29, 1978) storms (Figure 9). Although the Ada mesocyclone was at times as intense as Del City, this energy was confined to midlevels, precluding tornado formation. In contrast, Del City generated two strong tornadoes.

Recently *Donaldson and Desrochers* [1990], with an expanded sample of 17 Oklahoma mesocyclonic storms, improved the ERKE technique by integrating the values from surface up to several kilometers in height. They also performed similar integrations for other mesocyclonic features, including velocity and shear. They found (Figure 10) that ERKE predicted the violent (F4) tornadoes superbly, with a minimum lead time of 28 min and no false alarms. For the strong tornadoes (F2 and F3), velocity as well as ERKE provided a more than adequate warning lead time to the first occurrence of a strong tornado in a storm. Little or no skill was achieved for identification of the so-called weak (F0 and F1) tornadoes.

Recently, National Weather Service offices with access to Doppler radars have based accurate tornado warnings on detecting mesocyclones [*Dunn*, 1990; *Burgess et al.*, 1991, *Burgess and Lemon*, 1991]. Mesocyclone signatures are the most often used Doppler radar inputs to tornado warnings because they are larger and deeper than TVSs or misocyclones and therefore can be seen throughout all of the velocity coverage range of modern Doppler radars (230 km or 125 nm).

There are limitations in using mesocyclone information in tornado warnings. First, it is indirect evidence of a tornado. As already mentioned, less than 50% of all mesocyclones produce verified tornadoes. Second, remember that Earth curvature (horizon limitation) places the center of the lowest-elevation radar beam at greater than 4 km height when the range is 200 km (100 nm). No information is available on the low-level character of the mesocyclone, the important portion near the ground. Third, beam averaging will reduce the mesocyclone peak rotational velocity and, perhaps, cause underestimates of mesocyclone parameters. For all of these reasons, it is generally best to correlate Doppler signatures with spotter reports whenever possible.

3.3. Misocyclone Signatures

Beginning with the late 1970s, there was realization that Doppler detectable circulations and tornadoes occurred with other storm types besides supercells. *Fujita* [1979] documented a tornado along the leading edge of a squall line with a bow-shaped echo. *Burgess and Donaldson* [1979] compared a supercell mesocyclone with a smaller, weaker circulation in a newly developing, multicell storm and found the two signatures different. *Forbes and Wakimoto* [1983], studying conventional radar data, suggested that tornadoes observed with a squall line came from small-scale circulations, perhaps formed in association with interacting downbursts and gust fronts.

Bluestein [1985] observed a developing squall line and saw a waterspoutlike tornado that he dubbed a "landspout." Doppler studies of these phenomena by *Wilson* [1986], *Brady and Szoke* [1989], and *Wakimoto and Wilson* [1989] identified small-scale, single-Doppler signatures preceding and surrounding these nonsupercell tornadoes (see Plate 1 for an example). Since the signatures are smaller than average mesocyclones with supercells, they have been called misocyclones (less than 4 km size in the scale classification of *Fujita* [1981]).

Misocyclones and associated tornadoes occur with multicell storms, with ordinary cells in squall lines, and with newly developing storms (sometimes just cumulus congestus) along low-level, wind shear boundaries [*Wilson*, 1986]. A time-height plot of a misocyclone (Figure 11) indicates that the circulation first develops before there is appreciable radar echo, is mainly confined to the boundary layer during

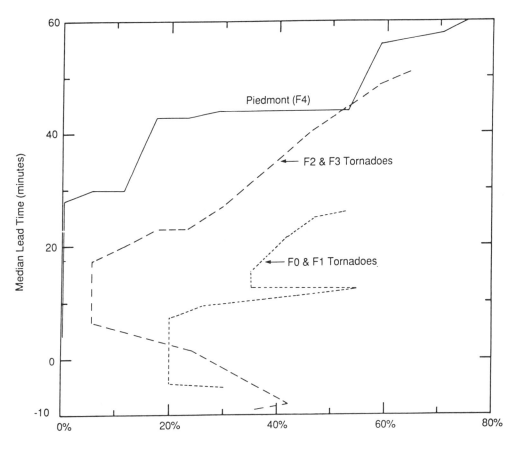

Fig. 10. Performance of excess rotational kinetic energy in warning of three intensity classes of tornadoes: violent (solid line), strong (long-dashed line), and weak (short-dashed line) [from *Donaldson and Desrochers*, 1990]. The ordinate gives median lead time, considering only the earliest tornado of that intensity in a storm. The abscissa shows combined errors expressed as the sum of false alarms and failed detections.

its lifetime, and has a relatively short lifetime. In general, these characteristics are much different from the supercell mesocyclone of Figure 5.

On the basis of Doppler radar data from Colorado, *Brady and Szoke* [1989] proposed a conceptual model of misocyclone tornado formation (Figure 12). Small vortices develop in the boundary layer along convergent wind-shift lines. New convective updrafts stretch and intensify the misocyclone. A tornado can result if the updrafts and the intermediate rotation are strong. As a strong thunderstorm is produced, the misocyclone weakens, and its diameter increases when precipitation-induced downdraft replaces updraft in the vortex vicinity.

Warnings of tornadoes associated with misocyclones are limited for several reasons. *Burgess and Donaldson* [1979] called attention to several instances of tornadoes associated with Oklahoma misocyclones. In each case the tornadoes formed during the early development phase of the storm, offering no opportunity, even under the best of viewing conditions, to verify rotation in the incipient misocyclone in time to provide a positive warning. In fact, in one case a tornado developed 5 min, before detection of a first radar echo! Furthermore, the small size of a misocyclone imposes a severe limit to the range at which the pattern of Doppler velocities might indicate the possibility of a tornado. Finally, there are indications that the most intense rotation of misocyclones is generally confined to heights just above the ground, and these most active parts of the phenomenon would be obscured by the Earth's curvature except at nearby ranges.

The possibility exists that certain large-scale features of the storm environment may provide some basis for a generalized warning of misocyclonic tornadoes. Some of the tornadoes occur in conditions of large thermal instability and near boundaries detected on synoptic surface maps. The Newkirk, Oklahoma, tornadoes of April 17, 1978, occurred in a storm developing rapidly at the intersection of a dry line and an overtaking cold front. Photographs presented by *Donaldson and Burgess* [1982] show that both boundaries were detected as clear-air echoes (Figure 13). The intersec-

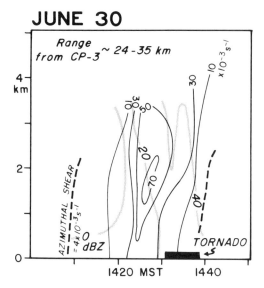

Fig. 11. Time-height plot of misocyclone shear on June 30, 1987. Gray lines are reflectivity; dashed line is threshold shear. The tornado is indicated by the solid bar at bottom. (From *Wakimoto and Wilson* [1989].

tion of the boundaries served as a preferred location for the strongest convective activity.

Wakimoto and Wilson [1989] reported on 17 misocyclones, none of them related to a supercell, during a Colorado field project in the summer of 1987. All of the vortices were located on radar-detected convergence boundaries, and eight of them occurred near the intersection of two or more boundaries, similar to the situation of the Newkirk, Oklahoma, storm. The misocyclone signatures were detected at ranges averaging 25 km, with a maximum range from the radar of 45 km (22 nm). Interestingly, there was some lead-time as the radar signatures preceded the observation of a visual vortex by an average of 14 min. The nearby locations and a radar capable of detecting weaker echoes may explain the lead-time differences from the Oklahoma observations.

Although misocyclone tornadoes are generally weak and short-lived, they are capable of inflicting injuries or fatalities, and deserve our best efforts to warn of their appearance. Unfortunately, the study by Wakimoto and Wilson suggests that a large minority of nonsupercell tornadoes may be undetectable by radar, with the detectable ones limited to a maximum range of around 50 km (25 nm), and with marginal warning times in many cases. However, these authors remind us that convergence boundaries, the spawning grounds of nonsupercell tornadoes, are often detected by radar to ranges of 100 km (50 nm) or more, and they suggest monitoring boundaries displaying strong horizontal shear, those with wavelike inflections, and intersections of boundaries as locations of possible tornadogenesis. We agree and offer the further suggestion that regions in boundaries with the greatest reflectivity may offer an alternative and perhaps more detectable means for indicating strong convergence.

3.4. *Conventional Radar*

We offer a short treatment of tornado warnings with conventional radar (reflectivity) signatures. This is justified because conventional radar is still the standard for most of the operational community and because proper use of Doppler radar involves combined interpretation of reflectivity as well as velocity signatures.

Sometimes the shape of reflectivity contours of radar echoes may provide a distinctive clue to the dynamics of processes occurring within storms (see Figure 4b). A hook-shaped appendage to a convective echo, the so-called tornado hook, has attracted the attention of forecasters since the early reports by *Stout and Huff* [1953] and *Fujita* [1958].

A number of years later, *Forbes* [1981] provided a definitive study of the reliability of the hook echo as a tornado signature. He found a moderate false alarm ratio but only a low probability of detection of a hook echo in a tornadic storm. Furthermore, the hook was absent three quarters of the lifetime of a tornadic storm and often made its first appearance after a tornado had already started. Forbes found somewhat better results with so-called "distinctive echoes." Clearly, hook echoes are not very helpful as a tornado warning indicator, though their presence may serve to reinforce the more accurate and timely predictors achievable through Doppler radar.

The bounded weak echo region (BWER) is another distinctive reflectivity feature that can be readily detected with conventional radar. The BWER, when observed for more than a brief interval, is a reliable indicator of a supercell, a storm type that produces the overwhelming majority of strong tornadoes and nearly all of the violent ones. In horizontal section the BWER is characterized by a region of diminished reflectivity entirely surrounded and capped by higher reflectivity. In vertical section, the highest echo top above the BWER marks the location of the intense, sustained updraft required for a supercell [*Browning and Ludlam*, 1962].

The value of the BWER as a forecasting aid in warning of severe and tornadic storms was fully appreciated by *Lemon* [1980] after his study of many Oklahoma storms. He devised an observing scheme to aid operational forecasters to rapidly identify storm types incorporating the BWER. His criteria, or variations thereof, based on three-dimensional reflectivity structure, are considered to be the most reliable indicators of tornado likelihood during the pre-Doppler operational era, at least in environments supportive of supercell formation.

3.5. *Estimated Skill*

An estimation of radar's potential skill in tornado warnings is given. These suggestions for operational performance of Doppler weather radar emanate from pseudo-operational tests performed over the last decade and include JDOP, Doplight (Doppler and lightning data application [*Forsyth et al.*, 1989]), CINDE (convective initiation and downburst experiment [*Wakimoto and Wilson* 1989]), and NEXRAD

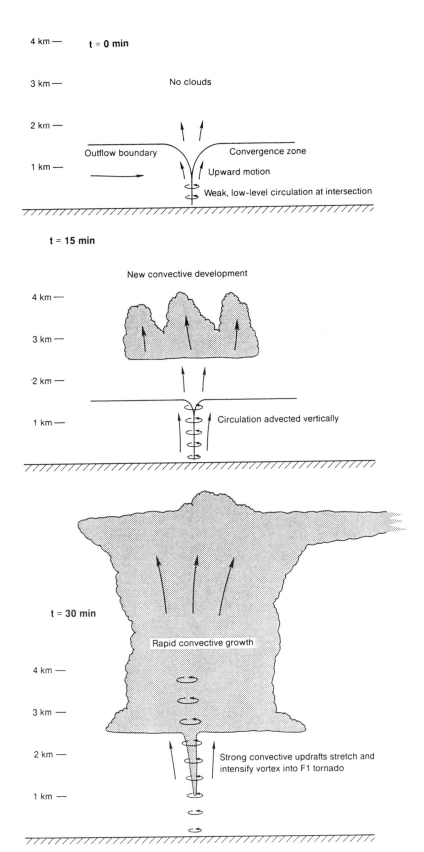

Fig. 12. Conceptual model of misocyclone and landspout tornado evolution. [from *Brady and Szoke*, 1989].

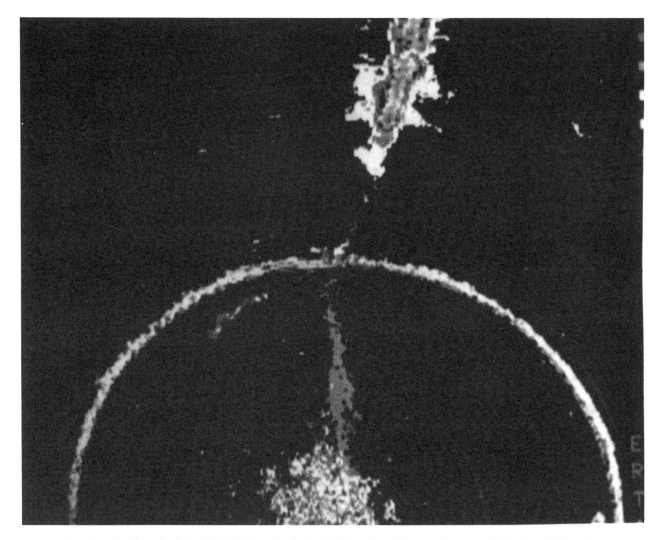

Fig. 13. Doppler relectivity at 1357 CST on April 17, 1978 [from *Donaldson and Burgess*, 1982]. One thin line (dry line) oriented north-south and a second thin line (cold front) oriented northeast-southwest are clear-air returns. Thunderstorm echoes are seen north of dry line/front intersection point. Range mark is 115 km.

IOT&E (initial operational test and evaluation, 1989; *Weyman and Clancy* [1989; also unpublished results, 1991]).

Radar of course is just one input into warning decision making, but the skill of that part of the warning process can be estimated. Also, warning skill will be discussed for operational radars (such as the WSR-88D), and skill scores will reflect values for the entire coverage range of the radars (assumed to be 230 km (125 nm)). Further, it is assumed that the radars meet network specifications (i.e., 10-cm wavelength, narrow beam width, high sensitivity, elaborate real-time processing, etc.). Finally, it should be known that better performance (higher skill) will be achieved for local use of the radars, such as protecting a nearby metropolitan area.

Table 1 lists estimated skill for the most commonly used radar signatures. The probability of detection (POD) is high for mesocyclone signatures because they can be seen over the entire radar coverage interval, but it is low for TVS and misocyclone signatures because they can only be seen in less than a quarter of the radar coverage area. POD is moderate for the hook echo because of its occasional, relatively long range of observation and close association with mesocyclones. The hook echo, however, has a high false alarm ratio

TABLE 1. Estimated Tornado Warning Skill Scores (Network Coverage of 125 nm)

Predictor	POD	FAR	CSI
Hook echo	M	H	L
TVS	L	L	L
Mesocyclone	H	M	M
Misocyclone	L	M	L

L, low; M, medium; H, high.

(FAR) owing to the high degree of subjectivity in correct identification. Mesocyclone and misocyclone signatures have moderate FARs because of the significant percentage of parent vortices that do not spawn tornadoes. Only the TVS, as the most direct indicator of a tornado, has a low FAR.

The resultant critical success index (CSI [*Donaldson et al.*, 1975]) for radar warning of tornadoes is low for all signatures except mesocyclones, where a moderate CSI is estimated. We feel that with today's radar technology and our current level of understanding a high CSI is unlikely except for violent tornadoes. Higher levels of success may be achieved through the combination of radar signatures with accurate and timely spotter reports.

Of course, it is disappointing that many tornadoes will go undetected, even with Doppler radar, but all misses should not be treated the same. Most tornadoes missed by Doppler radar are likely to be weak, small, and short-lived (F0-F1 on the Fujita scale). Strong tornadoes (F2-F3) and, particularly, violent tornadoes (F4-F5) seem to require a Doppler detectable signature for their existence. Also, a potential benefit of widespread Doppler use will be a reduction in FARs for tornado warnings. Some measures of pre-Doppler skill [e.g., *Kelly and Schaefer*, 1982] indicate FARs as high as 80%.

4. Automated Radar Algorithms

Algorithms for automated analysis and detection of severe weather phenomena such as TVSs and mesocyclones can significantly assist the forecaster if they are reliable, but these algorithms are likely to be dismissed if they are generally perceived to be ineffective [*Walker and Heideman*, 1989]. Reliability is utmost. Technically, an algorithm must be sufficiently robust to address the problem at hand while its computation load and memory requirements are kept to a minimum. An algorithm must have predictable response to permit operation in real time.

Hennington and Burgess [1981] developed a viable approach for automated vortex detection that has been adopted in general form in most subsequent TVS and mesocyclone algorithms. In the Rankine combined vortex used to model a mesocyclone, the relative velocity along the radar beam changes from one sign to the other across the mesocyclone core, and therefore the sign of the velocity gradient remains constant across the core. Segments where the sign of the velocity gradient is constant are referred to as pattern vectors (Figure 14). Since the mesocyclone is essentially a rotational phenomenon, pattern vectors are best defined in the azimuthal direction.

The beginning and ending points of a pattern vector represent relative maxima of incoming and outgoing velocities. For a mesocyclone or TVS the velocity maxima are located at the core radius. Since we are interested only in defining the core, it is sufficient to retain just the pattern vector endpoints and velocity maxima. Pattern vector analysis techniques therefore represent an efficient and concise methodology for extraction of well-behaved phenomena.

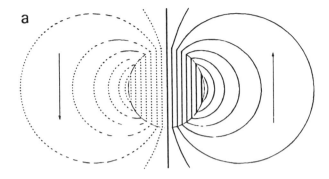
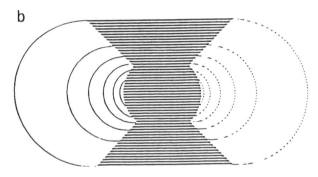

Fig. 14. (*a*) Single-Doppler mesocyclone couplet and (*b*) companion pattern vectors.

The basic steps of the mesocyclone and TVS algorithms are the identification of pattern vectors, the association of vectors from the same elevation scan into two-dimensional features, the association of features from differing elevations, and the evaluation of features for meteorological significance. Although the general frameworks of the algorithms to be discussed are similar, the intermediate steps are quite distinct. The following subsections will address some of the primary considerations and differences among the approaches.

4.1. Mesocyclone Algorithms

The ideas presented by *Hennington and Burgess* [1981] were implemented into a working mesocyclone algorithm by *Zrnić et al.* [1985b], and the technique was provisionally selected for use in NEXRAD. At the heart of this algorithm is the establishment of criteria to help restrict the detection of pattern vectors to mesocyclones, which has positive implications for memory requirements and processing speed. Thresholds of shear and momentum (length times velocity, which is more precisely momentum per unit mass) are used to identify likely mesocyclone-associated pattern vectors.

Pattern vectors from the same elevation scan are associated into features based on a general proximity condition. Feature parameters like size, shear, and momentum are determined from the collective values of all the pattern vectors that compose a feature. Values are momentum weighted to bias toward the greatest values of a feature. Features satisfying a shape and size requirement are identified as mesocyclonelike, having already satisfied pattern vector criteria for mesocyclones. Features from different elevations are associated together into the same feature if they overlap. Finally, a mesocyclone is identified when two or more elevations of a vertical feature contain a mesocyclonelike feature.

In a new algorithm developed by *Desrochers* [1991], thresholds of shear and velocity difference are used for pattern vector retention. This is done to overcome the size restriction imposed by momentum thresholding, permitting concurrent detection of TVSs.

Following the lead of *Wieler* [1986] in an earlier mesocyclone algorithm, resolution corrections are applied to the pattern vector thresholds in order to range normalize the detection criteria. This approach is based on work by *Brown and Lemon* [1976] illustrating how a Rankine combined vortex is viewed with varying radar resolution. Resolution corrections for vortex size and rotational velocity can be applied if the radar beam width is no greater than the mesocyclone core radius. As resolution degrades further the necessary correction becomes ambiguous. For an average size mesocyclone (core diameter of 5 km), range normalization is possible to ranges of about 150 km (75 nm) assuming illumination by a 1° beam.

Mesocyclone parameters such as size and rotational velocity are determined from a weighted average of pattern vectors by using only selected pattern vectors centered around the peak velocities of a feature. The technique allows the degree of rotation and divergence associated with any particular mesocyclone signature to be determined.

Three-dimensional (3-D) features are constructed by finding the "best" combination of features from differing elevations. This process considers the distance between features as well as their energy in ERKE. The objective is to identify the most energetic 3-D features. A 3-D feature is allowed to slope as much as 45° to account for actual tilt and to account for feature movement during the volume scan period. A tracking routine associates features from consecutive volume scans. Mesocyclones are identified by applying the classification criteria first proposed during JDOP.

One significant task that remains for NEXRAD is the regionalization of algorithms. This is especially true for mesocyclone detection. The NEXRAD provisional algorithm and the Desrochers' algorithm were designed entirely around Oklahoma-type supercell storms. Characteristics of shear and vertical extent are likely to vary significantly from region to region.

One example of site sensitivity is Ontario, Canada. *Joe* [1989] has found that the NEXRAD mesocyclone algorithm with the prescribed Oklahoma thresholds is not sensitive to the general Canadian mesocyclone. Pattern vector thresholds of shear and momentum were reduced by as much as 75% of their Oklahoma value to detect these phenomena.

Given this, Canadian mesocyclones appear to provide similar lead times to their American counterparts, 20 or more minutes from the time of first mesocyclone detection to incipient tornado formation.

4.2. *TVS Algorithms*

The provisional TVS algorithm for NEXRAD [*Forsyth*, 1984] was designed to operate in conjunction with the NEXRAD mesocyclone algorithm. Any detected mesocyclones are reexamined for regions of large, TVS-like shear (0.05 s^{-1}). When strong, rotationally indicative shear is detected at multiple elevations of a mesocyclone feature, a TVS is declared. Currently, the technique detects only cyclonically rotating features.

A shortcoming of the current NEXRAD TVS algorithm is its restriction of TVS detection to supercell-type tornadoes. Nonsupercell tornadoes, like those commonly observed in Colorado [*Wilson*, 1986; *Wakimoto and Wilson*, 1989], will escape detection with such a technique. *Albers* [1991] has subsequently modified the NEXRAD mesocyclone algorithm to detect these phenomena. For example, pattern vector thresholds of shear have been increased while those of momentum have been reduced, and an additional requirement for velocity difference across the pattern vector has been added to accept the pattern vectors of weak TVSs that momentum thresholding would eliminate. These changes have resulted in detection of some Colorado tornadoes, particularly at close range. Such results attest to the inherent tunableness of the original NEXRAD design.

A new TVS algorithm is under development at NSSL [*Vasiloff*, 1991; this volume] for possible inclusion in the Terminal Doppler Weather Radar Program (TDWR). Pattern vectors are utilized, but only adjacent azimuths are considered. Range-consecutive pairs of gates are required to meet certain velocity difference thresholds, and three elevation angles must be vertically associated to verify a TVS. No alignment with a mesocyclone signature is necessary for identification.

The Vasiloff algorithm has been tested on tornadoes from different type storms, occurring in different parts of the country. Results thus far are somewhat encouraging in that both nonsupercell tornadoes (gust front tornadoes, landspouts, etc.) and supercell tornadoes have been detected. However, as expected, detectability decreases as a function of reduced tornado size and range from the radar. Also, rapid scanning is necessary to detect short-lived tornadoes, and even then, some are missed.

4.3. *Tornado Prediction Algorithms*

Algorithms can reduce the tedium of radar monitoring and serve to call attention to developing situations. They can also provide useful quantitative assessments as well. One such application is the use of numerical output of the mesocyclone algorithm for tornado prediction. ERKE, a most promising identifier of tornadic mesocyclones, can be

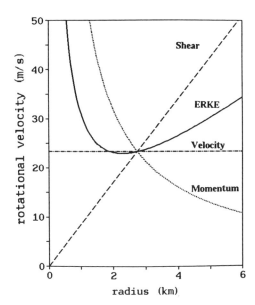

Fig. 15. The relationships among some mesocyclone evaluators using mean values of mature mesocyclone intensity.

readily obtained from the standard output of the mesocyclone algorithm.

Besides the initial performance capability for tornado prediction demonstrated thus far [*Donaldson and Desrochers*, 1990], ERKE has certain design attributes that make it attractive for algorithmic use. ERKE is conceived as a single-parameter function that relates tornado likelihood to one measure of energy. It is a continuous function that evolves gradually in magnitude between tornadic and nontornadic mesocyclones. It is readily tunable to varying storm structure. A comparison of ERKE with velocity, shear, and momentum is shown in Figure 15 for a typical mature mesocyclone as defined by *Burgess et al.* [1982].

Recently, the ERKE evaluation technique has been automated in the *Desrochers* [1991] mesocyclone algorithm. *Desrochers and Donaldson* [1992] tested it as a tornado probability algorithm on 23 Oklahoma mesocyclonic storms. They found the ERKE magnitude of tornadic mesocyclones distinctive from other mesocyclones at the 99.9% level of significance. In this limited sample, ERKE provided a median 27 min lead time for the first strong tornado (F2-F3) of a storm with no errors. ERKE performance statistics (Figure 16) show that there is a trade off between lead times above 27 min and false alarms. What is not shown in the figure is that the threshold range of ERKE necessary to provide positive lead times for tornadoes with small errors is large enough so that this algorithm should not be overly sensitive to viewing angle variations.

5. Future Directions

The timing of Tornado Symposium III in 1991 finds the weather radar community on the verge of a new era, just as it was with the beginning of the Doppler radar research era at the time of Tornado Symposium II. As described in the preceding sections, the implementation of the NEXRAD network will bring vast changes to the use of operational radar in tornado warnings. The new WSR-88D radars will feature rapid, three-dimensional scanning and will bring powerful new signatures and algorithms [*Alberty et al.*, 1991]. Plate 2 gives examples of useful WSR-88D products from a recent outbreak of violent tornadoes in Oklahoma and Kansas (April 26, 1991).

The advent of the NEXRAD network will aid warning forecasters greatly, but certain problem areas must be overcome to achieve maximum potential. Algorithm improvements are needed in several areas. Adaptation of algorithms to regional and seasonal differences in storm and tornado climatologies is needed. Overall algorithm immaturity, including the inability to detect consistently the existence of all hazards, will guide the near-future direction of applied research and development.

In the longer term, an increase in our understanding of thunderstorm morphology and tornado formation processes are essential to improved data interpretation and new algorithm creation. New conceptual models of tornadic weather systems are expected from analysis of WSR-88D data.

NEXRAD will revolutionize our nowcasting capability, but this technology is not singularly sufficient to provide for all our detection and warning needs. Warning/forecasting workstations that combine radar outputs with other weather data are needed. Because we know that the surrounding environment is related to storm type and tornado potential, diagnosis of a storm's mesoscale environment will lead to earlier and more reliable warnings.

Improvements in interaction between warning forecasters and the preparedness community (emergency management officials, storm spotters, the media, etc.) will result in better warnings. For example, there is no plan for Civil Defense

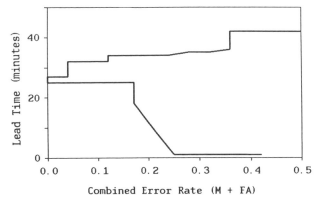

Fig. 16. Median lead time to the first strong tornado provided by excess rotational kinetic energy compared to errors composed of misses (M) and false alarms (FA).

directors to view the same radar displays that forecasters use, and forecasters can neither see spotter locations on their radar displays nor participate in directing the spotters to needed areas. Similar problems relate to interactions with the broadcast community. Radar is just a portion of the total warning system necessary to adequately protect the public; all portions of the system must function together if lives are to be saved.

Acknowledgments. Contribution by Ralph Donaldson to this manuscript was supported by the United States Air Force, Phillips Laboratory Geophysics Directorate, under contract F19628-90-C-0088. We thank Jim Wilson, Arthur Witt, and Steve Vasiloff for manuscript review and Joan Kimpel for graphics preparation.

References

Albers, S. C., The detection of isolated tornadic vortex signatures with NEXRAD algorithms, in *Preprints, 25th International Conference on Radar Meteorology*, pp. 123–126, American Meteorological Society, Boston, Mass., 1991.

Alberty, R. A., T. Crum, and F. Toepfer, The NEXRAD program: Past, present and future; a 1991 perspective, in *Preprints, 25th International Conference on Radar Meteorology*, pp. 1–8, American Meteorological Society, Boston, Mass., 1991.

Armstrong, G. M., and R. J. Donaldson, Jr., Plan shear indicator for real-time Doppler radar identification of hazardous storm winds, *J. Appl. Meteorol.*, 8, 376–383, 1969.

Atlas, D., Radar analysis of severe storms, *Meteorol. Monogr.*, 5, 177–220, 1963.

Bluestein, H. B., The formation of a "landspout" in a "broken line" squall line in Oklahoma, in *Preprints, 14th Conference on Severe Local Storms*, pp. 267–270, American Meteorological Society, Boston, Mass., 1985.

Brady, R. H., and E. J. Szoke, A case study of nonmesocyclone tornado development in northeast Colorado: Similarities to waterspout formation, *Mon. Weather Rev.*, 117, 843–856, 1989.

Brown, R. A., and L. R. Lemon, Single-Doppler radar vortex recognition. Part II: Tornadic vortex signatures, in *Preprints, 17th Radar Meteorological Conference*, pp. 104–109, American Meteorological Society, Boston, Mass., 1976.

Brown R. A., L. R. Lemon, and D. W. Burgess, Tornado detection by pulsed Doppler radar, *Mon. Weather Rev.*, 106, 29–38, 1978.

Browning, K. A., and F. H. Ludlam, Airflow in convective storms, *Q. J. R. Meteorol. Soc.*, 88, 117–135, 1962.

Burgess, D. W., Single-Doppler radar vortex recognition: Part I: Mesocyclone signatures, in *Preprints, 17th Conference on Radar Meteorology*, pp. 97–103, American Meteorological Society, Boston, Mass., 1976.

Burgess, D. W., and R. J. Donaldson, Jr., Contrasting tornadic storm types, in *Preprints, 11th Conference on Severe Local Storms*, pp. 189–192, American Meteorological Society, Boston, Mass., 1979.

Burgess, D. W., and L. R. Lemon, Severe thunderstorm detection by radar, in *Radar in Meteorology*, edited by D. Atlas, pp. 619–647, American Meteorological Society, Boston, Mass., 1990.

Burgess, D. W., and L. R. Lemon, Characteristics of mesocyclones detected during a NEXRAD test, in *Preprints, 25th International Conference on Radar Meteorology*, pp. 39–42, American Meteorological Society, Boston, Mass., 1991.

Burgess, D. W., L. R. Lemon, and R. A. Brown, Tornado characteristics revealed by Doppler radar, *Geophys. Res. Lett.*, 2, 183–184, 1975.

Burgess, D. W., R. J. Donaldson, Jr., T. Sieland, and J. Hinkelman, Final report on the Joint Doppler Operational Project (JDOP). Part I: Meteorological Applications. *NOAA Tech. Memo. ERL NSSL-86*, 84 pp., Natl. Oceanic and Atmos. Admin., Boulder, Colo., 1979, (Available as *NTIS PB80-107/88/AS* from Natl. Tech. Inf. Serv., Springfield, Va.)

Burgess, D. W., V. T. Wood, and R. A. Brown, Mesocyclone evolution statistics, in *Preprints, 12th Conference on Severe Local Storms*, pp. 422–424, American Meteorological Society, Boston, Mass., 1982.

Burgess, D. W., D. L. Andra, Jr., and W. F. Bunting, The Stillwater, Oklahoma tornado of May 15, 1990: A case study of the use of radar in the modernized National Weather Service, in *Preprints, 25th International Conference on Radar Meteorology*, pp. 51–54, American Meteorological Society, Boston, Mass., 1991.

Desrochers, P. R., Automated mesocyclone detection and tornado forecasting, *Tech. Rep. 91-2051*, 168 pp., Phillips Lab., Hanscom Air Force Base, Md., 1991.

Desrochers, P. R., and R. J. Donaldson, Jr., Automatic tornado prediction with an improved mesocyclone detection algorithm, *Weather Forecasting*, in press, 1992.

Desrochers, P. R., R. J. Donaldson, Jr., and D. W. Burgess, Mesocyclone rotational kinetic energy as a discriminator for tornadic and non-tornadic types, in *Preprints, 23rd Conference on Radar Meteorology*, pp. 1–4, American Meteorological Society, Boston, Mass., 1986.

Donaldson, R. J., Jr., Vortex signature recognition by a Doppler radar, *J. Appl. Meteorol.*, 9, 661–670, 1970.

Donaldson, R. J., Jr., Observations of the Union City tornadic storm by plan shear indicator, *Mon. Weather Rev.*, 106, 39–47, 1978.

Donaldson, R. J., Jr., and D. W. Burgess, Results of the Joint Doppler Operational Project, in *Proceedings of the NEXRAD Doppler Radar Symposium*, CIMMS, University of Oklahoma, Norman, pp. 102–123, 1982.

Donaldson, R. J., Jr., and P. R. Desrochers, Doppler radar estimates of the rotational kinetic energy of mesocyclones, in *Preprints, 14th Conference on Severe Local Storms*, pp. 52–55, American Meteorological Society, Boston, Mass., 1985.

Donaldson, R. J., Jr., and P. R. Desrochers, Improvement of tornado warnings by Doppler radar measurement of mesocyclone rotational kinetic energy, *Weather Forecasting*, 5, 247–258, 1990.

Donaldson, R. J., Jr., G. M. Armstrong, A. C. Chmela, and M. J. Kraus, Doppler radar investigation of air flow and shear within severe thunderstorms, in *Preprints, 6th Conference on Severe Local Storms*, pp. 146–154, American Meteorological Society, Boston, Mass., 1969.

Donaldson, R. J., Jr., R. M. Dyer, and M. J. Kraus, An objective evaluator of techniques for predicting severe weather events, in *Preprints, 9th Conference on Severe Local Storms*, pp. 321–326, American Meteorological Society, Boston, Mass., 1975.

Doviak, R. J., and D. S. Zrnić, *Doppler Radar and Weather Observations*, 458 pp., Academic, San Diego, Calif., 1984.

Dunn, L. B., Two examples of operational tornado warnings using Doppler radar data, *Bull. Am. Meteorol. Soc.*, 71, 145–153, 1990.

Forbes, G. S., On the reliability of hook echoes as tornado indicators, *Mon. Weather Rev.*, 109, 1457–1466, 1981.

Forbes, G. S., and R. M. Wakimoto, A concentrated outbreak of tornadoes, downbursts and microbursts, and implications regarding vortex classification, *Mon. Weather Rev.*, 111, 220–235, 1983.

Forsyth, D. E., TVS detection algorithm description, NEXRAD Algorithm Report, NEXRAD Joint Sys. Prog. Office, Washington, D. C., Dec. 1984.

Forsyth, D. E., D. W. Burgess, M. H. Jain, and L. E. Mooney, DOPLIGHT '87: Application of Doppler radar technology in a National Weather Service Office, in *Preprints, 24th Conference on Radar Meteorology*, pp. 198–202, American Meteorological Society, Boston, Mass., 1989.

Fujita, T. T., Mesoanalysis of the Illinois tornadoes of 9 April 1953, *J. Meteorol.*, 15, 288–296, 1958.

Fujita, T. T., Objectives, operations, and results of Project NIM-

ROD, in *Preprints, 11th Conference on Severe Local Storms*, pp. 259–266, American Meteorological Society, Boston, Mass., 1979.

Fujita, T. T., Tornadoes and downbursts in the context of generalized planetary scales, *J. Atmos. Sci.*, *38*, 1511–1534, 1981.

Hennington, L. D., and D. W. Burgess, Automated recognition of mesocyclones from single-Doppler data, in *Preprints, 20th Conference on Radar Meteorology*, pp. 704–706, American Meteorological Society, Boston, Mass., 1981.

Joe, P. I., The King City mesocyclone detection algorithm, report, 46 pp., Atmos. Environ. Serv. of Can., ARPP, Toronto, Ont., 1989.

Kelly, D. L., and J. T. Schaefer, Implications of severe local storm warning and verification, in *Preprints, 12th Conference on Severe Local Storms*, pp. 459–462, American Meteorological Society, Boston, Mass., 1982.

Kraus, M. J., Doppler radar observations of the Brookline, MA tornado of 9 August 1972, *Bull. Am. Meteorol. Soc.*, *54*, 519–524, 1973.

Lemon, L. R., Severe thunderstorm radar identification techniques and warning criteria, *NOAA Tech. Memo. NWS NSSFC-3*, 60 pp., Natl. Oceanic and Atmos. Admin., Boulder, Colo., 1980. (Available as *NTIS PB-273049* from Natl. Tech. Inf. Serv., Springfield, Va.)

Petit, P. E., Doppler radar detection capabilities at Montgomery, Alabama, 1982–1988, *Natl. Weather Dig.*, *15*, 23–28, 1990.

Stout, G. E., and F. A. Huff, Radar records Illinois tornadogenesis, *Bull. Am. Meteorol. Soc.*, *34*, 281–284, 1953.

Vasiloff, S. V., The TDWR tornadic vortex signature detection algorithm, in *Preprints, 4th International Conference on Aviation and Weather Systems*, pp. J43–J48, American Meteorological Society, Boston, Mass., 1991.

Vasiloff, S. V., Single-Doppler radar study of a variety of tornado types, this volume.

Wakimoto, R. M., and J. W. Wilson, Non-supercell tornadoes, *Mon. Weather Rev.*, *117*, 1113–1140, 1989.

Walker, D. C., and K. F. Heidman, DARE-I evaluation: Forecaster's assessment and use of the NEXRAD algorithm products during the 1987 and 1988 warm seasons, *NOAA Tech. Rep. ERL FSL 3*, 22 pp., Natl. Oceanic and Atmos. Admin., Boulder, Colo., 1989.

Weyman, J. C., and K. I. Clancy, NEXRAD initial operational test and evaluation phase II (IOT&E2), final report, *AFOTEC Rep. 0167*, 68 pp., Air Force Test and Evaluation Center, Kirtland Air Force Base, N. M., 1989.

Wieler, J. G., Real-time automated detection of mesocyclones and tornadic vortex signatures, *J. Atmos. Oceanic Technol.*, *3*, 98–113, 1986.

Wilson, J. W., Tornadogenesis by nonprecipitation-induced wind shear lines, *Mon. Weather Rev.*, *114*, 270–284, 1986.

Wilson, J. W., R. E. Carbone, H. Baynton, and R. J. Serafin, Operational applications of meteorological Doppler radar, *Bull. Am. Meteorol. Soc.*, *61*, 1154–1168, 1980.

Wood, V. T., and R. A. Brown, Single Doppler velocity signatures: An atlas of patterns in clear air, widespread, precipitation, and convective storms, *NOAA Tech. Memo. ERL NSSL-95*, 71 pp., Natl. Oceanic and Atmos. Admin., Boulder, Colo., 1983. (Available as *NTIS PB84155779* from Natl. Tech. Inf. Serv., Springfield, Va.)

Zrnić, D. S., D. W. Burgess, and L. D. Hennington, Doppler spectra and estimated windspeed of a violent tornado, *J. Clim. Appl. Meteorol.*, *24*, 1068–1081, 1985a.

Zrnić, D. S., D. W. Burgess, and L. D. Hennington, Automatic detection of mesocyclone shear with Doppler radar, *J. Atmos Ocean. Technol.*, *2*, 425–438, 1985b.

Single-Doppler Radar Study of a Variety of Tornado Types

STEVEN V. VASILOFF

National Severe Storms Laboratory, Norman, Oklahoma 73069

1. INTRODUCTION

Tornadoes occur in a spectrum of sizes and intensities [e.g., *Forbes and Wakimoto*, 1983; *Doswell and Burgess*, this volume]. While the largest and most violent garner the most attention, the predominance of tornadoes are weak to moderate in intensity. Herein, Doppler radar velocity signatures of tornadoes having a wide variety of sizes and intensities are examined. Attention is focused on radar characteristics of each tornado and their relationship to the tornado's visual appearance and associated damage.

Background material is provided which discusses formation mechanisms for various tornado types. Following presentation of the observations, attributes of the Doppler signatures will be related to their detectability by a TVS detection algorithm being developed at the National Severe Storms Laboratory [*Vasiloff*, 1991].

2. BACKGROUND

Tornado characteristics can be attributed largely to how they are formed. Near the low end of the spectrum are small spin-ups along thunderstorm outflow boundaries. These spin-ups are often referred to as gust front tornadoes (or "gustnadoes") and can cause F1 damage [*Fujita*, 1979; 1981]. Their vertical extent is limited to the lowest 1–2 km. In a dual-Doppler radar study of the Del City, Oklahoma tornado, *Brandes* [1978] showed that the signature of a gustnado was limited to the lowest 2 km and had no closed circulation field.

Higher up on the spectrum are waterspouts and waterspoutlike tornadoes occurring on the High Plains, sometimes referred to as "landspouts." *Wakimoto and Wilson* [1989] and *Brady and Szoke* [1989] document a number of Colorado landspouts, some of which were rated F2. They showed that these vortices form when preexisting vorticity along surface boundaries becomes entrained into and stretched by thunderstorm updrafts.

The largest and most violent tornadoes are produced by supercell thunderstorms [*Browning*, 1964]. Strong midlevel rotation in the updraft regions of these thunderstorms (i.e., mesocyclone) often precedes tornado formation by more than 30 min [*Burgess*, 1976]. *Brown et al.* [1978] found that the tornadic circulation for the Union City, Oklahoma, tornado formed at midlevels and lowered with time.

When a tornado is probed by a scanning pulsed Doppler radar, a tornadic vortex signature (TVS) may be revealed. A TVS is characterized by velocity extrema of opposite signs (after removal of a motion vector) at adjacent azimuths with the resolution volumes at the same range as the tornado. In studies of supercell tornadoes, *Brown et al.* [1978] found that the theoretical peaks in a TVS were one beamwidth apart. This is primarily the result of the tornado diameter being smaller than the radar beamwidth (see *Burgess et al.* [1991] for a review of Doppler limitations). *Burgess and Lemon* [1990] required that the TVS shear exceed the surrounding mesocyclonic shear by a factor of 5. For Colorado landspouts, *Wakimoto and Wilson* [1989] defined a TVS as approximately 6 times the surrounding "misocyclonic" shear. For the purposes of this study, we follow the *Brown et al.* [1978] definition of a TVS, which includes time and height continuity, but do not require that the shear be of any certain value.

3. OBSERVATIONS

The observations presented in this section are representative of a larger set of data that has been studied. Following the discussion in section 1, the data have been grouped into three categories: gust front tornadoes (or "gustnadoes"), landspouts, and supercell tornadoes. For reasons to be shown later, the supercell tornadoes are further divided into brief supercell and long-lived supercell tornadoes.

The Tornado: Its Structure, Dynamics, Prediction, and Hazards.
Geophysical Monograph 79
This paper is not subject to U.S. copyright. Published in 1993 by the American Geophysical Union.

Fig. 1. Photograph of two gust front tornadoes along a thunderstorm outflow boundary in west Oklahoma on August 21, 1979 (courtesy of G. Moore).

3.1. *Gustnadoes*

Three tornadoes were associated with a brief supercell storm that occurred on May 8, 1986. The first one was a small spin-up along a gust front on the right flank of the storm. A brief touchdown occurred near 2301 UTC in The Village near Oklahoma City, 25 km from the radar. The debris cloud and visual funnel appearance were very similar to the vortices shown in Figure 1 (D. W. Burgess, personal communication, 1991). Only minor tree damage resulted, thus the tornado's F0 rating. The time-height profile of the Doppler signature is shown in Figure 2. Data are from the Cimarron Doppler radar, one of two research Dopplers operated by the National Severe Storms Laboratory (NSSL). The TVS was very short-lived and shallow, extending to only 1.5 km above ground level (AGL). The maximum Δv (velocity difference between two adjacent gates at the same range) is 30 m s^{-1}, near the time of the observed funnel. A constant-elevation angle scan (also referred to as a plan position indicator (PPI)) of Doppler velocities at 1.1° through the signature at 2303 UTC is shown in Plate 1. The Doppler velocity signature of the tornado can be seen at about 70° azimuth and 24 km range. Although the velocities are relatively weak, the pattern is readily discernible in the velocity field.

Two other gustnadoes occurred in central Oklahoma on May 3, 1991. The evolution of the Doppler velocity signature of one of them is shown in Figure 3. The data are from NSSL's Norman Doppler. This particular gustnado formed near Blanchard, Oklahoma, about 20 km SW of Norman. Maximum Δvs were less than 25 m s^{-1}, although a damage survey revealed F1 damage, which *Fujita* [1981] translates into a wind speed estimate of 33–50 m s^{-1}. Unfortunately, timing on the damage for this event could not be ascertained during the damage survey. Placement of the damage path arises from when the signature was near the affected area and is in keeping with the timing observed for the gustnado

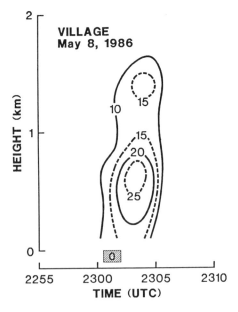

Fig. 2. Time-height plot of Δvs (in meters per second) from a gust front tornado that occurred in The Village on May 8, 1986. Contour intervals are every 5 m s^{-1}. The hatched bar along the bottom indicates times that the tornado was observed and contains damage intensities based on the Fujita scale.

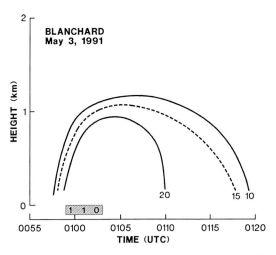

Fig. 3. As in Figure 2 except for the Blanchard gustnado of May 3, 1990. The placement of the hatched bar is approximate.

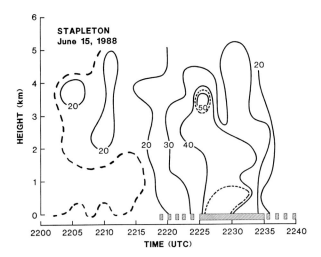

Fig. 4. As in Figure 2 except for the Stapleton tornado of June 15, 1988. Fifteen and 45 m s^{-1} contours have been added. The dashed hatched bar indicates times when only a funnel cloud was observed. Damage intensities as a function of time are not available.

that occurred in The Village (i. e., near the large Δv gradient at the beginning of the TVS). Since the low-level scans were separated by several minutes, the maximum winds in the gustnado may not have been sampled. Doppler velocity PPIs (not shown) of the May 3 events reveal signatures very similar in horizontal size and intensity to the signature in The Village.

3.2. Landspouts

At least four landspouts formed along colliding outflow boundaries near Denver, Colorado, on June 15, 1988. Data are from the Massachusetts Institute of Technology Lincoln Laboratory's FL-2 Doppler radar, which was part of the Terminal Doppler Weather Radar (TDWR) 1988 real-time demonstration [*Turnbull et al.*, 1989].

The Stapleton tornado (rated F2) caused temporary evacuation of nonessential personnel from the control tower at Stapleton international airport. The tornado touched down about 20 km from FL-2 and moved slowly to within 17 km of the radar before it dissipated. Maximum Δvs for this tornado (Figure 4) exceeded 45 m s^{-1} at the lowest levels, and a 60 m s^{-1} Δv was observed at midlevels. A circulation aloft was detected by the radar nearly 20 min before the tornado began. Weaker rotation near the surface was also detected (as shown by the 15 m s^{-1} contour). Note that the Δv gradient near the surface increased when the funnel came in contact with the ground. Furthermore, it appears that the stronger Δvs develop downward with time. This is contrary to most of the landspouts studied by *Wakimoto and Wilson* [1989], where the vorticity developed from the ground upward.

A photograph taken at 2223 UTC (Figure 5), when the Δvs were weaker, shows a funnel whose appearance did not change much over the lifetime of the event. When a video tape of the entire event was viewed, it often was difficult to distinguish between what was a funnel and what was a tornado; i.e., visual appearances are not indicative of damage intensity levels.

The second tornado occurred in the south part of Denver, about 22 km from FL-2. This tornado was rated F3, unusually high for a landspout. The TVS time-height profile for this tornado (Figure 6) is very similar to the Stapleton tornado's with a double maximum, one near the surface and another directly above, although Δvs aloft were 5–10 m s^{-1} less. However, in contrast to the Stapleton tornado signature, the stronger Δvs at low levels appear to build upward with time.

Plate 2 shows a Doppler velocity display at 2223 UTC. The Stapleton signature is WNW of the radar, while the Denver signature is nearly due west of the radar. Both signatures are easily discerned and similar in size and intensity to the signature in The Village.

3.3. Brief-Supercell Tornadoes

The May 8, 1986, Edmond, Oklahoma, tornado (rated F3) began just after the gustnado described previously dissipated. The tornado formed about 30 km from the NSSL Cimarron radar and moved toward the ENE. Unlike the gustnadoes and landspouts, the Edmond tornado was associated with a midlevel mesocyclone. The parent storm was a short-lived supercell which had a well-defined hook echo in the reflectivity field. The tornado formed along a low-level convergence boundary, away from the mesocyclone core. With time, the mesocyclone intensified while the tornado seemed to migrate toward its center, with the TVS blending in with the mesocyclone circulation; this is not a typical (at least for the small sample in the literature) supercell-tornado sequence of events.

The time-height profile of the Edmond TVS (Figure 7) is very similar to the landspout profiles. There is a simultaneous occurrence of a Δv maximum at midlevels above a

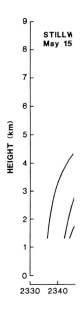

maximum
detailed
maximum
damage.
A photo
phase is
intense. T
signature
much sm

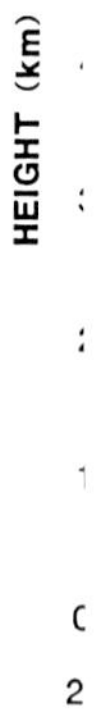

Fig. 6. A
1988.

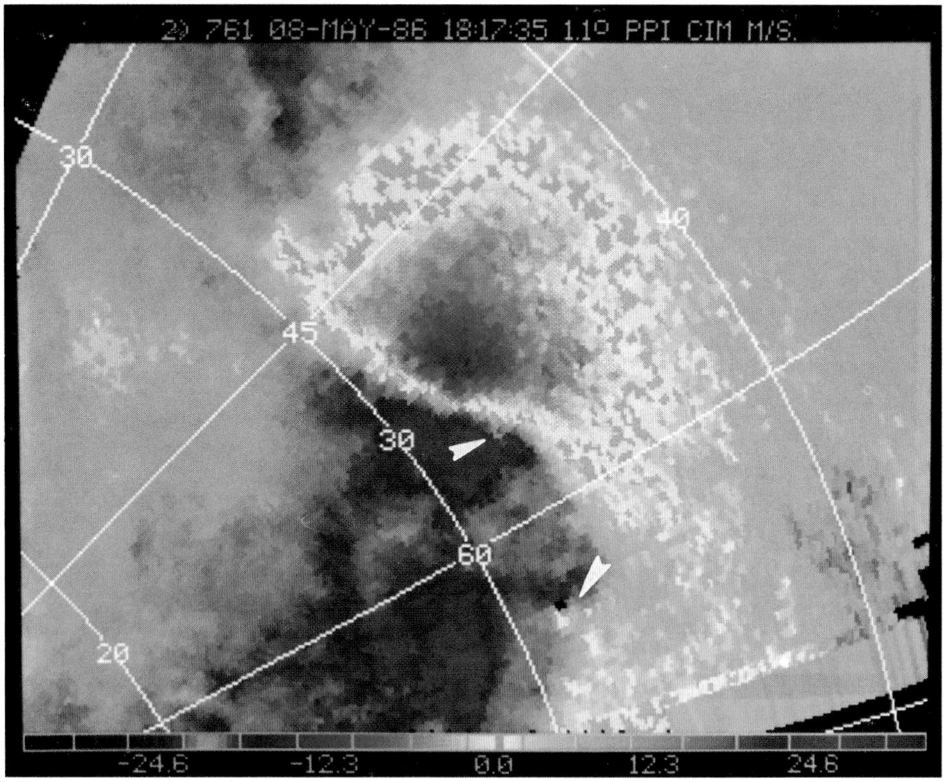

Plate 3. As in Plate 1 except for the Edmond tornado. Elevation angle is 1.1°.

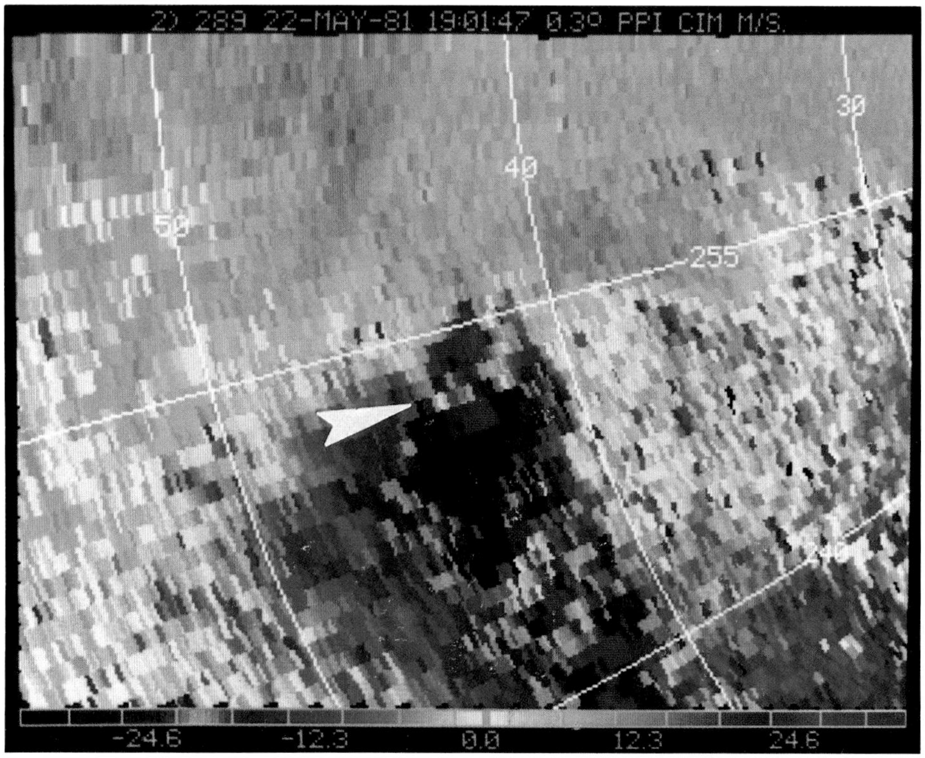

Plate 4. As in Plate 1 except for the Binger signature. Elevation angle is 0.3°.

the NSSL [*Vasiloff*, 1991]. As discussed in detail by *Burgess et al.* [this volume], resolution of a TVS by a Doppler radar is primarily limited by the tornado's aspect ratio with respect to the radar and horizontal and vertical beam spacing. The TVS algorithm requires at least three consecutive (in range) pairs of velocity extrema (so-called "pattern vectors") in order to ensure that a coherent feature is being detected. In addition, three different elevation angles are used in the vertical association of the individual features, where each feature must be within 2.5 km of the one below.

The TVS algorithm was tested on all of the above cases. Scan-by-scan results of several of them are given by *Vasiloff* [1991]. The algorithm had no problem detecting the largest signatures. However, as expected, detectability decreased as a function of size of the tornado. For example, the Edmond tornado had strong Δvs, but the number of pattern vectors was below algorithm thresholds and was not consistently detected. In contrast, the gustnado that occurred in The Village was detected, most likely because it was closer to the radar. The algorithm also performed well on the landspout signatures. However, the Δvs probably would not have met the algorithm threshold (nominally 17 m s^{-1}) if the tornadoes had been much farther away.

Another aspect of tornado detectability is the longevity of the tornado. Short-lived events can be missed if time intervals between volume scans are large. Recall that the gustnado signature in The Village lasted for only 5 min. The signature would have been missed if a 10-min interval between scans had been used.

5. SUMMARY AND DISCUSSION

Single-Doppler radar signatures from a wide variety of tornadoes have been examined. The tornadoes range from brief F0 gustnadoes to High Plains nonsupercell tornadoes to violent F4 tornadoes associated with long-lived supercell storms. Only the gustnadoes were clearly confined to the boundary layer. At the other end of the spectrum the long-lived supercell tornadoes had signatures which first appeared throughout a 3- to 4-km depth above the boundary layer before the tornadoes began doing damage. In addition, low-level to midlevel convergence was observed in the mesocyclone region just before the formation of the TVS.

Differences between brief supercell and nonsupercell tornadoes were not as apparent. Both types had a circulation maximum at low levels concurrent with a separate maximum aloft. While it is apparent that a continuum of tornado sizes and intensities exists, relationships among mechanisms that produce them are poorly understood.

Characteristics of the TVSs were used to assess their detectability by a TVS detection algorithm being developed at the NSSL. It was found that small tornadoes, even though they may produce up to F3 damage, may not be detected even at relatively close ranges (30 km or so), while large tornadoes can be detected at much longer ranges. Thus tornado detection by the algorithm is dependent largely on the distance between the radar and the tornado, as well as the size of the tornado. Work on the TVS algorithm continues, and techniques for better and more timely detection of weak tornadoes are being investigated.

Acknowledgments. The author is grateful to Don Burgess, who reviewed this paper and provided invaluable consultation during the progress of this study. Rodger Brown, Chuck Doswell, and Mike Eilts also reviewed the manuscript and provided helpful suggestions. Joan Kimpel drafted the figures. Doppler radar data used in this study are from the National Severe Storms Laboratory and the Massachusetts Institute of Technology Lincoln Laboratory under sponsorship from the Federal Aviation Administration.

REFERENCES

Brady, R. H., and E. J. Szoke, A case study of nonmesocyclone tornado development in northeast Colorado: Similarities to waterspout formation, *Mon. Weather Rev.*, *117*, 843–856, 1989.

Brandes, E. A., Mesocyclone evolution and tornado genesis: Some observations, *Mon. Weather Rev.*, *106*, 995–1011, 1978.

Brown, R. A., L. R. Lemon, and D. W. Burgess, Tornado detection by pulsed Doppler radar, *Mon. Weather Rev.*, *106*, 29–38, 1978.

Browning, K. A., Airflow and precipitation trajectories within severe local storms which travel to the right of the winds, *J. Atmos. Sci.*, *21*, 634–639, 1964.

Burgess, D. W., Single-Doppler radar vortex recognition: Part I: Mesocyclone signatures, in *Preprints, 17th Conference on Radar Meteorology*, pp. 97–103, American Meteorological Society, Boston, Mass., 1976.

Burgess, D. W., and R. J. Donaldson, Jr., Contrasting tornadic storm types, in *Preprints, 11th Conference on Severe Local Storms*, pp. 189–192, American Meteorological Society, Boston, Mass., 1979.

Burgess, D. W., and L. R. Lemon, Severe thunderstorm detection by radar, in *Radar in Meteorology*, edited by D. Atlas, chap. 30a, pp. 619–647, American Meteorological Society, Boston, Mass., 1990.

Burgess, D. W., R. J. Donaldson, Jr., and P. R. Desrochers, Tornado detection and warning by radar, this volume.

Doswell, C. A., and D. W. Burgess, Tornadoes and tornadic storms: A review of conceptual models, this volume.

Forbes, G. S., and R. M. Wakimoto, A concentrated outbreak of tornadoes, downbursts and microbursts, and implications regarding vortex classification, *Mon. Weather Rev.*, *111*, 220–235, 1983.

Fujita, T. T., Objectives, operations, and results of Project NIMROD, in *Preprints, 11th Conference on Severe Local Storms*, pp. 259–266, American Meteorological Society, Boston, Mass., 1979.

Fujita, T. T., Tornadoes and downbursts in the context of generalized planetary scales, *J. Atmos. Sci.*, *38*, 1511–1534, 1981.

Lemon, L. R., D. W. Burgess, and L. D. Hennington, A tornado extending to extreme heights as revealed by Doppler radar, in *Preprints, 12th Conference on Severe Local Storms*, pp. 430–432, American Meteorological Society, Boston, Mass., 1982.

Turnbull, D., J. McCarthy, J. Evans, and D. Zrnić, The FAA Terminal Doppler Weather Radar (TDWR) program, in *Preprints, 3rd International Conference on the Aviation Weather System*, pp. 414–419, American Meteorological Society, Boston, Mass., 1989.

Vasiloff, S. V., The TDWR tornadic vortex signature detection algorithm, in *Preprints, 4th International Conference on Aviation Weather Systems*, pp. J43–J48, American Meteorological Society, Boston, Mass., 1991.

Wakimoto, R. M., and J. W. Wilson, Non-supercell tornadoes, *Mon. Weather Rev.*, *117*, 1113–1140, 1989.

Zrnić, D. S., D. W. Burgess, and L. D. Hennington, Doppler spectra and estimated windspeed of a violent tornado, *J. Clim. Appl. Meteorol.*, *24*, 1068–1081, 1985.

Radar Signatures and Severe Weather Forecasting

PAUL JOE

King Weather Radar Research Station, Atmospheric Environment Service of Canada, Downsview, Ontario, Canada M3H 5T4

MIKE LEDUC

Ontario Weather Centre, Toronto AMF, Ontario, Canada L5P 1B1

1. INTRODUCTION

The central United States is well known throughout the world for the extreme intensity and frequency of occurrence of severe thunderstorms. This makes it an ideal location for the study of the most severe of storms with hail, tornadoes, gust fronts, and microbursts. Many insights have been gained and conceptual models have been derived from studies in these regions [e.g., *Donaldson*, 1965; *Browning*, 1982; *Ludlam*, 1980; *Burgess et al.*, 1982; *Burgess and Lemon*, 1990; *Fujita*, 1981].

The development of forecast techniques in other severe weather regimes has largely been based on these results [*Atmospheric Environment Service of Canada*, 1982; *Doswell*, 1982, 1985]. While it is expected that all regions will have severe weather, these other locales cannot match the climatology of the central United States in intensity nor frequency [*Newark*, 1984]. Different watch and warning requirements and constraints are imposed on weather forecasting programs. Therefore, while the physics and detection techniques may be similar, the application of the techniques will vary.

Since the last tornado symposium, Doppler radars have been implemented in several operational environments for the forecasting of tornadoes and other severe weather. This paper reviews the progress made at the Ontario Weather Centre (OWC) in Canada in the forecasting of these events after the implementation of Doppler capability at the King Radar Station [*Crozier et al.*, 1991]. Improvements in the forecast program were expected after successful demonstrations in the Joint Doppler Operational Project (JDOP) (1979) program. Such was the case at the OWC. However, the reasons for the improvement cannot be singularly attributed to the outputs of the Doppler radar. The implementation of Doppler capability has led to a significant improvement in the process of producing the severe forecast and warning. This involves a feedback loop.

The forecast process begins with an analysis of the synoptic data and leads to an operative conceptual model or hypothesis. Doppler radial velocity data presented in real time allow the forecaster to verify or discount the hypothesis. In either case, the radar data reinforce or negate the operative conceptual model, which may lead to a change in the warning or forecast. Effective use of the Doppler radar forces the forecaster to think to terms of wind fields on various spatial scales. Wind information is important to the forecaster because it is fundamental to the understanding of weather systems.

Now that the forecaster is thinking along these lines, there are benefits in the warning of storms that may be outside the effective Doppler coverage area. The forecaster is sensitized to various potential severe weather scenarios beyond Doppler range based on observations within the coverage area. The information gleamed from a limited region has been extrapolated to a larger area.

A review of the climatology of tornadoes and severe storms within range of the Doppler radar coverage of southern Ontario will emphasize the variety of storm structures that produce tornadoes in this area. If we are to do an effective forecasting job for this type of environment, we need to understand this variety.

Two examples of severe storms will be presented to illustrate the interactive process of severe weather forecasting described above. In one case, a weak tornado developed in a situation where the prestorm conceptual model indicated tornadoes were unlikely. Real time Doppler data allowed for revision of the model in time to affect the forecast process. The second case describes a "classic" tornado outbreak situation where radar data reinforced the expected scenario.

The Tornado: Its Structure, Dynamics, Prediction, and Hazards.
Geophysical Monograph 79
Published in 1993 by the American Geophysical Union.

Fig. 1. Tornado incidence map for North America, 1953–1979 [after *Newark*, 1984]. Thick solid line represents 4.0, dashed line is for 2.0, medium solid line is for 0.8, dotted line is for 0.4, and the thin lines are for less than or equal to 0.2 incidences per 100^2 km^2 per year.

2. Climatology

Figure 1 shows the climatology of tornado incidence per 100^2 km^2 per year for North America [after *Newark*, 1984]. In southern Ontario the incidence is about one quarter of that in Oklahoma. Generally, the intensities are much lower. Tornado incidence is shown since it is a phenomenon that is rare but memorable. Table 1 lists the severe weather events of a typical year (1986) within the Doppler coverage area of the King radar (110-km radius) [*Leduc and Joe*, 1987]. Notice that only one event is listed as a confirmed tornado though several did display mesoscale rotation. Four different types of Doppler signatures are indicated.

Though the incidence and intensities are lower than that in Oklahoma, there are over six million people who live under the Doppler radar coverage area of the King radar. Thus any severe weather event can potentially have grave consequences. In this environment the focus for the severe weather program at the Ontario Weather Centre, which has the watch and warning responsibility for this region, is on the "marginal" and stronger events.

3. Role of Algorithms

The complexity of the radial velocity images has hastened the development of automated algorithms for the detection of severe weather. Algorithms participate in the forecast process as a tool for the analysis and interpretation of the Doppler radar data.

Initial tests with the mesocyclone algorithm using threshold values appropriate for Oklahoma storms, suggested by *Zrnić et al.* [1985], filtered out the weak tornadic mesocyclone signatures that could be visually identified in the imagery. This should not be too surprising in that the data sets used to tune the algorithms were biased by research projects looking at the severest storms. At this stage, tuning the algorithm is subjective.

The algorithm has been detuned for operational use at the OWC to achieve high probability of detection for the "marginal" tornadic storms. Mesocyclones have been detected for over an hour preceding tornado touchdown [*Joe and Crozier*, 1988]. Meaningful statistics have not been established due to the rarity of the tornadic events.

4. Examples of the Forecast Feedback Process

In this section, two cases are presented to illustrate the feedback process in severe weather forecasting.

TABLE 1. Storms in 1986 Within Doppler Range (110 km) of King City Radar

Date	Time, UT	Reported Damage	Lead Time of Warning Before First Occurrence of Severe Weather, min	Lead Time of Doppler Signature Before First Occurrence of Severe Weather, min	Doppler Radar Signature
May 6	1930	wind	0	20	descending jet
June 16	1830	two wind events	60	40	rotation couplet descending jet
June 29	1830	wind	0	20–30	jet
July 13	2300	wind and hail	0	0	convergence
July 18	2100	wind and hail	30	40	jet
Aug. 1	2030	hail	120	90	convergence
Aug. 1	2400	hail	0	30	convergence
Aug. 2	1820	hail	70	60	convergence
Aug. 8	2400	wind	0	0	rotation
Aug. 15	2000	floods	40	70	rotation
Sept. 29	2330	F0 tornado	0	20	rotation

4.1. *September 17, 1988, F0 Tornado*

A weak tornado briefly touched down in a highly populated area northeast of the Pearson International Airport in Toronto on the Afternoon of September 17, 1988. Twelve hours before the event, a moderately active warm front moved through Southern Ontario.

The severe weather assessment (Figure 2) indicated (1) a low-level moisture axis pushed into the region with dew points of 19°C, (2) a 85-knot (42 m/s) high-level westerly jet, (3) a 30–40 knot (15–20 m/s) curving low-level west to southwesterly jet, and (4) a sharp vorticity trough at 500 mbar.

The modified 1200 UT Flint sounding indicated a potential buoyant energy of 1700 J/kg, and the average wind shear between the surface and 4 km was 9 m/s/km (Figure 3). Observed surface temperatures of 24°C and dew points of 19°C would break the inversion and produce clouds to 11 km. According the *Rasmussen and Wilhemson* [1983], this

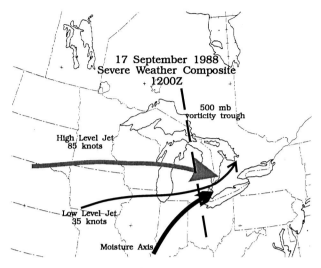

Fig. 2. Composite map for 1200 UT September 17, 1988, summarizing the features conductive for severe weather.

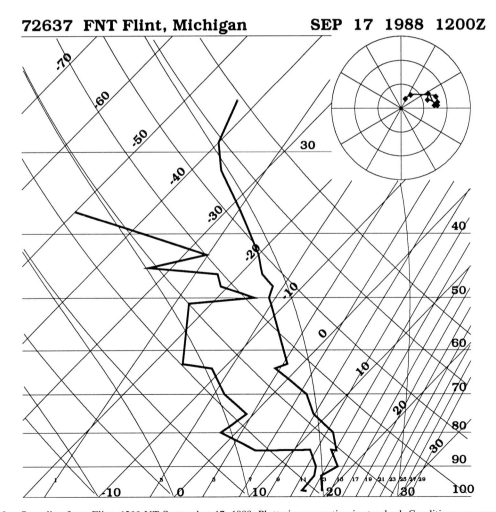

Fig. 3. Sounding from Flint, 1200 UT September 17, 1988. Plottwig convention is standard. Conditions are marginal for severe weather.

would place the environmental conditions as marginal for severe storm development.

The conventional radar observations indicated a few small isolated cells, none of which appeared severe. Echo tops were mostly in the 4- to 6-km range. However, the tornado-producing storm showed a small area of 50-dBZ echo, and the echo tops showed rapid development to 8–10 km in a 10-min period between 2020 and 2030 UT which was still below the height of the tropopause. Only three cloud to ground lightning strikes were detected. At this point, the storm track veered to the right, but this motion was attributed to steering of the storm by the winds at the 8- to 10-km height rather than to propagation effects [*Klemp*, 1977]. The strongest reflectivities were observed at a height of 3 km. Deep vertical development and high reflectivities aloft (40 dBZ at a height of >5.5 km), consistent with severe storms, were never observed [*Donaldson*, 1961]. Given this assessment, the storms observed on radar were assumed to be nonsevere.

The automated detection algorithm identified a tornadic mesocyclone beginning at 2030 and at 2040 UT, based on shear (>10 m/s) and momentum (>100 km m/s) thresholds (Plate 1 and Figures 4 and 5). At 2050 UT the classification of the mesocyclone was degraded to nontornadic status due to low momentum. The tornado touched down at 2120 UT. The classification is tentative and not statistically defensible due to a low number of samples.

This appears to be a situation where a tornado warning would never be issued. However, other data were available to modify this assumption. Storm relative helicity is a relatively new concept that has shown promise in discriminating tornadoes from nontornadoes [*Davies-Jones*, 1984; *Davies-Jones et al.*, 1990]. A storm motion of 290° at 10 m/s was deduced from radar data. Using the 1200 UT Flint

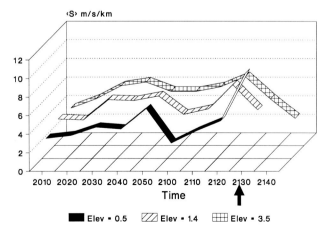

Fig. 4. Time evolution of shear for the September 17, 1988, F0 tornadic mesocyclone. Arrow indicates tornado dissipation time. The algorithm has been tuned to provide high probability of detection.

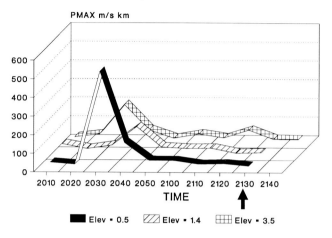

Fig. 5. Time evolution of momentum corresponding to Figure 4.

hodograph, the storm relative helicity values were around 150 m^2 s^2, which is marginal for tornado development. However, beginning at 2030 UT, a strong low level jet of 20–25 m/s was detected on Doppler radar. Using this wind field implies a storm relative helicity of around 300 m^2 s^2, well above the threshold for tornadoes. Other cases in southern Ontario [*Murphy*, 1991] have produced weak tornadoes with marginal dynamics but with storm relative helicity values just above 150 m^2 s^2.

The three pieces of radar information (cell motion, automated mesocyclone identification, and wind profiling) support and are consistent with each other in providing credence for tornado potential. This reassessment could not have been done without the Doppler radar information. This evidence is sufficient to alter the hypothesized conceptual model. The forecaster can take advantage of this to reevaluate the severe weather situation. A new model of a "weak tornado" case has been established. The type of warning message to issue in this case remains a problem. Further discussion is beyond the scope of this paper.

4.2. *May 31, 1985, F2–F4 Tornadoes*

On the afternoon of May 31, 1985, a cold front moved through southern Ontario triggering a series of strong to violent tornadoes. On the morning of May 31, a 98.5-kPa low-pressure center tracked across upper Michigan and moved to the north of Lake Huron by evening. A very sharp cold front trailed southward from the low-pressure center. Soundings east of the cold front showed warm moist air overlain by cool dry air, a classic "loaded gun" sounding. Potential buoyant energy was 3000 J/kg, and the 0- to 4-km shear was 13 m/s/km (Figure 6). The wind profile showed strong shear with southerly winds (180° and 10 knots (5 m/s)) at the surface and a strong westerly jet (200 knots (100 m/s)) at 10 km. The cold front and a sharp upper trough provided the triggering mechanism to release the instability. In short,

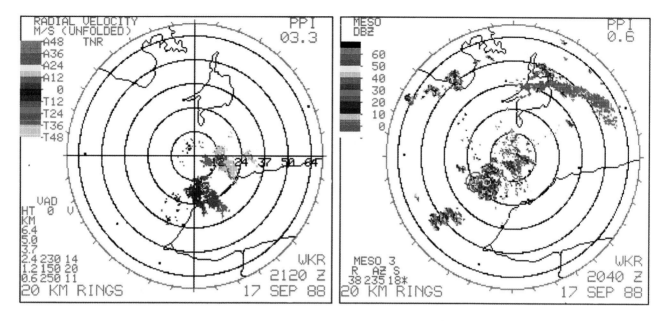

Plate 1. Radial velocity image (0.5° plan position indicator (PPI)) for 2120 UT September 17, 1988, showing a rotational signature at range 30 km and azimuth 190° and the output from the automated mesocyclone detection algorithm for 2040 UT showing a detected feature that was classified as tornadic (indicated by the asterisk).

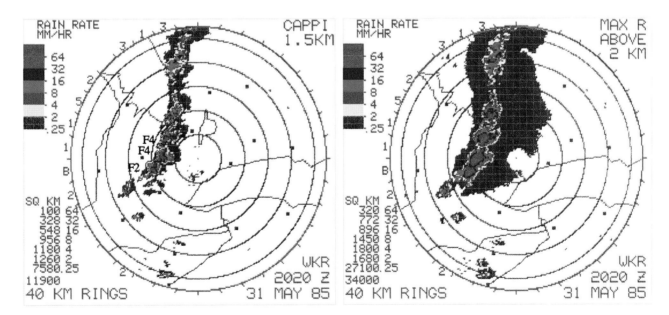

Plate 2. 1.5-km constant altitude plan position indicator (CAPPI) and a MAX R chart (maximum reflectivity above 2 km) showing hook echoes and weak echo regions for 2020 UT May 31, 1985. Areas where the MAX R shows intense echo above weaker echoes on the 1.5-km CAPPI are indicative of weak echo regions. The Fujita scale is used to indicate the intensity of the three tornadoes at this time.

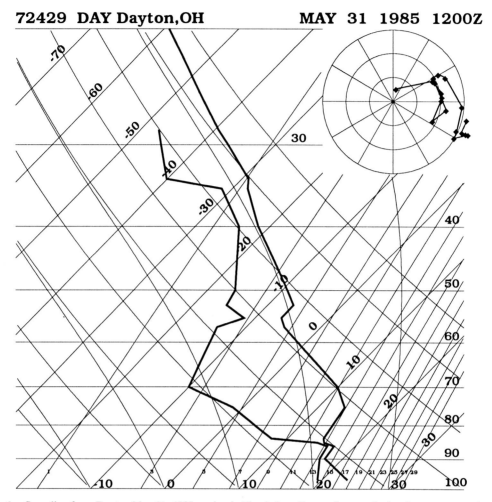

Fig. 6. Sounding from Dayton May 31, 1985, a classic "loaded gun" sounding conducive for severe weather.

all the necessary classic factors [*Fawbush et al.*, 1951; *Fawbush and Miller*, 1953] were present for tornadic severe storms (Figure 7).

The first thunderstorm cells developed at 1740 UT, and by 1820 UT, severe thunderstorm warnings were issued that included the statement "remember that some thunderstorms produce tornadoes." In this case, the radar data were consistent with the forecast scenario and confirmed the conceptual model. Table 2 shows the sequence of warnings and severe weather events. Each tornado was preceded by a warning of its imminence.

As the events unfolded, the conventional radar imagery exhibited all the classic reflectivity signatures of tornadic severe storms [*Donaldson*, 1965; *Forbes*, 1981; *Burgess and Lemon*, 1990] (high reflectivity throughout the core of the storm, large reflectivity gradients, hook echoes, high echo tops, weak echo region, large areal extent); this feedback to the forecaster continually reinforced the conceptual model (Plate 2).

In some respects, this case was "easy"; everything fit into

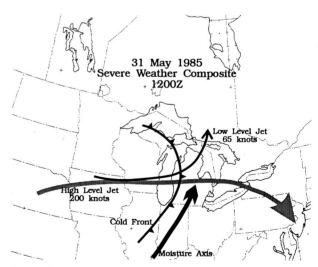

Fig. 7. Severe weather composite for 1200 UT May 31, 1985, showing the relevant synoptic features.

TABLE 2. Time of Warnings and Tornado Touchdown

County	Issue Time of Severe Thunderstorm Warning, UT	Issue Time of Tornado Warning, UT	Time of Actual Event (Tornado), UT	Town
Northern Bruce	1825		1900	Rush Cove
Northern Wellington	1915		2015	Arthur
Dufferin	1954		2028	Grand Valley
			2045	Orangeville
Southern Grey	1915		2017	Corbetton
Southern Simcoe	1954	2100	2118	Holland Landing
Northern Simcoe	1954		2100	Barrie
Northern York	2053	2100	2125	Holt
Northern Durham	2053	2120	2140	Wagner Lake
Southern Victoria	2120	2120	2205	Reaboro
Southern Peterborough	2205	2205	2220	Cavan
			2225	Birdsall
Southern Hastings	2225	2225	2235	Minto

a consistent picture, and all the signs pointed to tornadic severe weather. All the various severe weather indicators were at the extreme end of the spectrum.

5. Discussion

This paper discussed and illustrated the change in the severe weather forecasting process at the Ontario Weather Centre/King Radar that has come with the introduction of Doppler radar technology. There has been a positive influence, and this is reflected in the improved warning performance in the past few years. Figure 8 shows the results for verifications of 8 years of watches and warnings. Doppler capability was included in the last 4 years of the verification assessment.

The probability of detection (POD) and reliability (complement of the false alarm ratio) for watches over the two periods has not changed. This is synoptic in nature, and radar does not play a role here. There is a consistent 20% improvement in these measures for warnings during the Doppler era. It is estimated that 10–15% of the cases were related to tornadoes. Surveys have not been consistently done to distinguish between tornado and straight-line wind damage.

What is interesting in these results is that improvements have been noted even in geographical locations outside Doppler coverage. The consensus is that exposure to Doppler data has allowed the forecaster to practically use his or her knowledge of storm structure. This has led to better use of non-Doppler data that in turn leads to better warnings. For example, hook or pendant echoes, inflow notches, and rear inflow notches are readily observed, analyzed, and interpreted when combined with the velocity information. The Doppler data have reinforced the reflectivity observations.

An attempt has been made to explain the improvements in terms of an enhancement in the forecast process due to a feedback loop. The radial velocity information is central to the process. At this time, the direct outputs of the Doppler radar, the algorithms and imagery, play an initiating or supporting role to the warning process. This information must be combined with other observations and conceptual models for effective use as illustrated in the examples. The implication is that successful fully automated warning system must consider all the available information.

While the focus of this paper has been the feedback process in the forecasting of tornadoes, the comments apply to other weather situations on a variety of scales as well.

6. Conclusion

Doppler radar has had a positive impact on the severe weather program at the Ontario Weather Centre. It provides wind information that is directly applicable by the severe weather forecaster for validation of hypothesis. A feedback loop is established that allows the forecaster to iterate to the

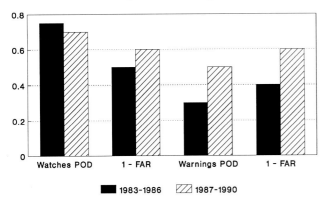

Fig. 8. Watch and warning verification statistics for 8 years (1983–1989).

correct forecast or warning. No other sensor provides information as effectively on this spatial and temporal scale. The feedback is most important in climatic regimes and population centers where severe weather of any intensity can have grave consequences.

Though the images can be difficult to interpret and the automated detection algorithms have yet to reach maturity, there is sufficient information to allow the forecaster to make creative use of his knowledge of storm structure to provide more accurate and timely warnings. Automated systems must consider all data sources and not just radar data to provide the most effective warnings.

Though there may be differences from one radar system to another, the major impact of these systems will be in the enhancement in knowledge of the user. The process of interpreting and using the Doppler imagery has led to a better understanding of mesoscale meteorology. This paper described the process by which this knowledge was and continues to be gained at the Ontario Weather Centre. Similar results can be attained elsewhere provided that the necessary training and education programs are in place. A development program would be a great benefit as well.

REFERENCES

Atmospheric Environment Service of Canada, Summer severe weather correspondence course, Training Branch, Toronto, Ont., 1982.

Browning, K. A., General circulation of middle-latitude thunderstorms, in *Thunderstorms*, vol. 2, edited by E. Kessler, pp. 211–248, National Oceanic and Atmospheric Administration, Boulder, Colo., 1982.

Burgess, D. W., and L. R. Lemon, Severe thunderstorm detection by radar, in *Radar in Meteorology*, edited by D. Atlas, pp. 619–647, American Meteorological Society, Boston, Mass., 1990.

Burgess, D. W., V. T. Wood, and R. A. Brown, Mesocyclone evolution statistics, in *Preprints, 12th Conference on Severe Local Storms*, pp. 422–424, American Meteorological Society, Boston, Mass., 1982.

Crozier, C. L., P. I. Joe, J. W. Scott, H. N. Herscovich, and T. R. Nichols, The King City operational Doppler radar: Development, all season applications and forecasting, *Atmos. Ocean*, 29, 479–516, 1991.

Davies-Jones, R. P., Streamwise vorticity: The origin of updraft rotation in supercell storms, *J. Atmos. Sci.*, 41, 2991–3006, 1984.

Davies-Jones, R. P., D. W. Burgess, and M. Foster, Test of helicity as a tornado forecasting parameter, in *Preprints, 16th Conference on Severe Local Storms*, pp. 588–592, American Meteorological Society, Boston, Mass., 1990.

Donaldson, R. J., Radar reflectivity profiles in thunderstorms, *J. Meteorol.*, 18, 292–305, 1961.

Donaldson, R. J., Jr., Methods for identifying severe thunderstorms by radar: A guide and bibliography, *Bull. Am. Meteorol. Soc.*, 46, 174–193, 1965.

Doswell, C. A., III, The operational meteorology of convective weather, vol. 1, Operational mesoanalysis, *NOAA Tech. Memo., NWS NSSFC-5*, 158 pp., Natl. Oceanic and Atmos. Admin., Boulder, Colo., 1982.

Doswell, C. A., III, The operational meteorology of convective weather, vol. 2, Storm scale analysis, *NOAA Tech. Memo., ERL ESG-15*, 240 pp., Natl. Oceanic and Atmos. Admin., Boulder Colo., 1985.

Fawbush, E. J., and R. C. Miller, Forecasting tornados, *Air Univ. Q. Rev.*, pp. 108–117, Maxwell Air Force Base, Ala., 1953.

Fawbush, E. J., R. C. Miller, and L. G. Starrett, An empirical method for forecasting tornado development, *Bull. Am. Meteorol. Soc.*, 32, 1–9, 1951.

Forbes, G. S., On the reliability of hook echoes as tornado indicators, *Mon. Weather Rev.*, 109, 1457–1466, 1981.

Fujita, T. T., Tornadoes and downbursts in the context of generalized planetary scales, *J. Atmos. Sci.*, 38, 1511–1534, 1981.

Joe, P., and C. L. Crozier, Evolution of mesocyclonic circulation in severe storms, paper presented at 10th International Cloud Physics Conference, International Association of Meteorology and Atmospheric Physics, Bad Homborg, Fed. Rep. of Germany, Aug. 15–20, 1988.

Klemp, J., Dynamics of tornadic thunderstorms, *Annu. Rev. Fluid Mech.*, 19, 369–402, 1987.

Leduc, M. J., and P. Joe, Use of a 5 cm Doppler radar to detect and forecast severe thunderstorms in a real time forecast operation, Proceedings Symposium on Mesoscale Analysis and Forecasting, *Eur. Space Agency Spec. Publ., ESA SP-282*, 49–54, 1987.

Ludlam, F., *Clouds and Storms: The Behavior and Effect of Water in the Atmosphere*, 405 pp., Pennsylvania State University Press, State College, 1980.

Murphy, B., Tornado and flash flood in southwestern Ontario on 31 May 1991, *Ontario Reg. Tech. Note, ORTN-91-3*, 8 pp., Atmos. Environ. Serv. of Canada, Toronto, Ont., 1991.

Newark, M. J., Canadian tornadoes, 1953–1979, *Atmos. Ocean*, 22, 343–353, 1984.

Rasmussen, E., and R. B. Wilhelmson, Relationship between storm characteristics and 1200 GMT hodographs, low-level shear, and stability, in *Preprints, 13th Conference on Severe Local Storms*, pp. 55–58, American Meteorological Society, Boston, Mass., 1983.

Zrnić, D. S., D. W. Burgess, and L. D. Hennington, Automatic detection of mesocyclonic shear with Doppler radar, *J. Atmos. Oceanic Technol.*, 24, 425–438, 1985.

The Use of Volumetric Radar Data to Identify Supercells: A Case Study of June 2, 1990

RON W. PRZYBYLINSKI

National Weather Service, NOAA, Saint Charles, Missouri 63304

JOHN T. SNOW AND ERNEST M. AGEE

Department of Earth and Atmospheric Science, Purdue University, West Lafayette, Indiana 47907

JOHN T. CURRAN

National Weather Service, NOAA, Indianapolis, Indiana 46241

1. INTRODUCTION

Conventional radar is used in National Weather Service offices to help issue timely warnings of severe weather (tornadoes, severe thunderstorms, and flash floods) to the public. Monitoring potentially severe or tornadic thunderstorms can become a significant challenge, particularly during a major tornadic outbreak. This was the situation during the evening of June 2, 1990, across much of Indiana when 38 tornadoes occurred over a period of 6 hours. Early in the outbreak, as many as seven supercells were identified simultaneously across parts of west central Indiana and extreme southeast Illinois. The rapid evolution and movement (greater than 25 m s^{-1}) of the individual radar echo structures further complicated an already difficult warning situation.

The purposes of this paper are twofold. The first part focuses upon the variation of radar echo structures associated with supercells which occurred during the evening of June 2, 1990. Both low-level and elevated presentations will be shown at select times during the evolution of the tornadic outbreak. Many of the examples are mature supercells.

The second part of this paper builds upon the current conceptual high-precipitation (HP) supercell models introduced by *Moller and Doswell* [1988] and *Moller et al.* [1990]. Two crude three-dimensional conceptual models of high-precipitation supercells are introduced. These models were derived from low-level and elevated plan position indicator (PPI) radar observations collected during the period from 1988 to the present across the mid-Mississippi and Ohio valley regions.

2. PRESENT KNOWLEDGE

During the early and middle 1960s, *Browning and Ludlam* [1962] and *Browning* [1964] introduced the term "supercell" to describe the quasi steady state structure attained by some storms during their intense phases. Supercell storms produce most, if not all, violent tornadoes and account for a high percentage of extremely large hail events. One important feature of the radar echo from this type of storm is a distinctly shaped region void of detectable echo which penetrates upward into the core of the storm beneath the storm's highest top. *Browning and Ludlam* [1962] referred to this region as an "echo free vault." *Chisholm* [1973] redefined this region as the "bounded weak echo region" (BWER) which is often observed with the main storm updraft near the right rear flank of the supercell. Depending upon radar resolution and range, the BWER may not always be detected; however, the weak echo region (WER) is often more identifiable. Other equally important radar echo features associated with the supercell include (1) the highest storm top is positioned over the BWER (or WER) and (2) a notch of weaker echo surrounded by a hook (pendant) echo signifying the location of possible updraft rotation. Research efforts conducted during the 1970s using Doppler radar have successfully identified the BWER as [*Burgess et al.*, 1977] a center of significant updraft and cyclonic rotation.

Lemon [1977, 1980] introduced a volumetric radar scanning method to aid operational forecasters in identifying

The Tornado: Its Structure, Dynamics, Prediction, and Hazards.
Geophysical Monograph 79
Copyright 1993 by the American Geophysical Union.

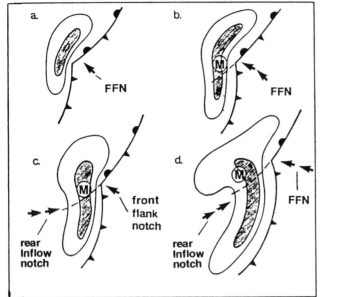
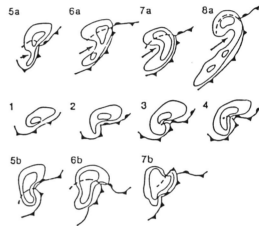

SCHEMATIC EVOLUTION OF HP SUPERCELL SHOWING DEVELOPMENT OF BOW ECHO STRUCTURE (DOSWELL, 1985).

COMPOSITE LIFE CYCLES OF HP STORMS THAT HAVE BEEN IDENTIFIED.

Fig. 1. Conceptual models and composite radar structures (life cycles) of high-precipitation supercells (adapted after *Doswell* [1985] and *Moller et al.* [1990]). M signifies probable location of mesocirculation. Arrows on right side of figure denote location of rear inflow jet.

classic supercells and multicell thunderstorms. The technique relies upon the identification of indicators of updraft strength, since the strength of the storm's updraft greatly influences the shape and appearance of the radar echo and type of severe weather events. *Forbes* [1981], using conventional radar reflectivity data from the April 3, 1974, super tornado outbreak over the Ohio valley region, examined the shape and reflectivity intensity of several tornadic and nontornadic storms. He classified a group of unique echoes as "distinctive echoes," suggesting that this group signified strong, persistent rotation and had the highest tornadic probability.

Recent studies completed by *Doswell et al.* [1990] (referred to hereinafter as D90) have shown that supercells can take on a variety of echo shapes and differing evolutions. They pointed out that there is a "spectrum of supercells" in which all supercells are neither completely unique nor completely identical. However, D90 have shown that the common denominator of all supercells is that they have updrafts with persistent rotation.

The spectrum of supercells can be subdivided into three basic categories: (1) classic (C) [*Browning*, 1964], (2) low-precipitation (LP) [*Bluestein and Parks*, 1983], and (3) high-precipitation (HP) [*Moller et al.*, 1990] (referred to hereinafter as M90). The conceptual model of HP supercells was recently introduced by M90 on the basis of radar and visual observations. M90 have shown that the key echo characteristics of HP supercells include (1) a weak reflectivity notch located along the forward flank (signifying the probable location of a rotating updraft or mesocyclone), (2) extensive precipitation (including torrential rainfall and hail) along the right rear flank, and (3) the mesocyclone embedded within significant precipitation.

Doswell [1985] (referred to hereinafter as D85) M90, D90, *Przybylinski* [1989] (referred to hereinafter as P89), and *Przybylinski et al.* [1990] (referred to hereinafter as P90) all have shown that HP storms can exhibit a number of echo shapes. Examples of radar images and echo evolution associated with HP storms are shown in Figure 1 (left side). Echo characteristics include (1) an overall echo that may take the form of a kidney-bean shaped configuration, (2) a hook echo that is exceptionally large (a broad hook echo filled with significant precipitation), (3) weak echo notches along the forward and rear flanks of the storm, (4) persistent low-level reflectivity gradients adjacent to the front flank notch (FFN), and (5) a spiral or "S" shaped echo structure. *Weaver and Nelson* [1982], *Foote and Frank* [1983], and *Nelson* [1987] have indicated that one variation of HP storms may contain multiple mesocyclones and BWERs. Their data set suggests that "hybrid" supercells (apparently the same as HP) may exhibit multicell traits (a storm having several high-reflectivity cores) and are often associated with widespread damaging hail events.

Studies by M90 also suggested that supercells follow a

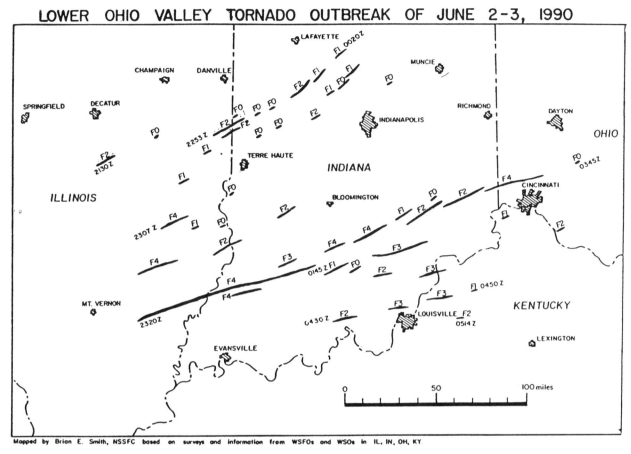

Fig. 2. Overall mapping of the damage caused by the June 2, 1990, storm. Dark solid contours represent tornado tracks. The damage survey was conducted by National Weather Service personnel from Indianapolis, Louisville, Evansville, Cincinnati, and Springfield. Data source is *Storm Data*.

variety of life cycles (right side of Figure 1). Some supercells follow the LP-C-HP (low-precipitation–classic–high-precipitation) lifecycle; a more common transition is the C to HP (classic to high-precipitation). Additionally, it is quite common for HP storms to evolve into a bow echo (BE) storm (HP-BE) with a rotating comma head. The rotating comma head echo documented by *Wakimoto* [1983], P89, and P90 can be a reflection of a strong mesocyclone. However, it is often noted during the dissipating stages of the HP storm. One of the unique characteristics of HP supercells is that the mesocyclone is embedded in significant precipitation for an extended period of time. This characteristic separates HP storms from classic supercells in which mesocyclones are wrapped up in significant precipitation for only a brief period during the dissipating stages.

The right-hand part of Figure 1 shows two different life cycles from the C-HP transition. The C-HP transition noted (2–4) may evolve either in a bowing echo structure (C-HP-BE, 2–8A life cycle) or, alternatively, as a new mesocyclone is forming on the storm's right flank (2–7B). M90 have noted that cyclic mesocyclone development can occur with either life cycle. This type of cyclic mesocyclone development is similar to cyclic mesocyclone evolution observed in classic supercells [*Burgess et al.*, 1982].

3. Radar Echo Patterns Identified on June 2, 1990: Illustrations of Classic and HP Supercells

The June 2, 1990, tornado outbreak over central and southern Indiana presents an excellent opportunity to show variations in the radar echo patterns of the supercells surveyed and their transitions within the classic phase and classic to HP storm phase. A map of the tornado tracks across central and southern Indiana is shown in Figure 2. Detailed surveys of the damage tracks and times of tornado occurrence were conducted by National Weather Service personnel (including two of the authors). The strongest tornadoes (F4 intensity) occurred across southern Indiana, where several supercells spawned families of tornadoes. Additional tornadoes (F2-F3 intensity) traveled across central sections of the state. Total damage exceeded 75 million dollars.

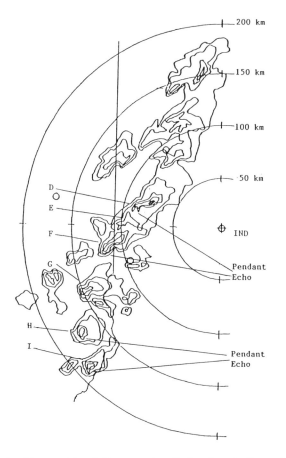
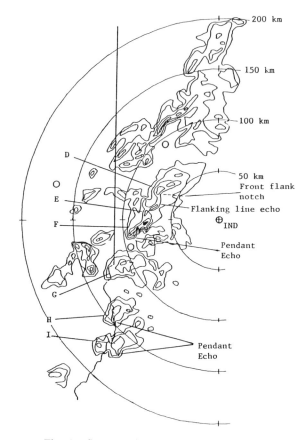

Fig. 3. Plan position indicator radar display from Indianapolis radar at 2328 UTC, June 2, 1990. Antenna elevation angle is 0.5°. Reflectivity contours are 18, 30, 41, and 46 dBZ. Shaded regions represent reflectivity values greater than 50 dBZ.

Fig. 4. Same as Figure 3 except for 2346 UTC.

Reflectivity data were collected from the WSR-74C (5.4 cm wavelength, 1.5° beam width) National Weather Service radar at Indianapolis, Indiana (IND). Both 16- and 35-mm photographic film records of PPI displays were collected for the study. Volumetric reflectivity data were collected at discrete times to determine storm structure. Owing to the large number of supercells present and rapid movement, only partial tilt sequences were accomplished. Lower and middle levels of storms were frequently examined to determine radar echo characteristics. When possible, the location of the maximum echo top was documented and compared to low-level and midlevel echo features.

Figure 3 shows a detailed view of the radar reflectivities from north central Indiana through extreme southeast Illinois at 2328 (all times UTC). A nearly continuous line of thunderstorms with embedded supercells extended across Indiana. Storm D spawned a strong F2 intensity tornado near the right rear flank of the cell at 2318 and produced a damage track of over 12 km in length. Between 2328 and 2346, storm D gradually followed the classic-HP (C-HP) transition in which the mesocirculation moved from the rear flank to the forward flank of the storm (Figures 3 and 4). A front flank notch evolved, and the mesocirculation was embedded in significant precipitation. A second strong F2 intensity formed near the front flank notch at 2346 and produced a continuous damage track of 18 km.

In contrast, storm F continued to show classic supercell characteristics in which the pendant echo, located near the right rear flank, was free of surrounding precipitation (Figure 3). Storm F spawned a series of weak intensity (F0-F1) tornadoes from 2315 through 2345.

Storms G, H, and I, also shown in Figure 3, were more isolated and exhibited variations of supercell structures. Storm G spawned its second F4 intensity tornado at 2307 before weakening considerably after 2328 as a weak convective cell near the southeast flank appeared to merge within the updraft region of the tornadic cell. Supercell H spawned an F4 intensity tornado at 2246 and a second tornado (F2 intensity) at 2345, while supercell I generated its first tornado after 0019.

Detailed low-level and elevated PPI scans of supercells D and F over west central Indiana at 2346 are shown in Figure 5. Storm D continued to exhibit HP storm characteristics from 2346 through its final stages. Characteristics included a front flank notch, indicating the probable updraft region of the cell, a mesocirculation embedded in significant rainfall,

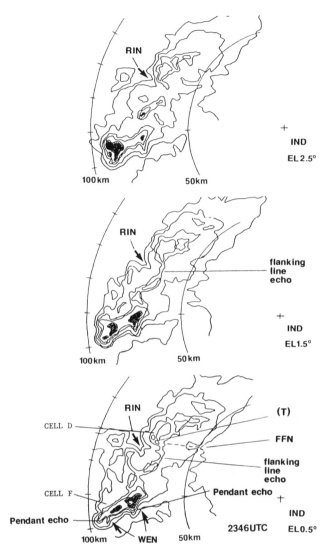

Fig. 5. Detailed low-level and elevated plan position indicator radar reflectivities from Indianapolis radar at 2346 UTC, June 2, 1990, of convective storms D and F. Reflectivity contours are 18, 30, 41, and 46 dBZ. Shaded regions represent reflectivity values greater than 50 dBZ. T signifies location of tornado. EL represents elevation angles during tilt sequence. RIN is rear inflow notch. WEN is weak echo notch. FFN is front flank notch.

and multiple high-reflectivity cores (multicell traits) downwind from the parent mesocirculation. Additional important echo features associated with storm D included a flanking line echo south of the parent circulation and a well-defined rear inflow notch (RIN) immediately upwind of the flanking line echo. Such echo features indicate the probable location of descending flow (rear flank downdraft) similar to radar signatures identified by D90 and M90.

In contrast, storm F continued to show nearly classic supercell characteristics. The two 50-dBZ cores identified within the larger echo mass suggest that storm F was composed of two supercells having distinct and separate updraft centers near the right rear flanks of each reflectivity core. The intense low-level reflectivity gradients, weak echo notches (WEN), and pendant echoes identified along the right flanks of each 50-dBZ core suggest the probable updraft region of each supercell.

If PPI reflectivity data were examined solely at 0.5° for storm F, the attention would have likely been focused on both pendant echo structures adjacent to each 50-dBZ reflectivity core for possible severe weather or tornadic activity. However, examination of PPI reflectivity data at subsequent elevation angles (1.5° and 2.5°) showed that the downwind (front flank) 50-dBZ reflectivity core had limited vertical depth compared to its upwind neighbor. The upwind 50-dBZ core extended well into the middle levels of the cell (2.5° scan) with well-defined echo overhang along the right flank. Such low-level and elevated reflectivity features suggest that the stronger of the two updraft centers was located near the right rear of the system. Having these additional data available, the warning forecaster would focus his attention on the right rear flank of storm F for possible tornadogenesis. Storm F spawned a strong F2 intensity tornado at 0030 approximately 50 km west-northwest of IND.

Examining PPI reflectivity data at two or more elevation angles to determine storm severity has significant advantages compared to viewing storms at only one elevation angle. This type of sampling format needs to be exercised frequently to observe changes in the storm evolution.

Detailed low-level and elevated scans of supercells H and I (Figures 6 and 7, respectively) illustrate some aspects of the vertical structures of these storms at 2346. Both supercells were isolated compared to their northern neighbors (D and F). Data sampled at 1.5° and 2.5° at a range of 150 km (storm position for cell H) represent heights of 5.0 and 7.5 km, respectively. The overall structure of supercell H suggests that the mesocirculation and tornado were wrapped in precipitation (near the southern flank of the VIP 3 echo). The RIN along the trailing edge of storm H at low-level and elevated scans indicates the probable location of descending flow similar to observations noted by D90, M90, and P89. The front flank notches, indicating potential updraft centers along the leading edge, are not well defined. The persistent vertical depth of the RIN and overall comma-shaped echo pattern noted at 2.5° suggested the presence of a mesocirculation which acted to redistribute the precipitation field (D90). At 2346, storm H was likely in the occlusion phase of a classic supercell evolution since a strong F2 intensity tornado having a 25-km damage track was occurring at this time. Supercell H later spawned a second F2 intensity tornado approximately 50 km southeast of Terre Haute, Indiana (HUF).

Further south, supercell I appeared to exhibit HP supercell characteristics in which a front flank notch was observed along the leading edge of the convective echo, and the mesocirculation outlined by the pendant echo appeared to be wrapped in significant precipitation (Figure 7). The RIN, signifying the location of descending flow, was not evident along the trailing edge as compared to storm H. However,

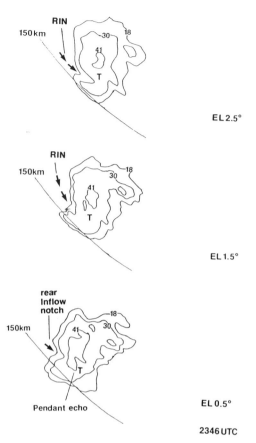

Fig. 6. Same as Figure 5 except for convective storm H.

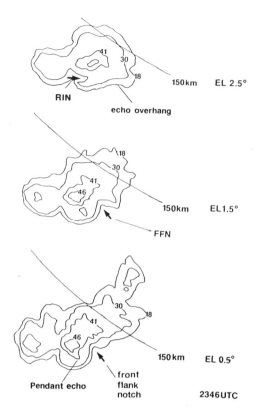

Fig. 7. Same as Figure 5 except for convective storm I.

the front flank notch downwind from the pendant echo was observed at low levels and capped by midlevel echo overhang. This echo feature appeared to be a reflection of the updraft region and is consistent with observations recorded by D85, D90, and M90. At 0019, approximately 35 min after this image, supercell I spawned its first of a family of eight tornadoes near Montgomery, Indiana, approximately 100 km north-northeast of Evansville, Indiana (EVV). The first tornado reached F3 intensity, while subsequent tornadoes ranged from F2 to F4.

A broad view of radar reflectivities of three large supercells (I, J, and K) at 0235 is shown in Figure 8. During the previous two hours each cell produced families of tornadoes varying from F1 through F4 intensity and continued to spawn tornadoes through 0500. The radar presentation showed that each storm exhibited multiple high-reflectivity cores, notches along the southern and eastern flanks, and an elongated echo structure. The storm locations aligned in a northeast-southwest orientation, suggesting each storm moved along the outflow boundary of its downwind neighbor. Essentially, storm K traveled along the outflow boundary produced by storm J, and storm J moved northeast along the outflow boundary generated by storm I. This type of storm movement is similar to that observed by *Weaver and Nelson* [1982] in which a hybrid supercell intersected and moved along a boundary generated by a downwind supercell. These boundaries act as a source of baroclinically generated horizontal vorticity, similar to numerical simulations shown by *Klemp* [1987] and observations documented by *Maddox et al.* [1980] and *Purdom and Sinclair* [1988].

The alignment of all three supercells and their boundaries also served to inhibit the transport of very unstable air northward across parts of central Indiana after 0100. The mapping of tornadic damage tracks also supports this claim since tornadic activity was absent east and just south of Indianapolis after 0115.

Detailed low-level and elevated reflectivity scans of storms J and K at 0235 illustrate some aspects of their vertical structure (Figure 9). Both storms revealed a general elongated echo structure with multiple 50-dBZ reflectivity cores similar to storms D and F and observations documented by *Nelson and Knight* [1987]. Notches linked to 50-dBZ cores and bounded by echo aloft along the southern and eastern flanks of each cell suggested the possible existence of multiple BWERs and WERs. At this time, an F3 intensity tornado (T) was located near the right rear flank of storm J's 50-dBZ core. The tornado produced a damage path of 54 km and had a life span of 42 min. It is noteworthy that storm J had two probable updraft centers, one near the forward flank capped by 50-dBZ core aloft and another near the right rear flank of the storm (near location of T).

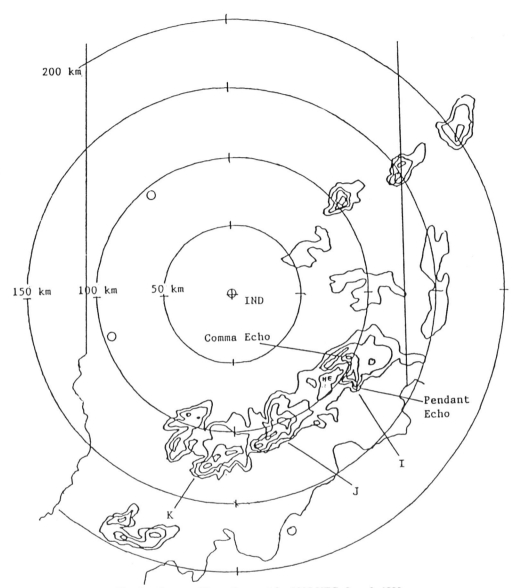

Fig. 8. Same as Figure 3 except for 0235 UTC, June 3, 1990.

Further upwind, storm K exhibited a weak echo notch along the forward flank of the southernmost 50-dBZ core, signifying a probable updraft center. The adjacent notch downwind of the first was probably linked to the downwind 50-dBZ core. Supercell K spawned a strong F2 intensity tornado at 0300 having a damage track of 9.3 km (approximately 50 km north northwest of Louisville) and a F3 intensity tornado at 0340 approximately 100 km northeast of Louisville.

Both convective storms J and K were variations of HP supercells in which each cell exhibited front flank notches (updraft centers), heavy precipitation along the right rear flank, and multiple 50-dBZ reflectivity cores within an elongated echo structure.

Figure 10 shows low-level and elevated scans of storm I at 0235. A small comma-shaped echo structure and pronounced RIN observed through 2.5° near the northern flank of the storm are reflections of a weakening circulation and probable descending flow. An F2 intensity tornado lifted near the comma head at approximately 0225. Near the right rear flank of storm I, the reflectivity structure at elevated scans suggests the probable presence of a BWER. A strong F2 intensity tornado (T) was located within the BWER at this time. The high reflectivities and reflectivity gradients surrounding the weaker echo within the center of the overall circular echo structure at all three scans appear to suggest the BWER evolving through a period of collapse. The reflectivity patterns suggest that the mesocyclone was sur-

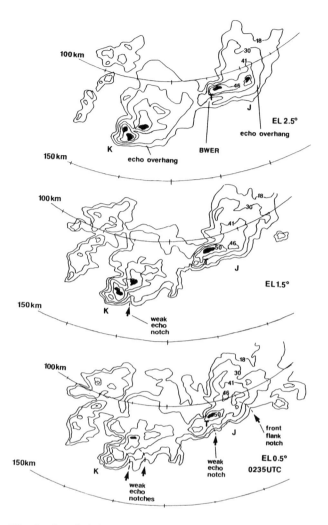

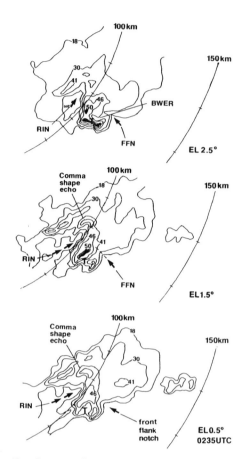

Fig. 9. Detailed low-level and elevated plan position indicator radar reflectivities from Indianapolis radar at 0235 UTC, June 3, 1990, of convective storms J and K. Reflectivity contours are 18, 30, 41, and 46 dBZ. Shaded regions represent reflectivity values greater than 50 dBZ. T signifies location of tornado. EL represents elevation angles during tilt sequence.

Fig. 10. Same as Figure 9 except for convective storm I.

rounded by significant precipitation. This claim was further supported by several witnesses who experienced heavy rainfall and hail with the tornado. The total damage track of this tornado was 47 km. Additionally, storm I spawned a stronger F4 intensity tornado at 0310. The reflectivity structures at low-level and elevated scans suggest that this storm was another variation of an HP supercell.

Table 1 reveals the phase of the identified tornadic storms. It is noteworthy that many of the tornadic storms evolved into the classic supercell category during the early part of their life cycle and gradually evolved into the HP storm phase during the later part (evening hours).

4. New Conceptual HP Storm Models

Conceptual models of HP storms presented by M90, D85, and D90 show evolutionary changes of the PPI radar echo at one elevation angle. In an attempt to show a crude three-dimensional view of the HP supercells, conceptual HP storm models are introduced in Figure 11. The models are based on a limited sample of 15 HP storms collected over the lower Ohio and mid-Mississippi valley regions during the period 1988–1991. The conceptual models are preliminary and continue to be tested and modified with updated radar observations.

TABLE 1. Storm Identification Letters and Type of Supercell Structure Observed During the June 2, 1990, Tornado Outbreak

Storm	Type of Supercell
D	C-HP
F	C
G	C
H	C-HP
I	C-HP-BE
J	C-HP
K	C-HP

C represents classic supercell structures. C-HP signifies transition from classic to high-precipitation supercell. C-HP-BE signifies transition from classic to high-precipitation supercell to bow echo structures.

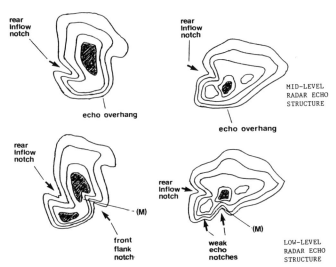

Fig. 11. Conceptual models of plan position indicator radar echo structures (at low and middle levels) of high-precipitation supercells during their mature phase. Contours are schematic reflectivities, with the shading showing high-reflectivity regions. M signifies probable location of mesocirculation. Storm motion is assumed to be to the right.

One model in Figure 11 (left side) shows that the parent mesocyclone and tornado are surrounded by high reflectivity (wrapped in significant rainfall and hail). During this phase, the overall echo pattern often takes the form of a large comma-shaped echo configuration. The notch observed along the forward flank of the storm is often bounded above by high reflectivities at midlevels. Such echo features suggest the presence of an updraft center along the forward flank of the storm. A singular weak echo notch along the trailing flank (RIN) is often observed within the lower and sometimes middle levels of the storm. This echo feature suggests that strong descending flow may be present. M90, P89, and P90 have documented the occurrence of damaging winds in the vicinity of this echo feature. Reflections of the parent circulation (mesocyclone) may at times be outlined by a spiral echo configuration. Additionally, more than one high-reflectivity core may be present within the overall storm structure suggesting the possible presence of multiple updraft centers. A data set of 10 storms has been documented to support this model.

Figure 11 (right side) shows the conceptual model of a second type of HP storm. This type of storm is composed of several high-reflectivity cores within an elongated echo structure. One or several weak echo notches, signifying the presence of updraft centers, are noted along the right side and forward flank of the storm; these notches are often capped by high reflectivites aloft. This type of storm structure is similar to structures documented by *Nelson and Knight* [1987] in which one or several weak echo regions are present. Only a small data set, comprising five storms, has been documented to support this model.

5. SUMMARY

A major tornado outbreak occurred across much of central and southern Indiana during the late afternoon and evening of June 2, 1990. Thirty-eight tornadoes were identified over a 6-hour period [*Storm Data*, 1990]. During the early part of the outbreak, as many as seven supercells were identified, some showing variations of classic supercell structures. However, during the latter stages of the outbreak, many of the storms evolved to HP supercell structures. The alignment of three supercells across southern Indiana suggested that each storm traveled along the outflow boundary of its downwind neighbor. This kind of storm configuration served to inhibit the transport of very unstable air northward across parts of central Indiana during the later part of the evening.

Two high-precipitation (HP) supercell conceptual models were derived from 15 cases. Several cases included volumetric reflectivity data. These models are additions to the current conceptual models presented by M88, M90, and D90. Both models revealed several important echo characteristics including (1) a broad or large hook echo at low levels along the right flank of the storm, (2) one or several weak echo notches, signifying an updraft center, bounded by high reflectivities aloft along the forward flank of the storm, (3) a singular rear inflow notch (RIN) having vertical depth along the trailing edge of the storm suggesting the presence of descending flow, and (4) a spiral echo structure at low and sometimes middle levels indicating a mesocirculation.

The conceptual models presented here and the models introduced by M88, M90, and D90 are being examined and tested with cases from other parts of the country. The testing includes the collection of new radar observations (including Doppler data) to determine if the models are consistent with data or if there is a need to alter the models in light of new evidence. The supporting evidence collected thus far has not been sufficient to conclusively validate the conceptual models.

The radar observations shown during the June 2, 1990, outbreak support the hypothesis that there are many variations within the classic and HP supercell themes. Identifying several supercells in real time may often be difficult due to the variety of echo structures and storm evolutions. However, persistent research utilizing volumetric reflectivity and Doppler velocity data will hopefully provide solutions for understanding the various supercell evolutions. The new WSR-88D radar systems will allow warning forecasters to survey the low-level and elevated reflectivity and velocity structures of several storms more efficiently than they can with today's methods. However, researchers and forecasters will face the challenge of properly interpreting these new data sets to produce the most accurate and timely warnings and forecasts.

Acknowledgments. The authors are grateful to Joseph T. Schaefer, Director NWSTC, Richard Livingston of Scientific Services Division (Central Region), Don Burgess of the Operational Support Facility, David Andra, WSFO Norman, and Fred Glass,

WSFO Saint Louis, for their many beneficial suggestions on improving the manuscript. The authors would also like to thank Al Moller for his review of the paper and acknowledge Steve Thomas (MIC/AM WSFO Saint Louis) for his encouragement in this study. Additional thanks goes to Jim Allsopp (WPM) WSFO Chicago and Phil Manuel and Jim Keene of WSO Cincinnati for their contributions. This paper is part of a COMET Partner's project between Purdue University and WSFO Indianapolis. This paper was partially funded by UCAR/COMET.

References

Bluestein, H. B., and C. R. Parks, A synoptic and photographic climatology of low-precipitation severe thunderstorms in the southern plains, *Mon. Weather Rev.*, *111*, 2034–2046, 1983.

Browning, K. A., Airflow and precipitation trajectories within severe local storms which travel to the right of the winds, *J. Atmos. Sci.*, *21*, 634–639, 1964.

Browning, K. A., and F. H. Ludlam, Airflow in convective storms, *Q. J. R. Meteorol. Soc.*, *88*, 117–135, 1962.

Burgess, D. W., R. A. Brown, L. R. Lemon, and C. R. Safford, Evolution of a tornadic thunderstorm, in *Preprints, 10th Conference on Severe Local Storms*, pp. 84–89, American Meteorological Society, Boston, Mass., 1977.

Burgess, D. W., V. T. Wood, and R. A. Brown, Mesocyclone evolution statistics, in *Preprints, 12th Conference on Severe Local Storms*, pp. 422–424, American Meteorological Society, Boston, Mass., 1982.

Chisholm, A. J., Alberta hailstorms, 1, Radar case studies and airflow models, *Meteor. Monogr.*, *14*, 1–36, 1973.

Doswell, C. A., III, The operational meteorology of convective weather, vol. 2, Storm scale analysis, *NOAA Tech. Memo. ERL ESG-15*, 240 pp., Natl. Oceanic and Atmos. Admin., Boulder, Colo., 1985.

Doswell, C. A., A. R. Moller, and R. W. Przybylinski, A unified set of conceptual models for variations on the supercell theme, in *Preprints, 16th Conference on Severe Local Storms*, pp. 40–45, American Meteorological Society, Boston, Mass., 1990.

Foote, G. B., and H. W. Frank, Case study of a hailstorm in Colorado, III, Airflow from triple-Doppler measurements, *J. Atmos. Sci.*, *40*, 686–707, 1983.

Forbes, G. S., On the reliability of hook echoes as tornado indicators, *Mon. Weather Rev.*, *109*, 1457–1466, 1981.

Klemp, J. B., Dynamics of tornadic thunderstorms, *Annu. Rev. Fluid Mech.*, *19*, 369–402, 1987.

Lemon, L. R., Severe thunderstorm evolution: Its use in a new technique for radar warnings, in *Preprints, 10th Conference on Severe Local Storms*, pp. 77–80, American Meteorological Society, Boston, Mass., 1977.

Lemon, L. R., Severe thunderstorm radar identification techniques and warning criteria, *NOAA Tech. Memo. NWS NSSFC-3*, 60 pp., Natl. Oceanic and Atmos. Admin., Boulder, Colo., 1980. (Available as NTIS PB-273049, Natl. Tech. Inf. Serv., Springfield, Va.)

Maddox, R. A., L. R. Hoxit, and C. F. Chappell, A study of tornadic thunderstorm interactions with thermal boundaries, *Mon. Weather Rev.*, *108*, 322–336, 1980.

Moller, A. R., and C. A. Doswell III, A proposed advanced storm spotter's training program, in *Preprints, 15th Conference on Severe Local Storms*, pp. 173–177, American Meteorological Society, Boston, Mass., 1988.

Moller, A. R., C A. Doswell, and R. W. Przybylinski, High-precipitation supercells: A conceptual model and documentation, in *Preprints, 16th Conference on Severe Local Storms*, pp. 52–57, American Meteorological Society, Boston, Mass., 1990.

Nelson, S. P., The hybrid multicell-supercell storm—An efficient hail producer, II, General characteristics and implications for hail growth, *J. Atmos. Sci.*, *44*, 2060–2073, 1987.

Nelson, S. P., and N. C. Knight, The hybrid multicellular-supercell storm—An efficient hail producer, 1, An archetypal example, *J. Atmos. Sci.*, *44*, 2042–2059, 1987.

Przybylinski, R. W., The Raleigh tornado—28 November, 1988, A radar overview, in *Preprints, 12th Conference on Weather Forecasting and Analysis*, pp. 186–191, American Meteorological Society, Boston, Mass., 1989.

Przybylinski, R. W., S. Runnels, P. Spoden, and S. Summy, The Allendale, Illinois tornado—7 January, 1989: One type of an HP supercell, in *Preprints, 16th Conference on Severe Local Storms*, pp. 516–521, American Meteorological Society, Boston, Mass., 1990.

Purdom, J. F. W., and P. C. Sinclair, Dynamics of convective scale interactions, in *Preprints, 15th Conference on Severe Local Storms*, pp. 354–359, American Meteorological Society, Boston, Mass., 1988.

Storm Data, Natl. Clim. Data Center, Natl. Oceanic and Atmos. Admin., Asheville, N. C., June 1990.

Wakimoto, R. M., The West Bend, Wisconsin storm of 4 April, 1981: A problem in operational meteorology, *J. Clim. Appl. Meteorol.*, *22*, 181–189, 1983.

Weaver, J. F., and S. P. Nelson, Multiscale aspects of thunderstorm gust fronts and their effects on subsequent storm development, *Mon. Weather Rev.*, *110*, 707–718, 1982.

Doppler Radar Identification of Nonsevere Thunderstorms That Have the Potential of Becoming Tornadic

RODGER A. BROWN

National Severe Storms Laboratory, NOAA, Norman, Oklahoma 73069

1. INTRODUCTION

Doppler weather radar has been an important research tool during the past 25 years in our efforts to understand the evolution of severe thunderstorms. Research starting in the late 1960s at the Air Force Cambridge Research Laboratories (now the Geophysics Directorate of U.S. Air Force's Phillips Laboratory) and in the early 1970s at the National Severe Storms Laboratory (NSSL) led to the identification of the Doppler velocity signature for a rotating updraft called a mesocyclone [e.g., *Donaldson*, 1970; *Brown et al.*, 1973]. Because the research results suggested that a mesocyclone is a precursor for tornado formation [e.g., *Burgess*, 1976], the Joint Doppler Operational Project (JDOP) was conducted by the National Weather Service, U.S. Air Force, and Federal Aviation Administration during 1977–1979 to test the operational significance of the research findings. The JDOP operation indicated that nearly all the storms with mesocyclone signatures produced surface damage, about one half of them produced tornadoes, and the existence of a mesocyclone signature permitted the lead time for tornado warnings to be increased by an average of 20 min [*Joint Doppler Operational Project Staff*, 1979]. Since the JDOP test also indicated other operational benefits for Doppler radar, the decision was made to procure a national network of Next Generation Weather Radars (NEXRAD) incorporating Doppler capability (now called WSR-88D).

In a single-Doppler radar study of 41 mesocyclones, *Burgess et al.* [1982] found that the mesocyclone within a supercell storm undergoes a distinct evolution. When a thunderstorm first becomes organized into the supercell stage, updraft rotation exists only at middle altitudes. However, as the storm reaches maturity, rotation is evident from near the ground to the upper portions of the storm. Tornadoes were produced during the mature stage of about 40% of the mesocyclones studied.

On a day with severe thunderstorm and tornado potential the forecaster ideally would use the WSR-88D to identify candidate severe storms before the formation of mesocyclones. In some instances, a precursor feature is evident in the Doppler radar data 30 min or more before the appearance of the mesocyclone/supercell storm. This feature is a vortex pair occurring at middle altitudes on the downstream flanks of a strong updraft, with cyclonic vorticity on the right forward flank and anticyclonic vorticity on the left forward flank. In this paper a controversial mechanism is proposed for the evolution of the vortex pair into a mesocyclone. Data then are presented in support of the proposed evolution process.

2. PROPOSED MECHANISM FOR MESOCYCLONE INITIATION

Given the presence of a middle-altitude vortex pair on the left and right forward flanks of a strong updraft, a mechanism is proposed to explain the transition of the cyclonic portion of the vortex pair into a mesocyclone. This mechanism, which incorporates portions of a hypothesis proposed by *Fujita and Grandoso* [1968], differs from the one that is currently accepted in the meteorological literature. Empirical evidence for the mechanism is based on the sole dual-Doppler radar study that provides detailed temporal resolution of the transition process [*Brown*, 1989].

The Fujita-Grandoso hypothesis starts with a developing storm that has a strong nonrotating updraft. A middle-altitude pair of counter-rotating vortices forms on the flanks of the updraft. As the expanding surface gust front moves beneath the vortex pair, convergence along the gust front forces a new updraft to form beneath both members of the pair. *Rotunno and Klemp* [1982] proposed that middle-altitude low pressure associated with each vorticity center produces an upward directed perturbation pressure gradient force that further enhances vertical motion at those locations.

In the more typical severe thunderstorm situation the significant updraft forms only on the right flank, in response to storm-relative, low-altitude flow that approaches the storm from the right. As the new updraft develops above the right flank gust front, it entrains the ambient vertical vorticity and starts to rotate cyclonically at middle altitudes as vorticity is stretched in the vertical. The process of discrete updraft propagation therefore is responsible for the storm's new updraft starting to rotate. Subsequent right flank updrafts likewise develop rotation through the entrainment of ambient vertical vorticity. Increasing precipitation produced by each new updraft blends with the decreasing precipitation from the previous dying updraft, resulting in a low- and middle-altitude radar reflectivity pattern that appears to move to the right in a continuous manner.

This hypothesis is distinctly different from the currently popular hypothesis that states that any strong updraft in an environment with low-altitude streamwise horizontal vorticity (that is, strong clockwise curvature in a storm-relative hodograph) should rotate because the horizontal vorticity is drawn into the updraft and tilted into the vertical [e.g., *Browning and Landry*, 1963; *Barnes*, 1970; *Rotunno*, 1981; *Davies-Jones*, 1984]. Data discussed in the next section indicate, however, that a strong nonrotating updraft can exist within an environment having low-altitude streamwise horizontal vorticity.

3. OBSERVATIONAL EVIDENCE

Dual-Doppler radar measurements permit the identification of updraft locations and strengths and of vorticity regions within a storm. Doppler radar evidence for the middle-altitude vortex pair being a precursor signature for mesocyclone initiation is presented for three Oklahoma supercell thunderstorms, one of which did not evolve beyond the organizing stage; dual-Doppler data were collected only during portions of the storms' life cycles. The storms are the Agawam hailstorm of June 6, 1979, [*Brown*, 1989] and the Billings hailstorm and Guthrie tornadic storm of April 26, 1984 [*Burgess and Curran*, 1985; *Bluestein and Woodall*, 1990].

Storm-relative hodographs for the environments within which the storms formed are shown in Figure 1. The environment on both days exhibited directional shear of over 90° in the lowest 3 km, minimal shear in the 4- to 6-km layer, and primarily speed shear above 6 km. With the storms moving from the southwest toward the northeast, the magnitude of the low-altitude (0–2 km), storm-relative flow approaching the right flanks of the storms was nearly twice as great on April 26, 1984, as on June 6, 1979. Thus one might expect stronger storms on April 26, 1984, owing to greater convergence along the storms' gust fronts and stronger perturbation pressure forces.

3.1. *Agawam, Oklahoma, Hailstorm of June 6, 1979*

The Agawam storm first appeared on radar at 1325 CST near Fort Sill as the right-hand member of a pair of adjacent storms. The storm moved to the northeast toward the NSSL

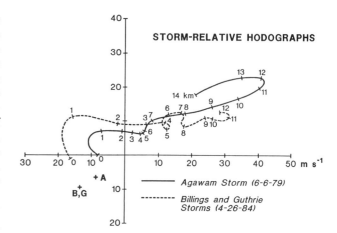

Fig. 1. Storm-relative hodographs for June 6, 1979 (solid) and April 26, 1984 (dashed). Plus signs indicate the origins of the respective ground-relative hodographs. The June 6, 1979, hodograph is a composite of the 1500 CST rawinsonde observations from Fort Sill and Norman, Oklahoma. The April 26, 1984, hodograph is a composite of the 1556 and 1806 CST rawinsonde observations from Chickasha, Oklahoma.

dual-Doppler area covered by the Norman radar and the Cimarron radar located 41 km to the northwest of Norman. Dual-Doppler data collection commenced at 1434 and continued through 1550. A storm intercept team reported hail 26 mm in diameter beneath the storm at 1535.

At 1434, a new nonrotating updraft (called U1) was just developing within the storm. A middle-altitude pair of vertical vorticity maxima was found on the right and left forward flanks of the updraft (Figure 2). As the updraft

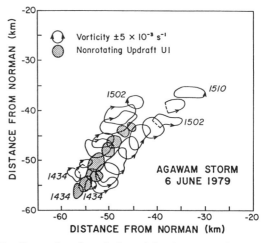

Fig. 2. Composite of vertical vorticity (contours of $\pm 5 \times 10^{-3}$ s^{-1}) relative to updraft U1 at a height of 5.5 km in the Agawam storm [after *Brown*, 1989]. Counterclockwise arrows indicate cyclonic vorticity, and clockwise arrows indicate anticyclonic vorticity. Vertical velocities ≥ 15 m s^{-1} within the updraft are shaded. The time interval between successive updraft and vorticity positions is 4 min.

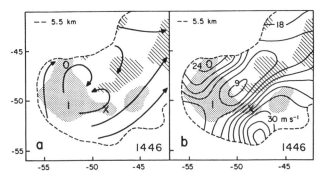

Fig. 3. (a) Storm-relative wind streamlines and (b) ground-relative wind speeds at 5.5-km height in the Agawam storm at 1446 CST [from *Brown*, 1989]. Dashed reflectivity outline is 20 dBZ. Updraft regions are stippled, and downdraft regions are hatched. Updraft U1 (labeled 1) is the mature nonrotating updraft, U0 is a dying left flank updraft, and UX is a new rotating right flank updraft.

moved to the northeast, the vortex pair remained on the forward flank of the updraft. Updraft U1 reached maximum strength (44 m s^{-1}) at 1446, after which time the pair moved slowly downstream relative to the updraft. By 1510 the updraft's maximum vertical velocity was less than 10 m s^{-1}, and the magnitude of vorticity within both members of the pair was dropping below 5×10^{-3} s^{-1}.

The middle-altitude flow field associated with the vortex pair at 1446 is shown in Figure 3. The basic storm-relative flow in Figure 3a diverged around updraft U1. The air closest to the updraft converged into the wake region behind the updraft where the associated downdraft was located. The ground-relative wind speeds in Figure 3b were a minimum in the wake region, being up to 12 m s^{-1} weaker than the undisturbed environmental flow of 19 m s^{-1} at the same height. The wind speeds on the lateral flanks of U1 were up to 7–18 m s^{-1} stronger than the undisturbed environmental flow. The wind speed pattern in Figure 3b is the same as the single Doppler velocity signature for a vortex pair when viewed from the southwest or northeast by a Doppler radar [e.g., *Brown and Wood*, 1991].

As indicated in Figure 3, both wind direction curvature and lateral speed shear contributed to anticyclonic vorticity in the vicinity of preexisting updraft U0 on the left flank of U1 and to cyclonic shear in the vicinity of new updraft UX on the right flank. The new updraft was a short-lived and relatively weak (maximum of 22 m s^{-1}) secondary updraft, but it had the distinction of being the first in a series of right flank updrafts to develop middle-altitude rotation.

The factors contributing to the initiation of updraft rotation are presented in composite form in Figure 4. Shown are the initiation locations for updrafts U1–U5 and UX relative to the right side of the surface gust front and to the pattern of cyclonic vertical vorticity at middle altitudes. Each updraft formed along the right side of the gust front where the low-altitude, storm-relative inflow (Figure 1) was essentially normal to the frontal surface, producing convergence. The location of updraft initiation along the gust front was beneath

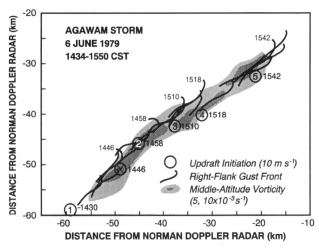

Fig. 4. Juxtaposition of the right edge of the surface gust front boundary (heavy curves), location of updraft initiation (circled numbers or letter), and time-space presentation of maximum cyclonic vertical vorticity at 5.5-km height (shaded region) in the Agawam storm between 1434 and 1550 CST. The spatial orientation of the vertical vorticity data was normal to the mean wind passing through the location of the strongest updraft at that time. Owing to the lack of dual-Doppler radar data at 1530 and 1534, outflow boundaries are missing and vorticity is interpolated for those two times. Times on the right are for updraft initiation, and times on the left represent the gust front locations at the times of updraft initiation.

the cyclonic member of the middle-altitude vorticity pair. Thus it appears that the interaction between the gust front and storm-relative inflow produced a zone of favorable convergence, whereas the upward directed pressure gradient force induced by low pressure at the vorticity center dictated where the updraft would form along the gust front [e.g., *Fujita and Grandoso*, 1968; *Rotunno and Klemp*, 1982]. As the updraft grew vertically, it entrained and stretched the surrounding vertical vorticity and started to rotate cyclonically at middle altitudes.

One can speculate that the Agawam storm did not develop beyond the organizing supercell stage because of the relatively rapid formation of new updrafts at 10- to 15-min intervals (compared to 40–50 min for the mature stage) that did not permit time for rotation to develop in the lower portions of the updraft. However, little is known about the organizing stage of supercell storms owing to the lack of other dual-Doppler radar case studies.

3.2. Billings and Guthrie, Oklahoma, Storms of April 26, 1984

At 1555 CST on April 26, 1984, four radar echoes appeared along the dry line in west central Oklahoma; these storms have been studied by *Burgess and Curran* [1985] and *Bluestein and Woodall* [1990], and additional information about them was obtained from photographs of the Norman Doppler velocity and reflectivity displays. Initially, these storms were low-precipitation (LP) storms because there was little visual precipitation below cloud base [*Bluestein and Woodall*, 1990]. The middle two storms were the stronger, and the single-Doppler velocity signature of a vortex pair (similar to the pattern in Figure 3b) quickly appeared at middle altitudes in both of them. At 1636, dual-Doppler analyses of the southern storm by *Bluestein and Woodall* [1990] revealed the presence of new right and left flank updrafts, which indicated that the storm was ready to split into right- and left-moving portions (beginning of the organizing supercell stage). The right flank updraft was rotating cyclonically at middle altitudes, and the left flank updraft was rotating anticyclonically. By that time, the updraft responsible for producing the vortex pair had decayed, and only the associated precipitation downdraft remained. Radar reflectivity patterns aloft in the northern storm indicated that it also was starting to split at that time. During and following the splitting process, the northern right-moving storm, called the Billings storm, produced damaging surface winds and hail as large as golf balls [*National Oceanic and Atmospheric Administration (NOAA)*, 1984]; after 1730, it moved beyond the first trip range from the Norman Doppler radar. The southern right-moving storm, called the Guthrie storm, produced damaging surface winds and hail up to baseballs and grapefruit in size. Intermittent damage from two tornadoes was reported from Guthrie northeastward to Stillwater between 1753 and 1906 [*NOAA*, 1984]. The important transition between the organizing and mature supercell stages taking place within the Guthrie storm between 1640 and 1720 could not be documented with dual-Doppler measurements because the storm was crossing the baseline passing through the Norman and Cimarron Doppler radar sites. Owing to the infrequent availability of dual-Doppler radar data, it was not possible to document the initiation and evolution of updrafts and their rotational characteristics within the Billings and Guthrie storms.

4. CONCLUDING DISCUSSION

On the basis of this limited survey of only three severe thunderstorms, it appears that the presence of a vortex pair at middle altitudes holds promise as an early indicator that a storm has the potential of becoming severe or tornadic. The sequence of events taking place within the three storms is summarized in Figure 5. A common feature is the middle-altitude vortex pair that appeared early in each storm's lifetime. There probably where sequential updrafts during the vortex pair stage, but neither dual-Doppler radar data (June 6, 1979) nor analyses from the literature (April 26, 1984) occurred at suitable times to document such an evolution.

The vortex pair appears to have played two roles in the development of a rotating updraft. First, low pressure associated with each middle-altitude vorticity center produced an upward directed pressure gradient force that enhanced updraft formation along the surface gust front beneath the vertical vorticity center [e.g., *Rotunno and Klemp*, 1982]. Second, the vertical vorticity field itself provided the vortic-

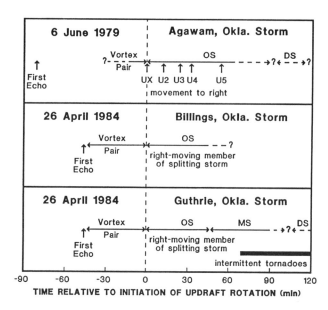

Fig. 5. Temporal variation of basic features within the Agawam, Billings, and Guthrie supercell thunderstorms. When known (solid lines) or estimated (dashed lines), the durations of the organizing (OS), mature (MS), and dissipating stages (DS) of the storms' mesocyclones are indicated.

ity that was entrained and stretched by the growing updraft to produce middle-altitude rotation [e.g., *Fujita and Grandoso*, 1968].

This explanation for the initiation of updraft rotation is in contrast with the popular explanation that employs the vertical tilting of low-altitude streamwise horizontal vorticity in the inflow air as it rises in the updraft [e.g., *Browning and Landry*, 1963; *Barnes*, 1970; *Rotunno*, 1981; *Davies-Jones*, 1984]. It can be noted in Figure 1 that the storm-relative winds in the lowest 3 km of the environment on both storm days were nearly parallel to horizontal vorticity vectors (normal to the hodograph curve, pointing to the left), in agreement with the popular explanation. However, not every strong updraft in these storms rotated; for example, the updrafts producing the vortex pairs did not rotate. The first updrafts that developed rotation appear to be those that were in the vorticity region of a member of the middle-altitude vortex pair on the lateral flanks of an existing nonrotating updraft. Thus vertical tilting of low-altitude streamwise horizontal vorticity does not appear to be the source of initial updraft rotation in these storms. However, streamwise vorticity may play some role during the mesocyclone's mature stage, when updraft rotation extends from near the ground to the upper portions of the storm.

The summary in Figure 5 indicates that the portions of mesocyclone/supercell evolution that could be documented fit the conceptual model of *Burgess et al.* [1982]. Initial updraft rotation occurred only at middle altitudes (organizing stage). The Guthrie storm was the only one to have a documented mature stage with rotation extending to the ground. Although all three storms produced damaging hail and the two on April 26, 1984, produced damaging winds, only the Guthrie storm produced tornadoes.

To properly interpret vertical vorticity patterns within a thunderstorm, it is important to know the relative locations of updrafts. Single-Doppler radar data provide two indications of updraft location, and both are situated in the upper portions of the storm. One is the single-Doppler velocity signature of divergence [e.g., *Brown and Wood*, 1991] near the top of the updraft. The other is the radar reflectivity maximum (upward bulge at storm top) that marks the presence of precipitation particles at the top of the updraft. During the vortex pair stage, the updraft is centered at the upwind end of the pair (Figure 2). During the mesocyclone/supercell stage, the updraft is coincident with a vorticity maximum, marking a rotating updraft.

The few single- or multiple-Doppler radar studies of supercell storms that exist provide only fragmentary documentation of the initiation and evolution of the supercell stage. The subset presented here suggests that the presence of a middle-altitude vortex pair is a necessary condition for the formation of updraft rotation and the transition of a storm into its supercell stage. However, a more systematic study of hundreds of severe and nonsevere thunderstorms is required before we will know how often a severe/tornadic storm is preceded by a vortex pair and how often a storm with a vortex pair evolves into a severe/tornadic storm.

Acknowledgments. Useful comments on the manuscript by Ralph Donaldson, Jr., Michael Eilts, David Stensrud, and an anonymous reviewer are most appreciated. Valuable editorial suggestions were provided by Christina Thomas. The figures were ably prepared by Joan Kimpel.

REFERENCES

Barnes, S. L., Some aspects of a severe, right-moving thunderstorm deduced from mesonetwork rawinsonde observations, *J. Atmos. Sci.*, *27*, 634–648, 1970.

Bluestein, H. B., and G. R. Woodall, Doppler-radar analysis of a low-precipitation severe storm, *Mon. Weather Rev.*, *118*, 1640–1664, 1990.

Brown, R. A., Initiation and propagation of thunderstorm mesocyclones, Ph.D. dissertation, 321 pp., Univ. of Okla., Norman, 1989.

Brown, R. A., and V. T. Wood, On the interpretation of single-Doppler velocity patterns within severe thunderstorms, *Weather Forecasting*, *6*, 32–48, 1991.

Brown, R. A., D. W. Burgess, and K. C. Crawford, Twin tornado cyclones within a severe thunderstorm: Single Doppler radar observations, *Weatherwise*, *26*, 63–69, 71, 1973.

Browning, K. A., and C. R. Landry, Airflow within a tornadic storm, in *Preprints, 10th Weather Radar Conference*, pp. 116–122, American Meteorological Soc., Boston, Mass., 1963.

Burgess, D. W., Single-Doppler radar vortex recognition: Part I: Mesocyclone signatures, in *Preprints, 17th Conference on Radar Meteorology*, pp. 97–103, American Meteorological Society, Boston, Mass., 1976.

Burgess, D. W., and E. B. Curran, The relationship of storm type to environment in Oklahoma on April 26, 1984, in *Preprints, 14th Conference on Severe Local Storms*, pp. 208–211, American Meteorological Society, Boston, Mass., 1985.

Burgess, D. W., V. T. Wood, and R. A. Brown, Mesocyclone evolution statistics, in *Preprints, 12th Conference on Severe Local Storms*, pp. 422–424, American Meteorological Society, Boston, Mass., 1982.

Davies-Jones, R. P., Streamwise vorticity: The origin of updraft rotation in supercell storms, *J. Atmos. Sci.*, *41*, 2991–3006, 1984.

Donaldson, R. J., Jr., Vortex signature recognition by Doppler radar, *J. Appl. Meteorol.*, *9*, 661–670, 1970.

Fujita, T., and H. Grandoso, Split of a thunderstorm into anticyclonic and cyclonic storms and their motion as determined from numerical model experiments, *J. Atmos. Sci.*, *25*, 416–439, 1968.

Joint Doppler Operational Project Staff, Final report on the Joint Doppler Operational Project (JDOP) 1976–1978, *NOAA Tech. Memo. ERL NSSL-86*, 84 pp., Natl. Severe Storms Lab., Norman, Okla., 1979. (Available as *NTIS PB80-107188/AS* from Natl. Tech. Inf. Serv., Springfield, Va.)

National Oceanic and Atmospheric Administration (NOAA), *Storm Data*, vol. 26, no. 4, pp. 50–51, Asheville, N. C., 1984.

Rotunno, R., On the evolution of thunderstorm rotation, *Mon. Weather Rev.*, *109*, 577–586, 1981.

Rotunno, R., and J. B. Klemp, The influence of the shear-induced pressure gradient on thunderstorm motion, *Mon. Weather Rev.*, *110*, 136–151, 1982.

An Examination of a Supercell in Mississippi Using a Tilt Sequence

DAVID A. IMY

National Weather Service Forecast Office, NOAA, Denver, Colorado 80239

KEVIN J. PENCE

National Weather Service Forecast Office, NOAA, Jackson, Mississippi 39288

1. INTRODUCTION

Most of the scientific community now recognize that a supercell is a storm that contains a deep persistent rotating updraft or a mesocyclone (i.e., vertical velocity correlated with vertical vorticity [*Doswell et al.*, 1990]). Most forecasters do not, however, have access to Doppler data to verify the existence of a mesocyclone in a storm. Operational forecasters have relied primarily on low-level reflectivity signatures such as hook echoes, "distinctive" echoes [*Forbes*, 1981], and notches [*Przybylinski et al.*, 1990] to infer the existence of a mesocyclone in a storm.

Relying on low-level reflectivity signatures for issuing tornado warnings, without investigating the vertical structure of a storm, many times results in missed tornadic events. The classic supercell is often identified on radar by the existence of a hook or distinctive echo in the lower portions of the storm. *Browning* [1964] found that the three-dimensional structure of the classic supercell was characterized with an extensive sloping overhang, a region of weak low-level reflectivity capped by the storm top (vault), and an intense hook-shaped echo surrounding the vault. The vault is now referred to as the bounded weak echo region (BWER) [*Chisholm and Renick*, 1972]. These features were noted with supercells in England as well as Oklahoma [*Browning*, 1963, 1965]. *Lemon* [1980] used the work of Browning and others to develop the Lemon Technique to examine not only the three-dimensional storm structure of supercells but all severe thunderstorms. The Lemon Technique uses the structural relationships among the low-level reflectivity gradient, midlevel overhang, and the location of the storm top as warning criteria to determine the storms potential for severe weather and/or tornadoes.

Using the low-level reflectivity field alone to determine if low-precipitation supercells (LP) [*Bluestein and Parks*, 1983] or high-precipitation supercells (HP) [*Moller et al.*, 1990] contain circulations is even less reliable than with the classic supercell. Hook echoes typically are not observed with these types of supercells. The LP supercell often exhibits relatively low reflectivities and usually lacks indications of a low-level circulation in the reflectivity field. *Przybylinski et al.* [1991] and *Doswell et al.* [1990] noted several variations in the low-level reflectivity echo shape associated with HP supercells. *Moller et al.* [1990] noted that the most commonly identified echo configurations associated with the HP supercell included the "comma," "kidney bean," spiral, and "fat hook" echoes.

National Weather Service forecasters in Mississippi [i.e., *Imy*, 1986] noted differences in the low-level reflectivity field associated with supercells in Mississippi from those described by Browning. In summary, hook echoes were rarely observed with these supercells. The storms usually were distinguished from other storms by their unusually large size (greater than 35 km in diameter). They contained an extensive area of high reflectivities (video integrator and processor (VIP 5), 50–56 dBZ). Most of the storms were not isolated but surrounded by light to moderate precipitation. Although most of these supercells in Mississippi generated hail, the hail was not much larger than 2.5 cm in diameter and normally affected only small areas. One supercell (February 28, 1987) spawned an F4 tornado, but a post storm survey found no one near or in the path of the tornado who observed hail prior to or after the tornadic event. Supercells in Mississippi seem to be more typically of the HP supercell variety, except they are not usually prolific hailers.

Although the low-level reflectivity field observed with these storms in Mississippi differed from those described by *Browning* [1963, 1965] and *Lemon* [1980], most of the storms unfortunately were not examined three-dimensionally. On

The Tornado: Its Structure, Dynamics, Prediction, and Hazards.
Geophysical Monograph 79
This paper is not subject to U.S. copyright. Published in 1993 by the American Geophysical Union.

March 12, 1986, however, a supercell in Mississippi was evaluated using a variation of the Lemon Technique. Although the low-level reflectivity characteristics resembled the HP supercell, the three-dimensional structure was similar to the classic supercell, with minor deviations. The need to use volumetric radar data to evaluate the potential severity of a storm is illustrated with this particular supercell. Also, an alternative tilt scan strategy is presented, since the full utilization of the Lemon Technique may at times be too time-consuming during the warning process.

2. THE SYNOPTIC/MESOSCALE ENVIRONMENT

At 1200 (all times UTC) on March 12, 1986, a long-wave midtropospheric trough over the central United States resulted in a strong southwesterly flow aloft over Mississippi. A series of short-wave troughs were rotating through the base of the long-wave trough and approached Mississippi from the southwest. The important low-level mesoscale feature was an outflow boundary/trough that was moving slowly eastward across Mississippi. The outflow boundary was generated by a line of thunderstorms that moved eastward from Louisiana and Arkansas during the night. This line contained numerous severe thunderstorms that yielded hail and wind damage along with a few tornadoes. The line of thunderstorms dissipated around 1200, but the outflow boundary persisted and became quasi-stationary northeast to southwest across Mississippi (Figure 1).

Surface observations in Mississippi and southern Louisiana together with radar data (that included a line-echo wave pattern, or LEWP), depicted a series of mesoscale low-pressure systems that moved rapidly northeastward along this boundary. As these surface perturbations neared reporting stations, the pressure dropped and the winds backed from the south to the southeast. Six tornadoes that developed along the boundary this day appeared to have occurred near or slightly north of these small areas of low pressure.

Moist low-level south to southeast winds were advecting air having 15°–18°C dew points northward ahead of the boundary, while dew points west of the surface trough lowered to around 12°C. The 1200 Jackson, Mississippi, sounding (not shown), with the boundary just east of the site, indicated that the atmosphere was moderately unstable with lifted indices around −6 and a convective available potential energy (CAPE) near 1500 m² s⁻² (or 1500 J kg⁻¹). The CAPE, or positive buoyant energy, is a measure of the positive energy located between the level of free convection (LFC) and equilibrium level (EL) on the sounding [*Rasmussen and Wilhelmson*, 1983]. In contrast to the Jackson sounding, a significant low-level inversion was present in southern Louisiana at Boothville (or ~120 km southeast of New Orleans). The low-level cap was 5.7°C, and without any other lifting, a surface temperature of 30°C was needed to break this cap; maximum temperatures reached only 26°C prior to the time of convection. This indicated that forced lifting via strong convergence, such as along the boundary, was necessary to break the cap.

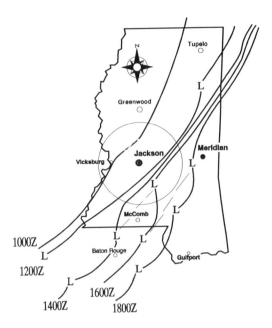

Fig. 1. Outflow boundary locations from 1000 to 1800 Z (UT), which served as the focus for the tornadic storms on March 12, 1986. The circle surrounding the radar represents the 92-km circle for references to Figures 3–8.

The 1200 Jackson, Mississippi, hodograph in the lower levels was unrepresentative of the environmental wind profile east of the boundary. The 1500 wind observation at McComb, Mississippi, was used to adjust the hodograph in the lower 2 km to better reflect the wind conditions on the east side of the boundary (Figure 2). The storm motion of 225° at 23 m s⁻¹ resulted in a strong storm-relative flow from the northeast in the lower 2 km. Southwest flow aloft, combined with the strong northeast storm-relative flow in the lower levels of the environment, indicated that the storm's updraft flank likely would be located on the southeast portion of the low-level precipitation echo.

The potential for mesocyclones existed as the helicity in the lowest 3 km was 173 m² s⁻², which exceeded the 150 m² s⁻² threshold given by *Davies-Jones et al.* [1990]. The positive shear was 9.3×10^{-3} s⁻¹ [*Davies*, 1989]. The *Johns et al.* [1990] relationship of positive shear and CAPE suggested that the potential existed for the development of strong to violent mesocyclone-related tornadoes.

3. THE SUPERCELL EVOLUTION

The storm of interest, on March 12, 1986, developed in south central Louisiana and moved northeastward into southwest Mississippi shortly after 1430. Two tornadoes (F2 and F1 intensity) were spawned with this storm between 1500 and 1700 in south central Mississippi, along with some hail and wind damage. The storm was examined almost continuously with a low-elevation angle (0.5°) using the WSR-57 located at Jackson, Mississippi. Occasionally, the

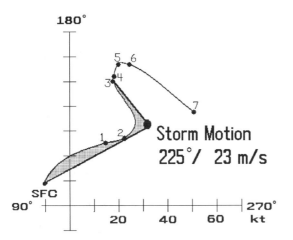

Fig. 2. Jackson 1200 hodograph adjusted in the lower 2 km to better reflect the wind profile along and east of the boundary. The helicity is 173 m^2 s^2.

operator elevated the antenna to examine the upper portions of the storm. Since this was performed during a warning situation, time did not allow for complete tilts. With the higher elevation tilt, the VIP 3 (41–46 dBZ) core (bright white) was used as the midlevel echo since the storm was surrounded by light to moderate precipitation. The storm top was not systematically located as described by *Lemon* [1980], since only one upper level tilt was used to examine the midlevels of the storm. We used the center point of the midlevel echo as the approximate location of the top. In Lemon's Technique it is not the height but the location of the top relative to the low-level reflectivity gradient that is important. If the midlevel overhang extends more than 6 km beyond the low-level reflectivity (as described by Lemon for severe thunderstorms), then it would indicate that the top is likely located on the updraft flank of the storm and not over the storm's core.

The low elevation angle slice of the storm, prior to the time it became severe, is illustrated in Figure 3 at 1455. A small appendage or pendant located on the storm's rear flank, which we will see later, was not associated with the updraft flank of the storm and therefore was not dynamically significant according to *Lemon* [1980]. Note also the band of precipitation to the southeast of this storm. The relationship between this band of precipitation and the development of the tornadic storm are unclear, but the band is notable in all of the low elevation slices used in this paper. The precipitation bands southeast and southwest of the echo resemble an "arch echo" found by *Smith and McKee* [1988].

The first elevated tilt of 8.3° at 1455 is presented in Figure 4 (all upper level tilts bisected the storm at 10 km above ground level, assuming standard atmospheric refraction con-

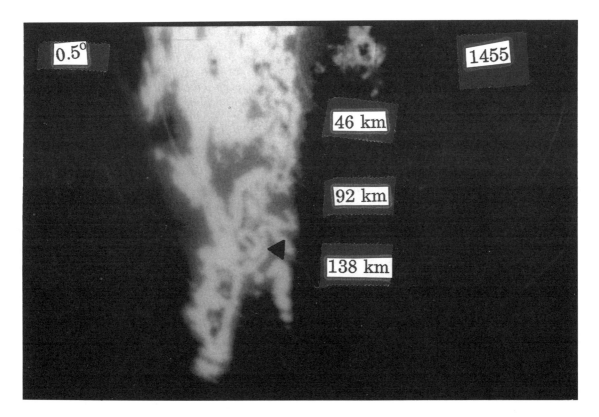

Fig. 3. Low level slice of storm located west of arrow at 1455.

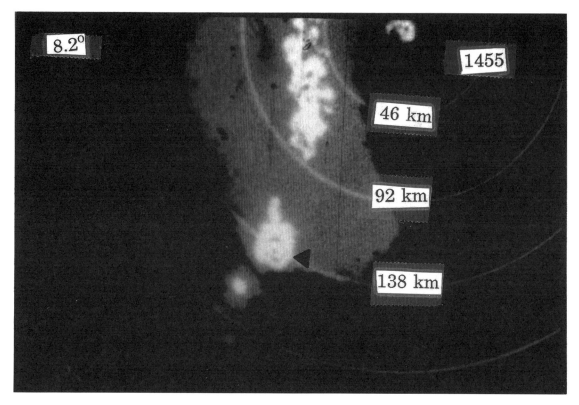

Fig. 4. Upper level slice of storm (8.3°) at 1455.

ditions). The midlevel core was almost vertical over the low level core. Both VIP 5 and 6 (≥50 dBZ) cores were evident even at this height. The VIP 3 midlevel echo (41–46 dBZ) failed to display a significant echo overhang. The height of the VIP 5 (and VIP 6) above 10 km, however, would warrant the issuance of a severe thunderstorm warning using Lemon's criteria.

In Figure 5, at 1519, the low-level echo configuration had evolved into a kidney bean shape, which, as mentioned earlier, may be suggestive of an HP supercell. A concavity was evident on the east side of the storm along with a tight reflectivity gradient, both suggestive of the storm's updraft flank. The kidney bean shape persisted for only about 5 min. An elevated cut was not available at 1519, but 2 min earlier, the higher tilt revealed an extensive midlevel overhang above the concavity on the east side of the storm (Figure 6). The development of the massive midlevel overhang since 1455 indicated that the updraft strength had increased significantly.

The combination of the tight low-level reflectivity gradient in the concavity, the expanse of the midlevel overhang, the mesoscale environment (storm moving along a preexisting boundary), moderate CAPE, and favorable helicity suggested that this storm had the potential to become tornadic. Once it appears that a storm is (or is becoming) tornadic, it is critical to pinpoint the tornadic region of the storm. With a large storm, such as this one, if a tornado warning was issued for the entire precipitation echo, several counties would have been overwarned. Locating the most dangerous portion of the storm is essential both for aiding spotters and warning those communities most threatened [Moller et al., 1990]. The combination of the low-level and high-level tilts indicated that the most dangerous portion of this storm was on the east side of the storm near the low-level concavity.

At 1525 an F2 tornado developed on the east side of the storm, just south of the concavity previously noted. The tornado was on the ground for approximately 15 min with a 10-km path length. The largest hail with the storm, 2 cm in diameter, occurred in the VIP 5 core north of the concavity.

At 1542 the low-level echo in Figure 7 had become elongated from north to south, but there were no hints of rotation. A channel of weak echo [Przybylinski and DeCaire, 1985] had developed on the back side of the storm, possibly indicative of the rear flank downdraft (RFD) [Moller et al., 1974]. The VIP 3 midlevel echo had a circular appearance at 1538 (Figure 8, closest available elevated tilt to the low tilt). No tornadoes ensued with the circular midlevel echo, but surface winds greater than 50 knots (26 m s^{-1}) occurred at the surface near this midlevel feature at 1545.

A small concavity was noted on the northeast side of the storm on the low elevation slice at 1605 (not shown). The high level tilt at 1604, however, revealed that the midlevel overhang was located on the east side of the storm. Thus this concavity probably was not dynamically significant, since it

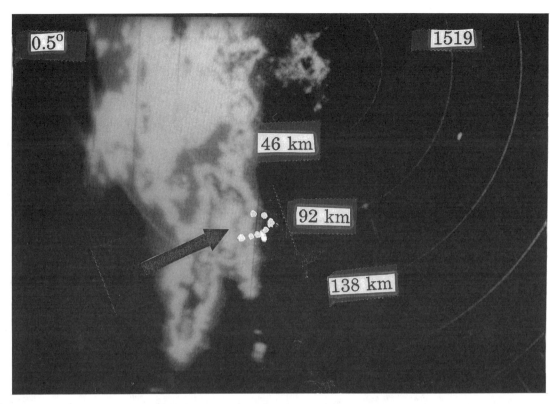

Fig. 5. Low-level slice of the storm at 1519. The midlevel overhang is depicted by the dotted line. The overhang extends approximately 10 nm east and northeast of the low-level reflectivity gradient in the concavity. A tornado initially touched down 6 min later. The arrow depicts the location of the tornado touchdown relative to the precipitation echo at this time.

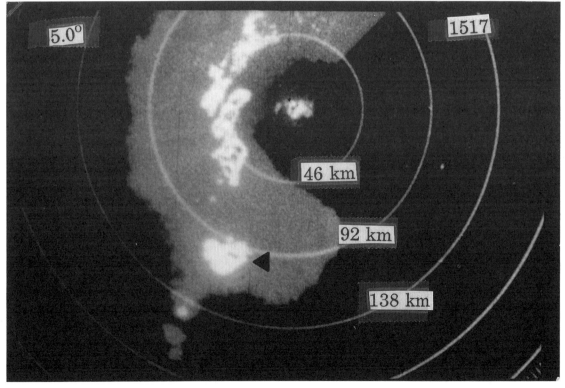

Fig. 6. Upper level slice of storm (5.0°) at 1517.

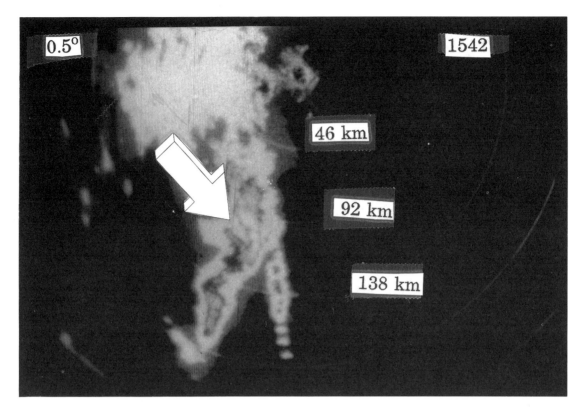

Fig. 7. Low-level slice of storm with a weak echo channel at the rear of storm (southeast of arrow) at 1542.

was not coupled with the storm's updraft. The overhang was associated with an intense updraft and low-level circulation located on the extreme southern portion of the precipitation echo. The overhang extended about 8 km northeast of the low-level reflectivity gradient. Also, the midlevel echo was in the shape of an "S," which *Doswell et al.* [1990] state can be an indication of circulation.

Without the use of an elevated tilt and a knowledge of the environmental wind profile, once again it would have been extremely difficult to determine the most menacing portion of this storm. A weak tornado (F1) developed on the southern portion of the storm with a 2-km path length. The low-level echo associated with this second tornado never assumed the kidney bean shape illustrating the variability of low-level echo configurations associated with supercells in Mississippi. Also, while the first tornado developed on the east side of the precipitation echo, the second tornado developed on the southern portion of the echo.

4. Conclusions

Strong storms in sheared environments must be analyzed frequently with at least one higher elevation angle to determine storm structural features suggestive of severe or tornadic storms. As illustrated in this case, forecasters should not rely solely on characteristic low-level configurations such as the hook or distinctive echoes for issuing severe thunderstorm or tornado warnings. Supercells in Mississippi, similar to the HP supercell, are characterized by continuously changing low-level echo configurations. Broad hooks may be observed, but waiting for these features to appear before issuing a tornado warning likely will result in small lead times or missed events.

The three-dimensional structural evolution of this supercell was similar to that described by *Browning* [1964] and *Lemon and Doswell* [1979] for supercells in the central plains, or the classic supercell. Although the low-level echo configurations does not resemble Browning's model for the classic supercell, this storm still displayed a vertical structure similar to the classic supercell. With or without Doppler data, only through a tilt sequence can most severe and tornadic thunderstorms be warned for effectively, regardless of the low-level echo shape. Any tilt technique, however, may become ineffective during the storm's collapse in the later stages of tornadic development, but tornado warnings should be considered throughout the life of a storm with a past history of tornadoes when Doppler information is not available [*Imy and Burgess*, 1991].

However, radar operators may feel that the Lemon Technique is too time-consuming for real-time warning operations. We propose the following methodology (as performed by the radar operator in this case) for analyzing storms using

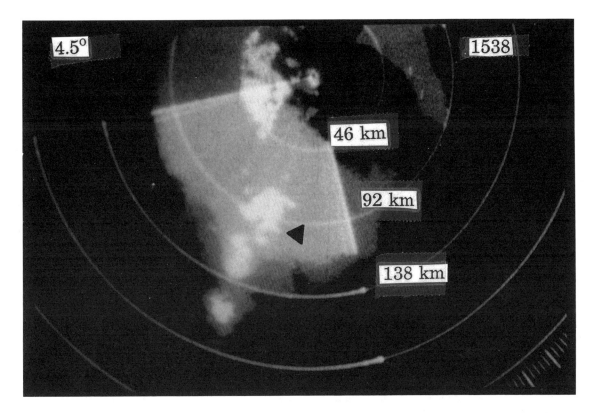

Fig. 8. Upper level slice of storm (4.5°) at 1538. Note the semicircle-shaped echo just south of the main midlevel echo.

conventional radar systems (with the antenna rotating continuously through both steps).

1. On the lowest elevation (usually 0.5°), contour the VIP 2 (30–41 dBZ) echo on the PPI scope, or the VIP 3 (41–46 dBZ) echo if VIP 2 (30–41 dBZ) precipitation surrounds the storm. Mark or highlight the strongest low-level reflectivity gradient if present.

2. Quickly elevate the antenna so the radar beam slices through the midlevel portion of the storm (5–12 km) [Lemon, 1980] using the VIP 3 (41–46 dBZ) contour.

After performing these steps, use Lemon's [1980] guidelines for determining if a warning is required and, if so, whether it should be for a severe thunderstorm or a tornado. This technique can be executed quickly (in less than a minute) yet provide critical information necessary to warn effectively for most severe storms, including the wide range of low-level echo configurations associated with supercells.

The Lemon Technique will be much easier to perform on the Weather Surveillance Radar-1988 Doppler (WSR-88D) system. The WSR-88D automatically makes one low-level slice (0.5°) and 8 or 13 elevated slices every 5 or 6 min. Up to four products can be displayed on one screen at the principal user processor (PUP) workstation at any one time [Alberty et al., 1991]. Depending on the distance of the storm from the radar, the operator could select three reflectivity slices depicting the lower, middle, and upper portions of the storm and an echo tops product in the fourth quadrant. On the other screen, velocity products at the same elevation angles as the reflectivity slices also could be displayed to help validate or nullify the existence of a mesocyclone. All of this can be done easily and repeatedly and in a few seconds through the use of a user function [Lemon et al., 1992]. The radar does the tedious work of collecting and displaying the data at various elevation angles, while the operator interprets the three-dimensional structure of storms to determine when and where warnings should be issued.

Acknowledgments. A special thanks to Les Lemon for his exceptionally helpful comments that significantly improved the content in this manuscript. We thank Bob Ponds, retired Senior Meteorological Technician, who examined this storm three-dimensionally. We thank Dan Purcell for preparing the radar pictures and Richard Armstrong for preparing one of the figures. Also, we thank the OSF staff and the anonymous reviewers for offering helpful comments that improved this manuscript.

References

Alberty, R. A., T. Crum, and F. Toepfer, The NEXRAD program: Past, present, and future; a 1991 perspective, in *Preprints, 25th International Conference on Radar Meteorology*, pp. 1–8, American Meteorological Society, Boston, Mass., 1991.

Bluestein, H. B., and C. R. Parks, A synoptic and photographic climatology of low-precipitation severe thunderstorms in the southern plains, *Mon. Weather Rev.*, *111*, 2034–2046, 1983.

Browning, K. A., Airflow and structure of a tornadic storm, *J. Atmos. Sci.*, *20*, 533–545, 1963.

Browning, K. A., Airflow and precipitation trajectories within severe local storms which travel to the right of the winds, *J. Atmos. Sci.*, *21*, 634–639, 1964.

Browning, K. A., Some inferences about the updraft within a severe local storm, *J. Atmos. Sci.*, *22*, 669–677, 1965.

Chisholm, A. J., and J. H. Renick, The kinematics of multicell and supercell Alberta hailstorms, Alberta Hail Studies, 1972, *Hail Stud., Rep. 72-2,24–31*, Res. Counc. of Alberta, Edmonton, 1972.

Davies, J. M., On the use of shear magnitudes and hodographs in tornado forecasting, in *Preprints, 12th Conference on Weather Analysis and Forecasting*, pp. 219–224, American Meteorological Society, Boston, Mass., 1989.

Davies-Jones, R. P., D. W. Burgess, and M. P. Foster, Test of helicity as a tornado forecast parameter, in *Preprints, 16th Conference on Severe Local Storms*, pp. 588–592, American Meteorological Society, Boston, Mass., 1990.

Doswell, C. A., III, A. R. Moller, and R. W. Przybylinski, A unified set of conceptual models for variations on the supercell theme, in *Preprints, 16th Conference on Severe Local Storms*, pp. 40–45, American Meteorological Society, Boston, Mass., 1990.

Forbes, G. S., Relationship between tornadoes and hook echoes on April 3, 1974, *Mon. Weather Rev.*, *109*, 1457–1466, 1981.

Imy, D. A., A modified supercell of November 31st–December 1st, *S. Reg. Headquarters Tech. Attachment, Mar. 25, 1986*, 6 pp., Natl. Weather Serv. S. Reg. Headquarters, Fort Worth, Tex., 1986.

Imy, D. A., and D. W. Burgess, The structural evolution of a tornadic supercell with a persistent mesocyclone, in *Preprints, 25th International Conference on Radar Meteorology*, pp. 408–411, American Meteorological Society, Boston, Mass., 1991.

Johns, R. H., J. M. Davies, and P. W. Leftwich, An examination of the relationship of 0–2 km AGL "positive" wind shear to potential buoyant energy in strong and violent tornado situations, in *Preprints, 16th Conference on Severe Local Storms*, pp. 593–598, American Meteorological Society, Boston, Mass., 1990.

Lemon, L. R., Severe thunderstorm radar identification techniques and warning criteria, *NOAA Tech. Memo. NWS NSSFC-3*, 60 pp., Natl. Oceanic and Atmos. Admin., 1980. (Available as *NTIS PB-273049*, Natl. Tech. Inf. Serv., Springfield, Va.)

Lemon, L. R., and C. A. Doswell III, Severe thunderstorm evolution and mesocyclone structure as related to tornadogenesis, *Mon. Weather Rev.*, *107*, 1184–1197, 1979.

Lemon, L. R., E. M. Quoetone, and L. J. Ruthi, WSR-88D: Effective operational applications of a high data rate, in *Preprints, Symposium on Weather Forecasting*, pp. 173–180, American Meteorological Society, Boston, Mass., 1992.

Moller, A. R., C. A. Doswell III, J. McGinley, S. Tegtmeier, and R. Zipser, Field observations of the Union City tornado in Oklahoma, *Weatherwise*, *27*, 68–77, 1974.

Moller, A. R., C. A. Doswell III, and R. W. Przybylinski, High-precipitation supercells: A conceptual model and documentation, in *Preprints, 16th Conference on Severe Local Storms*, pp. 52–57, American Meteorological Society, Boston, Mass., 1990.

Przybylinski, R. W., and D. M. DeCaire, Radar signatures associated with the derecho, a type of mesoscale convective system, in *Preprints, 14th Conference on Severe Local Storms*, pp. 228–231, American Meteorological Society, Boston, Mass., 1985.

Przybylinski, R. W., S. Runnels, P. Soden, and S. Summy, The Allendale, Illinois tornado—January 7, 1989: One type of an HP supercell, in *Preprints, 16th Conference on Severe Local Storms*, pp. 516–521, American Meteorological Society, Boston, Mass., 1990.

Przybylinski, R. W., J. T. Snow, E. M. Agee, and J. T. Curran, The use of volumetric radar data to identify supercells: A case study of June 2, 1990, this volume.

Rasmussen, E. N., and R. B. Wilhelmson, Relationships between storm characteristics and 1200 GMT hodographs, low-level shear, and instability, in *Preprints, 13th Conference on Severe Local Storms*, pp. J5–J8, American Meteorological Society, Boston, Mass., 1983.

Smith, C., and R. McKee, Arch echoes and maxi tornadoes, *S. Reg. Headquarters Tech. Attachment, Apr. 12, 1988*, 5 pp., Natl. Weather Serv. S. Reg. Headquarters, Fort Worth, Tex., 1988.

Satellite Observations of Tornadic Thunderstorms

JAMES F. W. PURDOM

Regional and Mesoscale Meteorology Branch, NOAA/NESDIS, Cooperative Institute for Research in the Atmosphere
Colorado State University, Fort Collins, 80523

1. INTRODUCTION

Satellite data provide valuable information concerning tornadic storms and the mesoscale environment in which they form. Below, selected aspects of the tornadic storm's mesoscale environment as observed using satellite data are discussed. That is followed by a discussion of features evident in satellite imagery that accompany tornadic storm development. Thoughts concerning the dynamic and thermodynamic importance of some of those features are discussed. In a few examples, other data are presented along with satellite data. However, while the author realizes the importance of other data sets in tornadic storm investigations, this paper focuses on uses of satellite data. The major emphasis is how to use satellite data to help the reader interpret why storms form, evolve, and take on characteristic appearances. A satellite image or a series of images represents on-going dynamic and thermodynamic processes in the atmosphere; what is important is recognizing those processes.

2. MESOSCALE ENVIRONMENT INFORMATION FROM SATELLITE DATA

Satellite data provide both quantitative and qualitative information concerning the mesoscale environment in which tornadic storms form. In the subsections below, both quantitative and qualitative (image interpretation) uses of satellite data are discussed.

2.1. Moisture and Water Vapor Data

The horizontal and vertical structure of moisture in the atmosphere, an important factor in storm development, exhibits significant mesoscale variability [*Doswell and Lemon*, 1979]. Satellite data, particularly from geostationary satellites, can play an important role in monitoring the horizontal variability of that constituent. That capability was first demonstrated by *Hillger and Vonder Haar* [1981] using polar orbiting satellite sounding data. They were able to extract moisture information at a much finer horizontal resolution than was possible using conventional data sources. Similar results have been demonstrated from geostationary altitudes using data from GOES-VAS [*Smith et al.*, 1984; *Hillger and Purdom*, 1990]. (GOES-VAS is a double acronym for Geostationary Operational Environmental Satellite–VISSR Atmospheric Sounder, with VISSR being visible and infrared spin scan radiometer.)

The characteristics of GOES-VAS sounding channels are shown in Table 1. For VAS the energy received in each channel originates in the region labeled "representative thickness." The amount of energy emitted in that region is a function of the concentration and temperature of the absorbing/emitting constituent, with clouds acting as a contaminant. For the window channels most of the energy originates at the Earth's surface, with a small amount of absorption and reemission by water vapor.

While the studies of Hillger and Smith cited above used all available data, simpler approaches using a limited number of VAS channels have been demonstrated [*Chesters et al.*, 1982; *Petersen and Mostek*, 1982]. For example, a product known as "split window" may be derived using the difference in energy between the 12.7-μm and 11.2-μm infrared (IR) channels. According to *Chesters et al.* [1983 p. 742]: "The VAS split window clearly differentiates those areas in which the water vapor extends over a deep layer and is more able to support convective cells from those areas in which the water vapor is confined to a shallow layer and is therefore less able to support convection." Split window products are useful over land during the afternoon when there are large differences in the signal between the two channels. This occurs with a high surface temperature, a

TABLE 1. GOES-VAS Channel Characteristics

Channel Central Wavelength, microns	Absorbing Constituent	Peak Level, mbar	Representative Thickness, mbar
14.7	CO_2	40	150–10
14.5	CO_2	70	200–30
14.3	CO_2	300	500–10
14.0	CO_2	450	800–300
13.3	CO_2	950	sfc–500
4.5	CO_2	850	sfc–500
12.7	H_2O	sfc	sfc–700
11.2	window	sfc	
7.3	H_2O	600	800–400
6.7	H_2O	450	700–250
4.4	CO_2	500	800–100
3.9	window	sfc	

The abbreviation sfc indicates surface.

steep lapse rate, and significant low-level moisture: ideal low-level thermodynamic conditions for intense convection. In the evening when the land cools (or over the ocean) the signal differential is small, and a meaningful split-window product is difficult to derive.

The 6.7-μm channel's data depict regions of middle level moisture and clouds. Distinct patterns of cool moist areas and warm dry areas are readily detected. Such features are related to areas of horizontal advection and vertical motion at both synoptic scales and mesoscales. When viewed in time lapse, they exhibit excellent spatial and temporal continuity. Strong baroclinic regions such as jet streams and vorticity maxima can often be identified in cloud free regions by the sharp moisture gradient detected in the 6.7-μm image [Weldon and Holmes, 1991].

2.2. Wind Field Information

Tornadic storm outbreaks are characterized by rapidly evolving storms embedded in an environment of strong vertical wind shear. Because of this, during outbreaks of severe weather the GOES satellite is placed in a rapid scan imaging mode (RISOP) to support operations at the National Severe Storm Forecast Center. During RISOP, 5-min interval imagery is scheduled for a portion of selected hours.

Although not done operationally, with RISOP imagery it is possible to compute high-resolution cloud motion vectors for both cumulus and cirrus [Lubich and Purdom, 1992]. When compared with winds from ground-based profilers, such cloud motion vectors were found to match within the noise level of the two systems [Dills and Smith, 1992]. Recently, Purdom and Weaver [1992] performed a detailed analysis of cloud motions for the April 26, 1991, Kansas/Oklahoma tornado outbreak. They found that (1) cirrus winds exhibited mesoscale detail as a jet streak moved into the outbreak area, (2) cirrus to cumulus level shear was greatest in the region of the tornadic storms, and (3) individual supercells blocked cirrus level flow.

2.3. Image Interpretation: Daytime Squall Line Development

Generally, organized convergence zones that trigger strong convection are detectable in satellite imagery prior to thunderstorm formation. Those convergence zones may be associated with dry lines, frontal zones, or areas of prefrontal convergence. The ability to locate areas of incipient squall line development was one of the earliest uses of geostationary satellite imagery in storm forecasting [Purdom, 1971].

Figure 1 presents a typical example of how squall line development appears in geostationary satellite imagery. In Figure 1 a well-defined convective line is located along a surface convergence zone. The convergence zone, analyzed in Figure 2, is moving into warm and moist surface air that extends from eastern Kansas into portions of Missouri and Iowa. Later, a squall line has developed along the northern portion of the convergence zone.

For the case shown in Figure 1, upper level dynamic support was strongest across the northern portion of the region. When 15-min interval satellite imagery was animated relative to the developing convective line, warm sector cumulus merged with the line to the north but not with the line to the south. This implies that stronger low-level convergence to the north aided earlier thunderstorm formation in that region.

Another example of squall line development is shown in Figure 3. In Figure 3a a bright convective line, located along a surface convergence zone, extends from north central Texas into southern Oklahoma. Smoke from range fires in west Texas extends into the rear of the convective line. To the east of the line skies are clear, becoming cloudy again in the vicinity of Tyler and Longview, Texas. Most of the cloudiness around Tyler is cumuliform with a distinct rope appearance. Such cloudiness is indicative of cumulus developing under a strong low-level inversion. The mechanism responsible for the capping inversion may be a "lid" of midlevel hot dry air that has been advected across the region [Carlson et al., 1983]. Figure 3b, taken a few hours after Figure 3a, shows a squall line across Texas and Oklahoma. About 100 miles (~161 km) to the east of that squall line the air is very stable. Abundant rope-appearing cumulus and the Longview, Texas, sounding (Figure 4) confirm that stability.

Is the Longview sounding representative of low levels immediately ahead of the squall line? If so, a very strong capping inversion must be removed for thunderstorms to form. Such stability may be overcome by (1) broad-scale vertical motion, (2) precipitation into the stable layer destroying the inversion through evaporative cooling, (3) differential advection at various levels, or (4) as described by Shapiro et al. [1985], strong localized vertical forcing along the surface convergence zone. Because of the proximity of the stable air to the squall line and the lack of midlevel clouds over the region, the first two mechanisms most likely did not play a significant role. Most likely, a combination of the latter two mechanisms came into play in the development of this squall line.

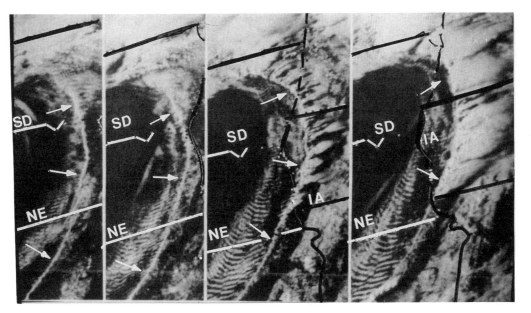

Fig. 1. Montage of GOES-East 1-km resolution visible images from left to right at 1300, 1400, 1600, and 1700 CST for June 14, 1976. Arrows point to organized cloudiness along the surface convergence line as well as the fully developed squall line.

2.4. Image Interpretation: Nighttime Squall Line Development

Squall lines may form at nighttime over the Great Plains of the United States. In many cases that development occurs with the northward advection of low-level moist air ahead of an approaching synoptic scale system. Because of a small amount of absorption and reemission by water vapor in the VAS 11.2-μm "window," such development may be monitored using that infrared channel's imagery. The following generalizations may be made concerning interpretation of 11.2-μm data: (1) for a moist low-level environment during the daytime, after the land surface has heated, temperatures will appear cooler than the actual surface temperature due to some absorption and emission by the overlying cooler moist air, (2) at nighttime, low-level moisture within a nocturnal inversion will cause temperatures to appear warmer than the colder underlying surface, (3) when the atmosphere is dry, day or night, 11.2-μm temperatures will be close to the true surface temperature. In addition, the surface will heat (cool) more rapidly during the day (night) in dry versus moist environments.

On the evening of May 13, 1985, strong nighttime thunderstorm activity developed in southwest Texas. That activity was triggered as a vorticity center and surface front moved out of New Mexico into a region of low-level moist air. In Figure 5a, thunderstorm activity extends from southern Kansas into central Texas. The black area at the southwest end of the anvil in north Texas is associated with the colder overshooting tops of a severe thunderstorm. That overshooting top region appears at the vertex of two cold plumes. That signature, an "enhanced V," has been corre-

lated with severe weather [Fujita, 1978; McCann, 1981; Heymsfield and Blackmer, 1988].

The air mass over northwest Texas and New Mexico is dry. In Figure 5a, relatively cold surface temperatures, with subtle variations due to terrain, are apparent. From the Rio Grande river into central Texas, temperatures appear much

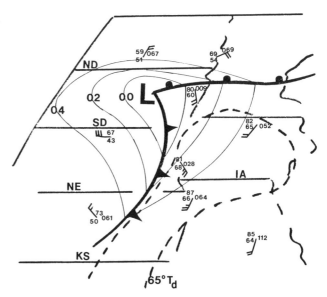

Fig. 2. Selected surface observations at noon CST on June 14, 1976, over the region shown in Figure 1. Analyzed are pressure (solid lines) and dew point temperature (dashed lines).

Fig. 3a. GOES-East 1-km resolution visible image at 1431 CST on March 11, 1988. Arrows point to initial cumulus formation in Oklahoma and north Texas where a squall line is beginning to form. Note the blowing smoke at A and the "rope" cumulus around Tyler (T) and Longview (G). b. GOES-East 1-km resolution visible image at 1731 CST on March 11, 1988. Arrows point to the squall line in eastern Oklahoma and Texas. Blowing smoke is still evident to the rear of the squall line, as is rope cumulus at Longview (G).

warmer (darker) than they do to the north and west. That dark region is a reflection of low-level moisture and perhaps low-level cloudiness at the height of the low-level inversion. During the evening this low-level warm and moist air advected north into the Midland, Texas, area. Another relatively warm region is apparent between the storms in north central Texas and the cooler ground in northwest Texas. The appearance of that region is the result of both warmer ground and low-level air that has been moistened by the earlier thunderstorm activity. The westward extent of that more moist air is determined by local terrain: the Palo Duro Canyon and region to the east is around 1000 feet (300 m) lower than the higher ground to the west (the Cap Rock escarpment).

Over the next several hours the low-level moisture's northward advance into the Midland area could be followed using animated GOES infrared imagery. In Figure 5b, upper level cloudiness associated with the vorticity center is evident in New Mexico. The crescent-shaped white area in southwest Texas locates the surface front. Figure 5c shows the squall line that formed as the front moved into the warm and moist environment to its east. Special ground-based radiometer data confirmed the rapid increase in moisture at Midland [*Parsons et al.*, 1986] during the time the dark region shown in Figure 5a moved across that area.

2.5. Image Interpretation: Mesoscale Convective Complexes

In the previous cases, anvil cirrus was seen blowing off to the east ahead of the squall line. This is typical for squall lines that develop under strong synoptic forcing where winds veer and increase to strong westerly with height. Another important type of highly organized mesoscale convective system, the mesoscale convective complex (MCC), forms with relatively weak synoptic scale forcing [*Maddox*, 1980]. MCCs appear to be a convectively driven weather system. With

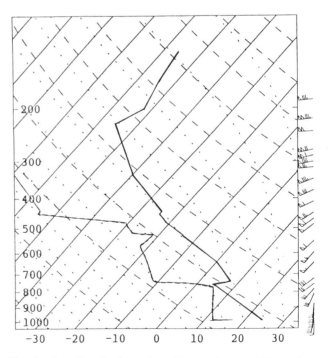

Fig. 4. Sounding for Longview, Texas, that would have been released near the time of Figure 3b. Note the strong capping inversion near 800 mbar.

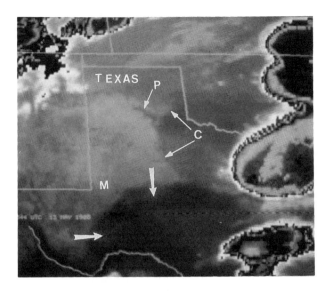

Fig. 5a. Enhanced GOES infrared image taken at 2144 CST on May 13, 1985. Dark to white represents warm to cold (−32°C); temperatures then become colder in 6°C increments. Arrows point to the low-level moist region referred to in the text. M, P, and C locate Midland, the Palo Duro Canyon, and Cap Rock escarpment, respectively. b. Enhanced imagery as in Figure 5a, 4 hours later. Arrows point to the surface front. c. Enhanced imagery as in Figure 5a, 6 hours later. Arrows point to a well-developed squall line.

Fig. 5c. As in Figure 5a, 6 hours later. Arrows point to a well-developed squall line.

mature MCCs, similar to tropical convection, there is often a reversal of vertical wind shear in the upper troposphere (relative to the moving MCC). The resulting appearance in a satellite image is active convection along the leading edge of the MCC with a large stratiform anvil area trailing behind.

This important spring and summertime convective weather system occurs most often over the central United States during the late evening and nighttime hours. With one type convective system that is similar to the MCC, the Derecho [*Johns and Hirt*. 1983], strong surface winds and gust front tornadoes are the primary type of severe activity. MCCs have been observed to interact with and modify the larger scale environment in which they are embedded. By influencing the larger-scale environment they affect downstream weather long after their demise.

3. STORM SCALE INFORMATION DETECTABLE IN SATELLITE DATA

In the previous section, aspects of the tornadic storm's mesoscale environment observed using satellite data were discussed. In the following section, features evident in satellite imagery that accompany individual tornadic storm development are discussed. At the end of the section, thoughts concerning those features are presented.

3.1. *Isolation of Preferred Areas for Tornadic Storm Development: Interaction With Boundaries*

Under proper dynamic forcing, tornadoes are spawned when squall line thunderstorms interact with an organized synoptic scale boundary, such as a warm front [*Miller*, 1972]. Another favored region for tornado activity exists where a strong thunderstorm and a low-level boundary due to previous convective activity interact [*Purdom*, 1976, 1990]. Usually, both synoptic and storm scale boundaries are detectable in satellite imagery. Such detection is especially aided by analysis of animated RISOP imagery [*Purdom and Weaver*, 1992]. When surface data are combined with satellite imagery, the ability to isolate boundaries is further enhanced [*Purdom*, 1990].

Fig. 5b. As in Figure 5a, 4 hours later. Arrows point to the surface front.

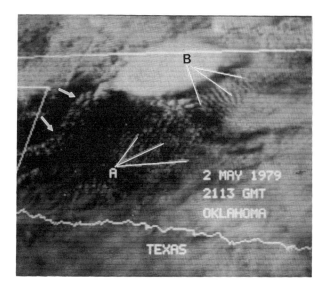

Fig. 6. GOES-East 1-km resolution visible image at 1513 CST on May 2, 1979. Wave clouds and low-level stratus predominate in the stable air mass (B), with cumulus filling the potentially unstable air (A). Arrows point to the convergence line referred to in the text.

Fig. 7. GOES-East 1-km resolution visible image at 1645 CST on June 13, 1976. Arrows at A point to the "well-defined" arc cloud line south of Chicago. The thunderstorm at B is tornadic. Another set of arrows points to arc cloud lines in western Indiana; they are a product of the storm to their east.

As mentioned above, prior convective activity may produce a low-level boundary that will focus tornado activity. Most often, that boundary appears as an arc cloud line [*Purdom*, 1976]; however, on occasion the boundary is located by a transition from low-level wave to cumulus cloudiness [*Beckman*, 1982]. In both cases, the boundary is located at the transition between warm sector air and boundary layer air that has been cooled and stabilized by the evaporation of rain. The reason for the difference in cloud appearance (wave/cumulus versus arc) has not been explained; however, the most likely cause is the difference in thermodynamic structure of the boundary layer air into which the rain fell and evaporated. Arc cloud lines prevail during afternoon hours, while wave/cumulus transition regions are mainly the result of nighttime and early morning thunderstorm activity.

On May 2, 1979, early morning thunderstorms helped stabilize the boundary layer in north central Oklahoma but not in southwest Oklahoma. By afternoon (Figure 6), wave clouds and low-level stratus predominate in the stable air mass, with cumulus filling the potentially unstable air in southwest Oklahoma. The intense thunderstorm activity in western Oklahoma developed where an organized convergence line merged with the boundary. Those thunderstorms produced tornadoes at Orienta and Lahoma, Oklahoma; downburst activity was also reported. Radar reflectivity and Doppler velocity images of the tornadic storms showed well defined hook echoes associated with strong cyclonic wind shears [*Purdom et al.*, 1982].

Arc cloud lines and their convective scale interactions are a natural part of the thunderstorm development and evolution process [*Purdom*, 1986]. Where an intense thunderstorm and an arc cloud line interact locates a favored region for tornado activity [*Purdom*, 1976, 1990]. On June 16, 1976, the stage for tornado activity around Chicago was set after thunderstorm activity moved through that area leaving behind a well-defined arc cloud line (Figure 7). Tornado activity was triggered where that arc cloud line merged with the thunderstorm in northeast Illinois. The satellite image in Figure 7 was taken while tornado activity was in progress near Chicago [*Fujita and Hjelmfelt*, 1977].

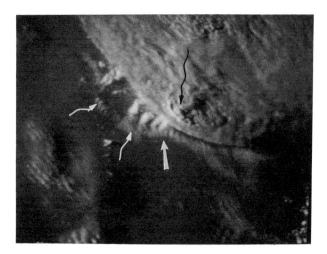

Fig. 8. GOES-West 1-km resolution visible image at 1816 CST on April 10, 1979. The view is into the back side of the Wichita Falls, Texas, tornadic storm. White wavy arrows point to the low-level arc cloud line, while the bold white arrow points to the location of the tower associated with the Wichita Falls tornado. The black wavy arrow points to the overshooting top above the tornadic region.

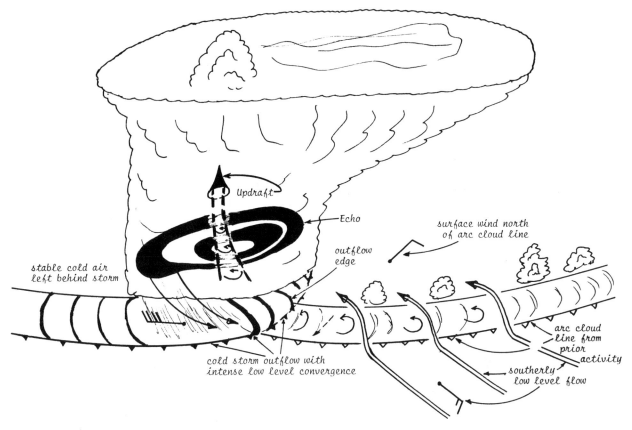

Fig. 9. Schematic of severe thunderstorm interacting with an outflow boundary from another storm, resulting in tornado development.

3.2. Satellite Observations of Supercells

Isolated supercells, apparently unassociated with a boundary produced by other convective activity, may produce strong tornadoes. The Wichita Falls, Texas, tornadic storm provides a good illustrative example of how a supercell thunderstorm appears in satellite imagery (Figure 8). On that day, anvil cirrus expanded rapidly eastward. Typical for such cases, masking by anvil cirrus and satellite viewing perspective combine to make it difficult to detect low-level storm features with GOES-East imagery. (GOES-East location is nominally over 75°W at the equator.) However, as can be seen in Figure 8, low-level features are easily detected with GOES-West imagery because of that satellite's improved viewing perspective. (GOES-West location is nominally over 135°W at the equator.)

The following conceptual model is given to help interpret Figure 8. Initial supercell development occurs in a well-mixed boundary layer. As the storm matures, a cold air mass is generated beneath its echo due to evaporation. As the storm evolves and moves away from its origin it leaves behind that stable dome of cold air: a new local boundary layer. In Figure 8 the thin curved arc cloud line behind the supercell encloses the new low-level boundary layer. That dome of cold air, subsiding in time, leads to clear skies behind the storm. New convective towers grow along the arc cloud line. They are especially intense where they join the right rear flank of the storm: at that location the supercell's updraft is regenerated. At that location in Figure 8, convective towers extend from near the surface into the overshooting top region at the rear of the anvil. The Wichita Falls tornado is on the ground beneath the eastern most of those towers. Evidence of rain-cooled air left behind by the supercells has been noted with other tornadic thunderstorms [*Purdom*, 1990; *Purdom and Weaver*, 1992].

In Figure 8 the towers at the rear of the storm extend upward into a well-defined region of overshooting tops. Using special 3-min-interval GOES-East imagery, *Anderson* [1982] found the overshooting tops associated with this storm were in cyclonic rotation. The National Severe Storm Laboratory's Doppler radar detected a well-defined cyclonic circulation with the Wichita Falls storm even though it was nearly 200 km from the radar site: this indicates the storm was in rotation to great depths. This author found evidence of cyclonic rotation at cloud top when viewing storm relative RISOP satellite imagery for the Plainfield, Illinois, tornadoes of August 28, 1990.

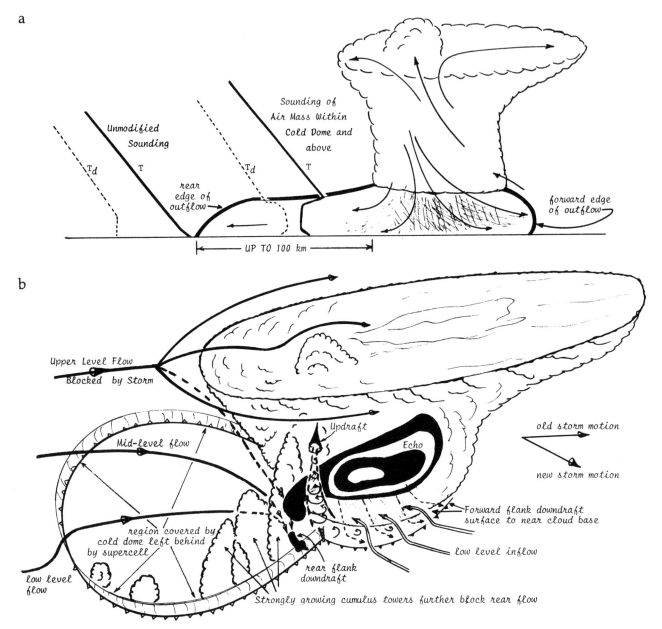

Fig. 10. Schematic illustration of (a) boundary layer modification due to cold dome left behind by thunderstorm and (b) supercell entering its tornadic phase. Both modified and unmodified soundings are similar above cold dome in middle and upper troposphere.

3.3. A Conceptual Model of Tornadic Storms

Severe thunderstorms interaction with a boundary: In case studies of tornadic storm development along larger mesoscale boundaries, *Maddox et al.* [1979] found that mesoscale convergence and cyclonic vorticity tended to maximize along the boundary. Using aircraft measurements, *Sinclair and Purdom* [1983] found low-level convergence, vertical wind shear, and an evolving solenoidal circulation concentrated along arc cloud lines.

On the basis of these observations the following conceptual model is presented for tornadic storm development after a severe thunderstorm interacts with a boundary (Figure 9). Along the boundary, strong vertical wind shear ($\delta v/\delta z$) is localized; this is due to the evolving solenoidal circulation and interaction of low-level southerly flow with the boundary. It is well known that severe thunderstorms have intense updrafts and downdrafts; across a very short distance strong convergence leads to intense vertical motion (a strong $\delta w/$

δx). Thus, when such a storm moves from the west into a preexisting low-level boundary, vorticity associated with the boundary is tilted in a cyclonic sense into the storm's updraft. Further, where the severe storm's downdraft interacts with the boundary, cyclonic boundary layer flow experiences intense convergence. Thus both convergence on existing vorticity and tilting of vorticity from one fluid plane to another results in a rapid increase in vertical vorticity and tornado genesis.

Is there a common link between the mechanism causing supercells to produce tornadoes and those described immediately above? Before answering that question, consider the development of the midlevel rear inflow jet, a critical component in supercell tornadic storm evolution. In their conceptual model of a supercell, *Lemon and Doswell* [1979] show upper level blocking by a supercell as necessary for the development of the midlevel rear inflow jet. As pointed out previously, such blocking has been observed using RISOP imagery. Furthermore, in Figure 10a the new boundary layer immediately to the rear of the storm, being shallow, cool, and stable, inhibits mixing between air near the surface with higher-momentum air aloft. Thus any forces acting to accelerate the air above the cold dome may do so without surface frictional and mixing effects. In laboratory experiments, *Simpson and Britter* [1980] found that when a density current moves into a fluid of lesser density the less dense fluid's flow accelerates and increases in velocity as it moves across the top of the density current. Upper level blocking coupled with the characteristics of the cool and stable low-level air to the rear of the storm should aid the development of a rear, midlevel inflow jet.

As shown in Figure 10b, the mature supercell continues generating a cold air mass beneath its echo, a forward flank downdraft. Along the boundary of the forward flank downdraft, the thermodynamic and dynamic characteristics of an arc cloud line exist: only amplified by proximity to the supercell. (For information concerning arc cloud line characteristics in relationship to the arc cloud line's source, see *Purdom* [1986].) The supercell, rotating cyclonically owing to updraft and environment interaction [*Rotunno and Klemp*, 1985], begins moving anomalously to the right. From early on, the dome of cold air left behind by the storm has been sinking and spreading out. Now, however, instead of being left behind by the supercell, it and the forward flank downdraft merge. At the merger location, very strong low-level convergence develops and intense convective towers grow. With the development of these new towers, the midlevel rear inflow jet is further blocked. This blocking leads to strong evaporation which aids in the development of an intense outflow along the back side of the flanking line. (According to *Moller et al.* [1974], ground intercept teams have observed cloud decks "apparently evaporated" and clearing of midlevel cloudiness at the rear of the flanking line.) That rear flank downdraft surges outward, intersecting the forward flank downdraft, where the same mechanisms discussed for a severe storm's interaction with a preexisting boundary lead to tornado genesis.

4. CONCLUSIONS

In the previous sections it was shown how satellite data may be used to help diagnose tornadic storms and their mesoscale environments. The data can be used to monitor the spatial variability of clouds and moisture from synoptic to thunderstorm scales; geostationary satellites add a vital temporal capability.

In discussing the mesoscale environment in which tornadic storms form, both quantitative and qualitative applications of satellite data were addressed. However, a number of issues remain (not solely satellite analysis issues). For example, how does instability combine with the vertical forcing within a convergence zone to produce thunderstorm activity? With satellite imagery, cloud development along a convergence zone may be monitored and cumulus flow relative to it may be measured. However, we do not know the convergence zone's width or depth (clear air returns from NEXRAD may allow analysis on a very restricted scale). That information is basic for performing the simplest of calculations for vertical motion at the scale on which triggering often occurs. Suppose we had the above information. What of the stability that must be overcome for storms to form? It is impractical to routinely release rawinsondes at frequent intervals and mesoscale spacings. Frequently, available information from satellite soundings may be able to provide such information. Even with such information, how do we use it?

Two types of tornadic storms were discussed: (1) intense thunderstorms that become tornadic after interacting with a low-level boundary due to another source and (2) supercells. In the former case, it was shown how satellite imagery could be used to help isolate the low-level boundary. For supercells, satellite viewing perspective was discussed with GOES-West often being preferred. A conceptual model was provided to help interpret supercell appearance in satellite imagery. That model explored the role of the modified boundary layer left in the wake of the storm. The conceptual model needs to be verified.

A common link was proposed for thunderstorms that become tornadic after interaction with a boundary due to another source and tornadic supercells. In both instances, storm/boundary interaction was proposed to be the key to tornado genesis. In the case of the supercell, the boundary was due to its forward flank downdraft. For both cases, convergence on preexisting vorticity and tilting of vorticity from one fluid plane to another was proposed to result in a rapid increase in vertical vorticity and tornado genesis. Is this common link correct? If so, how do we characterize boundaries and then use that information?

Acknowledgments. Portions of this work were supported by NOAA grant NA85RAH05045. My thanks to the reviewers for pointing out areas for clarification.

REFERENCES

Anderson, C. E., Dramatic development of thunderstorm circulation associated with the Wichita Falls tornado as revealed by satellite imagery, in *Proceedings of the 12th Conference on Severe Local Storms*, pp. 493–498, American Meteorological Society, Boston, Mass., 1982.

Beckman, S. K., Relationship between cloud bands in satellite imagery and severe weather, in *Proceedings of the 12th Conference on Severe Local Storms*, pp. 483–486, American Meteorological Society, Boston, Mass., 1982.

Carlson, T. N., S. G. Benjamin, G. S. Forbes, and Y. F. Li, Elevated mixed layers in the regional severe storm environment: Conceptual model and case studies, *Mon. Weather Rev.*, *111*, 1453–1473, 1983.

Chesters, D., L. W. Uccellini, and A. Mostek, VISSR Atmospheric Sounder (VAS) simulation experiment for a severe storm environment, *Mon. Weather Rev.*, *110*, 198–216, 1982.

Chesters, D., L. W. Uccellini, and W. D. Robinson, Low-level water vapor fields from the VISSR atmospheric sounder (VAS) "split window" channels, *J. Clin. Appl. Meteorol.*, *22*, 725–743, 1983.

Dills, P. N., and S. B. Smith, Comparison of profiler and satellite cloud-tracked winds, in *Proceedings of the 6th Conference on Satellite Meteorology and Oceanography*, pp. 155–158, American Meteorological Society, Boston, Mass., 1992.

Doswell, C. A., III, and L. R. Lemon, An operational evaluation of certain kinematic and thermodynamic parameters associated with severe thunderstorm environments, in *Proceedings of the 11th Conference on Severe Local Storms*, pp. 397–402, American Meteorological Society, Boston, Mass., 1979.

Fujita, T. T., Manual of downburst identification for Project NIMROD, *SMRP Res. Pap. 156*, 103 pp., Dep. Geophys. Sci., Univ. of Chicago, Chicago, Ill., 1978.

Fujita, T. T., and M. R. Hjelmfelt, Mesoanalysis of record Chicago rainstorm using radar, satellite, and raingauge data, in *Proceedings of the 10th Conference on Severe Local Storms*, pp. 65–72, American Meteorological Society, Boston, Mass., 1977.

Heymsfield, G. M., and R. H. Blackmer, Jr., Satellite-observed characteristics of midwest severe thunderstorm anvils, *Mon. Weather Rev.*, *116*, 2200–2224, 1988.

Hillger, D. W., and J. F. W. Purdom, Clustering of satellite radiances to enhance mesoscale meteorological retrievals, *J. Appl. Meteorol.*, *29*, 1344–1351, 1990.

Hillger, D. W., and T. H. Vonder Haar, Retrieval and use of high-resolution moisture and stability fields from NIMBUS 6 HIRS radiances in pre-convective situations, *Mon. Weather Rev.*, *109*, 1788–1806, 1981.

Johns, R. H., and W. D. Hirt, The derecho—A severe weather producing convective system, in *Proceedings of the 13th Conference on Severe Local Storms*, pp. 178–181, American Meteorological Society, Boston, Mass., 1983.

Lemon, L. R., and C. A. Doswell III, Severe thunderstorm evolution and mesoscale structure as related to tornadogenesis, *Mon. Weather Rev.*, *107*, 1184–1197, 1979.

Lubich, D. A., and J. F. W. Purdom, The use of cloud relative animation in the analysis of satellite data, in *Proceedings of the 6th Conference on Satellite Meteorology and Oceanography*, pp. 118–119, American Meteorological Society, Boston, Mass., 1992.

Maddox, R. A., Mesoscale convective complexes, *Bull. Am. Meteorol. Soc.*, *61*, 1374–1387, 1980.

Maddox, R. A., L. R. Hoxit, and C. F. Chappell, A study of thermal boundary/severe thunderstorm interactions, in *Proceedings of the 11th Conference on Severe Local Storms*, pp. 403–410, American Meteorological Society, Boston, Mass., 1979.

McCann, D. W., The enhanced V, a satellite observable severe storm signature, *NOAA Tech. Memo. NWS NSSFC-4*, 31 pp., Natl. Severe Storm Forecast Center, Kansas City, Mo., 1981.

Miller, R. C., Notes on analysis and severe-storm forecasting procedures of the Air Force Global Weather Central, *AWS Tech. Rep. 200 (rev)*, 190 pp., Air Weather Serv., Offutt, Air Force Base, Nebr., 1972.

Moller, A., C. Doswell III, J. McGinley, S. Tegtmeier, and R. Zipser, Field observations of the Union City tornado in Oklahoma, *Weatherwise*, *27*, 68–77, 1974.

Parsons, D. B., R. A. Hardesty, and M. A. Shapiro, The mesoscale structure of the dryline in the formation of deep convection, in *Proceedings of the International Conference on Monsoon and Mesoscale Meteorology*, pp. 117–122, American Meteorological Society, Boston, Mass., 1986.

Petersen, R. A., and A. Mostek, The use of VAS moisture channels in delineating regions of potential convective instability, in *Proceedings of the 12th Conference on Severe Local Storms*, pp. 168–171, American Meteorological Society, Boston, Mass., 1982.

Purdom, J. F. W., Satellite imagery and severe weather warnings, in *Proceedings of the 7th Conference on Severe Local Storms*, pp. 120–127, American Meteorological Society, Boston, Mass., 1971.

Purdom, J. F. W., Some uses of high resolution GOES imagery in the mesoscale forecasting of convection and its behavior, *Mon. Weather Rev.*, *104*, 1474–1483, 1976.

Purdom, J. F. W., Convective scale interaction: Arc cloud lines and the development and evolution of deep convection, *Atmos. Sci. Pap. 408*, 197 pp., Dep. of Atmos. Sci., Colo. State Univ., Fort Collins, 1986.

Purdom, J. F. W., Convective scale weather analysis and forecasting, in *Weather Satellites: Systems, Data, and Environmental Applications*, edited by P. Krishna Rao, chap VII-8, pp. 285–304, American Meteorological Society, Boston, Mass., 1990.

Purdom, J. F. W., and J. F. Weaver, Analysis of rapid scan satellite imagery to diagnose tornadic storms and the environment in which they form, in *Proceedings of the 6th Conference on Satellite Meteorology and Oceanography*, pp. J25–J30, American Meteorological Society, Boston, Mass., 1992.

Purdom, J. F. W., R. N. Green, and H. A. Parker, Integration of satellite and radar data for short range forecasting and storm diagnostic studies, in *Proceedings of the 9th Conference on Weather Forecasting and Analysis*, pp. 51–55, American Meteorological Society, Boston, Mass., 1982.

Rotunno, R., and J. Klemp, On the rotation and propagation of simulated supercell thunderstorms, *J. Atmos. Sci.*, *42*, 271–292, 1985.

Shapiro, M. A., T. Hampel, D. Rotzoll, and F. Mosher, The frontal hydraulic head: A micro-c scale (-1 km) triggering mechanism for mesoconvective weather systems, *Mon. Weather Rev.*, *113*, 1166–1183, 1985.

Simpson, J. E., and R. E. Britter, A laboratory model of an atmospheric mesofront, *Q. J. R. Meteorol. Soc.*, *106*, 485–500, 1980.

Sinclair, P. C., and J. F. W. Purdom, The genesis and development of deep convective storms, *CIRA Pap. 1*, 56 pp., 0737-5352, Cooperative Inst. for Res. in the Atmos., Colo. State Univ., Fort Collins.

Smith, W. L., et al., Nowcasting—Advances with MCIDAS III, in *Proceedings of the Nowcasting II Symposium*, pp. 433–438, European Space Agency, Kerplerlaan, Netherlands, 1984.

Weldon, R. B., and S. J. Holmes, Water vapor imagery: Interpretation and applications to weather analysis and forecasting, *NOAA Tech. Rep. NESDIS 57*, 213 pp., Natl. Oceanic and Atmos. Admin., Boulder, Colo., 1991.

Discussion

RON ALBERTY, SESSION CHAIR

National Weather Service

PAPER D1

Presenter, Don Burgess, National Severe Storms Laboratory [*Burgess et al.*, this volume, Tornado detection and warning by radar]

(Jim Wilson, National Center for Atmospheric Research) Don, I heard you say something about the percentage of mesocyclones that produces tornadoes. I've heard totally confusing numbers from several sources. Would you clarify it for us? What is the number?

(Burgess) It's a changing number from the samples we've seen. It changes with location and even with time period at the same location. The first work of NSSL was analyzing data after the fact, no real-time displays. We used the WSR-57 radar to search for big, ugly-looking storms, and then we pointed the Doppler radar at them. We found a strong relationship between mesocyclones and tornadoes because we were looking at the most classic supercells. After we began getting real-time displays, we found that there were a lot more weaker circulations out there that weren't classical features. Therefore the percentage of tornadic mesocyclones dropped. During the JDOP Experiment, where we were trying to look at all storms, we got something close to 50% of the mesocyclones producing tornadoes. Those were years with many significant supercells. Since that time, some years have had fewer supercells, and the percentages for certain periods have been as low as 30%. So a lot of mesocyclones don't produce tornadoes. It also depends on your threshold: How low a magnitude of shear or velocity, or whatever threshold you use, are you willing to call a circulation that you think is important? I am going to say, with the thresholds that I am familiar with and advocate, somewhere between 30 and 50% of the mesocyclones will produce tornadoes.

(Chair) We were operating the WSR-88D during last night's storms, and we saw several mesocyclones, one down along the Red River approximately 100 nm from Norman. I believe one or more of the mesocyclones produced tornadoes, but I don't know if they all did.

(Ian Harris, STX Corporation) You briefly mentioned the importance of a parameter called ERKE. I wonder if you or Ralph Donaldson could explain a little more about that?

(Donaldson) We like ERKE because it works very well for the most intense tornadoes. It works quite well as Don showed for the F2 and stronger tornadoes. Why it works, we are not too sure. Part of it is because we are in the proper domain of where observations have shown that mesocyclones later produce tornadoes. What it represents is the energy left after you subtract the velocity required to maintain an average mesocyclone. Therefore ERKE represents some kind of a surplus available for providing a tornado.

PAPER D2

Presenter, Steve Vasiloff, National Severe Storms Laboratory [*Vasiloff*, this volume, Single-Doppler radar study of a variety of tornado types]

(Howie Bluestein, University of Oklahoma) In the case in which it appears as if there is a strong TVS aloft and maybe one near the ground, is it possible that the vortex it tilted?

(Vasiloff) In the Stillwater case, there was no TVS aloft. I found the strongest shear at the lowest levels and high, but not maximum, shear above that. But by no means did the TVS extend to higher height than the 3.2° elevation you saw. The circulation aloft was more likely associated with the mesocyclone, and I'm not sure about the tilt.

(Jim Wilson, National Center for Atmospheric Research) In Colorado tornadoes, there is no evidence that we ever see anything build down. It's always ground up. You were showing something a little bit different.

(Vasiloff) There were four different signatures for that June day in Denver. I picked out the one that was most different to cause discussion, but it did appear to build down. We can compare data later if you like.

The Tornado: Its Structure, Dynamics, Prediction, and Hazards.
Geophysical Monograph 79
This paper is not subject to U.S. copyright. Published in 1993 by the American Geophysical Union.

PAPER D3

Presenter, Mike Leduc, Atmospheric Environment Service of Canada [*Joe and Leduc*, this volume, Radar signatures and severe weather forecasting]

(Arnold Court, California) I was wondering why one of your algorithms wasn't in terms of radians per second instead of shear.

(Leduc) Paul, can you answer that?

(Paul Joe, Atmospheric Environment Service of Canada) The nature of the Doppler estimates does not lend themselves to the use of that measure.

(Jeff Waldstreicher, National Weather Service, Eastern Region) I was wondering if you looked at the storm north of Lake Ontario and if you subtracted out storm motion? You had a storm moving away from the radars, and movement may have prevented seeing flow back toward the radar.

(Leduc) I haven't looked at that. Have you, Paul?

(Joe) Yes, I have, and when you remove storm motion, you do get a symmetric pattern, flow away and toward the radar. It is important to remember in real-time detection, if you don't remove storm motion, you need to look for change in velocity, not necessarily a change from outbound to inbound. Our algorithm uses velocity change, not necessarily crossing 0 radial velocity.

PAPER D6

Presenter, Dave Imy, National Weather Service, Operational Support Facility [*Imy and Pence*, this volume, An examination of a supercell in Mississippi using a tilt sequence]

(Keith Brewster, University of Oklahoma) You mentioned doing RHIs or vertical cross sections which you can do on NEXRAD. Do you have any recommendations on the orientation of cross sections that would be the best?

(Imy) If you know where your overhang is, the extent of overhang and the direction, you should set the cross section to go right through them. In fact, we have done just that, made cuts where we believe there to be an overhang, and we have seen bounded-weak-echo regions on NEXRAD cross-section products. For our study case with the WSR-57, looking for such features is nearly impossible.

Laboratory Models of Tornadoes

CHRISTOPHER R. CHURCH

Department of Aeronautics, Miami University, Oxford, Ohio 45056

JOHN T. SNOW

Department of Earth and Atmospheric Sciences, Purdue University, West Lafayette, Indiana 47907

1. INTRODUCTION

Nature provides many examples of intense but small-scale atmospheric vortices, the most devastating being tornadoes. Other small vortices include waterspouts, fire whirls, dust devils, and steam devils. Several aspects of small-scale atmospheric vortex flows are of concern to the atmospheric scientist, namely: determination of their kinematic structure, understanding of their formation and dynamics, identification of the factors that control their intensities, and application of new knowledge and insights in ways that will provide greater protection for society from the hazards of these phenomena. Although some of the vortex types listed above occur more frequently and are more readily available for observation than tornadoes, all small-scale vortices are inherently infrequent, short-lived phenomena; it has been expedient for some scientists to simulate tornadolike flows in the laboratory. This laboratory work constitutes a small but significant part of the overall tornado research effort.

Swirling flows are also studied in several fields of engineering. The resulting literature on experimental work is voluminous, and although only a small fraction is pertinent to the meteorological phenomenon, it should not be overlooked. An example of this material is the investigation of vortex breakdown (a phenomenon described below) by *Escudier* [1982]. *Maxworthy* [1982] provides an entry bibliography to some of this work.

Tornadolike vortices are created in a tornado vortex chamber (TVC). From necessity the background flow conditions are considerably simplified, as it does not seem feasible to incorporate in a laboratory model all the factors (dynamic, thermodynamic, and microphysical) that exist in the thunderstorm environment. A compelling reason for taking the laboratory approach is that it brings the classic attributes of experiments in physical science, namely, precision, control, and repeatability, to a complex and transient atmospheric phenomenon.

At the preceding Tornado Symposium, *Davies-Jones* [1976] gave a critical review of various TVCs that had been developed up to that time. His discussion of vortex core dynamics was based on available experimental results. The general purpose of the present paper is to review the progress that has been made since the 1976 symposium. It deals with several aspects of tornado vortex simulation and, in greater detail, with the contributions to the understanding of vortex flows that have resulted from the more recent laboratory experiments. The reader should be aware of other recent review papers on this subject: *Maxworthy* [1982], *Snow* [1982], and *Wilkins* [1988].

2. TORNADO VORTEX SIMULATION

The purpose of the laboratory apparatus is to create flows that, to the extent that it is possible, are geometrically and dynamically similar to natural flows, in order that meaningful observations and physical measurements can be made. By injecting visualizing material (for example, smoke and fog) into the surface boundary layer, one is able to observe details of the inflow and the core flow. The most important measurement is that of velocity, although measurements of static pressure have also proven useful in developing insights into vortex core dynamics. The accuracy of the laboratory data depends upon factors pertaining to the measurement technique and also upon the validity of the modeling technique, that is, on details in the design of the TVC. All TVCs incorporate boundaries and walls that have no counterparts in nature. These cause the development of secondary flows that may alter the vortex flows in some important ways. *Davies-Jones'* [1976] review of TVCs describes apparatus of many different types and sizes, some with air, others with water as the working fluid. He concluded that the most

The Tornado: Its Structure, Dynamics, Prediction, and Hazards.
Geophysical Monograph 79
Copyright 1993 by the American Geophysical Union.

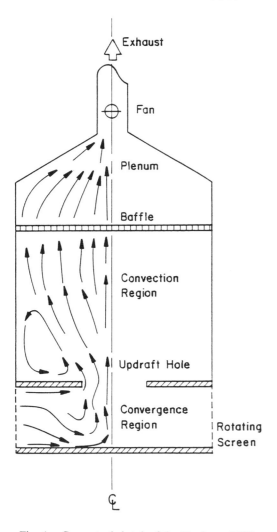

Fig. 1. Conceptual sketch of the Ward-type TVC.

appropriate apparatus for tornado vortex modeling was the TVC developed by *Ward* [1972].

2.1. The Ward-Type TVC

In this apparatus a swirling flow is created by injecting vorticity into a sink flow, so that an intense vortex forms on the centerline of a rotating upflow. Angular momentum is continuously supplied to maintain the vortex. A schematic form of the apparatus is shown in Figure 1. Its principal elements are as follows: (1) a means of supplying starting vorticity and then circulation to a radial inflow; in the original Ward apparatus this was provided by a rotating cylindrical meshwire, a feature that was also adopted in the Purdue TVC I [*Church et al.*, 1979]; (2) convergence and upflow created by an exhaust fan mounted at the top; and (3) a flow-straightening baffle placed upstream of the fan to provide a "free" downstream boundary condition on the swirling flow. (If a flow-straightening baffle followed by a large plenum is not used, the core diameter of the vortices that form are scaled by the diameter of the exhaust hole (see, e.g., *Wan and Chang* [1972]). The flow-straightening baffle also ensures that vorticity generated by the exhaust fan does not feed back into the swirling flow.)

The resultant axisymmetric flow is a representation of a tornado forming at the center of a mesocyclone. The updraft diameter is identified with that of the tornado mesocyclone. Examination of the streamline sketches in Figure 1 shows that a secondary circulating flow exists in the lower corner of the convection region, but this is judged to exert a negligible influence on the flow near the centerline.

In their analysis of swirling flows of this type, *Lewellen* [1962] and *Davies-Jones* [1973] showed that a complete set of governing equations in terms of the dimensionless circulation and stream function involve three key characteristic parameters:

$$Re_r = \frac{Q}{\nu}$$

$$S = \frac{r_0 \Gamma}{2Qh}$$

$$a = \frac{h}{r_0},$$

where Re_r is the radial Reynolds number, S is the swirl ratio, and a is the internal aspect ratio. Here Q is the volume flow rate per axial length; ν, the kinematic viscosity; r_0, the radius of the updraft; h, the depth of the inflow; and Γ, the circulation at r_0.

Swirl ratio is a particularly significant parameter in that it has been found to determine the particular vortex configuration, that is, one-cell versus two-cell, single versus multiple vortex flow.

External geometric parameters formed from dimensionless combinations of characteristic radial and axial dimensions are significant in characterizing the flow. These are r_s/r_0, r_w/r_0, and l/h, where r_s is the radius of the confluence/convergence region, r_w is the radius of the convection region, and l is the depth of the convection region.

For all these dimensionless groupings except one, a laboratory TVC can match the expected atmospheric values [*Church and Snow*, 1979]. The exception is the radial Reynolds number, which, when computed using the molecular kinematic viscosity for air, is several orders of magnitude greater for atmospheric flows. Fortunately, for flows over smooth surfaces, the dependence of flow characteristics on Reynolds number appears to be weak. For flows over rough surfaces, the dependence on Reynolds number has not been clearly established.

2.2. Current Status of Working TVCs

In comparing TVCs now being used to explore tornadolike flows, we find some common characteristics: (1) all are variants of the original Ward apparatus, and (2) all employ

TABLE 1. Tornado Vortex Chamber Characteristics

Principal Investigator/Location	Maximum Updraft Radius, m	Maximum Flow Rate, $m^3 \, m^{-1}$	No. of Vanes	Vane Spacing/Chord Length Ratio
Snow/PU	0.76	2.00	112	0.26
Wilkins/OU	0.61	1.56	NA	...
Monji/KU	0.38	0.55	48	0.5
Church/MU	0.26	0.57	18	0.92

PU, Purdue University; OU, University of Oklahoma; KU, Kyoto University; MU, Miami University. NA, not applicable.

systems of vanes in place of the rotating meshwire to provide circulation. (Not all TVCs have been built as research tools. During the early 1980s, *Kimpel* [1981] developed a Ward-type TVC for education and demonstration purposes. This TVC was housed in the OMNIPLEX, a science and art museum in Oklahoma City. It was viewed by a large number of people during its years of operation, and it probably represented the finest visual display of tornado vortex characteristics ever made available to the general public.) Table 1 compares some of the principal dimensional parameters of various chambers. Maximum updraft radius is particularly significant since the radius of the vortex core scales directly with it. Typically, vortex core radii vary from 3% to 50% of updraft radius, depending on swirl ratio. Tangential velocities in the vortex core scale directly with updraft velocity, which in turn is a function of the flow rate.

A second-generation TVC at Purdue University is described by *Snow and Lund* [1988] and illustrated in Figure 2. In addition to the use of vanes and larger dimensions, the main differences from TVC I are (1) an antiturbulence screen surrounding the inflow and a confluence zone of varying depth (the purpose of these will be made clear later), (2) a movable flow-straightening baffle in order to vary the aspect ratio of the convection zone, and (3) a laser Doppler velocimeter (LDV) for making nonintrusive velocity measurements.

The original Ward TVC at the University of Oklahoma was modified by *Rothfusz* [1986] in order to produce vortices from tilting of the horizontal vorticity present in a helical flow. This was accomplished by replacing the whole of the confluence/convergence region with a three-tier system of vanes. All the vanes in each tier were mutually parallel, but the vane angle from one tier to the next veered with height. This flow had zero circulation about the vertical, a feature that distinguishes this physical arrangement from all others. This resulted in a complex flow pattern (see Figure 3). Alternate quadrants had cyclonic and anticyclonic inflow. A mesocyclonelike zone of rotation developed with a smaller-scale tornadolike vortex at its center. This vortex appeared more turbulent than those produced in other simulators.

The TVC at Kyoto University [*Monji*, 1985] is illustrated in Figure 4. An earlier version of this apparatus [*Mitsuta and Monji*, 1984] incorporated a system of four fans operating in an annular region surrounding the inflow. These supplied

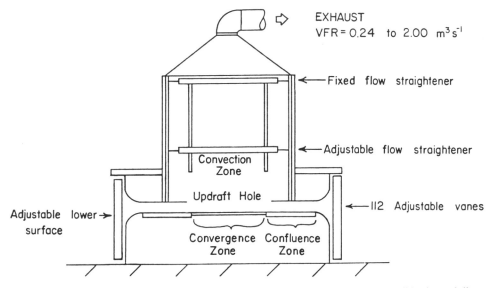

Fig. 2. Cross section of TVC II at Purdue University illustrating major components. Overall horizontal dimension is approximately 5 m. From *Snow and Lund* [1988].

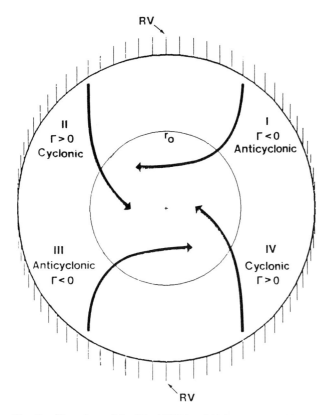

Fig. 3. Plan view of the Ward TVC at Oklahoma as modified by Rothfusz showing one tier of vanes. Air parcels entering particular quadrants follow trajectories that are either cyclonic or anticyclonic, as indicated by the arrows. From *Rothfusz* [1986].

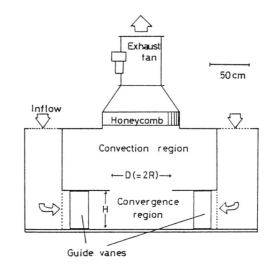

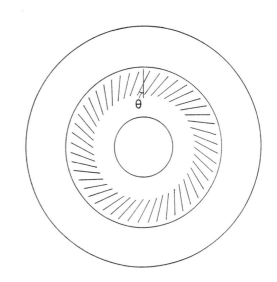

Fig. 4. Cross section and plan view of the TVC at Kyoto University. From *Monji* [1985].

circulation but produced turbulent vortices at all swirl ratios. It was found preferable for later experiments to configure the apparatus as is now illustrated. As in the Purdue TVC II, this employs a relatively large number of flat vanes that can be set at different angles. The annular vertical inflow feature of the original version is retained; this may help to reduce the disturbing effects of turbulent motions in the laboratory air.

Distinguishing features of the apparatus developed by Church at Miami University in Ohio are (1) aerodynamically shaped vanes that resemble turbine blades, (2) a large continuously variable iris diaphragm, and (3) termination of the vortex in a horizontally divergent flow. Flow is induced by 12 fans arranged around the top in a quasicylindrical configuration. The iris allows swirl ratio to be changed by varying r_0, eliminating laborious and time-consuming adjustment of the vanes. By this means a wide range of vortex flow configurations can be displayed in a matter of seconds. This investigator paid particular attention to producing stable, clearly defined vortices in a chamber of compact dimensions. In order to accomplish this, quite elaborate modifications to the lower working surface were required, as illustrated in Figure 5. The effectiveness of these features will be discussed below.

2.3. *Practical Aspects*

We consider briefly some of the experimental problems that confront investigators of laboratory vortex flows and some design aspects of tornado vortex simulators that have proven useful in overcoming these problems. These remarks should be helpful to potential designers and users of future TVCs.

Flow Stabilization. In fluid flow measurements, probes are placed at fixed locations, thus providing an Eulerian description of the flow. The amplitude of "vortex wander" (seemingly random motions about the mean position of the point of contact of the vortex core with the lower surface) can easily be on the same order of magnitude as the core radius. Given the large gradients of velocity and pressure

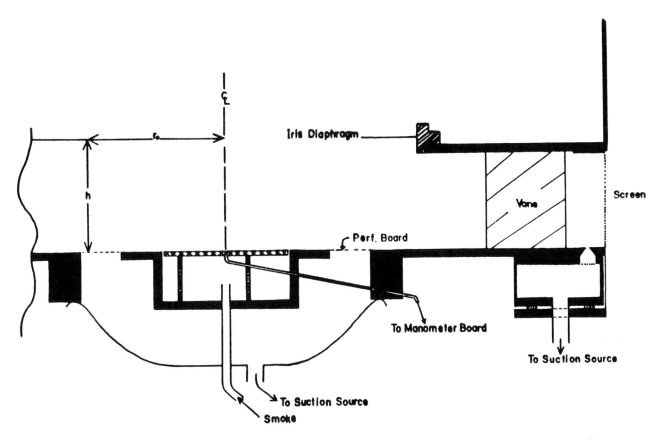

Fig. 5. Cross section of a portion of the inflow region of the TVC at Miami University. Distance from centerline to screen is 1.2 m.

that are present, measurements in such unsteady flows are difficult to interpret. While some degree of vortex wander may be an inherent property of a vortex core normal to a flat surface, two external causes of vortex wander have been identified: (1) turbulence in the air entering the TVC and (2) flow separation at the lower edge of the inflow.

To address the first, an antiturbulence screen consisting of a loosely woven fabric material surrounding the inflow has proven effective in reducing the strength of eddies drawn into the apparatus. Various schemes have been tried to address the second problem. The Purdue TVC II employs a variable depth inflow geometry to keep the boundary layer attached, although this feature also reduces the effective swirl angle in the chamber. The Miami TVC incorporates a circular gap placed close to the antiturbulence screen and connected to a source of suction; however, this has not produced any noticeable difference in the quality of the flow.

If circulation and flow rate are not steady, the swirl ratio will vary. The consequent variations in flow configuration will be reflected in measurements of velocity and pressure. This problem can be greatly reduced by using a system of vanes in place of a rotating meshwire. In the latter case, variations in circulation and flow rate are independent of one another and may thereby cause relatively large variations in S. It has been found for the vane-type system that variations in circulation largely compensate for variations in flow rate, keeping the ratio Γ/Q very nearly constant and so minimizing variations in S.

Church attempted to determine the optimum number, shape, and spacing of vanes required to produce a high swirl flow without high turbulence and concluded that (1) curved vanes are preferable to flat ones, (2) curved plates are just as effective as thick airfoil shapes, (3) it is preferable to use fewer vanes with longer chord length, and (4) the optimum vane spacing to chord length (s/c) is about 1.

The size of the chamber turns out to be a limiting factor in the quality of the flow, particularly at the highest swirl configurations. It has proven much more difficult to produce stable multiple-vortex patterns in the smaller Miami TVC ($r_0 = 0.26$ m) than in earlier experiments with the Purdue TVC I ($r_0 = 0.61$ m). In both TVCs, some means of extracting the turbulent inflowing air close to the surface of the convergence zone has proven very effective in improving the stability of multiple-vortex flows.

Instrumentation for Velocity Measurements. It is required to make measurements on flows that are inherently three dimensional and to derive information on mean velocities, turbulence intensities, and shear stresses. Many of the

TABLE 2. Recommended Dimensional Relationships for a Ward TVC

Dimensional Quantity	Functional Relationship
Overall radius of inflow, m	$3.5\, r_0$
Depth of inflow, m	$0.7\, r_0^{1.1}$
Depth of convection region, m	$3 r_0^{0.5}$
Diameter of convection region, m	$4 r_0$
Volume flow rate, m³ s⁻¹	r_0^3
Number of curved vanes	12–18
Vane turning angle, deg	60
Spacing between vanes, m	r_0
Vane chord length, m	r_0

flow features of interest occupy very small volumes and contain large gradients of velocity. Until now most velocity measurements have been made using intrusive probes. However, experience has shown that these can cause problems. For example, the probe and its supporting structure (perhaps large in size relative to the flow feature of interest) can perturb the flow. Because of the spiraling streamlines near the core, it is easy for the probe to be in the spreading wake of its own support. Further, when placed in a region that has a large velocity gradient, the sensing element on the probe may cut across a range of velocities, thereby resulting in a loss of spatial resolution. The magnitudes of the resulting errors are generally not known, and so the validity of the data becomes a subject for speculation. Such uncertainties can be avoided or greatly reduced by resorting to a nonintrusive technique such as laser Doppler velocimetry.

Recommended Design Dimensions for a TVC. For greater clarity and measurement accuracy the vortex core should be as large as possible. As mentioned above, the core radius is proportional to the size of the updraft hole, so the radius of the updraft (r_0) is the parameter of greatest importance in establishing the dimensions of the vortex. It also turns out that all of the significant dimensions of the TVC can be expressed as functions of r_0. These empirically derived relationships are presented in Table 2. The functional relationships shown were derived with economy in mind, that is, producing the largest vortices in relation to the overall size of TVC, and therefore these yield the minimum recommended dimensions for a given updraft radius and vortex core size. Experience has shown that it is very difficult to develop the wide range of tornadolike vortex configurations with updraft radii smaller than about 0.2 m. The general recommendation is that r_0 be of the order of 1 m.

3. Experimental Results

Investigators have developed insights into vortex dynamics from visual observations of laboratory vortices and have made measurements of velocity and pressure. In this section we review the principal aspects of such experiments and discuss them where appropriate with reference to field observations and to numerical simulations of laboratory vortices.

3.1. Visual Observations

Over the years there has developed a greater general awareness of the complexity of tornadic flows, owing in part to the results of laboratory simulation but also because of increased skill in intercepting and photographing them. *Ward* [1972] was the first to observe the wide range of vortex flow phenomena to be found in axisymmetric swirling flows in the TVC, and by now many other experimentalists have documented them, the most recent being *Monji* [1985]. *Church and Snow* [1979] described the kinematics of the different vortex flow configurations and presented photographs that showed striking visual similarities between laboratory vortex configurations and actual tornadoes. Various investigators have now concluded that the swirl ratio is the dominant parameter that governs the structure of the vortex core. The point of transition from one flow configuration to another depends not only on S but also on radial Reynolds number; however, the Reynolds number dependence has been found to be slight at the higher flow rates. A summary description of the evolution of the vortex core and the development of these vortex configurations as a function of increasing S now follows.

Vortex Core Formation. At very low values of swirl ($S \lesssim 0.1$), no concentrated vortex core exists at the surface. This is because of a ringlike zone of separated flow that forms in the inflow on the lower surface. It prevents low-level angular momentum from approaching the region close to the centerline. As S increases, a core zone develops aloft and builds downward. For $S \approx 0.1$, a concentrated core makes contact with the surface.

Intensification and Breakdown. For $S \gtrsim 0.1$, the flow is characterized by the following: (1) a thin inflow boundary layer in which the depth decreases with decreasing radius, (2) a corner region where the horizontal inflow turns into the vertical direction, and (3) a cylindrical column of upflow containing a rotational core and extending upward for a considerable distance.

The core radius increases gradually with height; its initial response to an increase in swirl is to contract. Vertical velocities (inside and outside the core) are everywhere upward with a vertical velocity maximum along the centerline; the vortex is thus said to exhibit a one-celled structure, and this general configuration is termed an end-wall vortex.

Except for the very small values of S, at some level in the convection zone the core structure undergoes a dramatic change due to vortex breakdown. An adverse axial pressure gradient in the core culminates in a free stagnation point which becomes the leading point of the vortex breakdown (VBD). There is an abrupt increase in core radius as upward moving fluid diverges and flows around an inner circulation termed the breakdown "bubble," which forms the boundary between a supercritical flow regime upstream and a subcritical flow regime downstream of that level. The inflow, the flow in the corner, and the supercritical core are often

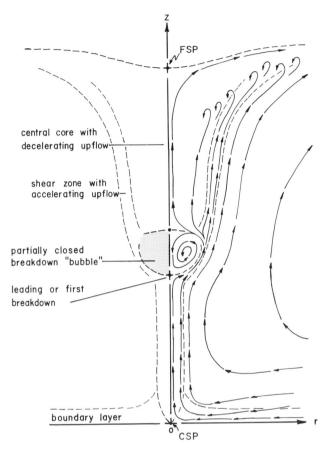

Fig. 6. Schematic of the flow at low swirl, showing the narrow laminar end-wall vortex, breakdown bubble, and downstream the enlarged turbulent core. From *Snow* [1982].

laminar in appearance, though this is not always the case. The subcritical flow downstream of the breakdown is almost always turbulent.

The significance of the criticality of the flow is that inertial waves are able to propagate upstream in a subcritical regime but not in a supercritical one. In a low-swirl one-cell vortex, this prevents filling in of the supercritical flow from above, and low central pressures and high velocities are sustained (otherwise, the pressure gradient force within the core, which is directed upstream, would promote a centerline downflow from aloft). The flow is turbulent immediately downstream of the breakdown bubble with a diminished and decelerating upflow near the axis and maximum vertical velocities in an annular region surrounding it.

The flow immediately downstream of the VBD is unsteady and has many variants, so verbal descriptions of it as well as physical measurements are challenging. The VBD features shown in Figure 6 are therefore a simplified representation of the flow. Various types of VBD have been photographed and discussed by *Sarpkaya* [1971], *Faler and Leibovich* [1977], and *Phillips* [1985]. At the time when VBDs were first being observed in TVCs, it was not clear whether they occurred in actual tornadoes. However, *Lugt* [1989] has since presented several examples of VBDs in atmospheric flows. From video tape of the funnel and debris cloud produced by the Minneapolis tornado of July 18, 1986, *Pauley and Snow* [1989] infer the presence of a vortex breakdown.

As the inner core flow continues to decelerate in the axial direction, another on-axis stagnation point occurs at some higher level. Downstream of this second stagnation point, vertical velocities are small. In some situations, an organized downflow is observed; in others, the inner core appears stagnant. *Mullen and Maxworthy* [1977] found a core structure similar to this in their dust devil model. On the other hand, *Pauley* [1989] did not find evidence of a mean upflow downstream of VBD for a wide range of flow conditions, suggesting that the total deceleration of the upflow and the second stagnation point had been incorporated in the structure of the breakdown bubble.

The far downstream core now appears broad and turbulent and superficially resembles the turbulent wake downstream of a bluff body placed in a straight-line flow. Vertical vorticity is concentrated within an annular region of upflow that surrounds a central downflow that contains comparatively little rotation. The concentric combination of upflow and downflow defines a two-celled vortex structure.

As the swirl ratio increases, the VBD moves upstream, that is, toward the lower surface, thereby reducing the depth of the supercritical region and causing the one-celled vortex to intensify. With sufficient swirl a condition is reached where the greatest vertical accelerations and largest pressure deficits are experienced in the surface layer (Figure 7). *Maxworthy* [1972] has termed this condition a "drowned vortex jump" (DVJ). In the laboratory the DVJ occurs at $S \approx 0.45$. It represents one of the most significant features of a tornadic flow because a tornado in this configuration may have its greatest impact on the surface.

Two-Celled Structure. As S is progressively increased, the leading breakdown point penetrates to the surface, followed by a general radial expansion of the core region and penetration of the downflow to the surface. The vortex has now become two celled over its entire length. Helical waves develop and propagate downstream in the annular cylindrical shear zone that surrounds a nearly quiescent inner core. As S increases further, the core radius increases, the shear zone thins out, and higher-order helical modes are excited. A more complex vortex configuration containing two intertwining spiral vortices emerges. At first these appear close together at the surface, as if forming a node, but additional swirl causes them to move apart and take up "antinodal" positions at the surface. They then resemble two separate tornadolike vortices, each spinning about its own axis and circulating about the centerline at the mean radius of the turbulent column. This occurs for $S \approx 1.0$. Further swirl produces higher-order multiple-vortex patterns; vortex cores containing up to six subsidiary vortices have been documented in the laboratory. These higher-order multiple-vortex flows are very difficult to maintain, and only at the lowest Reynolds numbers, because in the TVC they are so

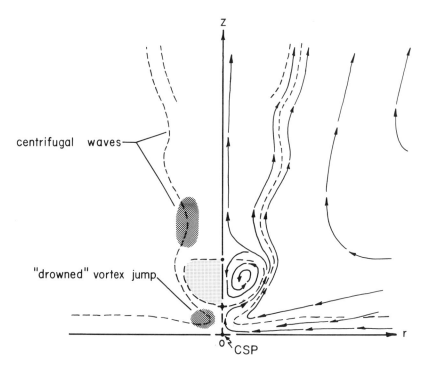

Fig. 7. Schematic of a one-celled flow showing the "drowned vortex jump" configuration, in which the highest velocities and greatest pressure falls are experienced at the surface. From *Snow* [1982].

closely spaced that they mutually interact and generate turbulence that causes them to break up. Tornadoes (and dust devils) have often displayed this type of multiple-vortex behavior (see, e.g., *Fujita* [1976]).

The complex nature of tornado damage can often be attributed to these types of flows. The combination of the translation of the parent thunderstorm and the rotation of subsidiary vortices about their own centers and about the centerline of the parent vortex leads to an inhomogeneous pattern of damage with a maximum impact at points where the velocity components add together most effectively and a reduced impact in nearby locations where the velocity components tend to cancel one another. At the present time, for a tornado formed in a flow of a given sink strength, it is not clear whether the DVJ configuration or the multiple-vortex configuration would produce the most violent effect on surface structures. This merits further investigation.

It would be incorrect to think that laboratory TVCs have been able to reproduce all of the vortex flow features that have been seen in nature. Over the past several years there has been a proliferation of automatic still and video cameras that has made it possible for members of the general public, perhaps sometimes injudiciously, to obtain excellent documentation of tornadoes. Some of the observed features serve to remind us that nature continues to hand out surprises in terms of new and unusual vortex features. It may be that the formation of some of these features depends on an absence of axial symmetry in the atmospheric flows, in which case they will not be produced in a conventional (axisymmetric) TVC.

3.2. Velocity Fields

Measurements With Hot Film Probes. The three-dimensional and turbulent nature of vortex flows makes velocity measurements difficult. In regions where the mean flow can be assumed to be two dimensional, effective measurements have been made using single hot film probes. By this means, *Baker and Church* [1979] measured the core radius and the average maximum core velocity (V_m) as a function of S for various flow rates. The mean updraft velocity (w) could be used as a scaling parameter, and results showed that the ratio V_m/w remained nearly constant at about 2.6 through a wide range of meteorologically important swirl ratios (including the VBD and multiple-vortex modes).

Baker [1981] further developed the technique of velocity measurement with hot film probes. By carefully combining single hot film measurements with a smoke filament tracing method, he was able to resolve three components of velocity at selected points. A detailed set of measurements was obtained for a vortex with a swirl ratio $S = 0.28$. This flow was one celled throughout the depth of the inflow layer. The VBD was located above the level of the updraft hole, in the lower part of the convection region. Some of the results are

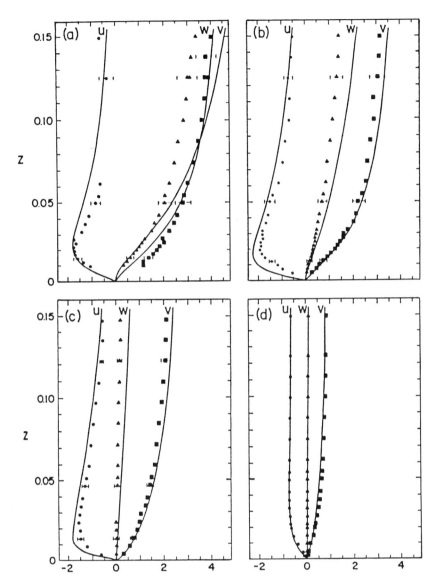

Fig. 8. Velocity profiles obtained by *Baker* [1981] for a vortex of $S = 0.28$ (data points) and the numerical simulation results (solid lines) of *Wilson and Rotunno* [1986] at four dimensionless radial positions: (a) $r^* = 0.0475$, (b) $r^* = 0.1025$, (c) $r^* = 0.2125$, and (d) $r^* = 0.75$. From *Wilson and Rotunno* [1986].

shown in Figure 8. (The points are Baker's data; the significance of the lines will be discussed under the next heading.)

Figure 8d shows data for a radial position far from the vortex core ($r^* = 0.75$), one where the flow is almost entirely horizontal (the vertical velocity increases from zero at the surface to some very small value at $z^* = 1$). Tangential velocities are small; radial velocities are even smaller. A weak maximum in the u profile can just be discerned at $z^* \approx 0.03$.

Moving nearer to the centerline, the velocity components in Figure 8c can be characterized thus: vertical velocities increase with height but are still small; tangential velocities have substantially increased; radial velocities have increased, though they are still smaller than the tangential velocities. Further, a significant "nose" has appeared in the vertical profile of radial velocities, with a peak at $z^* \approx 0.025$. Now the inflow boundary layer has a two-layer structure. A theoretical treatment by *Burggraf et al.* [1971] described this type of structure. They suggested that the boundary layer comprises an inner viscous sublayer and an outer inviscid layer.

Figure 8b shows a continuation of the trend toward higher values of u, v, and w. The thickness of the viscous sublayer has now diminished; the peak in the u profile now occurs at $z^* \approx 0.02$. At this location (about three core radii from the centerline) the strong radial jet achieves its maximum speed.

Close to the core, in the turning region (Figure 8a) the radial velocities become reduced. Over a wide range of heights the vertical and tangential velocities have quite similar magnitudes. Baker also measured vertical velocities along the centerline and showed that w attained its maximum value at $z^* \approx 0.25$ and diminished gradually with height toward the level of the VBD.

In reviewing the maximum values of u, v, and w that occurred in the flow, the following conclusions are drawn: (1) u_{max} was found at $z^* \approx 0.02$, $r^* \approx 0.1$; (2) v_{max} was found at $z^* \approx 0.25$, $r^* \approx 0.05$; (3) w_{max} was found at $z^* \approx 0.25$, $r^* = 0$; and (4) the quantities $u_{max} : v_{max} : w_{max}$ were in the approximate ratio $1:2:4$, indicating that one-celled vortex flows contain comparably large values of velocity in all three coordinate directions.

Numerical Simulations. In view of the difficulties involved in trying to simulate tornadoes numerically, a number of investigators have instead addressed numerical simulations of the flows in the Ward TVC. For a more complete review of these studies, see *Wilkins* [1988]. In this paper we focus on the results of *Wilson and Rotunno* [1986]. Of particular significance is their determination of the velocity field in a vortex flow that was identical to the one studied by Baker. The solid lines in Figure 8 show the results of their numerical simulation. The main differences are found in the vertical velocity field, where it appears that the numerical simulation may have predicted a somewhat larger core radius than was found experimentally. However, in general the numerical results match the laboratory data remarkably well. This independent validation of the experimental work tends to undermine the reservations expressed earlier about the validity of measurements made with intrusive velocity probes. It would be useful to extend the numerical simulation technique to include vortex flows at other swirl ratios, in order to see if the velocity components continue to be related in the same way.

From an analysis of the terms in the momentum equations, Wilson and Rotunno categorized the flow in the convergence zone as shown in Figure 9. The hyperbolic dashed line divides an outer irrotational flow from the inner rotational region. Thus we see the concentration of horizontal vorticity in a viscous surface sublayer and in an effectively inviscid layer above it. Vertical vorticity is concentrated in a cylindrical viscous subcore that is concentric with the centerline and in the effectively inviscid region that surrounds the viscous subcore.

3.3. Pressure Fields

Measurements With Pressure Probes. As static pressure is omnidirectional, it can be measured with comparative ease using a single pressure probe, although an investigator still has to deal with the problem of fluctuating signals. Useful insights into vortex core dynamics have been obtained from static pressure measurements made on the surface and along the central axis. Surface pressure fields as a function of swirl ratio were mapped by *Snow et al.* [1980]. They installed a

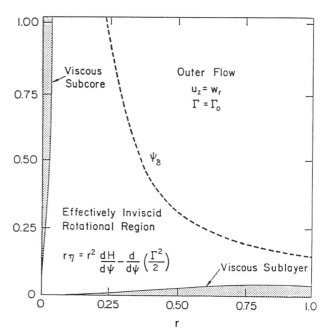

Fig. 9. Cross section of the convergence region showing the four principal vortex flow domains deduced by *Wilson and Rotunno* [1986]: irrotational outer flow, effectively inviscid rotational inner flow, viscous surface sublayer, and viscous subcore around the centerline. From *Wilson and Rotunno* [1986].

surface static port in a radial track. A sensitive transducer measured the static pressure deficit, that is, the difference between the surface static pressure at the various radial positions and the surface static pressure at the edge of the TVC. A sample of their results is shown in Figure 10. Principal conclusions were as follows:

1. In one-celled vortices the radial pressure gradients were generally of larger magnitude, and the central region of greatest pressure deficit, the "pressure well," was more closely confined to the centerline than in two-celled flows.

2. The magnitude of the central pressure deficit was sometimes larger in one-celled vortices than in the two-celled flows.

3. Radial variations of surface static pressure inside the cores of fully developed two-celled vortices were generally small in comparison with the magnitudes of the maximum core pressure deficits.

Pauley et al. [1982] examined how the central surface pressure deficits vary as a function of S and also presented measurements of the steady state surface pressure fields in multiple-vortex flows. Figure 11 is an example of how the central surface pressure deficit responds to changes in swirl ratio. The pressure values have been nondimensionalized with ρw^2 (the flow force per unit area). The symbols in Figure 11 identify a vortex configuration for each corresponding range of swirl ratios as one-celled laminar (L) or two-celled turbulent (T) flow containing two fully developed multiple vortices (2), transitions from two to three multiple vortices (2–3), and three multiple vortices (3). Figure 11

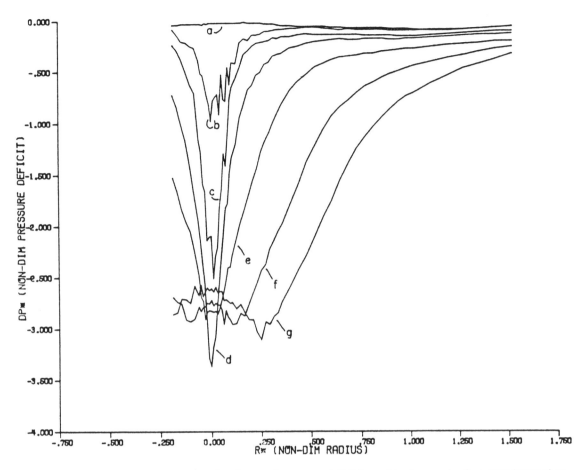

Fig. 10. Radial profiles of time-averaged surface static pressure deficit for swirl ratios ranging from zero (curve a) to 1.79 (curve g). From *Snow et al.* [1980].

presents the data in two ways: the time-dependent extrema of the fluctuating signal (solid line) and the time-averaged pressure values (dash-dot line). There are clearly substantial differences between the two quantities. The results show that vortex flow fields contain very large central pressure deficits. For example, they are larger by an order of magnitude or more than the pressure coefficients that are traditionally found in wind tunnel flows. The most prominent feature of these data is close to the point of transition from a laminar to a turbulent vortex and is attributed to the development of the DVJ. The very large values of central pressure deficit associated with this configuration ($\Delta P^* \approx 40$) were also found for a variety of flow rates and inflow geometries.

Church and Snow [1985] examined the axial distribution of pressure along the centerline in a series of nine one-celled and two-celled vortices over swirl ratios ranging from $S = 0.18$ to $S = 0.70$. The results are presented in Figures 12 and 13 and are summarized thus:

1. In one-celled vortices the central pressure deficit increased from a small value at the surface to a maximum (i.e., a pressure minimum) at $z^* \approx 0.3$ and diminished gradually with height above this level. At this level, ΔP^* increased rapidly with S, and an approximately cubic dependence on S was found. The general form of this relationship could be accounted for in terms of the similarity theory of *Burggraf et al.* [1971].

2. The largest value of pressure deficit anywhere in the flow was $\Delta P^* \approx 60$ (curve e); this occurred just upstream of the VBD with the VBD located at $z^* \approx 0.3$. The difference in pressure between this point and the central surface point can be used to derive a value for the maximum axial velocity (w_{max}) in terms of the mean updraft velocity (w). Thus $w_{max} = 10.5w$ at $z^* \approx 0.3$ for $S = 0.31$.

3. The largest value of pressure deficit at the surface was $\Delta P^* = 38$ (curve g), closely matching the corresponding value of *Pauley et al.* [1982] for the DVJ configuration under identical flow conditions.

4. Downstream of the VBD in the two-celled region of the vortex the pressure deficits became relatively small, with $\Delta P^* \approx 10$ typically and diminishing gradually with height. The static pressure profiles above $z^* \approx 0.3$ did not contain as much detail as they did below that level.

Pauley [1989] made additional axial pressure measurements, focusing on two-celled vortices. He argued that in the

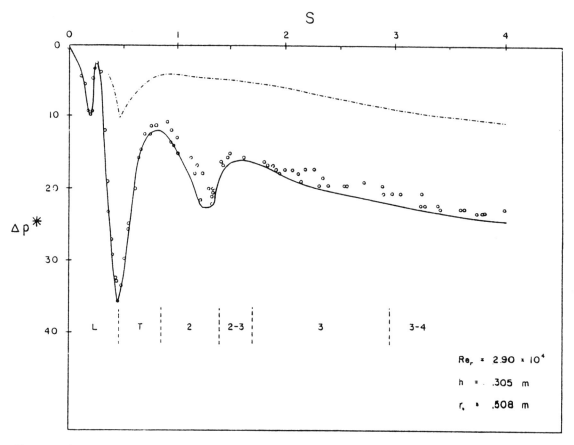

Fig. 11. Dimensionless central surface static pressure deficit versus swirl ratio, showing time-averaged data (dash-dot curve) and time-dependent maximum values (solid line). From *Pauley et al.* [1982].

turbulent flow downstream of the VBD, his time-averaged data captured the character of the axial pressure gradients better than the time-dependent measurements of *Church and Snow* [1985]. Pauley's values of ΔP^* were typically about one half of those of Church and Snow, with significantly better spatial resolution. Figure 14 shows an example of his data. From this it was possible to deduce the character of the downflow and to see that the strongest downflows occurred at middle levels in the two-celled flows. Pauley also concluded that (1) turbulent stresses are an important factor in helping to maintain low pressures and high velocities near the surface and (2) numerical simulations of tornadoes would be improved if they were to include turbulence modeling.

Applications to Tornadoes. It is tempting to try to use the laboratory pressure measurements to infer directly the minimum pressures in tornadoes, but there are serious difficulties with this. It is probable that the floor of the TVC (smooth in these experiments) was not as "rough" as the Earth's surface. *Pauley* [1980] demonstrated that roughening the flow with a covering of carpet caused a substantial reduction in the measured values of pressure. The effective viscosity of the fluid is also a problem. Theory implies that pressure deficits vary inversely with viscosity. In the atmosphere the effective viscosity is likely to be some orders of magnitude greater than the kinematic viscosity and also may vary in a complex manner. *Church and Snow* [1985] concluded that in spite of the relatively large pressure deficits that were measured in laboratory vortices, surface pressure deficits in tornadoes are expected to be small. So far there is no evidence to the contrary. It seems that the most effective way of inferring pressures inside tornadoes would be by carrying out integrations on the derived velocity fields.

The experimental pressure (and velocity) data were used by *Fiedler and Rotunno* [1986] in their development of a theory for maximum wind speeds in tornadoes. The theory addressed four physical quantities: (1) the maximum tangential velocity (v_{max}) in the supercritical flow, (2) the maximum axial velocity (w_{max}) in the supercritical flow, (3) the maximum tangential velocity (V_{sub}) in the subcritical flow (i.e., downstream of the VBD), and (4) the maximum swirl ratio possible (S^*) to maintain an end-wall vortex (i.e., the DVJ configuration). The following relationships were found from theory:

$$v_{max} \approx 1.7 v_{sub}$$

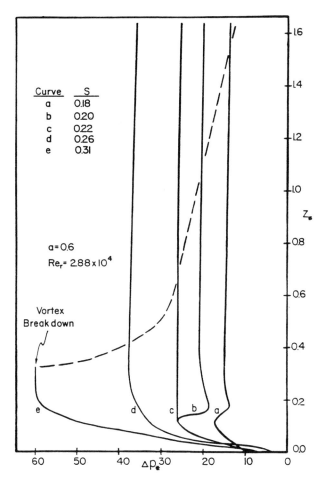

Fig. 12. Dimensionless central pressure deficit versus dimensionless height for five end-wall vortices for the range of swirl ratio $0.18 \leq S \leq 0.31$. (As was noted by *Pauley* [1989], owing to an error in measuring volume flow rate, the ΔP^* values here and in Figure 13 should be increased by about 30%.) From *Church and Snow* [1985].

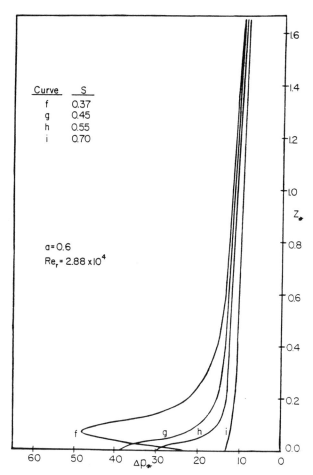

Fig. 13. Dimensionless central pressure deficit versus dimensionless height for four vortices for the range of swirl ratio $0.37 < S < 0.70$. Curve g ($S = 0.45$) is for the DVJ configuration. From *Church and Snow* [1985].

$$w_{max} \approx 2.1 v_{max}$$

$$S^* = 0.43.$$

It was demonstrated that the velocities that had been derived theoretically were in very good agreement with the corresponding experimental values. The maximum swirl ratio was also close to the experimental value of 0.45. Fiedler and Rotunno applied their results to the problem of the "thermodynamic speed limit," that is, that the maximum possible wind speeds in a saturated atmospheric column containing a Rankine vortex have been shown by thermodynamic argument to be ≈ 65 m s^{-1}. This is much smaller than the wind speeds in some tornadoes. By taking the 65 m s^{-1} limit as the value of v_{sub}, they thus obtained values for $v_{max} \approx 110$ m s^{-1} and $w_{max} \approx 220$ m s^{-1}. While the latter figure appears to be somewhat high, these are more in line with what damage analyses suggest for maximum wind speeds in tornadoes.

3.4. Surface Roughness

The issue of surface roughness adds one more dimension of complexity to the investigation of tornado dynamics. It is clear from the results of several investigators that the nature of the surface over which vortices form has a significant impact on their flow structure and intensity. Basic questions that have been addressed concern how the swirl ratio, core radius, velocities, and pressures are affected by changes in the roughness of the surface. To a certain extent, laboratory experiments, field observations, and numerical simulations have provided answers, but there is still scope for much more work. The results of previous efforts are summarized in Table 3. In some cases there appears to be good agreement; in others there appear to be conflicting conclusions.

Effect of Roughness on Swirl Ratios. There seems to be a consensus among investigators as to the effect on swirl ratio of increasing surface roughness: higher swirl vortices tend to form more readily over smooth surfaces. Increasing the roughness causes an increased frictional dissipation in

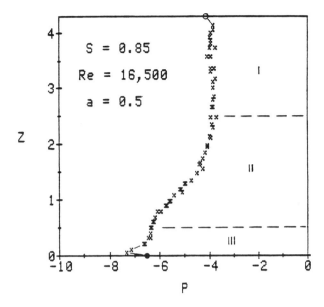

Fig. 14. Dimensionless central pressure deficit versus dimensionless height. Time-averaged data for a two-celled vortex ($S = 0.85$). Strongest downflow occurs in region II. From *Pauley* [1989].

the surface layer, causing transition to a lower swirl configuration. Alternatively, a greater angular momentum input is needed to maintain the same vortex configuration over a rough surface. The experimental work of *Dessens* [1972], *Wilkins et al.* [1975], *Leslie* [1977], and *Rostek and Snow* [1985], as well as the numerical study of *Ayad and Cermak* [1980], all support this conclusion.

The preceding sounds straightforward, but the following discussion illustrates the need to delve more deeply in order to provide substantive answers to practical questions. A typical question might be: if a tornado moves from smooth ground to rough ground, is it going to produce more or less damage? Field observations are conflicting. For a tornado in Indiana, *Baker et al.* [1982] found that after the tornado had entered a grove of trees the damage increased with depth of penetration. For tornadoes in France, *Dessens and Snow* [1989] also concluded that those occurring over forested areas had a greater than average intensity, path width, and path length. On the other hand, *Monji and Wang* [1989] showed that a tornado in Japan after having caused light damage in an urban (rough) area moved over smoother terrain and caused much heavier damage. There is no simple way to resolve this issue. Consider then the following points:

TABLE 3. Summary of Effects of Increasing Surface Roughness

Reference	Comments	$z_0 \times 10^4$ m	Results
Dessens [1972]	laboratory measurement	0, 20	Smaller azimuthal, larger vertical velocities; transition from 2-cell to 1-cell state; core radius larger and more turbulent.
Bode et al. [1975]	numerical study	...	Smaller azimuthal, larger vertical velocities; increased core radius.
Blechman [1975]	field observation	...	Transition to lower swirl state; increased inflow velocity.
Wilkins et al. [1975]	water tank	...	Increased convergence and vertical velocity; larger core radius; less intense vortices.
Leslie [1977]	laboratory measurement	0, 8.5	Flow more turbulent; effective swirl reduction of ~30%.
Leslie [1979]	laboratory measurement	...	Core radius decreased at small aspect ratios, increased at large aspect ratios.
Lewellen and Sheng [1979]	numerical study	...	Increased radial and vertical velocities.
Ayad and Cermak [1980]	numerical study	...	Transition from 2-cell to 1-cell state.
Nakamura and Nakama [1980]	laboratory measurement	...	Core radius increased.
Baker [1981]	field observation	...	Damage increasing with penetration after entering grove of trees.
Diamond and Wilkins [1984]	laboratory measurement	0, 34	Core radius decreased; effect of translation was to increase effective swirl and core radius.
Rostek and Snow [1985]	laboratory measurement	0, 0.5, 2.4, 3.6	Effective swirl reduced; thin ropelike vortex for medium roughness at $S = 0.1$; VBD over roughest surfaces did not penetrate to surface, owing to separated flow region; rough surface enhances high-pressure ring; little variation in core radius.
Monji and Wang [1989]	laboratory measurement	1.5, 5.6, 15	Smoke-filled core radius increases; thin ropelike vortex extends to surface in roughest case; location of maximum winds moves upward; surface winds decrease; zone of high radial winds broadens and deepens; zone of high tangential winds lifts and shrinks.
	field observation	...	Considerable increase in damage density and path width upon leaving urban (rough) area.
Dessens and Snow [1989]	field observation	...	Transition to lower swirl state. Tornadoes over wooded areas are more intense and have greater than average path widths and lengths.

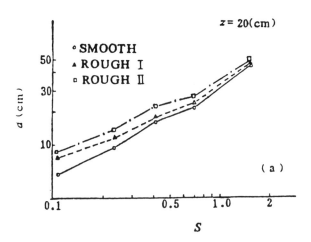
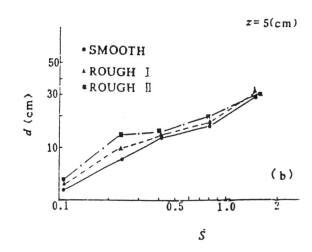

Fig. 15. Diameter of smoke-filled vortex core versus swirl ratio for surfaces of different roughness and for heights of 5 and 20 cm. Inflow depth: 47 cm. From *Monji and Wang* [1989].

1. If a vortex is one celled, roughening the surface is going to cause it to move further away (deintensify) from the DVJ configuration, thereby reducing near-surface wind speed.

2. If a vortex is two celled, roughening the surface is going to cause it to move closer to the DVJ configuration, thereby increasing near-surface winds.

3. It is to be expected that a rough surface creates higher radial velocities and lower tangential velocities than a smooth surface.

Does increased roughness cause the total velocity vector in the surface layer to increase or decrease? The answer is not clear. To summarize, the answer to an apparently simple question is evidently complex and may ultimately require a knowledge of how much swirl there is and how much the surface roughness changes in a given situation. These factors require more careful scrutiny in future laboratory work.

Effect of Roughness on Core Radius. A majority of the laboratory experiments and the numerical studies of *Bode et al.* [1975] and *Lewellen and Sheng* [1979] showed an increase in core radius with increasing surface roughness. The fact that some investigators found the opposite effect indicates that there is some degree of complexity to this issue also; the result may depend not only on how core radius is measured, but where it is measured. The most recent data are those of *Monji and Wang* [1989]. They used small wooden cubes as roughness elements and worked with three surfaces: smooth, rough I, and rough II (the latter being the roughest). They measured the diameters of smoke-filled cores at two levels as a function of swirl ratio. The results are shown in Figure 15. Both levels show a systematic increase in the diameter of the smoke-filled core with increasing roughness for all values of S, although the effect is less pronounced at the higher values (i.e., $S > 1$). Monji and Wang also measured horizontal velocities using a hot wire velocity probe and a minivane which enabled them to resolve the tangential and radial components. The tangential velocity data shown in Figure 16 provide another way of determining core radius. Here the trend is not so clear: It appears that the core radius for the rough I surface is larger than either the smooth or rough II surface, which is puzzling.

Effect of Roughness on Velocities. In a numerical study,

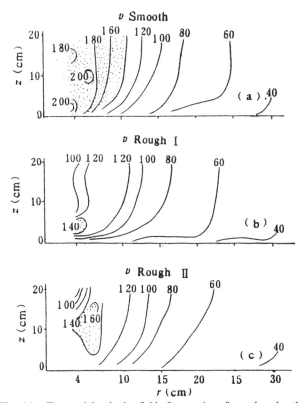

Fig. 16. Tangential velocity fields for vortices formed under the same background flow conditions over surfaces of different roughness. From *Monji and Wang* [1989].

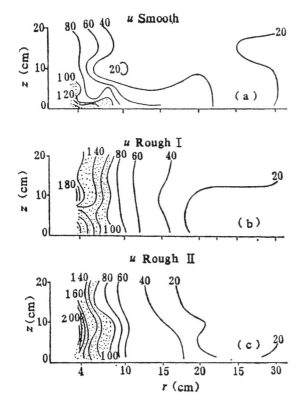

Fig. 17. Same as Figure 16, except radial velocity fields. From *Monji and Wang* [1989].

Lewellen and Sheng [1979] showed that the following occurred with increasing roughness: (1) an increase in the maximum radial and vertical velocities (nondimensionalized by the maximum tangential velocity), (2) an increase in the ratio of the height to radius at which the maximum tangential velocity occurred, (3) a decrease in the maximum tangential velocity, and (4) an increase in the radius of maximum tangential velocity. The experimental results of Monji and Wang (Figures 15, 16, and 17) are generally consistent with the conclusions of Lewellen and Sheng. The increase in radial velocities with roughness is striking; the details of the trends in the tangential velocity field appear less systematic. Monji and Wang also showed that there was little variation in the magnitude of the horizontal velocity vector (sum of radial and tangential components) with roughness. The main effect was that the maximum horizontal velocity was found close to the surface in the smooth case, and with increasing roughness the positions of the velocity maximum were found at progressively higher levels. The radial positions of the velocity maximum were relatively unchanged. It is not known at what swirl ratios these measurements were made, and they do not seem to address the complex nature of the issue as it was represented earlier in section 3.4. To conclude, the effect of increasing roughness is to increase the frictional dissipation of angular momentum. This results in an increased convergence accompanied by higher radial and vertical velocities and reduced tangential velocities.

Effect of Roughness on Pressures. As mentioned above, *Pauley* [1980] briefly examined the effect of surface roughness on the magnitudes of surface pressures. A more systematic series of measurements was performed by *Rostek and Snow* [1985]. The entire flow of the TVC was covered with a pegboard in which up to 5500 roughness elements (wooden pegs) were installed. The densities of the pegs determined the roughness characteristics of the surface. Roughness lengths were determined using separate peg board test surfaces placed in a wind tunnel.

Experiments were performed using one smooth surface and four of different roughness. The general characters of the swirling flows made visible by smoke injection were observed as a function of increasing roughness. Radial profiles of surface static pressure deficit were obtained for each of the surfaces. The general conclusions were as follows:

1. For a given input of angular momentum, a vortex over a rough surface has a structure and an associated pressure deficit profile similar to that of a vortex over a smooth surface with a smaller input of angular momentum.

2. Surface roughness has a significant impact on one stage in the evolution of the vortex, namely, the extension of the concentrated core downward to make contact with the surface. The evolution of the concentrated core and the descent of the breakdown bubble as seen over smooth surfaces becomes less distinct when observed over rough surfaces. Instead, a smaller subcore reached down through the thicker boundary layer and made contact with a localized surface inflow.

3. In order to establish multiple-vortex flows over rough surfaces, a greater angular momentum input is required.

4. In comparing pressure profiles of smooth and rough surfaces, a region of nearly zero radial pressure gradient within the core and surrounding the central pressure well was observed for rough surfaces.

This last may be seen in Figures 18 and 19 by comparing the pressure profiles for the smoothest and the least rough of the surfaces; the feature of interest is located between 1- and 7-cm radius. This feature may be compared with the notch or step observed within the core region of the 1962 Newton, Kansas, tornado [*Ward*, 1964]. It has also been seen in other barograph records of tornadoes. No generally accepted physical explanation for its appearance has yet been developed. (This feature should not be confused with the ring of higher pressure surrounding the central core that occurs for low swirl vortices.)

Several workers have now investigated the effect of varying the surface roughness of the lower boundary and have found that this exerts a significant impact on vortex dynamics. It is necessary to ask whether the roughened laboratory surfaces were valid representations of the Earth's surface characteristics. Table 3 lists the roughness lengths (z_0) that were determined for the experimental studies. It has been estimated that the ratio of the depth of the atmospheric

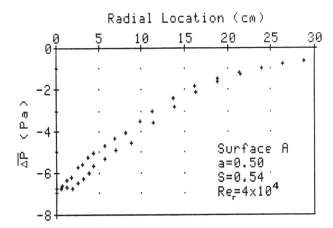

Fig. 18. Radial profile of time-averaged surface static pressure for a vortex flow ($S = 0.54$) over a smooth surface (surface A). From *Rostek and Snow* [1985].

boundary layer to the TVC boundary layer falls in the range 3000 to 8000. An upper bound for the roughness length of the Earth's surface is about 3 m [*Arya*, 1988], z_0 for most "rough" terrain being of the order of 0.1–1 m. From an examination of the experimental roughness lengths in Table 3 it would appear that a number of the experiments were performed over surfaces of extreme roughness, and consequently the effect of roughness may have been overstated in some of the results.

Another question concerns the means of preparing the rough surface. Some investigators used carpet of convenient thicknesses, but carpet is a nonstandard working material. Others followed the accepted wind tunnel practice of using regularly spaced arrays of roughness elements. This places a constraint on the spatial resolution of the measurements. It is concluded that other types of rough surface may be better

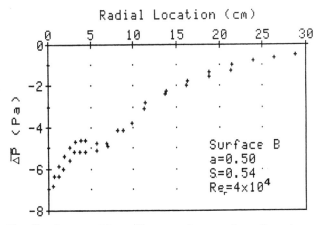

Fig. 19. Same as Figure 18 except for a surface of roughness length 5×10^{-5} m. Surface B was the least rough of the prepared surfaces. Profile is virtually identical to that of Figure 18 for radial positions greater than 7 cm. High-pressure feature of interest is located between 1 and 7 cm. From *Rostek and Snow* [1985].

suited to studies of swirling flows. For example, the application of various grades of sandpaper is an idea which merits further consideration in future laboratory experiments. From the foregoing discussion it is evident that further work is needed to clarify the issues pertaining to surface roughness.

4. SUMMARY AND SUGGESTIONS FOR FUTURE WORK

The past decade and a half have seen significant progress in the development of our understanding of the dynamics of tornadic flows. Laboratory modeling has played an important role in this progress. Tornado vortex chambers have demonstrated the range and complexity of atmospheric vortex configurations. Greater skill in documenting tornado events has led to a general confirmation of the similarity between vortex flows observed in nature and those observed in the laboratory. Physical measurements made in the laboratory have provided a significant amount of information on velocity and pressure fields. Numerical simulations of TVC flows using the latest generation of computers have demonstrated the feasibility of developing computer models that capture the principal features of the laboratory vortices and which lead to consideration of issues of practical concern, such as maximum intensities in actual tornadoes. Tornado intercept teams now seem to be on the threshold of providing physical measurements on tornadoes, which could then be used to help validate numerical models of the full-scale flows.

Concerning the future of laboratory measurements, we offer the following suggestions:

1. The LDV technique, although more difficult to implement than other technologies, is the only one capable of measuring the flow without perturbing it. LDV measurements on vortices will provide measurements of the mean values of three components of velocity at different swirl ratios. These data should overlap and extend those that have been obtained previously by other methods. From measurements of the fluctuating components it will be possible to determine turbulence intensities and Reynolds stresses, which will lead to a more complete empirical evaluation of the relative importance of the terms in the momentum equations.

2. Questions remain concerning surface roughness effects. Although much of the data obtained so far seem consistent with the basic physics, there is still no clear understanding of how the maximum near-surface winds depend on roughness. It is therefore desirable to continue the type of measurements outlined in suggestion 1 above using surfaces of suitable roughness.

3. As the dependence of flow characteristics on Reynolds number, especially for roughened surfaces, has not been established, it would be interesting to vary the kinematic viscosity of the working fluid. This would allow one to determine how the thickness of the boundary layers, magnitudes of velocity and pressure, and vortex transition points vary with viscosity. This could be done by setting up a

closed circulation system and, in place of air, establishing a flow of a different gas, for example, carbon dioxide. (Carbon dioxide has approximately one-half the kinematic viscosity of air.) Such an undertaking would lend itself best to the smaller TVC which would require less elaborate modifications to develop a closed system.

4. A question which has never been addressed directly through experiment is how does a tornado interact with the electrical environment of a thunderstorm. Although infrequent, numerous documented cases of luminous glows and other evidence of electrical discharges associated with tornado funnels have now been compiled. Somehow, electric fields develop that are close to breakdown strength: orders of magnitude greater than what is normally found near the surface under a thunderstorm. The surface of the Earth is a plentiful source of ions. One suggested experiment would be to determine how a vortex flow can influence an ionized surface layer and whether it could locally cause a sufficient amplification of the background electric field to produce glowlike discharges.

Finally, it should be borne in mind that in a laboratory experiment it takes an investigator of the order of 1000 s to make a single measurement. A supercomputer capable of one billion operations per second is able to generate velocity profiles far more rapidly than the experimentalist. The main question is the validity of the model. It is evident that for maximum economy, laboratory measurements must be made sparingly, an essential role of the laboratory work being to provide information that can be used to fine tune numerical models of swirling flows.

References

Arya, S. P., *Introduction to Micrometeorology*, 307 pp., Academic, San Diego, Calif., 1988.

Ayad, S. S., and J. E. Cermak, A turbulence model for tornado-like swirling flows, in *Vortex Flows*, edited by W. L. Swift, P. S. Barna, and C. Dalton, 171 pp., American Society of Mechanical Engineers, New York, 1980.

Baker, D., E. Agee, G. Baker, and R. Pauley, The Rush County, Indiana, tornado of 9 July 1980, in *Preprints, 12th Conference on Severe Local Storms*, pp. 379–392, American Meteorological Society, Boston, Mass., 1982.

Baker, G. L., Boundary layers in laminar vortex flows, Ph.D. thesis, 143 pp., Purdue Univ., West Lafayette, Ind., 1981.

Baker, G. L., and C. R. Church, Measurements of core radii and peak velocities in modeled atmospheric vortices, *J. Atmos. Sci.*, 36, 2413–2424, 1979.

Blechman, J. B., The Wisconsin tornado event of April 21, 1974: Observations and theory of secondary vortices, in *Preprints, 9th Conference on Severe Local Storms*, pp. 344–349, American Meteorological Society, Boston, Mass., 1975.

Bode, L., L. M. Leslie, and R. K. Smith, Numerical study of boundary effects on concentrated vortices with application to tornadoes and waterspouts, *J. R. Meteorol. Soc.*, 101, 313–324, 1975.

Burggraf, O. R., K. Stewartson, and R. Belcher, Boundary layer induced by a potential vortex, *Phys. Fluids*, 14, 1821–1833, 1971.

Church, C. R., and J. T. Snow, The dynamics of natural tornadoes as inferred from laboratory simulations, *J. Rech. Atmos.*, 12, 111–133, 1979.

Church, C. R., and J. T. Snow, Measurements of axial pressures in tornado-like vortices, *J. Atmos. Sci.*, 42, 576–582, 1985.

Church, C. R., J. T. Snow, G. L. Baker, and E. M. Agee, Characteristics of tornado-like vortices as a function of swirl ratio: A laboratory investigation, *J. Atmos. Sci.*, 36, 1755–1776, 1979.

Davies-Jones, R. P., The dependence of core radius on swirl ratio in a tornado simulator, *J. Atmos. Sci.*, 30, 1427–1430, 1973.

Davies-Jones, R. P., Laboratory simulations of tornadoes, in *Proceedings of the Symposium on Tornadoes: Assessment of Knowledge and Implications of Man*, pp. 151–173, American Meteorological Society, Boston, Mass., 1976.

Dessens, J., Influence of ground roughness on tornadoes: A laboratory simulation, *J. Appl. Meteorol.*, 11, 72–75, 1972.

Dessens, J., and J. T. Snow, Tornadoes in France, *Weather Forecasting*, 4, 110–132, 1989.

Diamond, C. J., and E. M. Wilkins, Translation effects on simulated tornadoes, *J. Atmos. Sci.*, 41, 2574–2580, 1984.

Escudier, M. P., Vortex breakdown and the criterion for its occurrence, in *Intense Atmospheric Vortices*, edited by L. Bengtsson and J. Lighthill, pp. 247–258, Springer-Verlag, New York, 1982.

Faler, J. H., and S. Leibovich, Disrupted states of vortex flow and vortex breakdown, *Phys. Fluids*, 20, 1385–1400, 1977.

Fiedler, B. H., and R. Rotunno, A theory for the maximum windspeeds in tornado-like vortices, *J. Atmos. Sci.*, 43, 2328–2340, 1986.

Fujita, T. T., Graphic examples of tornadoes, *Bull. Am. Meteorol. Soc.*, 57, 401–412, 1976.

Kimpel, J. F., An interactive weather exhibit at OMNIPLEX, *Bull. Am. Meteorol. Soc.*, 62, 1219–1223, 1981.

Leslie, F. W., Surface roughness effects on suction vortex formation: A laboratory simulation, *J. Atmos. Sci.*, 34, 1022–1027, 1977.

Leslie, F. W., The dependence of the maximum tangential velocity on swirl ratio in a tornado simulator, in *Preprints, 11th Conference on Severe Local Storms*, pp. 361–366, American Meteorological Society, Boston, Mass., 1979.

Lewellen, W. S., A solution for three-dimensional vortex flows with strong circulation, *J. Fluid Mech.*, 14, 420–432, 1962.

Lewellen, W. S., and Y. P. Sheng, Influence of surface conditions on tornado wind distributions, in *Preprints, 11th Conference on Severe Local Storms*, pp. 375–381, American Meteorological Society, Boston, Mass., 1979.

Lugt, H. J., Vortex breakdown in atmospheric columnar vortices, *Bull. Am. Meteorol. Soc.*, 70, 1526–1537, 1989.

Maxworthy, T., On the structure of concentrated, columnar vortices, *Astronaut. Acta*, 17, 363–374, 1972.

Maxworthy, T., The laboratory modeling of atmospheric vortices: A critical review, in *Intense Atmospheric Vortices*, edited by L. Bengtsson and J. Lighthill, pp. 229–246, Springer-Verlag, New York, 1982.

Mitsuta, Y., and N. Monji, Development of a laboratory simulator for small-scale atmospheric vortices, *Nat. Disaster Sci.*, 6, 43–54, 1984.

Monji, N., A laboratory investigation of the structure of multiple vortices, *J. Meteorol. Soc. Jpn.*, 63, 703–713, 1985.

Monji, N., and Y. Wang, A laboratory investigation of the characteristics of tornado-like vortices over various rough surfaces, *Acta Meteorol. Sin.*, 3, 506–515, 1989.

Mullen, J. B., and T. Maxworthy, A laboratory model of dust devil vortices, *Dyn. Atmos. Oceans*, 1, 181–214, 1977.

Nakamura, I., and N. Nakama, Some features of a tornado-like vortex core (in Japanese), *Summary 29*, pp. 39–47, Philos. Dep., Univ. of the Ryukyus, Okinawa, Japan, 1980.

Pauley, R. L., Laboratory measurements of surface pressure minima in simulated tornado-like vortices, M.S. thesis, 91 pp., Purdue Univ., West Lafayette, Ind., 1980.

Pauley, R. L., Laboratory measurements of axial pressures in two-celled tornado-like vortices, *J. Atmos. Sci.*, 46, 3392–3399, 1989.

Pauley, R. L., and J. T. Snow, On the kinematics and dynamics of the 18 July 1986 Minneapolis tornado, *Mon. Weather Rev.*, *116*, 2731–2736, 1989.

Pauley, R. L., C. R. Church, and J. T. Snow, Measurements of maximum surface pressure deficits in modeled atmospheric vortices, *J. Atmos. Sci.*, *39*, 368–377, 1982.

Phillips, W. R. C., On vortex boundary layers, *Proc. R. Soc. London, Ser. A*, *400*, 253–261, 1985.

Rostek, W. R., and J. T. Snow, Surface roughness effects on tornado-like vortices, in *Preprints, 14th Conference on Severe Local Storms*, pp. 252–255, American Meteorological Society, Boston, Mass., 1985.

Rothfusz, L. P., A mesocyclone and tornado-like vortex generated by the tilting of horizontal vorticity: Preliminary results of a laboratory simulation, *J. Atmos. Sci.*, *43*, 2677–2682, 1986.

Sarpkaya, T., On stationary and traveling vortex breakdowns, *J. Fluid Mech.*, *45*, 545–559, 1971.

Snow, J. T., A review of recent advances in tornado vortex dynamics, *Rev. Geophys.*, *20*, 953–964, 1982.

Snow, J. T., and D. E. Lund, A second generation tornado vortex chamber at Purdue University, in *Preprints, 15th Conference on Severe Local Storms*, pp. 323–326, American Meteorological Society, Boston, Mass., 1988.

Snow, J. T., C. R. Church, and B. J. Barnhart, An investigation of the surface pressure fields beneath simulated tornado cyclones, *J. Atmos. Sci.*, *37*, 1013–1026, 1980.

Wan, C. A., and C. C. Chang, Measurements of the velocity field in a simulated tornado-like vortex using a three-dimensional velocity probe, *J. Atmos. Sci.*, *29*, 116–127, 1972.

Ward, N. B., The Newton, Kansas tornado cyclone of May 24, 1962, in *Preprints, 11th Weather Radar Conference*, pp. 410–415, American Meteorological Society, Boston, Mass., 1964.

Ward, N. B., The exploration of certain features of tornado dynamics using a laboratory model, *J. Atmos. Sci.*, *29*, 1194–1204, 1972.

Wilkins, E. M., Influence of Neil Ward's simulator on tornado research, *CIMMS Rep. 87*, 64 pp., Univ. of Okla., Norman, 1988.

Wilkins, E. M., Y. Sasaki, and H. L. Johnson, Surface friction effects on thermal convection in a rotating fluid: A laboratory simulation, *Mon. Weather Rev.*, *103*, 305–317, 1975.

Wilson, T., and R. Rotunno, A numerical simulation of a laminar end-wall vortex and boundary layer, *Phys. Fluids*, *24*, 3993–4005, 1986.

Laser Doppler Velocimeter Measurements in Tornadolike Vortices

DONALD E. LUND AND JOHN T. SNOW

Department of Earth and Atmospheric Sciences, Purdue University, West Lafayette, Indiana 47907

1. INTRODUCTION

Using data recently gathered within the Purdue University tornado vortex chamber II (PU TVC II) using a state-of-the-art laser Doppler velocimeter (LDV), we present and discuss radial and vertical profiles of measured radial and tangential velocity components and derived vertical velocity component. A mean offset of the vortex from the centerline of the tornado vortex chamber (TVC) required a coordinate transformation of the measured data from a "chamber" reference frame to a "vortex" reference frame. Further, the impact of small vortex translations (vortex wander) on the data set is examined through the use of a simple mathematical model of the data collection process.

2. LABORATORY APPARATUS

Following *Ward* [1972], the PU TVC II is configured to give geometric and dynamic similarity to tornadolike flows (Figure 1). The portion of the lower chamber beneath the dividing surface is referred to as the confluence zone, and the region below the updraft hole is called the convergence zone. The upper chamber is termed the convection zone. In the PU TVC II, a mechanically driven throughflow is deflected by an assembly of vanes at the outer perimeter of the lower chamber. The vortex initially forms along the centerline through the convergence of vorticity and is maintained by a steady inflow of angular momentum. See *Snow and Lund* [1987] for more details concerning the PU TVC II.

Figure 1 depicts schematically the major components of the LDV as positioned in the PU TVC II. In this system the output of a He-Ne laser is transmitted via a fiber optic cable to an optical system that splits the single input beam. The two resultant beams are focused to cross in the working volume of the TVC. To facilitate detection of flow reversal, one of the beams is frequency shifted by an acousto-optic cell. A Doppler shift arises as a result of the relative motion of a scattering particle and two plane waves (one from each beam) that approach the beam crossing (the "measuring volume") from different directions. The velocity component in the plane of the beams and normal to the optical axis is detected as a beat frequency, or frequency difference, produced by the heterodyne mixing on a photodetector of backscattered light from particle interactions with each plane wave in the measuring volume [*Adrian*, 1983]. The optics are shown in Figure 1 configured to measure the radial velocity component. The entire assembly is rotated 90° to measure the tangential velocity component. A precision traversing mechanism ($r \pm 0.0005\ cm$, $z \pm 0.0025$ cm) allows positioning of the measuring volume anywhere in an rz plane cutting the convergence and confluence zones and the lower portion of the convection zone.

3. MEASUREMENT PROGRAM

3.1. Configuration of the Apparatus

The measurement region of the TVC was configured for the data set presented here with an updraft hole of radius $r_0 = 20.395$ cm and an inflow depth of $h = 30.383$ cm. The swirl ratio, the dimensionless parameter which characterizes the nature of the flow in a TVC [*Davies-Jones*, 1973] by the ratio of the mean swirl velocity v_0 to the mean throughflow velocity u_0, both measured at r_0, was $S = r_0 v_0 / 2hu_0 = 0.222$, a moderate value. The volume flow rate per unit axial depth was $Q = 2\pi r_0 u_0 = 2.35$ m^2 s^{-1}. Thus the radial Reynolds number, $Re_r = r_0 u_0/\nu$, was 2.46×10^4, where $\nu = 1.52 \times 10^{-5}$ m^2 s^{-1} is kinematic viscosity. These conditions resulted in a turbulent single-celled vortex in the convergence zone and lower one half of the convection zone (the only portions of the vortex that were visually examined). No vortex breakdown was observed; however, past experience suggests that for conditions such as these a breakdown was likely present in the upper one half of the convection zone.

Radial and tangential velocity components were measured

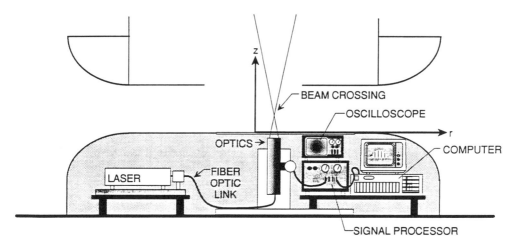

Fig. 1. Cross-sectional view of the Purdue University tornado vortex chamber II showing the arrangement of the major components of the laser Doppler velocimeter in the work space beneath the "floor" of the chamber.

in the convergence zone along zenith lines at radii of 5, 10, 15, and $r_0 = 20.395$ cm and along a chamber radius at elevations of 1, 3, 6, 15, and $h = 30.383$ cm. Measurements consisted of a minimum of 384 individual realizations accumulated according to particle arrival rates within the measuring volume. Velocities presented here are mean values of histograms constructed in this fashion.

3.2. Radial Profiles of Velocity Components

Figures 2 and 3 show typical radial profiles of mean radial and tangential velocity components, in this case for an elevation of 15 cm ≈ $h/2$, i.e., at midheight. In these figures and subsequent discussion, lowercase variables indicate a vortex chamber or absolute reference frame centered in the updraft hole, while uppercase variables denote a reference frame centered on the vortex axis. The peculiar pattern in the measured data of Figure 2 for $|r| < 4$ cm (as opposed to what might be anticipated for a corner flow) is to be noted. One possible cause for this peculiarity is an offset in the long-time mean position of the vortex with respect to the center of the updraft hole. This would place the line of measurement (LOM) along a chord of the vortex polar reference frame. Consequently, near the axis of the vortex, the velocity component along the LOM would be highly influenced by the tangential component of the vortex velocity field.

To determine and correct for this offset, a coordinate transformation technique was applied that assumes the magnitude of the mean horizontal velocity vector, $|\mathbf{V}_h|$, was constant on a circle of fixed radius R with respect to the mean vortex position. For each elevation, all measured (u, v) pairs along corresponding radial profiles were used to construct a plot of $|\mathbf{V}_h|$ versus position, r, along the LOM, where u and v are the velocity components measured along and normal to the LOM, respectively. Figure 4 shows the coordinate transformation geometry; (u, v) pairs determine values of θ_1 and θ_2, while corresponding positions along the LOM yield r_1 and r_2. These four parameters, together with the assumed circular symmetry, uniquely determine α, β, dr, l, and R. The measured velocity components may then be transformed to an orthogonal pair in the vortex reference frame through $|\mathbf{V}_h|$ and α and the position reassigned the value R.

Typical results of the transformation are also depicted in Figures 2 and 3, where $dr = 0.369$ cm and $l = -0.115$ cm. In Figure 2 the transformation has the effect of moving each data point upward and to the left, shifting the points toward what might be expected for corner flow, while in Figure 3, little change is noted. Careful inspection suggests that away from the axis the effect of this transformation is minimal and that near the axis, for $R < 0$, the radial component data are undercorrected, while for $R > 0$, they are overcorrected. The coordinate transformation has little impact on v since a mean offset along the LOM simply results in a shift of origin; a shift normal to the LOM (but within the core) results in a greater vortex tangential velocity, $V(R)$, at the location of the measurement, but measuring only the component of $V(R)$ normal to the LOM has a compensating effect.

This transformation fails near the core since it assumes the vortex is steady. In practice, the vortex constantly wanders about the mean offset position. We discuss the effect of wander on the near-axis measurements in section 4.

At first glance, the profile of Figure 3 appears similar to a classical Rankine-combined vortex (an irrotational outer flow surrounding a core in solid rotation). However, a least squares fit to the circulation field implied by these midheight data indicates that for $R > 1$ or $2R_c$ (where R_c is the core radius, here estimated to be about 1.08 cm), the outer flow is proportional to $r^{-0.63}$ rather than r^{-1}. This reflects the three-dimensional, viscous nature of the underlying swirling corner flow. (We will adopt this midheight value for the core radius as a reference length scale in what follows.)

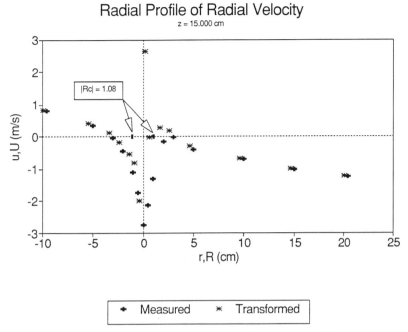

Fig. 2. Radial profile at 15-cm elevation of measured radial velocity component (pluses, lowercase variables) and results of coordinate transformation (asterisks, uppercase variables).

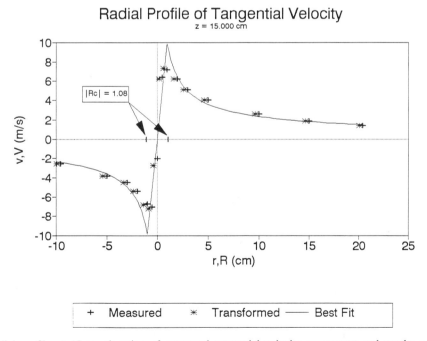

Fig. 3. Radial profile at 15-cm elevation of measured tangential velocity component and results of coordinate transformation. Symbols and notation as in Figure 2. Solid line represents the radial distribution of tangential component, $V = 27.1\Gamma_\infty R (R \leq R_c)$, $V = 4\Gamma_\infty (R_\infty/R)^n (R_c \leq R \leq R_\infty)$, resulting from a "best fit" to the circulation field of the transformed data. For outer radius R_∞ taken to be 25 cm, the best fit gives $R_c = 1.08$ cm, $n = 0.63$, and far-field circulation $\Gamma_\infty = 0.329$ m^2 s^{-1}. (We adopt this midheight value of R_c as our reference length scale.)

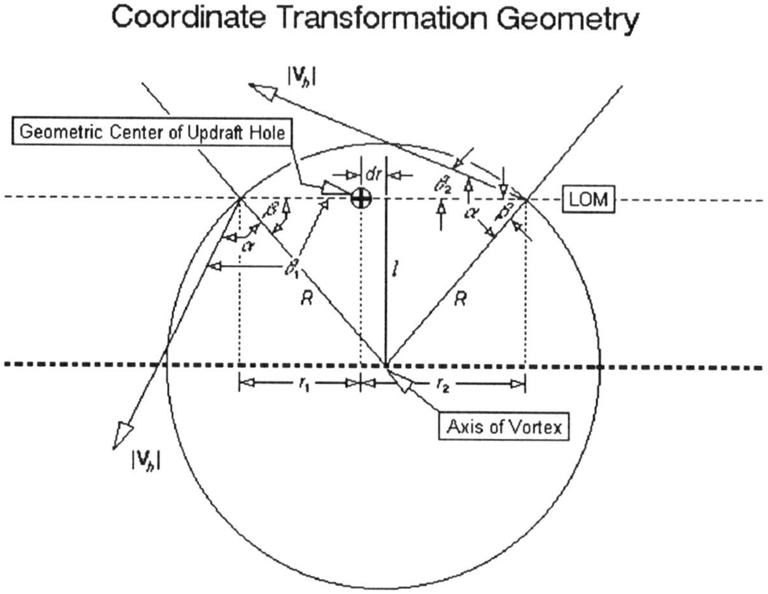

Coordinate Transformation Geometry

Fig. 4. Schematic depiction of coordinate transformation geometry. Measured (u, v) pairs provide θ_1, θ_2, r_1, and r_2. These parameters and circular symmetry uniquely determine α, β, dr, l, and R. Measured pairs can then be transformed through $|\mathbf{V}_h|$ and α and reassigned to the position R.

3.3. Vertical Profiles of Velocity Components in the Outer Flow

Figure 5 shows vertical profiles of radial velocity component at nominal radii of 5, 10, and 20 cm; a profile at nominal 15-cm radius was similar to the 10-cm profile and has been omitted for clarity. The overall flow field is one of acceleration below 5 cm and deceleration above 5 cm. Estimates of radial convective accelerations at the level of the "nose" of the developing surface inflow jet ($z \sim 1$ cm) give values of $4.7g$ at $R \sim 7.5$ cm, and decelerations (where the flow turns upward) were about $8.5g$ at $R \sim 2.3$ cm. The flow in this jet attains a peak measured speed of about 2.75 m s^{-1}. The location of this maximum is in agreement with *Baker*'s [1981] results, which indicate a peak velocity of about 0.86 m s^{-1} at $(r, z) \sim (4, 1)$ cm. The higher velocity in the present case is consistent with the larger Reynolds number (2.46×10^4 versus Baker's 1.22×10^4) and the smaller value of h (30 cm versus Baker's 41 cm).

The near-constant inflow speed in the vicinity of $z \sim 5$ cm marks the level where frictional effects have become small enough to allow cyclostrophic balance to establish itself. A fluid element can approach the axis with nearly constant speed while the opposing centrifugal and pressure gradient forces continue to build. This is consistent with the radial momentum balances of *Wilson and Rotunno*'s [1986] numerical experiment. Above 5 cm the radial momentum begins to decrease with decreasing radius, gradually at lower elevations, then more rapidly as the flow turns to exit the convergence zone through the updraft hole. The presence of a strong radial inflow jet associated with the upper dividing surface is prominent at a radius of 20 cm and still evident at a radius of 10 cm.

Vertical profiles of tangential velocity component (Figure 6) show an overall increase in V with decreasing R at all levels. The no-slip boundary condition associated with the underside of the upper dividing surface is manifest in the profile at $R \sim 20$ cm. That the largest values of tangential velocity component appear near $z \sim 30$ cm for $R = 5$, 10, and 15 cm reflects the convergence of the main body of the flow into the updraft hole. Fluid elements "spin up" as they approach the axis owing to conservation of angular momentum.

The oscillatory pattern in the profile at $R \sim 5$ cm seems too systematic to attribute to normal scatter. There is an indication of oscillations in the profile at $R \sim 10$ cm as well. A similar pattern is evident nearer the axis in the Table 1 data extracted from radial profiles. The V' in Table 1 are the measured maxima at the indicated levels. These observations possibly reflect the presence of an inertial standing wave in the core. Similar phenomena have been observed by *Church et al.* [1979] and modeled numerically by *Rotunno* [1979]. *Snow and Lund* [1989] describe inertial motions with a simple dynamical model.

A plot of a vertical profile of dimensionless tangential velocity component at $r \sim 20$ cm is shown in Figure 7. The friction velocity, $v^* = (\tau_w/\rho)^{1/2}$, is used to normalize z and v to produce the "inner variables" z^+ and v^+. The profile exhibits a logarithmic character for $100 < z^+ < 300$ and a linear behavior for $z^+ < 10$. *Phillips and Khoo* [1987] also observed this behavior in their LDV measurements of a

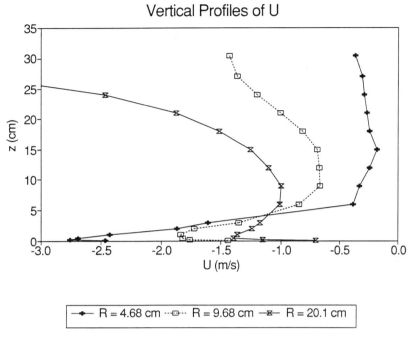

Fig. 5. Vertical profiles of radial velocity component showing the development of a low-level inflow jet between surface and 5 cm. Note that radial component magnitude increases to the left; the negative sign denotes inflow.

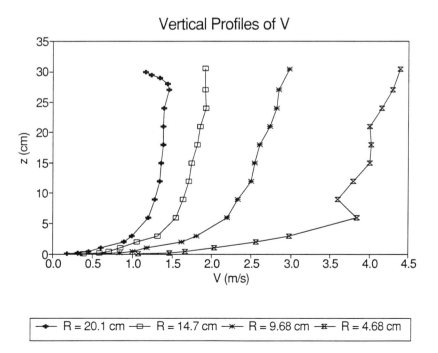

Fig. 6. Corresponding vertical profiles of tangential velocity component for the situation in Figure 5.

TABLE 1. Measured Maxima of Velocity Component Normal to LOM

Z, cm	V', m s^{-1}	R, cm
1.00	9.49	0.626
3.00	8.53	0.141
6.00	8.84	0.300
15.00	7.16	0.642
30.38	7.58	1.122

vortex generated in a rotating tank of water. *White* [1974, p. 473] describes this structure as a universal characteristic of all turbulent boundary layer data. This figure shows that credible near-wall velocity measurements with the LDV were attained well into the linear sublayer, down to $z^+ \sim 2.5$ (here equal to 1.65 mm).

3.4. *Estimates of the Vertical Component of Velocity*

Radial profiles of vertical velocity component (Figure 8) were derived from the measured radial velocity component via an integrated continuity equation. In recognition of the difficulty in resolving the profiles near the core, these data were selectively reduced, eliminating "bad" points by inspection prior to being used for computations. Advantage was again taken of circular symmetry by reflecting points for $R < 0$ about both coordinate axes, thus enhancing the data set.

The weak vertical velocities near the surface at outer radii first increase gradually with decreasing R, then more rapidly near the core where an axial jet "erupts" from the boundary layer. Vertical velocities nearly quadruple between 1 and 3 cm near the axis and reach a maximum of about 10.5 m s^{-1} at $z = h$. These findings are in qualitative agreement with *Baker*'s [1981] measured vertical velocities. In both cases the jet appears to erupt inside a radius of about 4 cm at 3-cm elevation. Baker's peak velocity approaches the somewhat lower value of about 4 m s^{-1}, whereas in the present case the vertical velocity at the same dimensionless height is at least 5.5 m s^{-1}. This is consistent with the higher Re_r and smaller r_0 in the present case. While direct comparisons with the LDV measurements of *Phillips and Khoo* [1987] are more difficult owing to differences in experimental apparatus, similar profile shapes exist at about 3-cm elevation, and the axial jet erupts from the boundary layer inside a radius of about 6 cm.

Rough estimates of vertical convective accelerations in the erupting jet yield values around $34g$. To put this in a different perspective, a fluid element moving through a divergence field of 1 m s^{-1} cm^{-1} at 3.3 m s^{-1} experiences a $34g$ acceleration. This occurs at a relatively low elevation of $z \sim 2R_c$, just above the entrance (at $z \sim R_c$) to the annular volume containing the greatest tangential speeds; see below.

The vertical velocity profile for $z = h$ reflects the presence of the upper boundary at outer radii, where the flow turns abruptly as it passes the edge of the updraft hole.

4. Assessment of the Impact of Vortex Wander

To better assess the impact of vortex wander on near-axis measurements, a numerical simulation of the measurement

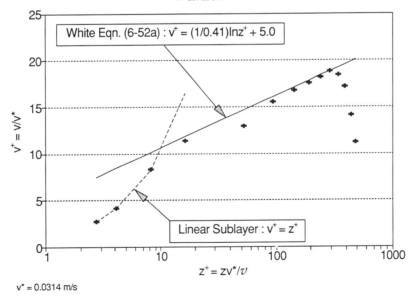

Fig. 7. Boundary layer profiles plotted in terms of "inner variables" [*White*, 1974], dimensionless height z^+ and dimensionless tangential velocity component v^+, showing the linear and logarithmic layers that are universally characteristic of turbulent boundary layer data.

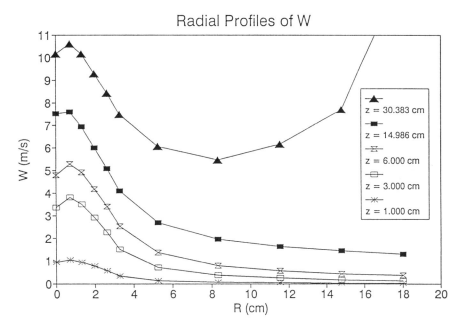

Fig. 8. Radial profiles of derived vertical velocity component. An edge effect associated with the upper surface is evident in the profile for $z = h$.

process has been developed. In this model, length measures are defined as in Figure 9. All quantities are taken to be dimensionless, with the core radius as the length scale and the actual tangential velocity maximum as the velocity scale. Using prescribed functions for the radial distribution of radial and tangential velocity components, the velocity components along and normal to the LOM (see Figure 9) can be expressed as

$$u(r) = u'(r) + \bar{u}(r) \quad (1)$$

$$v(r) = v'(r) + \bar{v}(r), \quad (2)$$

where $u(r)$ and $v(r)$ are measured radial and tangential velocity components, respectively. Quantities $u'(r)$ and $v'(r)$ are the contributions from actual radial and tangential velocities, $U(R)$ and $V(R)$, respectively, and are given by

$$u'(r) = |U(R)| \cos(\alpha) \quad (3)$$

$$v'(r) = |V(R)| \sin(\beta). \quad (4)$$

Finally, $\bar{u}(r)$, the contribution to the measured radial velocity component from $V(R)$, and $\bar{v}(r)$, the contribution to the tangential component from $U(R)$, are given by

$$\bar{u}(r) = |V(R)| \cos(\beta) \quad (5)$$

$$\bar{v}(r) = |U(R)| \sin(\alpha). \quad (6)$$

For this analysis, radial velocities were taken to be proportional to the square root of the radius and normalized to preserve the ratio $u(r_0)/v_{\max}$ in Figures 2 and 3. The tangential velocity field was taken to be a Rankine-combined vortex.

The first numerical experiment considered a stationary vortex with the offset normal to the LOM having a typical value of $l = -0.500$. The resulting radial velocity profiles are shown in Figure 10. The model-measured $u(r)$ closely resemble the profile of measured data in Figure 2. The

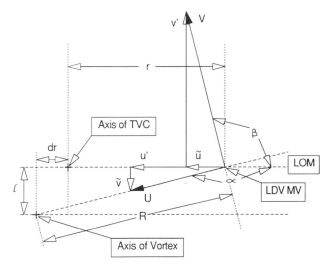

Fig. 9. Schematic depiction of the geometry for an instantaneous measurement of the velocity component along or normal to the LOM. The measured velocity components u and v are each comprised of contributions from both actual velocity components U and V. LDV MV denotes the location of the measuring volume.

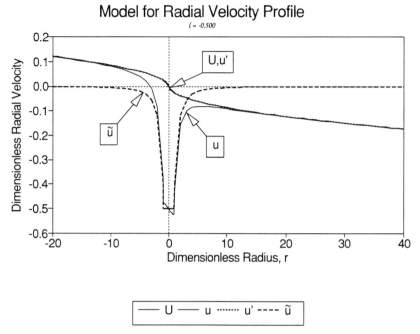

Fig. 10. Mathematical model for the profile of radial velocity component through a stationary vortex with a mean offset normal to the LOM of $l = -0.500$.

component velocity u' is virtually identical to $U(r)$, suggesting from (1) that a reasonable estimate of U can be obtained from $u' = u - \bar{u}$.

Previous comments with regard to the general insensitivity of the tangential velocity profile to a small offset normal to the LOM are borne out in Figure 11. Tangential velocities V, v, and v' were so nearly identical that they have been presented in terms of absolute error $[v(r) - V(r)]$ and $[v'(r) - V(r)]$. Inside the core, for both r and $R < 1$, $[v'(r) - V(r)] = 0$. From (6) and Figure 9 one sees that $\bar{v}$ is small everywhere since for large r, $\sin(\alpha) \to 0$ and when $|\sin(\alpha)| \to 1$ as $r \to 0$, $U \to 0$. Hence $\bar{v}$ is plotted as a velocity component in Figure 11. The sign of $\bar{v}$ remains the same along the entire profile (compare Figure 9) and hence in this case increases sensor error for $r > 0$ and decreases it for $r < 0$. Errors are everywhere below 5% except in the range $1 < |r| < 2$. With the exception of an underestimate of $V_{\max}$ by about 20%, $v(r)$ is a good estimate of $V(r)$. To summarize, u' is a good approximation of U, which implies that we must subtract the contribution $\bar{u}$ from the measured velocity u, and the measured velocity v is a good estimate of V with the exception of $V_{\max}$.

In a second experiment, the vortex was allowed to "wander" in a random fashion within ranges of offsets dr and l. These ranges were chosen to yield profiles that resembled those of Figures 2 and 3. At each of 17 radii corresponding roughly to the locations of actual measurements, histograms of u and v were constructed by generating sequences of 384 vortex positions. Radial profiles were then plotted using the mean values. Since the LDV measures only one velocity component at a time, the model is run once for each velocity profile. The sample size of 384 represents the minimum used for each laboratory measurement. Since LDV measurements are made on the basis of particle arrival rates at the measuring volume, this was considered a reasonable simulation of the data collection process.

Mean values of the histograms are plotted in Figure 12, along with the prescribed functions for U and V. The most striking feature is the gross underestimate of $V_{\max}$ (here more than 50%) that results from the inherent averaging induced by vortex wander. Companion experiments indicate that this is somewhat dependent on profile shape and appears to be around 25% for more realistic profiles.

In spite of wander-induced averaging, high velocities were still observed in the vortex in which the data of Figures 2 and 3 were gathered. This flow had an observed maximum tangential velocity component of 9.49 m s^{-1} at $(R, z) = (0.626, 1.000)$ cm. The above arguments would suggest that the actual speed was somewhat higher than this, perhaps as high as 11.9 m s^{-1}.

5. SUMMARY AND CONCLUSIONS

The juxtaposition of a tornadolike vortex with the axis of a TVC affects profiles of velocity components gathered in a "chamber" reference frame. A straightforward coordinate transformation to recover actual tangential and radial profiles meets with limited success. For the vortex considered here, radial velocity component data collected outside of $|r| \sim 5$ cm were affected little by the degree of offset and motion

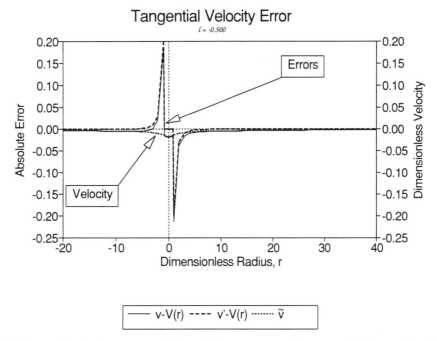

Fig. 11. Model tangential velocity components for a stationary vortex with a mean offset were virtually the same except for values of V_{max} and are presented in terms of error for clarity. Actual values were retained for $\tilde{v}$ owing to their small magnitude. Here $l = -0.500$.

of the vortex with respect to the TVC centerline, while tangential component data were only significantly affected (albeit dramatically) for $1 \text{ cm} \leq |r| \leq 2 \text{ cm}$.

Vertical and radial profiles of radial and tangential velocity components reveal characteristic boundary layer and vortex flow features. A series of vertical profiles of radial velocity show the rapid development of a near-surface inflow jet. Vertical profiles of tangential velocity suggest inertial oscil-

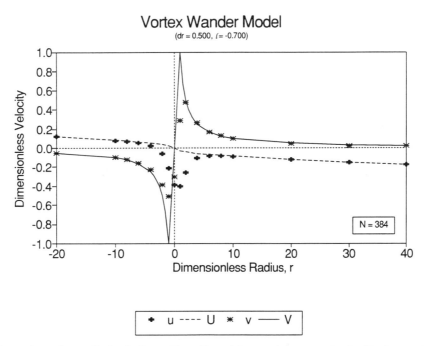

Fig. 12. Comparison of prescribed velocity profiles with model-generated measured velocities for a vortex in random motion with a specified degree of wander. Here $dr = 0.500$ and $l = -0.700$.

lations in the core, while radial profiles exhibit the irrotational nature of the outer flow and the solid-body rotation of the core. Vertical profiles in dimensionless inner variables substantiate the capability of near-wall measurements even into the linear sublayer. Radial profiles of vertical velocities derived through continuity compare qualitatively to measurements by *Baker* [1981] and *Phillips and Khoo* [1987].

Rough calculations at the location of the maximum in tangential velocity (9.49 m s^{-1} at $(R, z) = (0.626, 1.000)$ cm) give a centrifugal acceleration of about $1470g$ with a corresponding pressure gradient, assuming cyclostrophic balance, of about 170 mbar m^{-1}. The region of greatest tangential velocity component (here taken to be speeds greater than 7.5 m s^{-1}) extended upward from the location of the maximum of 9.49 m s^{-1}. As is shown in Table 1, a secondary maximum of 8.84 m s^{-1} occurs at (0.300, 6.000) cm; this maximum is accompanied by a centrifugal acceleration of about $2655g$ and a cyclostrophic pressure gradient of 310 mbar m^{-1}. Maxima in centrifugal acceleration and pressure gradient from observed data occur at a height of 3.00 cm where these quantities have values of about $5260g$ and 614 mbar m^{-1}, respectively.

The greatest tangential speeds were found in an annular volume of small radius and small radial width but significant vertical extent (and coincident with the axial jet), with large centrifugal accelerations and largest pressure gradients occurring well aloft (up to a height of about $5.6R_c$). This finding suggests that a single-cell, F1 or F2 intensity tornado (the type modeled here) may contain a larger region of potentially damaging wind than previously thought. (Recall that intensities and hence wind speeds in such tornadoes are inferred from surface structural damage.) Two-cell tornadoes may attain F4 or F5 ratings simply because this region is broader and comes closer to the surface, exposing more structures to high wind speeds. The laboratory observations of *Church et al.* [1979] also seem to show such structure.

The occurrence of $34g$ convective accelerations at low levels (here at $2R_c$) in the erupting jet suggests (in a respect after *Fiedler and Rotunno* [1986]) that some of the destructive force in tornado vortices of the type modeled here lies within the near-surface axial jet.

A simple numerical investigation of the impact of wander on the measured speeds has shown that u' is a good approximation of U, implying that the contribution $\bar{u}$ must be subtracted from the measured velocity u and that the measured velocity v is a good estimate of V with the exception of $V_{\max}$.

In conclusion, the feasibility of LDV measurements in tornadolike flows has been demonstrated. We have shown that LDV measurements can be used to obtain quantitative information about critical flow structure such as the near-surface jet in the radial component of velocity in the vortex boundary layer. Measurements have been made deeper into the near-surface core region than ever before. However, vortex wander precludes accurate measurements of the innermost core. It appears that a separate, simultaneous measurement of the wander may be necessary. Possibilities include pressure measurements or video imagery. Parallel mathematical modeling of the data acquisition process also appears to be a necessity both to clearly understand what is being measured and to develop techniques to compensate for wander.

Acknowledgments. A special thanks to P. J. Smith and W. L. Wood for their support while one of us (J.T.S.) participated in Operations Desert Shield/Desert Storm. R. L. Walko provided a number of useful ideas in support of this work. C. R. Church and R. Rotunno provided useful remarks on the initial manuscript. This research was supported by the National Science Foundation through grant ATM 8703846.

REFERENCES

Adrian, R. J., Laser velocimetry, in *Fluid Mechanics Measurements*, edited by R. J. Goldstein, chap. 5, pp. 155–244, Hemisphere, Washington, D. C., 1983.

Baker, G. L., Boundary layers in laminar vortex flows, Ph.D. dissertation, Purdue Univ., 143 pp., West Lafayette, Ind., 1981.

Church, C. R., J. T. Snow, G. L. Baker, and E. M. Agee, Characteristics of tornado-like vortices as a function of swirl ratio: A laboratory investigation, *J. Atmos. Sci.*, 36, 1755–1776, 1979.

Davies-Jones, R. P., The dependence of core radius on swirl ratio in a tornado simulator, *J. Atmos. Sci.*, 30, 1427–1430, 1973.

Fiedler, B. H., and R. Rotunno, A theory for the maximum wind speeds in tornado-like vortices, *J. Atmos. Sci.*, 43, 2328–2340, 1986.

Phillips, W. R. C., and B. C. Khoo, The boundary layer beneath a Rankine-like vortex, *Proc. R. Soc. London, Ser. A*, 411, 177–192, 1987.

Rotunno, R., A study in tornado-like vortex dynamics, *J. Atmos. Sci.*, 36, 140–155, 1979.

Snow, J. T., and D. E. Lund, A second generation tornado vortex chamber at Purdue University, in *Preprints, 15th Conference on Severe Local Storms*, pp. 323–326, American Meteorological Society, Boston, Mass., 1987.

Snow, J. T., and D. E. Lund, Inertial motions in analytical vortex models, *J. Atmos. Sci.*, 46, 3605–3610, 1989.

Ward, N. B., The explanation of certain features of tornado dynamics using a laboratory model, *J. Atmos. Sci.*, 49, 1194–1204, 1972.

White, F. M., *Viscous Fluid Flow*, 725 pp., McGraw-Hill, New York, 1974.

Wilson, T., and R. Rotunno, Numerical simulation of a laminar end-wall vortex and boundary layer, *Phys. Fluids*, 29, 3993–4005, 1986.

Vortex Formation From a Helical Inflow Tornado Vortex Simulator

JAMES G. LaDUE

Cooperative Institute for Mesoscale Meteorological Studies, Norman, Oklahoma 73019

1. INTRODUCTION

The University of Oklahoma tornado vortex chamber (TVC) was recently reconfigured from the *Rothfusz* [1985] version with the goal of improving the horizontal homogeneity of the helical inflow. The three tiers of guide vanes used in the Rothfusz TVC were replaced with a set of flat boxes stacked on top of each other. Each box was able to be oriented in any direction allowing for a variety of shear profiles to be created.

Using the new configuration of the TVC inflow layer, the objectives of this paper are to (1) establish the flow parameters that produce the most intense tornadolike vortex, (2) report on significant changes to vortex morphology due to changing inflow shear profiles, (3) show similarities and differences of the strongest vortex morphology to the vortex produced by the Rothfusz TVC and to that of observed and modeled supercell tornadic thunderstorms, and (4) offer suggestions for an improved helical inflow TVC based on current observations.

The goal of tornado simulation by physical means is to produce a vortex in a laboratory environment and to do so from a flow field with characteristics of the real atmosphere. Most varieties of vortex simulators generate vorticity in a fluid and then concentrate that vorticity by some means to produce a vortex [*Davies-Jones*, 1976]. The Ward-type tornado simulator, known as the tornado vortex chamber, which is to be discussed here, more closely represents the formation and stretching of vorticity within the tornadic region of a thunderstorm [*Ward*, 1972].

Based on observations of tornado-producing thunderstorms, *Ward* [1972] produced a TVC with a shallow inflow layer depth (h) compared with the radius (r_0) of the updraft region such that the ratio h/r_0, is of the order of unity. The inflow layer allowed air to flow into the interior region of the TVC with a symmetrical vertical circulation situated in such a way that the updraft fan assembly concentrated the circulation into a vortex. Either a rotating screen or a set of vanes around the perimeter of the inflow layer provided the source of vertical circulation. A baffle was placed below the updraft fan assembly to distribute the pressure minimum over a wider area than the vortex as is the case in the upper regions of a thunderstorm. These simulator characteristics allowed the Ward-type TVC to produce vortices that exhibited many similarities to actual tornadoes [*Church et al.*, 1979].

For investigating the structure of the tornadolike vortex, the classical Ward-type TVC is satisfactory. However, most major tornadoes are produced by supercell thunderstorms which derive their vertical vorticity from the horizontal helicity in the environment [*Lilly*, 1986; *Davies-Jones*, 1984].

To simulate tornadoes in such highly helical environments, *Rothfusz* [1985] used the inflow layer design mentioned previously to produce a positive helical inflow with no net vertical circulation. Any vertical circulation that was created within the TVC was found to come from the tilting of the ambient horizontal vorticity in the flow. Rothfusz was successful in creating a small-scale version of a mesocyclone with an embedded tornadolike vortex. Even though *Rothfusz* [1985] was able to create a strong vortex connected to the ground with the new inflow configuration, there were still inherent problems with the design. The configuration did not allow the inflow layer to have horizontally homogeneous helicity, and thus a new configuration was designed.

2. METHODOLOGY

Presently, the configuration of the inflow layer has similar dimensional measurements as that of the original Ward TVC as shown by a side view schematic in Figure 1 and listed in Table 1, although the meanings of some nondimensional numbers [*Davies-Jones*, 1973] are different. For instance, the formulation of the swirl ratio becomes unclear because there is no symmetrical vertical circulation being introduced into the TVC. The relative changes in the swirl ratio can be brought about in the current TVC configuration by adding or subtracting tiers.

Three types of airflow through each tier are possible in the new inflow configuration shown in Figure 2. The first is a

---

The Tornado: Its Structure, Dynamics, Prediction, and Hazards.
Geophysical Monograph 79
This paper is not subject to U.S. copyright. Published in 1993 by the American Geophysical Union.

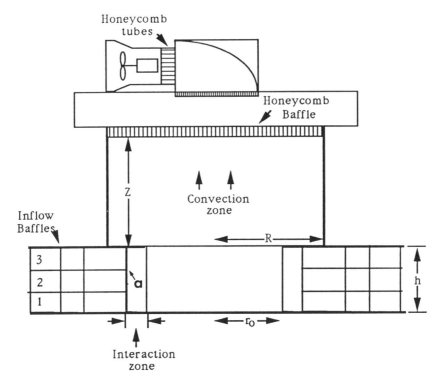

Fig. 1. A side view schematic of the latest University of Oklahoma helical flow TVC. The accompanying dimensions of the variables portrayed are listed in Table 1.

full-sectional airflow, where air is allowed in from both ends of each tier. The second configuration is a half-sectional inflow where one end of each tier is blocked. The third occurs with the ability to enhance the inflow with blower fans mounted at the "downstream" end of each tier. Each tier has the option of being pivoted about a vertical axis such that vertically veering inflow can be achieved. The angle between successive tier orientations is defined as the angle β. As one final option, the addition of filters at one or both ends of each tier can induce many permutations of vertical speed shear.

When the tier boxes veer with height, helicity is created as the vorticity along the top and bottom boundaries between

TABLE 1. Dimensional Measurements for the Current University of Oklahoma Tornado Vortex Chamber

Variable	Value
r_0	0.6096 m
R	0.9100 m
Z	0.9100 m
Inflow tier dimensions	
For each tier	
h_i	0.1780 m
w	2.2900 m
Four-tier system	
h	0.7100 m
Area of entry ($=4 \cdot h_i \cdot w$)	1.625 m^2
Three-tier system	
h	0.5300 m
Area of entry ($=3 \cdot h_i \cdot w$)	1.218 m^2
Area of updraft, πr_0	1.167 m^2
Re ($=1.4 \times 10^4$ at inflow)	0.6 m s^{-1}
Laminar viscosity, $Q/Re\ h$	1.45×10^{-5} m^2 s^{-1}
Aspect ratio a ($=h/r_0$)	
Four-tier system	1.16
Three-tier system	0.87

The measurements for the four- and three-tier system correspond to the number of tiers used in the inflow layer. The variable Q stands for the volume flow rate, and Re is the Reynolds number. All other variables are measurements which are labeled in Figure 1.

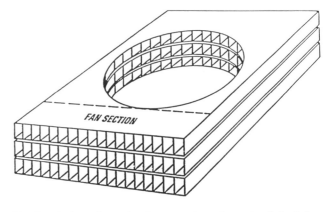

Fig. 2. A three-dimensional perspective schematic of the inflow layer accompanying the helical flow TVC at the University of Oklahoma oriented unidirectionally (most experimentation was done with four tiers).

the tiers reaches a common edge at the beginning of the interaction zone. An example is illustrated by Figure 3: horizontal vorticity $\vec{\zeta}_1$ is created near the ceiling of tier one by velocity $\vec{v}_1$, while $\vec{\zeta}_2$ is created by $\vec{v}_2$ along the floor of tier two. At the edge of the interaction zone (labeled "a" in Figure 1) the two vorticity vectors interact constructively to create $\vec{\zeta}_t$ in a narrow vertical strip parallel to the mean velocity vector between the two tiers. That narrow strip of vorticity eventually diffuses vertically within the interaction zone to fill in the layer above and below the tier one/tier two interface. The width of the interaction zone determines to what degree the vorticity diffuses into the spaces between the tier interfaces.

The purpose of the experimentation using the current TVC configurations is to determine whether a mesocyclone with a concentrated vortex can be produced. Experimentation using combinations of varying updraft strength, inflow layer fans, and even mechanical downdrafts is carried out to determine the configuration that produces strong cyclonic vortices. Once realistic results are obtained, observations of the internal flow are documented and compared with the flow observed in the Rothfusz TVC. Comparisons of the flow structure are also made with observational and numerical studies of tornadic supercells.

Using the TVC shear layer configuration that produces the strongest cyclonic vortex as a standard, additional experiments were conducted by employing variations in the shear layer profile while keeping the β angle constant. These shear tests explored the changes of the vortex character and strength to changes in the inflow layer velocity in individual layers. Seven experiments of significance will be reported in this paper. A seventh experiment will be discussed as part of the sensitivity studies with one of the tiers removed to make a three-tier inflow system and thus lowering the aspect ratio. It was hoped that lowering the aspect ratio (three-tier aspect ratio of 0.87) would increase the swirl ratio enough to

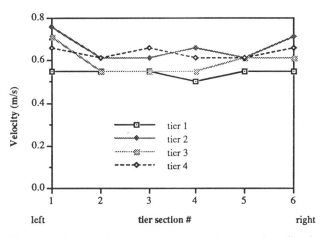

Fig. 4. Inflow velocity profile across the entrance region of each tier corresponding to half-sectional inflow used in the experimentation. The width of each tier was divided into six sections labeled across the bottom.

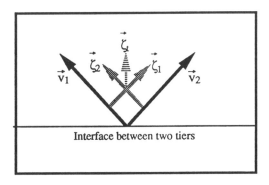

Fig. 3. Schematic of the edge of the interface between two tiers showing the generation of helicity between two tiers. The subscript 1 (2) refers to the quantity in the lower (upper) tier. The variable $\vec{v}$ is the velocity in the middle of each tier, while $\vec{\zeta}$ is the vorticity generated along the surface of a tier (see text for more details). Variable $\vec{\zeta}_t$ corresponds to the total vorticity vector produced by adding the other two vorticity components (the length of the vectors is not to scale).

produce multiple vortices within the rotation [*Davies-Jones*, 1973; *Church et al.*, 1979].

3. RESULTS

Experimentation has shown that a single-cell, end-wall vortex formed with the current configuration of the inflow layer. In each experiment conducted where rotation occurred, a counterrotating couplet was observed. Cyclonic rotation was observed to the right of the anticyclonic rotation with respect to the inflow layer shear vector in all cases where rotation was observed.

The major problem discovered in many of the experiments has been that at times, the anticyclonic rotation was observed to be stronger than expected. Anticyclonic rotation dominated the cyclonic rotation in all of the TVC inflow configurations when the TVC inflow was facing the laboratory room entrance. Even when the TVC inflow layer was oriented to minimize the anticyclonic bias caused by the room, there were still some unexpectedly strong anticyclonic vortices with some of the shear layer profile experiments. The additional anticyclonic bias was found to be the fault of the TVC itself when a series of hot film anemometer measurements were taken in equidistant intervals along the entrance of each tier to measure the consistency of the inflow velocities. Tiers two and three had considerably higher relative inflow velocities near their left edge shown in Figure 4. A visual flow analysis using smoke tracers seemed to indicate that the airflow entering the inflow region of the middle two tiers was blocked and channeled around the midsection of each tier. The channeling observed in the middle two tiers was caused by the partial obstruction by the corners of the surrounding adjacent tiers. This effect only occurred with a veering shear layer profile. Under unidirectional shear flow, no significant net anticyclonic shear was

TABLE 2. Experiments Completed as Part of the Vortex Sensitivity Experiments Whose Hodographs Appear in Figures 5 Through 7

Experiment	Filter Arrangement	No. of Tiers	Total Helicity, $m^2 s^{-2}$	Figure No.	Remarks
1	no filters	4	0.42	5a	dominant cyclonic vortex (strongest case)
2	3, 4	4	0.50	5b	dominent cyclonic rotation (no concentrated vortex)
3	1	4	0.49	6a	dominant anticyclonic vortex
4	1, 2, 4	4	0.45	6b	dominent anticyclonic rotation (no concentrated vortex)
5	1, 4	4	0.63	6c	dominant anticyclonic vortex (weaker than experiment 3)
6	2, 3	4	0.58	7a	no observed rotation
7	2, 3, 4	4	0.63	7b	no observed rotation
8	no filters	3	0.70	5c	dominant cyclonic vortex (wider vortex than experiment 1)

The β angle is the same for all experiments. The column under the filter arrangement lists the tiers filtered, 1 (4) being the lowest (highest) tier.

introduced into the TVC as evidenced by the lack of rotation observed in the interior. Even with the anticyclonic bias, further experimentation was considered possible.

Table 2 provides a description of the shear profile experiments and the experiment with the lower aspect ratio. The hodographs of the experiments which produced dominant cyclonic rotation are shown in Figure 5. Dominant anticyclonic vortices occur with the hodographs shown in Figure 6, while the hodographs in Figure 7 occurred with no observed rotation. For each of the experiments, helicity is calculated for comparative reasons by taking 2 times the area underneath the hodograph for the entire inflow layer.

In the set of shear tests completed, concentrated cyclonic vortices were produced when the flow in the lowest two tiers was left unobstructed. Even with the negative velocity shear profile from tiers two to three shown in Figure 5b, cyclonic vortices formed. However, when the inflow velocities of tier one relative to tiers two or three were dropped, allowing the strongest horizontal anticyclonic shear to dominate the flow inside the TVC, a strong anticyclonic vortex was created. Lowering the negative shear in the TVC to the lower half of the shear layer (Figure 7) resulted in no rotation of the flow.

In summary, a slowly changing profile of velocity and horizontal vorticity throughout the inflow layer (Figures 5a and 5b) produces the strongest cyclonic vortex. However, strong velocity and vorticity in the lowest two tiers is more important than in any other set of tiers in determining the flow morphology in the TVC. Lowering the inflow velocity in the lowest two tiers as shown in Figure 6 enhances the strength of the anticyclonic vortex since the available horizontal helicity in the lower half of the inflow layer fails to match the horizontal anticyclonic shear in the middle of the inflow layer. Lowering the positive horizontal vorticity in the lowest two tiers as shown in Figure 7 results in no organized rotation. Note that the helicity for the entire inflow layer shows no significant trend in determining the strength or type of rotation observed in the TVC.

Another part to the experiments listed in Table 2 was the lowering of the aspect ratio of the TVC in the hopes that there would be a transition from single cell to multiple vortex. The same unfiltered, half-sectional inflow that produced the strongest cyclonic vortex (Figure 5a) was also used in the lower aspect ratio test. Lowering the aspect ratio also meant lowering the volume of the inflow area since one tier was removed. As a result, inflow velocities increased along with the observed helicity (Figure 5b). Two effects on the vortex were noted when the aspect ratio was lowered. First the vortex diameter appeared larger in the low aspect ratio case. Second, the vortex exhibited many disruptions, turbulent breakdowns (not related to the breakdown bubble), and more rapid cycloidal motions. *Davies-Jones* [1973] observed that increases in vortex diameter occurred with increasing swirl ratios. A similar effect may be happening here. As for the second effect, the sudden increase of the inflow shear from $1.3 s^{-1}$ to $2.2 s^{-1}$ while holding the updraft strength constant may have led to a more turbulent inflow leading to a less organized vortex. No sign of multiple vortices was observed.

The aspect ratio threshold needed for vortex breakdown may need to be much smaller in a helical TVC than in the standard Ward TVC. This is an expected departure from axisymmetric simulators where radially consistent angular momentum produces circulations occupying the entire area of the updraft core. A helical flow TVC, where no net vertical angular momentum is introduced into the inflow, implies that there are equal amounts of positive and negative angular momentum flux. As a result, in the present experimentation, the radius of the circulation is only 1/3 that of r_0, which means the aspect ratio of the disturbance is 3 times higher than the TVC aspect ratio.

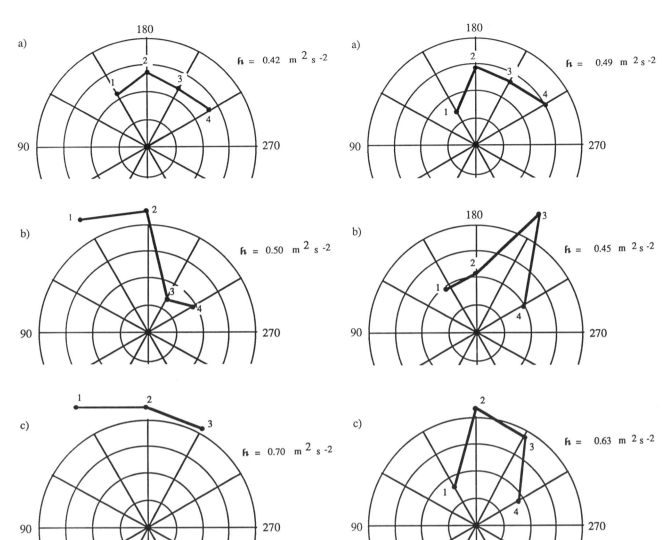

Fig. 5. Hodographs corresponding to (a) half-sectional unfiltered inflow, (b) half-sectional inflow single air filters on tiers 3 and 4, and (c) half-sectional unfiltered inflow with three tiers. The values h correspond to the helicity over the inflow layer depth. Labels upon the hodograph correspond to the tier number. The directional labels on the major axis of the hodograph are merely used as a convenient frame of reference. The maximum radius of each hodograph is 1.0 m s^{-1}.

Fig. 6. Similar to Figure 5 except (a) single filter on tier 1, (b) single filters on tiers 1, 2, and 4, and (c) single filters on tiers 1 and 4.

With both the high and low aspect ratio flow fields within the TVC, there are many similarities between the present TVC inflow version and that of *Rothfusz* [1985] and *Rothfusz and Lilly* [1989]. The flow on the right side of the vortex as seen in Figures 8a and 9a reacts in a similar way to the flow on the right side of the Rothfusz vortex by becoming drawn into the vortex at ground level. On the left side of the vortex, the flow turns anticyclonically. However, as in the Rothfusz case, the surface air flow is forcefully lifted by a convergence boundary formed by the opposing flow and never becomes entrained into the cyclonic circulation. The photograph in Figure 8b shows the cyclonic vortex to the left and a deep layer of smoke on the right associated with the lifted anticyclonic flow (labeled A).

Significant differences in the internal flow field between the present version and the Rothfusz version of the TVC also exist. Since the most productive cyclonic vortices come from half-sectional inflow (Rothfusz used a full-sectional inflow), there is a single region of anticyclonic flow (Figure 9a), whereas the Rothfusz version has two regions, one on each side of the cyclonic vortex (Figure 9b). The source of the surface convergence line in the present version is the high stagnation pressure as the flow meets the opposite wall while the opposing airstreams in the Rothfusz version force the surface convergence. One unique feature of the present TVC is the existence of a closed anticyclonic circulation left

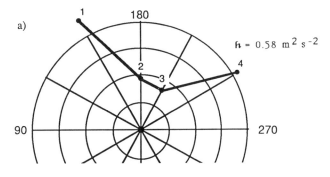

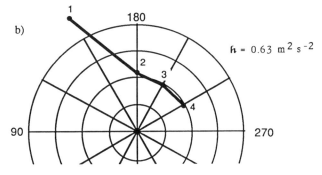

Fig. 7. Similar to Figure 5 except (a) single filters on tiers 2 and 3 and (b) single filters on tiers 2, 3, and 4.

of the cyclonic vortex with respect to the tier one inflow vector. This anticyclonic circulation may be enhanced by the anticyclonic bias of the TVC. Another feature unique to the present TVC is the evidence of a downdraft on the rear side of the cyclonic vortex revealed by the observation of smoke tracers. The downward moving air is observed to reach the floor to the left of the vortex and then become part of the opposing flow which forms the convergence boundary. There are no such observations of downward air motions near the vortex in the Rothfusz version of the TVC [Rothfusz, 1985].

In the relatively steady state of the vortex (well after vortex initiation), the source of the vertical vorticity at the lowest levels also appears similar to the scenario proposed by Rothfusz and Lilly [1989]. The nearly linear flow with positive angular momentum is suppressed and allowed to feed rotation to the vortex while the negative angular momentum is forcefully lifted out and away from the vortex by the strong convergence line. Vertical vorticity is also supplied by the tilting of boundary layer vortex lines contained in the flow reaching the vortex.

Now the question remains whether the simulated vortex flow structure is analogous to established observations and numerical simulations of the flow morphology in a typical tornadic supercell. A numerical simulation schematic from Klemp and Rotunno [1982] and a dual-Doppler analysis by Brandes [1984] of the low-level flow around a tornadic supercell are shown in Figures 10a and 10b, respectively, for a basis of comparison of the observed surface-simulated vortex. Several similarities are apparent upon inspection (the surface inflow vector is used as a frame of reference unless otherwise noted), including (1) the presence of a rear flank gust front left of the surface cyclonic vortex labeled RF, (2) a region of positive (negative) angular momentum to the right (left) of the vortex where only the flow on the right side is entrained into the vortex, (3) anticyclonic curvature left of the vortex, maximized on the left side of the rear flank gust front (labeled A), and (4) kidney-shaped updraft conforming to the shape of the rear flank gust front.

The presence of the anticyclonic circulation in Figure 8a would seem to be a manifestation of the anticyclonic bias discussed earlier. However, there were observations that indicated more than the anticyclonic bias was responsible for this circulation. First, the experimentation showed that no rotation was observed in the unidirectional flow control test. Such evidence suggests that under directional shear conditions, an underlying region of anticyclonic flow would be necessary in order for the anticyclonic bias to produce an observable circulation center. Second, the cyclonic and anticyclonic circulation centers always remained in the same relative positions to each other. This fact held true even when the anticyclonic was stronger than the cyclonic vortex. The anticyclonic rotation always remained left of the cyclonic vortex. Then, as mentioned before, anticyclonic flow was observed by dual-Doppler radar (labeled A in Figure 10b) by Brandes [1984] in this same region.

Both visual observations (right side of the photograph in Figure 8b) inside the TVC and inspection of Figure 10b indicate that this anticyclonic flow is located on the edge of a region of positive vertical velocity. This is the region relative to the cyclonic tornado where anticyclonic tornadoes have also been observed to form by Brown and Knupp [1980]. It is possible that the anticyclonic rotation observed is due to the stretching of the anticyclonic vorticity by a strong gradient of positive vertical velocity found along the tail end of the rear flank gust front. The anticyclonic bias inherent in the present TVC acts to enhance the strength of the circulation.

4. DISCUSSION

A number of performance tests were made upon a newly reconfigured helical flow TVC consisting of three and four independently movable inflow boxes. This reconfiguration of the inflow layer was considered to be an improvement over the circular guide vane inflow of the Rothfusz TVC version in that no negative helical flow regions would be introduced into the TVC from the inflow layer.

The results were rather mixed because of the inadvertent horizontal anticyclonic shear that was introduced into the TVC inflow box construction and from the laboratory room. Even with the anticyclonic bias, meaningful results have been achieved with the present TVC. A few important results are summarized below:

1. Strong vortices develop only when there exists a convergence layer as deep as the inflow layer. The half-sectional

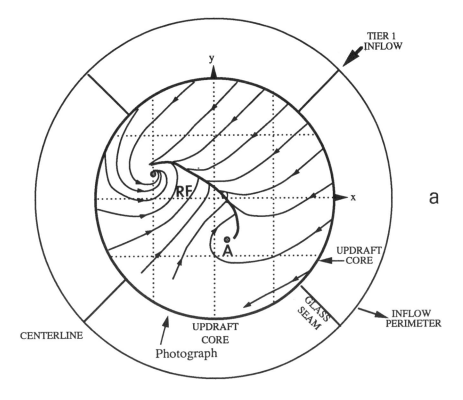

Fig. 8. (a) Surface streamline analysis of the flow structure corresponding to the hodograph in Figure 5a. The thick line corresponds to the convergence boundary and the small stippled circles represent the average location of the rotation centers. The label RF marks the rear flank boundary, while A marks the anticyclonic rotation. (b) Accompanying photograph taken from the TVC exterior at a location labeled "photograph" in Figure 8a. The end-wall vortex can be seen to the left. The label A marks the region over the anticyclonic rotation shown in Figure 8a.

inflow layer allowed the opposing wall to the inflow layer to create a stagnation region and a deep convergence boundary.

2. The shear profile experiments suggested that the velocity and vorticity of the lowest two tiers influenced the vortex morphology more than any other layer. Additionally, a slowly varying profile of velocity and vorticity produced the strongest cyclonic vortex, not just high helicity.

3. The TVC vortex morphology showed many similari-

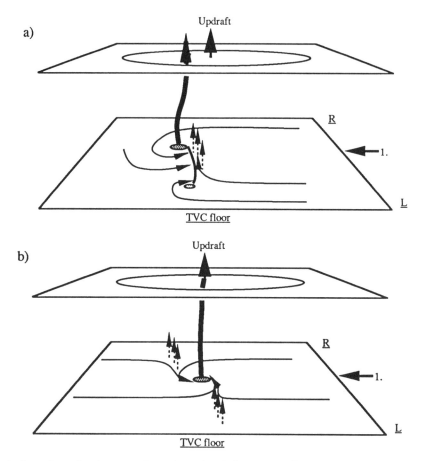

Fig. 9. (a) A three-dimensional perspective schematic of the current flow morphology corresponding to the hodograph in Figure 5a. The thickest line indicates the primary vortex location, while the line bridging the two vortex centers marked by the two stippled circles indicates a convergence boundary. The dotted arrows indicate strong updraft regions outside vortex centers. (b) Similar to Figure 9a except valid for the Rothfusz full-sectional inflow case.

ties to observational and numerical studies of classical tornadic supercells including a rear flank gust front and the presence of anticyclonic curvature in a region known to produce real anticyclonic tornadoes.

4. The dynamics maintaining the vortex are similar to the dynamics of the Rothfusz TVC. Flow on the right side of the vortex containing positive angular momentum was allowed to reach the vortex while the negative angular momentum flow was removed vertically by the rear flank gust front. The flow reaching the vortex also allows a continuous supply of boundary layer vorticity to be tilted up into the updraft, as is the case with axisymmetric Ward-type simulators.

The capability of the present TVC to produce flow structure similar to that of a tornadic supercell indicates that further investigations will be useful; however, the inflow layer of a future helical flow TVC will have to be redesigned to improve the homogeneity of the vertical shear profile. First, the entrance to each inflow box must be unobstructed, either from the exterior room geometry or from adjacent boxes. Currently, the corners of the tiers obstruct the inflow to the middle of adjacent tiers creating the anticyclonic bias.

Second, the width of the interaction zone labeled in Figure 1 will have to be increased such that the diffusion of intertier boundary layer vorticity results in more complete smoothing of the inflow layer vorticity profile at the edge of the updraft hole. Observations now indicate that there are regions of high and low horizontal vorticity as the flow passes inside of the interaction zone.

A future helical flow TVC will also need the capability of having independently controllable inflow and outflow flow velocities without changing the aspect ratio or the use of extra fan assemblies in the inflow layer. Increasing the updraft in the present inflow layer configuration leads to increasing the shear layer velocity when the aspect ratio is held constant. Similarly, decreasing the aspect ratio reduces the inflow volume and thereby increases the inflow velocity given the same updraft velocity. The simplest way of keeping the updraft velocity, the aspect ratio, and the inflow velocity independent from each other will be to have inflow layer tiers with adjustable widths (therefore an adjustable inflow volume). In such a configuration, the aspect ratio will be able to be lowered without a corresponding change in the

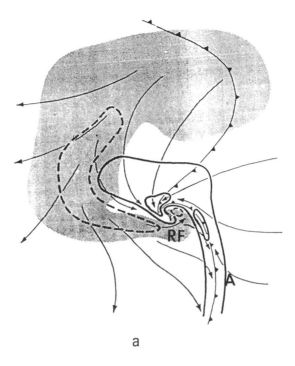

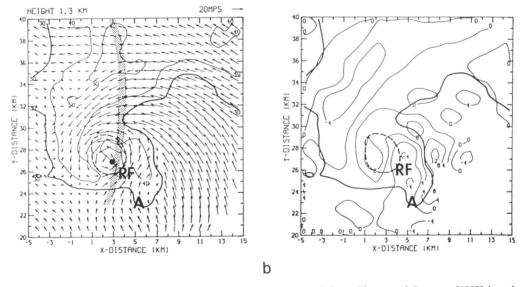

Fig. 10. (a) A schematic of the surface flow in a tornadic supercell from *Klemp and Rotunno* [1982] based on numerical simulations. The solid (dashed) lines represent positive (negative) vertical velocity. Flow arrows represent storm relative streamlines, and the shaded region represents the surface rainwater field. A vertical vorticity maximum is labeled with a small T. (b) Low-level wind field of a tornadic thunderstorm near Del City, Oklahoma, from a dual-Doppler analysis. Left panel shows storm relative horizontal wind with contours of radar reflectivity (in dBz). Right panel shows vertical velocity (in meters per second). The heavy dashed line represents the mesocyclone [from *Brandes*, 1984]. The labels RF and A mark similar regions described in Figure 8.

ratio of the inflow layer to the updraft velocity. That means the inflow layer helicity could be held constant in an experiment of changing updraft velocity or aspect ratio. Such an increase in flow control could help determine if a helical flow TVC can produce multiple vortices.

Acknowledgments. Throughout this work, I greatly appreciate the patience and help of my immediate adviser, Eugene Wilkins, for his insight and support. Lans Rothfusz helped me in getting familiar with the laboratory equipment, and Meta Sienkiewicz helped with the myriad of computer-related problems. I am also grateful to Sue Weygandt for helping with the figures. This work was conducted under National Science Foundation grant ATM 8914434.

REFERENCES

Brandes, E. A., Vertical vorticity generation and mesocyclone sustenance in tornadic thunderstorms: The observational evidence, *Mon. Weather Rev.*, *112*, 2253–2269, 1984.

Brown, J. M., and K. R. Knupp, The Iowa cyclonic-anticyclonic tornado and its parent thunderstorm, *Mon. Weather Rev.*, *108*, 1626–1646, 1980.

Church, C. R., J. T. Snow, G. L. Baker, and E. M. Agee, Characteristics of tornado-like vortices as a function of swirl ratio: A laboratory investigation, *J. Atmos. Sci.*, *36*, 1755–1776, 1979.

Davies-Jones, R. P., The dependence of core radius on swirl ratio in a tornado simulator, *J. Atmos. Sci.*, *30*, 1427–1430, 1973.

Davies-Jones, R. P., Laboratory simulation of tornadoes, in *Proceedings of the Symposium on Tornadoes: Assessment of Knowledge and Implications for Man*, pp. 151–174, American Meteorological Society, Boston, Mass., 1976.

Davies-Jones, R. P., Streamwise vorticity: The origin of rotation in supercell storms, *J. Atmos. Sci.*, *41*, 2991–3006, 1984.

Klemp, J. B., and R. Rotunno, A study of the tornadic region within a supercell thunderstorm, *J. Atmos. Sci.*, *40*, 359–377, 1982.

Lilly, D. K., On the structure, energetics and propagation of rotating storms, II, Helicity and storm stabilization, *J. Atmos. Sci.*, *43*, 113–125, 1986.

Rothfusz, L. P., A mesocyclone and tornado-like vortex generated by the tilting of horizontal vorticity: A laboratory simulation, M.S. thesis, 106 pp., Sch. of Meteorol., Univ. of Okla., Norman, 1985.

Rothfusz, L. P., and D. K. Lilly, Quantitative and theoretical analysis of an experimental helical vortex, *J. Atmos. Sci.*, *46*, 2265–2279, 1989.

Ward, N. B., The exploration of certain features of tornado dynamics using a laboratory model, *J. Atmos. Sci.*, *29*, 1194–1204, 1972.

Discussion

GENE WILKINS, SESSION CHAIR

University of Oklahoma

PAPER E1

Presenter, Chris Church, University of Miami, Ohio [*Church and Snow*, this volume, Laboratory models of tornadoes]

(Brian Fiedler, University of Oklahoma) When Rich [Rotunno] and I estimated that the maximum vertical velocity would be roughly ten times the average external [vertical] velocity [*J. Atmos. Sci.*, *43*, p. 2337, 1986], we did not intend that to be applied to thunderstorms because of the different outflow conditions; that really only applies to the Purdue simulator. We think that ratio is too high in the simulator. In my paper [A4], that ratio was 2, not 10. Could you comment on your new simulator, where there is divergence aloft instead of the fan sucking the air out vertically? What is the ratio of maximum vertical velocity to mean updraft velocity in that simulator? Is it the same as in the Purdue chamber?

(Church) I haven't measured it.

(Fiedler) Do you have the same Δp^* [nondimensional central pressure deficits] as in the Purdue chamber?

(Church) They are very similar. But you must appreciate that it takes a long time to make these measurements, and one does not routinely compare one's own chamber with other ones by repeating experiments. So I don't have very many substantive measurements for comparison.

(Bruce Morton, Monash University) I couldn't really pass over that challenge to me. It is quite simple. The set of vanes are oriented so that the vanes are at a constant angle to the radials. When the flow is turned on, each vane sheds a starting vortex, which has positive vorticity, and there is negative circulation around the vane. Thereafter, the negative circulation continues around the vane and the contribution to positive circulation continues in the center. Positive vorticity is generated on the outer side of the vane, negative vorticity on the inner side, at equal rates. So the circulation around the whole set of vanes is zero, as it ought to be because there is negative circulation around each vane and positive circulation at the center. But the circulation around a circuit inside the vanes is positive, of course; otherwise, there wouldn't be a vortex.

(Church) In the Rothfusz experiments that I was referring to, the vanes are set up in a very different way. The inflow layer was divided into three layers. In the lowest layer, the vanes are all parallel, e.g., oriented in the north-south direction. The middle and highest layers are similar, but the direction of the vanes veers with height between the layers, e.g., NE-SW in the middle layer and E-W in the top layer. Each layer produces no circulation. Nevertheless, vortices form. Basically, it is helical flow, but I don't know why concentrated vortices are produced.

PAPER E2

Presenter, Don Lund, Purdue University [*Lund and Snow*, this volume, Laser Doppler velocimeter measurements in tornadolike vortices]

(Brian Fiedler, University of Oklahoma) The question keeps coming up of whether or not the ratio of maximum tornado wind speed to thermodynamic speed limit still applies when the lower boundary layer is turbulent. Could an experiment be performed, using a rough lower surface, to see if axial jets in the core still occur with a turbulent boundary layer?

(Lund) Snow did some work with Wayne Rostek [see *Preprints*, 14th Conf. Severe Local Storms, 1985] using roughness elements in the chamber. Are you familiar with this work?

(Fiedler) Did he find axial jets in the core?

(Lund) Unfortunately, I am not familiar with it. LDV [laser Doppler velocimeter] work with roughness elements in the chamber would be problematic because of having to see through the roughness elements.

(Rich Rotunno, National Center for Atmospheric Research) Aren't the vortex wander and the helicoidal waves a disappointment? I thought that the motivation for the new apparatus was to get a vortex that was stable and vertical. Are the vortex wander and the helicoidal wave on the core different or the same phenomena?

(Lund) I would like to see if the apparent helicoidal structure is repeatable. Regarding the wander, the data in Figure 2, which are averages of lots of samples, look very symmetric about the origin. So there is still a question whether wander is really present or whether there is a steady state offset in the vortex position [from the axis of the apparatus]. I haven't determined that yet.

(Rotunno) If the bottom part of the vortex wandered and the top part didn't, there would be a helical structure. So the wander and the helical structure would be connected.

Paper E3

Presenter, Jim LaDue, University of Oklahoma [*LaDue*, this volume, Vortex formation from a helical inflow tornado vortex simulator]

(Bob Walko, Colorado State University) In your successful experiment, for each tier of ducts, did you let air flow in from opposite sides of the apparatus, or did you let air in on one side and draw air out on the opposite side?

(LaDue) No, we didn't draw air out. We created a stagnation pressure on the opposite side from the inflow, so that we produced a convergence layer that was as deep as the inflow layer and that was steady underneath the updraft.

(Walko) What was the condition at the opposite side from the side where most of the flow was coming in?

(LaDue) At each tier level, flow would come in from one side, the opposite side was blocked. There was no air passing in or out from the opposite side, the air exited upwards.

(Chair) Bob Walko, by the way, is responsible for much of the construction of this apparatus during the time he was at the University of Oklahoma.

(Rich Rotunno, National Center for Atmospheric Research) I have a comment. It seems to me that it is more difficult to explain what's going on in that experiment than to explain the tornado.

(LaDue) That's true. There are too many boundaries and walls. But there is some significance to our experiments because there was no organized rotation when all the inflow was from one direction or when the inflow speeds decreased markedly with height. Organized rotation occurred only for favorable shears.

(Chair) Jim has tried different speed and directional shears, and has drawn hodographs, so that the simulator results may be interpreted in terms of hodograph shape. He has much more information.

(Bruce Morton, Monash University) I don't yet fully understand what is happening. Is the mean inflow velocity the same or different in the successive layers?

(LaDue) It depends on what experiment we're doing, but . . .

(Morton) But you should always measure that so that you know what is happening. You can try to explain the vortex without worrying about all the boundaries until later. But I believe that it is essential to know how the inflow profile varies from layer to layer.

(LaDue) That's what the hodographs try to show, but they are not totally accurate. The velocity varies from left to right in each tier, and sometimes this leads to vorticity of the wrong sign or unwanted vorticity. The tiers have to be constructed very carefully to ensure that the velocity is more or less uniform in each tier.

(Chair) Jim has tried many different vertical profiles of velocity. He doesn't have time to present all of them.

A Review of Tornado Observations

HOWARD B. BLUESTEIN

School of Meteorology, University of Oklahoma, Norman, Oklahoma 73019

JOSEPH H. GOLDEN[1]

National Oceanic and Atmospheric Administration, Office of the Chief Scientist, Washington, D. C. 20235

1. INTRODUCTION

During the past 15 years following Tornado Symposium II, our understanding of the basic morphology of tornadoes and their parent storms has advanced significantly. The fragmentary observations obtained by the National Severe Storms Laboratory (NSSL)–organized Tornado Intercept Project [*Golden and Morgan*, 1972; *Davies-Jones*, 1988] have evolved to a three-dimensional, time-variant conceptual model. There has been a synthesis of more sophisticated (i.e., well trained) eyewitness and storm-chaser observations, including mobile soundings and various new portable instrumentation with fixed Doppler radar data.

Our ideas concerning tornadogenesis mechanisms have tended to focus on the wealth of observations taken over the southern Great Plains but have also evolved with increasing field experiments in diverse locations, such as Alabama, Florida, Colorado, and the U.S. West Coast. Indeed, there has been a reexamination of how we should define a tornado [*Forbes and Wakimoto*, 1983] in relation to the primary source of vorticity in the parent storm system. Mechanisms involving features other than the mesocyclone (visualized by the presence of a "wall cloud" or other rotating cloud base) appear to be operative in certain situations, e.g., along gust fronts and other types of convergence zones or boundaries. We shall attempt here to summarize these new findings regarding tornado observations but shall exclude climatology and fixed Doppler radar observations of storm-scale features, which are addressed elsewhere in this volume.

---

[1]Now at NOAA, Oceanic and Atmospheric Research, Silver Spring, Maryland.

The Tornado: Its Structure, Dynamics, Prediction, and Hazards.
Geophysical Monograph 79
Copyright 1993 by the American Geophysical Union.

2. VISUAL OBSERVATIONS

Many of the early visual eyewitness accounts of tornadoes were well summarized by *Flora* [1953]. In fact, the difficulty of Flora's accounts is that some of them were obviously made under great psychological stress by observers and must, therefore, be viewed with skepticism. One is struck by the vivid account of one eyewitness who survived the overhead passage of a large tornado and was able to peer upward into the heart of the funnel cloud from his basement as the remainder of the house was disintegrated and blown away. This observer, and several others in similar encounters, reported seeing intense electrical discharges inside the funnel walls. These accounts, while fascinating, have not been replicated by trained storm chase teams over the past two decades; moreover, they fed controversy regarding the possible role of magnetic anomalies and/or lightning in tornadogenesis and waterspouts which persists to this day [e.g., *Davies-Jones and Golden*, 1975a, b, c; *Colgate*, 1975; *Rossow*, 1969, 1970; *Vonnegut*, 1960]. These eyewitness credibility problems are also addressed by *Grazulis* [this volume] in assessing the 100-year climatological record of tornadoes in this country. Nevertheless, technological advances have occurred since Tornado Symposium II that have permitted more comprehensive studies of intracloud and cloud-to-ground lightning in tornadic storms [*MacGorman*, this volume; *MacGorman and Nielsen*, 1991]. Arguably, the single most comprehensive set of eyewitness photographs and accounts were those assembled and diagnosed in a study by *Fujita* [1960] on the June 20, 1957, Fargo, North Dakota, tornadoes.

2.1. NSSL Tornado Intercept Project

The advent of the NSSL Tornado Intercept Project in the early 1970s [*Golden and Morgan*, 1972] led to confirmation of Fujita's earlier Fargo model in a number of cases. One of the best photographed tornadic storms in the early days of

Fig. 1. Same rotating storm as Plate 1, looking NW-NNE from light aircraft. Note large tilt of bell-shaped cloud toward the right and tornado pendant from flanking line merging, with relatively shallow cloud tops, into main updraft. (Courtesy of D. Ray Booker.)

the Intercept Project occurred 1 hour before sunset on April 30, 1972, in extreme western Oklahoma. The small tornado that it produced lasted only a few minutes (Plate 1) and remained over open farm country [*Golden and Morgan*, 1972]. However, several weeks later it was discovered that the same visually rotating, bell-shaped cumulonimbus was photographed by two aircraft flying at different altitudes (just below cloud base (Figure 1) and by a U-2 flying at 18 km (Figure 2)). Unfortunately, NSSL Doppler data were not available that day, and the storm would have been beyond the effective range anyway. Conventional WSR-57 radar reflectivities depicted a very large, intense (radar reflectivity in excess of 60 dBz) storm, with a possible hook echo at the time and location of the tornado.

The Union City, Oklahoma, tornado of May 24, 1973, has to date provided the best overall data set for synthesizing Doppler radar observations with the evolving three-dimensional visual storm and tornado structures. The storm morphology and strategy used to intercept it on the ground are given by *Golden* [1976]. The Union City case led to the development of a life cycle hypothesis for tornadoes and is illustrated in Figure 3, after *Golden and Purcell* [1978a]. Finally, *Golden and Purcell* [1978b] presented photogrammetric wind speed estimates derived from high-quality films of the Union City tornado. These results clearly indicated that even though in later stages of its life cycle the tornado shrinks, it retains destructive wind speeds (up to 65 m s^{-1}).

2.2. Tornadoes and Rough Terrain

Plates 2a, 2b, and 2c show typical examples of tornadoes observed in very high mountainous terrain, elevated mountain valleys, and at the foot of Pikes Peak. The latter produced F1/F2 damage in Manitou Springs, Colorado, even though the visible funnel never extended more than halfway to the ground. This is a typical feature of tornadoes in mountainous or elevated terrain, and we believe that it is primarily due to somewhat drier environments in the low levels [*Szoke et al.*, 1984].

While destructive tornadoes appear to be still rare in mountainous terrain, *Fujita* [1989] documented a very large and >100-km path length tornado that devastated a heavily forested area at over 3 km above mean sea level (MSL) from NE Utah and climbed the Continental Divide into the Grand Teton National Park, Wyoming, before weakening. *Fujita* [1974] found that a few of the tornadoes in the 1974 Jumbo Outbreak developed in valleys and climbed substantial mountain ridges of up to 0.5-km elevation, maintaining continuous destructive paths on both sides of the mountain.

There continues to be some controversy regarding the possibility of strong tornadoes affecting urban areas. *Dessens* [1972] was the first to suggest that tornado formation could be affected by surface roughness, and he based this inference on laboratory simulations. *Blechman* [1975] studied the formation of a multivortex tornado and found good

Fig. 2. Photo of same rotating storm as Plate 1 and Figure 1 taken from overflying U-2 jet about 1 hour earlier from 18-km altitude. Note overshooting cloud tops, striated midlevel cloud bands, and flanking line entering main cumulonimbus from center left. View NE-E. (Courtesy of J. T. Lee.)

correlations of suction vortex formation with increasing surface roughness. Later, *Elsom and Meaden* [1982] used long-term storm records over the greater London area to infer that the urban "heat island effect," coupled with greater surface roughness, accounted for the minimum of "weak" tornadoes in the climatic record over London. Nevertheless, significant tornadoes have been documented in the Lubbock, Texas, Kalamazoo, Michigan, and Denver, Colorado, urban areas. (See *Szoke et al.* [1984] and *Szoke and Rotunno* [this volume] for the Denver area cases.)

2.3. A Conceptual Model for Great Plains Supercell Tornadoes

We summarize in Plate 3 the overall structure of a tornado-spawning supercell over the southern Great Plains, as seen by an observer positioned to the southeast of the storm. This is meant to be a conceptual model and will not apply to most tornadic storms in the High Plains of Colorado and Wyoming and in the southeast U.S. Note that many of the cloud features first documented by Fujita in the Fargo tornado appear in this model, such as the rotating, precipitation-free base of the wall cloud, which may be associated with one or more pendant tornado funnels during its lifetime. In addition, smaller, short-lived tornadoes may form beneath the flanking line to the left of the wall cloud in Plate 4.

There may also be a "tail cloud" and a "feeder band" spiraling into the wall cloud from the northeast. Doppler analyses presented elsewhere in this volume indicate that the "wall cloud" is usually the cloud base reflection of an organizing mesocyclone aloft within the supercell storm. Rarely, the mesocyclone becomes so intense that the wall cloud itself lowers to the ground, although this was the case in the Binger, Oklahoma, tornado studied by *Zrnic et al.* [1985].

Variants of the conceptual model in Plate 4 exist for other parts of the country but have not yet been documented in the literature. In fact, there is still some lively debate about the conjectured preeminent role of supercells in tornadogenesis (see *Burgess and Lemon* [1990] and *Golden et al.* [1990] for detailed arguments). More recently, there have been low-precipitation (LP) and high-precipitation (HP) supercell storms documented by chase teams (see, for example, *Bluestein and Woodall* [1990]); generally only the latter are important tornado producers, although several cases of LP storms with tornadoes have been documented by chase teams recently over eastern Colorado, Wyoming, and Texas (E. W. McCaul and D. Blanchard, personal communication, 1990). On the other hand, tornadoes in the southeast U.S. often occur late at night and are frequently ensheathed by heavy rain curtains (e.g., Huntsville, Alabama [*Goodman*

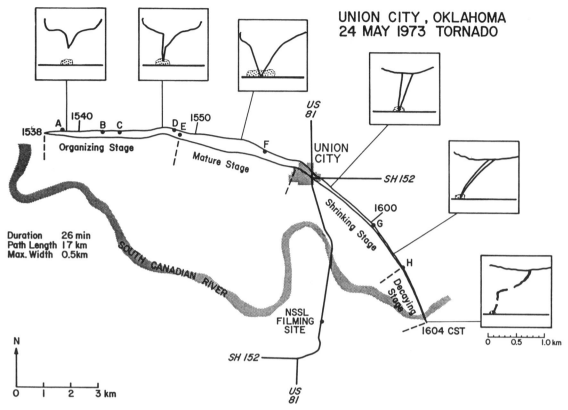

Fig. 3. Synthesis of the May 24, 1973, tornado damage track through Union City, Oklahoma, with insets depicting tornado morphological changes with life cycle stages along track (After *Golden and Purcell* [1978b].)

and Knupp, 1990; McCaul, 1987]). Wall clouds are occasionally observed with these tornadoes but may exhibit little or no rotation in early stages.

3. WATERSPOUTS AND THEIR LINK TO TORNADOES

Our understanding of waterspout structure, wind speeds, and behavior prior to the late 1960s was based largely upon mariner's tales of chance encounters at sea [e.g., *Hurd*, 1950; *Gordon*, 1951]. That waterspouts pose a threat to structures situated near their paths is beyond doubt [*Golden*, 1973; *Fujita et al.*, 1972; *Macky*, 1953]. Many of the damaging tornadoes affecting the central and eastern Gulf Coast during the late fall and early spring originate over the northern Gulf of Mexico as intense waterspouts. Some of these do not have hook echoes or other classic, distinguishing characteristics on radar. Moreover, many of the tornadoes along the U.S. West Coast originate offshore as waterspouts in the presence of an intense upper cold low. *Golden* [1973] noted that these events tend to be small and short lived, with minor damage.

3.1. *The Waterspout Life Cycle*

During the late 1960s, *Rossow* [1970] used an instrumented Navy Grumman S2E aircraft to study waterspout developments in the Florida Keys. He wanted to test the hypothesis that waterspouts and their land-based cousins, tornadoes, are electrically driven. However, he found from several sets of measurements that "atmospheric vortices over water can exist without an electric field or current of appreciable magnitude." Lightning strokes were indicated by the instruments while near waterspouts on four occasions, but in only one instance could the strokes have been in the parent cloud containing the waterspout.

Golden [1974a, b] followed up on Rossow's pioneering studies and was able to document a repetitive waterspout life cycle, from close-range aircraft observations of over 100 events in the 1969 summer season over the lower Florida Keys. *Church et al.* [1973] developed a trailing-wire probe which they towed through Florida Keys waterspouts from a light aircraft. However, the complicated interactions between the probe itself and the flow field in and around the waterspout funnel made reduction of the pressure and temperature data extremely difficult. The double-peaked temperature anomalies around the waterspout core appear to be systematically warm, as depicted in Figure 4, a three-dimensional conceptual model based on 3 years of aircraft observations and tracer experiments [*Golden*, 1971, 1974a].

During the latter part of the 1970s, further advances were made on our knowledge of the internal flow and thermal features of waterspouts. *Leverson et al.* [1977] used an AT-6 acrobatic aircraft equipped with a gust probe to penetrate waterspouts just below cloud base. Their data and findings (based on a limited sample of small- to medium-sized funnels in the Florida Keys) show that in the waterspout funnel the following features exist:

Waterspout Vortex Structure Near Cloud Base

1. A core of rising motion of ~5–10 m s^{-1}.
2. A horizontal circular flow field (helical upward) that is broader in weaker waterspouts and spatially concentrated in more intense waterspouts.
3. A central core region of ~0.3°K warmer than ambient temperatures.
4. A core pressure deficit of the order of ~1–10 mbar, depending on waterspout intensity.

One example of the data *Leverson et al.* [1977] obtained on tangential and vertical wind components, temperature, and pressure anomalies is given in Figure 5. Note that all of the cases sampled by the AT-6 were relatively small waterspouts. No data were obtained above parent cloud base or within the waterspout's spray vortex. An important new finding was the asymmetric structure of the waterspout's circulation around the funnel (noted in the spray vortex of an intense anticyclonic waterspout by *Golden* [1974a]) and its large lateral radius of influence. Large upward vertical velocities peak just outside the visible funnel, with likely weak descent in the core.

Schwiesow [1981] and *Schwiesow et al.* [1981] used an airborne infrared Doppler lidar to probe waterspout features in the subcloud layer. Two of the more interesting radial velocity power spectra obtained through waterspout funnels are shown in Figures 6a and 6b. We note the relatively close

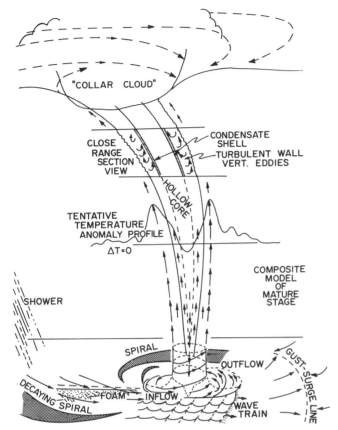

Fig. 4. Composited conceptual model of mature stage of waterspout life cycle [after *Golden*, 1971, 1974a, c]. Vertical scale greatly exaggerated.

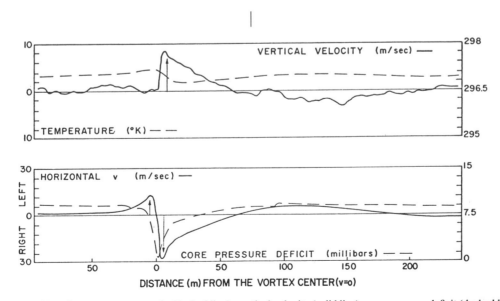

Fig. 5. Profiles of temperature anomaly (dashed line), vertical velocity (solid line), core pressure deficit (dashed line), and tangential velocity V_r (solid line) for AT-6 penetration through an anticyclonic waterspout. Aircraft altitude was about 645 m and funnel diameter at penetration time was about 15–20 m (midsized case). (After *Leverson et al.* [1977].)

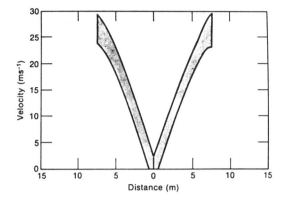

a

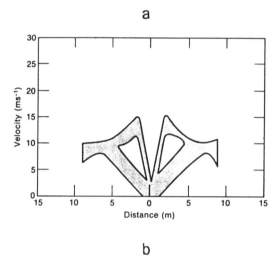

b

Fig. 6. (a) Waterspout velocity spectra measured by airborne Doppler lidar at midlevel (460 m), after *Schwiesow* [1981]. The simple, early solid rotation velocity structure shows the highest tangential velocity observed in the experiment and the usual transition to reduced angular velocity near the edge of the condensation funnel. The funnel core diameter was approximately half that near cloud base. Translation inflow was 4 m s^{-1}. (b) Same as Figure 6a except for a different waterspout at an altitude of 675 m. This was a double-walled funnel with sufficient condensation to mark the flow between and outside the two major shells.

fit of the first, most intense example to a classical Rankine-combined profile and the double peak in velocity of the second (a rare, double-walled funnel).

Holle and Maier [1980] reported a tornado over the Florida peninsula which was associated with invigorated convection as two gust fronts intersected. *Simpson et al.* [1985] reported on an investigation of six waterspout events documented during the 1974 GATE project. Triggering of intensification to funnel stage appears in some cases to be effected by intersecting convergent features, with an invigorated cumulus tower overhead.

Comparing tropical with severe continental funnel events, *Simpson et al.* [1985, 1986] noted that there is now some evidence that the parent clouds have some important features in common. The severe midwestern events have preconditioning processes that are mechanistically different, stronger, and larger in scale and extend through the depth of the troposphere, as commonly do the parent clouds. *Simpson et al.* [1991] extended this work to a well-documented anticyclonic waterspout over the Great Salt Lake. The cloud-scale modeling study confirmed earlier GATE results that a superadiabatic lapse rate, together with moderate shear, needs to be collocated in the layer in which the updraft increases upward, in order for strong cumulus-scale vortices to have maximum intensities at the lowest levels. However, these ingredients had different root causes in the Great Salt Lake case and GATE waterspouts. A high-resolution axisymmetric vortex model [*Howells and Smith*, 1983] was used to investigate whether an intense vortex could develop within a 5- to 15-min interval, when the model cumulus vortices were used as initial conditions [*Dietachmeyer*, 1987]. *Simpson et al.* [1991] found that with the strong Salt Lake cumulus vortex, an intense vortex with maximum tangential velocity of 35 m s^{-1} developed in nearly all the experiments they ran.

3.2. Waterspout Linkages to Tornadoes

Golden [1971, 1973, 1974b, c] has presented evidence to show that larger waterspouts have many similarities to weak-to-moderate tornadoes. *Golden* [1971] documented a large tornadic waterspout which was spawned by a line of thunderstorms and pendant from a "wall cloud." In addition, several intense waterspouts have been observed that moved onshore, or crossed over islands, inflicting considerable damage [*Maier and Brandli*, 1973; *Golden and Sabones*, 1991]. Both types of convective vortices have preexisting cloud-scale circulations, which if large enough and close enough to a weather radar may appear as a hook echo or "spike" in the case of large waterspouts.

The first case of a Doppler-observed mesocyclone associated with a large, damaging waterspout in the Cape Kennedy, Florida, area was documented by *Golden and Sabones* [1991]. In that case, the mesocyclone was only detected in the lower levels of the parent thunderstorm, at the southwest end of a line. The case of preexisting mesocyclones for at least the supercell-spawned class of tornadoes is compelling; *Golden* [1974c] has used ground-based time lapse photography to document precedent circulations at cloud base which subsequently were associated with waterspout formation.

During the period since the last Tornado Symposium, there have been increasing numbers of tornadoes documented near and east of the Rocky Mountains. In fact, tornado incidence over the state of Colorado has increased more than 100% in the decade of the 1980s as compared with the 1970s (F. Ostby, personal communication, 1990). Many of the increased sightings have occurred along the Front Range corridor of the Rockies from Denver north and northeast in conjunction with field experiments conducted by Programs for Regional Observing and Forecasting Services (PROFS) and the National Center for Atmospheric Research (NCAR).

Zipser and Golden [1979] studied a mini-outbreak of three tornadoes northeast of Denver, Colorado, on August 14, 1977. They documented several atypical features of the parent cloud systems; indeed, the tornadoes were pendant from rapidly growing cumulus congestus clouds on the northwest flank of strong thunderstorms but 5–10 km distant from any significant precipitation. Two of the three tornadoes were documented photographically, as shown in Plates 4a and 4b and were of large size and long duration. A pronounced dust band was a conspicuous, although transient, feature during the mature stage of the largest tornado (Plate 2b) and is reminiscent of a similar feature documented in the Great Bend, Kansas, tornado by *Golden and Purcell* [1977]. Aspects of the tornadoes which could cause public confusion were noted, such as the disproportionately short condensation funnel from high-based cumulus clouds (Plate 4a).

Indeed, recent field studies of tornadoes in the High Plains of eastern Colorado have highlighted the fact that many tornadoes develop without an associated mesocyclone [*Wakimoto and Wilson*, 1989]. *Wilson* [1986] described the formation of several small tornadoes (sometimes termed "gustnadoes"), not associated with mesocyclones, along boundary layer convergence lines in northeast Colorado. On the basis of a study of the evolution of the July 26, 1985, Erie, Colorado, tornado, *Brady and Szoke* [1989] developed the schematic model shown in Figure 7. They believe that this represents the sequence of formation of a class of tornadoes which they term "landspouts," after *Bluestein* [1985], who observed a similar vortex development in Oklahoma during benign synoptic conditions. Landspouts are weaker, boundary-layer-forced tornadoes whose initial circulations form along or near the intersection of mesoscale boundaries and then develop up to cloud base with time. *Brady and Szoke* [1989] assert that the processes depicted in Figure 7 "may account for a significant percent of all confirmed F0 and F1 tornadoes, and that processes involved in misocyclone tornadogenesis may be crucial to the formation of mesocyclone tornadoes." (According to *Fujita* [1981], misocyclones range from 40 m to 4 km in diameter.)

Hess and Spillane [1990] presented striking photographs of two mature waterspouts in the Australian Gulf of Carpentaria and noted that they exhibited mirror images of most of the key features documented by *Golden* [1974a, b, c] during the five stages of the Florida Keys waterspout life cycle. These Australian waterspouts were produced from a warm top cumulus cloud line, with rain showers present, and converging gust fronts may have played a role in vortex formation.

Finally, we conclude by noting the strong linkages between waterspouts and tornadoes found by *Golden and Purcell* [1978b], after their detailed photogrammetric and damage analyses of the Union City, Oklahoma, tornado. They were able for the first time to define a tornado life cycle, depicted for the Union City case in Figure 3, which closely parallels that documented earlier by *Golden* [1974a,

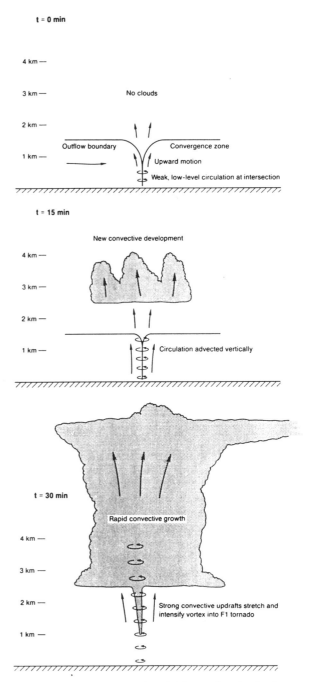

Fig. 7. Schematic conceptual model illustrating the development of the Erie, Colorado, tornado studied by *Brady and Szoke* [1989]. They hypothesized that this basic scenario appears to be valid for most nonmesocyclone tornadoes.

c] for Florida Keys waterspouts. The two vortices appear to be similar in most respects, with differences primarily relating to parent cloud development (both appear to require vigorous convection) and the near environment (tornadoes seem to require a stronger shear profile and attendant helicity; both have high low-level lapse rates). Wind profilers and

NEXRAD Doppler radars may offer the prospect of adequate sampling of a variety of tornadic storm environments.

4. Photogrammetric Wind Speed Estimates

There have been only a few new cases of photogrammetry applied to eyewitness tornado films since the last Tornado Symposium. This appears to be largely due to the sudden shifting of interest on the tornado wind speed problem by previously supportive federal agencies to seismic and other issues. *Golden* [1976] summarized the spectrum of photogrammetric analyses of tornado films up to that time and noted that this technique gives conservative estimates of a tornado's wind components in the plane normal to the line of sight of the camera.

The assignment of radius is difficult because of typical asymmetric flow fields in the tornado's debris cloud. These problems are discussed in detail by *Golden and Purcell* [1978a]. These results taken together imply that highest wind speeds are in the tornado's boundary layer, below 50 m above ground level (AGL), and generally in the range 75–95 m s^{-1}. However, we should emphasize that this is precisely where there are the fewest data points in previous photogrammetric tornado studies, owing to the frequent obscuration by intervening shrubs and/or buildings. Figures 8a and 8b show velocity estimates and boundary layer streamlines obtained by *Golden and Purcell* [1978a] for the Union City tornado, by photogrammetric analysis of cloud tags and debris elements, in the tornado's decay stage.

4.1. *Flow Asymmetries*

It became increasingly apparent during the mid-1970s that a significant class of tornadoes exists that exhibit "suction vortices" at some stage in their life cycle [*Fujita*, 1970]. Moreover, there are other forms of asymmetric flow structures in waterspouts and tornadoes, including orbiting upward jets at low levels [*Golden*, 1974a, c] and accelerating radial inflow along trailing dust bands [*Golden and Purcell*, 1977]. The latter authors documented for the first time quantitatively the existence of such a dust band, as well as important upward nonhydrostatic accelerations in the dense debris cloud of the Great Bend, Kansas, tornado, as illustrated in Plate 5. *Lee* [1981] also presented new results on two tornadoes that exhibited suction vortices during a significant portion of their lifetimes: Seymour, Texas, and Orienta, Oklahoma. Highest net velocities measured photogrammetrically from the movies (tracking dust parcels and cloud tags) were about 90 m s^{-1} at 30 m AGL for the Seymour case and 76 m s^{-1} at 50 m AGL for the translation speed of a suction vortex in the Orienta case.

The highest wind speed measured photogrammetrically thus far by NSSL is the 95 m s^{-1} net velocity measured at 200 m AGL in the Xenia, Ohio, tornado [*Golden*, 1976]. Speeds almost as large were measured at much lower heights (15–50 m) in movies taken by the Tornado Intercept Project (see Table 1, after *Lee et al.* [1981]). Note that wind speeds measured photogrammetrically may be somewhat lower than actual maximum wind speeds because (1) the velocity component along the camera axis is not measured, (2) differences may exist between tracer motions (size of several centimeters to a few meters over time periods of 2–10 s) and actual air motions, (3) the locus of maximum winds may lie within an opaque dust column or in a region devoid of tracers, and (4) the tangential wind velocity around suction vortex axes often cannot be resolved.

Pauley and Snow [1988] photogrammetrically analyzed the slowly translating Minneapolis, Minnesota, tornado and documented vortex breakdown, a precessing vortex core, and a suction vortex. Their conjectures for the structure of this tornado, which was videotaped in great detail from a TV news helicopter, are given in Figure 9. It illustrates a supercritical flow at the surface, a vortex breakdown a short distance aloft, and a region of subcritical flow containing axial downflow just above the breakdown.

Forbes [1979] also made some observations of relationships between tornado structure, underlying surface, and tornado appearance for the Parker, Indiana, tornado of April 3, 1974, and the Cabot, Arizona, tornado of March 29, 1976. He proposed that a "poor man's swirl ratio"

$$S = V_{max}/W_{max}$$

may be useful in classifying single-vortex and multivortex tornadoes. Multivortex tornadoes were observed for a range of S approximately 1.2 to 2.5. *Forbes* [1978] found that the Parker, Indiana, tornado probably represents the extreme (after the Xenia, Ohio, case) in development of suction vortices. He was able to photogrammetrically analyze motions in the suction vortices themselves and found maximum net velocities of up to 125 m s^{-1} aloft and maximum vertical velocity of up to 36 m s^{-1} (at 324-m elevation) in a suction vortex. It should be noted that there are different suction vortex configurations. *Agee et al.* [1977] documented a tornado near West Lafayette, Indiana, in which suction vortices were confined to the surface boundary layer beneath a single condensation funnel; on the other hand, suction vortices in a tornado at Parker, Indiana, extended to cloud base.

Finally, *Rasmussen* [1982] and *Rasmussen and Peterson* [1982] performed a digitally based photogrammetric analysis, including an assessment of errors, on the Tulia, Texas, outbreak tornado of May 28, 1980. *Campbell et al.* [1983] also applied the same methodology to the Lakeview, Texas, tornado of May 28, 1980 (in the same cyclic outbreak); radial-vertical cross sections of the three wind components for that tornado at its most intense stage are shown in Figures 10a, 10b, and 10c. Principal findings were (1) the Lakeview tornado had a highly asymmetric velocity distribution like that of an occluded mesoscale frontal system; (2) inflow was confined below about 50 m AGL; (3) core flow oscillations possibly indicate the up-and-down translation of an upper stagnation point in the lowest 300 m (compare with Figure 4), and core flow reversal and transition to a two-cell structure apparently occurred above 300 m; and (4) vortex

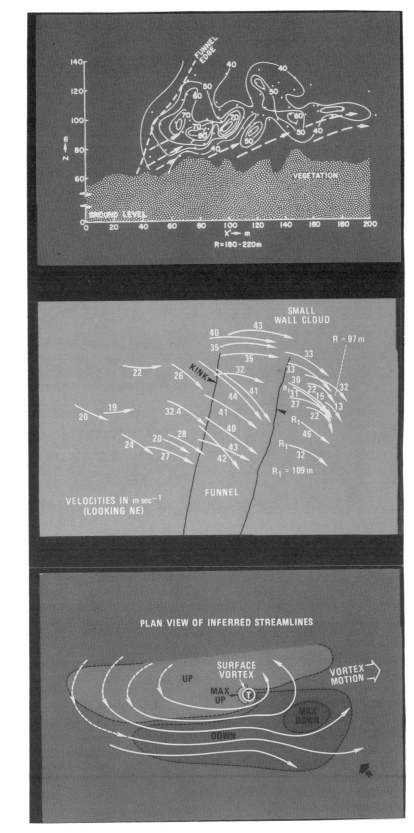

Fig. 8. (a) Photogrammetric wind speeds from Union City, Oklahoma, tornado in meters per second showing cloud tag velocity estimates around upper portion of funnel during tornado's decay stage. (b) Inferred horizontal streamflow and vertical velocity distribution around tornado's base circulation in decay stage. (By J. H. Golden and D. Purcell; see *Lee et al.* [1981] for other details.)

TABLE 1. Tornado Movies Analyzed by NSSL, After *Lee et al.* [1981]

Tornado	Date	Taken By	Maximum Measured Wind Speed, m s^{-1}	Occurred at Height, m	Official F Scale Damage Rating of Tornado	Reference
Kailua Kona, Hawaii (tornadic waterspout)	Jan. 28, 1971	eyewitness	56	40	...	*Zipser* [1976]
Union City, Oklahoma	May 24, 1973	NSSL	80	90	F4*	*Golden and Purcell* [1978a]
Salina, Kansas	Sept. 25, 1973	eyewitness	59	45	F5	*Zipser* [1976]
Xenia, Ohio	April 3, 1974	eyewitness	95	200	F5	J. Golden and D. Purcell (unpublished)
Great Bend, Kansas	Aug. 30, 1974	eyewitness	85	80	F2	*Golden and Purcell* [1977]
Alva, Oklahoma (anticyclonic tornado)	June 6, 1975	private chase team	<60?	?	F1	unpublished
Seymour, Texas	April 10, 1979	NSSL	77	15	F2	R. P. Davies-Jones [*Lee et al.*, 1981]
Seymour, Texas	April 10, 1979	G. Moore	90	30	F2	R. P. Davies-Jones [*Lee et al.*, 1981]
Orienta, Oklahoma	May 2, 1979	OU	76	50	F2	R. P. Davies-Jones [*Lee et al.*, 1981]

*Should have been F5 according to *Davies-Jones et al.* [1978].

translation speed decreased markedly as the tornado intensified and increased as it weakened.

4.2. *Probable Maximum Wind Speeds and Uncertainties*

The highest wind speeds obtained so far by photogrammetric studies are in the tornado's boundary layer, below z = 50 m AGL, and generally within the range of 75–95 m s^{-1}. (However, most data points in these studies lie above this layer.) The highest wind speeds tend to be closely associated with transient features like suction vortices, dust bands, etc. There are, at times, large upward nonhydrostatic accelerations, as well as horizontal (inflow) accelerations. Highest wind speed estimates so far (about 125 m s^{-1}) are well above the surface [e.g., *Forbes*, 1978] and somewhat higher than fixed-site Doppler estimates (90 m s^{-1} for the 1981 Binger, Oklahoma, case [*Zrnic et al.*, 1985]). Unresolved issues requiring further research include the frequency, role, and causes of suction vortices and the role of horizontal and vertical wind components in producing tornado damage.

5. Wind Estimates From Tornado Damage Surveys

Some of the earliest assessments of tornado severity from aerial photographs of damage patterns were made by *Van Tassel* [1955] and *Prosser* [1964]. Moreover, *Van Tassel* [1955] published a remarkable photograph of cycloidal patterns (loops) in the damage path of a tornado, which he attributed to scratch marks on the ground made by large debris elements. *Fujita et al.* [1970], however, suggested that similar cycloidal patterns in the damage to grain fields from one of the 1965 Palm Sunday tornadoes was due to the presence of one or more "suction spots" which rotated around the tornado's core. By using aerial photographs to make estimates of damage-loop parameters, *Fujita et al.* [1970] calculated ground speeds of 72 to 78 m s^{-1} for one tornado.

The systematic, quantitative assessment of tornado wind speeds from surface and aerial damage surveys gained momentum from the maxitornado at Lubbock, Texas, in May 1970. *Golden* [1976] gives details on methods developed by the wind engineering community to study modes of failure to engineered structures for objective estimates of

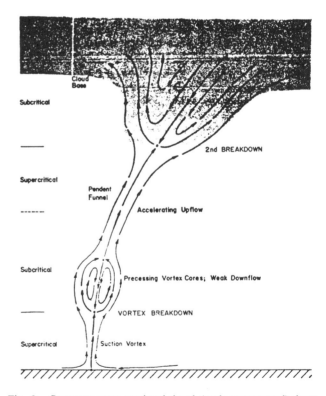

Fig. 9. Summary cross-sectional sketch (scale exaggerated) showing conjectured meridional flow structure of Minneapolis, Minnesota, tornado [after *Pauley and Snow*, 1988]. This is a composite and does not necessarily represent the tornado's flow or other features at any single instant.

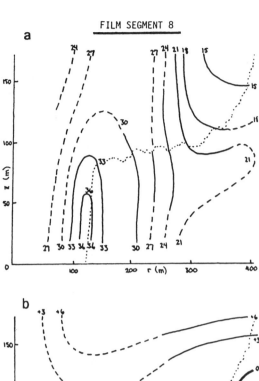
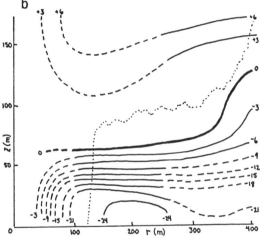
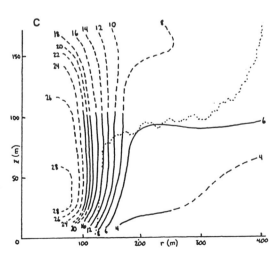

wind speeds that caused the failure (damage). Both wind engineers and Fujita and colleagues independently studied the Lubbock damage. From the Lubbock case and subsequent tornadoes, *Fujita* [1971] and *Fujita and Pearson* [1973] developed the F scale, which subjectively ranks tornadoes according to their damage severity (discrete wind speed range from each of six categories), path length, and path width. Concerns developed regarding the application of the F scale by untrained personnel and especially its use for residential damage where there may be great local inhomogeneity in the quality of construction and materials used.

From all tornadoes surveyed up to 1977, Texas Tech wind engineers claim [*Minor et al.*, 1977] that "no conclusive evidence can be found that ground level wind speeds exceed 112 m s^{-1}, and most building damage is caused by winds in the 34–56 m s^{-1} range." They also found that atmospheric pressure changes in tornadoes play only a minor role in the damaging mechanism. In particular, they suggested that the discrete wind speed ranges assigned to the F3-F4-F5 categories should be lowered somewhat and should overlap [*Minor et al.*, 1977; *Marshall et al.*, 1983, 1985; *Marshall*, 1985].

The F scale is now a routine part of local National Weather Service office tornado reporting to the national storm data archives; Fujita has cautioned, however, that the assignment of an F scale rating to a given tornado event is dependent on observer training and experience and that these ratings have a possible range of error of ±1 F number. This is especially true when the tornado in question has moved over open terrain; for example, the Seymour, Texas, tornado had photogrammetric wind speed estimates of 77 m s^{-1} at 15 m AGL, at a point in its track over open country where the damage rating was F0 [*Lee*, 1981].

It is imperative that damage surveys be continued, not only for verification with research programs on severe storms, but for operational verification as we enter the NEXRAD era. There will be many hybrid severe weather phenomena, including convective vortex phenomena over a wide spectrum, which require surface "ground truth." *Doswell and Burgess* [1988] argue convincingly for the continued need for quick-response damage surveys following severe storm events. Ideally, teams should be composed of at least one meteorologist and a wind engineer and should be dispatched within 24 hours after the event if at all possible, owing to the rapid cleanup of damage. The process needs to be both interdisciplinary and institutionalized; we should be prepared for many surprises in the NEXRAD/wind profiler era. Moreover, there is an urgent research need for independent wind estimates at various points and times

Fig. 10. (Opposite) Vertical-radial cross sections, relative to Lakeview, Texas, tornado's center at lower left of (*a*) tangential velocity component, (*b*) radial component, and (*c*) vertical component, all in meters per second. Obtained by photogrammetric analyses of film segments, after *Campbell et al.* [1983]. Outline of main tornado debris cloud is dotted.

along the same tornado's path, using remote sensors, quickly deployed in situ sensors, photogrammetry, and damage surveys. *Fujita et al.* [1976] made this point eloquently at the last Tornado Symposium, and progress in this area has been slow up to recent years.

6. IN SITU MEASUREMENTS

Although the wind field may be mapped above the ground remotely with Doppler radars (see section 7), pressure, temperature, and water vapor content cannot be similarly measured with ease. Furthermore, even Doppler radar cannot measure the wind field directly at the surface. During the late 1970s, several instruments were developed which could be brought to bear on tornadic storms by intercept crews, so that the number of direct measurements could be increased substantially over that obtained by the chance passage of a tornado or its parent wall cloud over a fixed instrument.

6.1. Serendipitous Surface Measurements

Occasionally, a tornado may pass over or close by a weather station and produce a pressure or wind trace, if the instruments survive and can be calibrated. *Flora* [1953] and *Davies-Jones and Kessler* [1974] have presented most of these records, which include eyewitness observations of wind or pressure extrema on anemometer or aneroid barometer dials. While pressure drops of up to about 200 mbar have been observed on aneroid barometers by private citizens, these pressure drop measurements are suspect. A 34-mbar pressure deficit was recorded in the 1962 Newton, Kansas, mesocyclone [*Ward*, 1964]. Some measurements of small pressure deficits may be underestimated owing to the slow response time and damping of the instruments.

The bulk of the data, in our opinion, indicates that the largest, credible pressure drop measurement to date in a tornado's core is about 100 mbar [*Davies-Jones and Kessler*, 1974]. There have been fewer reliable wind records that have survived direct hits by tornadoes. *Fujita et al.* [1970] found one anemometer trace that clearly shows the effects of two tornado passages: the first caused peak gusts to 65 m s^{-1} as the tornado center passed just north of the station; the second had gusts only to near hurricane force ($\gtrsim$30 m s^{-1}) about 1 hour later.

Tornadoes have passed close to or directly over NSSL mesonet surface sites on more than one occasion, but the instruments have usually been destroyed or severely damaged. *Barnes* [1978] studied an extensive outbreak of severe storms that passed directly through the NSSL mesonetwork on April 30, 1970; one of the tornadoes passed close by a mesonet site and produced peak gusts to 46 m s^{-1}. Another NSSL mesonet site was completely destroyed by a tornado near Fort Cobb, Oklahoma, on May 20, 1977. However, it has never been possible to determine whether or not a tornado's core region passed directly over the recording instrument in these serendipitous cases.

Fig. 11. TOTO being tested in a wind tunnel at Texas A&M University in March 1983. The strip charts and electronic circuitry are contained inside the cylinder; the wind, pressure, temperature, and corona discharge instruments are mounted on the boom at the top. (Photograph copyright by H. Bluestein.)

6.2. TOTO

In 1980, *Bedard and Ramzy* [1983] developed a 181.6-kg (400 lb) portable instrument package (Figure 11) called TOTO (Totable Tornado Observatory), which was appropriately named after Dorothy's dog, who, along with Dorothy, was carried away by a tornado in C. F. L. Baum's book *The Wizard of Oz*. TOTO was developed further by J. Carter, S. Frederickson, and E. Kelly at NSSL. The device used "hardened" sensors that had been used to measure wind shear and severe downslope winds. It was designed to be mounted in a pickup truck and deployed in 30 s or less. Wind speed, wind direction, temperature, static pressure, and corona were recorded on strip charts. TOTO was first tested in Colorado during the summer of 1980.

Intercept teams from the University of Oklahoma (OU) traveled in two vehicles, the parent vehicle and TOTO's pickup truck, to storms in Oklahoma and nearby parts of Texas during the spring seasons of 1981–1983, in an attempt to place TOTO directly in the path of tornadoes (Figure 12),

Fig. 12. Photograph of a tornado on May 11, 1982, over Altus Air Force Base, Oklahoma, to the southwest, another funnel cloud developing to the west, and TOTO deployed in the foreground. The tornado dissipated and did not pass over TOTO; the funnel cloud to the right developed into a large tornado that moved to the north-northwest for over an hour and also never passed over TOTO. (Photograph by K. Hodson.)

and then to recover the instrument package for analysis back home [*Bluestein*, 1983a, b]. Teams from NSSL attempted to do the same during 1984 and 1985 [*Burgess et al.*, 1985].

Pressure perturbation deficits of 2–5 mbar were found within 1.5 km of the path of tornadoes and under wall clouds. The highest wind speeds (gusts to 36 m s^{-1}) were measured on May 22, 1981, near Cordell, Oklahoma, in the rear flank downdraft, 2.3 km from a tornado. The traces of pressure, wind speed, wind direction, and temperature for a case in which TOTO was located in the path of a tornado are shown in Figure 13. The maximum instantaneous rate of pressure fall was 16 mbar min^{-1}; the maximum wind gust, however, was only 30 m s^{-1}.

TOTO, which is no longer being used, is currently at National Oceanic and Atmospheric Administration Headquarters in Washington, D. C. It turned out to be extremely difficult to place TOTO directly in the path of tornadoes, owing to the short-lived nature of most tornadoes and the coarseness of the network of good roads. Furthermore, wind tunnel tests indicated that TOTO's tip-over wind speed is only approximately 50 m s^{-1}, which is less than the maximum wind speed in some tornadoes. If TOTO is ever used again, the tip-over speed will have to be increased by rapid anchoring of TOTO or by increasing TOTO's diameter.

6.3. *Turtles*

Ten portable instrument packages, which are much smaller and lighter than TOTO, were developed at OU and first tested in the spring of 1986 [*Brock et al.*, 1987] in Oklahoma and Texas. Owing to their appearance, the instruments were named "Turtles" (Figure 14). Only pressure and temperature can be measured; they are recorded, and the Turtles are retrieved and read later, back home. The probability of making a direct measurement in a tornado is increased by deploying a number of instruments, such as the Turtle, in series along roadways rather than by deploying only one, such as TOTO. Furthermore, a small-scale network can be deployed in the vicinity of the tornado by several storm-intercept vehicles, with the objective of mapping out the larger-scale features around the tornado.

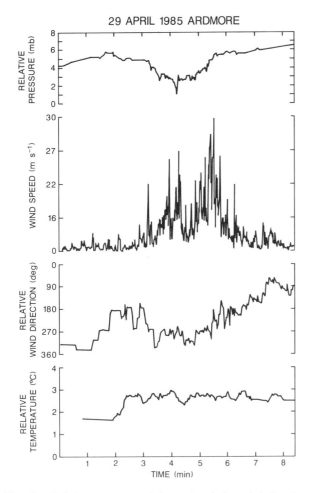

Fig. 13. Relative pressure, wind speed, relative wind direction, and relative temperature as a function of time on April 29, 1985, as recorded by TOTO almost directly in the path of an F2 tornado near Ardmore, Oklahoma. (Courtesy of R. P. Davies-Jones.)

6.4. *Portable Rawinsondes*

With the advent of portable radiosonde units in the early 1980s, efforts were undertaken to release soundings into the wall cloud of tornadic and potentially tornadic storms and, just outside these storms, to determine the "environmental" profile of temperature and water vapor content. Such soundings are useful to numerical modelers who try to simulate tornadic storms and to theoreticians who try to understand why certain environments are conducive to tornadogenesis. Mobile ballooning increases substantially the likelihood of obtaining an in-cloud or tornado proximity sounding, because the probability that a tornadic wall cloud will pass near a fixed sounding site at launch time is extremely low.

The first attempted sounding release occurred in a tornadic storm on April 26, 1984, in central Oklahoma [*Bluestein and Woodall*, 1990]; however, owing to underinflation, the sounding skimmed across a wheat field toward the wall cloud of the storm. The first successful release of soundings both into the wall cloud of a tornadic supercell and in the storm's environment (Figure 15) occurred in the Canadian, Texas, storm of May 7, 1986 [*Bluestein et al.*, 1988]. Winds were computed using an optical theodolite; an updraft of almost 50 m s^{-1} was estimated and compared to parcel theory calculations, which were in near agreement. Other soundings have been successfully released into wall clouds [*Bluestein et al.*, 1989, 1990*a*] and close to growing storms [*Bluestein et al.*, 1990*b*].

The first attempts to measure both the thermodynamic structure and the wind profile inside severe storms began in 1987, when NSSL had NCAR's Cross-Chain LORAN Atmospheric Sounding System (CLASS) [*Lauritsen et al.*, 1987] installed in one of their storm intercept vehicles [*Rust et al.*, 1990; *Rust and Marshall*, 1989]. They named their mobile version of CLASS "M-CLASS." (NCAR now has its own M-CLASS unit, which is used in field experiments.) On May 31, 1987, NSSL successfully launched a CLASS sounding in eastern Oklahoma into the wall cloud of a storm after a tornado had dissipated [*Rust*, 1990]. The advantage of CLASS is that the winds inside the storm can be determined

On May 2, 1988, a Turtle was placed under a dissipating tornado by OU students near Reagan, Oklahoma. Unfortunately, the Turtle was apparently picked up and tampered with; no useful data were recovered (J. LaDue, personal communication, 1988). There were a number of Turtle deployments in 1989 and 1991 (J. LaDue and M. Shafer, personal communication, 1991). On May 26, 1991, four Turtles were deployed around a tornadic storm near Mooreland, Oklahoma. One recorded a pressure perturbation deficit of 4 mbar, approximately 1.6 km from a tornado. This is consistent with TOTO's measurements. Based on the TOTO and Turtle measurements to date, it seems that one must be much closer than 1 km from a tornado to observe pressure deficits in excess of 5 mbar. Temperature anomalies within a tornado's core region at the ground have not yet been measured. It is anticipated that the Turtles will continue to be used at OU.

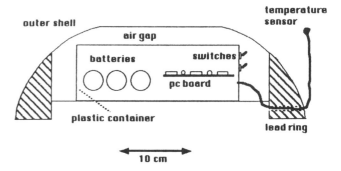

Fig. 14. Schematic cross section, drawn to scale, of the Turtle. The plastic container that contains the electronic circuits and batteries can be removed quickly from the rest of the protective pod. (From *Brock et al.* [1987].)

from tracking using LORAN. The main problems involve high noise levels inside electrically active storms and "locking on" distant LORAN stations.

6.5. Instrumented Rockets

Colgate [1982] designed and constructed small, lightweight (less than 1 kg) rockets which he proposed to launch from a light aircraft into tornado funnels. Miniaturized electronics were packaged to measure pressure, temperature, ionization and electric field variations along a rocket trajectory penetrating a tornado funnel. Colgate believed that such a tornado rocket probe should fly close to Mach I in order to penetrate the tornado funnel and not have its trajectory severely perturbed by the tornado's flow field.

Figure 16 shows the rockets mounted on the wing rack of the private aircraft. Colgate flew for three spring storm seasons in Oklahoma and adjacent states, guided by forecasts and radar information relayed in real time from NSSL. Colgate was able after several unsuccessful flights to intercept tornadoes and launch his rockets into a few of them;

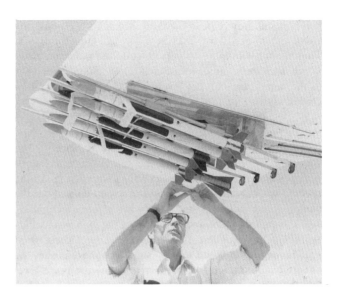

Fig. 16. S. A. Colgate inspecting small tornado rocket probes mounted on wing rack of his Cessna light aircraft, prior to a mission.

unfortunately, the light weight and fragility of the rockets themselves, mandated by Federal Aviation Administration regulations, proved to be an insurmountable obstacle. Many of the rockets became soaked with water during the aircraft's maneuvers around the tornado in heavy precipitation and misfired, missing their intended target. Moreover, in 1982 Colgate's aircraft was caught in an estimated 63 m s^{-1} inflow into a tornado near cloud base, experienced extreme turbulence, and was forced to make an emergency landing in an open field.

One of the tornadoes probed by Colgate's rockets is shown in Plate 6; the path of the rockets was documented by three different photographic systems mounted on the aircraft. Colgate noted that the "pilot load with this much equipment has proved excessive." A potentially more promising technique for launching instrumented, constant-volume balloons into a tornado was proposed by *Grant* [1971], using a research aircraft to drop the self-inflating balloon package ahead of the tornado's path. In the light of hindsight, this approach deserves reevaluation, perhaps with field trials on waterspouts.

7. Doppler Radar Measurements

The safest way to obtain wind measurements in tornadoes is to probe them remotely by Doppler radar. The Doppler radar measures the line-of-sight (radial) component of the wind in the presence of precipitation or cloud droplets.

The main problem in obtaining good Doppler data sets is that the tornadoes must be relatively close to the radar, so that the spreading of the radar beam with range does not compromise the spatial resolution of the measurements and so that measurements can be made relatively close to the ground, where we lack data in other techniques. For exam-

Fig. 15. Photograph of an OU faculty member (G. Lesins) and graduate student (E. W. McCaul, Jr.) releasing a radiosonde into a developing supercell in western Oklahoma on May 20, 1985. (Photograph copyright by H. Bluestein.)

ple, a 1° half-power beamwidth at 20 km corresponds to a cross section of 350 m, which is wider than most tornadoes. Thus the tornado must be very close to the radar, an event which is unlikely. However, even if the tornado is very close to the radar, scans at low levels might be contaminated by ground clutter.

When the radar field of view encompasses all or most of the tornado, spectra (backscattered power versus wind speed) of the radial wind component are computed. The spectra allow one to estimate the maximum wind speed within the tornado and, by examining the shape of the spectra, learn, with some assumptions, something about the nature of the radial profile of the wind [Zrnic and Doviak, 1975]. Doppler radar wind estimates of the maximum winds are likely to be underestimates because the highest wind speeds may be confined to very small volumes, so that the energy backscattered to the radar is below the noise level and therefore cannot be detected. When the radar field of view encompasses a small portion of the tornado, the mean velocities within azimuth and range bins are computed. These Doppler radar wind estimates are also likely to be underestimates owing to averaging within the radar volume. The ideal situation is one in which the tornado is close enough to two radars to obtain a dual-Doppler analysis on a scale small enough to resolve the wind speeds within the tornado vortex, an event which is not likely.

Unfortunately, the maximum unambiguous velocity of typical pulsed Doppler radars operating at wavelengths of 3, 5, and 10 cm, with pulse repetition frequencies short enough to obtain reasonably long unambiguous ranges, is small in comparison with the maximum wind speeds expected in tornadoes. For example, the maximum unambiguous velocity for a maximum unambiguous range of 100 km at 10 cm (3 cm) is only 37.5 m s^{-1} (11.3 m s^{-1}) (see equation (7.4) of Doviak and Zrnic [1984]). To attain maximum unambiguous velocities in excess of 100 m s^{-1} at 10 cm, the maximum unambiguous range must be less than 37.5 km; since thunderstorms are typically of the order of 30 km or more in width, the signal may be contaminated with range folding.

7.1. Fixed-Site Doppler Radar Observations of Supercell Tornadoes

J. Q. Brantley of Cornell Aeronautical Laboratory first suggested in 1957 that Doppler radar could be used to probe tornadoes [Brantley, 1957]. The first Doppler radar measurements of wind speeds in a funnel cloud were made by the U.S. Weather Bureau on June 10, 1958, at a range of 41 km in a supercell storm that shortly thereafter produced a tornado that hit El Dorado, Kansas [Smith and Holmes, 1961]. The 3-cm, continuous wave (CW) radar, obtained from the U.S. Navy, yielded wind spectra that did not discriminate between approaching and receding velocities. The maximum velocities detected were 92 m s^{-1}. (Range folding is not a problem in a CW radar; however, no range information is obtained.)

NSSL in the 1970s and early 1980s made a systematic attempt to obtain tornadic wind spectra with a fixed 10-cm pulsed Doppler radar located in Norman, Oklahoma. In all of their successful measurements, the parent storm of the tornado had supercell characteristics. Spectra were first obtained on May 24, 1973, in a tornado that struck Union City, Oklahoma, at a range of approximately 50 km [Zrnic and Doviak, 1975]. For the first time the sense of wind direction along radials was determined. With corrections for velocity folding (aliasing) and simulated spectra fitted to the observations, maximum wind speeds of 72 m s^{-1} were estimated [Zrnic et al., 1977]. In a tornado that hit Stillwater, Oklahoma, on June 13, 1975, at a range of approximately 100 km, maximum wind speeds of 92 m s^{-1} were estimated using the same techniques used to obtain the wind speed estimates in the Union City tornado.

NSSL obtained the first tornadic wind spectra uncontaminated by velocity folding in the Del City, Oklahoma, tornado on May 20, 1977, at a range of 35–40 km, with the radar operating at a high pulse repetition frequency (PRF). Combined with simulated spectra, Zrnic and Istok [1980] reported maximum wind speeds of 65 m s^{-1}. Unaliased spectra were obtained in the high-PRF mode in the Binger, Oklahoma, tornado on May 22, 1981, at a range of 30 km [Zrnic et al., 1985]. Maximum wind speeds of 90 m s^{-1} were estimated (Figure 17).

7.2. Fixed-Site Doppler Radar Observations of Nonsupercell Tornadoes

In eastern Colorado during the late spring and early summer, tornadoes pendant from parent thunderstorms that are not supercells ("gustnadoes" and "landspouts") are more common than supercell tornadoes in Oklahoma. Consequently, between 1984 and 1988 there have been more than two dozen serendipitous measurements at ranges of 15–45 km in these tornadoes during various field experiments using 3-cm, 5-cm, and 10-cm Doppler radars, which were operated by NCAR and NOAA. Since the tornadoes were relatively small in diameter, most of the observations were of the tornadoes' parent vortex; consequently most of the wind measurements are probably underestimates.

Maximum single-Doppler wind speed estimates (from mean data, not from spectra) range from 16 to 57 m s^{-1} [Wilson, 1986; Wakimoto and Martner, 1989; Brady and Szoke, 1989; Wakimoto and Wilson, 1989]. Photographs of a tornado near Denver, Colorado, on July 2, 1987, superimposed upon analyses of radial wind data (Figure 18), have for the first time revealed the relationship between the wind field in a tornado and the appearance of its condensation funnel [Wakimoto and Martner, 1989].

The first dual-Doppler measurements of a tornado were also made in the tornado near Denver on July 2, 1987 [Uttal et al., 1989], using two pulsed 3-cm radars operated by NOAA's Wave Propagation Laboratory (WPL) at a range of

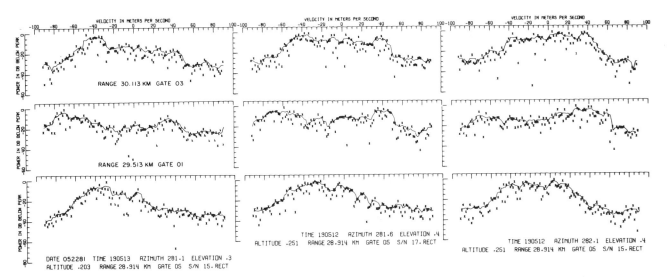

Fig. 17. Spectra for three azimuths and three range gates in the Binger, Oklahoma, tornado on May 22, 1981. The solid line is a running average of data points (crosses). The ranges of the spectra are for volumes 28–30 km from the Norman Doppler radar. The altitude of the volumes is about 200 m AGL. Power in decibels below the peak power is plotted on the ordinate; the radial wind velocity in meters per second is plotted on the abscissa. (From *Zrnic et al.* [1985]; courtesy of D. Zrnic.)

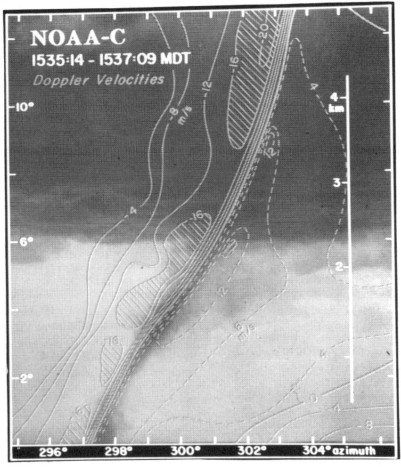

Fig. 18. Photograph of a tornado on July 2, 1987, near Denver, Colorado, superimposed on a vertical cross section of radial wind velocities in meters per second, with solid (dashed) contours representing flow out from (into) the plane of the cross section, which passes through the center of a hook echo associated with the storm. Elevation angle plotted on the ordinate; azimuth from the radar plotted on the abscissa; the height above ground (kilometers) indicated on the right. (Courtesy of R. Wakimoto; from *Wakimoto and Martner* [1989].)

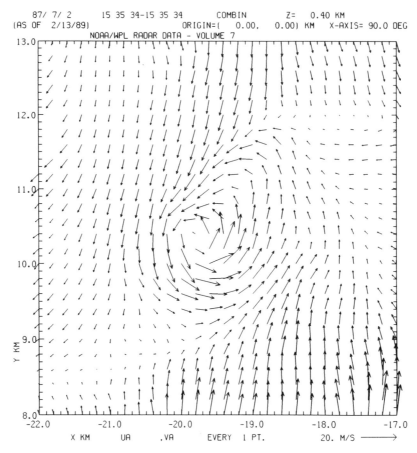

Fig. 19. Dual-Doppler analysis of the wind field in the tornado near Denver, Colorado, on July 2, 1987, at 1535 MDT. Abscissa (ordinate) is distance in kilometers east (north) of the NOAA-C radar. (Courtesy of B. Martner.)

approximately 20 km. The vortex is clearly resolved at a height of 400 m above the ground (Figure 19). Although the parent vortex is well resolved, details about the vortex are limited by the 200-m resolution. Dual-Doppler analyses (Figure 20) were also obtained in other tornadoes near Denver on June 15, 1988, based upon data from the Massachusetts Institute of Technology's Lincoln Laboratory 10-cm pulsed radar and the University of North Dakota's 5-cm pulsed radar at ranges of approximately 20 km [*Roberts and Wilson*, 1989; *Wilson and Roberts*, 1990].

7.3. *Portable Doppler Radar Observations of Supercell Tornadoes*

Zrnic et al. [1985] suggested that higher-resolution wind data in tornadoes could be obtained with a portable radar, which could be transported by storm intercept teams to locations close to tornadoes. A portable Doppler radar is useful also because the number of data sets obtained can be significantly increased by going to the tornadoes, rather than by waiting for the tornadoes to come to the radar; furthermore, the sensitivity to the highest wind speeds can also be increased by getting closer to the tornado. The portable Doppler radar can also probe volumes closer to the ground than distant radars. Simultaneous visual documentation is also possible with a portable radar; videotapes can be photogrammetrically analyzed to obtain complementary azimuthal wind speed estimates.

In 1986, T. Morton of Texas Instruments contacted H. Bluestein and suggested that a portable radar he had been working on at the Los Alamos National Laboratory might be useful for tornado research. The portable, low-power CW Doppler radar (Figure 21) was modified in 1987 for meteorological use [*Bluestein and Unruh*, 1989]. The radar was similar to the *Smith and Holmes* [1961] radar, except that approaching and receding velocities could be discriminated. During the summer of 1988 the radar was modified so that range information could be obtained using an FM-CW processor [*Strauch*, 1976]. Details on the technical aspects of the radar and its signal processing and results are given by *Bluestein and Unruh* [1991, this volume].

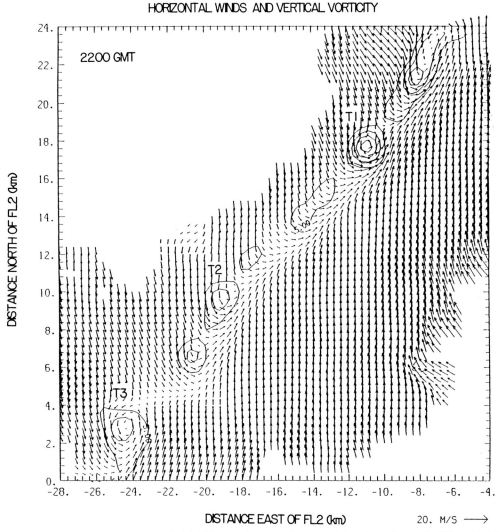

Fig. 20. Dual-Doppler analysis of the wind field in tornadoes near Denver, Colorado, on June 15, 1988, at 2200 UTC. T_1, T_2, and T_3 are locations of three tornadoes that occurred; note that not all vortices produced tornadoes. Vertical component of vorticity contoured at intervals of 5×10^{-3} s^{-1}. Abscissa (ordinate) distance in kilometers east (north) of the Massachusetts Institute of Technology Lincoln Laboratory radar. (Courtesy of R. Roberts.)

The portable radar was first used in CW mode to probe wall clouds (Figure 22). On May 2, 1987, a distant high-based funnel cloud was probed; however, its field of view was only a tiny fraction of the volume illuminated by the radar. The first tornado to appear in the field of view of the radar occurred briefly, without adequate video documentation, on May 25, 1987, in the northern Texas Panhandle; wind speeds as high as 60 m s^{-1} or higher were recorded. On May 13, 1989, the complete life history of a tornado in Texas was captured in the field of view of the radar (Figure 23). Unfortunately, both the CW and FM-CW data sets were contaminated as a result of an incorrectly set bias in a preamplifier, which had been recently added to allow for FM-CW processing. An FM-CW data set was first recorded in a wall cloud on June 6, 1989, near Floydada, Texas. Subsequent CW data sets, some FM-CW data sets, and some simultaneous videotapes were recorded for many wall clouds and for one tornado in 1990 and five tornadoes in 1991. A preliminary inspection of the data indicates wind speeds in the CW mode as high as 120–125 m s^{-1} [*Bluestein and Unruh*, this volume] in a tornado on April 26, 1991, in north central Oklahoma; the radar was 1.2 km away from the tornado's damage path, which was 0.8 km wide. This might be the first recorded data set in an F5 tornado [*Fujita*, 1981] by a Doppler radar. The results from the 1991 data are forthcoming.

Fig. 21. Photograph of University of Oklahoma graduate student G. Martin operating the first version of the Los Alamos National Laboratory portable Doppler radar in 1987. (From *Bluestein and Unruh* [1989].)

8. THE FUTURE OUTLOOK

An impressive array of mobile instruments was developed in the 1980s. Instruments now in existence need to be improved. Other devices need to be developed.

8.1. Turtles

The Turtles need to be developed further, with the addition of water vapor sensors and telemetering capability back to the intercept vehicle. The locations of deployed Turtles need to be determined from navigation aids, and the most promising technology for doing this accurately is with the GPS satellite system.

8.2. The Remotely Piloted Vehicle

The development of a remotely piloted vehicle (RPV) began in the fall of 1987 by K. Bergey and his aerospace and mechanical engineering and electrical engineering students at OU (K. Bergey and J. LaDue, personal communication, 1988). The objective of the RPV is to fly around a tornado and telemeter back thermodynamic measurements.

To date, two planes, with 1.5- and 2.4-m wing spans, have been developed. One is equipped with a real-time video camera and can be piloted remotely. It takes 2 min to launch the RPV from an intercept vehicle. Experience needs to be acquired in launching the RPV near storms. The RPV should be developed further to have a navigation aid location system and meteorological instrumentation. It has the potential to provide valuable thermodynamic and wind information near tornadoes.

8.3. Portable Doppler Radar

Improvements are suggested for the existing portable CW/FM-CW radar by *Bluestein and Unruh* [this volume]. In addition, technology has advanced to the point at which it may be possible to develop a portable, low-power, pulsed Doppler radar.

In order to improve the sensitivity of a portable radar to a tornado composed of cloud droplets and small precipitation particles, millimeter-wavelength radars need to be developed [*Pasqualucci et al.*, 1983; *Hobbs et al.*, 1985; *Lhermitte*, 1987]. A portable 35-GHz CW Doppler radar already exists at the Los Alamos National Laboratory (W. Unruh, personal communication, 1991) and needs to be tested; if tests are successful, the radar should be modified to have FM-CW capability.

8.4. Portable and Airborne Doppler Lidar

Schwiesow et al. [1981] and *Schwiesow* [1981] used an airborne 10.6-μm CW Doppler lidar from WPL in the Florida Keys in August and September 1976 to obtain radial wind spectra, with 0.75-m azimuthal resolution, of waterspout winds integrated over range. *McCaul et al.* [1987] used NASA's airborne, fore-and-aft scanning, pulsed Doppler lidar to obtain analyses of the horizontal wind field in gust fronts and around growing cumulus congestus clouds. Pulsed Doppler lidars from WPL and NASA were used in Colorado in 1982 to obtain dual-Doppler analyses of gust fronts, with 320-m range resolution [*Rothermel et al.*, 1985]. Based on the aforementioned experiments, it appears that a Doppler lidar mounted on an aircraft or made portable enough to be mounted in an intercept vehicle can be used to obtain wind measurements, with fine azimuthal resolution, in the clear air regions of tornadoes and in condensation funnels. NCAR is developing an airborne infrared lidar system (NAILS) that will have 100-m range resolution and a range of 7.5–10 km [*Schwiesow*, 1987; *Schwiesow et al.*, 1989; R. L. Schwiesow, personal communication, 1990]. NASA Marshall Space Flight Center, in cooperation with WPL and the Jet Propulsion Laboratory, is also tentatively planning to develop a new airborne Doppler lidar system (J. Rothermel, personal communication, 1991).

The problems in designing a lidar system that is suitable for probing the wind field of a tornado are formidable.

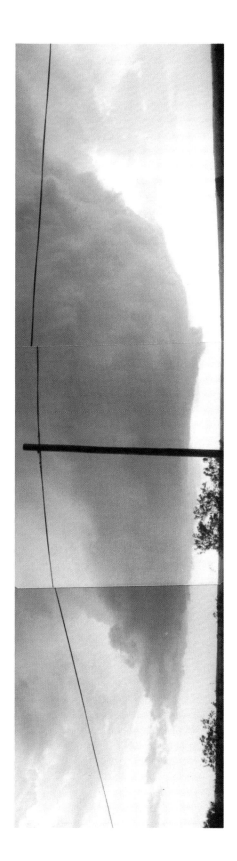
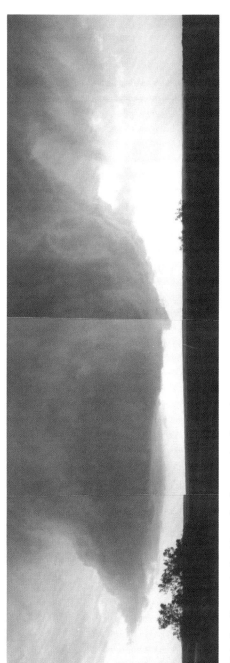

Plate 1. Composited wide-angle photograph of rotating, tornadogenetic cumulonimbus cloud intercepted by NSSL crew in extreme southwestern Oklahoma, shortly after 1900 CST, April 30, 1972 (see *Golden and Morgan* [1972]; photo by first author). View looking SW-NW.

Plate 2a. Tornado in mountainous terrain: over Basalt Mountain (10,000 feet MSL) in Colorado at 1300 MDT on July 12, 1982. (Photo by S. Jones from 16 km away.)

Plate 2b. Tornado in mountainous terrain 4 km south, over Alamosa Valley in Colorado (near the New Mexico border) at 1115 MDT on July 6, 1979, looking SSE. (Photo by R. Alexander and courtesy of R. F. Abbey.)

Plate 2c. Destructive tornado over Manitou Springs, Colorado, at the foot of Pikes Peak (elevation 2 km) on June 24, 1979. Funnel never extended more than 30–40% downward to surface from parent cloud base.

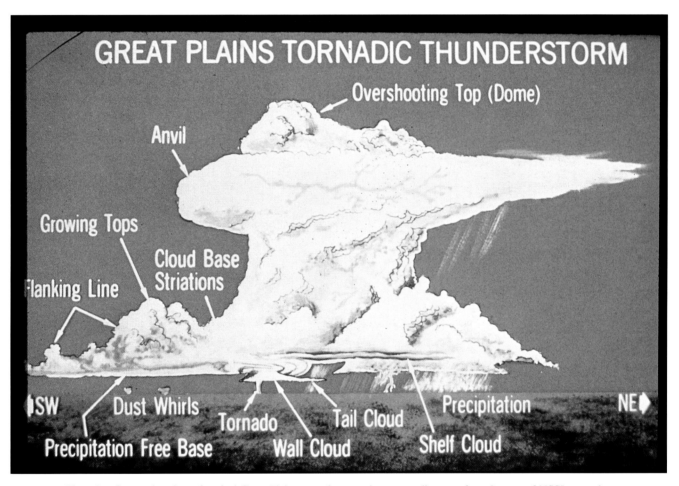

Plate 3. Composite view of typical Great Plains tornado-spawning supercell storm, from 8 years of NSSL tornado intercept observations (produced circa 1980 by J. H. Golden). Horizontal scale is compressed, and all features shown cannot necessarily be seen simultaneously from a single location.

Plate 4a. Early stage of intensifying tornado near Bennett, Colorado, on August 14, 1977, part of a minioutbreak studied by *Zipser and Golden* [1979]. Note short condensation funnel, narrow debris cloud and tilt of vortex toward the east.

Plate 4b. Same tornado, viewed at close range on opposite side of Plate 4a, 8 min later, toward the SSW. Note lack of visible funnel and trailing dust band. (Courtesy of M. White, Colorado State Patrol, and Ms. Copeland and Ms. Saylor of Bennett, Colorado.)

Fig. 22. Photograph of W. Unruh operating the radar and probing the wind field in a wall cloud west of Morton, Texas, at 2357 UTC on May 30, 1988. Looking to the northwest. (From *Bluestein and Unruh* [1989].)

Although the azimuthal resolution of a lidar is excellent, the range resolution is limited by the pulse length. The lidar system must be sensitive enough to yield a strong enough backscattered signal from aerosol and other tiny scatterers that an aircraft or storm intercept vehicle can be positioned at a safe distance. Furthermore, in an airborne system, the location and attitude of the aircraft must be known with high accuracy.

9. Summary

Tornadoes have been observed since the last Tornado Symposium in seasons and locales where once they were considered extremely unlikely (e.g., the destructive Teton-Yellowstone case); indeed, there has been a major shift in tornado incidence to states farther west of the Mississippi River from the decade of the 1970s to the 1980s (F. Ostby, personal communication, 1990). Many larger tornadoes undergo a life cycle similar to the waterspout life cycle; however, there are notable differences between the classic supercell tornadic thunderstorms, which contain mesocyclones, and the nonmesocyclone tornadic thunderstorms that produce "landspouts" and "gustnadoes." There is some difference of opinion among researchers concerning the resemblance of the latter class of tornadoes to the supercell-spawned variety. For example, we have noted the contrast of the *Simpson et al.* [1991] waterspout results and the "updraft and downdraft mechanism" with the more predominant updraft-driven mechanism found in the work of *Brady and Szoke* [1989] and *Wakimoto and Wilson* [1989].

The highest wind speed estimates in tornadoes are around 90 m s^{-1}, with some preliminary estimates as high as 120–130 m s^{-1}. Technology has advanced to the point at which we can further refine these estimates using radars and lidars mounted on mobile platforms and using advanced photogrammetric techniques that make use of more than one camera. The thermodynamic structure of tornadoes may be mapped with sensors aboard RPVs.

New instruments need to be tested on vortex phenomena such as dust devils and waterspouts, which are much more common and easy to intercept than tornadoes. The instruments can then be used to make measurements in and near tornadoes; data from them can be integrated with information from damage surveys and photogrammetric analyses to yield a clearer picture of the dynamics of the tornado vortex. Data will provide better boundary conditions for high-resolution vortex models.

The resolution of current wind speed and minimum pressure controversies will have not only a scientific payoff, but an economic payoff as well. Damage to structures can be minimized by constructing new buildings that are strong enough to withstand tornadic, but not any greater, wind speeds.

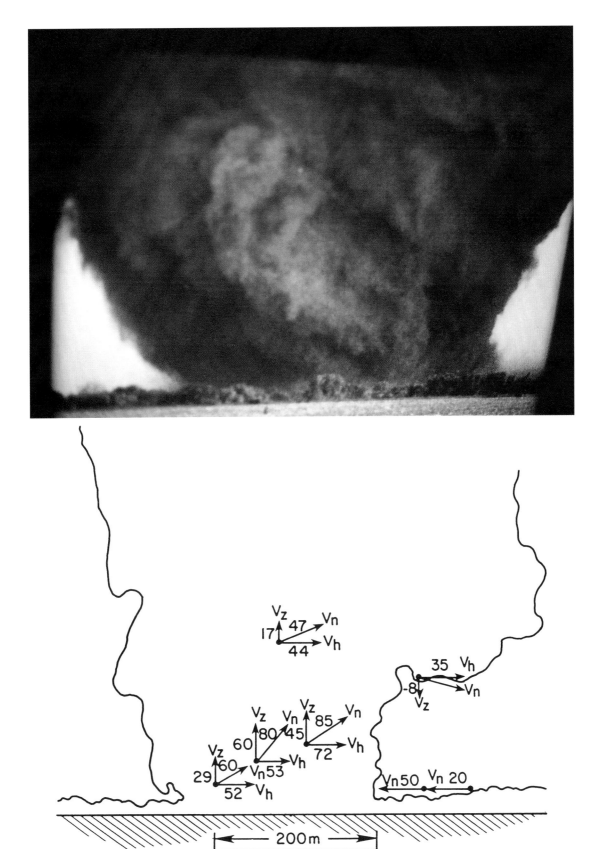

Plate 5. (*a*) Photograph of Great Bend, Kansas, tornado's dense, turbulent debris cloud with trailing dust band. (*b*) Scaled outline of same tornado's dust column with representative net horizontal and vertical velocity vectors (units of meters per second) superimposed. Velocities derived by photogrammetric tracking and analysis as described by *Golden and Purcell* [1977]. Note accelerations implied vertically and along trailing dust band, from the right.

348 REVIEW OF TORNADO OBSERVATIONS

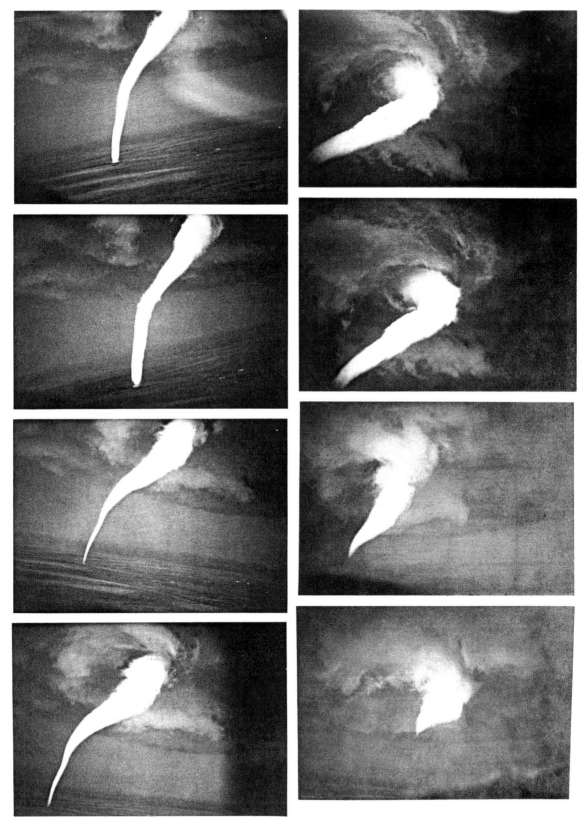

Plate 6. Example of tornado investigated by S. A. Colgate over western Oklahoma. View taken by him from aircraft preparing to launch instrumented rocket. Note nearby precipitation, which impaired rocket's durability.

Fig. 23. Photograph of University of Oklahoma graduate students S. Hrebenach, G. Martin, and S. Contorno, attempting to probe a tornado to the west at 2329 UTC on May 13, 1989, west of Hodges, Texas. (Photograph copyright by H. Bluestein.)

Acknowledgment. Part of this paper was supported by National Science Foundation grant ATM-8902594.

REFERENCES

Agee, E. M., J. T. Snow, F. S. Nickerson, P. R. Clare, C. R. Church, and L. A. Schaal, An observational study of the West Lafayette, Indiana, tornado of 20 March 1976, *Mon. Weather Rev.*, *105*, 893–907, 1977.

Barnes, S. L., Oklahoma thunderstorms on 29–30 April 1970, Part I, Morphology of a tornadic storm, *Mon. Weather Rev.*, *106*, 673–684, 1978.

Bedard, A. J., Jr., and C. Ramzy, Surface meteorological observations in severe thunderstorms, Part I, Design details of TOTO, *J. Appl. Meteorol.*, *22*, 911–918, 1983.

Blechman, J. B., The Wisconsin tornado event of April 21, 1974: Observations and theory of secondary vortices, in *Preprints, 9th Conference on Severe Local Storms*, pp. 267–270, American Meteorological Society, Boston, Mass., 1975.

Bluestein, H. B., Surface meteorological observations in severe thunderstorms, Part II, Field experiments with TOTO, *J. Appl. Meteorol.*, *22*, 919–930, 1983a.

Bluestein, H. B., Measurements in the vicinity of severe thunderstorms and tornadoes with TOTO: 1982–1983 results, in *Preprints, 13th Conference on Severe Local Storms*, pp. 89–92, American Meteorological Society, Boston, Mass., 1983b.

Bluestein, H. B., The formation of a "landspout" in a "broken-line" squall line in Oklahoma, in *Preprints, 14th Conference on Severe Local Storms*, pp. 267–270, American Meteorological Society, Boston, Mass., 1985.

Bluestein, H. B., and W. P. Unruh, Observations of the wind field in tornadoes, funnel clouds, and wall clouds with a portable Doppler radar, *Bull. Am. Meteorol. Soc.*, *70*, cover, 1514–1525, 1989.

Bluestein, H. B., and W. P. Unruh, On the measurement of wind speeds in tornadoes with a portable CW/FM-CW Doppler radar, in *Preprints, 25th International Conference on Radar Meteorology*, pp. 848–851, American Meteorological Society, Boston, Mass., 1991.

Bluestein, H. B., and W. P. Unruh, On the use of a portable FM-CW Doppler radar for tornado research, this volume.

Bluestein, H. B., and G. R. Woodall, Doppler-radar analysis of a low-precipitation severe storm, *Mon. Weather Rev.*, *118*, 1640–1664, 1990.

Bluestein, H. B., E. W. McCaul, Jr., G. P. Byrd, and G. R. Woodall, Mobile sounding observations of a tornadic storm near the dryline: The Canadian, Texas storm of 7 May 1986, *Mon. Weather Rev.*, *116*, 1790–1804, 1988.

Bluestein, H. B., E. W. McCaul, Jr., G. P. Byrd, G. R. Woodall, G. Martin, S. Keighton, and L. Showell, Mobile sounding observations of a thunderstorm near the dryline: The Gruver, Texas storm complex of 25 May 1987, *Mon. Weather Rev.*, *117*, 244–250, 1989.

Bluestein, H. B., E. W. McCaul, Jr., G. P. Byrd, and R. L. Walko, Thermodynamic measurements under a wall cloud, *Mon. Weather Rev.*, *118*, 794–799, 1990a.

Bluestein, H. B., E. W. McCaul, Jr., G. P. Byrd, R. L. Walko, and R. Davies-Jones, An observational study of splitting convective clouds, *Mon. Weather Rev.*, *118*, 1359–1370, 1990b.

Brady, R. H., and E. J. Szoke, A case study of nonmesocyclone tornado development in northeast Colorado: Similarities to waterspout formation, *Mon. Weather Rev.*, *117*, 843–856, 1989.

Brantley, J. Q., Some weather observations with a continuous-wave Doppler radar, in *Proceedings of the 6th Weather Radar Conference*, pp. 297–306, American Meteorological Society, Blue Hill Meteorological Observatory, Geophysics Research Directorate of the Air Force Cambridge Research Center, Massachusetts Institute of Technology, Boston and Cambridge, Mass., 1957.

Brock, F. V., G. Lesins, and R. L. Walko, Measurement of pressure and air temperature near severe thunderstorms: An inexpensive and portable instrument, in *Extended Abstracts, 6th Symposium on Meteorological Observations and Instrumentation*, pp. 320–323, American Meteorological Society, Boston, Mass., 1987.

Burgess, D. W., and L. R. Lemon, Severe thunderstorm detection by radar, in *Radar in Meteorology*, edited by D. Atlas, pp. 619–647, American Meteorological Society, Boston, Mass., 1990.

Burgess, D. W., S. V. Vasiloff, R. P. Davies-Jones, D. S. Zrnic, and S. E. Frederickson, Recent NSSL work on windspeed measurement in tornadoes, in *Proceedings, 5th U.S. National Conference on Wind Engineering*, pp. 1A-53–1A-60, Texas Tech University, Lubbock, 1985.

Campbell, B. D., E. N. Rasmussen, and R. E. Peterson, Kinematic analysis of the Lakeview, TX tornado, in *Preprints, 13th Conference on Severe Local Storms*, pp. 62–65, American Meteorological Society, Boston, Mass., 1983.

Church, C. R., C. M. Ehresman, and J. H. Golden, Instrumentation for probing waterspouts, in *Preprints, 8th Conference on Severe Local Storms*, pp. 169–172, American Meteorological Society, Boston, Mass., 1973.

Colgate, S., Comment on "On the relation of electrical activity to tornadoes" by R. P. Davies-Jones and J. H. Golden, *J. Geophys. Res.*, *80*, 4556, 1975.

Colgate, S., Small rocket tornado probe, in *Preprints, 12th Conference on Severe Local Storms*, pp. 396–400, American Meteorological Society, Boston, Mass., 1982.

Davies-Jones, R. P., Tornado dynamics, in *Thunderstorms—A Social, Scientific and Technological Documentary*, vol. 2, *Thunderstorm Morphology and Dynamics*, 3rd ed., edited by E. Kessler, 412 pp., University of Oklahoma Press, Norman, 1988.

Davies-Jones, R. P., and J. H. Golden, On the relation of electrical activity to tornadoes, *J. Geophys. Res.*, *80*, 1614–1616, 1975a.

Davies-Jones, R. P., and J. H. Golden, Reply, *J. Geophys. Res.*, *80*, 4557–4558, 1975b.

Davies-Jones, R. P., and J. H. Golden, Reply, *J. Geophys. Res.*, *80*, 4561–4562, 1975c.

Davies-Jones, R. P., and E. Kessler, Tornadoes, in *Weather and Climate Modification*, edited by W. N. Hess, pp. 552–595, John Wiley, New York, 1974.

Davies-Jones, R. P., D. W. Burgess, L. R. Lemon, and D. Purcell, Interpretation of surface marks and debris patterns from the 24 May 1973 Union City, Oklahoma, tornado, *Mon. Weather Rev.*, *106*, 12–21, 1978.

Dessens, J., Influence of ground roughness on tornadoes: A laboratory simulation, *J. Appl. Meteorol.*, *11*, 72–75, 1972.

Dietachmeyer, G., Numerical modelling of intense atmospheric vortices, Ph.D. dissertation, Dep. of Math., Monash Univ., Clayton, Victoria, Australia, 1987.

Doswell, C. A., and D. W. Burgess, On some issues of U.S. tornado climatology, *Mon. Weather Rev.*, *116*, 495–501, 1988.

Doviak, R. J., and D. S. Zrnic, *Doppler Radar and Weather Observations*, 458 pp., Academic, San Diego, Calif., 1984.

Elsom, D. M., and G. T. Meaden, Suppression and dissipation of weak tornadoes in metropolitan areas: A case study of greater London, *Mon. Weather Rev.*, *110*, 745–756, 1982.

Flora, S. R., *Tornadoes of the United States*, 194 pp., University of Oklahoma Press, Norman, 1953.

Forbes, G. S., Three scales of motion associated with tornadoes, *NUREG/CR-0363*, final report, 359 pp., U.S. Nucl. Reg. Comm., Washington, D. C., 1978. (Available as *NTIS PB-288 291* from Natl. Tech. Inf. Serv., Springfield, Va.)

Forbes, G. S., Observations of relationships between tornado structure, underlying surface, and tornado-appearance, in *Preprints, 11th Conference on Severe Local Storms*, pp. 351–356, American Meteorological Society, Boston, Mass., 1979.

Forbes, G. S., and R. M. Wakimoto, A concentrated outbreak of tornadoes, downbursts, and microbursts, and implications regarding vortex classification, *Mon. Weather Rev.*, *111*, 220–235, 1983.

Fujita, T. T., Mother cloud of the Fargo tornadoes of 20 June 1957, in *Cumulus Dynamics*, pp. 175–177, Pergamon, New York, 1960.

Fujita, T. T., The Lubbock tornadoes: A study of suction spots, *Weatherwise*, *23*, 160–173, 1970.

Fujita, T. T., Proposed mechanism of suction spots accompanied by tornadoes, in *Preprints, 7th Conference on Severe Local Storms*, pp. 208–213, American Meteorological Society, Boston, Mass., 1971.

Fujita, T. T., Jumbo outbreak of 3 April 1974, *Weatherwise*, *27*, 116–126, 1974.

Fujita, T. T., Tornadoes and downburst in the context of generalized planetary scales, *J. Atmos. Sci.*, *38*, 1511–1534, 1981.

Fujita, T. T., The Teton-Yellowstone tornado of 21 July 1987, *Mon. Weather Rev.*, *117*, 1913–1940, 1989.

Fujita, T. T., and A. D. Pearson, Results of FPP classification of 1971 and 1972 tornadoes, in *Preprints, 8th Conference on Severe Local Storms*, pp. 142–145, American Meteorological Society, Boston, Mass., 1973.

Fujita, T. T., D. L. Bradbury, and C. F. Van Thullenar, Palm Sunday tornadoes of April 11, 1965, *Mon. Weather Rev.*, *98*, 29–69, 1970.

Fujita, T. T., K. Watanabe, K. Tsuchiya, and M. Schimada, Typhoon-associated tornadoes in Japan and new evidence of suction vortices in a tornado near Tokyo, *J. Meteorol. Soc. Jpn.*, *50*, 431–453, 1972.

Fujita, T. T., A. D. Pearson, G. S. Forbes, T. A. Umenhofer, E. W. Pearl, and J. J. Tecson, Photogrammetric analyses of tornadoes, in *Proceedings, Symposium on Tornadoes*, edited by R. E. Peterson, pp. 43–88, Institute for Disaster Research, Texas Tech University, Lubbock, 1976.

Golden, J. H., Waterspouts and tornadoes over south Florida, *Mon. Weather Rev.*, *99*, 146–154, 1971.

Golden, J. H., Some statistical aspects of waterspout formation, *Weatherwise*, *26*, 108–117, 1973.

Golden, J. H., The life-cycle of Florida Keys' waterspouts, I, *J. Appl. Meteorol.*, *13*, 676–692, 1974a.

Golden, J. H., The life-cycle of Florida Keys' waterspouts, II, *J. Appl. Meteorol.*, *13*, 693–709, 1974b.

Golden, J. H., Life-cycle of Florida Keys' waterspouts, *NOAA Tech. Memo.*, *ERL-NSSL-70*, 147 pp., 1974c.

Golden, J. H., An assessment of wind speeds in tornadoes, in *Proceedings, Symposium on Tornadoes*, edited by R. E. Peterson, pp. 5–42, Institute for Disaster Research, Texas Tech University, Lubbock, 1976.

Golden, J. H., and B. J. Morgan, The NSSL/Notre Dame tornado intercept program, spring 1972, *Bull. Am. Meteorol. Soc.*, *53*, 1178–1179, 1972.

Golden, J. H., and D. Purcell, Photogrammetric velocities for the Great Bend, Kansas tornado of 30 August 1974: Accelerations and asymmetries, *Mon. Weather Rev.*, *105*, 485–492, 1977.

Golden, J. H., and D. Purcell, Airflow characteristics around the Union City tornado, *Mon. Weather Rev.*, *106*, 22–28, 1978a.

Golden, J. H., and D. Purcell, Life-cycle of the Union City, OK tornado and comparison with waterspouts, *Mon. Weather Rev.*, *106*, 3–11, 1978b.

Golden, J. H., and M. E. Sabones, Tornadic-waterspout formation near intersecting boundaries, in *Preprints, 25th International Conference on Radar Meteorology*, pp. 420–423, American Meteorological Society, Boston, Mass., 1991.

Golden, J. H., et al., Severe storm detection: Panel report, in *Radar in Meteorology*, edited by D. Atlas, pp. 648–656, American Meteorological Society, Boston, Mass., 1990.

Goodman, S. J., and K. R. Knupp, Tornadogenesis via squall line and supercell interaction revisited: The 15 November, 1989 Huntsville tornado, in *Preprints, 16th Conference on Severe Local Storms*, pp. 566–571, American Meteorological Society, Boston, Mass., 1990.

Gordon, A. H., Waterspouts, *Weather*, *6*, 364–371, 1951.

Grant, F. C., Proposed technique for launching instrumented balloons into tornadoes, *NASA Tech. Note*, *TN D-6503*, 21 pp., 1971.

Grazulis, T. P., J. T. Schaefer, and R. F. Abbey, Jr., Advances in tornado climatology, hazards, and risk assessment since Tornado Symposium II, this volume.

Hess, G. D., and K. T. Spillane, Waterspouts in the Gulf of Carpentaria, *Aust. Meteorol. Mag.*, *38*, 173–179, 1990.

Hobbs, P. V., N. T. Funk, R. R. Weiss, Sr., J. D. Locatelli, and K. R. Biswas, Evaluation of a 35 GHz radar for cloud physics research, *J. Atmos. Oceanic Technol.*, *2*, 35–48, 1985.

Holle, R. L., and M. W. Maier, Tornado formation from downdraft interaction in the FACE mesonetwork, *Mon. Weather Rev.*, *108*, 991–1009, 1980.

Howells, P., and R. K. Smith, Numerical simulation of tornado-like vortices, Part I, Vortex evolution, *Geophys. Astrophys. Fluid Dyn.*, *27*, 253–284, 1983.

Hurd, W. E., Some phases of waterspout behavior, *Weatherwise*, *3*, 75–82, 1950.

Lauritsen, D., Z. Malekmadani, C. Morel, and R. McBeth, The Cross-Chain LORAN Atmospheric Sounding System (CLASS), in *Preprints, 6th Symposium on Meteorological Observations and Instrumentation*, pp. 340–343, American Meteorological Society, Boston, Mass., 1987.

Lee, J. T., D. S. Zrnic, R. P. Davies-Jones, and J. H. Golden, Summary of AEC-ERDA-NRC supported research at NSSL 1973-1979, *NOAA Tech. Memo., ERL NSSL-90*, 93 pp., 1981.

Leverson, V. H., P. C. Sinclair, and J. H. Golden, Waterspout wind, temperature and pressure structure deduced from aircraft measurements, *Mon. Weather Rev., 105*, 725-733, 1977.

Lhermitte, R., A 94-GHz Doppler radar for cloud observations, *J. Atmos. Oceanic Technol., 4*, 36-48, 1987.

MacGorman, D. R., Lightning in tornadic storms: A review, this volume.

MacGorman, D. R., and K. E. Nielsen, Cloud-to-ground lightning in a tornadic storm on 8 May 1986, *Mon. Weather Rev., 119*, 1557-1574, 1991.

Macky, W. A., The Easter tornadoes at Bermuda, *Weatherwise, 6*, 7475, 1953.

Maier, M. W., and H. W. Brandli, Simultaneous observations of a tornado, waterspout, and funnel cloud, *Weather, 28*, 322-327, 1973.

Marshall, T. P., Damage analysis of the Mesquite, Texas tornado, in *Proceedings, 5th U.S. National Conference on Wind Engineering*, pp. 4B-11–4B-18, Texas Tech University, Lubbock, 1985.

Marshall, T. P., J. R. McDonald, and K. C. Mehta, Utilization of load and resistance statistics in a wind speed assessment, *Rep. 670*, 91 pp., Inst. for Disaster Res., Tex. Tech Univ., Lubbock, 1983.

Marshall, T. P., J. R. McDonald, and K. C. Mehta, Performance of structures during Hurricane Alicia and the Athens tornado, in *Proceedings, DOE Natural Phenomena Hazards Mitigation Conference*, pp. 140-149, U.S. Department of Energy, Washington, D. C., 1985.

McCaul, E., Observations of the Hurricane "Danny" tornado outbreak of 16 August 1985, *Mon. Weather Rev., 115*, 1206-1223, 1987.

McCaul, E. W., Jr., H. B. Bluestein, and R. J. Doviak, Airborne Doppler lidar observations of convective phenomena in Oklahoma, *J. Atmos. Oceanic Technol., 4*, 479-497, 1987.

Minor, J. E., J. R. McDonald, and K. C. Mehta, The tornado: An engineering-oriented perspective, *NSSL Tech. Memo. ERL NSSL-82*, Natl. Severe Storms Lab., Natl. Oceanic and Atmos. Admin., Norman, Okla., 1977.

Pasqualucci, F., B. W. Bartram, R. A. Kropfli, and W. R. Moninger, A millimeter-wavelength dual-polarization Doppler radar for cloud and precipitation studies, *J. Clim. Appl. Meteorol., 22*, 758-765, 1983.

Pauley, R. L., and J. T. Snow, On the kinematics and dynamics of the 18 July 1986 Minneapolis tornado, *Mon. Weather Rev., 116*, 2731-2736, 1988.

Prosser, N. E., Aerial photographs of a tornado path in Nebraska, May 5, 1964, *Mon. Weather Rev., 92*, 593-598, 1964.

Rasmussen, E. N., The Tulia outbreak storm: Mesoscale evolution and photogrammatic analysis, M.S. thesis, 180 pp., Dep. of Geosci., Tex. Tech Univ., Lubbock, 1982.

Rasmussen, E. N., and R. E. Peterson, Tornado photogrammetry: Determination of windspeeds and assessment of errors, report, 56 pp., Inst. for Disaster Res., Tex. Tech Univ., Lubbock, 1982.

Roberts, R. D., and J. W. Wilson, Multiple Doppler radar analyses of the 15 June 1988 Denver tornadoes, in *Preprints, 24th Conference on Radar Meteorology*, pp. 142-145, American Meteorological Society, Boston, Mass., 1989.

Rossow, V. J., On the electrical nature of waterspouts, in *Preprints, 6th Conference on Severe Local Storms*, pp. 182-187, American Meteorological Society, Boston, Mass., 1969.

Rossow, V. J., Observations of waterspouts and their parent clouds, *NASA Tech. Note, D-5854*, 63 pp., 1970.

Rothermel, J., C. Kessinger, and D. L. Davis, Dual-Doppler lidar measurement of winds in the JAWS experiment, *J. Atmos. Oceanic Technol., 2*, 138-147, 1985.

Rust, W. D., Severe storm electricity research with a mobile laboratory and mobile ballooning, in *Preprints, 16th Conference on Severe Local Storms*, pp. J31-J35, American Meteorological Society, Boston, Mass., 1990.

Rust, W. D., and T. C. Marshall, Mobile, high-wind, balloon-launching apparatus, *J. Atmos. Oceanic Technol., 6*, 215-217, 1989.

Rust, W. D., R. Davies-Jones, D. W. Burgess, R. A. Maddox, L. C. Showell, T. C. Marshall, and D. K. Lauritsen, Testing of a mobile version of a cross-chain LORAN atmospheric sounding system (M-CLASS), *Bull. Am. Meteorol. Soc., 71*, 173-180, 1990.

Schwiesow, R. L., Horizontal velocity structure in waterspouts, *J. Appl. Meteorol., 20*, 349-360, 1981.

Schwiesow, R. L., The NCAR airborne infrared lidar system (NAILS) design and operation, *NCAR Tech. Note NCAR/TN-291+IA*, 38 pp., Atmos. Tech. Div. Natl. Center for Atmos. Res., Boulder, Colo., 1987.

Schwiesow, R. L., R. E. Cupp, P. C. Sinclair, and R. F. Abbey, Jr., Waterspout velocity measurements by airborne Doppler lidar, *J. Appl. Meteorol., 20*, 341-348, 1981.

Schwiesow, R. L., A. J. Weinheimer, and V. M. Glover, Applications and development of a compact, airborne Doppler lidar for atmospheric measurements, in *Proceedings, 5th Conference on Coherent Laser Radar: Technology and Applications*, pp. 20-24, Society of Photo Optical Instrumentation Engineers, Bellingham, Wash., 1989.

Simpson, J., M. C. McCumber, and J. H. Golden, Tropical waterspouts and tornadoes, in *Preprints, 14th Conference on Severe Local Storms*, pp. 284-288, American Meteorological Society, Boston, Mass., 1985.

Simpson, J., B. R. Morton, M. C. McCumber, and R. S. Penc, Observations and mechanisms of GATE waterspouts, *J. Atmos. Sci., 43*, 753-782, 1986.

Simpson, J., G. Roff, B. R. Morton, K. Labas, G. Dietachmeyer, M. McCumber, and R. Penc, The Great Salt Lake waterspout, *Mon. Weather Rev., 119*, 2741-2770, 1991.

Smith, R. L., and D. W. Holmes, Use of Doppler radar in meteorological observations, *Mon. Weather Rev., 89*, 1-7, 1961.

Strauch, R. G., Theory and application of the FM-CW Doppler radar, Ph.D. thesis, 97 pp., Dep. of Electr. Eng., Univ. of Colo., Boulder, 1976.

Szoke, E. J., and R. Rotunno, A comparison of surface observations and visual tornado characteristics for the June 15, 1988, Denver tornado outbreak, this volume.

Szoke, E. J., M. L. Weisman, J. M. Brown, F. Caracena, and T. W. Schlatter, A subsynoptic analysis of the Denver tornado of 3 June 1981, *Mon. Weather Rev., 112*, 790-808, 1984.

Uttal, T., B. E. Martner, B. W. Orr, and R. M. Wakimoto, High resolution dual-Doppler radar measurements of a tornado, in *Preprints, 24th Conference on Radar Meteorology*, pp. 62-65, American Meteorological Society, Boston, Mass., 1989.

Van Tassel, E. L., The North Platte Valley tornado outbreak of June 27, 1955, *Mon. Weather Rev., 83*, 255-264, 1955.

Vonnegut, B., Electrical theory of tornadoes, *J. Geophys. Res., 65*, 203-212, 1960.

Wakimoto, R. M., and B. E. Martner, Photogrammetric/radar analysis of the 2 July tornado during CINDE, in *Preprints, 24th Conference on Radar Meteorology*, pp. 58-61, American Meteorological Society, Boston, Mass., 1989.

Wakimoto, R. M., and J. W. Wilson, Non-supercell tornadoes, *Mon. Weather Rev., 117*, 1113-1140, 1989.

Ward, N. B., The Newton, Kansas tornado cyclone of May 24, 1962, in *Preprints, 11th Weather Radar Conference*, pp. 410-415, American Meteorological Society, Boston, Mass., 1964.

Wilson, J. W., Tornadogenesis by non-precipitation-induced wind shear lines, *Mon. Weather Rev., 114*, 270-284, 1986.

Wilson, J. W., and R. D. Roberts, Vorticity evolution of a non-supercell tornado on 15 June 1988 near Denver, in *Preprints, 16th Conference on Severe Local Storms*, pp. 479-484, American Meteorological Society, Boston, Mass., 1990.

Zipser, R. A., Photogrammetric studies of a Kansas tornado and a Hawaiian tornadic-waterspout, M.S. thesis, 75 pp., Dep. of Meteorol., Univ. of Okla., Norman, 1976.

Zipser, E. J., and J. H. Golden, A summertime tornado outbreak in Colorado: Mesoscale environment and structural features, *Mon. Weather Rev.*, *108*, 1328–1342, 1979.

Zrnic, D. S., and R. J. Doviak, Velocity spectra of vortices scanned with a pulse-Doppler radar, *J. Appl. Meteorol.*, *14*, 1531–1539, 1975.

Zrnic, D. S., and M. Istok, Wind speeds in two tornadic storms and a tornado, deduced from Doppler spectra, *J. Appl. Meteorol.*, *19*, 1405–1415, 1980.

Zrnic, D. S., R. J. Doviak, and D. W. Burgess, Probing tornadoes with a pulse Doppler radar, *Q. J. R. Meteorol. Soc.*, *103*, 707–720, 1977.

Zrnic, D. S., D. W. Burgess, and L. Hennington, Doppler spectra and estimated windspeed of a violent tornado, *J. Clim. Appl. Meteorol.*, *24*, 1068–1081, 1985.

A Comparison of Surface Observations and Visual Tornado Characteristics for the June 15, 1988, Denver Tornado Outbreak

E. J. SZOKE

Forecast Systems Laboratory, NOAA, and National Center for Atmospheric Research, Boulder, Colorado 80307

R. ROTUNNO

National Center for Atmospheric Research, Boulder, Colorado 80307

1. INTRODUCTION

On June 15, 1988, four tornadoes developed in and near Denver, Colorado, along the Denver Convergence-Vorticity Zone (DCVZ) (or Denver Cyclone [*Szoke et al.*, 1984]), after its intersection with two thunderstorm outflow boundaries. Although the tornadoes formed in proximity to each other in time and space, the two observed by the authors were distinctly different in appearance. One was quite typical of tornadoes in this area, with the vortex mainly defined by dust, whereas the other had an unusually large condensation funnel. In this paper the question of why the tornadoes were so visibly different is examined using two networks of nearby automated surface stations. Our hypothesis is that the visual differences were primarily the result of local variations in low-level moisture. Before examining the June 15 case, however, in keeping with the theme of the Tornado Symposium, tornado observational and research highlights from northeastern Colorado during the past decade will be summarized.

2. REVIEW OF COLORADO TORNADO OBSERVATIONS AND RESEARCH DURING THE PAST DECADE

In this brief review we focus on observational research, concentrating on eastern Colorado. A number of field experiments conducted in northeastern Colorado in the 1980s have provided the data to increase our understanding of convection in general and tornadogenesis in particular.

One data set that has provided valuable information about the environment of storms is that generated by a mesonetwork of up to 22 automated surface stations (the current configuration is shown in Figure 1). The network was put into place in 1980 by the Program for Regional Observing and Forecasting Services (PROFS) of the National Oceanic and Atmospheric Administration (NOAA) (now part of NOAA's Forecast Systems Laboratory (FSL)) to test the effect such data might have on improving short-range forecasts. The mesonet is noteworthy in that it has provided a long-term set of dense observations. Additional observations from Doppler radar(s), storm intercept teams, special soundings, aircraft, and finer-scale mesonets have been obtained during various PROFS forecast exercises and a number of experiments including the Joint Airport Weather Studies (JAWS) [*McCarthy et al.*, 1983] in 1982 and the Convective Initiation and Downburst Experiment (CINDE) in 1987 [*Wilson et al.*, 1988].

One of the first discoveries noted when the PROFS mesonet data became available was the formation of a zone of convergence, the DCVZ (see Figure 1), that typically developed near the Denver area when the ambient flow farther east on the plains was from the south or southeast. There had been speculation that such a feature existed before the advent of the mesonet observations. For example, *Zipser and Golden* [1979] noted mesoscale convergence and cyclonic vorticity based on Surface Aviation Observation (SAO) stations Denver (DEN) and Limon (LIC) before an August 1977 tornado outbreak east of Denver. Noteworthy among their observations was that the storms producing the tornadoes were of the ordinary cell type and did not display the hook echo (from the WSR-57 10-cm wavelength National Weather Service (NWS) radar at LIC) characteristic of more organized supercellular storms. These storms were growing "explosively," with new development on the western edges, just before the tornadoes were observed. The tornado formation during the rapid vertical development phase of the

The Tornado: Its Structure, Dynamics, Prediction, and Hazards.
Geophysical Monograph 79
Copyright 1993 by the American Geophysical Union.

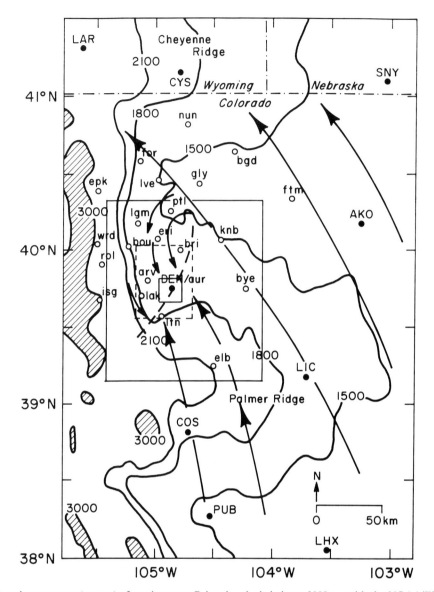

Fig. 1. Elevation contours (meters) of northeastern Colorado, shaded above 3000 m, with the NOAA/FSL mesonet stations (open circles), SAO stations (closed circles), schematic flow for the DCVZ (with "mean" location shown by the dashed line), analysis areas for Figures 10a–10f (two solid boxes), and area in Figure 4 (dashed box).

parent thunderstorm was also noted by *Golden* [1978] for the May 30, 1976, tornado just east of Denver.

The importance of the DCVZ for the initiation of convection and its possible relationship to tornadic development was documented by *Szoke et al.* [1984] in their study of the June 3, 1981, Denver area tornadoes, which caused about 15 million dollars in damage. They speculated that the Denver area thunderstorm became tornadic at least in part because of low-level vertical vorticity along the DCVZ.

Subsequent tornado events have been observed with Doppler radar and other supplementary instruments, allowing for a more complete picture of the tornadogenesis process for these types of storms. Using independent data sets, *Brady and Szoke* [1989] and *Wakimoto and Wilson* [1989] documented tornado development and proposed similar tornado mechanisms. Brady and Szoke studied the July 26, 1985, tornado that developed near Erie ("eri" in Figure 1), Colorado, within 20 km of a Doppler radar. They found that low-level circulations of ≥ 5 m s^{-1} differential velocity over an $\sim$500-m diameter were evident along the DCVZ before any clouds had formed along the zone. Extremely rapid thunderstorm development occurred along the DCVZ after a thunderstorm outflow boundary from weak thunderstorms over the foothills to the west collided with the DCVZ. Doppler data showed that the tornado formed in association with an intensified low-level circulation, presumably owing to stretching of air columns in the vertical beneath the vigorous convection that developed after the intersection. The tornado began before any precipitation had fallen from the storm, and no midlevel rotation (down to the maximum radar observable scale for their case, an ~300-m by 350-m radar gate) in the cloud preceded the tornado. Brady and Szoke made a detailed comparison with the life cycle of a waterspout, as documented by *Golden* [1974*a*, *b*]. Because of the similarities in the formation mechanisms they used the name "landspout" to describe this type of tornado. A schematic from Brady and Szoke of the typical life cycle of the landspout is shown in Figure 2. The landspout terminology was first used by *Bluestein* [1985] to document an atypical Oklahoma tornado that formed under conditions of weak vertical wind shear. Bluestein observed the landspout under "a line of rapidly growing towers along a cold front."

Wakimoto and Wilson [1989] used data collected in 1987 during CINDE to document the single-Doppler radar and visual characteristics of 27 vortices, many of which were tornadoes. Their schematic for the nonsupercell tornado is shown in Figure 3. Thus two independent studies documented tornadogenesis in association with the stretching of low-level vertically oriented cyclonic vorticity located along a boundary (or intersecting boundaries) beneath a rapidly developing parent cloud. In both studies a midlevel mesocyclone was not present before tornado formation.

Evidence that tornadoes develop elsewhere from storms without mesocyclones is found not only from Bluestein's observations, but also from the Oklahoma observations of *Burgess and Donaldson* [1979]. Additionally, Brady and Szoke speculated that areas like Florida might be especially susceptible to nonmesocyclone or landspout tornadogenesis, because of the common occurrence of suitable low-level boundaries (the sea breeze and Lake Okeechobee breeze fronts).

A vorticity analysis using multiple-Doppler data of one of the CINDE cases by *Wilczak et al.* [1992] suggests a somewhat more complex picture for the July 2, 1987, tornado. They found that the tornado developed from both vertical stretching and the tilting of environmental horizontal vorticity after the collision of a thunderstorm outflow boundary with a stationary boundary. The environmental horizontal vorticity was apparently enhanced in the vicinity of the collision. This storm appears similar to two of the tornadoes in the May 1984 case studied by *Wilson* [1986], where the tornadoes were produced from more organized storms than those associated with the nonmesocyclone type tornadoes noted earlier, but with the boundaries needed for tornadogenesis already present, rather than produced by the storm itself (as in the typical evolution of a supercell tornado).

Wilson also studied three smaller tornadoes that occurred during the May 1984 case, of the type often referred to as "gustnadoes." His study may be the first Doppler radar documentation of the gustnado phenomenon, which is a relatively common type of weak tornado that occurs in Colorado and elsewhere. The term gustnado refers to a type

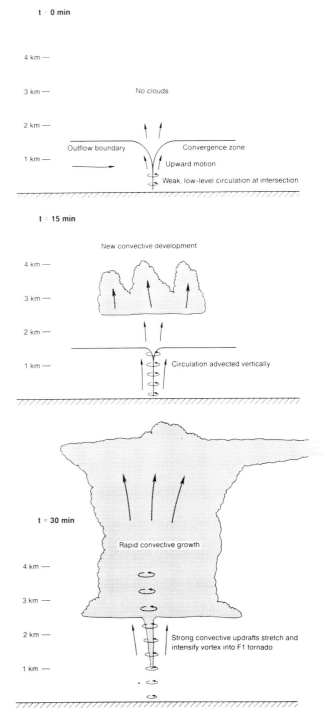

Fig. 2. Schematic of the typical landspout, from *Brady and Szoke* [1989].

of tornado that forms along the gust front from a thunderstorm and is removed from the parent storm and any associated mesocyclone, if present. *Forbes and Wakimoto* [1983] discuss the range of vortices that they believe fit the

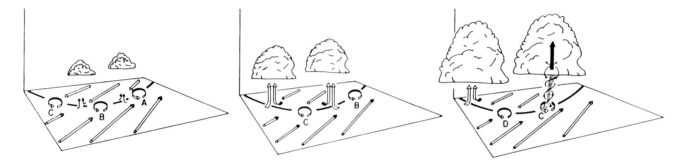

Fig. 3. Schematic of tornadogenesis from nonsupercell storms, from *Wakimoto and Wilson* [1989].

classification of a tornado and suggest including the gustnado because it is associated with a thunderstorm. The stretching mechanism responsible for the landspout or nonmesocyclone tornado also occurs with the gustnado, but the lack of a vigorous updraft over the low-level vortex (in many cases) prevents the gustnado from becoming very strong and could be considered a distinguishing factor when compared with the landspout or nonsupercell tornado. Whatever the terminology used, the important point is that these tornadoes are distinguished from supercell tornadoes by the lack of a mesocyclone in the parent thunderstorm.

The discussion above has emphasized the occurrence of tornadoes from storms that do not possess the typical characteristics of a tornadic supercell. Although these types of tornadoes account for many of those observed in Colorado, especially closer to the Denver area, long-lived supercell storms do occur in Colorado, as documented for the 1972 Fleming storm [*Browning and Foote*, 1976] and the Limon tornado of June 6, 1990. No detailed study has been done to indicate the frequency of such supercells in Colorado or the percentage that produce tornadoes. The above studies suggest that a variety of tornadic storm types exist in Colorado, from the supercell to the nonsupercell or landspout type, with a type of "hybrid" storm between these extremes that might possess a midlevel mesocyclone but would not develop a tornado without the proper preexisting boundary configuration and interaction.

3. THE JUNE 15, 1988, TORNADOES

The four tornadoes that developed in and near the Denver area on June 15, 1988, were the most damaging in Denver since the June 3, 1981, tornadoes [*Szoke et al.*, 1984]. The two tornadoes that touched down in Denver itself were classified as F2, with one having a small area of F3 damage (Table 1).

The day was characterized by a DCVZ, and although the tornadoes eventually developed in the area where the convergence zone was located, they formed only after two thunderstorm outflow boundaries intersected at the location of the DCVZ. A detailed multiple-Doppler study of this case is under way [*Roberts and Wilson*, 1989; *Wilson and Roberts*, 1990], and preliminary results indicate that the storms did not possess a midlevel mesocyclone before the tornadoes formed. Instead, the rotation associated with the tornadoes began at the surface, intensified, and built upward with time beneath the storm updrafts, similar to the cases documented earlier. In light of this Doppler study and its implications for tornadogenesis with the circulation intensifying from the surface upward, our assumption is that the condensation funnels are composed of low-level air rising upward. Thus we feel that surface-based values of moisture are most relevant for this case when discussing the funnel characteristics.

In the remainder of this section we concentrate on the visual contrast between the tornadoes and examine mesonet data to discern the reason for the variation in appearance.

Visual Observations

When the authors arrived in Denver (at location L1 in Figure 4) at about 2200 UTC (1600 MDT; all subsequent times will be UTC), a north-northeast to south-southwest expanse of relatively uniform cloud bases extended to either side of our location. Initially, a few large raindrops fell at location L1, but these soon ended, and we saw the first

TABLE 1. Summary of the June 15, 1988, Denver Area Tornadoes

Tornado	Time, UTC	Fujita Rating	Damage	Visual Characteristics
T1	2205–2213	F1	open terrain	dust column, tiny condensation funnel
T2	2215–2236	F2	old neighborhood with tall trees	huge condensation funnel
T3	2218–2235	F2–F3	commercial area	moderate size condensation funnel
T4	2220–2233	F1	open terrain	huge dust column

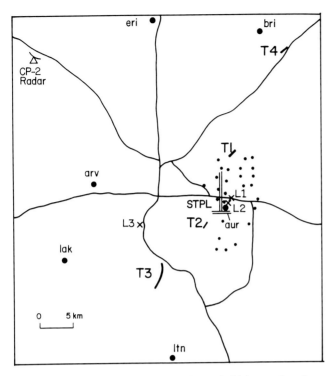

Fig. 4. Map of the Denver area showing main highways, locations of tornadoes (T1–T4), photographer locations (L1–L3), and mesonet stations (large circles for the larger-scale mesonet and smaller circles for the more dense mesonet).

tornado form to the north (T1 in Figure 4). Figure 5 shows this tornado at ~2208 from L1. The characteristics of the tornado are fairly classic for this area: a tiny condensation funnel extends from a flat cloud base, and the tornado itself is defined by a column of dust extending from the ground to cloud base. As with most of the tornadoes in the Zipser and Golden case, the tornado was at the extreme west edge of the storm. T1 passed mainly over an open area with very few structures.

Within minutes after the first tornado had ended, another funnel cloud appeared a few kilometers to the south. Unlike the first tornado, the condensation funnel associated with this tornado (T2) continued to grow, and by ~2220 it had become very large with condensation nearly to the ground. A photograph of this tornado at ~2223 from L2 is shown in Figure 6. During the time of this photograph, occasional wisps of condensation cloud were seen at and near the ground, with an area of clear air remaining between the ground and the lowest extent of the funnel. In this case, no dust cloud was observed because the tornado was passing over an older residential area noted for its boulevards with very large trees. The increased roughness in this area may have reduced the actual surface damage, in the manner discussed by *Elsom and Meaden* [1982].

While T2 was on the ground, a third tornado (T3) developed 7 km to the southwest. This tornado was not visible from L2, but a photograph taken from L3 at ~2225 is shown in Figure 7. Although the condensation funnel associated with T3 is much larger than that associated with T1, it is considerably smaller than T2's funnel. Occasionally, a very narrow additional condensation funnel was visible in a portion of the area beneath the main funnel, faintly visible in the original photograph. A debris cloud is apparent in the video of this tornado as it passed over a commercial area of south Denver, causing F2 and some F3 damage. In the video for T2 and T3, it is clear that the rotation in the condensation funnel of T3 is considerably more rapid than the slowly turning rotation in the larger funnel with T2 (at a comparative vertical location midway between the ground and cloud base).

Finally, a fourth tornado (T4, Figure 8) formed minutes after T2 and T3 had developed, just to the southeast of the town of Brighton ("bri"). It touched down in an open field and was marked by an impressive cloud of dust that extended to cloud base. Little if any condensation cloud was evident with T4, so visually it resembled T1.

In summary, the appearance of the two northern tornadoes (T1 and T4) was similar to the "typical" nonmesocyclone tornado of this area; they had very small (if any) condensation funnels and were made visible by dust clouds. In contrast, T2 had one of the largest condensation funnels observed in this area, whereas the southernmost tornado, T3, had a somewhat smaller condensation funnel.

Overview of Conditions

The overall synoptic scenario was like that described by *Doswell* [1980] for multiday severe weather events over the High Plains. The event was preceded by a cold front passage from the north early on June 13 that began several days of thunderstorms and severe weather in eastern Colorado.

The Denver sounding at 1200 had a conditionally unstable lapse rate, with a potential buoyancy (CAPE) of 1154 J kg^{-1} and a Lifted Index of -5.5°C for predicted afternoon conditions (the actual prestorm dewpoint in the Denver area was higher than the value used for this calculation). An estimate from profiler winds of the hodograph for Denver at the time of the tornadoes is shown in Figure 9. The combination of increased southerly flow, following the passage of the outflow boundary from the south, and midlevel west to northwest flow created ~20 m s^{-1} of shear in the lowest 5 km. Wind profiler observations from sites northeast of DEN indicated stronger midlevel northwest winds that increased during the day and, coupled with increasing low-level southeasterly flow, might explain the existence of a long-lived supercellular storm in extreme northeastern Colorado at the time of the Denver area tornadoes.

With ambient southeasterly flow, a DCVZ had developed by late morning, approximately north-south across the eastern portion of Denver. Convection began by late morning over the higher terrain along the Front Range. By noon (1800 UTC), storms had moved out onto the plains and developed eastward along outflow boundaries. These early storms tended to weaken as they moved east away from the higher

358 TORNADO CHARACTERISTICS

Fig. 5. Photograph of T1 at ~2208, looking north from location L1. (Photograph by E. Szoke.)

terrain because temperatures had not yet warmed to convective levels over the plains.

As noted by *Purdom and Weaver* [1990], who studied this event using visible satellite data, the evolution of development on this day was rather complex. By 1900 some cells had begun to form along the DCVZ boundary, but the intersection of an outflow boundary with the DCVZ failed to initiate further growth. This has been found in other cases

Fig. 6. Photograph of T2 at ~2223, looking south from location L2. (Photograph by E. Szoke.)

Fig. 7. Photograph of T3 at ~2225, looking south-southeast from location L3. (Photograph by S. McGinniss of the Colorado State Patrol.)

(see, for example, *Szoke and Brady* [1989]) and reflects surface temperatures too low to overcome the capping inversion even with the added lift provided by a boundary collision. Purdom and Weaver indicated that by 2000, one main outflow boundary had become organized from about Platteville ("ptl") south into the Denver area, with the strongest push to the south toward the Palmer Divide. Northward moving outflow from storms that had developed over the western portion of the higher terrain of the Palmer Divide intersected the southward moving outflow and the southern portion of the DCVZ to initiate a new thunderstorm just southeast of Denver around 2000. This storm grew strongly after 2030 and produced golf ball size hail east of Littleton ("ltn") at 2115. The storm developed a midlevel

Fig. 8. Photograph of T4 at ~2225, taken off of video from a television station helicopter. (Photograph courtesy of KMGH-TV, Channel 7, Denver.)

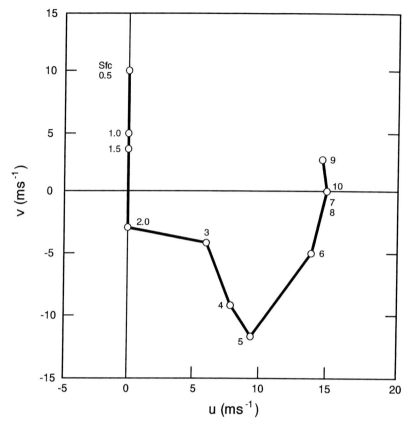

Fig. 9. Hodograph (heights in kilometers AGL) for DEN at 2200, using the DEN surface observation, DEN profiler winds through 4 km, and PTL profiler winds from 5 to 10 km.

mesocyclone (as observed by the Mile High 10-cm wavelength Doppler radar [*Pratte et al.*, 1991], located 15 km northeast of DEN) and visually had a well-defined lowered base (wall cloud) by 2100. However, the storm quickly died as it moved east into the cold air behind the northward moving outflow.

By contrast, a strong thunderstorm that had formed well west of Fort Collins (FCL) over the mountains maintained its intensity as it moved eastward. This storm had a midlevel mesocyclone as early as 1945 as it was just coming off the foothills. Of importance to events in the Denver area was a very organized strong outflow produced by the storm that moved southward.

What followed between 2130 and 2200 is an ideal type of boundary collision that could generate low-level cyclonic vortices and lead to a tornado outbreak of this type. The detailed observations follow in the next section, but in essence the two outflow boundaries intersected at the location of what remained of the DCVZ, in an area of full sunshine and possibly deeper moisture (owing to the DCVZ). Additionally, the boundaries intersected in such a manner (not head on) as to maximize the cyclonic shear in and north of the Denver area along the intersection.

Detailed Surface Observations

In addition to the 5-min data from the mesonetwork shown in Figure 1, a much smaller-scale mesonetwork of surface stations was in place around Stapleton Airport, as shown in Figure 4. This network was established to make detailed 1-min observations of surface parameters for an experiment to measure and predict downburst/microburst winds. Although individual pressure observations from the mesonet were somewhat suspect, the temperature and moisture measurements were accurate. Therefore temperature and relative humidity are presented in the following figures.

In terms of the tornado intensities, it is difficult to be certain of the true maximum tornado winds. Although *Roberts and Wilson* [1989] and *Wilson and Roberts* [1990] have examined the tornadoes using several Doppler radars at close range, the maximum resolution of the data is ~250 by 250 m, so that the actual intensity of the tornadic winds cannot be determined. Their vorticity estimates indicated maximum values of 200 ($\times 10^{-3}$ s^{-1}) for T1, 250 for T2, and 125 for T3 (T4 was not studied). It is possible that the smaller values for T3, which actually had the highest F rating, were caused by a narrower vortex not being suitably resolved.

Spectra data, if available, would provide a more certain estimate of the actual winds in the tornadoes. Photogrammetric estimates might also be possible, although these have not been attempted.

The F ratings in Table 1 would indicate that T2 and T3, the tornadoes with the larger condensation funnels, were stronger than T1 and T4, which had funnels composed mainly of dust. If the F ratings were in fact representative of the tornado intensities, then the differences in the visual appearance could, at least in part, be accounted for by T2 and T3 being stronger tornadoes, given that stronger rotation would lead to a lower pressure deficit in the tornado and a comparatively longer condensation funnel, other factors being equal. However, as noted in the earlier discussion and summarized in Table 1, T1 and T4 passed over vastly different terrain, with virtually no structures that could be damaged. This makes the F scale a less than satisfactory means of comparing intensities for this case (a more general problem discussed by *Grazulis* [this volume]). The Doppler studies cited above suggest that the tornadoes are more comparable in intensity than the F scale ratings would indicate, but given the maximum resolution of the radar data, one cannot really determine by what degree the actual tornadic winds differ. We believe there is uncertainty regarding the comparative strengths of the tornadoes, but there is insufficient evidence to be certain of the differences. Regardless of the possible variations in intensity, the appearance of the funnels is consistent with the variations in the low-level moisture that will now be presented and therefore consistent with our hypothesis that such moisture variations can account for the funnel appearances.

Figures 10a through 10f present the mesonet data from the two networks for three representative times during the tornado outbreak. Images for the same times from the Mile High Doppler radar, located ~15 km northeast of DEN, are displayed in Plates 1a and 1b (0.7° elevation angle scans) and Plate 1c (4.0° scan). The first time shown in Figures 10a and 10b is 2205, when the first tornado, T1, had touched down. It is apparent from the analyses that at the time of the first tornado the air was rather dry at low levels, with relative humidity (RH) values less than 50% north of Denver. Additionally, T1 occurs at the very western edge of the parent cloud, away from any moist low-level air. The more detailed mesonet data for 2205 (Figure 10b) show the highest RH well to the southeast of T1. This moisture had increased rapidly from values more uniformly between 45% and 50% before 2200, as rain began to fall from the eastern portion of the developing line.

During the 2218–2220 time period (Figures 10c and 10d), three tornadoes were in progress (T1 had ended at 2213). By this time the strongest low-level reflectivity (Plate 1b) consisted of a line of several cells extending from east of "bri" southward to eastern Denver county, consistent with the higher RH located in the same area that would be the result of downdrafts and raining out of the initial cell in the line associated with T1. At higher levels, greater reflectivity extended into southern Denver and over T3, indicating active updraft in this area. Similar to the earlier tornado, T1, tornado T4 also occurred at the western edge of the echo and the northern end of the line (Plate 1b). From Figure 10c it is clear that dry northerly flow was present just to the west of T4, from the observation at "bri," located 3 km to the west-northwest, which reported a northerly wind and an RH of only 40% at 2220. We hypothesize that the visual appearance of T4 (Figure 8), composed mainly of a dust cloud, resulted from the tornado occurring at the western edge of the echo, in close proximity to the dry, low-level northerly flow.

By contrast, tornado T2 with the huge condensation funnel is seen from Figure 10c to be located closer to more moist air. T2 was ~3 km west of a broad area of ≥60% RH, while the earlier T1 had been surrounded by drier low-level air (see Figures 10a and 10b). Analysis of the more detailed mesonet in Figure 10d indicates two separate areas of higher (≥60%) RH in the larger area shown in Figure 10c to the east of T2. The area of higher RH northeast of STPL was the result of rain from the cell associated with the first tornado, T1. The higher RH just to the east of T2 was from the echo in the line of cells that was directly associated with T2, and this area of moisture increased during the lifetime of the tornado, as seen in Figure 10f at 2230. Note also from Figure 10f the strong easterly flow from the very moist air toward T2, with the RH at the station just to the east of T2 having increased to 96%, with an easterly wind of 11 m s^{-1}.

The detailed mesonet's western edge is just near the location of T2, so it does not resolve the drier northwesterly flow to the west side of the intersecting boundaries where the tornado had formed. Single-Doppler velocities though indicate that the stronger northwesterly flow, which was also dry as indicated from Figures 10c and 10e, was at least 5–10 km west of T2 between 2215 and 2230. The low-level radar reflectivity did indicate that T2 also formed at the western edge of the echo (Plate 1b), although not as much at the extreme western edge as for T1 and T4. Our hypothesis for the appearance of T2 is that the higher low-level moisture values just to the east of T2 were in large part responsible for the unusually large condensation funnel observed.

The conditions associated with T3 were somewhat more complex. The condensation funnel associated with T3 was relatively large for this area and considerably different in appearance than T1 and T4. However, except during a short period near the end of its lifetime, the condensation funnel of T3 was much narrower and did not extend as near to the ground as for T2. Like T2, T3 formed in an area of much higher low-level moisture than T1 and T4. Unfortunately, the detailed mesonet is more than 6 km to the northeast of T3. However, examining the less dense mesonet data (Figure 10e) together with the Doppler radar data indicates that three different airflows were probably involved in the low-level air feeding the tornadic storm: (1) moist air to the east and northeast of the storm, some of which was also involved in T2; (2) drier northerly flow to the west that had surged southward; and (3) a surge of relatively warm and dry air from the southwest. Thus T3 essentially evolved at a triple-

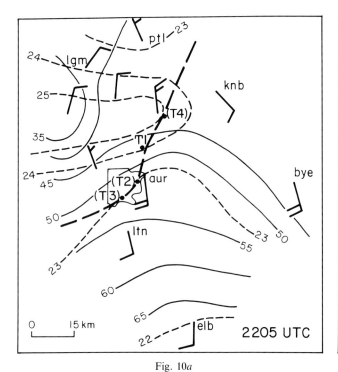

Fig. 10a

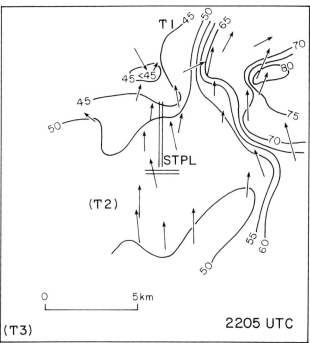

Fig. 10b

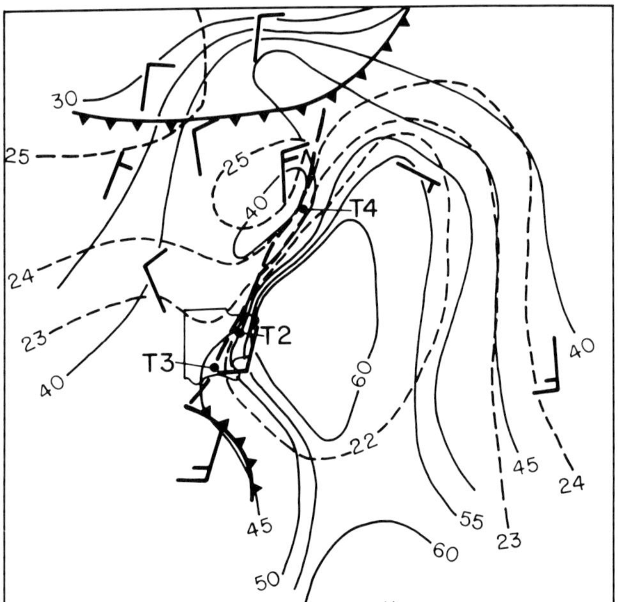

Fig. 10c

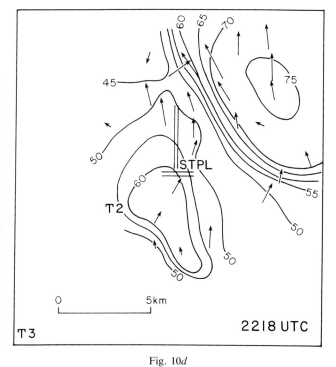

Fig. 10d

Fig. 10. Analyses of relative humidity (percent) for the two mesonets. Also shown for the larger-scale mesonet (Figures 10a, 10c, and 10e) is a temperature analysis, with values reduced to the elevation of DEN (degrees Celsius, short dashes) wind barbs (full wind barb, 5 m s^{-1}; short barb, 2.5 m s^{-1}), and boundaries. Arrows are displayed for the winds with the smaller-scale mesonet (Figures 10b, 10d, and 10f).

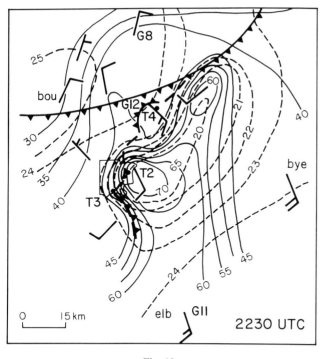
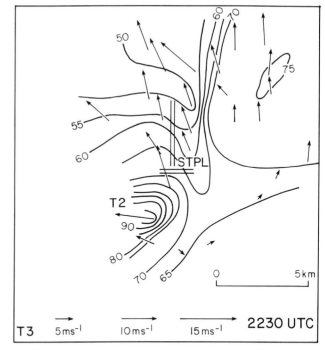

Fig. 10e

Fig. 10f

Fig. 10. (continued)

point intersection, which might explain why, at least from the video, it appears to have the most intense vortex. Interestingly, this triple-point configuration resembles the schematic for the low-level boundaries with a supercell tornadic storm [*Lemon and Doswell*, 1979], although in our case the boundaries evolved from the interaction of nearby events, rather than an evolution of the storm itself. The source of the dry air from the southwest was a weak downburst from a separate echo that had moved east from the mountains. From the video and Doppler radar data it appears that the increase in the size of the visible condensation funnel toward the end of the lifetime of T3 occurred as more precipitation began falling from the cloud line.

4. SUMMARY AND CONCLUSIONS

Studies of the June 15, 1988, tornado outbreak in the Denver area indicate this is a good example of tornadogenesis resulting from the intersection of two thunderstorm outflow boundaries with a stationary boundary. The intersection occurred in such a manner as to maximize the cyclonic shear across the intersection, yielding a number of small-scale centers of vertical vorticity [*Wilson and Roberts*, 1990]. Some of these were associated with tornadogenesis as they were stretched in the vertical beneath vigorously growing storms initiated by the colliding boundaries. This scenario for nonsupercell tornadogenesis is consistent with that documented by *Wakimoto and Wilson* [1989] and *Brady and Szoke* [1989] from studies of other Colorado events.

The differing appearances of the tornadoes for the June 15 case were related to the low-level moisture present near the tornado circulations using detailed mesonet observations. The observations are consistent with the appearance of the tornadoes and support the idea that variations in the low-level moisture could be responsible for both the enormous funnel that occurred with T2 and the lack of condensation funnels with T1 and T4. The wide variations in conditions over such a small area and the apparent effects on the visual appearance of the tornadoes raise some interesting questions regarding spotter observations of tornadoes. The size of the visible funnel cloud is quite often used as an indication of tornado intensity when observing tornadoes from a distance. Our observations suggest that this may not always be a reliable indicator. Whether our results for this type of tornadogenesis apply to more supercellular storms is open to question, but certainly variations in the boundary layer moisture occur. Additionally, visual observations of tornadoes associated with heavy-precipitation supercells alone might be considerably different than those of the same strength tornado associated with a low-precipitation storm [*Doswell et al.*, 1990].

Considerable progress has been made in the observation and understanding of tornadoes of this type over the past decade, primarily from observations obtained through a number of field programs. A good conceptual model of this type of tornadogenesis and the antecedent conditions to be observed now exists [*Brady and Szoke*, 1989; *Wakimoto and*

364 TORNADO CHARACTERISTICS

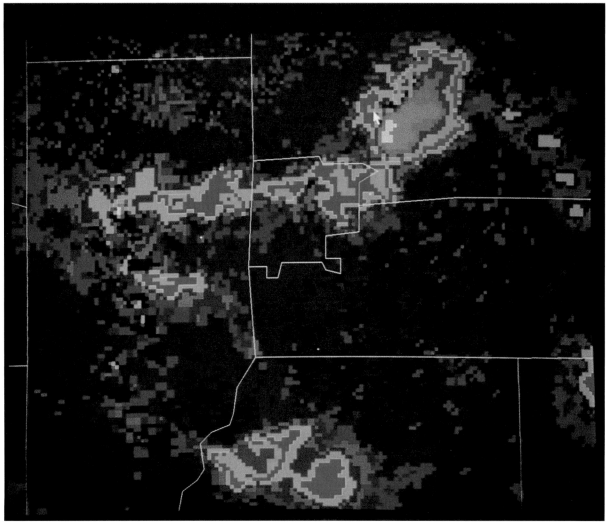

Plate 1a

Plate 1. Doppler radar reflectivity images from Mile High radar for the same times as in Figure 10. Elevation angles are 1.2° for Plates 1a and 1b and 4.0° for Plate 1c. The white arrow marks the location of (a) T1, (b) T4, and (c) T3. The county outlines are shown for reference to the Denver county outline in Figures 10a, 10c, and 10e.

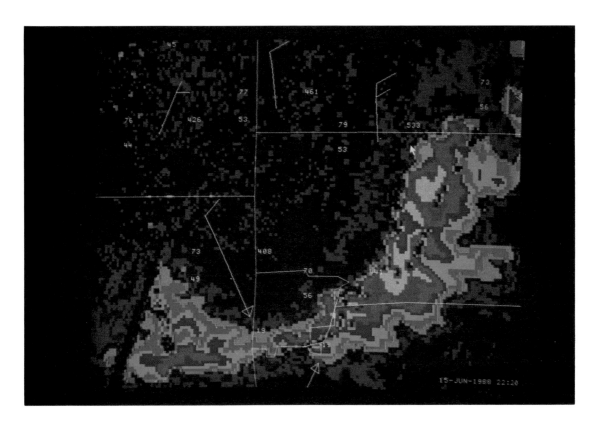

Plate 1b

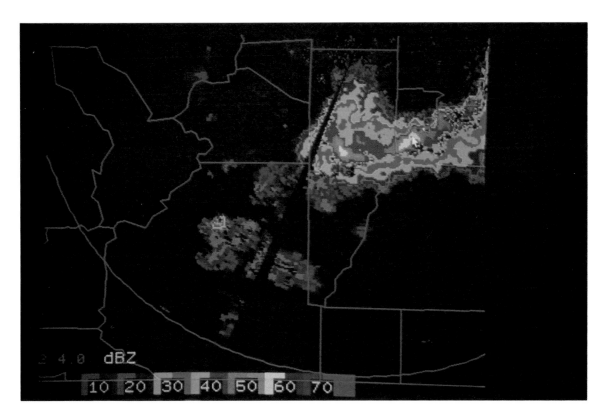

Plate 1c

Wilson, 1989]. Nonetheless, observation of these events over the years from the forecasting standpoint indicates that there is often a fine distinction between the production of tornadoes from the type of interaction documented here and just the formation of thunderstorms. A number of the thunderstorms forming along the DCVZ will develop funnel clouds that do not become tornadoes. Further documentation of a number of cases over a long period would be useful in determining the distinctions between the three outcomes: ordinary thunderstorms, funnels only, and tornadoes. It is an important distinction from the public's perspective, since unlike conditions associated with supercell storms, in many cases the tornado is the only form of severe weather that occurs with these nonmesocyclone type storms.

Acknowledgments. Thanks to Rita Roberts (NCAR) for useful discussions regarding Roberts and Wilson's Doppler analyses of the June 15 tornadoes and for help in getting the smaller-scale mesonet data. The authors appreciate helpful reviews of this manuscript by Morris Weisman (NCAR), Christopher Church, and an anonymous reviewer. The National Center for Atmospheric Research is sponsored by the National Science Foundation.

REFERENCES

Bluestein, H. B., The formation of a "landspout" in a "broken-line" squall line in Oklahoma, in *Preprints, 14th Conference on Severe Local Storms*, pp. 267–270, American Meteorological Society, Boston, Mass., 1985.

Brady, R. H., and E. J. Szoke, A case study of nonmesocyclone tornado development in northeast Colorado: Similarities to waterspout formation, *Mon. Weather Rev., 117*, 843–856, 1989.

Browning, K. A., and G. B. Foote, Airflow and hail growth in supercell storms and some implications for hail suppression, *Q. J. R. Meteorol. Soc., 102*, 499–533, 1976.

Burgess, D. W., and R. J. Donaldson, Contrasting tornadic storm types, in *Preprints, 11th Conference on Severe Local Storms*, pp. 189–192, American Meteorological Society, Boston, Mass., 1979.

Doswell, C. A., Synoptic-scale environments associated with High Plains severe thunderstorms, *Bull. Am. Meteorol. Soc., 61*, 1388–1400, 1980.

Doswell, C. A., A. R. Moller, and R. Przybylinski, A unified set of conceptual models for variations of the supercell theme, in *Preprints, 16th Conference on Severe Local Storms*, pp. 40–45, American Meteorological Society, Boston, Mass., 1990.

Elsom, D. M., and G. T. Meaden, Suppression and dissipation of weak tornadoes in metropolitan areas: A case study of greater London, *Mon. Weather Rev., 110*, 745–756, 1982.

Forbes, G. S., and R. M. Wakimoto, A concentrated outbreak of tornadoes, downbursts and microbursts, and implications regarding vortex classification, *Mon. Weather Rev., 111*, 220–235, 1983.

Golden, J. H., The life cycle of Florida Keys' waterspouts, *J. Appl. Meteorol., 13*, 676–692, 1974a.

Golden, J. H., Scale-interactions for the waterspout life cycle, *J. Appl. Meteorol., 13*, 693–709, 1974b.

Golden, J. H., Picture of the month: Jet aircraft flying through a Denver tornado?, *Mon. Weather Rev., 106*, 575–578, 1978.

Grazulis, T. P., A 110-year perspective of significant tornadoes, this volume.

Lemon, L. R., and C. A. Doswell, Severe thunderstorm evolution and mesocyclone structure as related to tornadogenesis, *Mon. Weather Rev., 107*, 1184–1197, 1979.

McCarthy, J., R. Roberts, and W. Schreiber, JAWS data collection, analysis highlights and microburst statistics, in *Preprints, 21st Conference on Radar Meteorology*, pp. 596–601, American Meteorological Society, Boston, Mass., 1983.

Pratte, J. F., J. H. Van Andel, D. G. Ferraro, R. W. Gagnon, S. M. Maher, and G. L. Blair, NCAR's Mile High meteorological radar, in *Preprints, 25th Conference on Radar Meteorology*, pp. 863–866, American Meteorological Society, Boston, Mass., 1991.

Purdom, J. F. W., and J. F. Weaver, A satellite perspective of the 15 June, 1988 tornado outbreak in Denver, Colorado, in *Preprints, 16th Conference on Severe Local Storms*, pp. 167–170, American Meteorological Society, Boston, Mass., 1990.

Roberts, R. D., and J. W. Wilson, Multiple Doppler radar analysis of the 15 June 1988 Denver tornado, in *Preprints, 24th Conference on Radar Meteorology*, pp. 142–145, American Meteorological Society, Boston, Mass., 1989.

Szoke, E. J., and R. H. Brady, Forecasting implications of the 26 July 1985 northeastern Colorado tornadic thunderstorm case, *Mon. Weather Rev., 117*, 1834–1860, 1989.

Szoke, E. J., M. L. Weisman, J. M. Brown, F. Caracena, and T. W. Schlatter, A subsynoptic analysis of the Denver tornadoes of 3 June 1981, *Mon. Weather Rev., 112*, 790–808, 1984.

Wakimoto, R. M., and J. W. Wilson, Non-supercell tornadoes, *Mon. Weather Rev., 117*, 1113–1140, 1989.

Wilczak, J. M., T. W. Christian, D. E. Wolfe, R. J. Zamora, and B. Stankov, Observations of a Colorado tornado, Part 1, Mesoscale environment and tornadogenesis, *Mon. Weather Rev., 120*, 497–520, 1992.

Wilson, J. W., Tornadogenesis by nonprecipitation induced wind shear lines, *Mon. Weather Rev., 114*, 270–284, 1986.

Wilson, J. W., and R. D. Roberts, Vorticity evolution of a non-supercell tornado on 15 June 1988 near Denver, in *Preprints, 16th Conference on Severe Local Storms*, pp. 479–484, American Meteorological Society, Boston, Mass., 1990.

Wilson, J. W., J. A. Moore, G. B. Foote, B. Martner, A. R. Rodi, T. Uttal, and J. M. Wilczak, Convection Initiation and Downburst Experiment, *Bull. Am. Meteorol. Soc., 69*, 1328–1348, 1988.

Zipser, E. J., and J. H. Golden, A summertime tornado outbreak in Colorado: Mesoscale environment and structural features, *Mon. Weather Rev., 107*, 1328–1342, 1979.

On the Use of a Portable FM-CW Doppler Radar for Tornado Research

HOWARD B. BLUESTEIN

School of Meteorology, Energy Center, University of Oklahoma, Norman, Oklahoma 73019

WESLEY P. UNRUH

Mechanical and Electronic Engineering Division, Los Alamos National Laboratory, Los Alamos, New Mexico 87545

1. INTRODUCTION

Although much has been learned about tornadoes in the last 15 years, we still do not know their maximum wind speeds, three-dimensional wind field, and source of vorticity, especially in supercell storms. Photogrammetric analysis of debris movies [e.g., *Golden and Purcell*, 1978] does not reveal the line-of-sight wind component and does not work inside opaque condensation funnels, where debris motion is hidden, or elsewhere where there is no debris. Serendipitous direct measurements by instruments are rare, and the instruments are often destroyed [*Davies-Jones*, 1986], while direct measurements made with instruments deliberately placed in the path of tornadoes are also rare and difficult to obtain [e.g., *Bluestein*, 1983]. Analysis of condensation funnel photographs [*Davies-Jones*, 1986] and nearby soundings [*Snow and Pauley*, 1984] based on the hydrostatic assumption has yielded wind speed estimates thought to be dubious [*Fiedler and Rotunno*, 1986], owing to strong vertical accelerations in the tornado boundary layer. Damage analysis [*Fujita*, 1981] requires that objects exist and are damaged and that we understand their response to wind [*Doswell and Burgess*, 1988].

Previous studies with fixed Doppler radars are reviewed by *Bluestein and Golden* [this volume]. Doppler radar measurements of tornadoes in supercells are not common because it is rare that tornadoes pass by close to them (i.e., within 20–50 km). The first measurements of wind speeds in a tornado with a radar were made by *Smith and Holmes* [1961]. The National Severe Storms Laboratory measured wind (power) spectra (backscattered power as a function of radial wind speed) with their 10-cm pulsed Doppler radar in only four supercell tornadoes from 1973 to 1981 [e.g., *Zrnic et al.*, 1985]. The maximum wind speeds found were around 90 m s^{-1}. Volume-averaged wind speeds have been estimated in several dozen nonsupercell tornadoes in Colorado [e.g., *Wakimoto and Wilson*, 1989]; however, volume-averaged wind speeds are less than the wind speeds determined from spectral shapes.

The use of a portable Doppler radar to make wind speed measurements in tornadoes was suggested by *Zrnic et al.* [1985]; they noted that higher spatial resolution could be obtained by transporting a radar to a location near a tornado. There are additional reasons for using a portable radar. Since the likelihood a tornado will pass near a fixed-site radar is very small, a portable radar can increase substantially the number of data sets collected. The strategy used to get near tornadoes in an intercept vehicle is given by *Bluestein* [1980]. When a radar is brought to within 3–5 km of a tornado, it is usually also possible to confirm the detection of the tornado and to map out photogrammetrically the azimuthal (cross line of sight) component of the wind if debris and cloud tags are trackable and to relate the winds to the condensation funnel, if it is present. The photogrammetrically determined wind analyses complement the radial (line of sight) measurements made by the radar. Nearby radar observations can also determine the wind field below cloud base, something that distant radars cannot do. Finally, we note that since the highest wind speeds in tornadoes are probably located in relatively small volumes, the radar's ability to measure the speeds of the fastest-moving scatterers is improved by bringing the radar as close as possible to the tornado.

2. PRINCIPLES OF FM-CW DOPPLER RADAR

The principles of FM–continuous wave (CW) Doppler radar are reviewed by *Skolnik* [1980], *Strauch* [1976], and *Doviak and Zrnic* [1984]. One way to modulate the transmitted frequency as a function of time is shown in Figure 1. The

---

The Tornado: Its Structure, Dynamics, Prediction, and Hazards.
Geophysical Monograph 79
Copyright 1993 by the American Geophysical Union.

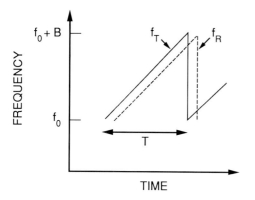

Fig. 1. Transmitted frequency f_T (solid line) and received frequency f_R (dashed line) as a function of time. The sweep width is B; the sweep repetition frequency is $1/T$. (Adapted from *Strauch* [1976].)

frequency is linearly ramped by B Hz, the sweep width, periodically every T s, the sweep repetition time; $1/T$ is the sweep repetition frequency. The latter is analogous to the pulse repetition frequency in pulsed radars. FM-CW Doppler radars are sometimes referred to as "chirped" radars, because birds make chirping sounds as the pitch they emit changes. The backscattered signal is mixed with the transmitted signal; only that part of the signal representing the difference frequency is retained.

If the radar target is stationary, then the signal is a periodic series of wave segments of identical phase (Figure 2a); the corresponding spectrum is composed of a series of equally spaced impulse functions, whose spacing is constrained to be the sweep repetition frequency (Figure 2b). The range to the target is obtained from the Doppler shift frequency of the impulse function having the greatest amplitude (i.e., from the beat frequency).

If the target is moving, then the phase of each wave segment is uniformly advanced or retarded from the phase of the previous one (Figure 3a). (Only for certain speeds will there be a phase change from segment to segment of integer multiples 2π.) It is this change in phase from segment to segment that allows one to estimate the Doppler velocity. The spectrum of the moving target (Figure 3b) is similar to the spectrum of the stationary target; it is equivalent to the stationary spectrum but shifted in frequency from the stationary spectrum due to the progressive phase shift.

In practice, one computes the spectrum from a finite series of ramps, not from an infinite series. The spectrum of a moving target computed from a series of N ramps is similar to the spectrum shown in Figure 3b, except that the impulse functions are broadened and there are some low-amplitude power fluctuations at frequencies in between the frequencies at which there are larger-amplitude peaks (not shown).

Like the pulsed Doppler radar, the FM-CW radar is susceptible to range folding and velocity aliasing. After *Strauch* [1976], we list the following important characteristics of a linearly modulated radar system:

1. The maximum unambiguous range is about $cT/4$.
2. The maximum unambiguous velocity is $\lambda/4T$, where λ is the wavelength.
3. The range resolution is $c/2B$.
4. The velocity resolution is $\lambda/2NT$.

For equivalent data record lengths, the sensitivity of an FM-CW Doppler radar is lower than that of a CW Doppler radar because the effective bandwidth required to recover

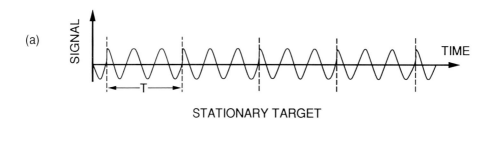

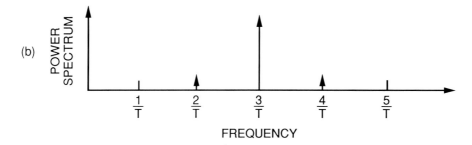

Fig. 2. (a) Received signal as a function of time for a stationary target. (b) Power spectrum for Figure 2a. (Adapted from *Strauch* [1976].)

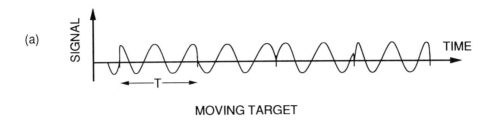

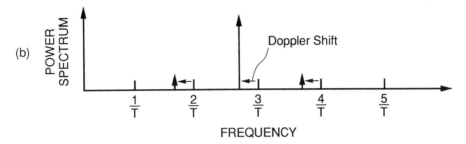

Fig. 3. As in Figure 2, but for a moving target. The spectrum is identical to that in Figure 2b, but for a moving target. (Adapted from *Strauch* [1976].)

both range and Doppler velocity is larger than that necessary to recover Doppler motion alone.

3. DESIGN CONSTRAINTS

The size of the radar places constraints on the power and wavelength. In order that the radar be portable, it must be small, the power requirements relatively low, and the antennas relatively small. Although higher power output is attainable using solid-state devices, it is prohibitively costly. The antennas cannot be too small, however, or the half-power beamwidth would be too large and the spatial resolution compromised. For a given size, an antenna for a smaller-wavelength radar can have a smaller beamwidth and increased sensitivity for small scatterers. However, if the wavelength is too short, attenuation by hydrometeors could be a problem. If the wavelength is relatively long, then attenuation is much less of a problem, but the antenna would have to be large to get a reasonably small beamwidth and high spatial resolution, and sensitivity to small scatterers is reduced.

The range folding and velocity aliasing characteristics of the radar also impose serious constraints. If the sweep repetition frequency is too low, the maximum unambiguous velocity is too low; if the sweep repetition frequency is too high, the maximum unambiguous range is too low.

Other constraints are imposed by the sweep width. If it is too low, the range resolution is poor; if it is too high, the detected bandwidth is high, and it becomes very expensive to record the data with a high-bandwidth recording device or to process the data in real time.

The radar and its recording system must have a wide enough dynamic range to make measurements of weak radar return for which the scatterers are relatively small and/or distant, yet not be saturated by ground clutter and strong scattering from nearby targets.

4. DESIGN OF THE LOS ALAMOS NATIONAL LABORATORY PORTABLE RADAR

The Los Alamos National Laboratory (LANL) radar is a solid-state, bistatic (i.e., having separate receiving and transmitting antennas) device [*Bluestein and Unruh*, 1989]. Detailed specifications of the radar are summarized in Table 1. A functional diagram of the radar is given in Figure 4. Originally a CW radar, it was modified in 1988 to have FM-CW capability so that range information could also be obtained. In FM-CW mode it operates with a sweep repetition frequency derived from the boresighted video camera. The FM-CW signals are recorded in VHS video format; it is therefore convenient to derive the FM modulation from the horizontal sync signals of the video field. FM-CW Doppler radars have been used before in hydrometeor studies [*Ligthart and Nieuwkerk*, 1989]; however, the LANL radar is the first FM-CW radar to be used for tornado research.

The bistatic radar has two parabolic antennas with linearly polarized feeds (Figure 5), separated by a septum to provide increased isolation; one is for transmitting, and the other is for receiving. Although individual antenna polarizations can be set to be horizontal or vertical, they cannot be adjusted in the field. We have used horizontal polarization for both antennas. Tapered feeds to the reflectors reduce sidelobes to over 20 dB down below the main beam and thus help to reduce near-field clutter. At 1.6 km (3 km) the azimuthal resolution is 140 m (262 m). For comparison we note that some fixed-site radars have beamwidths as low as 0.8°, for

TABLE 1. Specifications of the LANL Portable Doppler Radar (1991)

	Specifications		
	General	CW	FM-CW
Wavelength/frequency	3 cm/10.5 GHz		
Power output	1 W maximum		
Dynamic range (before recording)	70 dB		
Antenna type	2 parabolic dishes		
Antenna half-power beamwidth	5°		
Weight (approximate)			
Radar-antenna package	<50 lb		
Case containing batteries/recorders	50 lb		
Setup time (approximate)	2–3 min		
Power requirements	two 12-V rechargeable batteries		
Modulation			
Sweep repetition frequency		NA*	15.750 kHz
Sweep width		NA	1.92 MHz
Maximum unambiguous range		NA	5 km
Maximum unambiguous velocity		±292 m s^{-1}	±115 m s^{-1}
Range resolution		NA	78 m
Velocity resolution		NA	1.8 m s^{-1}
Data recording		Stereo audio HD channels of video recorder; stereo channels of hi-fi audio recorder	video channel of video recorder
Signal monitoring		earphones	video monitor
Boresighted video monitoring and recording		video monitor, video channel of video recorder	none

*Not applicable.

which the azimuthal resolution is 140 m (262 m) at 10 km (18.8 km) range.

Backscattered return is amplified directly at 10 GHz by a low-noise microwave preamplifier. This device sets the overall noise figure of the radar. The microwave detectors are conventional homodyne in-phase/quadrature (I/Q) mixers. The low-level signals are amplified at baseband (0–5 MHz) by matched operational amplifiers. High-pass filters further reduce the effects of ground clutter.

The radar and antennas are built into one package, which fits into an altitude/azimuth mounting tripod and is pointed by hand. Batteries provide over 2 hours of operation in the field. The radar can be instantly switched between CW mode and FM-CW mode with a switch accessible to the operator. A boresighted video camera for visual documentation is mounted on the radar. An umbilical cord of several cables from the radar is connected to terminals on a separate case (Figure 6), which contains the batteries, a videotape recorder and monitor, an audiotape recorder, and meters that monitor the state of the batteries. The whole system is easily set up and dismantled by three people. One sets up the tripod, while the other two mount the radar-antenna package onto the tripod. One then prepares the package, while the other two set out the case and connect the case to the radar. Care must be taken to deploy the radar away from structures (e.g., trees and houses) that might produce excessive ground clutter.

A dedicated broadband 90° audio phase shift circuit allows us to separate, in real time, the CW Doppler signals from the I/Q mixers into two broadband (0–20 kHz) audio signals, one of which contains the approaching (negative) and the other the receding (positive) Doppler shifts. These two audio signals are recorded directly. The effective channel separation is greater than 50 dB, which provides Doppler signals uncontaminated by opposing motion in the same field of view. These audio signals are monitored in a set of stereo headphones worn by the operator; approaching targets are heard in one ear only, receding in the other. With no radar return, only the steady hiss due to the noise floor of the radar is heard on both channels. The pitch of the sound from an actual Doppler-shifted return is usually proportional to the line-of-sight wind speed. Ground clutter from waving wheat, grass, and tree leaves has a low-frequency, random sound; birds and passing cars that appear in the antenna sidelobes have nearly pure tones that come and go rather quickly. Although it is somewhat of an art to distinguish and interpret in the field the meteorologically significant sounds from the extraneous ones, one soon learns to make good use of this real-time audio output of the radar as an aid in pointing the device in CW mode. One can, for example, locate the center of a tornado vortex and the region of highest Doppler velocities by scanning across the funnel cloud or debris cloud and listening to the stereo audio signal.

The power output of the radar is adjusted manually by the operator so that the recorded signal is not saturated. The power is not recorded, however, so that quantitative reflec-

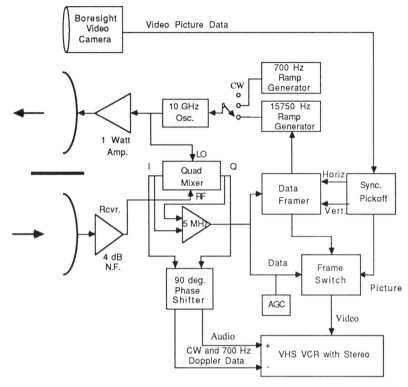

Fig. 4. Functional block diagram of the LANL portable Doppler radar. (The 700-Hz modulation capability is for another application.)

tivity measurements are not possible. In CW mode, boresighted video of the tornado is simultaneously recorded on the video channel of the videotape, with the analog I and Q signals stored on the so-called audio "HD" channels of the VHS format tape. A separate, very high quality audio recorder is used, as well, to provide redundancy (in case the video recorder or tape fails) and improved signal-to-noise recovery of the data.

In the FM-CW mode the wideband radar signal is stored on the video channel of the videotape. Coincidental voice documentation from a microphone mounted on the radar is recorded on the audio channels of the videotape and backed up on the audiotape. The quality of the video signal representing the FM-CW return is viewed on the video monitor. Since the dynamic range of the video cassette recorder (VCR) is only about 30–40 dB, which is at least 30 dB less than that of the radar, one must be very careful not to produce a saturated signal, which appears as an unstable, unsynchronized picture. Ground clutter can easily saturate the recording system, even with the high-pass filters in the radar.

Scattering return in the FM-CW mode is formatted to be recorded as VHS video fields. Each horizontal scan line of the video field represents the recovered radar signal from a single ramp of the periodic frequency sweep. For a single, stationary target, the return from each ramp is a single sinusoid whose frequency is proportional to range. The resulting visual pattern on the monitor appears as a pattern of vertical lines or bars whose separation on the screen is inversely proportional to range. If the target is moving, the progressive phase shift resulting from the motion leads to a tilt of this pattern of lines on the screen, the direction of the tilt being determined by the sense of the motion. Range folding is seen as an abrupt change in the periodicity of this pattern. Ground clutter, resulting from scattering close to the radar, is seen as an overall smooth variation in pattern brightness across the video field. The radar power level is adjusted to keep this strong signal component from saturating the overall recorded signal. Complex targets, in which all types of motion are present, result in complex visual video patterns. Such complexity appears in video data from a tornado (Figure 7). The real-time appearance of these video patterns is thus a powerful diagnostic of the actual scattering characteristics being measured.

Our scan strategy is to scan back and forth across the tornado (Figure 5) as near to the ground as we can without saturating the radar with ground clutter. We scan only in the CW mode, so that we have a record on videotape of where the radar is viewing. We switch into the FM-CW mode only when the radar is stationary. Usually we have to turn the radar power down to prevent saturating due to ground clutter and then turn it back up when in CW mode. We frequently alternate between CW mode and FM-CW mode. In CW mode we have a nearly unlimited unambiguous

Fig. 5. Photograph of University of Oklahoma (OU) graduate student J. LaDue operating the LANL radar, with OU graduate student D. Speheger (left) and OU undergraduate student H. Stein (right) attending to the radar box, while collecting data for a tornado in north central Oklahoma on April 12, 1991. (Photograph by H. B. Bluestein.)

velocity (292 m s^{-1}) but get no range information; in the FM-CW mode we do get range information but are susceptible to range folding, velocity aliasing, and more susceptible to ground clutter contamination. When the tornado is beyond the maximum unambiguous range or does not produce a satisfactory signal in the FM-CW mode, we record data only in the CW mode. We scan at higher-elevation angles only after we have obtained what we judge to be a good data set at as low an elevation angle as possible. Typically, the elevation angle is around 10°–20°. At these low-elevation angles the error added to the horizontal Doppler velocity estimates is less than 5 m s^{-1}. If there is no tornado but there is a rotating wall cloud, then we scan at high-elevation angle near the base of the wall cloud to search for signs of an incipient tornado circulation.

5. DATA PROCESSING

The recorded video data from the FM-CW mode is visually examined, frame by frame, to identify segments of data judged to be the best available. The desired field (one half of the interleaved frame) from this videotape record is captured for analysis by computer hardware and software that digitizes a single field of the record. The result consists of concatenated time series of the 252 horizontal lines of the video field, each digitized at 128 points, which forms the primary data set of 32,256 points used by the computer. From this stored data set, a segment of at least 16,384 points (128 × N, where N is usually around 200), which is judged to have the least contamination, is chosen for analysis.

Stationary ground clutter is removed from this time series by a time average that leaves only time-varying signals in the record. This is accomplished by averaging together the 128 separate returns from individual ramps and subtracting this composite average from each of the ramp returns. The subtracted signal represents the return from targets that are essentially stationary over the elapsed time of 128 contiguous ramps (~8 ms). We have found this procedure to be highly effective.

Spectral estimation of this corrected time series is obtained by using a slight modification of the Welch procedure [*Welch*, 1967] to compute the discrete Fourier transform and at the same time obtain optimum noise reduction. The total time series is divided into three overlapping segments of 16,000 points each. Each of these uniformly overlapping time series segments has a Hanning window applied before the fast Fourier transform is computed. Range normalization by a factor of $1/R^2$ is applied in the transform process. The resulting spectral densities are averaged together to provide a spectral estimation record of 8192 points. This 8000 record represents 64 range bins, each with 128 unambiguous Doppler velocity points ranging from −115 m s^{-1} to +115 m s^{-1}, with velocity resolution of 1.8 m s^{-1}. Data in the first half and last half of the first and last range bins, respectively, are contaminated by folding and are not used.

6. PRELIMINARY RESULTS

The CW wind spectrum (average of four obtained over a 0.5- to 1-s period) measured looking toward a tornado (Figure 8) in the northern Texas Panhandle on May 31, 1990, is shown in Figure 9. The highest wind velocities estimated appear to be between ±80 and ±90 m s^{-1}; they are indicated at the points where the spectrum falls off to the noise level. These wind velocities are consistent with a National Weather Service (NWS) damage survey estimate (courtesy of S. Cooper, NWS, Amarillo, Texas) of F3 damage. The width of the tornado was estimated photogrammetrically to be approximately 200–300 m in diameter; this compares to the antenna half-power beamwidth of slightly under 1000 m.

The tornado shown in Figure 8 was too far away to obtain acceptable FM-CW data. However, when the rotating wall cloud associated with the remains of the storm's circulation came closer, we were able to obtain a data set looking at the

Fig. 6. Photograph of the electronic equipment inside the radar box: audio recorder (rear left), batteries (rear center), video recorder (front left), meters (center front), and video monitor (far right). (Photograph by H. B. Bluestein.)

southern portion, where the rear flank downdraft is located. The FM-CW spectra (Figure 10) indicated return from approximately 2.3–3 km in range; this suggests that the wall cloud was about 700 m in diameter. Approaching wind speeds around 35 m s^{-1} were indicated. This estimate is consistent with gusts experienced from the southwest as the wall cloud passed by overhead and 6-cm hail fell.

During the spring of 1991, five tornado data sets were

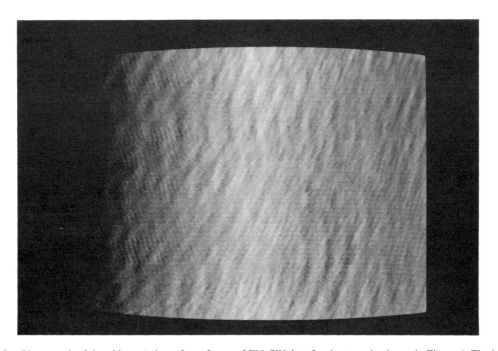

Fig. 7. Photograph of the video monitor of one frame of FM-CW data for the tornado shown in Figure 4. The image is dominated by lines that slope upward to the right, which indicate motion away from the radar. The spacing between the lines is inversely proportional to the range of the targets. The darker area on the left-hand side of the video frame is due to nearby ground clutter.

Fig. 8. Photograph looking west at a tornado located east of Spearman, Texas, at 1909 CDT on May 31, 1990. (Photograph by H. B. Bluestein.)

obtained. Of these, two included FM-CW data and excellent video documentation. A table of all tornado data sets collected is shown in Table 2. Wind speeds at or in excess of 70 m s^{-1} were common. The Doppler wind speeds in a tornado that passed by at 1.6-km range were estimated as high as 120–125 m s^{-1}. These are the highest wind speeds ever measured in a tornado by Doppler radar and verify the existence of F5 wind speeds. Detailed results of analyses of the 1991 CW and FM-CW data are forthcoming.

7. Suggested Future Improvements

The greatest problem associated with the LANL portable FM-CW Doppler radar is ground clutter contamination. It could be reduced by spectrally analyzing the data in real time, to make use of the full 70-dB dynamic range of the radar and then recording the spectra. Currently, the system is limited by the 30- to 40-dB dynamic range of the video recorder and the lack of real-time signal processing.

The radar could be made more versatile if the operator had the ability to change sweep repetition frequency and sweep width at the flick of a switch, so that the resolution and the folding and aliasing characteristics could be adapted to each specific case. The problem of how to record the data and display them on a video monitor would have to be solved. The "housekeeping" data would also have to be recorded.

Finally, if the beamwidth of the antenna could be reduced, without increasing the size of the antenna to the point at which the radar system is no longer portable, the spatial resolution of the radar could be much improved. A shorter-wavelength radar with an antenna having the same dimensions as that of the LANL radar would have increased resolution; it would also be more sensitive to smaller targets such as cloud droplets in a tornado condensation funnel.

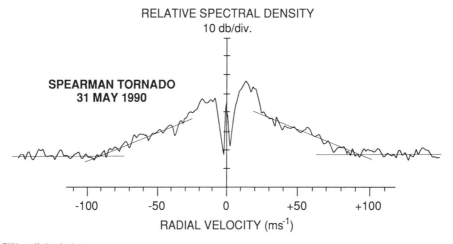

Fig. 9. CW radial wind (power) spectrum of the tornado shown in Figure 8, several minutes later. Noise level is indicated by horizontal lines; linear dropoff in logarithm of spectral density indicated by a sloping solid line.

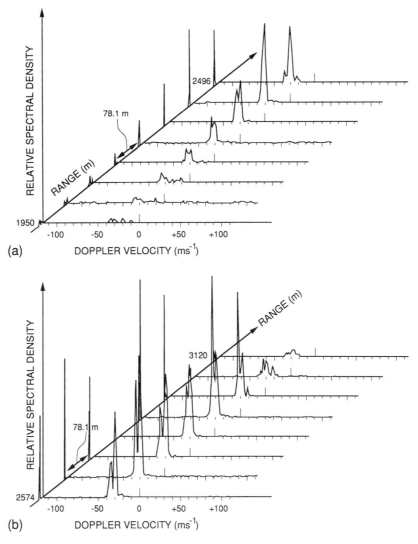

Fig. 10. FM-CW radial wind (power) spectra of the southern portion of the wall cloud associated with the tornado in Figure 8, which had dissipated 10–20 min earlier. Each spectrum is plotted as a function of range every 78 m from (a) 1950–2496 m and (b) 2574–3120 m. The ordinate is the relative spectral density plotted on a logarithmic scale.

TABLE 2. Tornado Data Sets Obtained With the LANL Portable Doppler Radar

Date	Approximate Location	Maximum Wind Speed, m s^{-1}	CW	FM-CW
May 25, 1987	Gruver, Tex.	60	yes	no
May 13, 1989	Hodges, Tex.	...	EM	EM
May 31, 1990	Spearman, Tex.	80–90	yes	WC
April 12, 1991	Enid, Okla.	75–80	yes	yes
April 12, 1991	Pond Creek, Okla.	55–60	yes	no
April 26, 1991	Red Rock, Okla.	120–125	yes*	no
May 16, 1991	Clearwater, Kans.	70–75	yes	yes
May 26, 1991	Mooreland, Okla.	100–105	yes*	no

EM, equipment malfunction; no useful data. WC, for wall cloud only.
*Receding spectrum only.

Since the number of tornado wind speed measurements made with the LANL portable radar over several years has far exceeded the number made with a fixed radar over many years, we believe that the portable radar should continue to be improved and used in future field experiments. An improved radar could provide the high resolution and sensitivity needed to map out the wind field in tornadoes. A Doppler lidar, which can sense aerosol motion in "clear air" outside a tornado debris cloud or condensation funnel could augment the portable Doppler radar, which can estimate the motion inside the debris cloud or condensation funnel. Finally, several portable Doppler radars and lidars could probe tornadoes from different viewing angles and thereby provide detailed information on the three-dimensional wind field.

Acknowledgments. This project has been funded by National Science Foundation grant ATM-8902594. M. Wolf and R. Bracht at LANL contributed to the design and modification of the radar. C. Doniec, one of LANL's summer students, was especially helpful in making the FM-CW modification. Recent storm intercept crews have included J. LaDue, H. Stein, D. Speheger, G. Martin, E. W. McCaul, Jr., S. Hrebenach, and S. Contorno. The National Severe Storms Laboratory, the National Weather Service in Norman, and students at the University of Oklahoma contributed nowcasting support. C. Church and two anonymous reviewers provided helpful comments on the manuscript.

REFERENCES

Bluestein, H. B., The University of Oklahoma Severe Storms Intercept Project—1979, *Bull. Am. Meteorol. Soc.*, *61*, 560–567, 1980.
Bluestein, H. B., Surface meteorological observations in severe thunderstorms, Part II, Field experiments with TOTO, *J. Clim. Appl. Meteorol.*, *22*, 919–930, 1983.
Bluestein, H. B., and J. H. Golden, A review of tornado observations, this volume.
Bluestein, H. B., and W. P. Unruh, Observations of the wind field in tornadoes, funnel clouds, and wall clouds with a portable Doppler radar, *Bull. Am. Meteorol. Soc.*, *70*, 1514–1525, 1989.
Davies-Jones, R., Tornado dynamics, in *Thunderstorm Morphology and Dynamics*, edited by E. Kessler, pp. 197–236, University of Oklahoma Press, Norman, 1986.
Doswell, C. A., and D. W. Burgess, On some issues of U.S. tornado climatology, *Mon. Weather Rev.*, *116*, 495–501, 1988.
Doviak, R. J., and D. S. Zrnic, *Doppler Radar and Weather Observations*, 458 pp., Academic, San Diego, Calif., 1984.
Fiedler, B. H., and R. Rotunno, A theory for the maximum windspeeds in tornado-like vortices, *J. Atmos. Sci.*, *43*, 2328–2340, 1986.
Fujita, T. T., Tornadoes and downbursts in the context of generalized planetary scales, *J. Atmos. Sci.*, *38*, 1511–1534, 1981.
Golden, J. H., and D. Purcell, Airflow characteristics around the Union City tornado, *Mon. Weather Rev.*, *106*, 22–28, 1978.
Ligthart, L. P., and L. R. Nieuwkerk, Studies of precipitation processes in the troposphere using an FM-CW radar, *J. Atmos. Oceanic Technol.*, *6*, 798–808, 1989.
Skolnik, M. I., *Introduction to Radar Systems*, 581 pp., McGraw-Hill, New York, 1980.
Smith, R. L., and D. W. Holmes, Use of Doppler radar in meteorological observations, *Mon. Weather Rev.*, *89*, 1–7, 1961.
Snow, J. T., and R. L. Pauley, On the thermodynamic method for estimating maximum tornado windspeeds, *J. Clim. Appl. Meteorol.*, *23*, 1465–1468, 1984.
Strauch, R. S., Theory and application of the FM-CW Doppler radar, Ph.D. thesis, 97 pp., Dep. of Electr. Eng., Univ. of Colo., Boulder, 1976.
Wakimoto, R. M., and J. W. Wilson, Non-supercell tornadoes, *Mon. Weather Rev.*, *117*, 1113–1140, 1989.
Welch, P. D., The use of the fast Fourier transform for the estimation of power spectra, *IEEE Trans. Audio Electroacoust.*, *AU15*, 70–73, 1967.
Zrnic, D. S., D. W. Burgess, and L. Hennington, Doppler spectra and estimated windspeed of a violent tornado, *J. Clim. Appl. Meteorol.*, *24*, 1068–1081, 1985.

Discussion

RON TAYLOR, SESSION CHAIR

National Science Foundation

PAPER F1

Presenters, Joe Golden, National Oceanic and Atmospheric Administration, Office of the Chief Scientist, and Howie Bluestein, University of Oklahoma [*Bluestein and Golden*, this volume, A review of tornado observations]

(Arnold Court, California State University) One lesson from the last symposium, 15 years ago, has not been learnt. Then we spent half a morning arguing about the meaning of wind speeds. The point was that a wind speed statement must include the volume and the time interval of the estimate. For example, an estimate might be for a cube with 10-m sides over a 1-min period. Before you compare wind speeds, you must decide what they represent. Joe, what do your estimates represent?

(Golden) The answer depends on the measuring technique. With a lidar, only the wind component along the beam is sampled. Photogrammetric estimates are conservative because the tracers range in size from dust particles to large chunks of rooftops. The same applies to Doppler radar measurements.

(Court) Yes, but you should at least indicate what volume and time interval is involved. Otherwise, people compare lidar measurements with the velocity of barn doors, which, incidentally, fly about 25% less than wind speed because they depend on relative air motion to keep them aloft.

(Golden) For the photogrammetric estimates, the sampling time is less than 10–20 seconds.

(Bluestein) For Doppler radar data, the volumes are typically several hundred meters on a side. The Doppler lidar has a very narrow, collimated beam, perhaps a few centimeters wide. The azimuthal resolution is tremendous, but the range resolution is not very good, around 150 m. In the future, the range resolution may be reduced by an order of magnitude.

PAPER F2

Presenter, Ed Szoke, Forecast Systems Laboratory and National Center for Atmospheric Research [*Szoke and Rotunno*, this volume, A comparison of surface observations and visual tornado characteristics for the June 15, 1988, Denver tornado outbreak]

(Jim Purdom, NOAA/NESDIS CIRA, Colorado State University) You said that the Denver Convergence Zone was associated with all of these tornadoes. It seemed to me that the Denver Convergence Zone did not play a role in the formation of the Denver tornadoes on this day. Also, we get other types of tornadoes in Colorado, not just ones in the narrow Denver Convergence Zone. One inflicted heavy damage on Limon, Colorado [on June 6, 1990]; this formed where a supercell intersected a boundary.

(Szoke) I meant to mention the Limon tornado. Yes, different types of tornadoes do occur in Colorado. I read your work with John Weaver on satellite analysis of the day of the Denver tornadoes. I believe that there was still a weak Denver Convergence Zone present in midafternoon just prior to tornado development and that two outflows collided right on the zone. Obviously, the presence of the outflows increased the threat of severe weather considerably. We have found that the probability of tornadoes forming on the zone increases when the zone is intersected by another boundary.

PAPER F3

Presenter, Al Bedard, Wave Propagation Laboratory, NOAA (A. Bedard, Tornadoes and the hierarchy of atmospheric vortices, not in this volume)

(Joe Golden, National Oceanic and Atmospheric Administration, Office of the Chief Scientist) I agree with your conclusions except the first one ["For a range of atmospheric vortices, tangential wind speed measurements indicate solid-body rotation within the core, $1/r$ variation outside the core"]. Although

The Tornado: Its Structure, Dynamics, Prediction, and Hazards.
Geophysical Monograph 79
Copyright 1993 by the American Geophysical Union.

some of my own data on large waterspouts do indicate this profile, I believe that we have to worry about asymmetries, such as suction vortices and bands of stronger inflow, which make the flow noncyclostrophic and depart from the Rankine combined profile. Many of the most destructive tornadoes have these asymmetries, so we need better models.

(Bedard) I agree, but I think that most vortices are feeding off a quasi-axisymmetric larger-scale circulation. The complexities that you mention are small-scale features.

Paper F4

Presenter, Howie Bluestein, University of Oklahoma [*Bluestein and Unruh*, this volume, On the use of a portable FM-CW Doppler radar for tornado research]

(Ed Szoke, Forecast Systems Laboratory and National Center for Atmospheric Research) How difficult would it be to save some velocity spectra when there is a tornado close to a fixed-base Doppler radar?

(Bluestein) Ed Brandes or Don Burgess should answer that.

(Don Burgess, National Severe Storms Laboratory) The WSR-88D is not a radar that collects time series. It estimates the mean velocity in the sampling volume and the spectral width in real time, and that is all the information you get. A different radar configuration is needed to collect time series. The National Severe Storms Laboratory has a research radar with this capability.

Design for Containment of Hazardous Materials

ROBERT C. MURRAY

Lawrence Livermore National Laboratory, Livermore, California 94550

JAMES R. MCDONALD

Texas Tech University, Lubbock, Texas 79409

1. INTRODUCTION

Use of deterministic design and evaluation criteria to meet probabilistic performance goals is the approach used by Department of Energy (DOE) facilities across the United States as given in UCRL-15910, "Design and Evaluation Guidelines for Department of Energy Facilities Subjected to Natural Phenomena Hazards" [Kennedy et al., 1990]. An overview of the approach is presented in this paper.

In this paper, three types of winds are discussed: extreme (straight), hurricane, and tornado. Extreme (straight) winds are nonrotating such as those found in thunderstorm gust fronts. Wind circulating around high- or low-pressure systems are rotational in a global sense, but they are considered "straight" winds in the context of this paper. Tornadoes and hurricanes both have rotating winds. The diameter of rotating winds in a small hurricane is considerably larger than the diameter of a large tornado. However, most tornado diameters are relatively large in comparison with the dimensions of typical buildings. It is estimated that the diameter of 80% of all tornadoes is greater than 300 ft.

(Metric equivalents, in general, will not be provided. However, the conversion factors are as follows: 1 in. = ~2.5 cm; 1 ft = ~30.5 cm; 1 mi = ~1.6 km; and 1 lb = ~0.5 kg.)

Wind pressures produced by extreme winds are studied in boundary layer wind tunnels. The results are generally considered reliable because they have been verified by selected full-scale measurements. Investigations of damage produced by extreme winds tend to support the wind tunnel findings. Although the rotating nature of hurricane and tornado winds cannot easily be duplicated in a wind tunnel, damage investigations suggest that pressures produced on enclosed buildings and other structures by extreme, hurricane, and tornado winds are so similar that it is almost impossible to look at damage to an individual structure and tell which type of wind produced it. Thus the approach for determining wind pressures on buildings and other structures proposed in this paper is considered to be independent of the type of wind storm. The recommended procedure is essentially the same for straight, hurricane, and tornado winds.

UCRL-15910 wind and tornado provisions use deterministic evaluation criteria with the hazard annual probability of exceedance specified to obtain design/evaluation wind speeds. The evaluation of response and structure or equipment capacities is performed using deterministic methods which are familiar to engineers. Wind speeds are developed from probabilistic hazard curves and are used to meet performance goals for various building occupancies.

2. PERFORMANCE GOALS

Performance goals may be expressed in terms of annual probability of exceedance of some level of damage. Levels of damage can include that damage beyond which occupants are endangered, beyond which hazardous materials cannot be confined, beyond which a facility cannot safely shut down, or beyond which a facility cannot continue its mission. The performance goal for the safety of advanced light water reactors is of the order of 1.0×10^{-5} annual probability of exceedance for external events induced core damage [Electric Power Research Institute, 1990]. Performance goals upon which design/evaluation guidelines are based for DOE facilities in UCRL-15910 are presented in Table 1.

For each performance goal there are separate wind and tornado design and evaluation criteria. DOE management categorize facilities or individual structures, systems, or components depending on the cost, mission importance, or hazard to people or the environment. For UCRL-15910, a

TABLE 1. Performance Goals for Each Usage Category

Usage Category	Performance Goal Description	Performance Goal (Annual Probability of Exceedance)*
General use	maintain occupant safety	$\sim 10^{-3}$ of the onset of major component damage to the extent that occupants are endangered
Important or low hazard	occupant safety, continued operation with minimal interruption	$\sim 5 \times 10^{-4}$ of component damage to the extent that the component cannot perform its function
Moderate hazard	occupant safety, continued function, confinement	$\sim 10^{-4}$ of component damage to the extent that the component cannot perform its function
High hazard	occupant safety, continued function, very high confidence of confinement	$\sim 10^{-5}$ of component damage to the extent that the component cannot perform its function
Commercial reactor†		

*Component refers to structure, equipment, or distribution system.
†Currently beyond the scope of UCRL-15910.

DOE natural phenomena hazards panel selected reasonable and achievable performance goals, bounded by current practice:

1. For ordinary facilities the performance goals are consistent with design according to conventional building code provisions.
2. For high-hazard facilities the performance goals are comparable to performance reached by nuclear power plants as measured by probabilistic risk assessments (PRAs). (Note that the UCRL-15910 performance goals are related to individual component behavior, while PRA results are in terms of core damage.)

3. Performance Goal Achievement

Structure/equipment performance is a function of (1) the likelihood of hazard occurrence and (2) the strength of the structure or equipment item. Therefore design and evaluation criteria have been developed to attain performance goals by (1) specification of hazard probability for definition of wind and tornado loadings and (2) specification of response evaluation methods, acceptance criteria, and design detailing requirements with controlled levels of conservatism. Acceptable performance can only be achieved by consistent specification of all design or evaluation criteria elements.

UCRL-15910 wind and tornado design and evaluation guidelines contain the following provisions: (1) lateral force provisions, (2) damage control provisions, (3) detailing provisions, and (4) quality assurance and peer review provisions.

Wind/tornado performance depends on the level of the hazard and on conservatism in the response evaluation and acceptance criteria. For example, a performance goal of 1×10^{-4} can be achieved either by (1) a conservative evaluation/acceptance approach for a more frequent hazard probability, such as 1×10^{-3}, or by (2) a median-centered evaluation/acceptance approach coupled with a less frequent 1×10^{-4} hazard probability. UCRL-15910 uses the former approach because conservative evaluation/acceptance approaches are well established, extensively documented, and commonly practiced.

4. Design and Evaluation Criteria for Wind Load

A uniform approach to wind load determination that is applicable to the design of new facilities and the evaluation of existing facilities has been developed for DOE. A uniform treatment of wind loads is recommended to accommodate extreme, hurricane, and tornado winds. Buildings or facilities are first assigned appropriate usage categories. Criteria are recommended such that the performance goals for each category can be achieved. Procedures according to American Society of Civil Engineers (ASCE) 7-88 (formerly ANSI A58.1) [*American Society of Civil Engineers*, 1990] are recommended for determining wind loads produced by straight, hurricane, and tornado winds. The extreme wind/tornado hazard models developed for DOE sites were used to establish site-specific criteria for each of some 25 DOE sites [*Coats and Murray*, 1985].

The performance goals established for general use and important or low-hazard usage categories are met by conventional building codes or standards [see *International Conference of Building Officials*, 1988]. These criteria do not account for the possibility of tornado winds, because wind speeds associated with extreme winds typically are greater than those for tornadoes at exceedance probabilities greater than approximately 1×10^{-4}. For this reason, tornado design criteria are specified only for buildings and facilities in moderate and high-hazard categories, where hazard exceedance probabilities are less than 1×10^{-4}.

The traditional approach for establishing tornado design

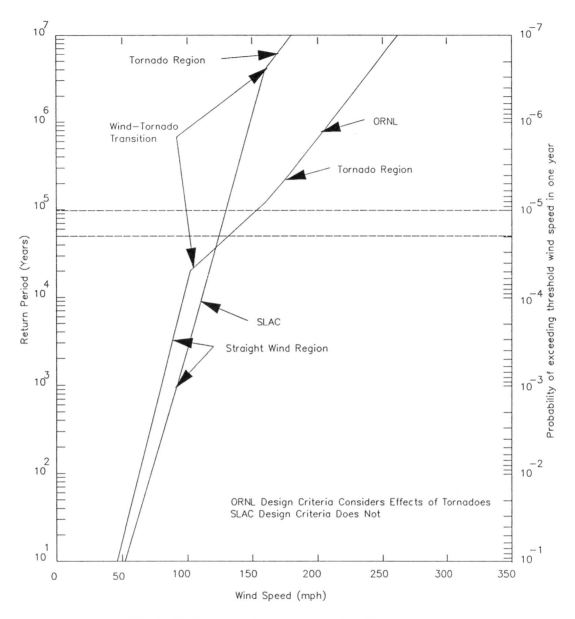

Fig. 1. Straight wind and tornado regions of wind hazard curves.

criteria is to select extremely low exceedance probabilities. For example, the exceedance probability for design of commercial nuclear power plants is 1×10^{-7} [*U.S. Nuclear Regulatory Commission*, 1974; *Kimura and Budnitz*, 1987; *Ravindra and Nafday*, 1990]. There are reasons for departing from this traditional approach. The low exceedance value for commercial nuclear power plants was established circa 1960, when very little was known about tornadoes from an engineering perspective. Much has been learned about tornadoes since that time. Use of a low-hazard probability is inconsistent with the practice relating to other natural hazards, such as earthquakes. There are many uncertainties in tornado hazard probability assessment, but they are not significantly greater than the uncertainties in earthquake probability assessment. The strongest argument against using low-probability criteria is that a relatively short period of record (40 years) must be extrapolated to extremely small exceedance probabilities. For these reasons, an alternative approach was used in the UCRL-15910 guidelines.

Establishment of Wind and Tornado Hazard Annual Probabilities

The rationale for establishing tornado criteria is described below. Figure 1 shows the tornado and straight wind hazard curves for two DOE sites (SLAC and ORNL). The wind speed at the intersection of the tornado and straight wind curves is defined for purposes of this discussion as the transition wind speed. An exceedance probability is associated with each transition wind speed. If the exceedance probability of the transition wind speed is less than 1×10^{-5} per year, tornadoes are not a viable threat to the site, because straight wind speeds are higher than tornadoes for a given return period. Thus from Figure 1, tornadoes need not be considered at SLAC, but they should be considered at ORNL.

Mean wind and tornado hazard curves were developed for each DOE site, along with the transition wind speed. Those sites with transition wind speed exceedance probabilities greater than 10^{-5} should be designed for tornadoes; others should be designed for extreme winds or hurricanes.

The tornado wind speed is obtained by selecting the wind speed associated with an exceedance probability of 2×10^{-5} per year. The value of 2×10^{-5} is the largest one that can be used and still represent a point on the tornado hazard curve. For example, the tornado wind speed for the ORNL is 130 mph (peak gust at 10 m).

A comparison of the slopes of the tornado hazard curves for the DOE sites reveals that the slopes are essentially the same even though the transition wind speeds are different. The criteria required to meet the performance goals of moderate and high-hazard facilities can be met by using multipliers that are equivalent to an importance factor in the ASCE 7-88 design procedure. The multipliers are specified in lieu of two different exceedance probabilities for moderate and high-hazard facilities. The value of the importance factor is selected to achieve lower probability of tornado damage for high-hazard facilities compared with moderate hazard facilities. The importance factors are then chosen to meet the performance goals.

In general, design criteria for each usage category include the following: (1) annual hazard exceedance probability, (2) importance factor, (3) wind-generated missile parameters for moderate and high-hazard facilities, and (4) tornado parameters for moderate and high-hazard facilities, if applicable. The criteria are formulated in such a way that a uniform approach for determining design wind loads, per ASCE 7-88, can be used for extreme, hurricane, and tornado winds.

In order to apply the ASCE 7-88 procedure, tornado gust wind speeds must be converted to fastest-mile speeds. Appropriate gust response factors and velocity pressure exposure coefficients are utilized in the process of determining wind loads. Appropriate exposure categories also are considered in the wind load calculations. Open terrain (exposure C) should always be assumed for tornado winds, regardless of the actual terrain conditions.

Criteria for Design of Facilities

The criteria presented herein are consistent with the performance goals described earlier. Buildings or facilities in each category have a different role and represent different levels of hazard to people and the environment. In addition, the degree of wind hazard varies geographically. Facilities in the same usage category, but at different geographical locations, will have different input loading specified to achieve the same performance goal.

The minimum wind design criteria for each of the four usage categories are summarized in Table 2. The recommended basic wind speeds for extreme wind, hurricanes, and tornadoes are contained in Table 3. All wind speeds are fastest mile. Minimum recommended basic wind speeds are also noted in the table.

General Use Facilities

The performance goals for general use facilities are consistent with objectives of ASCE 7-88 building class I, ordinary structures. The wind force resisting structural system should not collapse under design load. Survival without collapse implies that occupants should be able to find an area of relative safety inside the building. Breach of the building envelope is acceptable, since confinement is not essential. Flow of air through the building and water damage are acceptable. Severe damage, including total loss, is acceptable, so long as the structure does not collapse.

The ASCE 7-88 standard calls for the basic wind speed to be based on an exceedance probability of 0.02 per year. The importance factor for this class of building is 1.0. For those sites within 100 miles (160 km) of the Gulf of Mexico or Atlantic coastlines, a slightly higher importance factor is recommended to account for hurricanes.

Terrain surrounding the facilities should be classified as exposure B, C, or D, as appropriate. Gust response factors and velocity pressure exposure factors should be used according to rules of the ASCE 7-88 procedures.

Wind pressures are calculated on the walls and roofs of enclosed buildings by appropriate pressure coefficients specified in the ASCE 7-88 standard. Openings, either of necessity or created by wind forces or missiles, result in internal pressures that can increase wind forces on components and cladding. The worst cases of combined internal and external pressures should be considered, as required by the standard.

Structures in the general use category may be designed by either allowable stress design (ASD) or strength design (SD) as appropriate for the material used in construction. Except when applicable codes provide otherwise, plausible load combinations shall be considered to determine the most unfavorable effect on the building, foundation, or structural member being considered. When using ASD methods, allowable stresses appropriate for the building material shall be used with the following combinations that involve wind:

$$DL + WL \tag{1}$$

$$0.75(DL + WL + LL) \tag{2}$$

where DL is dead load, LL is live load, and WL is wind load. When using SD methods, the following load combinations that involve wind are recommended:

$$0.9DL + 1.3WL \tag{3}$$

TABLE 2. Summary of Minimum Wind Design Criteria

Building Category	General Use	Important or Low Hazard	Moderate Hazard	High Hazard
			Wind	
Annual probability of exceedance	2×10^{-2}	2×10^{-2}	1×10^{-3}	1×10^{-4}
Importance factor*	1.0 (1.05)	1.07 (1.11)	1.0 (1.05)	1.0 (1.05)
Missile criteria			2 × 4 timber plank 15 lb at 50 mph (horiz.); maximum height 30 ft	2 × 4 timber plank 15 lb at 50 mph (horiz.); maximum height 50 ft
			Tornado	
Annual hazard probability of exceedance			2×10^{-5}	2×10^{-5}
Importance factor			$I = 1.0$	$I = 1.35$
APC†			40 psf at 20 psf/s	125 psf at 50 psf/s
Missile criteria			2 × 4 timber plank 15 lb at 100 mph (horiz.); maximum height 150 ft; 70 mph (vert.)	2 × 4 timber plank 15 lb at 150 mph (horiz.); maximum height 200 ft; 100 mph
			3-in. diameter standard steel pipe, 75 lb at 50 mph (horiz.); maximum height 75 ft; 35 mph (vert.)	3-in. diameter standard steel pipe, 75 lb at 75 mph (horiz.); maximum height 100 ft; 50 mph (vert.)
				3000-lb automobile at 25 mph, rolls and tumbles

*The first value represents the importance factor for sites 100 miles (160 km) or more inland. The value in parentheses should be used at the coastline. Linear interpolation between the two values should be used for sites within 100 miles (160 km) of the coastline to account for hurricanes.

†Atmospheric pressure change.

$$1.2DL + 0.5LL + 1.3WL \qquad (4)$$

The SD method requires that the strength provided be greater than or equal to the strength required to carry the factored loads. Appropriate strength reduction factors must be applied to the nominal strength of the material being used.

Important or Low-Hazard Facilities

Important or low-hazard facilities are equivalent to essential facilities (class II), as defined in ASCE 7-88. The structure's main wind force resisting structural systems shall not collapse at design wind speeds. Complete integrity of the building envelope is not required because no significant quantities of toxic or radioactive materials are present. However, breach of the building envelope may not be acceptable if wind or water interferes with the facility function. If loss of facility function is caused by water damage to sensitive equipment, collapsed interior partitions, or excessive damage to HVAC ducts and equipment, then loss of cladding and missile perforation at the design wind speeds must be prevented.

An annual wind speed exceedance probability of 0.02 is specified, but the importance factor for important or low-hazard category structures is 1.07. For those sites located within 100 miles (160 km) of the Gulf of Mexico or Atlantic coastlines, a slightly higher importance factor is used to account for hurricane winds.

Once the design wind speeds are established and the importance factors applied, the determination of wind loads on important or low-hazard category structures is identical to that described for general use category structures. Facilities in this category may be designed by ASD or SD methods, as appropriate, for the construction material. The load combinations described for general use structures are the same for important or low-hazard structures. However, greater attention should be paid to connections and anchorages for main members and components, such that the integrity of the structure is maintained [see *McDonald*, 1988].

Moderate Hazard Facilities

The performance goal for moderate hazard facilities requires more rigorous criteria than is provided by standard building codes. For some DOE sites, tornadoes must be considered.

Extreme Winds and Hurricanes. For those sites where tornadoes are not a viable threat, the recommended basic wind speed is based on an annual exceedance probability of 1×10^{-3}. The importance factor is 1.0. For those sites located within 100 miles (160 km) of the Gulf of Mexico or Atlantic coastlines, a slightly higher importance factor is specified to account for hurricanes.

Once the basic wind speeds are established and the importance factors applied, determination of moderate hazard category wind loads is identical to that described for the

TABLE 3. Recommended Basic Wind Speeds (in Miles per Hour) for DOE Sites

DOE Project Sites	Building Category/Fastest-Mile Wind Speeds at 10-m Height					
	General Use/Wind	Important or Low Hazard/Wind	Moderate Hazard/		High Hazard/	
			Wind	Tornado	Wind	Tornado
Bendix Plant, Mo.	72	72	⋯	144	⋯	144
Los Alamos National Laboratory, N. M.	77	77	93	⋯	107	⋯
Mound Laboratory, Ohio	73	73	⋯	136	⋯	136
Pantex Plant, Tex.	78	78	⋯	132	⋯	132
Rocky Flats Plant, Colo.	109	109	138	⋯†	161	⋯†
Sandia National Laboratories, Albuquerque, N. M.	78	78	93	⋯	107	⋯
Sandia National Laboratories, Livermore, Calif.	72	72	96	⋯	113	⋯
Pinellas Plant, Fla.	93	93	130	⋯	150	⋯
Argonne National Laboratory—East, Ill.	70*	70*	⋯	142	⋯	142
Argonne National Laboratory—West, Idaho	70*	70*	83	⋯	95	⋯
Brookhaven National Laboratory, N. Y.	70*	70*	⋯	95‡	⋯	95‡
Princeton Plasma Physics Laboratory, N. J.	70*	70*	⋯	103	⋯	103
Idaho National Engineering Laboratory, Idaho	70*	70*	84	⋯	95	⋯
Feed Materials Production Center, Ohio	70*	70*	⋯	139	⋯	139
Oak Ridge National Laboratory, X-10, K-25, and Y-12, Tenn.	70*	70*	⋯	113	⋯	113
Paduach Gaseous Diffusion Plant, K. Y.	70*	70*	⋯	144	⋯	144
Portsmouth Gaseous Diffusion Plant, Ohio	70*	70*	⋯	110	⋯	110
Nevada Test Site, Nev.	72	72	87	⋯	100	⋯
Hanford Project Site, Wash.	70*	70*	80*	⋯	90*	⋯
Lawrence Berkeley Laboratory, Calif.	72	72	95	⋯	111	⋯
Lawrence Livermore National Laboratory, Calif.	72	72	96	⋯	113	⋯
Lawrence Livermore National Laboratory, Site 300, Calif.	80	80	104	⋯	125	⋯
Energy Technology and Engineering Center, Calif.	70*	70*	⋯	95‡	⋯	95‡
Stanford Linear Accelerator Center, Calif.	72	72	95	⋯	112	⋯
Savannah River Plant, S. C.	78	78	⋯	137	⋯	137

*Minimum extreme wind speed.
†Although extreme winds govern at Rocky Flats, it is recommended that facilities be designed for the tornado missile criteria. APC need not be considered.
‡Minimum tornado speed.

general use category. Facilities in this category may be designed by ASD or SD methods, as appropriate, for the material being used in construction. Plausible load combinations shall be considered to determine the most unfavorable effect on the building, foundation, or structural member being considered. When using ASD, allowable stresses appropriate for the building material shall be used with the following load combinations:

$$0.9(DL + WL) \quad (5)$$

$$0.67(DL + WL + LL) \quad (6)$$

The SD load combinations recommended for the moderate hazard category are

$$DL + 1.3WL \quad (7)$$

$$1.1DL + 0.5LL + 1.2WL \quad (8)$$

Greater attention should be paid to connections and anchorages for main members and components, such that the integrity of the structure is maintained [see *McDonald*, 1988].

A minimum missile criterion is specified to account for objects or debris that could be picked up by extreme winds, hurricane winds, or weak tornadoes. A 2 × 4-in. timber plank weighing 15 lb is the specified missile. Its impact speed is 50 miles per hour (mph) at a maximum height of 30 ft above ground level. The missile will break glass; it will perforate sheet metal siding, wood siding up to 3/4 in. thick, or form board. The missile could pass through a window or a weak exterior wall and cause personal injury or damage to interior contents of a building. The specified missile will not perforate unreinforced masonry or brick veneer walls or other more substantial walls.

Tornadoes. For those sites requiring design for tornadoes, the criteria are based on site-specific studies. The basic wind speed is associated with an annual hazard probability of exceedance of 2×10^{-5}. The wind speed obtained from the tornado hazard model is converted to fastest mile. The importance factor for the moderate hazard category is 1.0.

With the wind speed converted to fastest-mile wind and an importance factor of 1.0, equations in the ASCE standard should be used to obtain design wind pressures on the structure. Exposure category C should always be used with tornado winds regardless of the actual terrain roughness. The velocity pressure exposure coefficient and the gust response factor are obtained from appropriate tables in the ASCE standard. External pressure coefficients are used to obtain tornado wind pressures on various surfaces of the structure. A distinction is made between the main wind force resisting system and components and cladding.

If the building is not specifically sealed to maintain an internal negative pressure for confinement of hazardous materials, or if openings greater than 1 ft^2 (~0.1 m^2) per 1000 ft^3 (~28 m^3) of volume are present, or if openings of this size can be created by missile perforation, then the effects of internal pressure should be considered according to ASCE procedures. If the building is sealed, then atmospheric pressure change (APC) pressures associated with the tornado should be considered.

APC pressure is half its maximum value at the radius of maximum wind speed in a tornado. Thus critical tornado loading will be one-half the maximum APC pressure plus the maximum tornado wind pressure. A loading condition of APC alone can occur on the roof of a buried tank or sand filter, if the roof is exposed at the ground surface. APC pressure always acts outward. The effect of rate of pressure change on ventilation systems should be analyzed to assure that it does not interrupt any function or processes carried out in the facility. Procedures and computer codes are available for such analyses.

Plausible load combinations shall be considered to determine the most unfavorable effect on the building, foundation, or structural member being considered. When using ASD methods, allowable stresses appropriate for the building materials shall be used with the following load combinations that involve tornado loading:

$$0.75(\text{DL} + W_t) \qquad (9)$$

$$0.63(\text{DL} + W_t + \text{LL}) \qquad (10)$$

The SD load combinations recommended for the moderate hazard category are

$$\text{DL} + W_t \qquad (11)$$

$$\text{DL} + \text{LL} + W_t \qquad (12)$$

where W_t is tornado loading, including APC, as appropriate.

Two missiles are specified as minimum criteria for this usage category. The 2 × 4 in. timber plank weighing 15 lb is assumed to travel in a horizontal direction at a speed up to 100 mph. The horizontal speed is effective up to a height of 150 ft above ground level. If carried to a great height by the tornado winds, the timber plank could achieve a terminal vertical speed of 70 mph in falling to the ground. The horizontal and vertical speeds are assumed to be uncoupled, and they should not be combined. The missile will perforate most conventional wall and roof cladding except reinforced masonry or concrete. The cells of concrete masonry walls must be filled with grout to prevent perforation by the timber missile. The second missile is a 3-in.-diameter standard steel pipe, which weighs 75 lb. It can achieve a horizontal impact speed of 50 mph and a vertical speed of 35 mph. Its horizontal speed could be effective to heights of 75 ft above ground level. The missile will perforate conventional metal siding, sandwich panels, wood and metal decking on roofs, and gypsum panels. In addition, it will perforate unreinforced concrete masonry and brick veneer walls, reinforced concrete masonry walls less than 8 in. thick, and reinforced concrete walls less than 6 in. thick. Although wind pressure, APC, and missile impact loads can act simultaneously in a tornado, the missile impact loads can be treated independently for design and evaluation purposes.

High-Hazard Facilities

The performance goal can be achieved for this category if the main wind force resisting members do not collapse, structural components do not fail, and the building envelope is not breached at the design wind loads. Loss of cladding, broken windows, collapsed doors, or significant missile perforations must be prevented. Strong air flow through the building or water damage cannot be tolerated.

Extreme Winds and Hurricanes. For those sites that do not require specific design for tornado resistance, the recommended basic wind speed is based on an annual hazard exceedance probability of 1×10^{-4}. The importance factor is 1.0. The wind speed is fastest mile at an anemometer height of 10 m above ground level. Once the basic wind speeds are established and the importance factors are applied, determination of high-hazard facility wind loads is identical to that described for the general use category. Facilities in this category may be designed by ASD or SD methods, as appropriate, for the material being used in construction. Recommended wind load combinations are the same as for moderate hazard facilities. Greater attention should be paid to connections and anchorages for main members and components, such that the integrity of the structure is maintained [see *McDonald*, 1988].

The missile criteria are the same as for the moderate hazard category, except that the maximum height achieved by the missile is 50 ft instead of 30 ft.

Tornadoes. For those sites requiring design for tornado resistance, the criteria are based on site-specific studies. The recommended basic wind speed is associated with an annual hazard probability of exceedance of 2×10^{-5} (the same as the moderate hazard category). The wind speed obtained from the tornado hazard model is converted to fastest mile. The importance factor for the high-hazard category is 1.35.

With the wind speed expressed as fastest mile and an importance factor of 1.35, ASCE 7-88 equations should be used to obtain design wind pressures on the structure. Exposure category C should always be used with tornado winds regardless of actual terrain roughness. The velocity pressure exposure coefficient and the gust response factor are obtained from appropriate tables in the ASCE standard. External pressure coefficients are used to obtain tornado wind pressures on various surfaces of the structure. A distinction is made between the main wind force resisting system and components and cladding in determining wind pressures.

TABLE 4. Ratio of Hazard Probability to Performance Goal

Usage Category	Performance Goals	Wind/Tornado Hazard Probability	Ratio of Hazard Probability to Performance Goal
Extreme Winds			
General use	10^{-3}	2×10^{-2}	20
Important or low hazard	5×10^{-4}	10^{-2}*	20
Moderate hazard	10^{-4}	10^{-3}	10
High hazard	10^{-5}	10^{-4}	10
Tornadoes			
Moderate hazard	10^{-4}	2×10^{-5}	<1
High hazard	10^{-5}	3×10^{-6}†	<1

*Here 2×10^{-2} with $I = 1.07 \sim 10^{-2}$.
†Here 2×10^{-5} with $I = 1.35 \sim 3 \times 10^{-6}$.

If the building is sealed to confine hazardous materials, the wind and APC load combinations specified for the moderate hazard usage category also should be used for this category. The effects of rate of pressure change on ventilating systems should be analyzed. Recommended tornado wind load combinations for moderate hazard facilities also apply to high-hazard facilities.

Three missiles are specified as minimum criteria for this usage category. The 2×4- in. timber plank weighing 15 lb is assumed to travel in a horizontal direction at speeds up to 150 mph. The horizontal missile is effective to a maximum height of 200 ft above ground level. If carried to a great height by the tornado winds, it could achieve a terminal speed in the vertical direction of 100 mph. The horizontal and vertical speeds are uncoupled and should not be combined. The missile will perforate most conventional wall and roof cladding except reinforced masonry and concrete. Each cell of the concrete masonry shall contain a $\frac{1}{2}$-in.-diameter rebar and be grouted to prevent perforation by the missile. The second missile is a 3-in.-diameter standard steel pipe, which weighs 75 lb. It can achieve a horizontal impact speed of 75 mph and a vertical speed of 50 mph. The horizontal speed could be effective at heights up to 100 ft above ground level. This missile will perforate unreinforced concrete masonry and brick veneer walls, reinforced concrete masonry walls less than 12 in. thick, and reinforced concrete walls less than 8 in. thick. The third missile is a 3000-lb automobile that is assumed to roll and tumble on the ground and achieve an impact speed of 25 mph. Impact of an automobile can cause excessive structural response to columns, walls, and frames. Impact analyses should be performed to determine specific effects. Collapse of columns, walls, or frames may lead to further progressive collapse.

Tornado Missile Impact Tests

Wall barrier specimens have been tested at the Tornado Missile Impact Facility at Texas Tech University. The facility has an air-activated tornado missile cannon capable of firing 2×4 timer planks weighing 12 lb up to 150 mph and 3-in.-diameter steel pipes weighing 75 lb at speeds to 75 mph. Wall barriers tested to date include reinforced concrete walls from 4 in. to 10 in. thick; 8-in. and 12-in. walls of reinforced concrete masonry units (CMU); two other masonry wall configurations consisting of an 8-in. CMU and a 4-in. clay brick veneer; and a 10-in. composite wall having two wythes of 4-in. clay brick.

The impact tests series was designed to determine the impact speed required to produce backface spall of each wall barrier. A set of 15 wall sections have been constructed and tested at this time. Preliminary findings suggest that all cells of CMU walls must be grouted to prevent missile penetration. Results of these impact tests as well as previous available test data will be compared with existing impact formulas. In particular, the Rotz Equation and the Modified NDRC formula are being examined. A new impact formula for CMU walls is being developed.

Comments on Load Combinations

The ratios of hazard probabilities to performance goal probabilities for the usage categories as shown in Table 4 are an approximate measure of the conservatism required in the design to achieve the performance goals. The most conservatism is needed in the response evaluation and acceptance criteria for design of general use and important or low-hazard facilities. Somewhat less conservatism is needed for moderate and high-hazard facilities. The hazard probabilities specified for tornadoes are less than the performance goal probabilities. Hence the performance goals are theoretically met with no added conservatism in the design.

Conservatism can be achieved in designs by specifying factors of safety for allowable stress design (ASD) and load factors for strength design (SD). Consistent with the ratios in Table 4, the loading conditions recommended for design for DOE facilities are summarized in Table 5.

Since the ratio of extreme wind hazard probability to performance goal probability for general use and important or low-hazard facilities are the largest, 20, designs for these categories should be the most conservative in terms of factors of safety for ASD and load factors SD. The recommended combinations are essentially those given in ASCE

TABLE 5. Summary of Recommended Wind and Tornado Load Combinations

Facility	General Use	Important or Low Hazard	Moderate Hazard	High Hazard
ASD				
Extreme winds	DL + WL	DL + WL	0.9(DL + WL)	0.9(DL + WL)
	0.75(DL + LL + WL)	0.75(DL + LL + WL)	0.67(DL + LL + WL)	0.67(DL + LL + WL)
Tornadoes			$0.75(DL + W_t)$	$0.75(DL + W_t)$
			$0.63(DL + LL + W_t)$	$0.63(DL + LL + W_t)$
SD				
Extreme winds	0.9DL + 1.3WL	0.9DL + 1.3WL	DL + 1.3WL	DL + 1.3WL
	1.2DL + 0.5LL + 1.3WL	1.2DL + 0.5LL + 1.3WL	1.1DL + 0.5LL + 1.2WL	1.1DL + 0.5LL + 1.2WL
Tornadoes			$DL + W_t$	$DL + W_t$
			$DL + LL + W_t$	$DL + LL + W_t$

ASD, allowable stress design; use allowable stress appropriate for building material SD, strength design; use ϕ factors appropriate for building material. DL, dead load; LL, live load; WL, extreme wind load; and W_t, tornado load, including APC if appropriate.

7-88 [*American Society of Civil Engineers*, 1990]. The recommended load combinations for moderate and high-hazard facilities are slightly less conservative than for the general use and important or low-hazard categories because the ratio of extreme wind hazard probability to performance goal probability is less. The load factor coefficients have been reduced by approximately 10%.

The tornado hazard probabilities for both moderate and high-hazard facilities are less than the performance goal probabilities. The tornado load combinations for both ASD and SD recognize that the performance goals are theoretically met and no added conservatism in the load combination is required. This approach is consistent with criteria for commercial nuclear power plants as given in ACI 349 [*American Concrete Institute*, 1985] for concrete and ANSI/AISC N690 [*American Institute of Steel Construction*, 1984] for steel.

Additional background material on wind and tornado effects can be found in the work of *McDonald* [1985a, b]. A commentary on the design criteria summarized there is also included in Appendix B of *Kennedy et al.* [1990].

5. Summary and Conclusions

UCRL-15910 is an example of deterministic design/evaluation criteria developed to achieve probabilistic performance goals. UCRL-15910 also covers seismic and flood as well as wind and tornado criteria as discussed in this paper. The criteria developed are consistent with the consensus standard ASCE 7-88 and required by DOE General Design Criteria [*U.S. Department of Energy*, 1989]. Conservatism is specified which is sufficient to achieve the performance goals. This conservatism increases from the general use to the high-hazard usage category.

The UCRL-15910 wind/tornado design and evaluation guidelines follow the philosophy of (1) gradual reduction in hazard annual exceedance probability and (2) gradual increase in conservatism of evaluation procedure as one goes from a general use to a high-hazard facility. Four separate sets of design/evaluation criteria have been presented in UCRL-15910, each with a different performance goal. In all these criteria, loading is selected from hazard curves on a probabilistic basis, but response evaluation methods and acceptable behavior limits are deterministic approaches with which design engineers are familiar.

References

American Concrete Institute, Code requirements for nuclear safety-related concrete structures (ACI 349-85) and commentary (ACI 349R-85), Detroit, Mich., 1985.

American Institute of Steel Construction, Nuclear facilities—Steel safety-related structuring for design, fabrication, and erection, *ANSI/AISC N690*, Chicago, Ill., 1984.

American Society of Civil Engineers, Minimum design loads for buildings and other structures, *ASCE 7-88* (formerly ANSI A58.1), New York, 1990.

Coats, D. W., and R. C. Murray, Natural phenomena hazards modeling project: Extreme wind/tornado hazard models for Department of Energy sites, *Rep. UCRL-53526 Rev. 1*, Lawrence Livermore Natl. Lab., Livermore, Calif., 1985.

Electric Power Research Institute, *Advanced Light Water Reactor Requirements Document*, vol. 1, ALWR Policy and Summary of Top-Tier Requirements, Palo Alto, Calif., 1990.

International Conference of Building Officials, *Uniform Building Code*, 1988 ed., Whittier, Calif., 1988.

Kennedy, R. P., S. A. Short, J. R. McDonald, M. W. McCann, R. C. Murray, and J. R. Hill, Design and evaluation guidelines for Department of Energy facilities subjected to natural phenomena hazards, *Rep. UCRL-15910*, Lawrence Livermore Natl. Lab., Livermore, Calif., 1990.

Kimura, C. Y., and R. J. Budnitz, Evaluation of external hazards to nuclear power plants in the United States, NUREG/CR-5042, *Rep. UCID-21223*, Lawrence Livermore Natl. Lab., Livermore, Calif., 1987.

McDonald, J. R., Extreme winds and tornadoes: An overview, *Rep. UCRL-15745*, Lawrence Livermore Natl. Lab., Livermore, Calif., 1985a.

McDonald, J. R., Extreme winds and tornadoes: Design and evaluation of buildings and structures, *Rep. UCRL-15747*, Lawrence Livermore Natl. Lab., Livermore, Calif., 1985b.

McDonald, J. R., Structural details for wind design, *Rep. UCRL-21131*, Lawrence Livermore Natl. Lab., Livermore, Calif., 1988.

Ravindra, M. K., and A. M. Nafday, State-of-the-art and current research activities in extreme winds relating to design and evaluation of nuclear power plants, *Rep. NUREG/CR-5497*, U.S. Nucl. Reg. Comm., Washington, D. C., 1990.

U.S. Department of Energy, General design criteria, *DOE Order 6430.1A*, Washington, D. C., 1989.

U.S. Nuclear Regulatory Commission, Regulatory guide 1.76, Washington, D. C., 1974.

State-of-the-Art and Current Research Activities in Extreme Winds Relating to Design and Evaluation of Nuclear Power Plants

M. K. RAVINDRA

EQE International, Irvine, California 92715

INTRODUCTION

The objective of this paper is to review results from recent and ongoing research projects on extreme winds and summarize the current state of the art as a background for the development of criteria for individual plant examination of nuclear power plants for external events (IPEEE). Past studies have shown that extreme winds from tornadoes, hurricanes, and extratropical storms may be an important external event for some nuclear power plants in the United States. The evolution of tornado design criteria in the nuclear industry is traced, starting with U.S. Nuclear Regulatory Commission (NRC) Regulatory Guide 1.76 and the Standard Review Plan (SRP) and concluding with American Nuclear Society (ANS) Standard 2.3. The review covers the hazard analysis for tornadoes including tornado missiles and straight winds. Fragility analysis of structures for extreme wind loading is discussed. The techniques used in the systems analysis and quantification are described along with a summary of the results from different probabilistic risk assessments. A study was performed for those nuclear power plants that conform to the current licensing criteria specified in the Standard Review Plan to show that these plants are likely to have a mean core damage frequency of less than 10^{-6}/yr. Finally, the procedure developed for extreme wind analysis in IPEEE is described.

Code of Federal Regulations (CFR) Title 10 part 50 appendix A criterion 2 requires that nuclear power plants be designed to withstand the effects of extreme winds such as tornadoes, hurricanes, and extratropical winds. These effects include wind pressures, pressure drop, and missile impacts. The procedures for design against extreme winds have been developed over the last 30 years and are described in regulatory guides (RG), the Standard Review Plan, and industry standards.

Extensive analytical and experimental research in the modeling of tornadoes, tornado missile trajectory, and impact effects on structural barriers was sponsored by the Nuclear Regulatory Commission and the Electric Power Research Institute (EPRI) in the 1970s in order to develop the design procedures that are included in the regulatory criteria. The wind speed data collected at over 130 weather stations were examined systematically to develop mean annual recurrence intervals for different wind speeds near nuclear power plant sites [*Simiu et al.*, 1979].

In the last 10 years, probabilistic risk assessment (PRA) studies of nuclear power plants have considered extreme winds as external event initiators of accidents and estimated the frequencies of such accidents. These studies included the wind hazard analysis, wind fragility evaluation of structures and equipment, and systems analysis of accident sequences. It was found that the contribution of extreme winds to the overall risk of nuclear power plants is highly plant specific and may not be insignificant [*Kimura and Budnitz*, 1987]. Similar conclusions were drawn from the studies performed by Sandia National Laboratories as part of Task Action Plan (TAP) A-45, "Decay Heat Removal Requirements."

In order to provide information that would be useful in preparing guidelines for extreme wind analyses within IPEEE, current information on wind design procedures, wind hazard, fragility, and systems analysis is needed. For this purpose, the NRC funded research at Lawrence Livermore National Laboratory (LLNL) to survey the state of the art in the treatment of extreme winds in nuclear power plant design and risk assessment [*Ravindra and Nafday*, 1990].

CURRENT NRC REQUIREMENTS

Title 10 of the Code of Federal Regulations (10 CFR) specifies, in general terms, the conditions and factors that must be considered in constructing, licensing, and operating a nuclear power plant and the regulatory process that must be followed in performing this function. The regulation of nuclear power plants for protection from high winds/tornado hazards is given in 10 CFR 50 (design) and 10 CFR 100 (siting).

The Tornado: Its Structure, Dynamics, Prediction, and Hazards.
Geophysical Monograph 79
Copyright 1993 by the American Geophysical Union.

Fig. 1. Tornado regionalization map of the United States.

The scope of regulation of nuclear power plant design and requirements for acceptability have evolved over the years. Initially, the General Design Criteria were adopted as the minimum requirements in developing the standards for detailed design. Safety guides issued in 1970 became part of the Regulatory Guide series in 1972. The specific regulatory documents of interest are the NRC Regulatory Guides 1.76 and 1.117. RG 1.76 defines three tornado regions in the United States based on the regional variations of tornado hazard adopted in WASH-1300 [*Markee et al.*, 1974] and specifies a design basis tornado (DBT) for each of these regions (Figure 1). DBT for each region is described in terms of the maximum wind speed, rotational wind speed, translational wind speed, radius of maximum rotational speed, and the rate of pressure drop. DBT for region I, covering all states east of the Rocky Mountains, has a maximum wind speed of 360 miles per hour (mph) (1 mi. = ~1.6 km). Region II covers the coastal western United States, and its DBT has a maximum wind speed of 300 mph. Region III covers the remaining western states. *Ramsdell and Andrews* [1986] have collected tornado strike data since the publication of WASH-1300 and updated the tornado regionalization map for the United States.

RG 1.117 lists the plant systems, structures, components, areas, etc., to be considered in the study of tornado risk to ensure (1) the integrity of the reactor coolant pressure boundary, (2) the capability to shut down the reactor and maintain it in a safe shutdown condition, and (3) the capability to prevent accidents that could result in potential off-site exposures that are a significant fraction of the guideline exposure of 10 CFR 100. The portions of structures, systems, or components whose continued function is not required but whose failure could reduce the functional capability of any plant feature included in the specified list are also required to be designed for the design basis tornado.

Detailed requirements for tornado loadings are covered in NRC Standard Review Plan section 3.3.2 and are essentially based on RG 1.76. Procedures based on American National Standard ANSI A 58.1 [*American National Standards Institute*, 1982] are accepted for transformation of DBT to actual loadings: the appropriate load combinations are also specified in the SRP. SRP section 3.3.1 gives the acceptance criteria for straight winds (i.e., hurricanes and extratropical storms) as speeds with 100-year mean return period, but this value is much less than the DBT wind speed. Therefore the design basis for the plants is invariably controlled by tornadoes for all regions in the United States. The frequency of hurricanes and extratropical storms exceeding DBT wind

TABLE 1. NRC Spectrum II Missiles

	Horizontal Strike Speed,* mph		
	Region I	Region II	Region III
Wood plank, 4 in. × 12 in. × 12 ft, weight 115 lb	185	157	130
Steel pipe, 6-in. diameter, schedule 40, 15 ft long, weight 285 lb	117	94	23
Steel rod, 1-in. diameter × 3 ft, weight 8 lb	114	89	86
Steel pipe, 12-in. diameter, schedule 40, 15 ft long, weight 750 lb	105	63	16
Utility pole, 13.5-in. diameter, 35 ft long, weight 1500 lb	123	107	58
Automobile, frontal area 28 ft^2 4000 lb	132	116	91

Metric conversions: 1 in. = ~2.5 cm; 1 ft = ~0.3 m; 1lb = ~0.45 kg.
*Vertical speeds should be taken as 70% of horizontal speed.

speeds is usually much less than 10^{-7}/yr. Section 3.5.1.4 of SRP describes the acceptance criteria for missiles generated by tornadoes, their impact speeds, and transformation to static loads on the structures. In 1977 the NRC adopted alternative missile criteria consisting of two sets of missiles denoted spectrum I and spectrum II. Spectrum I missiles consisted of a rigid slug and a 2 in. × 4 in. timber plank. Spectrum II missiles are more conservative and are listed in Table 1. (Metric conversion rates for inches, feet, and pounds are provided in a footnote to Table 1.) The total tornado load is calculated using a number of load combinations which include tornado wind load, differential pressure load, and missile load. Load combinations (for wind and tornado loads with other loads) and structural design criteria are specified in SRP section 3.8.

An industry standard, ANSI/ANS 2.3 [*American National Standards Institute/American Nuclear Society*, 1983], has been developed which specifies guidelines to determine the wind velocity, atmospheric pressure changes, missile type, and impact speed that result from tornadoes, hurricanes, and other extreme winds to be used in nuclear plant design. This standard provides maps of tornado wind speeds corresponding to annual exceedance probabilities of 10^{-7}, 10^{-6}, and 10^{-5}. Each map defines three tornado intensity regions. Records from 1916–1975 are used in the tornado hazard assessments. The methodology by *Fujita* [1978] which accounts for gradations of damage across and along the path length, topographical features, and biases in reporting occurrences and intensity of tornadoes was used in developing these maps. A spectrum of missiles was selected by the American Nuclear Society Working Group ANS 2.3 to cover the range of characteristics associated with all potential objects that can be propelled by a tornado. The two missiles, a 750-lb-wide flange steel beam and a 4000-lb automobile, have been documented to be actual tornado-generated projectiles. The beam controls the design of wall thickness to prevent perforation and scabbing (i.e., back face damage and disintegration of concrete), whereas the automobile determines the design for structural response of wall panels, columns, or frames.

TORNADO MISSILE RISK

Missiles generated by tornadoes may cause damage if they impact safety-related structures or exposed equipment. Equipment located within the structures is vulnerable to damage from missile penetration of protective structures, secondary impact from spalled concrete, or missiles entering through openings. Damage depends on the mass, terminal velocity, deformation characteristics, and angle of impact of the missiles. Historically, the licensing criteria specify sufficient thickness of concrete to prevent perforation, spalling (i.e., front face damage and disintegration of concrete), or scabbing of the barriers in the event of missile impact. Several formulas have been suggested for determining wall thicknesses required to prevent scabbing on the basis of missile impact tests. From the tests conducted by Bechtel [*Vassallo*, 1975], EPRI [*Stephenson*, 1977], and Stone and Webster [*Jankov et al.*, 1976] and using different types of missiles, it was concluded that the concrete thickness required to prevent damage is much less than 12 in. for all missiles except for the 12-in.-diameter steel pipe missile (a barrier thickness of 18 in. is required). However, examples of 12-in.-diameter pipe missiles have never been observed as documented in several damage studies [*Abbey and Fujita*, 1975]. SRP section 3.5.3 gives the minimum acceptable wall and roof thicknesses to protect against local damage from tornado missiles (Table 2).

Damage from tornado missiles results only as the culmination of a sequence of low-probability events. This sequence starts with the tornado strike in the plant vicinity and includes missile injection (from available objects) and transport, missile impact, and barrier damage. Owing to analytical complexity of the tornado missile injection, transport, and impact analyses, the probability of strike is computed by simulation as described by *Twisdale and Dunn* [1981], *Johnson et al.* [1977], and *Reinhold and Ellingwood* [1982]. These

TABLE 2. Minimum Acceptable Barrier Thickness for Local Damage Protection Against Tornado-Generated Missiles

Region	Concrete Strength, psi	Wall Thickness, in.	Roof Thickness, in.
Region I	3000	23	18
	4000	20	16
	5000	18	14
Region II	3000	16	13
	4000	14	11
	5000	13	10
Region III	3000	<6	<6
	4000	<6	<6
	5000	<6	<6

TABLE 3. Extreme Wind Core Damage Frequencies From Plant-Specific PRA

Plant Name	Tornado Frequency (Any Size), $(mi^2/yr)^{-1}$	Straight Wind Core Damage Frequency, yr^{-1}	Tornado Core Damage Frequency, yr^{-1}	Total Core Damage Frequency, yr^{-1}
Indian Point 2	1.00E − 04*	3.60E − 05	<E − 07	3.60E − 05
Indian Point 3	1.00E − 04	1.30E − 06	<E − 07	1.30E − 06
Limerick 1 and 2	1.13E − 04	9.00E − 09	<E − 08	9.00E − 09
Millstone 3	1.87E − 04	low	<E − 07	<E − 07
Oconee 3	2.50E − 04	low	<E − 09	<E − 09
Seabrook 1 and 2	1.26E − 03	<3.89E − 08	2.06E − 09	2.06E − 09
Zoin 1 and 2	1.00E − 03	NA	<E − 08	NA

NA, not available.
*Read, for example, 1.00E − 04 as 1.00×10^{-4}.

studies were based on computer simulations involving mathematical modeling of the geometrical distribution of potential missiles, tornado movement, tornado wind field, missile injection into the wind field, aerodynamic flights of missiles, and the missile interaction with the plant. The results from these studies show that the three parameters important for the analysis of tornado missile risk are (1) tornado characteristics such as wind speed, radius, and pressure drop; (2) number and locations of potential tornado missiles; and (3) resistance of plant barriers to missile strikes.

The study by *Twisdale and Dunn* [1981] simulated tornado missile histories for two hypothetical plants utilizing the standard spectrum of missiles. Their results indicate that the annual probability of missile strike is less than 10^{-7} in NRC RG region I. The probability of back face scabbing was found to be several orders of magnitude lower. Conservatively assuming that scabbing damage results in core damage, the thickness of reinforced concrete walls required is obtained as less than 8 in. for region A, the region of highest intensity according to the tornado region classification by Twisdale. This region is entirely contained in region I of NRC RG 1.76. Barrier thicknesses (Table 2) for all nuclear plants in the United States designed to SRP criteria are much larger than 8 in. Therefore it is concluded that the plants designed to SRP criteria for missile strikes have a mean core damage frequency of substantially less than 10^{-7}/yr. In a recent paper, *Twisdale* [1988] has provided further examples on tornado missile risk analysis.

Components inside the reinforced concrete buildings still may be vulnerable from missiles entering through openings such as vents, ducts, doorways, and roll-up doors. The probability of damage to these components depends on the number and size of these openings and the location of critical components inside the building. In some nuclear power plants constructed in the 1960s and early 1970s, it is possible to find safety-related components outdoors or in non–category I structures. Potential for missile generation from failure of any nonsafety structures such as metal sided structures, masonry walls, and steel and concrete stacks also needs examination.

EXTREME WIND RISK

Winds due to tornadoes, hurricanes, and straight winds can cause failure of structures due to direct pressure loading. Failure modes may be local (e.g., wall panel or roof) or global (e.g., failure of building shear walls or buckling collapse of tanks and stacks). The probability of failure induced by wind loading for any structure is evaluated by integrating (the procedure is called "convolution") the extreme wind hazard probability distribution with the fragility of the structure for the specified failure mode. Wind hazard denotes the annual probability of exceedance of a hazard parameter, such as wind velocity. Wind fragility is defined as the conditional probability of failure of a structure for a given wind velocity. In case the wind hazard and the wind fragility are fully known, there will be a single hazard curve and a single fragility curve; the result of the convolution is a single value of the annual frequency of failure. If there is uncertainty in the hazard, that is represented by a discrete set of hazard curves with subjective weights attached to them. Similarly, a family of fragility curves may be specified. The convolution is done by taking one hazard curve and one fragility curve at a time, and the resulting frequency of failure carries a subjective weight equal to the product of the subjective weights assigned to these specific hazard and fragility curves. The result of convolving the family of hazard curves with the family of fragility curves is a subjective probability distribution of the annual frequency of failure. This convolution could be accomplished by numerical integration or by simulation techniques.

Extreme wind PRAs have been conducted for seven nuclear power plants. Table 3, extracted from a report by *Kimura and Budnitz* [1987], summarizes the tornado frequencies, the tornado strike frequencies, the straight wind core damage frequencies, the tornado wind/missile core damage frequencies, and the total straight wind/tornado core damage frequencies for these sites. A study of these previous PRAs shows that core damage frequency from straight winds/tornadoes is much lower than 10^{-6}/yr except for Indian Point 2. The frequency at this plant was estimated to be higher because of a unique plant feature; that is, the postulated collapse of the unit 1 superheater stack was judged to affect the unit 2 diesel generator building and/or control building. Abbreviated PRAs have been conducted for five nuclear power plants under the NRC's program for the resolution of generic safety issue A-45 (decay heat removal) [Reed and Ferrell, 1987]. The core damage frequencies due to high winds alone, tornado winds, and/or missiles and their combination were estimated. The core damage frequency computations were based on the assumption of loss of off-site power. All these core damage frequencies are generally higher than those reported in Table 1 because in almost every plant, the risk was dominated by non–category I structures interacting with safety-related systems. It is to be noted that in the above studies, resistance to wind loading for reinforced concrete structures was considered to be adequate, and wind-loading-induced failure of these structures was eliminated on the basis of its negligible contribution to the frequency of core damage.

It is shown in the following that the structures designed to SRP requirements have negligible contribution ($<10^{-6}$/yr) to the core damage frequency. Core damage frequency computations were performed for two sites in tornado region I (region of highest tornado hazard), representing extremely high tornado hazard (plant A, Cooper, Nebraska) and another site in the region with substantial contribution to the hazard curves from the hurricanes (plant B, St. Lucie, Florida). The plant A site is exposed to potential winds generated by tornadoes and extratropical storms, whereas the plant B site experiences hurricanes, tornadoes, and extratropical storms. The hazard data specified for these sites for both tornado and straight winds are available in the TAP A-45 reports [Reed and Ferrell, 1987].

For the structures designed to SRP requirements, fragility curves can be developed on the basis of the plant design criteria. The purpose of wind pressure fragility analysis is to develop a relationship between the conditional probability of failure of a structure and the wind velocity. For example, if the wind velocity is V, and the response (e.g., force) due to this wind velocity at a specified location is R, the structure's capacity to withstand wind force is a random variable, C. The fragility of the structure is then defined as

$$f = \text{Probability } \{C \leq R\} \quad (1)$$

If the structure capacity is modeled as a lognormally distributed random variable with median C_m and logarithmic standard deviation β, then the fragility is given by

$$f = \Phi\left[\frac{\ln(C_m/R)}{\beta}\right] \quad (2)$$

where $\Phi(\cdot)$ is the standard Gaussian cumulative distribution function; for different values of wind speed the fragility f is calculated, and a curve called "fragility curve" is obtained by plotting f versus v (or a factor of safety with respect to DBT) (Figure 2).

Development of the fragility is done in terms of the factor of safety, defined as the capacity divided by the response associated with design basis loads from extreme winds. The structure's capacity to withstand wind load effects is formulated in terms of the variables influencing the wind velocity necessary to cause failure. For example, if the design wind speed is V_d, the structure capacity C can be expressed as

$$C = V_d F_W F_S \quad (3)$$

where F_W is the safety factor relating the design wind pressure to the actual wind pressure on the structure and F_S is the safety factor relating the actual capacity of the structure to the calculated capacity.

Each of these safety factors, say F, can be modeled as

$$F = F_m e_R e_U \quad (4)$$

where F_m is the median safety factor, e_R is a random variable reflecting the inherent randomness in the safety factor, and e_U is a random variable reflecting the uncertainty in the calculation of F_m. Both e_R and e_U are assumed to be lognormally distributed with logarithmic standard deviations β_R and β_U. Using the median safety factors F_{Wm} and F_{Sm}, logarithmic standard deviations due to randomness $\beta_{R,W}$ and $\beta_{R,S}$, and logarithmic standard deviations due to uncertainty $\beta_{U,W}$ and $\beta_{U,S}$, the median structural capacity C_m, its variability due to randomness $\beta_{R,C}$ and uncertainty in the median $\beta_{U,C}$ can be computed.

The variables influencing the factor of safety for external wind pressure are structural strength, structural modeling, and determination of the wind pressure distribution including effects of adjacent structures and terrain. The factor of safety due to structural strength refers to the actual strength compared with the stress limitations imposed by the original design acceptance criteria. The factor of safety due to modeling refers to the accuracy of analytical prediction of the loads and stresses. The factor of safety due to determination of wind pressure distribution includes the variables associated with wind pressure coefficients, gust factors, etc. This factor also includes the consideration of shielding and channeling effects of other structures at the site together with local site grade changes in determining median-centered wind loads on the structure.

Lognormal distribution is usually adopted for modeling structural fragilities. From the knowledge of median wind capacity C_m and variability due to randomness $\beta_{R,C}$ and variability due to uncertainty $\beta_{U,C}$ in terms of their logarithmic standard deviations, fragility of the structure, f', at a wind speed V for any nonexceedance probability level Q can be derived by using the formulation given in the *PRA Procedures Guide* [1983]:

$$f' = \Phi\left[\frac{\ln(V/C_m) + \beta_{U,C}\Phi^{-1}(Q)}{\beta_{R,C}}\right] \quad (5)$$

where $Q = P[f \leq f'|V]$ is the subjective probability ("degree of belief") that the conditional probability f is less

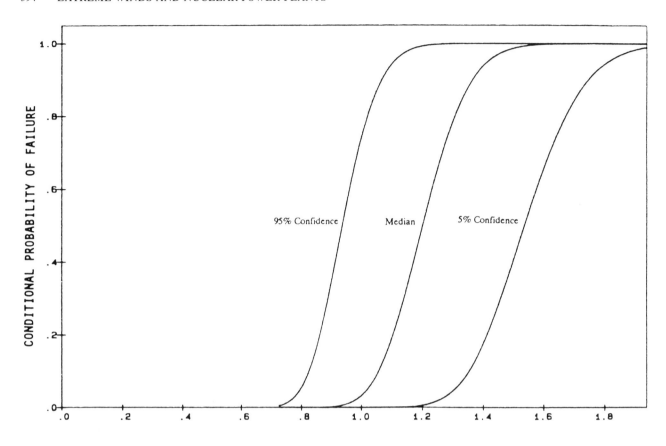

FACTOR OF SAFETY WITH RESPECT TO DESIGN BASIS TORNADO

Fig. 2. Fragility curves for structures designed to SRP criteria.

than f' given a wind speed V and where $\Phi^{-1}(\cdot)$ is the inverse of the standard Gaussian cumulative distribution function. Note that Q is the sum of the probabilities assigned to all the fragilities less than f'. In this formulation, both the inherent randomness and the uncertainty are explicitly represented, and this leads to a family of fragility curves for the specified failure mode.

A structure which has been designed for a region I design basis tornado is expected to have a wind-resisting capacity larger than 360 mph because of conservatisms inherent in design codes, material strength specifications, assumed failure modes, etc. The median capacity and the variability associated with the calculated median capacity are estimated by accounting for these conservatisms as explained below.

The main source of conservatism in the capacity is due to the fact that the nominal steel yield strengths and the nominal concrete strengths specified by the designer are much lower than the median values. This conservatism factor in nominal yield strength to median strength has been estimated to be 1.2 [*Ravindra and Banon*, 1988]. Variability in wind pressure is a function of variabilities in material yield strength, pressure coefficient, and modeling of structures for wind load analysis. The coefficient of variation in material yield strength, β_{my}, was estimated to be 0.15 [*Galambos and Ravindra*, 1978; *Mirza and MacGregor*, 1979; *Mirza et al.*, 1979].

The pressure coefficient relates the induced pressure on wall panels to maximum wind velocity. Induced pressure on a wall panel is a function of structural shape as well as the location on a wall panel. Therefore there is some uncertainty associated with the pressure coefficient. There are no data on tornado wind pressures to derive the uncertainties in pressure coefficient and modeling. Since the physical phenomenon of induced pressure due to tornadoes and straight winds can realistically be considered to be similar, the values estimated by *Ellingwood* [1978] for straight winds are used, i.e., uncertainty in pressure coefficient (β_{pc}) of 0.15 and uncertainty in wind modeling (β_{wm}) of 0.05.

Next, it is assumed that the variability in wind pressure (e_p) can be modeled as the product of random variables representing variability in pressure coefficient (e_{pc}), wind modeling (e_{wm}), and material yield strength (e_{my}):

TABLE 4. Annual Extreme Wind Core Damage Frequencies for Two Nuclear Power Plants in RG 1.76 Region I

	Tornado	Straight Wind	Combined Tornado and Straight Wind
Plant A	3.20E − 07*	8.88E − 14	3.20E − 07
Plant B	1.39E − 08	2.65E − 09	1.65E − 08

*Read, for example, 3.20E − 07 as 3.20×10^{-7}.

$$e_p = e_{pc} e_{wm} e_{my} \qquad (6)$$

Since the uncertainties associated with these variables are essentially independent variables, they may be combined to find the overall or combined uncertainty β_p by

$$\beta_p^2 = \beta_{pc}^2 + \beta_{wm}^2 + \beta_{my}^2 \qquad (7)$$

where β_{pc} is the lognormal standard deviation for the pressure coefficient, β_{wm} is the lognormal standard deviation for wind modeling, and β_{my} is the lognormal standard deviation associated with material yield strength.

The logarithmic standard deviation for the wind pressure (β_p) is calculated as 0.21. Since the calculated wind pressure is proportional to the square of wind velocity, logarithmic standard deviation (β_R) of wind velocity is 0.1. The uncertainty in structural modeling (β_U) is estimated as 0.15 [Ravindra and Banon, 1988]. Thus the wind capacity of reinforced concrete structures defined in terms of design basis tornado has a median capacity (V_m) of 1.2, β_R of 0.1, and β_U of 0.15. Using these computed values, a family of fragility curves can be obtained. Figure 2 shows fragility curves for reinforced concrete structures designed to SRP requirements.

It is conservatively assumed that the failure of a structure results in failure of all safety-related equipment inside the structure and leads to subsequent core damage. The core damage frequency is evaluated by the convolution of the family of fragility curves for the structure and hazard curves for tornado and straight winds for the two sites, A and B. The mean frequencies obtained are shown in Table 4. The mean frequencies for combined tornado and straight winds are approximated by adding their individual frequencies. The values of core damage frequencies from extreme winds for both the sites are substantially lower than 10^{-6}/yr.

To summarize, the analysis described above has indicated that the frequency of failure under straight wind and tornado loading for the structures designed to SRP criteria for structures housing the safety-related equipment is less than 10^{-6}/yr.

INDIVIDUAL PLANT EXAMINATION OF EXTERNAL EVENTS

Based on the Policy Statement on Severe Accidents, the NRC has requested that the licensee of each nuclear power plant perform an individual plant examination. This plant examination systematically looks for vulnerabilities to severe accidents and cost-effective safety improvements that reduce or eliminate the important vulnerabilities. Extreme winds (called high winds) are identified as an external event that needs to be included specifically in the IPEEE. A progressive screening approach has been developed [Nuclear Regulatory Commission, 1991; Prassinos et al., 1989] to identify potential vulnerabilities from extreme winds at nuclear power plants. Figure 3 shows a flow chart of this approach. The steps shown in the figure represent a series of analyses in increasing level of detail, effort, and resolution. However, the licensee may choose to bypass one or more of the optional steps so long as the 1975 Standard Review Plan criteria are met and the potential vulnerabilities are either identified or demonstrated to be insignificant. This procedure is based on the study results discussed above: current SRP requirements assure that the frequency of tornado and other extreme wind damage from pressure loading and missile impacts is acceptably low (i.e., $<10^{-6}$/yr). The major steps are as follows:

1. Review the site-specific wind hazard data and licens-

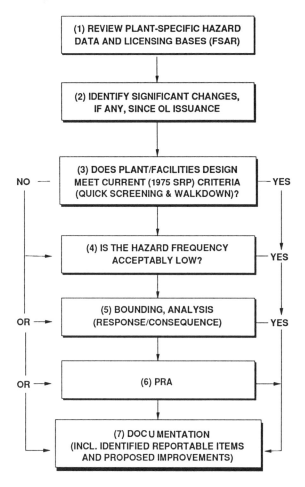

Fig. 3. NRC-recommended IPEEE approach for extreme winds.

ing bases, including how the specific hazard issues were resolved at the licensing stage.

2. Identify significant changes since the issuance of the operating license; in the case of an extreme wind event, these changes could be new sources of tornado missiles or construction of structures vulnerable to extreme winds.

3. Determine if the plant meets the Standard Review Plan criteria. As part of this evaluation, a confirmatory walkdown of the plant should be performed. This walkdown would concentrate on outdoor facilities (steel structures, unreinforced masonry walls, steel tanks, etc.) that could be affected by extreme winds and missiles. If this evaluation indicates that the plant conforms to the SRP criteria and the walkdown reveals no potential vulnerabilities, it is judged that the contribution from the extreme wind hazard to core damage frequency is less than 10^{-6}/yr and that the IPEEE screening criterion is met.

If the SRP criteria are not met, the licensee should take one or more of the optional steps given below to further evaluate the extreme wind event.

4. Determine if the extreme wind hazard frequency is acceptably low. If the original design basis does not meet the current regulatory requirements, the licensee may choose to demonstrate that the frequency of exceeding the original design basis is sufficiently low (that is, less than 10^{-5}/yr) and that the conditional core damage frequency from winds exceeding the original design basis is judged to be less than 10^{-1}/yr. If the original design basis hazard multiplied by the conditional core damage frequency is not sufficiently low (that is, less than the screening criterion of 10^{-6}/yr), additional analysis may be needed.

5. Perform a bounding analysis: this analysis is intended to provide a conservative calculation showing that either the hazard would not result in core damage or the core damage frequency is below the reporting criterion of 10^{-6}/yr. The level of detail is that level needed to demonstrate this point.

6. Perform a probabilistic risk assessment: if options 4 and 5 are not able to screen out the extreme wind event, a PRA consisting of hazard analysis, fragility evaluation, plant systems and accident sequence analysis, and core damage frequency estimation should be performed. If the core damage frequency is less than 10^{-6}/yr, the event need not be considered further.

Summary and Conclusions

This paper has reviewed the results from recent and ongoing research projects on extreme winds with the objective of summarizing the current state of the art as a background for the development of criteria for extreme wind analysis in IPEEE. The evolution of tornado design criteria in the nuclear industry is traced starting with RG 1.76 and the Standard Review Plan and concluding with ANS 2.3. The review has covered the hazard analysis for tornado including tornado missiles and straight winds. Fragility analysis of structures subjected to extreme wind effects is discussed. The techniques used in the systems analysis and quantification are described along with a summary of the results from different PRAs.

The review showed that the risk of tornado missile damage for nuclear power plants designed according to the Standard Review Plan requirements is acceptably low (i.e., less than 10^{-6}/yr). In addition, case studies described in the paper have indicated that the frequency of wind-induced damage of structures designed to the Standard Review Plan criteria is also less than 10^{-6}/yr. Results from past PRAs have indicated that the major contribution to core damage frequency comes from failures of non-safety-related items which may endanger the safety-related structures by collapsing on them or by missile impact. Therefore the procedures developed for the wind portion of IPEEE require a confirmatory walkdown of the plant to search for vulnerabilities to extreme wind even when the plant design is considered to have met the SRP requirements.

References

Abbey, R. F., and T. T. Fujita, Use of tornado path lengths and gradations of damage to assess tornado intensity probabilities, in *Preprints of the Ninth Conference on Severe Local Storms*, American Meteorological Society, Boston, Oct. 1975.

American National Standards Institute, Minimum design loads for building and other structures, *ANSI A58.1-1982*, New York, March 1982.

American National Standards Institute/American Nuclear Society, Standard for estimating tornado and extreme wind characteristics at nuclear power sites, *ANSI/ANS 2.3*, Am. Nucl. Soc., La Grange Park, Ill., 1983.

Ellingwood, B., Reliability basis of load and resistance factors for reinforced concrete design, Natl. Bur. of Stand., Washington, D. C., Feb. 1978.

Fujita, T. T., Workbook of tornadoes and high winds for engineering applications, *SMRP Pap. 165*, Univ. of Chicago, Chicago, Ill., 1978.

Galambos, T. V., and M. K. Ravindra, Properties of steel for use in LRFD, *J. Struct. Div. Am. Soc. Civ. Eng.*, *104*, 1459–1468, 1978.

Jankov, Z. D., J. A. Shanahan, and M. P. White, Missile tests of quarter-scale reinforced concrete barriers, paper presented at the Symposium on Tornadoes, Assessment of Knowledge and Implications for Man, Tex. Tech Univ., Lubbock, June 1976.

Johnson, B., et al., Simulation of tornado missile hazards to the Pilgrim 2 nuclear thermal generating station, draft report, Science Applications, Inc., Sunnyvale, Calif., Nov. 1977.

Kimura, C. Y., and R. J. Budnitz, Evaluation of external hazards to nuclear power plants in the United States, *NUREG/CR-5042*, Lawrence Livermore Natl. Lab., Livermore, Calif., Dec. 1987.

Markee, E. H., J. G. Beckerley, and K. E. Sanders, Technical basis for interim regional tornado criteria, *WASH-1300*, U.S. Gov. Print. Off., Washington, D. C., May 1974.

Mirza, S. A., and J. G. MacGregor, Variability of mechanical properties of reinforcing bars, *J. Struct. Div. Am. Soc. Civ. Eng.*, *105*, 921–937, 1979.

Mirza, S. A., M. Hatzinikolas, and J. G. MacGregor, Statistical descriptions of strength of concrete, *J. Struct. Div. Am. Soc. Civ. Eng.*, *105*, 1021–1037, 1979.

Nuclear Regulatory Commission, Procedural and submittal guidance for the individual plant examination of external events (IPEEE) for severe accident vulnerabilities, *Final Rep. NUREG-1407*, June 1991.

PRA Procedures Guide, Analysis of external events, chap. 10, *NUREG/CR-2300*, vol. 2, American Nuclear Society and Nuclear Regulatory Commission, Washington, D. C., Jan. 1983.

Prassinos, P. G., et al., Individual plant examination for external events: Guidance and procedures, draft, *NUREG/CR-5259*, Lawrence Livermore Natl. Lab., Livermore, California, March 1989.

Ramsdell, J. V., and G. L. Andrews, Tornado climatology of the contiguous United States, *NUREG/CR-4461, PNL-5697*, Pac. Northwest Lab., Richland, Wash., March 1986.

Ravindra, M. K., and H. Banon, External event scoping quantification of LaSalle unit 2 nuclear power plant: Risk methods integration and evaluation program (RMIEP), *NUREG/CR-4832*, Sandia Natl. Lab., Albuquerque, N. M., Jan. 1988.

Ravindra, M. K., and A. M. Nafday, State-of-the-art and current research activities in extreme winds relating to design and evaluation of nuclear power plants, *NUREG/CR-5497, UCID-21933*, Lawrence Livermore Natl. Lab., Livermore, Calif., May 1990.

Reed, J. W., and W. L. Ferrell, Extreme wind analyses for five nuclear power plants, Appendix G, in *Shutdown Decay Heat Removal Analysis of Nuclear Power Plants*, *NUREG/CR-4713, 4767, 4458, 4448 and 4762*, Sandia National Laboratories, Albuquerque, N. M., 1987.

Reinhold, T. A., and B. Ellingwood, Tornado damage risk assessment, *NUREG/CR-2944*, Nucl. Reg. Comm., Washington, D. C., 1982.

Simiu, E., M. J. Changery, and J. E. Filliben, Extreme wind speeds at 129 stations in the contiguous U.S., *NBS Build. Sci. Ser. 118*, Natl. Bur. of Stand., Washington, D. C., 1979.

Stephenson, A. E., Full scale tornado missile impact tests, *Final Rep. NP-440*, Electric Power Res. Inst., Palo Alto, Calif., July 1977.

Twisdale, L. A., Probability of facility damage from extreme wind effects, *J. Struct. Div. Am. Soc. Civ. Eng.*, *114*, 2190–2209, 1988.

Twisdale, L. A., and W. L. Dunn, Tornado missile simulation and design methodology, *NP-2005*, vol. 1, Research Triangle Inst., Research Triangle Park, N. C., Aug. 1981.

Vassallo, F. A., Missile impact testing of reinforced concrete panels, *HC-5609-D-1*, report prepared for Bechtel Corp., Calspan Corp., Buffalo, N. Y., Jan. 1975.

Wind/Tornado Design Criteria Development to Achieve Required Probabilistic Performance Goals

DOROTHY S. NG

Lawrence Livermore National Laboratory, Livermore, California 94550

1. INTRODUCTION

The U.S. Department of Energy (DOE) has established specific requirements for the design of structures, systems, and components (SSCs) of critical facilities for resisting all types of internal and external natural hazard events. The Lawrence Livermore National Laboratory (LLNL) was chartered to develop design guidelines for natural hazards utilizing the best technical knowledge and engineering judgment to assure that the design of the SSCs conform to performance goals set forth by DOE.

This report documents the strategy employed to develop wind/tornado hazard design criteria for a critical facility to withstand loading induced by the wind/tornado hazard. These guidelines were developed by a Wind/Tornado Working Group (WTWG) at LLNL composed of six experts who are knowledgeable in the fields of structural engineering, wind/tornado engineering, and meteorology. Utilizing their best technical knowledge and judgment in the wind/tornado field, they met and discussed the methodologies and reviewed available data. A review of the available wind/tornado hazard model for the site, structural response evaluation methods, and conservative acceptance criteria led to a proposed design criteria that has a high probability of achieving the required performance goals.

2. DESIGN REQUIREMENTS

The primary considerations of the design of the critical facility are public safety, worker safety, and environmental protection. The DOE design requirements document serves as the basis for design of the facilities. The contractors may propose methods which offer improvements over the specified requirements. However, if such methods are proposed, the contractor is required to demonstrate the validity of resulting improvements.

The pertinent wind/tornado design requirements are summarized below:

1. The design of facilities shall encompass annual safety performance goals of (1) 5×10^{-7}/yr probability for an early fatality to an average individual assumed to be located within 1 mile (~1.6 km) of the reactor facility control perimeter; (2) 2×10^{-6}/yr probability for a long-term fatality to plant workers or a member of the general public located within 10 miles (~16 km) of the reactor facility, but outside of the control perimeter; (3) 1×10^{-6}/yr probability for a large release of radioactive materials from a reactor accident; and (4) 1×10^{-5}/yr probability for core damage.

2. DOE encourages the contractors to incorporate the knowledge and experience over many years in nuclear reactor design accumulated.

3. DOE requires the contractors to make use of past regulatory compliance in commercial reactor design or demonstrate improvement over regulatory provisions.

The design requirements document stated that external events shall be addressed proactively in the design process in the context of the large-release goal. To investigate the achievement of the large-release goal requires additional consideration of containment failure modes, namely, direct, bypass, and penetration failures. It is more realistic to show the achievement of the core damage goal as an interim step, by wind/tornado risk assessment. The WTWG expects that the mean large-release risk frequency will be at least an order of magnitude lower than the mean core damage risk frequency because it is unlikely that wind and missile effects will significantly contribute to breaching containment. Therefore achieving the core damage goal is believed to ensure achievement of the large-release goal.

3. DESIGN GUIDELINES DEVELOPMENT STRATEGY

To ensure the safety of the plants, DOE requires that facility design meet stringent safety goals and should incorporate reactor experience and compliance with regulatory requirements in facility design. An effective approach to satisfy these requirements is to employ a panel of experts who have technical knowledge, experience, and judgment as a result of years of research and service in this area. Thus the WTWG was established for the development of wind/tornado design guidelines. The WTWG consists of a group of six nationally recognized technical experts; they are briefly described below in alphabetical order:

Robert F. Abbey, Jr., is a well-known meteorologist who is currently serving as the director of Marine Meteorology Research, Office of Naval Research. He specializes in probabilistic and statistical analyses of the occurrence of natural phenomena.

W. Lynn Beason is associate professor in the Civil Engineering Department of Texas A&M University. His expertise is in wind-borne missile research, tornado risk analysis, and structural design.

T. Theodore Fujita is a world-known tornado assessment expert. Fujita developed the Fujita scale (F scale) for estimating the relative intensity of tornadoes and the DAPPLE method for estimating the probability of tornado occurrence.

Dale C. Perry, chairman of the WTWG, is the head of the Department of Construction Science of Texas A&M University. His expertise is in structural design and wind engineering. He has over 20 years of experience in postdisaster investigations.

John W. Reed is an associate at Jack R. Benjamin & Associates, Inc., California. He has 25 years of experience in nuclear reactor design and probabilistic risk assessment (PRA) for seismic and wind hazards. Reed performed the preliminary wind speed PRA for this project.

Lawrence A. Twisdale, Jr., is the senior vice president of Applied Research Associates, Inc., Southeast Division in North Carolina. He specializes in tornado missile simulation and design methodology for tornado hazard assessment. Twisdale performed the preliminary missile criteria PRA for this project.

Additionally, James McDonald, a professor and researcher at Texas Tech University, is a member of the senior review group for this project; he provides guidance to the WTWG and reviews reports produced by the group. McDonald specializes in wind hazard assessment and missile impact research.

Shortly after formation of the WTWG and prior to their first meeting, the existing wind/tornado hazard assessment reports, Nuclear Regulatory Commission (NRC) provisions, design standards, and other references available for the site of interest were distributed to the WTWG. The methodologies and data utilized in existing regional and site-specific hazard assessments were compared and discussed by the WTWG in a series of meetings. Then, using their cumulative technical knowledge and judgment in the wind/tornado field, the WTWG planned the development of wind/tornado design guidelines.

The six basic goals of the WTWG strategy are as follows: (1) to build in conservatism by reducing wind/tornado goals to 1/10 of total project goals; (2) to specify design wind load specifications based on site-specific hazard frequency curves; (3) to specify design missile load specifications based on DOE guidelines and the Electric Power Research Institute (EPRI) NP-2005 report; (4) to define conservative structural response evaluation and acceptance criteria to substantiate the design guidelines; (5) to assure the conformance of design guidelines with safety goals by performing preliminary probabilistic risk assessments; and (6) to make use of nuclear industry standards, NRC provisions, and national codes and standards as benchmarks and references.

3.1. *Developing Wind Load Specifications Based on Site-Specific Hazard Frequency Curves*

In the 1970s the NRC divided the U.S. into three geographical regions and recommended design basis tornado (DBT) wind speeds for each region in NRC Regulatory Guide 1.76 [*U.S. Nuclear Regulatory Commission*, 1974]. These DBT wind speeds are believed to be quite conservative, resulting from coarse assumptions and problems with the regional data sets employed.

More recently, the *Office of Nuclear Reactor Regulation* [1988] position paper divided the U.S. into four regions and recommended DBT wind speeds based on regional analyses. Recommendations based on regional analyses cannot provide a realistic portrayal of the risk associated with specific sites within the region.

In 1979, two leading wind engineering experts, Fujita and McDonald, performed independent site-specific wind/tornado hazard assessments for 26 DOE facilities. The site of interest was among them. These analyses utilized advanced methodologies. Two of the most important improvements involved the use of wind speed gradation, both across the width and along the length of tornado damage paths. In addition, techniques were introduced to allow adjustments to be made to tornado data sets to account for unreported tornado events. Based on these hazard assessment results, *Kennedy et al.* [1989], in their DOE-UCRL-15910 report, provided design and evaluation guidelines for hazard frequencies up to 2×10^{-5}/yr.

In 1985, a third site-specific hazard assessment was performed by Twisdale for the site of interest. This analysis used computerized simulation techniques to estimate hazard probabilities and accounted for building size effects.

In summary, wind assessment technology has evolved from overly conservative regional analyses to realistic site-specific assessments. In this process, coarse assumptions have been refined by application of accumulated knowledge appropriate to tornado hazard assessment. In addition, during the ensuing years since the introduction of NRC Regulatory Guide 1.76, more complete tornado occurrence data have been recorded, allowing more accurate assessments. The resulting site-specific analyses, which take these advancements into account, are improvements over earlier regional analyses. Therefore the WTWG opted to incorporate site-specific hazard curves in the design guideline development.

The use of site-specific hazard frequency curves in design guidelines development reflects the accumulated knowledge and experience in wind engineering. The recommended design guidance presented in this paper is a major improvement to NRC provisions. The conformance of these proposed design guidelines for design was confirmed by preliminary PRAs described later in this paper.

It must be noted in wind hazard assessment that limitations of tornado intensity and occurrence data, F scale wind speed assumptions, and analysis processes incorporated in the hazard models result in assessment uncertainties.

3.2. Developing Missile Load Specifications Based on DOE Guidelines and EPRI Tornado Missile Simulation Report

The *Office of Nuclear Reactor Regulation* [1981] Standard Review Plan (SRP), NUREG-0800, recommends a conservative spectrum of tornado-generated missiles and corresponding velocity fractions. There is no regional distinction in the SRP missile types and sizes. However, the missile impact velocities vary on the basis of regional DBT wind speeds. DOE-UCRL-15910 [*Kennedy et al.*, 1989] also recommended missile spectra for various categories of facilities based in part on results of missile research performed at Texas Tech University. The missile spectra were developed for risk frequencies up to 2×10^{-5}/yr.

The Electric Power Research Institute sponsored development of a probabilistic tornado missile risk methodology to assess potential missile hazard for nuclear power plants. Results of this effort are presented in the EPRI NP-2005 report [*Twisdale and Dunn*, 1981]. Missile impact velocities were estimated on the basis of computer simulations of potential missile populations at nuclear power plants and computer-generated missile trajectories. The velocity fractions for steel pipes generated by EPRI are slightly larger than SRP velocity fractions.

In summary, state-of-the-art missile simulation technology incorporating relevant data bases and newly developed experimental results is an improvement over the technology used for NRC missile impact estimates in the 1970s. In the judgment of the WTWG, the recommended missile spectrum and the maximum impact velocities guidelines should be formulated by modification of DOE UCRL-15910 guidelines. The modification consists of incorporation of additional missiles and extrapolation from the EPRI tornado missile simulation results.

3.3. Assuring Conformance of Design Guidelines by Performing Preliminary Probabilistic Risk Assessment

Probabilistic risk assessments were performed to estimate annual damage risk frequencies for the facilities based on the recommended design guidelines. These wind damage risk frequencies are measures of achievement of the performance goals. To ensure the conformance of the design with the required performance goals, independent preliminary PRAs were performed to evaluate both wind speed and missile impact.

4. DESIGN GUIDELINES TECHNICAL SUPPORTING BASES

The objective of the design guidelines development is to recommend guidance for the design of the facility to meet the safety goals set forth by DOE. The basic philosophy for design guidelines development is to select low wind load specifications. The design goal is met by applying conservative response evaluation and acceptance criteria.

Core damage guidelines were developed for the structures, systems, and components required to bring a plant to a safe shutdown and to maintain this condition following a wind or tornado event. Production guidelines were developed for SSCs which cannot be cost effectively repaired or replaced within 90 days following a wind or tornado event. Ordinary SSCs, which are classified as low-hazard SSCs, must satisfy only UCRL-15910 wind/tornado design guidelines.

4.1. Wind Load Specification

At the work group meetings the WTWG held long discussions to establish the proper direction of wind/tornado design guidelines. Based upon references provided to the working group and the combined judgment of the hazard assessment and the risk assessment analysts, the following major factors were taken into account: (1) conservatism embedded in methodologies of assessment, (2) uncertainties identified in the assessment, (3) site-specific wind speeds in the geographical areas of interest, (4) site-specific wind speeds which achieve risk goals, and (5) acceptable wind speeds by the consensus of WTWG members.

After careful study of the hazard frequency curves and the examination of preliminary risk calculations, through an iterative process, the WTWG selected a tornado fastest-mile wind speed of 180 miles per hour (mph) (1 mi. = ~1.6 km), which corresponds to a probability of occurrence of 3×10^{6}/yr, and it achieves the core damage goal of 1×10^{-6}/yr.

A preliminary risk analysis on the production event indicates that straight wind controls the design. Using the straight wind model, a fastest-mile straight wind speed of 130 mph was selected. This wind speed corresponds to an occurrence probability of 2×10^{-4}/yr.

4.2. Missile Load Specification

After evaluating the SRP provisions, EPRI missile simulation results, and DOE-UCRL-15910 missile spectra, the WTWG formulated missile guidelines based on the following supporting bases: (1) selection of missile types that have been observed in past tornado damage investigations, (2) modification of the DOE-UCRL-15910 to extend the risk frequency from 2×10^{-5}/yr to 1×10^{-6}/yr, and (3) extrapolation of the EPRI tornado missile simulation results.

As a result of this effort, the WTWG added an 8-in.-diameter steel pipe to the DOE UCRL-15910 missile spectrum to extend the risk frequency to 1×10^{-6}/yr. The horizontal impact velocity for the 8-in. steel pipe was obtained by extrapolating EPRI results for a 6-in.-diameter

TABLE 1. Wind-Borne Missiles Criteria

Missile	Missile Criteria	Safety	Production
2 in. × 4 in. timber plank 12 ft long, 15 lb	horizontal impact speed, mph	150	115
	effective height, ft	200	200
	vertical impact speed, mph	100	75
3-in.-diameter standard steel pipe, 10 ft long, 75 lb	horizontal impact speed, mph	120	90
	effective height, ft	150	150
	vertical impact speed, mph	80	60
8-in.-diameter standard steel pipe, 15 ft long, 430 lb	horizontal impact speed, mph	100	75
	effective height, ft	75	30
	vertical impact speed, mph	70	50
Automobile, 3000 lb	rolls/tumbles along ground, mph	35	25
	effective height, ft	30	30
	vertical impact speed, mph	20	15

Metric conversions: 1 in. = ~2.5 cm; 1 ft = ~0.3 m; 1 lb = ~0.45 kg.
Horizontal and vertical missile speeds are uncoupled and should not be combined. Specific missiles represent a class of debris. Effective heights are elevations above ground level.

steel pipe at the 90 percentile. This extrapolation included conversion from a maximum velocity of 300 mph to 200 mph. The vertical impact velocity was set at 2/3 of the horizontal impact velocity by the WTWG.

Table 1 presents spectra of potential missiles with recommended impact velocities and corresponding effective heights (metric conversions are provided in a footnote). The barrier thickness shall be determined by equations given by *Twisdale and Dunn* [1981]. The minimum thickness of reinforced concrete barriers is 12 in. for all elevations below 75 feet, and 8 in. for all elevations above 75 feet. The minimum thickness of steel barriers is 1/2 in. for all elevations.

4.3. Structural Response Evaluation and Acceptance Levels

The WTWG incorporated conservative structural response evaluation methods and acceptance levels as follows:
1. Account for the fact that the critical facility is a high-hazard facility by increasing the importance factor to 1.35.
2. Use exposure C for tornado-controlled designs by assuming that the facilities are located in open terrain with obstruction having heights generally less than 30 feet. This assumption accounts for uncertainties in the characteristics of tornadoes.
3. Account for the variability in structural resistance properties by using a load factor of 1.3 on design wind pressures.
4. Select the velocity pressure exposure coefficient and the gust factor corresponding to the height of structures for both tornado and straight wind controlled designs.
5. Account for the uncertainty in missile impacts by following the American Concrete Institute (ACI) 349-85 recommended 20% increase in calculated barrier thickness.
6. Since a conservative missile spectrum and maximum tornado-generated impact velocities were selected for barrier design, the load factor of 1.3 does not apply to missile loads.

The SSCs shall be designed for the most severe total wind load (W_t) weight. This is a combination of wind pressure (W_w), atmospheric pressure change (APC) (W_p), and missile impact loads (W_m). The load factor applies only to wind pressure W_w and APC pressure W_p, but not to missile loads. In addition, the APC pressure load is half of its maximum value at the radius of maximum wind speed in a tornado; therefore, only $0.5W_p$ is combined with wind pressure and/or missile impact, as given below:

$$W_t = 1.3W_w$$

$$W_t = 1.3W_p$$

$$W_t = W_m$$

$$W_t = 1.3W_w + W_m$$

$$W_t = 1.3W_w + 0.5 \times 1.3W_p$$

$$W_t = 1.3W_w + 0.5 \times 1.3W_p + W_m$$

In the case of events controlled by straight wind, the APC pressure load does not exist, and the load combinations are as follows:

$$W_t = 1.3W_w$$

$$W_t = W_m$$

$$W_t = 1.3W_w + W_m$$

4.4. Risk Assessment for Wind Speeds and Missile Criteria

The preliminary PRAs were performed during the conceptual design stage; hence, component and plant details were not available. This limits the applicability of the preliminary PRAs to the final design. In addition, the results must be considered to be preliminary owing to the following simplifications which were built into the wind speed risk analyses:

1. The preliminary PRA analysis was simplified by using the results from previous PRAs to obtain an approximate but useful estimate of the expected damage frequency.

2. The preliminary mean damage frequency, which was used, represents damage within the 60–90% probability range.

3. Mean damage fragility curves used in the analysis were inferred on the basis of knowledge of the rules and procedures used in typical plant design processes. A more rigorous analysis would include performing detailed fragility analysis for each component and combining the resulting mean component fragility curves through Boolean logic.

The tornado wind speeds of 180 mph and a straight wind of 130 mph, corresponding to 3×10^{-6} and 2×10^{-4} annual hazard probabilities of occurrence, produce damage frequencies of 1×10^{-6} and 5×10^{-5} for core damage and production events, respectively.

For missile criteria risk analyses the simplifications are as follows:

1. Evaluation of missile damage is limited to exterior barriers for core damage and production events.

2. Estimation of impact probabilities for vulnerable areas, such as doorways and vent openings, is not included in the analysis.

3. Failures of plant stacks due to high winds are not considered.

4. Missile generation contributions resulting from the collapse and/or failure of tall structures are not included.

5. Nonlinear relationships between the numbers of missiles, wind speed, facility layout, and damage probabilities are not considered.

For core damage events, reinforced concrete barriers were assumed to be damaged if the barriers suffered backface scabbing or worse from any single missile impact. As a first approximation, it was assumed that these minimum barrier thicknesses controlled plant design.

The sum of scabbing damage risk frequencies for minimum design requirements is 1.1×10^{-6}/yr. This value is slightly higher than the core damage goal of 1×10^{-6}/yr. However, this damage criterion is very conservative, because even if backface scabbing occurs, a core damage event is not very likely to occur. The WTWG believes that the core damage risk frequency is at least 1 to 2 orders of magnitude less than the scabbing damage risk frequency. Therefore the WTWG believes that the recommended missile criteria conform to the core damage goal by a significant margin.

At an impact velocity of 35 mph, which is the core damage design guideline, the exceedance probability for automobile impact is 3.7×10^{-7}/yr; thus, the automobile results meet the core damage performance goal.

5. SUMMARY OF DESIGN GUIDELINES DEVELOPMENT

LLNL evaluated NRC provisions and acknowledged the excess conservatism embedded in their supporting analysis. Therefore design guidelines which deviate from NRC provisions are developed on the basis of site-specific hazard assessments and EPRI report results for the project. These credible design guidelines were developed on the basis of technological advancements and past experience in wind engineering. The load specifications reflect updates and enhancements to NRC regulations. The justification and supporting bases for these updates and enhancements are provided in this paper. The proposed structural response evaluation and acceptance criteria follow industry codes and standards with minor modifications. The general design guidelines recommended in this paper are believed to closely comply with DOE orders.

Acknowledgment. This work was performed under the auspices of the U.S. Department of Energy by the Lawrence Livermore National Laboratory under contract W-7405-Eng-48.

REFERENCES

Kennedy, R. P., S. A. Short, J. R. McDonald, M. W. McCann, and R. C. Murray, Design and evaluation guidelines for DOE facilities subjected to natural phenomena hazards, *UCRL-15910*, Lawrence Livermore Natl. Lab., Livermore, Calif., Oct. 1989.

Office of Nuclear Reactor Regulation, Standard Review Plan for the review of safety analysis reports for nuclear power plants, *NUREG-0800* (formerly NUREG-75/087), rev. 2, U.S. Nucl. Reg. Comm., Washington, D. C., July 1981.

Office of Nuclear Reactor Regulation, Safety evaluation by the Office of Nuclear Reactor Regulation of recommended modification to the R.G. 1.76 tornado design basis for the ALWR, NRR position paper, U.S. Nucl. Reg. Comm., Washington, D. C., March 1988.

Twisdale, L. A., and W. L. Dunn, *Tornado Missile Simulation and Design Methodology*, vol. 1, *Simulation Methodology, Design Applications, and TORMIS Computer Code*, EPRI NP-2005, Electric Power Research Institute, Palo Alto, Calif., Aug. 1981.

U.S. Nuclear Regulatory Commission, Design basis tornado for nuclear power plants, NRC Reg. Guide 1.76, Washington, D. C., April 1974.

Discussion

BOB MURRAY, SESSION CHAIR

Lawrence Livermore National Laboratory

PAPER G1

Presenter, Bob Murray, Lawrence Livermore National Laboratory [*Murray and McDonald*, this volume, Design for containment of hazardous materials]

(Unknown) I assume for straight-line winds you take the gust response factor from the UBC or ASCE manual?

(Murray) No, use the ASCE standard. The concept is built around the ANCI 58.1, now ASCE 788. You use the rules in there to go from the wind speeds to the pressure coefficients. We've imposed the importance factors to use, and we've varied the importance factors if you're at a coastline where a hurricane could be the threat, and let that slide as you go inland 100 miles.

(Unknown) Do you take the same gust response factor for a tornado?

(Murray) No gust response factors are used for tornadoes.

(Joe Golden, National Oceanic and Atmospheric Administration, Office of the Chief Scientist) I have two questions. First, is the infamous document, the so-called New-Reg 1200, still enforced for nuclear power plants?

(Murray) I think the information from that is incorporated in *Regulatory Guide 176*, which provides the three tornado zones for the country, and a standard review plan, which is what the NRC uses to review a design.

(Golden) My urgent recommendation is that the community represented at this symposium take a careful look at that document to see if the criteria in the document are still valid. Based on what we've heard so far and will hear tomorrow, I don't believe they are. They may have been the best results of their time, but that was 15 or 20 years ago.

(Murray) The whole panel agrees with that also.

(Golden) My second question concerns looking at the tornado incidence in the west over the last two decades. Ostby has shown that there has been a significant increase in western states' tornadoes during that period, increases amounting up to 300% in some cases. You seem to be ignoring these tornadoes and worrying only about the straight-line winds.

(Murray) You are correct. As a matter of fact, the next speaker lives in a California city that just recently had a tornado. It's part of the element of continually reviewing this information and learning from new storm data that we are doing now. One of our southern California sites, Santa Suzanna, has tornadoes as a design criterion.

PAPER G2

Presenter, Ravi Ravindra, EQE Engineering [*Ravindra*, this volume, State-of-the-art and current research activities in extreme winds relating to design and evaluation of nuclear power plants]

(Arnold Court, California) Any probabilistic risk assessment must involve first and foremost the expected or desired lifetime of the structure. And because the longer you want it to last, the greater the risk that it will be damaged, so you start first with the lifetime. Do you assume that these plants are built to last 10 years, 500 years, or what?

(Ravindra) The nuclear power plants are licensed to operate for 40 years. There is currently a program to extend the license period by another 20 years, so you can assume a lifetime of about 60 years. The probabilistic risk assessment that is done today is being done about 20 years into the operation of the plant. So the risk numbers that you see are typically over the life of the plant, or you can talk about per year values.

(Court) So what is the calculated risk? 10^{-5} of failure within the next 40 years?

(Ravindra) Yes, a very low risk. Because meteorologists have done a good job in saying how much wind to expect, to design for, the probabilistic risk assessments show that the risk in somewhere between 10^{-4} and 10^{-6}, depending upon the site.

(Court) I have another question about missiles. Are they assumed to fall end on, or broadside?

(Ravindra) In the design, it is assumed to fall end on. In the risk analysis, like Twisdale is going to talk about, the impact is randomly oriented. Sometimes it could be end on; other times it could be broadside.

(Roger Tanner, National Climatic Data Center) Are there studies that take into consideration, particularly in the fragility area, the temperature in any of the three tornado regions? It could affect tensile strength of the reinforced steel or concrete. Also, what about soil conditions.

(Ravindra) The temperature effects are considered mainly for the internals of the nuclear power plant. There is concern about how the construction would perform over the years. The embrittlement is considered in terms of the structure, but the effect is minimal.

(Tanner) What about moisture?

(Ravindra) It is minimal in the sense of the aging we are involved in. We are looking at ways to inspect the structures to see if there is some deterioration as a result of moisture or temperature. But from what we've seen the effect is minimal. This is because the structures are pretty beefy and the power plants are well maintained by law.

(Joe Golden, National Oceanic and Atmospheric Administration, Office of the Chief Scientist) Missiles in tornadoes are becoming more complicated than we used to think. The slide I'm having projected shows the aftermath of the Huntsville, Alabama, tornado. What you see there are about eight automobiles that have been wrapped together by the tornado. How would your model handle something like that?

(Ravindra) The plants are designed to withstand automobile impact, based on speed and weight. The estimates are conservative, so they could withstand additions from multiple automobiles. Also, the plants are typically located in isolated areas, where you wouldn't expect so many cars to impact on them at one time.

(Golden) That's not true. In Amarillo, Texas, they have a lot of automobiles and there's a plant just to the north. Nor is it true at Rocky Flats, Colorado, where there is a main highway that goes right in front of the plant.

(Ravindra) When doing a risk analysis for a nuclear power plant, one would consider two stages. Where the plant is being built, and a second stage when it is operating. Then there is the strength of construction material to consider. We do consider all of those things in the analysis.

(Golden) I would like to ask a second question. In your hazard analysis, do you use *Storm Data*? I mean, do you look at the storm statistics?

(Ravindra) Yes.

(Golden) You should know that *Storm Data* is in a very fragile state. There are some important considerations that everyone in this room needs to understand when they use *Storm Data*. I think we need to get the community together to support *Storm Data* and to improve it.

(Ravindra) I fully concur with you. Perhaps the engineers and meteorologists have not had enough interaction.

Paper G4
Presenter, Jim Hill, Department of Energy (J. Hill, not in this volume, Natural hazard losses: A Department of Energy Perspective—Injury and property damage experience from natural phenomena hazards)

(Don Burgess, National Severe Storms Laboratory) Did I understand you correctly that DOE has a Doppler radar at Amarillo, Texas?

(Hill) No, not a Doppler, lightning detection equipment. Also we have equipment to detect potential electrical gradient. We find high gradients from wind and dust as well as thunderstorms. They have sensitive explosives at that plant they do not like to operate at times of high potential gradient. They have had problems with farmers plowing their fields during strong winds and producing dangerous, high potential gradients. They shut the plant down when the potential gradient exceeds a certain amount, no matter what the cause. I could add a comment about weather prediction at our sites. At least five of the sites have very sophisticated and complex weather forecasting for local weather. For example, when a tornado touched down at the Savannah River Plant last March, people were sheltered. There was some damage but no serious structural damage. Also, if there is any kind of release of hazardous materials, then we can do tracking and postulate potential trajectories for the release of materials.

Paper G5
Presenter, Larry Twisdale, Applied Research Associates (L. A. Twisdale, Jr., not in this volume, Safety assessment of critical facilities for tornado effects)

(Arnold Court, California) The winds that you are using, are those fastest mile?

(Twisdale) No, those are gusts.

(Court) One second, 10 seconds, what?

(Twisdale) Damage-producing gusts averaged over, we would guess, several seconds.

(Court) Two-second gusts?

(Twisdale) No, I wouldn't say two seconds. I would just call them damage-producing gusts. We don't know what the averaging times are, since tornado wind speeds are based on damage produced at the ground. So it would vary, structure by structure. Most engineers would say it's just damage-producing gusts.

(Jim Hill, Department of Energy) The tornado model that you used, was that the NRC model or was that a site-developed model?

(Twisdale) That was a model we developed as part of an EPRI program. We reviewed three or four models that contained the basic physics, and we added a tornado boundary layer and inflow. We essentially synthesized the model from Fujita and a few others. We developed probability distributions for four or five key parameters that were best guesses at the time. We also put in suction vortices. They were added in 1979 or 1980 because we were concerned that they could lead to more missile effects. What we found was that if you had a mini vortex with the peak speed the same as in the parent vortex, you could use the parent vortex model and get essentially the same results.

(Hill) What was the maximum wind speed that you used?

(Twisdale) We tracked the hazard curve, so it's a risk assessment.

(Hill) I understand, but what would be the upper limit on your curve?

(Twisdale) The curve goes to infinity. The curve that I developed goes asymptotic at about 300 mph. And it has a slightly downward slope, which means that as you get to higher and higher risk, you would expect less incremental increase in wind speeds.

(Hill) For the example you showed, what was the return period of the wind that caused the damage to the facility.

(Twisdale) All of them contributed, and the range from F2 to F4 dominated the prediction of failure frequencies.

(Rudy Engleman) I'm a little surprised that you have so little risk involved with operator error. In view of Three Mile Island and Chernobyl and other incidents, I've sort of come to the conclusion that operator error is the biggest problem.

(Twisdale) That's an interesting comment. We used inputs from the plant on operator error. However, we did a study where we assumed perfectly reliable and perfectly unreliable operators. There was little difference for the sequence of tornado-initiated events.

(Engleman) This is just for tornado-initiated events?

(Twisdale) Yes, any damage that shows up is just because you've got the tornado.

(Ravi Ravindra, EQE Engineering) I don't want to leave the impression that the older plants will go without evaluation and identification. The main purpose of the IPEEE is to look for vulnerabilities, and it calls for a detailed walk-down through the plant. Some of the things we have learned over the last 20 to 30 years of tornado design and that we've seen in real tornado damage can be applied. If vulnerabilities exist, the IPEEE will correct them in older operating plants. In new plants, obviously they don't exist because we have taken care of them in the design.

(Twisdale) I agree with you except that I think that identifying these vulnerabilities, when we've done so few analyses, is heavily dependent on the individual. There are a lot of cases where, at least at plants we've looked at, some things have cropped up that don't necessarily show up from the walk-down. Sometimes it's good to run through the numbers.

Advances in Tornado Climatology, Hazards, and Risk Assessment Since Tornado Symposium II

THOMAS P. GRAZULIS

St. Johnsbury, Vermont 05819

JOSEPH T. SCHAEFER

Scientific Services Division, National Weather Service Central Region, Kansas City, Missouri 64106

ROBERT F. ABBEY, JR.

Office of Naval Research, Arlington, Virginia 22217

1. RISK ANALYSIS

1.1. Introduction

At Tornado Symposium II, R. F. Abbey, Jr., of the Nuclear Regulatory Commission (NRC), presented an outline of the history of efforts to advance the science of tornado climatology and risk assessment. This presentation summarized the relatively slow progress in the field prior to 1971. However, a few advancements in the years just prior to that 1976 Symposium drastically changed the course of tornado research and permanently altered our discussions of tornadoes in general. One was the development of the Fujita scale of tornado intensity [*Fujita*, 1971]. Of equal importance was the decision by the National Weather Service (NWS) and National Severe Storms Forecast Center (NSSFC) Director, A. Pearson [*Fujita and Pearson*, 1973], to embrace it as the official government-certified intensity classification system for tornadoes. Despite its shortcomings the Fujita scale became a much needed point of focus for discussions and an integral part of nearly all modern tornado climatology and risk assessment.

The introduction of the Fujita scale was almost coincidental with the entry of the NRC into tornado hazard research. The NRC, under the guidance of R. Abbey, Jr., provided a source of funding for projects that ranged from historical climatology to computer simulation of tornado wind fields.

In addition, the NRC established specific quantitative goals for tornado climatology. Prior to this time, as illustrated by *Court* [1970], most climatological studies involved the creation of tornado occurrence maps, with little or no separation of events by destructive potential. The NRC required an estimate of the actual maximum wind speed values that could be expected from tornadoes for different levels of probability, particularly 10^{-7}. The first numerical models that could be used to create wind speed probability maps from climatological data were developed prior to Tornado Symposium II. This was followed by years of refinement, analysis, discussion of limitations, and the development of alternative models. With NRC support, tornado climatology and risk analysis advanced steadily for a decade. However, by 1984 only a few scattered projects in risk analysis were still underway, and those were largely efforts to solidify the gains of the previous decade.

1.2. Probability Map Sets

The first full set of wind speed probability maps was presented by *Abbey* [1976] at Tornado Symposium II. Since then, three separate sets of such maps have been created. Figure 1 reproduces two maps from Tornado Symposium II, based on a model by *Abbey and Fujita* [1975]. These eventually evolved into the more refined set shown in Figure 2 [from *Tecson and Fujita*, 1985], which used the Fujita/University of Chicago (UC) tornado data base. A model (using the same UC data base) was developed by J. McDonald at Texas Tech. Charts are shown in Figure 3 as they appeared in the work of *McDonald and Allen* [1981].

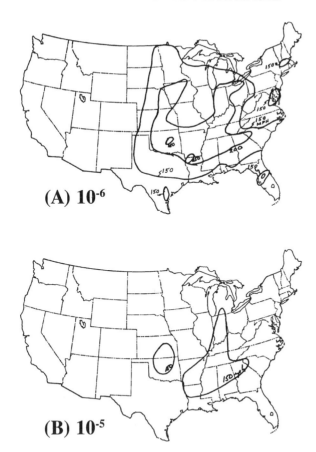

Fig. 1. Wind speed probability charts based on the model by *Abbey and Fujita* [1975] and presented at Tornado Symposium II: (*a*) occurrence probability of 10^{-6} and (*b*) occurrence probability of 10^{-5}.

An alternate model, using the NSSFC data base, was developed by J. Schaefer and colleagues at the NWS offices in Kansas City, Missouri. It does not deal directly with specific wind speeds, but rather gives the probability that tornadoes greater than each individual F scale category will occur. Some maps are shown in Figures 4–7 as they appeared in the work of *Schaefer et al.* [1986].

There were four major efforts in tornado risk analysis during this period, all fostered (at least in their early stages) by the NRC. These will be analyzed in a bit more detail.

1.3. *University of Chicago Model*

The most comprehensive study on tornado risk was headed by T. T. Fujita at the University of Chicago. That work culminated in that of *Tecson and Fujita* [1985] and *Fujita* [1987]. Much of the basis of the model for the Fujita/UC effort was detailed by the *Fujita* [1978] and by *Abbey and Fujita* [1979]. This rigorous treatment of the "design tornado" addresses such concepts as the wind speed distribution in suction vortices. Figures 8, 9, and 10 reproduce diagrams from *Fujita* [1978] dealing with wind distribution in suction vortices. They illustrate a small part of the Fujita-Abbey attempt to deal with the horizontal and vertical winds fields in tornadoes.

At the heart of this model is the damage area per path length (DAPPL) method for risk computation. This method is centered around the concept that each tornado produces a damage area that includes a portion of all F scale levels from the estimated maximum intensity down. The F scale velocity lines, or isovels, for a model F-3 tornado are shown in Figure 11 [from *Fujita*, 1978].

A difficulty in implementing this model is the assignment of appropriate areas for each F scale damage. Since full-path photographic aerial surveys rarely are performed for tornadoes, it was necessary to develop a model using data collected from the most thoroughly surveyed population of tornadoes that were available. That was, and is, the spectacular outbreak of April 3–4, 1974.

From extensive analysis of the 148 tornado paths on April 3–4, 1974, empirical formulas were developed that related the width of the tornado track and its maximum intensity to the distribution of damage by F scale within the track. For instance, a 400-yd-wide (366.2 m wide) F-5 tornado might contain a 5-yd-wide (4.6 m wide) swath of F-5 intensity. The remaining 395 yd (361.6 m) would be distributed as 7 yd (6.4 m) of F-4, 17 yd (15.6 m) of F-3, 41 yd (37.5 m) of F-2, 97 yd (88.8 m) of F-1, and 233 yd (213.3 m) of F-0. In reality, of course, the F-5 damage swaths would be larger and scattered along the track. There has been some discussion as to whether the tornado population of April 3–4, 1974, actually is typical of all tornadoes in all areas of the country. It has been suggested that perhaps these tornadoes have a higher percentage of F-3, F-4, and F-5 damage areas than the general population of tornadoes. Since only a few of the most destructive tornadoes each year ever get a full photographic survey, there is no alternative to using this outbreak, and no conclusive data exist to resolve any debate on the subject.

Path length is an important parameter in the DAPPL method. Extensive efforts were made to produce models to correct path lengths for various topographic and population-based factors. Indices for six parameters that might affect path length reports were created. The United States was divided into 13,690 subboxes, each 15 min by 15 min. A value for each parameter was assigned to each box from 1970 census data and U.S. Geological Survey (USGS) 1:250,000 scale maps. This was a tedious process, because census data are organized by county, not 15° box. There were some efforts at UC to adjust tornado paths for the percentage of water or forest area, slope, road separation, number of communities, and population. *Tecson et al.* [1983] presented maps for population and topography indices, and these are reproduced in Figure 12.

In the end the area in each subbox affected by tornado damage of a given F category was computed by taking the sum of the products obtained by multiplying the adjusted path length for each F scale greater than or equal to the one

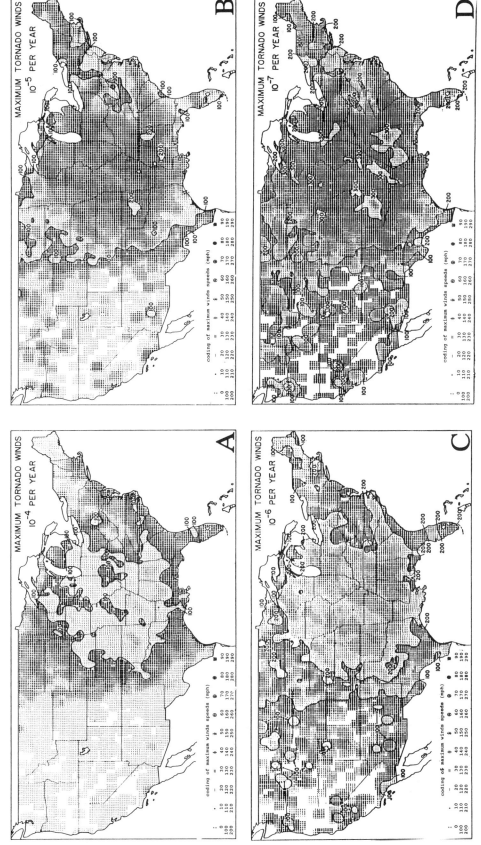

Fig. 2. Wind speed probability charts based on the DAPPL model as presented by *Tecson and Fujita* [1985]: (*a*) occurrence probability of 10^{-4}, (*b*) occurrence probability of 10^{-5}, (*c*) occurrence probability of 10^{-6}, and (*d*) occurrence probability of 10^{-7}

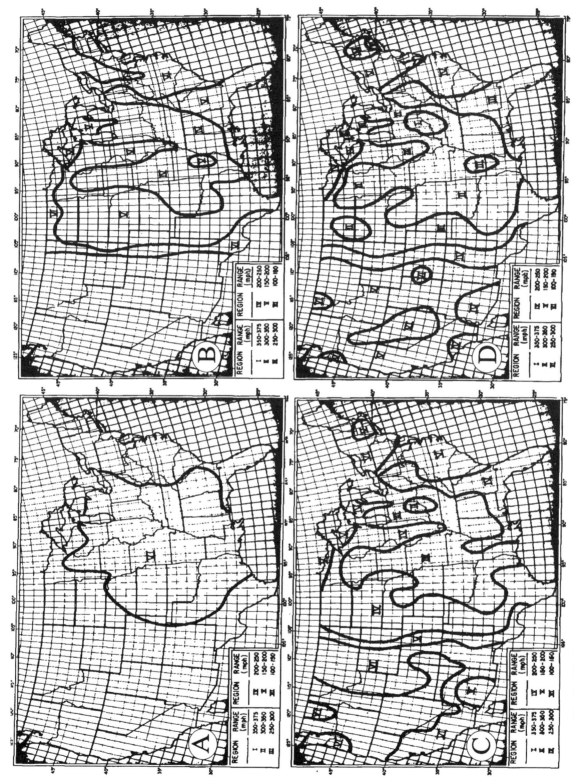

Fig. 3. Wind speed probability charts based on the model of McDonald presented by *McDonald and Allen* [1981]: (*a*) occurrence probability of 10^{-4}, (*b*) occurrence probability of 10^{-5}, (*c*) occurrence probability of 10^{-6}, and (*d*) occurrence probability of 10^{-7}.

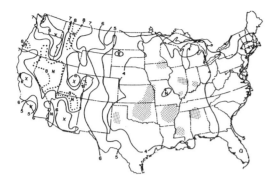

Fig. 4. Probable annual tornado hazard. Contours are labeled in negative powers of 10 per year (i.e., 4 indicates 10^{-4}). Maxima are denoted by "X," minima by "N." Stippled area has a hazard greater than 4×10^{-3} per year. Dotted line is 10^{-8} or less.

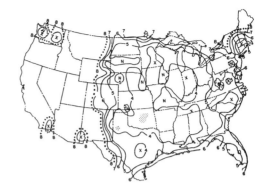

Fig. 6. Probable annual tornado hazard from F-3 or greater tornadoes; legend as in Figure 4.

of interest by the appropriate mean path width, and the DAPPL ratio. This approximation requires that path length not be correlated with path width, and that the DAPPL numbers obtained from the outbreak of April 3–4, 1974, be universally true.

1.4. NWS/NSSFC

The research effort at the NWS office in Kansas City, Missouri, analyzed tornadoes documented in the National Severe Storms Forecast Center (NSSFC) data base. Statistics considered were occurrence, affected area, and specific F scale category rather than specific wind speeds. These efforts were headed largely by J. T. Schaefer. The resulting probability maps were shown in previous figures as published by *Schaefer et al.* [1986]. Here the hazard probability is obtained by summing the individual damage areas of each tornado of intensity greater than the desired F scale threshold that was reported within localized quasi-homogeneous regions. This tornado-affected area is then normalized by the size of the region and the duration of the data base.

Implicit in this analysis is the assumption that all significant information about a tornado is reported. If no F scale is reported, then the tornado produced negligible damage. Similarly, if the path length or path width is missing from the report, then the tornado had minimal ground contact or was extremely narrow and impacted very little area. Caution must be exercised when interpreting this chart in areas such as the Dakotas or Colorado, where many tornado records contain no path width estimates and the observed damage area data may tend to underestimate the true hazard.

The NWS office in Kansas City used the same empirical methods to produce other charts which can be termed risk or hazard maps. Figure 13, reprinted in many different government publications, shows the frequency of occurrence of all tornadoes within a 10,000-mi^2 (25,900 km^2) box. Figure 14 [from *Schaefer et al.*, 1980] shows the isopleths for the area that will contain 1 mi^2 (2.59 km^2) of tornado damage in 100 years.

1.5. Texas Tech University

The Texas Tech University group, headed by J. McDonald, divided the United States into the 14 regions. Empirical tornado area-intensity relationships were determined in each of these "global" regions. This was an effort to address the concept that tornado characteristics are not geographically uniform. For instance, tornadoes in Florida have different

Fig. 5. Probable annual tornado hazard from F-2 or greater tornadoes; legend as in Figure 4.

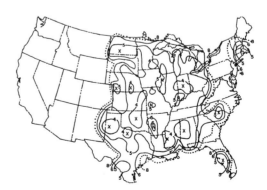

Fig. 7. Probable annual tornado hazard from F-4 and F-5 tornadoes; legend as in Figure 4.

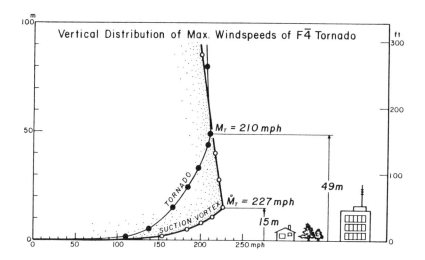

Fig. 8. The maximum wind speeds inside an F-4 tornado with an embedded suction vortex as modeled by *Fujita* [1978].

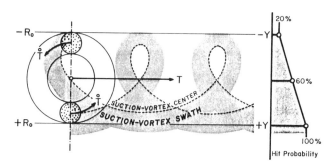

Fig. 9. The "hit probability" across the path of a tornado with an embedded suction vortex as modeled by *Fujita* [1978].

area-intensity characteristics than those in Texas and Oklahoma.

The Texas Tech methodology used empirically derived occurrence-intensity relationships for local regions (3° squares) containing the 1° square of interest. Thus the probability of experiencing or exceeding winds in any F scale interval could be determined for any point.

Texas Tech determined that at least 6500 tornadoes were probably missing from the earlier years of the NSSFC tornado data set (1950–1972). These 6500 "missing" events were distributed across the United States and included in the probability calculations. The distribution is shown in Figure 15. The placement of these unreported tornadoes was based on studies of reporting efficiency as determined by *LaGreca* [1980].

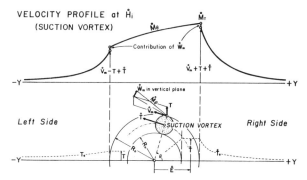

Fig. 10. Velocity profile across a tornado with an embedded suction vortex as modeled by *Fujita* [1978].

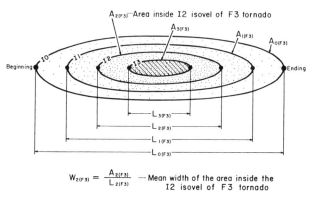

Fig. 11. A schematic of the F scale distribution of damage across the swath of an F-3 tornado as modeled by *Fujita* [1978].

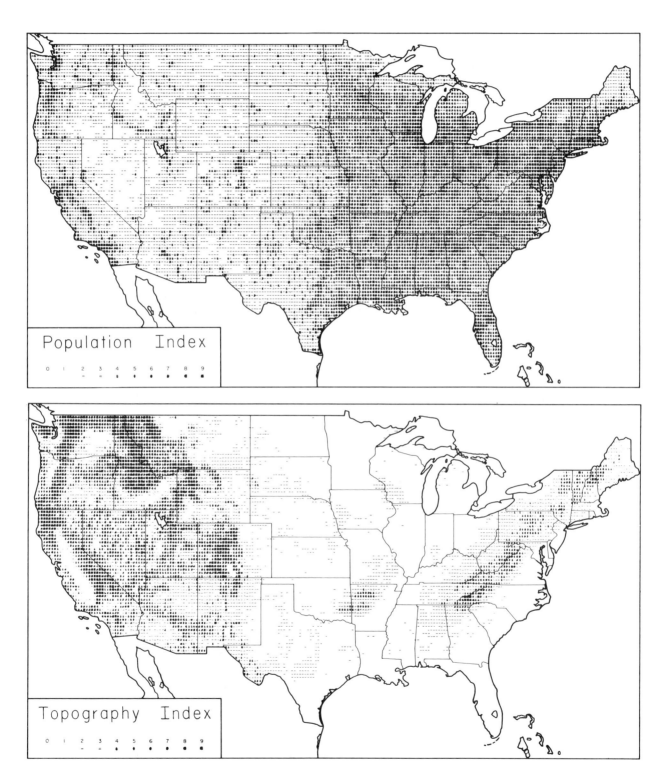

Fig. 12. Charts of Fujita's indices for (top) population and (bottom) topography as presented by *Tecson et al.* [1983].

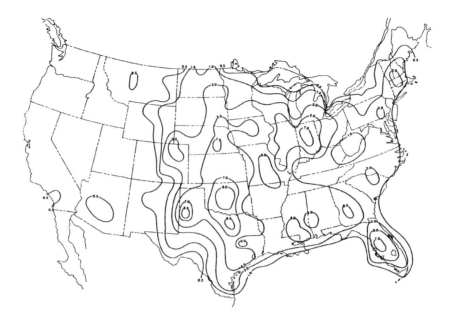

Fig. 13. Average annual tornado incidence per 10,000 mi^2 (25,900 km^2), as published by *National Oceanic and Atmospheric Administration* [1982].

1.6. *Twisdale/Electric Power Research Institute*

Twisdale [1978, 1982, 1983], of the Electric Power Research Institute (EPRI), used a rigorous statistical approach to identify probable sources of error in the NSSFC tornado data set and to deal with them on a regional basis. The sources of error he identified were unreported events, unreported F scale, direct misclassification of F scale, and misclassification of F scale due to the randomness of damage. He synthesized his data into four broad regions and gives an analysis of occurrence rate, intensity, path length, path width, and direction of tornado movement for each.

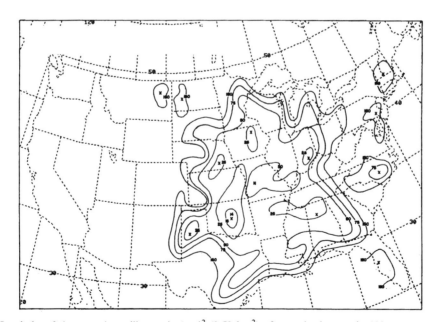

Fig. 14. Isopleths of the area that will contain 1 mi^2 (2.59 km^2) of tornado damage in 100 years as published by *Schaefer et al.* [1980].

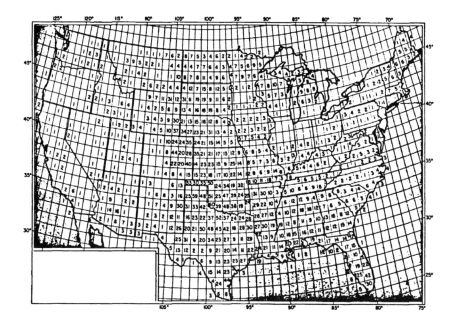

Fig. 15. The number of unreported tornadoes, 1950–1979, per 1° latitude/longitude square as estimated by *McDonald and Allen* [1981].

Twisdale [1978] suggested an alternative set of wind speeds to be associated with the Fujita scale. Twisdale's and Fujita's values are compared in Table 1. The debate as to just what wind speeds are needed to cause just what kind of damage also continues. Any adjustments to the Fujita scale, if and when needed, would come by either altering descriptions of the kinds of damage ascribed to each wind speed range or changing the wind speeds assigned to each of the six levels.

1.7. Fujita-McDonald Comparison

Despite differing methodologies, the Fujita and McDonald site-specific 10^{-7} wind speed probabilities agree fairly well across much of the United States. Differences are usually less than 50 mi/hr (mph) (22.4 m s^{-1}) at any site. *Coats and Murray* [1985], in a report by Lawrence Livermore National Laboratory (LLNL) for the Department of Energy, compared the Fujita and McDonald models for 10^{-7} wind speed probability at 25 sites.

These sites are listed in Table 2 and ordered by the difference between the wind speed estimates produced by the two models. The exact reasons for the differences are complex and a product of the assumptions in the models and the way the data are handled.

The largest differences are in the Ohio Valley area of the April 3–4, 1974, outbreak. This may be because the Fujita/DAPPL model uses data extending back to 1916, while the McDonald/Texas Tech model uses data beginning in 1950. *Grazulis* [1991] noted that the 1916–1949 period was less active for violent tornadoes in the Ohio Valley. Grazulis also suggested that the wind speed may be exaggerated at the Pinellas, Florida, site. The data there are heavily influenced by the violent, 135-mi-long (217-km long), Gulf-to-Atlantic tornado track of April 4, 1966. This entire track was given an F-4 rating based on the destruction of a single home in a rural community in central Florida. The event was probably a tornado family which was not broken into individual family members.

Perhaps the west coast of Florida, the Florida Panhandle, central Florida, and the hurricane-prone areas of southeast Florida should all be treated as independent subregions. While other states may need to be treated in this way, the meteorological factors that underlie the long-term tornado risk are not as readily identifiable as they are in Florida.

Difficulties associated with analysis of long-track tornadoes and tornado families were addressed by *Doswell and Burgess* [1988]. Their reanalysis of the historic Woodward Tornado of April 9, 1947, is shown in Figure 16. The paper is a summary and discussion of some of the many difficult issues in tornado climatology that remain unresolved.

TABLE 1. Comparison of Wind Speeds

Scale	Fujita Range, mph	Twisdale Range, mph
0	40–72	40–73
1	73–112	73–103
2	113–157	103–135
3	158–206	135–168
4	207–260	168–209
5	261–318	209–277

TABLE 2. Site-Specific 10^{-7} Wind Speed Probabilities

Site	State	Fujita (A),* mph	McDonald (B),† mph	B − A,‡ mph
Mound	Ohio	283	364	81
Oak Ridge	Tennessee	261	340	79
FMPC	Ohio	287	364	77
Portsmouth	Ohio	257	330	73
Paducah	Kentucky	289	340	51
Argonne, east	Illinois	276	318	42
Pantex	Texas	271	297	26
Rocky Flats	Colorado	204	228	24
Pinellas	Florida	268	244	−24
Princeton	New Jersey	229	210	−19
Site 300 (LLNL)	California	164	182	18
Bendix/Kansas City	Missouri	292	310	18
Stanford	California	182	165	−17
Sandia, Albuquerque	New Mexico	175	191	16
Nevada Test Site	Nevada	152	136	−16
Brookhaven	New York	224	215	−9
Los Alamos	New Mexico	183	190	7
ETEC	California	179	174	−5
Berkeley	California	168	165	−3
Hanford	Washington	179	177	−2
Sandia, Livermore	California	164	165	1
Argonne, west	California	185	184	−1
Idaho (inel)	Idaho	185	184	−1
Livermore (LLNL)	California	164	165	1
Savannah River	South Carolina	283	283	0

*The 10^{-7} probability maximum wind according to the Fujita model.
†The 10^{-7} probability maximum wind according to the McDonald model.
‡The difference between the two models.

1.8. Engineering Studies

There have been many important wind-engineering studies which potentially can impact tornado climatology but were not, in themselves, climatological studies. The most significant of these was probably by *Minor et al.* [1977], which broadened the discussion of what kinds of damage occur at various wind speeds. The concept that structural failure of homes and other buildings (resulting in complete "destruction") can take place under both lower F scale wind (F-2) and higher F scale winds (F-4) continues to evolve. The irregular means by which meteorologists become aware of engineering concept topics may affect climatology by creating state-to-state inconsistencies in the application of the Fujita scale.

2. CLIMATOLOGY

2.1. General Climatology

Wilson and Kelly [1977] examined the distribution of tornadoes by day of the week and noted a statistically significant (10–20%) lower number of tornadoes on Saturday. This study concluded that Saturday tornadoes were less likely to be reported by newspapers. Newspapers alert NWS personnel to the occurrence of about 10% of tornadoes, which probably would not have been reported otherwise.

Grazulis [1991] also noted difficulty in locating newspaper descriptions for tornadoes that occurred on Saturday.

It seems likely that these unreported tornadoes would be weak. However, the Saturday minimum is present at even the higher F scale levels. This may be due to a lack of impressive newspaper damage photographs, resulting in a tendency for Saturday tornadoes to be rated lower. This hebdomadal distribution does not hold for tornado outbreaks. Two of the 10 largest 1-day tornado outbreaks occurred on Saturday (the lowest tornado day). In contrast, Thursday and Friday (the two most likely tornado days) accounted for zero and one of the largest outbreaks, respectively. *Tecson et al.* [1979] noted a similar day-of-the-week distribution but found it less significant. Statistical studies on tornado path length and path width were done on the UC data set by *Tecson et al.* [1979], and on the NSSFC tape by *Kelly et al.* [1978] and by *Schaefer et al.* [1986].

The paper by *Kelly et al.* [1978] contains detailed analysis of tornado occurrences for each F scale category. Among the findings were that violent (F-4–F-5) tornadoes occurred at all times of the day or night. Weak (F-0–F-1) and strong (F-2–F-3) tornadoes showed more diurnal trends. Studies were done of tornadoes in three geographical regions. Figure 17 shows clearly that the southeast has the greatest frequency of morning tornadoes.

Table 3 provides some comparative path length and path

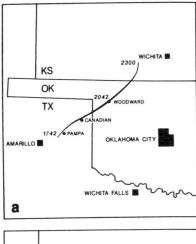

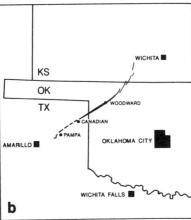

Fig. 16. Path of the April 9, 1947, tornado which struck Woodward, Oklahoma, (a) as originally classified and (b) as reclassified, on the basis of a review of the data. Tornado occurrence times shown in Figure 16a are in central standard time. The dashed segment in Figure 16b corresponds to what was in 1947, and still is, a very sparsely populated region within which path continuity cannot be determined definitely [after *Doswell and Burgess*, 1988].

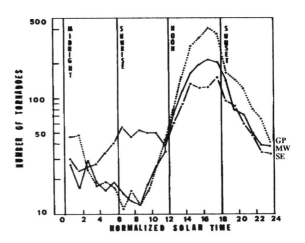

Fig. 17. The diurnal distribution of tornado occurrence time (normalized with respect to sunrise/sunset) in the Great Plains (GP, dotted line), midwest (MW, solid line), and southeast (SE, dashed line) [after *Kelly et al.*, 1978].

width values for UC data [from *Tecson et al.*, 1979] and for NSSFC data [from *Schaefer et al.*, 1986]. A comparison of the NSSFC mean shows a shorter path length than for UC mean F-2–F-5 tornadoes. This is possibly due to the lower path length for many tornadoes in the 1916–1949, pre-NSSFC period, covered by UC data. More complete surveys in "modern" times (i.e., since 1953) tend to find that actual path lengths are often longer than initially reported in the press. The 1916–1949 shorter path lengths may have been, in some cases, only the length of the path that produced obvious damage. The UC path lengths for F-0–F-1 are longer than the NSSFC path lengths. This may be due to the nonlisting of minor, short path length tornado in the 1916–1949 period. Another factor may be that the UC data set (column 1) gives poorly documented tornadoes a default path length of 0.5 mi (0.8 km). The calculations from the

TABLE 3. Comparative Path Length and Path Width Data

F Scale	1916–1978 UC Mean PL, mi	1950–1982 NSSFC Mean PL, mi	1950–1982 NSSFC Median PL, mi	1950–1978 NSSFC Mean PW, yd	1950–1978 NSSFC Median PW, yd
F-0	1.41	1.11	0.30	46	17
F-1	2.94	2.59	0.98	93	47
F-2	5.56	5.66	2.19	167	99
F-3	10.25	12.08	6.76	290	180
F-4	18.66	22.42	13.80	432	297
F-5	28.55	34.17	23.44	616	496
All	4.67	4.40	0.98	128	48

Abbreviations are as follows: PL, path length; PW, path width. One mile equals 1.6 km, and 1 yd equals 0.9 m.

Fig. 18. The distribution of 372 tornadoes associated with major winter (December–February) tornado outbreaks between 1950 and 1977 [after *Galway and Pearson*, 1979].

NSSFC tape used only tornadoes that had a path length and a path width recorded. The NSSFC data set for the 1950–1982 period had 22,840 tornadoes, of which 14,243 had both a path length and a path width.

It is worth noting the difference between the mean and the median values. Relatively few very long track tornadoes skew the distribution and boost the mean value so much that the median is a more representative figure for tornadoes, in general. Path lengths are in miles, path widths are in yards.

Some conversion from kilometers to miles and miles to yards was done from the original papers.

Galway and Pearson [1979] studied winter tornado outbreaks, finding they were often accompanied by heavy snow or ice conditions to the northwest of the track of the synoptic low. Winter outbreaks were comparable in violence to spring outbreaks but caused an even higher number of deaths than spring outbreaks, despite average or above average watch and warning times. This is possibly due to the unexpected nature of the events at that time of year. Winter outbreak tornadoes are mapped in Figure 18.

Galway's [1983] climatology of killer tornadoes found total death statistics dominated by F-4/F-5 tornadoes; however, half of all killer tornadoes caused only a single death. He also found a clear relationship between long-track tornadoes and high death totals for individual tornadoes.

An apparent natural division of the United States into eastern and western groups of tornadoes was revealed by *Tecson et al.* [1982]. The two large groups of tornadoes are separated by a slanted axis (along a relative minimum of activity) that extends from upper Michigan and northern Wisconsin through the Ozark mountains to southeast Texas. A plot of total tornado path length, from west to east, across the United States is shown in Figure 19. The central minimum is apparent. This central minimum was also located by *Grazulis and Abbey* [1983] in a documentation of F-4 and F-5 tornadoes, as seen in Figure 20. It was concluded in both papers that population may not be the sole reason for the existence of the relative minimum along the dividing axis.

Changnon [1982] looked at trends in tornado days, rural population, and the number of days with tornado deaths in Illinois since 1916. He noted that the increase in tornado

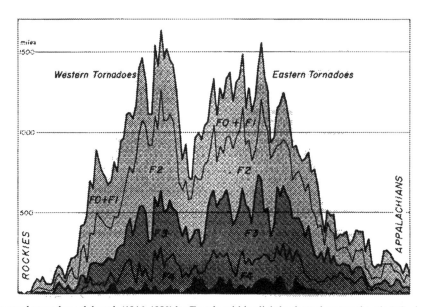

Fig. 19. Summed tornado path length (1916–1980) by F scale within slightly slanted narrow bands (15 min of longitude wide) stretch from 30°N to 42°N [after *Tecson et al.*, 1982].

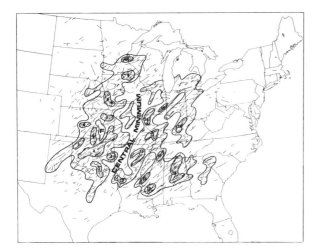

Fig. 20. A map of violent (F-4 and F-5) tornadoes from *Grazulis and Abbey* [1983]. Isopleths are drawn for 40 and 80 path mi per 1000 mi^2 (1 mi = 1.6 km). Twenty subregional maxima are enclosed by 80-mi isopleths. A central minimum is so labeled, and a relative minimum in the sparsely populated Flint Hills area of Kansas and Oklahoma is marked with an "F."

the many potential pitfalls that await the tornado climatologist are quite valuable. He also noted how easily tornado statistics can be twisted and misinterpreted. The "user beware" notice he advocated, albeit tongue in cheek, is very applicable to most tornado statistics.

2.2. The Annual Number of Tornadoes

Eshelman and Stanford [1977], in a study of Iowa tornadoes for the year 1974, reported that funnel touchdowns could be confirmed 81 times in Iowa during 1974. This is 3 times as many tornadoes as the officially recorded 27. Similar drastic differences in the number of actual, compared to the number of recorded, tornadoes were found in the statistics compiled by *Snider* [1977] on Michigan tornadoes.

Schaefer and Galway [1982] divided counties into four groups based on population density. They noted that the tornado distribution was rather uniform across the categories and that on a nationwide basis there was little statistical evidence for population bias, except in counties with extremely high populations. This lack of a strong statistical bias for most counties runs contrary to the standard logic that few tornado reports would come from counties with few people. This lack of population bias is most evident along the western plains from Oklahoma through Kansas, to the Dakotas (see Figure 21).

Grazulis [1991] found that in the moderately populated counties of eastern Kansas, about 50% of tornadoes hit buildings. Oddly enough, he found that same percentage (50%) of the tornadoes in the sparsely populated rural counties of western Kansas also hit buildings. Because of the variance in building density, this strongly hints that there is

numbers since 1916 took place while rural population was dropping and that the increase was due to a gradual increase in awareness. It was concluded that at least some of the large increase in tornado numbers from 1971 to 1980 was due to the reporting of various nontornadic windstorms as tornadoes. Many of the trends arise from changes in society, rather than changes in the meteorology. His cautions about

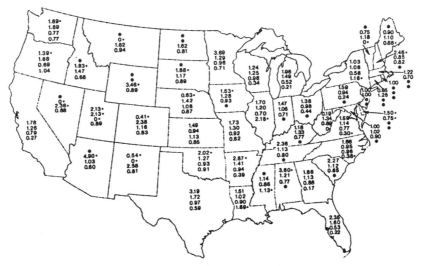

Fig. 21. Apparent population bias by county in the number of reported tornadoes, 1950–1979. Top number is for counties with a 1970 population density greater than or equal to 250 mi^2; place is for population density 50 to 249 mi^2; penultimate number is for population density of 10 to 49 mi^2; bottom number is for counties with population density of less than 10 mi^2. (One mile equals 1.6 km. Asterisk indicates no counties in that category; a dot after number indicates less than two counties in category.) No bias indicated by 1.00; numbers greater than 1.00 indicate overreporting of tornadoes; numbers less than 1.00 indicate underreporting (see *Schaefer and Galway* [1982] for details).

TABLE 4. Comparison of Number of Tornadoes Reported in Urban Counties Containing State Capital to Statewide Tornado Counts

City, State (State Area)*	County (Area)*	Tornadoes Recorded in County†	Tonadoes Recorded in State†	Tornadoes Estimated in State‡	Tornadoes Recorded in State,§%
Little Rock, Arkansas (52,078)	Pulaski (767)	30	597	2037	29
Indianapolis, Indiana (35,932)	Marion (396)	19	557	1724	32
Oklahoma City, Oklahoma (68,655)	Oklahoma (708)	48	1515	4655	33
Des Moines, Iowa (55,965)	Polk (582)	26	934	2501	37
Topeka, Kansas (81,778)	Shawnee (549)	18	1105	2681	41
Springfield, Illinois (55,645)	Sangamon (785)	29	785	1863	42
Jackson, Mississippi (47,233)	Hinds (875)	29	686	1565	44
Columbus, Ohio (41,004)	Franklin (982)	13	441	982	45
Austin, Texas (262,017)	Travis (989)	31	3861	8213	47
Madison, Wisconsin (54,426)	Dane (1205)	17	599	768	78

*Areas given in square miles; 1 mi^2 = 2.6 km^2.
†NSSFC data for 1959–1989.
‡Based on a simple proportion, the number of tornadoes that would have been reported in the state if the capital county were representative.
§The percentage of tornadoes actually reported, compared to the estimate.

a higher percentage of unreported tornadoes in western Kansas than in eastern Kansas. He thus concluded that tornado numbers actually increased as one moved westward across Kansas. This increase in tornado numbers compensated for the drop in population, thus yielding the statistical lack of a population bias in reported tornadoes.

This same situation occurs in Wisconsin. A cursory glance at the damage codes on the NSSFC data base shows that virtually all of the counties in the heavily forested areas of northeast Wisconsin report high damage figures for all tornadoes. A check of *Storm Data* and newspapers indicates that this is because the tornadoes hit one or more buildings. It is highly unlikely that 100% of all tornadoes would hit a building in the sparsely populated counties of northern Wisconsin. It should be noted that *Grazulis* [1991] studied only Wisconsin and Kansas, and only to the extent that it provided a hint at explaining the lack of a strong population bias found by *Schaefer and Galway* [1982].

Grazulis [1991] compared the number of tornadoes reported in urban counties containing the state capital to statewide tornado counts (Table 4). Other capitals, such as Nashville, St. Paul, or Lansing, and large cities, such as Birmingham, Minneapolis, St. Louis, Chicago, Omaha, and Dallas, could be used to generate even more drastic numbers. On the basis of these results, *Grazulis* [1991] suggested that about 2000 tornadoes go unreported each year. He noted that most of these "missing" tornadoes probably fall into three categories: (1) tornadoes on the western plains and in other very rural areas that were unseen and thus unreported; (2) poorly formed tornadoes that were not recognizable as such from a distance and hit nothing except perhaps tree tops or open fields; (3) gust front and other shear zone tornadoes that have a convergent rotation but may not be connected to the base of a thunderstorm, and yet they are typically reported as tornadoes by the general public in urban areas.

Forbes and Wakimoto [1983] studied an outbreak of 18 tornadoes across corn fields near Springfield, Illinois, on August 6, 1977 (see Figure 22), and discussed the larger problem of which vortices should be officially called tornadoes. Many of these August 6, 1977, tornadoes were gust front or other shear zone vortices. That study hints at the possibility that some large areas of thunderstorm wind damage may be riddled with small tornadolike vortices. Category 3 contains mostly weak tornadoes and probably accounts for most of the 2000 "missing" events.

Fujita [1987] presents a wide variety of climatological maps, as derived from the UC data tape, through 1985. Figures 23, 24, 25, and 26 are a few samples of the over 120 figures in that volume, which displays the great variability in tornado activity.

2.3. State Climatologies

Tornado climatologies have been compiled for many states. Because of the number of them, only a few of the more notable are cited here. In California, *Hales* [1983] found several areas of concentrated tornado activity. He suggested that the relative maximum in the number of tornadoes occurring in the Los Angeles area is caused by enhanced convergence induced by local topography. There is a southeast to northwest curvature in both the coastline and an adjacent mountain barrier that is positioned somewhat inland, rather than right along the coast.

Speheger et al. [1990] noted significant differences in the number of reports of Indiana tornadoes, funnel clouds, and thunderstorm winds between the 1980s and previous decades. The number of funnel clouds and tornado reports were decreasing while the number of downburst/microburst wind reports were increasing. These trends were thought to be, at least in part, due to changes in reporting procedures and to changing attitudes toward event classification within the publication *Storm Data*.

Some aspects of the unique and little studied tornado

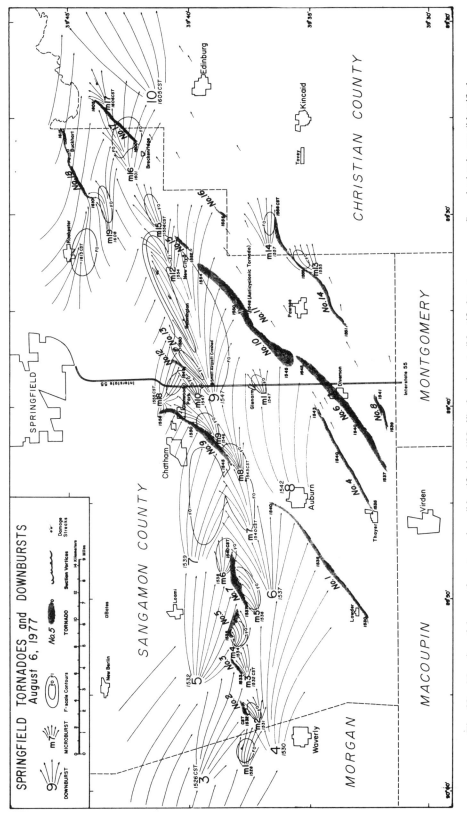

Fig. 22. Mapping of damage from tornadoes (identified by "No."), downbursts (identified by large numeral), and microbursts (identified by "m") on August 6, 1977, in central Illinois. "Streamlines" of damage and F scale contours are mapped [after Forbes and Wakimoto, 1983].

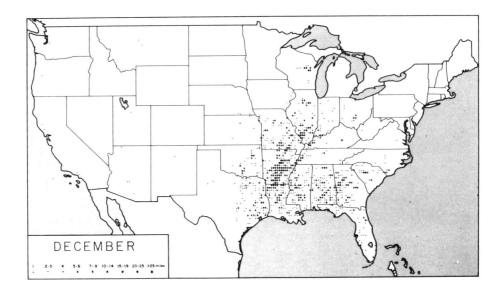

Fig. 23. Chart of the summed path length per 15-min box of December tornadoes, 1916–1985, constructed from the UC data tape [after *Fujita*, 1987].

climatology of Florida were noted by *Anthony* [1988]. Among the findings were that an unusually high proportion (42%) of severe reports in that state were tornadic.

3. THE FUTURE OF TORNADO CLIMATOLOGY AND RISK ASSESSMENT

For the foreseeable future, studies in the tornado climatology of individual states and regions will continue. However, the next "great leap," similar to those in the 1970s, is probably in the distant future. Such a leap could be a national mesocyclone climatology, based on NEXRAD Doppler radar data. The mesocyclone data might be used as a basis for adjusting traditional climatological data for less populated areas. Another significant advancement may come from the use of portable Doppler radar. Such units can be brought to within 1 mi (1.6 km) of even the most intense tornadoes. These instruments may drastically change our

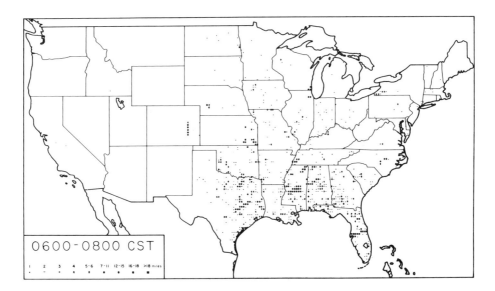

Fig. 24. Chart of the summed path length per 15-min box of tornadoes, 1916–1985, that occurred between 0600 and 0800 CST constructed from the UC data tape [after *Fujita*, 1987]. The anomaly in Colorado arises from the tornadoes of November 4, 1922, while the one in western Nebraska is a reflection of the events of May 8, 1927.

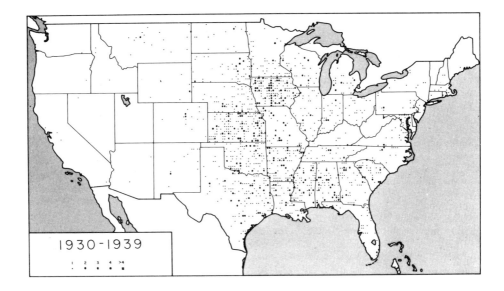

Fig. 25. Chart of the summed tornadoes per 15-min box that occurred during the 1930s constructed from the UC data tape [after *Fujita*, 1987]. Note that there was more tornado activity in Iowa than Indiana during this decade.

perceptions of tornado wind speeds and thus require a rethinking of current risk analysis models. In the meantime it is likely that traditional methods of tornado climatology will produce only minor refinements in our present overall picture. This period would also be an ideal time to address some of the questions concerning tornado documentation and use of the Fujita scale that were brought forth by *Doswell and Burgess* [1988].

REFERENCES

Abbey, R. F., Jr., Risk probabilities associated with tornado wind speeds, paper presented at Symposium on Tornadoes: Assessment of Knowledge and Implications for Man, Tex. Tech Univ., Lubbock, Tex., June 22–24, 1976.

Abbey, R. F., Jr., and T. T. Fujita, Use of tornado path lengths and gradations of damage to assess tornado intensity probabilities, in *Preprints, Ninth Conference on Severe Local Storms*, pp. 286–293, American Meteorological Society, Boston, Mass., 1975.

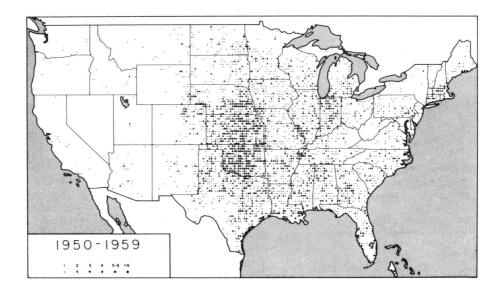

Fig. 26. Chart of the summed tornadoes per 15-min box that occurred during the 1950s constructed from the UC data tape [after *Fujita*, 1987]. Note the contrast in the relative amount of activity in Iowa and Indiana between this decade and the 1930s (Figure 25).

Abbey, R. F., Jr., and T. T. Fujita, The DAPPL method for computing tornado hazard probabilities: Refinements and theoretical considerations, in *Preprints, 11th Conference on Severe Local Storms*, pp. 241–248, American Meteorological Society, Boston, Mass., 1979.

Anthony, R., Tornado/severe thunderstorm climatology for the southeastern United States, in *Preprints, 15th Conference on Severe Local Storms*, pp. 511–516, American Meteorological Society, Boston, Mass., 1988.

Changnon, S. A., Trends in tornado frequencies: Fact or fallacy, in *Preprints, 12th Conference on Severe Local Storms*, pp. 42–44, American Meteorological Society, Boston, Mass., 1982.

Coats, D. W., and R. C. Murray, Natural phenomena hazards modeling project; extreme wind/tornado hazard models for department of energy sites, *Rep. UCRL-535266*, Lawrence Livermore Natl. Lab., Livermore, Calif., 1985.

Court, A., Tornado incidence maps, *Tech. Memo. ERLTM-NSSL 49*, 76 pp., Natl. Severe Storms Lab., Environ. Sci. Serv. Admin., Norman, Okla., 1970.

Doswell, C. A., III, and D. W. Burgess, Some issues of United States tornado climatology, *Mon. Weather Rev.*, *116*, 495–501, 1988.

Eshelman, S., and J. L. Stanford, Tornadoes, funnel clouds and thunderstorm damage in Iowa during 1974, *Iowa State J. Res.*, *51*, 327–361, 1977.

Forbes, G. S., and R. M. Wakimoto, A concentrated outbreak of tornadoes, downbursts, and microbursts, and implications regarding vortex classification, *Mon. Weather Rev.*, *111*, 220–235, 1983.

Fujita, T. T., Proposed characterization of tornadoes and hurricanes by area and intensity, *Res. Pap. 91*, Satell. and Mesometeorol. Res. Proj., Univ. of Chicago, Chicago, Ill., 1971.

Fujita, T. T., Workbook of tornadoes and high winds for engineering application, *Res. Pap. 165*, Satell. and Mesometeorol. Res. Proj., Univ. of Chicago, Chicago, Ill., 1978.

Fujita, T. T., U.S. tornadoes, part 1, 70 year statistics, report, 122 pp., Satell. and Mesometeorol. Res. Proj., Univ. of Chicago, Chicago, Ill., 1987.

Fujita, T. T., and A. D. Pearson, Results of FPP classification of 1971 and 1972 tornadoes, in *Preprints, Eighth Conference on Severe Local Storms*, pp. 142–145, American Meteorological Society, Boston, Mass., 1973.

Galway, J. G., and A. D. Pearson, Winter tornado outbreaks, in *Preprints, 11th Conference on Severe Local Storms*, pp. 1–6, American Meteorological Society, Boston, Mass., 1979.

Grazulis, T. P., *Significant Tornadoes, 1880–1989*, vol. I and II, 970 pp., Tornado Project, St. Johnsbury, Vt., 1991.

Grazulis, T. P., and R. F. Abbey, Jr., 103 years of violent tornadoes ... patterns of serendipity, population, and mesoscale topography, in *Preprints, 13th Conference on Severe Local Storms*, pp. 124–127, American Meteorological Society, Boston, Mass., 1983.

Hales, J. E., Synoptic features associated with Los Angeles tornado occurrences, in *Preprints, 13th Conference on Severe Local Storms*, pp. 132–135, American Meteorological Society, Boston, Mass., 1983.

Kelly, D. L., J. T. Schaefer, R. P. McNulty, C. A. Doswell III, and R. F. Abbey, Jr., An augmented tornado climatology, *Mon. Weather Rev.*, *106*, 1172–1183, 1978.

LaGreca, K. W., Factors affecting tornado hazard probabilities, Master's thesis, Tex. Tech Univ., Lubbock, 1980.

McDonald, J. R., and B. S. Allen, Regionalization of tornado hazard probabilities, *Preprint 81-540*, 16 pp., Am. Soc. of Civ. Eng., St. Louis, Mo., 1981.

Minor, J. E., J. R. McDonald, and K. C. Mehta, The tornado; an engineering-oriented perspective, *Tech. Memo. ERL-NSSL-82*, 192 pp., Natl. Oceanic and Atmos. Admin., Boulder, Colo., 1977.

National Oceanic and Atmospheric Administration, Tornado safety, surviving nature's most violent storms, *Rep. NOAA/PA 82001*, Boulder, Colo., Jan. 1982.

Schaefer, J. T., and J. G. Galway, Population biases in the tornado climatology, in *Preprints, 12th Conference on Severe Local Storms*, pp. 51–54, American Meteorological Society, Boston, Mass., 1982.

Schaefer, J. T., D. L. Kelly, and R. F. Abbey, Jr., Tornado track characteristics and hazard probabilities, in *Wind Engineering*, edited by J. E. Cermak, pp. 95–110, Pergamon, New York, 1980.

Schaefer, J. T., D. L. Kelly, and R. F. Abbey, A minimum assumption tornado hazard probability model, *J. Clim. Appl. Meteorol.*, *25*, 1934–1945, 1986.

Snider, C. R., A look at Michigan tornado statistics, *Mon. Weather Rev.*, *105*, 1341–1342, 1977.

Speheger, D. A., D. J. Shellberg, J. R. Gibbons, J. A. DeToro, J. T. Snow, and T. P. Grazulis, A climatology of severe thunderstorm events in Indiana, in *Preprints, 16th Conference on Severe Local Storms*, pp. 18–23, American Meteorological Society, Boston, Mass., 1990.

Tecson, J. J., and T. T. Fujita, Automated mapping of maximum tornado wind speeds in the United States as a function of occurrence probabilities, in *Preprints, 14th Conference on Severe Local Storms*, pp. 21–24, American Meteorological Society, Boston, Mass., 1985.

Tecson, J. J., T. T. Fujita, and R. F. Abbey, Jr., Statistics of the U.S. tornadoes based on the DAPPL tornado tape, in *Preprints, 11th Conference on Severe Local Storms*, pp. 227–230, American Meteorological Society, Boston, Mass., 1979.

Tecson, J. J., T. T. Fujita, and R. F. Abbey, Jr., Climatologic mapping of U.S. tornadoes during 1916–1980, in *Preprints, 12th Conference on Severe Local Storms*, pp. 38–41, American Meteorological Society, Boston, Mass., 1982.

Tecson, J. J., T. T. Fujita, and R. F. Abbey, Jr., Statistical analyses of U.S. tornadoes based on the geographic distribution of population, community, and other parameters, in *Preprints, 13th Conference on Severe Local Storms*, pp. 120–123, American Meteorological Society, Boston, Mass., 1983.

Twisdale, L. A., Tornado data characterization and wind speed risk, *J. Struct. Div.*, *104*(ST10), 1978.

Twisdale, L. A., Regional tornado data base and error analysis, in *Preprints, 12th Conference on Severe Local Storms*, pp. 45–50, American Meteorological Society, Boston, Mass., 1982.

Twisdale, L. A., Probabilistic analysis of tornado wind risks, *J. Struct. Eng.*, *109*(2), 1983.

Wilson, L. F., and D. Kelly, Tornado climatology by day of the week, in *Preprints, 10th Conference on Severe Local Storms*, pp. 194–198, American Meteorological Society, Boston, Mass., 1977.

Comparative Description of Tornadoes in France and the United States

JEAN DESSENS

Laboratoire d'Aérologie, Université Paul Sabatier, 65300 Campistrous, France

JOHN T. SNOW

Department of Earth and Atmospheric Sciences, Purdue University, West Lafayette, Indiana 47907

1. INTRODUCTION

In comparison with the Great Plains of the United States, severe tornadoes are rare in France. Statistics based on significant (strong (F-2 and F-3) and violent (F-4 and F-5) [see *Hales*, 1988]) tornadoes that have occurred in France in the past three decades, however, show that this country averages two such tornadoes a year. By combining these observations with information from the scientific literature about events that occurred before the modern period, it has been possible to establish a tornado climatology for France based on 107 significant tornadoes. The main data of this climatology, recently published by *Dessens and Snow* [1989], are summarized here and compared with American tornadoes.

2. SUMMARIZED CLIMATOLOGY

2.1. Geographical Distributions

Figure 1 shows the geographical location of the 107 significant tornadoes. These are distributed in intensity as follows: 50 F-2, 44 F-3, 11 F-4, and two F-5. These events are concentrated in the northwestern part of the country, with a small secondary concentration in the south, around the Mediterranean shore.

It is clear from the mapping in Figure 2 that the distribution in winter is different from that in summer; during the cold months of November through March, tornadoes occur mainly in the two coastal zones, while during the warm months of April through October, they occur in the interior. This is very similar to the distributions observed (on a larger scale) in the United States, where the region of maximum frequencies for tornadoes moves from the lower Mississippi in winter to the central part of the United States in summer [see *Fujita*, 1978, Figures 5.7–5.12]. In France the wintertime concentration in the coastal zones probably reflects greater availability of both latent heat energy and (to a lesser extent) kinetic energy (winds are stronger along the coasts in winter) in these regions, while the summertime events in the interior reflect the stronger surface heating that takes place there.

2.2. Diurnal and Seasonal Distributions

Figure 3 shows hourly and monthly distributions of the 103 significant tornadoes for which time of touchdown is known. The central portion of this figure is a plot of day and time of occurrence of each of the events. The histogram along the ordinate gives frequency of occurrence by month, while that along the abscissa gives frequency of occurrence by hour of the day.

June and August are the months with the greatest number of tornadoes. The interval 1600 to 1700 UTC is the one in which occurrence has been most frequent, with a secondary maximum between 1800 and 1900 UTC. The mean time for all tornadoes is 1512 UTC (UTC and Sun time are not very different, since the Greenwich meridian crosses over France). The mean time for "winter" tornadoes is 1322 UTC, while the mean time for "summer" tornadoes is 1546 UTC, a difference of 144 min.

These results can be compared with the diurnal distribution of U.S. tornadoes, as summarized in by *Abbey* [1976, Table 2]. While the monthly distribution is different (in the United States, May and June are the months with the greatest activity), the same trend is present for the seasonal mean times of occurrence: summer tornadoes occur later in the afternoon, with a difference of 51 min between the mean times of occurrence for winter and summer events.

Winter events are more scattered throughout the day than

The Tornado: Its Structure, Dynamics, Prediction, and Hazards.
Geophysical Monograph 79
Copyright 1993 by the American Geophysical Union.

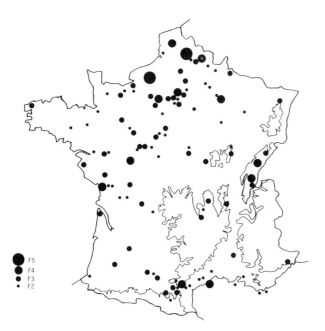

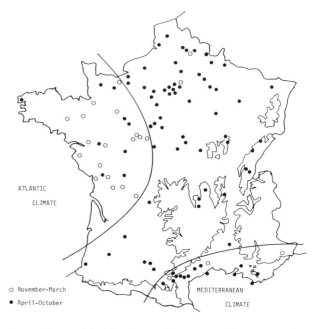

Fig. 1. Geographical locations of 107 significant tornadoes that occurred in France in the period 1680–1988. Greatest reported intensity is indicated by the size of the plotted circle. Light lines outline the 500-m elevation contour and so show the Pyrenees Mountains in the south, the Alps Mountains to the east, and the Massif Central in the southeast.

Fig. 2. Geographical locations with corresponding season of occurrence. The two curved lines divide France into an interior region and the Atlantic and Mediterranean coastal regions. Topography is as in Figure 1.

the summer events, which are concentrated in the afternoon. For the 24 tornadoes in the November through March period, only nine (37%) occurred between solar noon and sunset. Of the 79 tornadoes in the April through October period, 65 (82%) occurred between solar noon and sunset. While the size of the winter sample is small, these percentages suggest that winter events are less dependent on destabilization due to solar heating. Related findings in the United States are the bimodal daily distribution in tornado occurrences in the southeastern states with a secondary peak near sunrise [see Kelly et al., 1978, Figure 12] and findings by Moller [1979] that "cool" season events in the southern Great Plains are more diurnally scattered than late spring events. The data for France shown in Figure 3 support Moller's conclusion that cool/winter season events are largely dependent on dynamic forcing.

2.3. Path Dimensions

Figure 4 is a scatter plot of path length L and maximum path width W for the 78 significant tornadoes for which both L and W were available. The regression between L and W (with a correlation coefficient of 0.53) is given by the following equation:

$$\log_{10}L = -0.60 + 0.57 \log_{10}W,$$

or

$$L = 0.25W^{0.57},$$

with L in kilometers and W in meters. The mean path length for significant events is 8.5 km, while their mean path width is 300 m (see Dessens and Snow [1989, Table 3] for more detail). (For a few "weak" (F-0 and F-1) tornadoes, path length (14) and path width (12) data are also available. If

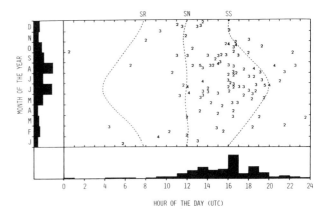

Fig. 3. Hourly and monthly distributions of tornadoes. Numbers refer to F scale intensities. The dashed curves give the times of sunrise (SR), solar noon (SN), and sunset (SS) at Paris (48°52′N, 2°20′E). The histograms bordering the ordinate and the abscissa show frequency of occurrences by month and time of day, respectively. (This figure corrects a small error in the corresponding figure of Dessens and Snow [1989].)

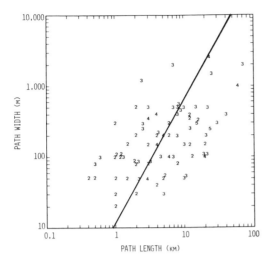

Fig. 4. A plot of the maximum reported path width versus the reported path length for each significant tornado for which both parameters are known. Numbers refer to F scale intensities. The solid line is the regression line (see text).

these data are merged with those for significant events, the mean path length becomes 7.8 km (based on 100 events) and the mean path width becomes 268 m (94 events).)

The mean path length and path width are different for winter and summer tornadoes:

Winter tornadoes

$$\bar{L} = 5.4 \text{ km}, \qquad \bar{W} = 190 \text{ m};$$

Summer tornadoes

$$\bar{L} = 9.5 \text{ km}, \qquad \bar{W} = 330 \text{ m}.$$

Another stratification can be made between "coastal" tornadoes, ones occurring less than 50 km inland from the Atlantic or from the Mediterranean, and those occurring in the "interior" of France:

Coastal tornadoes

$$\bar{L} = 5.5 \text{ km}, \qquad \bar{W} = 154 \text{ m};$$

Interior tornadoes

$$\bar{L} = 10.3 \text{ km}, \qquad \bar{W} = 350 \text{ m}.$$

Using data for all tornadoes (weak, strong, and violent), *Thom* [1963], *Reed* [1971], and *Howe* [1974] have computed tornado path sizes for U.S. tornadoes, finding a mean path width of around 150 m. An analysis by *Fujita and Pearson* [1973, Figure 2] of path statistics by intensity suggests that U.S. significant (strong and violent) tornadoes have a mean path width of around 300 m.

It appears that in the mean, significant French tornadoes have path widths similar to their American counterparts. If data on all tornadoes occurring in France were available, the path statistics would probably shift toward those for American events; the limited sample of data for F-0 and F-1 events supports this. All French tornadoes (winter and summer) have mean path sizes very similar to the tornadoes of Arkansas-Tennessee [see *Howe*, 1974, Figure 2].

3. RISK PROBABILITY

Path area A was computed for the 78 significant events for which both L and W were available. The mean value of this area, 4.0 km^2, allows an estimation of the mean risk probability of tornado occurrence for France. During the well-surveyed modern period (1960 through 1988), there were 58 tornadoes of intensity F-2 or greater (about two per year). Therefore on a national basis the mean area exposed to significant tornadoes is about 8 km^2 per year. From the ratio of this area with that for France (549,000 km^2), the probability of a given "point" being struck by a significant tornado is about 1.5×10^{-5} per year, and the return period is about 70,000 years. (Inclusion of the 10 weak events for which both path length and width data are available modifies the mean path area value to 3.58 km^2 but does not significantly change the overall conclusion.)

When compared with the United States, the mean annual frequency of tornadoes per area of 1° squares [*Thom*, 1963, Figure 4], together with the mean tornado path area by region [*Howe*, 1974, Figure 1] or per 10,000 mi^2 (25,900 km^2) [*Reed*, 1971, Figure 6], shows that the risk is about 100 times smaller in France than in Oklahoma, or about the same as in the northeast of the United States.

The geographical distributions of Figure 1 shows that the level of risk is not the same in all regions of France. Along the Atlantic the tornado frequency is relatively high, but the mean path area struck by a tornado in this region is low. For example, a computation made for a region near Nantes gives a mean path area of 0.6 km^2 per tornado. The risk probability for this area is 4×10^{-6} per year.

4. PARTICULAR FEATURES RELATED TO GROUND CONFIGURATION

In France, topography and surface characteristics change markedly from one region to another, often over short distances. This allows computation of mean values of F scale, path length, and maximum path width for significant events by landscape type. A very marked result appears concerning mean path dimensions of tornadoes over wooded areas as contrasted to those over more open areas:

"Forest" tornadoes

$$\bar{L} = 17.1 \text{ km}, \qquad \bar{W} = 850 \text{ m};$$

Other tornadoes

$$\bar{L} = 6.8 \text{ km}, \qquad \bar{W} = 180 \text{ m}.$$

We suggest that this may reflect an effect of surface roughness on tornadoes. Several laboratory simulations [e.g., *Wilkins et al.*, 1975; *Dessens*, 1972; *Diamond*, 1982;

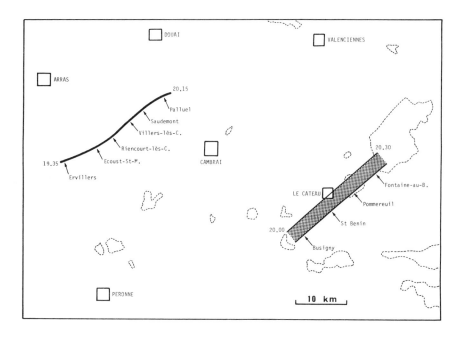

Fig. 5. Paths of the June 27, 1967, tornadoes across the departments of Nord and Pas-de-Calais. Locations of villages where significant structural damage occurred and times of touchdown and termination are given. The dashed lines outline densely forested areas. The stippled rectangle denotes the broad path of the tornado through partly forested regions.

Diamond and Wilkins, 1984] and numerical models [e.g., *Bode et al.*, 1975] suggest that an increase of core radius with increasing surface roughness is to be expected. The increase of mean path length for tornadoes over a forest may be a consequence of the increase of the mean path width, as there is a moderately strong direct relationship between the two dimensions (see section 2).

Another particular feature related to surface characteristics is that several reports support the concept of occurrence of suction vortices when a tornado goes from a rough area (forest) to a smooth one (e.g., bare ground). A few well-documented cases illustrating these observations will be briefly related.

4.1. June 27, 1967, Palluel (Pas-de-Calais) and Pommereuil (Nord)

Two companion violent tornadoes killed eight people and injured 80 in the north of France on June 27, 1967. The nearly parallel tornado paths were separated by about 35 km (Figure 5). The east tornado touched down while the west tornado was in action. The eastern tornado was rated F-4, and the western tornado probably reached F-5 intensity. Seventeen houses were razed to the ground, cars were lifted up and hurled in fields or over houses, grass blades and tree branches were stuck into tree trunks, large pieces of wood and tree trunks were stuck into the ground, and tombs were opened at Riencourt, and so on by the F-5 event.

The western tornado path was only 250 m wide, while the eastern tornado was nearly 10 times larger. This difference correlates with the nature of the ground, which is completely bare for the western path but partly wooded for the eastern one.

4.2. September 20, 1973, Sancy-les-Provins (Seine-et-Marne)

This tornado system occurred about 80 km east of Paris. It initially was a large ($\simeq$2000 m diameter) F-2 tornado, moving through a forested area from Moret-sur-Loing to Rampillon (Figure 6). After coming out of the forest, it continued for a few kilometers to near Vanville and Maison-Rouge as two vortices moving along parallel tracks separated by about 1 km. As it continued on across the plain for 10 km toward Neuvy, it took the form of an F-3 tornado, leaving a 200-m-wide damage path across bare ground.

4.3. A Tornado Alley in the Jura Mountains?

A local "tornado alley" [*Gallimore and Lettau*, 1970] appears to exist north of Geneva, along the France-Switzerland border. In this region the southwest-northeast foldings of the Jura mountains provide a mesoscale pattern of significant relief, in some places with as much as 500-m differences in elevation between adjacent crests and valleys.

On August 19, 1890, an F-4 tornado touched down between Oyonnax and Saint-Claude (Figure 7) and then remained on the ground for 1 hour to Croy (Switzerland). This tornado was well documented by *Bourgeat* [1890] for the French portion of its track and by *Gauthier* [1890] for the

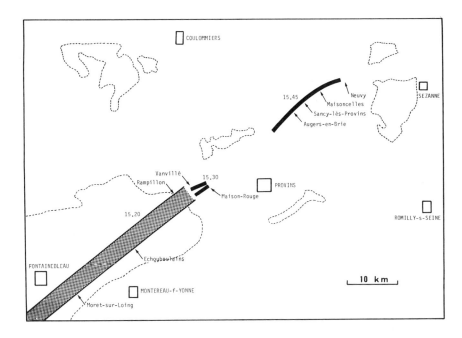

Fig. 6. Damage tracks of the tornadoes of September 20, 1973, across the department of Seine-et-Marne. Locations of villages where significant structural damage occurred and reported times of tornado passage at a few locations are given. The dashed lines again outline densely forested areas. Here the stippled rectangle shows the broad path of the tornado through a wholly forested region.

Swiss portion. Reexamination of their descriptions is of interest now that we understand better the physics of tornadoes. For example, *Gauthier* [1890, p. 419] reports, "The ground configuration transformed the main vortex into strong gales or collateral and secondary tornadoes, only on the right-hand side. The longest of these ramifications (Bois d'Amont-Carroz) was 3 km long and went 1 km eastwards."

On August 26, 1971, an F-4 tornado retraced almost

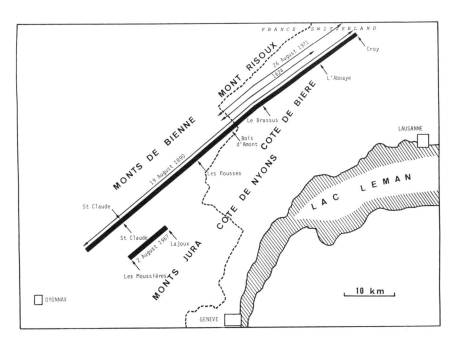

Fig. 7. Mapping of selected damage tracks in the "tornado alley" of the Jura Mountains.

exactly the Swiss portion of this 1890 tornado [*Piaget*, 1976]. The replication was so perfect that a reorganization of the 1971 tornado into several vortices was observed along the same part of the path where this had been observed in the 1890 event, above the flat, bare ground of the valley of Joux, between the wooded areas of Bois d'Amont and Le Brassus [see *Piaget*, 1976, Figure 27].

The tornadoes of 1890 and 1971 both traveled at 60 km hr^{-1}, indicative of strong mid and upper level winds. In fact, the wind velocity was 30 m s^{-1} at 300 hPa and 55 m s^{-1} at the tropopause in the jet stream over the tornado area on August 16, 1971 [*Piaget*, 1976].

Three other violent storms, in the years 1624, 1768, and 1842, have devastated the same valley. All are suspected to have been tornadoes. The event in the year 1624 is sufficiently well documented to allow a detailed mapping of the storm path (Figure 7). In 1967 an F-3 tornado struck the village of Lajoux, 6 km to the right of this "tornado alley" (also shown in Figure 7). On several other occasions, narrow swaths have been reported in the forests of the same area; they are also thought to be due to tornadoes. These observations suggest that low-level wind flow modified through channeling by the mountains might provide a locally favorable wind shear for tornadogenesis [*Nuss*, 1986].

5. Meteorological Conditions

Synoptic situations on days of significant tornadic events have been classified according to surface and 500-hPa flow regimes. Generally, these days may be identified as falling into either a summer or a winter pattern.

For tornado-producing configurations in summer (roughly June–August), the 500-hPa-height field typically shows a closed low either over the Bay of Biscay (to the west and south, between France and Spain) or farther west in the Atlantic and high pressure centered over the Sahara. As a consequence of the circulation around these two centers, there is midlevel flow from the southwest of about 20 m s^{-1} over France. In the surface pressure field, there is a weak low over France. A weak, slow moving frontal system is sometimes present just west of the tornado area. Surface winds are usually weak. These conditions are illustrated in the hodographs in Figure 8.

Back trajectories over isentropic surfaces computed for two cases illustrate the summer tornado-producing configuration (Figure 9). These trajectories show that the low-level inflow was Mediterranean air moving very slowly into central France from the south. Trajectories at high levels show a much faster flow over the Iberian Peninsula coming from the mid-Atlantic. This turns to the northeast to flow over France. This situation is roughly similar to that observed in the Great Plains of the United States (see, for example, the tornado-producing synoptic pattern shown by *Pearson* [1976, Figure 8]).

In winter (roughly the period October–March) the 500-hPa pattern for a tornado-producing situation typically shows a strong midlevel flow (with wind speeds of 20–40 m s^{-1}) from

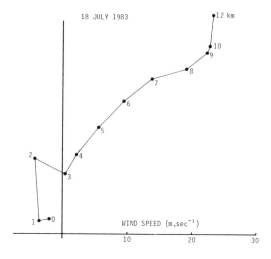

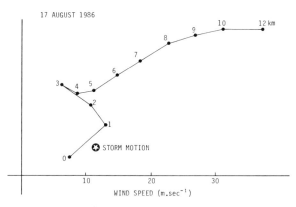

Fig. 8. Hodographs for (top) July 18, 1983, and (bottom) August 17, 1986. Both taken at Bordeaux in southwest France at 1200 UTC on the days indicated. For the July 18, 1983, event this location is 70 km to the southeast of and 6 hours before the tornado. For the August 17, 1986, event this location is 350 km to the southwest of and 6 hours before the tornado; it is almost exactly upwind of the touchdown point. Storm motion is shown for August 17, 1986; unfortunately, these data are not available for July 18, 1983.

the west over France. This is associated with a strong trough containing a deep primary low to the northwest of the British Isles. In the surface pressure field, there is a deep low in the same area, with a well-defined cold front west of the tornado region. Strong westerly winds extend down to near the surface, so that there is little directional shear in the lower troposphere.

6. Summary and Conclusions

An examination of historical and modern accounts spanning more than 300 years has shown that a few square kilometers of France are struck by tornadoes each year. While the effects of these occurrences have not usually been sufficient to cause a national calamity, losses in life and property and the number of injuries have occasionally been substantial. This explains the interest of several national

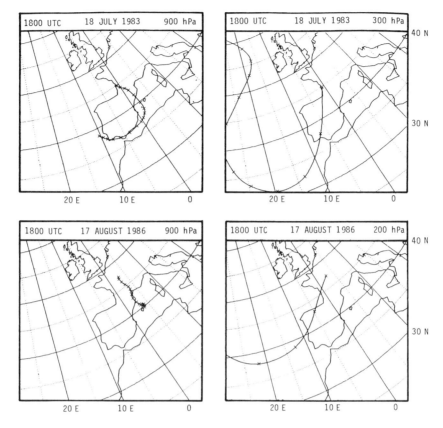

Fig. 9. (Top) Air parcel trajectory arriving at the (left) 900-hPa and (right) 300-hPa levels over Bordeaux during the F-3 tornado outbreak of July 18, 1983, and (bottom) air parcel trajectory arriving at the (left) 900-hPa and (right) 200-hPa levels over La Charité during the F-3 tornado outbreak of August 17, 1986. Spacing between two crosses along a trajectory represents the distance traveled in 6 hours.

agencies, such as Délégation aux Risques Majeurs and Electricité de France, and of many of the French people.

To summarize, France is somewhat like a miniature of the United States. This is particularly true when considering the synoptic conditions supporting tornadogenesis: in France, tornadoes result from the interaction of slow moving, low-level Mediterranean air with an upper level cold Atlantic flow, whereas in the United States, the interaction occurs between upper level are masses of Pacific origin and low-level northward incursions of tropical air from the Gulf of Mexico. This situation probably explains why France is one of the very few European countries where "strong" and "violent" tornadoes are observed from time to time.

Why are tornadoes some 15 times less frequent in France than in the United States? A possible explanation is that there is a requisite "gestation time" for tornado generation in a thunderstorm. In the central United States the relatively uniform extensive topography of the Great Plains frequently allows quasi-steady atmospheric conditions. Strong thunderstorms can often maintain their intensity for periods of time sufficient for tornadoes to form. In France, with diverse topography and mesoclimates in a relatively small region, such conditions occur only rarely. Probably the best United States analog to French conditions is found in the New England area, which does indeed have recurrence rates and intensity statistics similar to those observed in France.

Acknowledgments. The authors wish to acknowledge the efforts of an anonymous reviewer and of Alan Moller. Moller in particular made numerous suggestions that led to several significant improvements in the content of the manuscript. His observation of the greater scatter to be found in the winter events displayed in Figure 3 was especially important.

REFERENCES

Abbey, R. F., Jr., Risk probabilities associated with tornado windspeeds, paper presented at Symposium on Tornadoes, Tex. Tech. Univ., Lubbock, Tex., June 22–24, 1976.

Bode, L., L. M. Leslie, and R. K. Smith, A numerical study of boundary effects on concentrated vortices with application to tornadoes and waterspouts, *Q. J. R. Meteorol. Soc.*, *101*, 313–324, 1975.

Bourgeat, M., Premieres observations sur le cyclone du 19 aôut dans le Jura, *C. R. Hebd. Seances Acad. Sci.*, *111*, 385–389, 1890.

Dessens, J., Influence of ground roughness on tornadoes: A laboratory simulation, *J. Appl. Meteorol.*, *11*, 72–75, 1972.

Dessens, J., and J. T. Snow, Tornadoes in France, *Weather Forecasting*, *4*, 110–132, 1989.

Diamond, C. J., Laboratory simulation of tornado-like vortices under the effects of translation, M.S. thesis, 75 pp., Univ. of Okla., Norman, 1982.

Diamond, C. J., and E. M. Wilkins, Translation effects on simulated tornadoes, *J. Atmos. Sci.*, *41*, 2574–2580, 1984.

Fujita, T. T., Workbook of tornadoes and high winds for engineering applications, *Res. Pap. 165*, 142 pp., Satell. and Mesometeorol. Res. Proj., Univ. of Chicago, Chicago, Ill., 1978.

Fujita, T. T., and A. Pearson, Results of FPP classification of 1971 and 1972 tornadoes, in *Preprints, Eighth Conference on Severe Local Storms*, pp. 142–145, American Meteorological Society, Boston, Mass., 1973.

Gallimore, R. G., and H. H. Lettau, Topographic influence on tornado tracks and frequencies in Wisconsin and Arkansas, *Wis. Acad. Sci. Arts Lett.*, *58*, 101–127, 1970.

Gauthier, L., La trombe-cyclone du 19 août 1890, *C. R. Hebd. Seances Acad. Sci.*, *111*, 417–421, 1890.

Hales, J., Improving the watch warning program through the use of significant event data, in *Preprints, 15th Conference on Severe Local Storms*, pp. 165–168, American Meteorological Society, Boston, Mass., 1988.

Howe, G. M., Tornado path sizes, *J. Appl. Meteorol.*, *13*, 343–347, 1974.

Kelly, D. L., J. T. Schaefer, R. P. McNulty, C. A. Doswell III, and R. F. Abbey, An augmented tornado climatology, *Mon. Weather Rev.*, *106*, 1172–1183, 1978.

Moller, A., Climatology and synoptic meteorology of southern plains tornado outbreaks, Masters thesis, 70 pp., Univ. of Okla., Norman, 1979.

Nuss, W. A., Observations of a mountain tornado, *Mon. Weather Rev.*, *114*, 233–237, 1986.

Pearson, A., Tornado prediction, paper presented at Symposium on Tornadoes, Tex. Tech. Univ., Lubbock, Tex., June 22–24, 1976.

Piaget, A., L'évolution orageuse au nord des Alpes et la tornade du Jura Vaudois du 26 août 1971, *Publ. 35, SZ ISSN 0080-7346*, 102 pp., L'Inst. Suisse de Météorol., Payerne, Switzerland, 1976.

Reed, J. W., Some averaged measures of tornado intensity based on fatality and damage reports, in *Preprints, Seventh Conference on Severe Local Storms*, pp. 187–193, American Meteorological Society, Boston, Mass., 1971.

Thom, H. C. S., Tornado probabilities, *Mon. Weather Rev.*, *91*, 730–736, 1963.

Wilkins, E. M., Y. Sasaki, and H. L. Johnson, Surface friction effects on thermal convection in a rotating fluid; a laboratory simulation, *Mon. Weather Rev.*, *103*, 305–317, 1975.

Tornadoes of China

XU ZIXIU, WANG PENGYUN, AND LIN XUEFANG

Academy of Meteorological Science, State Meteorological Administration, Beijing, China

1. INTRODUCTION

While the most important severe convective weather phenomenon in China is hail, the occasional tornadoes that do occur can inflict considerable damage. For example, on March 26, 1967, a family of 13 tornadoes occurred near Shanghai and North Zhejiang province. Along the path of this tornado outbreak, more than 10,000 homes were damaged. Several steel high-voltage power transmission towers, including some built to withstand Beaufort force 12 winds, were blown down or severely damaged.

Tornadoes have been recorded in China as far back as 1000 years, but the total number of such reported events is rather small, indicating that many events have gone unreported. In 1949 many new weather stations were established all over China, and tornado reporting in considerably greater detail has resulted. In this paper we shall present information based on the records from 26 provinces and municipalities of China (Xinjing, Tibet, Guangdong, and Taiwan provinces are not included owing to lack of data) for the period 1971–1981. This data set provides an introduction to the climatological tornado distribution over China. We also shall present some brief case examples to illustrate the general synoptic and mesoscale regimes under which tornadoes occur in China. Finally, some radar and visual characteristics of tornadoes in China will be presented.

2. GEOGRAPHICAL DISTRIBUTION AND CLIMATIC CHARACTERISTICS

China is characterized by diverse and complex terrain. In broad terms, high plateaus are characteristic of the west, whereas low plains are typical in the east. The western plateaus include the Qinghai-Xizhang plateau (with typical mean sea level (msl) elevations of over 4000 m, and its Mount Qomulangma is known as the "roof of the world"), the Inner Mongolia plateau, the Yellow Earth (or "loess") plateau, and the Yunnan-Guizhou plateau (with typical msl elevations of 1000–2000 m). Near the eastern coasts, there are several major plain regions, including (from north to south) the Northeast China Plain, the North China Plain, and the plains of the middle and lower reaches of the Yangtze River. All of these plains are typified by msl elevations of less than 50 m.

During the 11-year period of record, from 1971 to 1981, it was noted that the region receiving hail most frequently was not coincident with the tornado frequency maximum (see Figure 1). Hail is more common in the western part of China, with its high plateaus, than in the eastern plains (see Figure 2). We note that in the United States, large-hail (≥ 2 cm) occurrences are most frequent in about the same area as the tornado frequency maxima [e.g., *Kessler and White*, 1981; *Kelly et al.*, 1985]. However, considerable hail that is below the arbitrary size threshold for "severe" hail occurs over the high plains of the United States, accounting for the results of *Changnon et al.* [1977], which show peak hail fall frequency over the high plains of Colorado, Wyoming, and Nebraska.

Tornadoes in China seem to occur most frequently over the plains of the middle and lower Yangtze River (Figure 3), especially in Jiangsu, Anhui, and Hunan provinces, with Jiangsu province having the highest frequency at 30 tornado days during the 11-year period. This averages out to 2.7 tornado days per year. In Jiangsu province, there was a total of 217 hail days during the same period, so hailstorm days were about 7 times more frequent than tornado days.

For the same period in Anhui province, there were 12 tornado days and 218 hailstorm days, making hail days about 18 times more frequent. In Hunan province the ratio reached about 30. Elsewhere in China the ratio is even larger. Some of the reasons for this might include low population densities in some of the provinces.

Generally speaking, the region of maximum tornado frequency in China is associated with the common presence of frontal boundaries and extratropical storms during the spring and summer. Moreover, the coastal plains have bountiful low-level moisture and a lack of terrain features to disrupt the flow of moisture and instability.

As can be seen in Table 1, tornadoes occur from April through August, with a peak frequency in July. Tornadoes

The Tornado: Its Structure, Dynamics, Prediction, and Hazards.
Geophysical Monograph 79
Copyright 1993 by the American Geophysical Union.

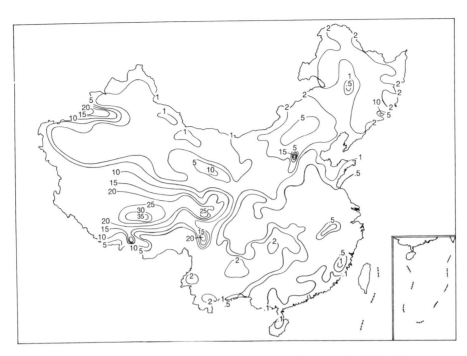

Fig. 1. Annual average number of hailstorm days in the period 1951–1980.

Fig. 2. Total number of tornado days and hailstorm days (in parentheses) in each province in the period 1971–1981.

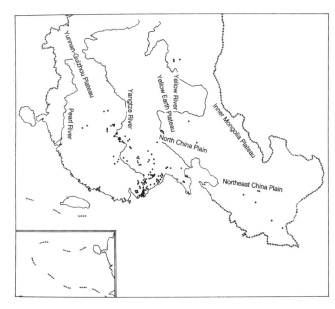

Fig. 3. Geographical distribution of tornadoes during the period 1971–1981.

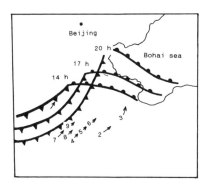

Fig. 4. Tornado tracks (small arrows) on August 26, 1987, in the North China Plain region.

are virtually nonexistent during the winter months. Similar to tornadoes in the United States [*Kelly et al.*, 1978] and elsewhere, most tornadoes in China occur between 1200 and 1800 LT.

3. Meteorological Conditions With Tornado Outbreaks

By analyzing the synoptic charts for 94 tornado days in China, we determined the following factors were common to the events: (1) potential instability, (2) moisture flux convergence at low levels, (3) a localized lifting mechanism (front, terrain, etc.), (4) upper and lower level jet streams, (5) strong vertical wind shear, and (6) potentially cold air in midlevels. These conditions are quite similar to those found for tornadoes elsewhere in the world. In China, these factors are brought together in different ways in different parts of the country, and the typical synoptic situations differ from place to place. Some basic patterns are as follows.

3.1. Middle and Lower Reaches of the Yangtze Valley

3.1.1. Southerly flow around a subtropical high. When a low-pressure trough is present with southerly flow around the western side of a subtropical high-pressure center, there often is a mesoscale system (either a closed low-pressure system or a convergence boundary) at the surface, with abundant moisture. Tornadic storms develop in association with such systems.

3.1.2. Stationary fronts. When a frontal boundary becomes stationary over the plains of the middle and lower Yangtze River, it may be associated with strong southerly flow (jetlike streams) at 850 and 700 hPa. Cyclonic waves can develop on the quasi-stationary front, leading to the development of tornadic storms.

3.1.3. Typhoons. When a weakening typhoon moves from Fujian province northward, the northern part of the circulation moves over the plains of the middle and lower Yangtze River. This puts a low-level cyclonic shear line over the area, having northeasterly flow converging with southeasterly winds, in the presence of considerable low-level moisture. When cold dry air at 700 hPa is coming from northern China at the same time, the stratification can become quite potentially unstable. In such cases, tornadoes of strong intensity can develop.

In the southern part of such a typhoon a convergence line having northwesterly and southwesterly winds can be present. If tornadoes occur with this feature, they tend to be weak. Tornadoes are quite uncommon in other quadrants of typhoons.

3.1.4. Extratropical cyclones. Some tornadoes occur within the warm sector of extratropical cyclones. In such situations it is common for the warm sector to have above-normal moisture content, with the moisture axis at 850 hPa ahead of the cold front, as in the work of *Miller* [1972]. In fact, these situations look very much like those described in "classical" tornado outbreaks in the United States, having

TABLE 1. Monthly Distribution of Tornado Days in China During the Period 1971–1981

Month	Tornado Days
January	0
February	1
March	3
April	14
May	14
June	16
July	27
August	14
September	3
October	2
November	0
December	0

Fig. 5. A series of photographs of the tornado near Laishui, Hebei province, on May 15, 1981, with times indicated in local standard time.

Fig. 5. (continued)

Fig. 6. Waterspout over the Yangtze River, photographed in 1933 near Shanghai.

vortex, it may move across the North China Plain. This trough often is associated with severe thunderstorms, especially when the trough interacts with a preexisting, low-level mesoscale boundary [Xu et al., 1977].

4. VISUAL ASPECTS OF TORNADOES

Tornadoes in China present a visual appearance that is quite comparable to that of their counterparts in the United States and elsewhere around the world. Figure 5 illustrates a tornado that developed near Laishui in Hebei province on May 15, 1981 [Chang, 1981]. A large thunderstorm was observed to be moving toward the northwestern part of Laishui at 1220 (all times local standard time). At 1245 a funnel cloud was observed. This funnel lengthened and then dissipated, but by 1249, dust and debris were observed below the funnel cloud. Four minutes later the funnel became more clearly visible, and the damage track width increased as the tornado intensified. From 1253 to 1257 the tornado attained its greatest intensity, with a damage track 50 m wide. By 1301 the tornado began to weaken, and the funnel diameter shrank into a ropelike appearance, with the tornado dissipating by 1310. Total track length was about 5 km. This visual evolution is quite similar to that described by Golden and Purcell [1978], although the tornado shown did not become as large during its mature stage as some tornadoes do in the United States.

dry, potentially cold air in midlevels, jet streams at both low and high levels, etc.

3.1.5. *Prefrontal squall lines.* When strong cold fronts come from northern China into the Yangtze River plains, tornadoes may occur with squall lines. Tornadoes in such cases arise most frequently when a preexisting mesoscale convergence boundary is found ahead of the approaching front.

3.2. *North China Plain*

The frequency of tornadoes in the North China Plain is much lower than in the middle and lower reaches of the Yangtze River, but when tornadoes do occur, they can cause serious damage. As before, the conditions for tornadoes arise in several different synoptic situations.

3.2.1. *Extratropical cyclones.* On August 26, 1987, nine tornadoes occurred in Hebei and Shandong provinces of the North China Plain (Figure 4). They developed in the warm sector of an extratropical cyclone associated with a strong closed low at 700 and 500 hPa [Zhao, 1990]. The situation was characterized by strong moisture flux convergence at low levels, substantial horizontal and vertical wind shears, and considerable instability.

3.2.2. *Mongolian cold vortices.* When a cold vortex aloft is present over Mongolia, daily thunderstorms often develop in the North China Plain. When a trough is found in the westerly jet stream along the southern part of the cold

Fig. 7. Funnel cloud near Haikou, Guangdong province.

Fig. 8. Funnel clouds offshore near Qingdao, Shandong province.

Waterspouts also are observed in China. A waterspout is essentially a type of tornado, only over water. Shown in Figure 6 is a waterspout on the Yangtze River not far from Shanghai in 1933. Noteworthy is the height to which the water spray was observed; one explanation for this can be found in the work of *Nalivikin* [1982], although there is no way to verify this explanation from the information available.

Of course, funnel clouds are observed in China. In Figure 7 an example from Haikou, Guangdong province, is shown, while twin funnels over the East China Sea (near Qingdao, Shandong province) are shown in Figure 8. In the latter example the funnels are coming from a cumulonimbus cloud; the two funnels persisted for about 10 min. The left funnel touched down briefly as a waterspout.

5. RADAR ASPECTS OF TORNADOES

Radar observations in China indicate that most tornadoes develop at the right rear portion of the radar-observed thunderstorm echoes, as in the United States. Those radar echoes may take on the "distinctive" forms [*Forbes*, 1981] of hooks, pendants, and S shapes.

As an example, on June 8, 1977, at 1520 (all times are local standard time), severe thunderstorms were observed by the Yongdeng county radar in Gansu province [*Zheng and Liu*, 1982]. The position of the radar echo and tornado at various times relative to the radar is shown in Figure 9. The storm developed over relatively flat terrain in a valley; at 1530 it was a part of an east-west oriented squall line about 200 km long, moving toward the southeast.

A hook-shaped echo with an apparent "eye" was observed at 1530 (all times are local standard) from the right rear quadrant of the parent echo (Figures 10a and 10b); such a configuration is obviously a clear indication of a supercell storm. The eye's diameter was less than 1 km and was in the center of the circular end of the hook, similar to *Fujita*'s [1965] observations. The eye was not visible at 1532 or 1535 without attenuation, but appears with 22-dB attenuation. Although its appearance was quite variable with time, the eye was maintained until 1542. Figure 10c reveals the presence of a bounded weak echo region (or vault) that extends to a height of about 6 km. The tornado began at about 1530 and dissipated around 1551, having traveled roughly 10 km. The tornado was on the ground during the time when the radar was revealing, at least intermittently, the eyelike feature at low levels.

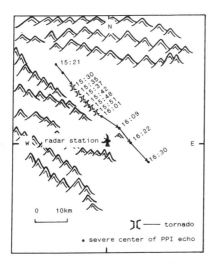

Fig. 9. Sketch of terrain, showing the track of the main radar echo that produced a tornado at the indicated local standard times near Yongdeng, Gansu province, on June 8, 1977.

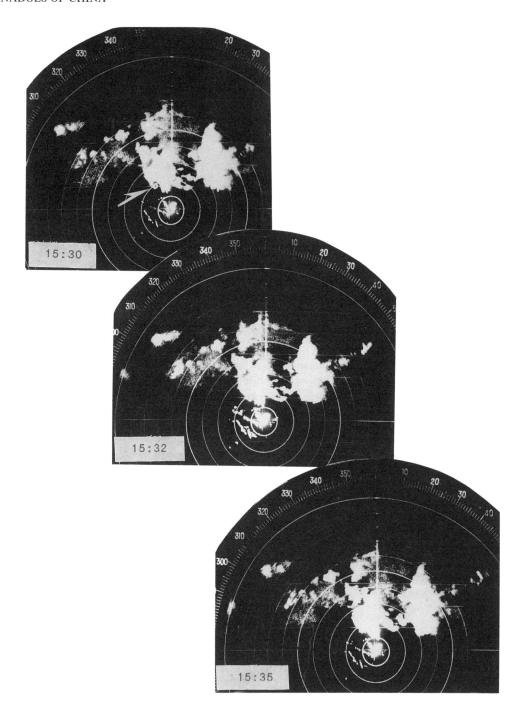

Fig. 10a. Plan position indicator (PPI) photographs taken from the 3-cm radar in Yongdeng county on June 8, 1977. The radar was operating at 1.6° elevation, and range circles are at 10-km intervals.

6. Discussion

The geographical distribution of tornado reports in China, showing most events on the relatively flat plains may be explained, at least in part, by the experiments of *Monji and Wang* [1989], showing that surface roughness can cause the wind speed maximum in a circulation to be higher above the surface than when the surface is smooth. In western China, where the surface roughness is large, the height of the maximum wind is higher than over smooth terrain.

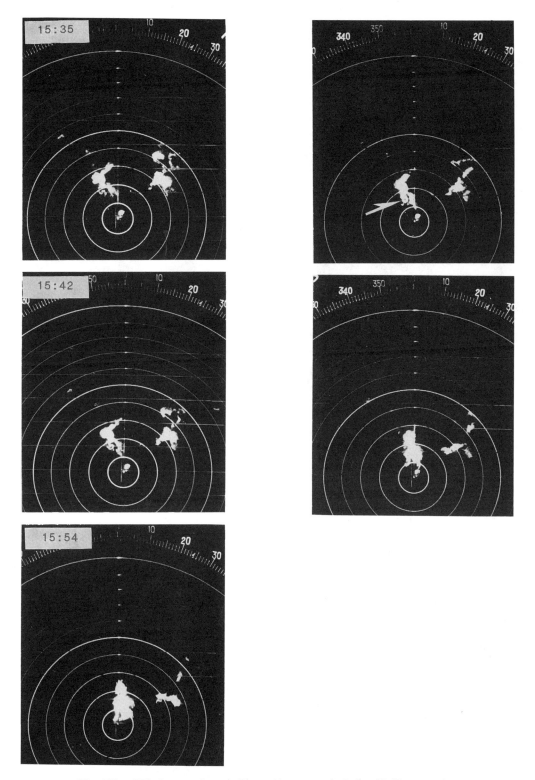

Fig. 10b. PPI photographs as in Figure 10a, except including 22-dB attenuation.

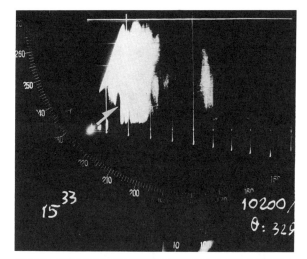

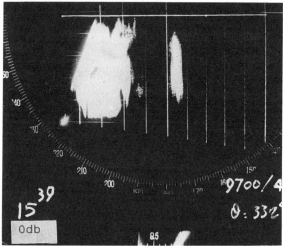

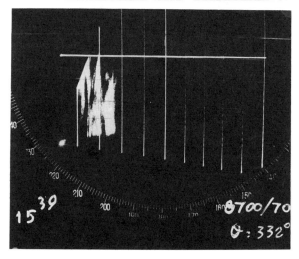

Fig. 10c. Range height indicator photographs for the events shown in Figures 10a and 10b across the hook-shaped echo, at (top) 1533, with no attenuation and azimuth angle of 329°, (middle) 1539, with no attentuation and azimuth of 332°, and (bottom) 1539, with 22-dB attenuation and azimuth of 332°.

On the other hand, in eastern China, the smooth terrain encourages the development of strong low-level circulations from which tornadoes may develop. It also appears that the tilting term in the vorticity equation may be particularly important for tornadogenesis in association with tropical cyclones [*Shen*, 1990].

However, owing to the lack of detailed information for tornadoes in China, such hypotheses can be offered only as speculative ideas. Many more observations and studies are needed to confirm them.

Acknowledgments. The authors are grateful to Charles A. Doswell III for his help in editing and revising this paper, as well as to the anonymous reviewers for their valuable comments. Joan Kimpel (Cooperative Institute for Mesoscale Meteorological Studies, Norman, Oklahoma) helped with some of the figures, for which we are grateful.

References

Chang, J. A case study of tornado in Laishui city, Hebei province (in Chinese), *Meteorol. Monogr.*, *11*, 32, 1981.

Changnon, S. A., Jr., et al., Hail suppression: Impacts and issues, *Final Rep. ERP75-00980*, Ill. State Water Surv., Urbana, 1977.

Forbes, G., On the reliability of hook echoes as tornado indicators, *Mon. Weather Rev.*, *109*, 1457–1466, 1981.

Fujita, T. T., Formation and steering mechanism of tornado cyclones and associated hook echoes, *Mon. Weather Rev.*, *93*, 67–78, 1965.

Golden, J. H., and D. Purcell, Life cycle of the Union City, Oklahoma tornado and comparison with waterspouts, *Mon. Weather Rev.*, *106*, 3–11, 1978.

Kelly, D. L., J. T. Schaefer, R. P. McNulty, C. A. Doswell III, and R. F. Abbey, Jr., An augmented tornado climatology, *Mon. Weather Rev.*, *106*, 1172–1183, 1978.

Kelly, D. L., J. T. Schaefer, and C. A. Doswell III, Climatology of nontornadic severe thunderstorm events in the United States, *Mon. Weather Rev.*, *113*, 1997–2014, 1985.

Kessler, E., and G. F. White, Thunderstorms in a social context, in *Thunderstorms: A Social, Scientific, and Technological Documentary*, vol. 1, *The Thunderstorm in Human Affairs*, edited by E. Kessler, pp. 1–22, U.S. Department of Commerce, Washington, D. C., 1981.

Miller, R. C., Notes on analysis and severe-storm forecasting procedures of the Air Force global weather central (revised), *Tech. Rep. 200*, 190 pp., Air Weather Serv., Scott Air Force Base, Ill., 1972.

Monji, N., and Y. K. Wang, A laboratory investigation of characteristics of tornado-like vortices over various rough surfaces, *Acta Meteorol. Sinica*, *3*, 506–513, 1989.

Nalivikin, D. V., *Hurricanes, Storms, and Tornadoes*, pp. 282–284, Amerind, New Delhi, India, 1982.

Shen, S., An analysis of general characteristics and genesis conditions of tornado in front of typhoon (in Chinese), *Meteorol. Monogr.*, *16*, 11–16, 1990.

Xu, Z., et al., The analyses of mesoscale weather processes in the summer in the Beijing-Tianji-Tangshan areas (in Chinese), in *Collected Papers in Radar Meteorology*, pp. 17–39, Institute of Central Meteorological Bureau, 1977.

Zhao, Y., An analysis of environmental conditions for generation of tornadoes over North China (in Chinese), *Meteorol. Monogr.*, *16*, 36–38, 1990.

Zhen, C., and D. Liu, An analysis of radar echoes of tornado, plateau (in Chinese), *Meteorol. Monogr.*, *9*, 95–98, 1982.

Seasonal Tornado Climatology for the Southeastern United States

LINDA PICKETT GARINGER AND KEVIN R. KNUPP

Atmospheric Science Program, University of Alabama, Huntsville, Alabama 35899

1. INTRODUCTION

Although the climate of the southeastern United States (referred to hereinafter as Southeast) is classified as humid (continental) subtropical, inspection of average annual and seasonal precipitation totals [*Shea*, 1984], thunderstorm frequency [*Court and Griffiths*, 1982], and the diurnal nature of precipitation and thunderstorms [*Wallace*, 1975] shows appreciable spatial variability within the region. For example, *Wallace* [1975] found a strong diurnal tendency in thunderstorm frequency during summer. Within the interior Southeast a prominent peak in thunderstorm frequency was found near 1500 local solar time (LST), while along the Gulf Coast and Florida Peninsula, thunderstorm frequency peaked near 1300–1400 LST. Such summertime variability can be related to mesoscale influences produced by land-ocean contrasts (e.g., the sea breeze).

Patterns of severe weather, including tornadoes, also are inhomogeneous in frequency and diurnal distribution, over the United States in general and the Southeast in particular. *Kelly et al.* [1978] conducted a comprehensive climatological survey for the period 1950–1976, in which the diurnal behavior of tornadoes for the three distinct regions (Great Plains, Midwest, and Southeast) was examined. (They defined the Southeast as consisting of the major portions of Mississippi, Alabama, and Georgia.) The diurnal behavior of tornadoes over the Southeast was found to differ from that of the other two regions in having a weak maximum in frequency in the early morning. All three regions showed a prominent middle to late afternoon maximum. *House* [1963] found relative patterns similar to those of *Kelly et al.* [1978] for the period 1916–1955. In an earlier study, *Skaggs* [1969] performed harmonic analysis of hourly tornado frequency, which displayed appreciable inhomogeneity in diurnal behavior over the Southeast.

In a preliminary tornado climatology which examined tornado and severe weather statistics for individual states within the Southeast region, *Anthony* [1988] also found appreciable inhomogeneity in both monthly and hourly tornado frequency. Anthony's analysis indicates that individual seasons display contrasting diurnal patterns in tornado frequency. Thus potentially confusing results may be obtained when hourly tornado frequency is composited over the entire year. This study extends the work of Anthony by examining and quantifying the diurnal distribution of tornadoes by season, over four relatively homogeneous subregions of the Southeast.

2. DATA AND ANALYSIS PROCEDURES

This analysis utilizes the National Severe Storms Forecast Center's (NSSFC) updated tornado data base covering the 39-year period 1950–1988. Data used include tornado location information, time, and intensity estimates. The limitations of these data have been summarized by *Kelly et al.* [1978, 1985] and *Schaefer et al.* [1986]. Some of the more noteworthy problems include population bias and geographical (state) variations in severe weather recognition and reporting procedure. For large samples we assume that random time errors will cancel. In addition, analysis biases may have been introduced by variations in reporting accuracy as a function of time of day (e.g., day versus night), at least for weak tornadoes. The analysis in subsequent sections was conducted on five subregions shown in Figure 1. The two Atlantic Coast subregions are combined in the analysis of monthly and hourly frequency in sections 3 and 4. Subregion boundary selection was based on several factors, including (1) the results of previous climatologies [e.g., *Anthony*, 1988], (2) geographical breakdown, and (3) an attempt to equalize the area within each region. However, since the areas within each subregion are not equal, the data were normalized to a standard unit area of 10^4 square statute miles (25,900 km^2).

All times have been converted from the standard time (EST or CST) recorded in the data log to local mean solar time (LST). All recorded times were discretized into 1-hour intervals beginning on the hour. Thus a given point in the time series represents 30 min past the hour. In the seasonal analyses, winter is defined as December, January, and

The Tornado: Its Structure, Dynamics, Prediction, and Hazards.
Geophysical Monograph 79
Copyright 1993 by the American Geophysical Union.

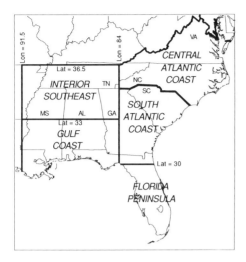

Fig. 1. Definition of subregional boundaries.

February; spring as March, April, and May; summer as June, July, and August; and fall as September, October, and November.

In section 4, harmonic analyses have been completed for hourly distributions (24 samples) of all tornadoes, subdivided by subregion and by season. In each case the data were first smoothed by an equal-weighted, three-point moving average to reduce high-frequency noise. A slow Fourier transform then was applied to obtain frequency components. The resulting equation,

$$Y = \sum A_p \cos(\phi_p + \omega_p t) \quad (1)$$

$$p = 1, 2, 3, \cdots, 10,$$

(where A is amplitude, p is the order of the harmonic, ϕ is phase angle, ω is frequency, and t is time) describes the total distribution. The variance of the pth harmonic is given by

$$r_p^2 = A_p^2/2. \quad (2)$$

The variance fraction, which indicates the importance of a single harmonic relative to the total, is defined as

$$VF = r_p^2 \bigg/ \sum r_{pi}^2, \quad (3)$$

where the summation is again over all 10 harmonics.

3. General Patterns

We first describe some of the general spatial and temporal inhomogeneities of tornado frequency that are apparent over the Southeast. Tracks of very strong to violent (F-3, F-4, and F-5) tornadoes during the 39-year period 1950–1988 are shown in Figure 2. This provides a good indication of the inhomogeneous nature of tornado distribution over the Southeast and shows a prevalence of very strong tornadoes

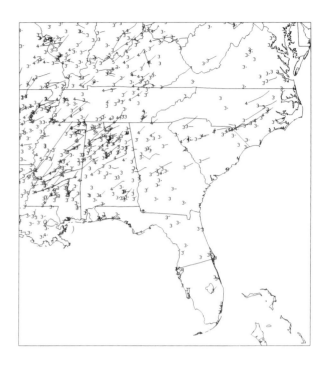

Fig. 2. F scale values and tracks (lines connecting beginning and ending points) of very strong to violent (F-3 and greater) tornadoes reported during the 39-year period 1950–1988.

over the Interior Southeast and Gulf Coast subregions. A relatively large number of long-track tornadoes occur over Mississippi, northern Alabama, and eastern North Carolina, as identified previously by *Kelly et al.* [1978].

The average (normalized) number of all reported tornadoes, and of F-3 and greater tornadoes, per 10^4 mi^2 (25,900 km^2) per year for each subregion is depicted in Figure 3. The Florida Peninsula has by far the greatest number of tornadoes, with an average of 12.1 per unit area per year, but very few (1.1%) are F-3 or stronger. By contrast, 11.6% of the

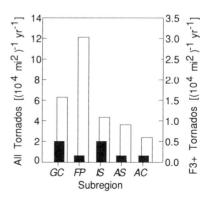

Fig. 3. Average annual number of all tornadoes (open histogram) and very strong to violent tornadoes (F-3 and greater; solid histogram) per 10,000 mi^2 (25,900 km^2), during 20 years (1969–1988) for each of the five subregions defined in Figure 1.

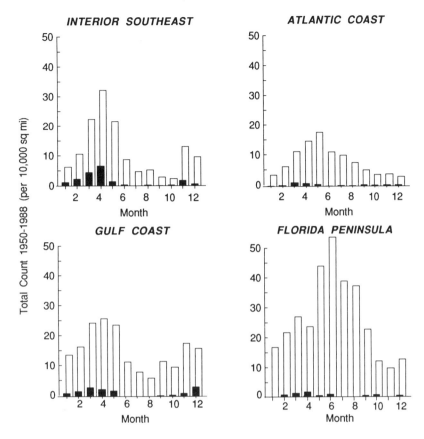

Fig. 4. Monthly distribution of area-normalized tornado frequency in each subregion for the period 1950–1988. The South and Central Atlantic Coast subregions are combined. Open columns represent all tornadoes, and solid columns represent F-3 and greater.

tornadoes in the Interior Southeast subregion are classified as F-3 or greater. The Gulf Coast is second in total number of occurrences, with an average of 6.3 tornadoes per unit area, 7.9% of which are classified as F-3 and greater. The Central and South Atlantic Coast subregions display similar numbers of total tornadoes, 3.64 and 2.42 events respectively, with F-3 and greater tornadoes accounting for 4.5% and 5.7% of the total. Because the Central and South Atlantic Coast areas are similar in number and relative distribution (monthly and hourly) of tornadic occurrences, they have been combined into a single Atlantic Coast subregion in the following sections.

Monthly distributions of tornadoes display dissimilarities for each subregion and are generally similar to those described on a statewide basis by *Anthony* [1988]. Spring produces the greatest number of tornadoes in the Gulf Coast and Interior Southeast subregions, with significant secondary peaks during November–December (Figure 4). The November–December peak in the Gulf Coast and Interior Southeast distributions is likely due to southward advance of the jet stream over moist air masses produced by the relatively warm adjacent Gulf waters.

The Atlantic Coast subregions display a maximum of activity somewhat later, from spring to early summer. The secondary peak during November–December is less significant both here and in the Florida Peninsula. The Florida Peninsula shows a maximum in July, although May, June, and August also display significantly more tornadoes than any individual monthly maximum in the other subregions. The monthly distribution of very strong to violent (F-3 and greater) tornadoes, also shown in Figure 4, indicates that the Gulf Coast region has a significant number of F-3+ events in December, in addition to the spring.

The general patterns described above require physical explanation. Assuming that F-3 and greater tornadoes originate from supercell storms, we suggest that Florida experiences only a few such occurrences during the cold season when upper level winds are strong. We hypothesize that further north and west, over the Interior Southeast and Gulf Coast subregions, environmental conditions for long-lived supercell storms, such as strong vertical shear of environmental wind [*Weisman and Klemp*, 1982] or high values of storm-relative helicity [*Davies-Jones et al.*, 1990], are more favorable.

The preponderance of tornadoes during summer over Florida (and to some extent over the Atlantic Coast regions) is likely associated with strong mesoscale forcing associated

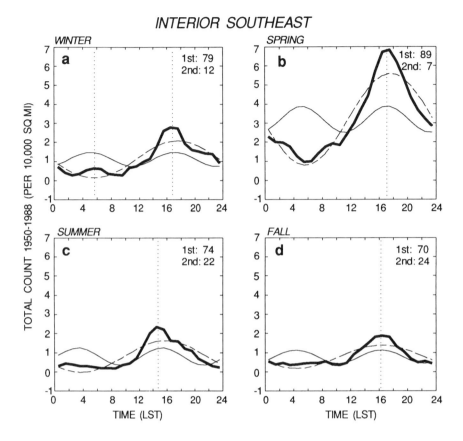

Fig. 5. Smoothed hourly distribution of all tornadoes (heavy curves) in the Interior Southeast subregion for each season. Dashed curves represent the first harmonic, and the light solid curves represent the second harmonic. The variance fractions from significant harmonics are indicated in the top right-hand corner of each panel. Vertical dotted lines show the approximate times of local maxima in the smoothed data.

with sea breeze circulations that frequently interact with other boundaries such as thunderstorm outflows [Purdom, 1982]. Such tornadoes originate from nonsupercell storms and are hence weak and short-lived [e.g., Holle and Maier, 1980]. It remains to be determined whether the mechanisms of tornadogenesis resemble those within nonsupercell storms which generate tornadoes over eastern Colorado [e.g., Wakimoto and Wilson, 1989].

4. DIURNAL BEHAVIOR OF TORNADOES

The seasonal (smoothed) hourly distributions of tornado frequency for each subregion are shown in Figures 5–8. In addition, those harmonic components, generally the first two, that have a variance fraction (VF) exceeding 5% are superimposed. In many of the distributions the peak in one or more harmonics coincides very closely with the data peak, while in others there is a significant offset due to skewed distributions and/or secondary maxima. The Interior Southeast region is the only one in which the time series data peak before the peak in the first harmonic. The other three subregions exhibit the opposite behavior. As expected from the results of Skaggs [1969], Wallace [1975], and Anthony [1988], the diurnal cycle (first harmonic) is dominant (52% to 95%) in all cases, but a large amount of variability exists on both a seasonal and an interregional basis.

With one exception (the winter distribution over the Florida Peninsula) all distributions peak during the middle to late afternoon hours. During spring and fall the southern subregions generally peak at times similar to a summer distribution, while maximum value times in the northern subregions more closely resemble a winter distribution (Table 1). The Gulf Coast exhibits significant secondary maxima in the morning (0700 to 0900) during all seasons and has the lowest interseasonal variability. The secondary morning maximum also was noted by Kelly et al. [1978] for a portion of the southeast United States consisting primarily of Mississippi, Alabama, and Georgia. This study indicates that the geographical area primarily responsible for the secondary morning maxima is enclosed by the Gulf Coast subregion. Some additional seasonal characteristics are described in the following sections.

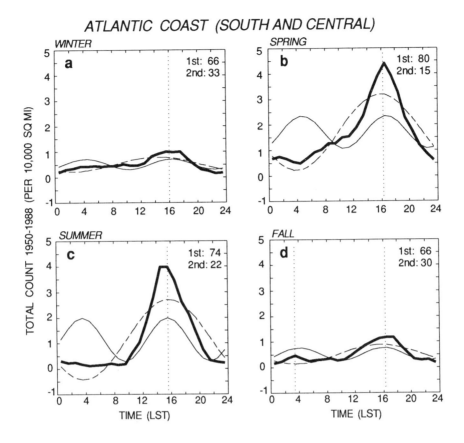

Fig. 6. As in Figure 5, for the combined (South and Central) Atlantic Coast subregions.

4.1. Winter Patterns

The winter season shows the most variability in the time series distributions among subregions. The Florida Peninsula distribution (Figure 8a) displays a very strong (95%) contribution from the first harmonic which reaches its maximum at ~1100. This late morning peak in the first harmonic is unique to the winter Florida Peninsula, as noted by *Anthony* [1988]. This prominent morning tornado maximum contrasts with the late afternoon (1700–1800 LST) cold-season thunderstorm frequency maximum in the first harmonic determined by *Wallace* [1975]. A physical explanation for this time difference between thunderstorm and tornado frequency is not obvious.

The Gulf Coast distribution (Figure 7a) has a primary maximum at ~1700 LST and a significant secondary maximum at ~0900. The contribution from the first harmonic is very weak (52%) when compared to the contributions from the first harmonic in other seasons and subregions. This is the only case in which the contribution from the third harmonic is significant (>5%). The peak in the third harmonic occurs at ~0900 and coincides with the secondary morning maximum in the smoothed time series.

The Interior Southeast distribution (Figure 5a) also displays two maxima, a primary peak at ~1700 and a weak secondary maximum at ~0530 LST. In this case, however, the first-harmonic contribution is stronger (79%), and that from the second (and third) is much less. Although the Atlantic Coast subregions (Figure 6a) show relatively little tornadic activity during the winter, a secondary morning peak is suggested.

4.2. Spring Patterns

In relative terms, spring is the most active season for all of the study area except the Florida Peninsula. All subregions, again with the exception of the Florida Peninsula, show data maxima at 1600–1700 LST. All subregions display skewed distributions in which tornado frequency increases slowly during the early morning hours (0600–0800) and then decreases more rapidly after the afternoon peak. An earlier maximum at ~1430 is seen in the Florida Peninsula distribution (Figure 8b). The Florida distribution seems to contain characteristics of both the summer and the winter distributions. The contribution from the first harmonic is relatively high (80–90%) in all subregions except the Florida Peninsula, which displays a moderate contribution from the first harmonic (74%) and a moderately strong contribution from the second harmonic (22%). The Gulf Coast (Figure 7b)

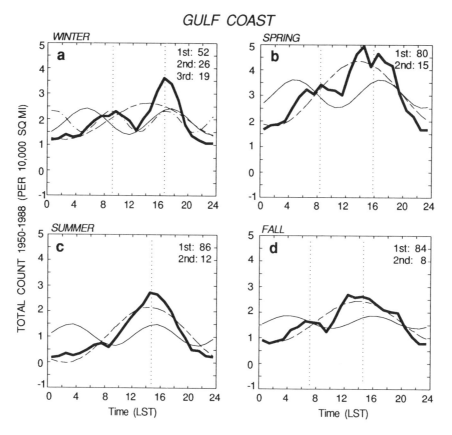

Fig. 7. As in Figure 5, for the Gulf Coast subregion, except that the third harmonic is shown as a dotted-dashed curve in Figure 7a.

again shows a prominent secondary maximum centered at ~0800.

4.3. Summer Patterns

The summer distributions are generally symmetric, with the exception of the Interior Southeast, which is skewed into the evening. The Florida Peninsula (Figure 8c) records a maximum in activity during summer, when the Gulf Coast (Figure 7c) subregion displays a relative minimum (Figure 4). In comparison to the other seasons, the secondary morning maximum of the Gulf Coast subregion is relatively weak. The time series for the Florida Peninsula and Atlantic Coast subregions are similarly symmetric in shape, although the Atlantic Coast exhibits lower frequency. The variance fractions and the harmonic phases are almost identical for the first two harmonics in these subregions. The increasing significance (from winter to summer) of the second harmonic in Florida is consistent with the work of *Brier and Simpson* [1969], who demonstrated a close relationship between cloud/rainfall patterns and the semidiurnal pressure oscillation (which has a very strong second harmonic) within tropical oceanic regimes. Although the relative influence of the semidiurnal tide would increase from winter to summer

as synoptic scale perturbations migrate northward, the Florida time series does not exhibit relative peaks at the times (~0600 and ~1800 LST) found by Brier and Simpson.

4.4. Fall Patterns

Fall generally is the least active time of year for the Interior Southeast, Atlantic Coast, and Florida Peninsula subregions (Figures 4, 5d, 6d, and 8d). As indicated in section 3, the Interior Southeast and Gulf Coast subregions display a significant secondary peak in monthly tornado distribution from late fall to early winter (Figure 4). The secondary morning maximum in the Gulf Coast subregion is again evident (Figure 7d). Contributions from the first harmonic vary from a weak 66% in the Atlantic Coast subregion to a strong 84% in the Gulf Coast subregion, while contributions from the second harmonic range from a strong 30% in the Atlantic Coast subregion to a very weak 8% in the Gulf Coast subregion.

5. SUMMARY

In general, the results emphasize the variability in tornado characteristics and temporal patterns from both regional and

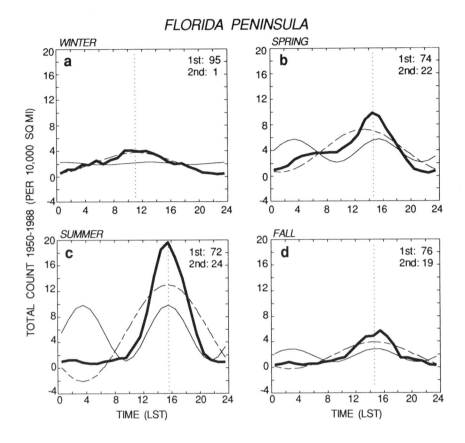

Fig. 8. As in Figure 5, for the Florida Peninsula subregion.

seasonal perspectives. Some of the specific results of this study are as follows:

1. Tornadoes occur most frequently per unit area over the Florida Peninsula subregion, followed by the Gulf Coast, Interior Southeast, South Atlantic, and Central Atlantic. Very strong to violent tornadoes (F-3 and stronger) occur most frequently in the Gulf Coast and Interior Southeast, followed by South Atlantic, North Atlantic, and finally the Florida Peninsula subregions.

2. The monthly distributions for the Florida Peninsula show peak tornado frequency during the summer. The other subregions display a maximum monthly frequency in the spring. Only two of the four subregions (Interior Southeast and Gulf Coast) show significant secondary peaks in late fall to early winter.

3. The hourly tornado frequency distributions show several significant seasonal and interregional differences. Winter, in particular, reveals that each subregion has unique distribution characteristics. The Florida Peninsula displays a morning (1100 LST) maximum in frequency, which is represented almost entirely by the first harmonic. In contrast, the Gulf Coast shows a significant contribution from the second (26%) and third (19%) harmonics during the winter. Only the Gulf Coast has a significant secondary peak during the morning hours (0800–0900 LST) in all seasons.

The observed seasonal and regional differences likely point to a multitude of environmental (mesoscale) controls on storm structure which are not understood fully. Further physical interpretation of these findings is clearly needed. A composite analysis of proximity soundings, combined with hourly wind profiler data or NEXRAD VAD winds, would provide additional insight on the physical mechanisms responsible for the observed patterns. In addition, synoptic scale compositing and analysis of selected case studies would likely shed insight on the physical processes responsible for the variation in observed hourly distributions.

TABLE 1. Approximate Time of Maximum Number of Tornadoes by Season and Subregion

	Interior Southeast	Atlantic Coast	Gulf Coast	Florida Peninsula
Winter	1645	1600	1630	1100
Spring	1700	1615	1545	1430
Summer	1445	1530	1445	1530
Fall	1615	1615	1430	1430

Acknowledgments. This research was sponsored by the National Oceanic and Atmospheric Administration through the Southeast Regional Climate Center. Tornado frequency data were provided by the National Severe Storms Forecast Center (NSSFC). The National Center for Atmospheric Research (NCAR) provided the graphics package used in this study.

References

Anthony, R., Tornado/severe thunderstorm climatology for the southeastern United States, in *Preprints, 15th Conference on Severe Local Storms*, pp. 511–516, American Meteorological Society, Boston, Mass., 1988.

Brier, G. W., and J. Simpson, Tropical cloudiness and rainfall related to pressure and tidal variations, *Q. J. R. Meteorol. Soc., 95*, 120–147, 1969.

Court, A., and J. F. Griffiths, Thunderstorm climatology, in *Thunderstorms: A Social, Scientific and Technological Documentary*, vol. 2, *Thunderstorm Morphology and Dynamics*, edited by E. Kessler, pp. 11–77, U.S. Department of Commerce, Washington, D. C., 1982.

Davies-Jones, R., D. Burgess, and M. Foster, Test of helicity as a tornado forecast parameter, in *Preprints, 16th Conference on Severe Local Storms*, pp. 588–592, American Meteorological Society, Boston, Mass., 1990.

Holle, R. L., and M. W. Maier, Tornado formation from downdraft interaction in the FACE mesonetwork, *Mon. Weather Rev., 108*, 1010–1028, 1980.

House, D. C., Forecasting tornadoes and severe thunderstorms. *Meteorol. Monogr., 5*(27), 141–156, 1963.

Kelly, D. L., J. T. Schaefer, R. P. McNulty, C. A. Doswell III, and R. F. Abbey, Jr., An augmented tornado climatology, *Mon. Weather Rev., 106*, 1172–1183, 1978.

Kelly, D. L., J. T. Schaefer, and C. A. Doswell III, Climatology of nontornadic severe thunderstorm events in the United States, *Mon. Weather Rev., 113*, 1997–2014, 1985.

Purdom, J. F. W., Subjective interpretation of geostationary satellite data for nowcasting, in *Nowcasting*, edited by K. A. Browning, pp. 149–166, Academic, San Diego, Calif., 1982.

Schaefer, J. T., D. L. Kelly, and R. F. Abbey, A minimum assumption tornado-hazard probability model, *J. Appl. Clim. Meteorol., 25*, 1934–1945, 1986.

Shea, D. J., The annual variation of precipitation over the United States and Canada, *Rep. NCAR/TN-243+STR*, 66 pp., Natl. Cent. for Atmos. Res., Boulder, Colo., 1984.

Skaggs, R. H., Analysis and regionalization of the diurnal distribution of tornadoes in the United States, *Mon. Weather Rev., 97*, 103–115, 1969.

Wakimoto, R. M., and J. W. Wilson, Non-supercell tornadoes, *Mon. Weather Rev., 116*, 1113–1140, 1989.

Wallace, J. M., Diurnal variations in precipitation and thunderstorm frequency over the conterminous United States, *Mon. Weather Rev., 103*, 406–419, 1975.

Weisman, M. L., and J. B. Klemp, The dependence of numerically simulated convective storms on vertical wind shear and buoyancy, *Mon. Weather Rev., 110*, 504–520, 1982.

Oregon Tornadoes: More Fact Than Fiction

GEORGE R. MILLER

National Weather Service Forecast Office, Portland, Oregon 97218

1. INTRODUCTION

Factual accounts of tornadoes in Oregon are rare. The U.S. Department of Commerce's Weather Bureau Annual Meteorological Summary With Comparative Data for Portland, Oregon, for the years 1922 and 1944 seems to verify this with two very bold statements. The 1922 publication on page 3 states, "Tornadoes are unknown." The 1944 publication on page 8 modifies this slightly by saying, "Well-developed tornadoes are unknown."

These accounts fit well with the opinion of the majority of Oregonians. They will tell you that they are happy to be living in this state, where storms like the ones in the Midwest do not occur. This paper will show that tornadoes do occur in Oregon, under atmospheric conditions that are similar to those that produce tornadoes elsewhere. A few of the atmospheric characteristics associated with Oregon tornadoes will be examined. The paper also offers the hypothesis that tornadoes occur in Oregon at a frequency greater than that perceived by the populace.

2. TORNADO CLIMATOLOGY

Oregon is definitely out of the tornado belt [*Flora*, 1953]. One study [*Hollifield*, 1990] shows that Oregon averaged one tornado and one tornado day per year for the period 1953–1989. That study indicates that the greatest number reported for 1 year was four in 1984.

There have been no recorded deaths from tornadoes this century in Oregon. (On April 5, 1972, a tornado touched down on the Oregon side of the Columbia River, moved across the river, and struck a shopping center and school, killing six people and injuring 300 in Vancouver, Washington.) Newspaper accounts, however, indicate that four persons were killed from tornadoes in eastern Oregon prior to 1900. Three of those deaths occurred in the eastern Oregon community of Long Creek on June 3, 1894. The fourth was in the small farming community of Lexington in north-central Oregon on June 14, 1888.

Tornadoes have been reported in Oregon in every month except January (Figure 1). But even in this month an unconfirmed report was received in 1989 from the coast. There are two distinct maxima, one in the spring and one in the fall. April, May, and June have the greatest number, which is similar to other sections of the country [*Kelly et al.*, 1978]. The incidence drops off markedly in July, the driest month in Oregon. The fall maximum is quite likely due to the migration southward of the jet stream and storm track. The total number, excluding the January 1989 report, is 44.

Tornadoes that occur in Oregon during the cool season (October through March) are confined to the area west of the Cascade Mountain Range which bisects the state north to south. During these months, near-surface air east of the Cascades is quite cool. Thus the instability that is necessary for convective activity is almost nonexistent. All reports of tornadoes in December and January have been from the coastal strip.

3. TORNADO INTENSITY

Fujita classifies tornadoes as F-0 to F-5 [*Fujita*, 1987]. His scheme is abbreviated here for those unfamiliar with it: F-0, winds 40–72 mph (18–32 m/s); F-1, winds 73–112 mph (33–50 m/s); F-2, winds 113–157 mph (51–70 m/s); F-3, winds 158–206 mph (71–92 m/s); F-4, winds 207–260 mph (93–116 m/s); and F-5, winds 261–318 mph (117–142 m/s). Using this classification, Oregon tornadoes are ranked in the F-0 or F-1 range.

Oregon tornadoes investigated over the last 5 years by the author all have been less than F-2. The tornado that touched down briefly in Oregon before crossing the Columbia River in 1972 was classified F-2 or F-3 in Washington, but no evidence exists that it reached this intensity in Oregon.

The Long Creek and Lexington tornadoes probably come the closest to F-2 strength in Oregon. A vivid, dramatic account of the former storm can be found in an article in *The Long Creek Eagle* [*Patterson*, 1894]. A similar account of the Lexington tornado can be found in a historical document

The Tornado: Its Structure, Dynamics, Prediction, and Hazards.
Geophysical Monograph 79
This paper is not subject to U.S. copyright. Published in 1993 by the American Geophysical Union.

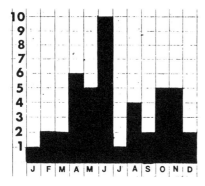

Fig. 1. Plot of tornadoes reported in Oregon since 1884. The January report was unconfirmed.

about Umatilla and Morrow counties [*Parsons and Shiach*, 1902]. It should be taken into consideration, when reading historical accounts of tornadoes from small newspapers, that there may be a bit of dramatization to the reports. Nevertheless, they underline the fact that tornadoes have caused destruction and killed people in Oregon.

4. METEOROLOGICAL FACTORS (SYNOPTIC SCALE) AND EXAMPLES

As of this writing, there have been no detailed case studies of tornado occurrences in Oregon. No single weather pattern stands out as the most favorable for tornado formation in Oregon. An unstable air mass with a strong jet and a vigorous shortwave trough are items for which to look. However, these features occur with regularity in Oregon as elsewhere. An important difference between these features associated with tornado events in Oregon and those in the Central Plains appears to be in the degree of instability. Large, negative lifted index values are virtually unheard of in Oregon, even with tornadoes.

The jet stream appears to be a factor for tornado formation in Oregon. Tornadoes have been reported with a strong northwesterly jet, a westerly jet, a south to southwesterly jet, and a cyclonically curved jet. However, despite the orientation of the jet, most tornadoes cases that were examined occurred in the left front quadrant of the jet maximum (jet streak). This coincides with the area where upward vertical motion would be expected with a jet streak circulation [*Uccellini and Johnson*, 1979].

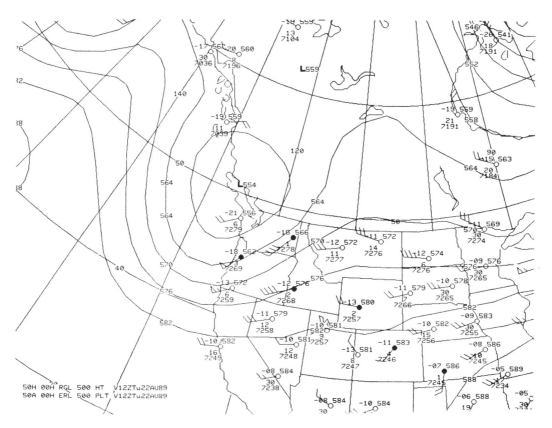

Fig. 2. The 500-mbar geopotential height analysis (60-m contours), 1200 UTC, August 22, 1989.

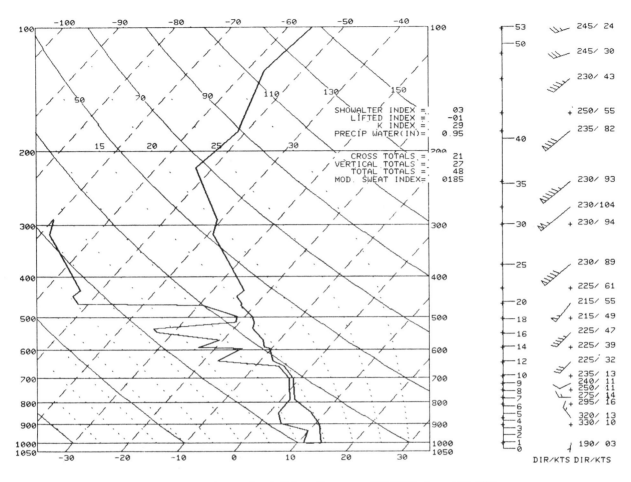

Fig. 3. Upper air sounding in Salem, Oregon, at 1200 UTC, August 22, 1989.

Bend, Oregon, Tornado, August 22, 1989

On August 22, 1989, at around 1 P.M. PDT (2000 UTC) a tornado was reported just east of Bend, Oregon, in the central portion of the state. This area is east of the Cascade Mountains and at an elevation of 3000–4000 feet (915–1220 m). An active shortwave trough (Figure 2) is entering the Pacific Northwest with relatively cool air aloft. The trough moves slowly through the area.

No soundings are available from east of the Cascades in Oregon. The winds at Salem (Figure 3) show backing with height below about 700 mbar, indicating cold air advection. Above 700 mbar, winds show considerable speed shear but little directional change with height. (The portion of the sounding below about 700 mbar is probably unrepresentative for the central part of the state, but the moderate lapse rate and considerable speed shear above 700 mbar suggest that the sounding in central Oregon may have been quite comparable to tornado soundings elsewhere.) The subsequent sounding from Salem at 0000 UTC on August 23, 1989 (not shown), revealed cooling of the order of 2°–5°C at most levels of the atmosphere, especially above 550 mbar.

Eugene, Oregon, Tornado, November 24, 1989

On November 24, 1989, a tornado was sighted just south of Eugene, Oregon, in the southern Willamette Valley, west of the Cascade Mountain Range. A jet maximum of 130 knots (65 m/s) (Figure 4) was just west of the Oregon/

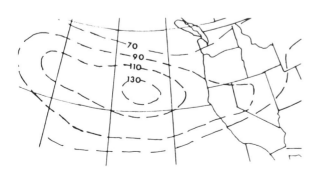

Fig. 4. The 300-mbar isotachs (knots; 1 knot = 0.5 m/s) at 1200 UTC, November 24, 1989.

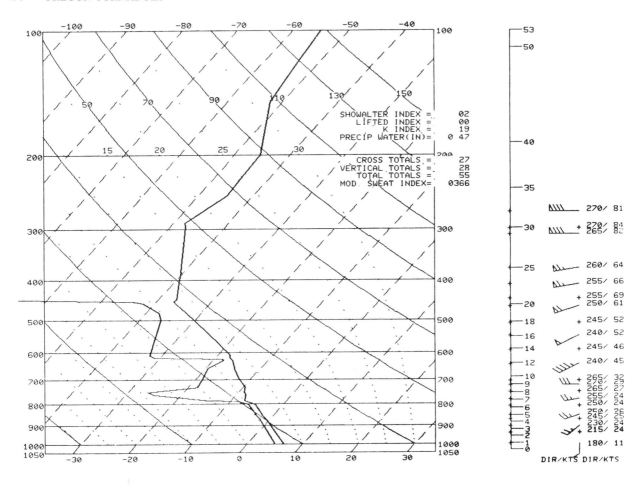

Fig. 5. Upper air sounding in Salem, Oregon, at 1200 UTC, November 24, 1989.

California border, and Oregon was in the left front quadrant.

The lifted index, as shown in Figure 5, is near zero with a strong jet above the tropopause. Although it is relatively shallow, a layer of steep lapse rate is present in midtroposphere (about 600 to 450 mbar), and the winds show veering off the surface. The entire sounding is cool relative to its midwestern counterparts, but it is not inconsistent with a shallow supercell. As *Braun and Monteverdi* [1991] have shown, supercells are not unheard of on the West Coast. The subsequent sounding taken at 0000 UTC on November 25, 1989 (not shown), revealed cooling above 600 mbar and a slight decrease in wind with little change in direction.

Vancouver, Washington, Tornado,
April 5, 1972

The pattern that spawned the destructive Vancouver, Washington, tornado on April 5, 1972, consisted of a south-southwesterly jet in excess of 110 knots (55 m/s) (Figure 6). The tornado occurred around 1 P.M. local time (2000 UTC)

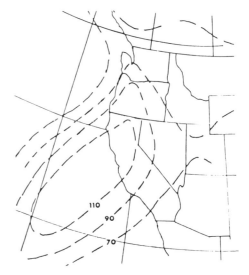

Fig. 6. The 300-mbar isotachs (knots; 1 knot = 0.5 m/s) for 1200 UTC, April 5, 1972.

in postfrontal conditions and was associated with an active squall line.

5. SUMMARY AND CONCLUSIONS

There have been documented accounts of tornadoes in Oregon with recorded deaths. Some have caused considerable damage, others slight damage. In the tornado cases considered, soundings taken before and after generally showed negative to only slightly positive lifted indices. For the May through September tornadoes the mid and upper atmospheric flow was generally from the southwest with a tropopause level of between 30,000 and 40,000 feet (9 to 11 km). October through March tornado cases show mostly a west to northwesterly flow.

Much work remains to be done regarding tornadoes in Oregon. There is a need for detailed studies about specific occurrences, including sounding and hodograph analyses, as well as constant level charts. Perhaps after this is accomplished, some definite ideas can be identified for anticipating tornado formation in Oregon.

Currently, the only facts about Oregon tornadoes are as follows: (1) They reach a maximum in June. (2) Tornadoes in the colder months of the year (October–March) are confined to the western sections of Oregon. (3) In the few cases that were superficially studied, Oregon was located in the left front quadrant of the jet stream.

It seems likely that reported occurrences of tornadoes in Oregon will increase. This is due, in part, to the publicity these storms are receiving in other parts of the United States. Another contributing factor is the increase in the number of residents from areas outside Oregon where tornadoes are more common.

As has been noted elsewhere, tornado reporting is always uneven. In states where the perception is that tornadoes are uncommon, it is typical that tornadoes often go unreported [e.g., *Doswell*, 1980]. Oregon appears to be typical in this regard, but the author is trying to change this perception. More occurrences should lead to a better understanding of the meteorological factors associated with tornado development in Oregon.

Acknowledgments. The author is grateful for the numerous suggestions and helpful comments provided by the reviewers, the staff in Scientific Services Division, Western Region, National Weather Service, and especially to C. A. Doswell III, National Severe Storms Laboratory, for his immeasurable assistance.

REFERENCES

Braun, S. A., and J. P. Monteverdi, An analysis of a mesocyclone-induced tornado occurrence in northern California, *Weather Forecasting*, 6, 13–31, 1991.

Doswell, C. A., III, Synoptic scale environments associated with High Plains severe thunderstorms, *Bull. Am. Meteorol. Soc.*, 60, 1388–1400, 1980.

Flora, S. D., *Tornadoes of the United States*, 194 pp., University of Oklahoma Press, Norman, 1953.

Fujita, T. T., U.S. tornadoes, part I, 70 year statistics, report, 120 pp., Satell. and Mesometeorol. Res. Proj., Dep. of the Geophys. Sci., Univ. of Chicago, Chicago, Ill., 1977.

Hollifield, J., National summary of tornadoes, 1989, *Storm Data Bull.*, 31, 1-10, Dec. 1990.

Kelly, D. L., J. T. Schaefer, R. P. McNulty, C. A. Doswell III, and R. F. Abbey, Jr., An augmented tornado climatology, *Mon. Weather Rev.*, 106, 1172–1183, 1978.

Parsons, W., and W. S. Shiach, *An Illustrated History of Umatilla County and Morrow County*, W. H. Lever, 1902.

Patterson, O. L., in *The Long Creek Eagle*, p. 1, Friday, June 8, 1894.

Uccellini, L. W., and D. R. Johnson, The coupling of upper and lower tropospheric jet streaks and implications for the development of severe convective storms, *Mon. Weather Rev.*, 107, 682–703, 1979.

The Stability of Climatological Tornado Data

JOSEPH T. SCHAEFER[1] AND RICHARD L. LIVINGSTON

Central Region Scientific Services Division, National Weather Service, Kansas City, Missouri 64106

FREDERICK P. OSTBY AND PRESTON W. LEFTWICH[2]

National Severe Storms Forecast Center, National Weather Service, Kansas City, Missouri 64106

1. INTRODUCTION

The National Severe Storms Forecast Center's (NSSFC) log of all tornadoes reported in the United States since January 1, 1950, contains the latitude/longitude of each tornado's beginning and ending points, its observed track length and average path width, the FPP indices as defined by *Fujita and Pearson* [1973], and an estimate of the dollar amount of damage caused. Many climatological-type studies [e.g., *McNulty et al.*, 1979; *Schaefer et al.*, 1980, 1986] have used this historical data set. However, none have questioned whether these studies fulfill the basic purpose of a climatology, the presentation of the characteristic state and behavior of the atmosphere [*Jagannathan et al.*, 1967].

To see the dependence of the apparent tornado climatology upon the period over which data were analyzed, one need only compare state-by-state tornado totals for the 1970s and 1980s (Figure 1). The marked differences between the state totals for the two 10-year periods indicate that as a general rule, tornado activity decreased east of the Mississippi River and increased west of it. Thus a 10-year tornado summary could be used to indicate either that Illinois ranks fifth among the states in the annual number of tornadoes (from the 1970–1979 data) or fifteenth (1980–1989 data).

Climatological data are usually presented in terms of "normals" which, in accordance with World Meteorological Organization (WMO) procedures, are averages over specified 30-year periods. However, the length of a statistically meaningful averaging period varies among weather parameters and from station to station [*Court*, 1968]. Hence an empirical exploratory data analysis approach is needed to describe the climate adequately [*Guttman*, 1989].

In this paper, tornado occurrence statistics over the coterminous United States for the 40-year period 1950–1989 are examined. First, variations in the annual number of tornadoes reported during the study period are analyzed. Then the spatial distribution of tornado occurrence frequency is computed for various averaging periods. Finally, it is shown that the tornado occurrence pattern does not stabilize until 35 years of data are considered. It is argued that this pattern corresponds to the U.S. tornado climatology.

2. TIME SERIES ANALYSIS

In the 40 years 1950–1989, confirmed tornado reports over the coterminous United States totaled 28,384. The minimum was 201 in 1950, while the maximum was 1102 in 1973. A graph of the annual tornado count (Figure 2) shows marked year-to-year fluctuations. The highest count in the first three years of the data base (260 in 1951) is only 62% of the lowest count of the remaining 37 years (421 in 1953). Thus a sizable number of tornadoes were most likely unreported before 1953.

While the 1953 count is also low, since it is comparable to the 463 reports received in 1963, it cannot be so easily flagged as being erroneous. The increase in the number of tornadoes reported in the early 1950s can be related to public awareness. The U.S. Air Force started alerting their facilities of potential tornadic weather in 1948, and the Weather Bureau started releasing public tornado forecasts in 1952 [*Galway*, 1993]. Further, tornadoes devastated Waco, Texas; Flint, Michigan; and Worcester, Massachusetts, in 1953, causing 114, 116, and 90 fatalities, respectively. (There is some disagreement as to the number of fatalities with the Flint, Michigan, tornado. The NSSFC data base, *Flora* [1954], and *Snider* [1975] each attribute 116 fatalities to this storm. However, *Linehan* [1957] only lists 115 deaths. *Grazulis* [1990] puts

---

[1] Now at National Weather Service Training Center, Kansas City, Missouri 64124.
[2] Now at Central Region Scientific Services Division, National Weather Service, Kansas City, Missouri 64106.

The Tornado: Its Structure, Dynamics, Prediction, and Hazards.
Geophysical Monograph 79
This paper is not subject to U.S. copyright. Published in 1993 by the American Geophysical Union.

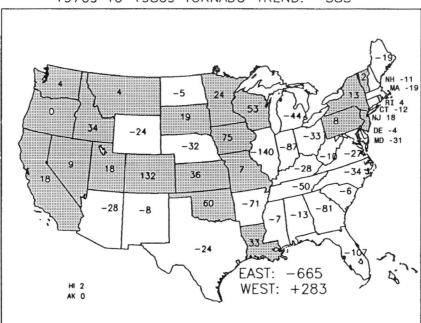

Fig. 1. The change in decadal tornado count by state between the 1970s and 1980s. A decrease in the number of reported tornadoes was recorded in the 1980s as compared to the 1970s. The national change does not equal the sum of the changes for the individual states because of tornadoes that crossed from one state into another.

the death toll for this storm at 115 and attributes another death to a second storm that occurred in the same area that night. The NSSFC data base shows no deaths with this second storm.)

To distinguish subtle features in the pattern underlying these data, a shape-preserving smoother developed by *Tukey* [1977] was applied to it. The resulting curve (superposed on Figure 2) shows that the increase in the annual number of tornadoes through the 1950s and 1960s did not persist through the 1970s and 1980s. In fact, quite possibly tornado activity has decreased since 1973. Perhaps this decreasing trend in tornado occurrences is related to procedural changes in the National Weather Service (NWS), which abolished its state climatologist positions during 1972 and 1973. After the demise of this program there were no longer individuals who were dedicated to the collection of information on the occurrence of severe thunderstorms.

One way to work around the apparent inflation in the annual number of reported tornadoes is to consider "tornado days" rather than actual tornadoes. *Court and Griffiths* [1985] showed that since 1916, the annual number of tornado days has increased much less than the number of tornadoes. Part of this is a reflection that there is a physical upper bound to the possible number of annual tornado days. Further, as noted by *Court* [1970, p. 39], "the nonmeteorologist is interested in the one tornado that may effect him, and hence the public is better served by tabulations and maps of tornado incidence rather than of the number of days on which they occur."

The vagaries of the annual tornado count can be seen in accumulated tornado occurrences for local areas (Figure 3). A rapid increase in population typically corresponds to a consistent growth in the rate that tornadoes are reported. An example of this is Hillsborough County in west central Florida. This county, which contains the city of Tampa, had a population increase of 156,695 between 1970 and 1980. The cumulative number of tornadoes in Hillsborough County since 1950 is concave downward through the mid-1980s, indicating that, in general, more tornadoes were reported each subsequent year. In the greater Houston area, Harris County, Texas, the 1970–1980 population increase was 667,632. The accumulated tornado curve generally parallels that of Hillsborough County, Florida, except for a distortion in 1983 when 22 tornadoes occurred as hurricane Alicia approached the Gulf Coast.

In contrast, the population of Kay County, Oklahoma, which contains Ponca City and is in "tornado alley," has been rather sluggish. It had a population decrease of 2251 during the 1960s and gained only 1065 people in the 1970s. The slope of the accumulated tornado curve was rather constant, indicating that the rate of reported tornado occurrence did not change substantially over the past four decades. Weld County, Colorado, around Greeley in the north central part of the state, has also remained rather unpopulated. Even though its population grew by 34,141 during the 1970s, the 1980 population density was still less than 31 people per square mile. Through 1978 most of the tornadoes reported in this county occurred in a few active years (1955,

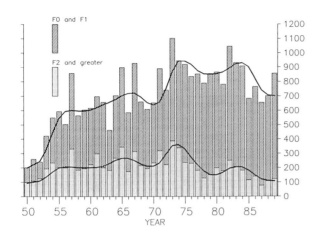

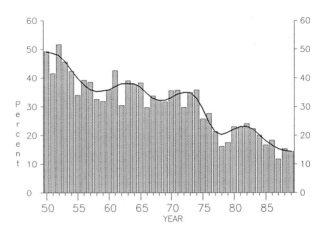

Fig. 2. The number of reported tornadoes in the United States by year. The solid lines superimposed over the bars indicate data smoothed by the Tukey filter.

Fig. 4. The ratio of strong and violent tornadoes (F-2 and greater) to the total number of reported tornadoes.

1957, 1958, and 1976). Then tornadoes started occurring extremely often but at a nearly constant rate. One possible explanation for this is the advent of the Prototype Regional Observing and Forecast Service (PROFS) program with its tornado "chasers" and its emphasis on providing ground truth over northeastern Colorado [*Beran and Little*, 1979].

It has been noted that population biases in tornado reporting are reduced when only stronger tornadoes are considered [*Kelly et al.*, 1978]. A time series for those tornadoes that are rated as strong and violent in the NSSFC log (F-2 and greater on the Fujita scale) was constructed (Figure 2). While the fluctuations are smaller, the Tukey-smoothed annual count of these more intense tornadoes has a shape similar to the one for the annual number of all reported tornadoes, regardless of intensity. However, the percentage of tornadoes that are F-2 or greater (Figure 4) has decreased with time. It has been proposed that the decrease in the annual number of tornadoes since 1973 was a result of the recognition of the microburst phenomena [*Fujita*, 1976] causing storms that had once been reported as weak tornadoes to be classified as straight-line severe thunderstorm winds. However, the con-

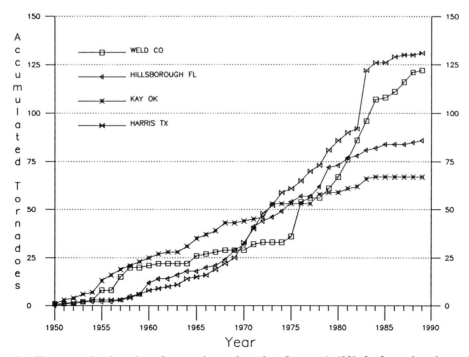

Fig. 3. The accumulated number of reported tornadoes since January 1, 1950, for four selected counties.

Fig. 5. Tornado frequency computed for four different but overlapping time periods: (a) 5 years (1960–1964), (b) 10 years (1960–1969), (c) 15 years (1955–1969), and (d) 20 years (1955–1974). Stippled areas represent four to six, eight to 10, and 12 to 14 tornadoes per 10,000 mi^2 (25,900 km^2) per year. Crosses mark the locations of relative maxima, and N marks the locations of relative minima.

tinued increase in the percentage of weak tornadoes (F-0 and F-1) through the 1980s counters this notion.

3. TIME DEPENDENCE OF SPATIAL DISTRIBUTION

To evaluate the geographic distribution of tornadoes, the annual frequency of tornado occurrence per unit area is needed. This parameter is computed by tabulating the number of tornado touchdowns in 2° latitude-longitude quadrangles and then normalizing by both area and time. Calculating the 2° "square" values at 1° latitude and longitude increments across the coterminous United States yields a light smoothing, equivalent to that achieved by passing a boxcar average over 1° data [Kelly et al., 1978].

However, as noted in the introduction, the period of the record, or "averaging period," plays a significant role in determining the apparent tornado climatology. Court [1970] published a compilation of 109 different maps and/or pseudomaps of tornado occurrence across the United States dating back to 1884. In his discussion of these charts he noted that the true distribution of tornadoes cannot be established since the various analyses were based on reports for differing periods. On the basis of the premise that "the most recent decade is a better estimator of the next one than the past half-century," Court [1970, p. 40] recommended that tornado frequency charts for 10-year periods be produced every 5 years. While this is a reasonable procedure for many meteorological variables, Figure 1 demonstrates that it is not valid for tornado data.

To illustrate the effect of the averaging period on apparent tornado frequency, first consider the chart produced from the tornado record from the 5-year period 1960–1964 (Figure 5a). This short record yields the rather classic picture of a "tornado alley" running north-northeastward from near Lubbock, Texas, through Oklahoma City, Oklahoma, and immediately west of Kansas City, Missouri, to the Mason City, Iowa, area. An area of enhanced tornado frequency (greater than six tornadoes per 10,000 mi^2 (25,900 km^2) per year) is also found in north central Indiana.

The averaging time is then increased by 5-year increments, still including the initial data. Adding the second 5 years, 1965–1969, expanded the data base to include the 1965 Palm Sunday tornado outbreak across Iowa, Wisconsin, Illinois, Michigan, and Indiana and the 118 tornadoes reported in south Texas between September 18 and September 23, 1967, while the circulation of Hurricane Beulah was affecting the area. (Estimates of the number of tornadoes

Fig. 6. Tornado frequency computed for four different 25-year periods: (a) 1950–1974, (b) 1955–1979, (c) 1960–1984, and (d) 1965–1989. Legend is as in Figure 5.

associated with Hurricane Beulah range from 47 [*Dye*, 1968] to 141 [*Novlan and Gray*, 1974]. Both of these events are reflected in the decadal frequency chart (Figure 5b).

The 15-year presentation created by including 1955 through 1959 in the analysis (Figure 5c) gives less prominence to Iowa and Minnesota. However, largely because of February outbreaks in both 1956 and 1959, the area from St. Louis, Missouri, across Illinois to north of Indianapolis, Indiana, now has an apparently higher frequency.

The addition of a fourth 5-year period still does not yield a stable pattern. The tornado distribution computed over the 20 years 1955–1974 (Figure 5d) is markedly different from the one obtained from the 15 years 1955–1969. Large portions of Mississippi, Alabama, and Georgia now have an apparent annual tornado frequency of greater than four tornadoes per 10,000 mi^2 (25,900 km^2). This arose because the early 1970s were extremely active in the southeast, with Georgia and Mississippi ranking in the top six states in tornado occurrences in both 1971 and 1972.

4. Consistency of Spatial Patterns

The 25-year tornado occurrence chart computed from 1950 through 1974 (Figure 6a) is quite similar to the 20-year one computed from 1955 through 1964. However, before one can argue that this pattern is the "real" tornado climatology for the coterminous United States, it is necessary to determine if the 25-year pattern is stable. Accordingly, four 25-year charts were constructed from the 40-year data set (Figure 6).

Marked interchart differences, over and above the population density increases in northeast Colorado and central Florida, appear. The 1950–1974 chart indicates a corridor of minimal activity running northward from Louisiana to northwest Iowa. In the 1955–1979 presentation this corridor becomes broken up and remains so in the later two charts. The zone of enhanced activity that extended northward from central Oklahoma through central Kansas in the 1955–1979 map is replaced by a local minimum in the 1965–1989 analysis.

Since the 25-year pattern is not stable, a longer data span is needed to define the tornado climate. Three presentations, based on a 30-year average time, were constructed (Figure 7). Significant changes between the charts are found in Kansas, Illinois, and Louisiana. Tornado alley, which is a definite feature in the 1950–1979 chart, does not appear in the 1960–1989 analysis. These charts illustrate that the usual climatological averaging period of 30 years is not long enough to yield a meaningful representation of the tornado occurrence pattern.

Fig. 7. Tornado frequency computed for three different 30-year periods: (a) 1950–1979, (b) 1955–1984, and (c) 1960–1989. Legend is as in Figure 5.

enough to yield a meaningful representation of the tornado occurrence pattern.

In contrast, the two 35-year climatologies (Figure 8) constructed from these data are quite consistent. Further, they are compatible with the 40-year one constructed using the entire data set (Figure 9). Obviously, part of the consistency can be attributed to the fact that all three of these presentations contain data from the same 30-year span (1955–1984). However, the first two 30-year displays (Fig-

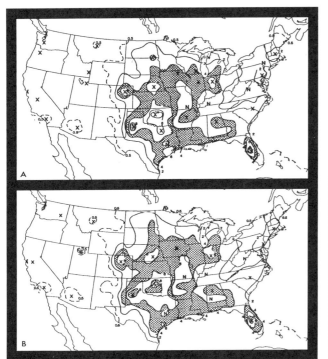

Fig. 8. Tornado frequency computed for two different 35-year periods: (a) 1950–1984 and (b) 1955–1989. Legend is as in Figure 5.

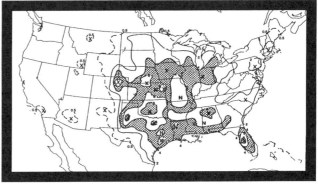

Fig. 9. Tornado frequency computed over 40 years, 1950–1989. Legend is as in Figure 5.

used to depict the tornado occurrence pattern across the contiguous United States.

5. U.S. Tornado Climatology

The salient feature of the tornado climatology of the United States is a general tornado-prone region between the Rockies and the Appalachians, generally south of 44°N, which contains embedded areas of enhanced activity from northern Texas through Oklahoma, and along a west-to-east corridor from northeastern Colorado through eastern Nebraska and central Iowa to north central Indiana. Areas of decreased activity in this area are zones from northeastern Missouri through central Arkansas, eastern Mississippi through central Alabama, and a small pocket in east central Texas.

The above procedures also were applied to the data set containing only strong and violent tornadoes (F-2 and greater). The geographic pattern again stabilized with the 35-year averaging time (Figure 10). Again, the 35-year charts are similar to the one constructed from the entire 40 years of the data set (Figure 11). The area most prone to strong and violent tornadoes is central Oklahoma, but small pockets of enhanced occurrence are also found in eastern Kansas, eastern Missouri, central Arkansas, southeastern Mississippi, northern Alabama, western Kentucky, north central Indiana, and north central Iowa.

However, since many of these areas are located around cities, these may not be real zones of enhanced strong and violent tornado activity, but rather reflections of population biases in the data and the fact that probability of obtaining a high rating in a damage-based tornado classification system increases with the number of damageable structures as well as with the strength of the storm. Indeed, the listing of individual "significant" tornadoes over the last century compiled by *Grazulis* [1990] indicates the many areas that historically have been susceptible to strong and violent tornadoes are not enhanced zones in the 40-year climatology. It is quite likely that the relative infrequency of strong and violent tornadoes (the NSSFC log only records 8271 of those storms in its 40 years of record) precludes determining a meaningful occurrence pattern of F-2 and greater tornadoes over the United States at the present time.

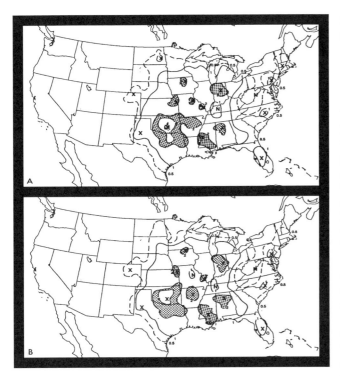

Fig. 10. Frequency of strong and violent tornadoes (F-2 and greater) computed over two 35-year periods: (*a*) 1950–1984 and (*b*) 1955–1989. Stippled areas represent two to three and four to five tornadoes per 10,000 mi^2 (25,900 km^2) per year. Crosses mark the locations of relative maxima, and N marks the locations of relative minima.

ures 7*a* and 7*b*) cover a common 25-year span (1955–1979) and do not show a similar consistency. It is hard to argue that having 86% of the time span in common (30 of 35 years) is significantly different than having 83% (25 of 30 years) in common. Thus it can be stated heuristically that a 35-year averaging time is necessary to obtain statistics that can be

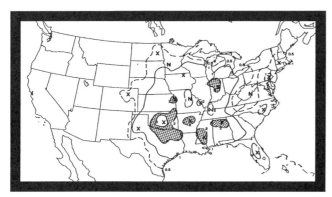

Fig. 11. Frequency of strong and violent tornadoes (F-2 and greater) computed over 40 years, 1950–1989. Legend is as in Figure 10.

Acknowledgments. We owe special gratitude to Leo Grenier and John Halmstad, who maintain the NSSFC tornado data base and who provided us with much detailed information about particular items discussed in this paper. The graphics were generated by Michael Manker, and the words were expertly edited and processed by Beverly Lambert. Joe Galway provided a very insightful review which markedly clarified the presentation.

References

Beran, D. W., and C. G. Little, Prototype Regional Observing and Forecast Service: From concept to implementation, *Natl. Weather Dig.*, *4*, 2–5, 1979.

Court, A., Climatic normals as predictors, 5, Conclusions, *Rep.*

AFCRL 69-0003, 74 pp., Air Force Cambridge Res. Lab., Bedford, Mass., 1968.

Court, A., Tornado incidence maps, *Tech. Memo. ERLTM-NSSL 49*, 76 pp., Natl. Severe Storms Lab., Environ. Sci. Serv. Admin., Norman, Okla., 1970.

Court, A., and J. F. Griffiths, Thunderstorm climatology, in *Thunderstorms: A Social, Scientific, and Technological Documentary*, vol. 2, *Thunderstorm Morphology and Dynamics*, 2nd ed., edited by E. Kessler, pp. 9–40, University of Oklahoma Press, Norman, Okla., 1985.

Dye, L. W., Tornado summary for 1967, *Weatherwise, 21*, 22–25, 1968.

Flora, S. D., *Tornadoes of the United States*, revised ed., 206 pp., University of Oklahoma Press, Norman, Okla., 1954.

Fujita, T. T., Spearhead echo and downburst near the approach end of a John F. Kennedy Airport runway, New York City, *Res. Pap. 137*, 51 pp., Satell. and Mesometeorol. Res. Proj., Univ. of Chicago, Chicago, Ill., 1976.

Fujita, T. T., and A. D. Pearson, Results of FPP classification of 1971 and 1972 tornadoes, in *Preprints, Eighth Conference on Severe Local Storms*, pp. 142–145, American Meteorological Society, Boston, Mass., 1973.

Galway, J. G., Early severe thunderstorm research by the U.S. Weather Bureau, *Weather Forecasting*, in press, 1993.

Grazulis, T. P., *Significant Tornadoes 1880–1989*, vol. II, *A Chronology of Events*, 368 pp., Environmental Films, St. Johnsbury, Vt., 1990.

Guttman, N. B., Statistical descriptors of climate, *Bull. Am. Meteorol. Soc., 70*, 602–607, 1989.

Jagannathan, P., R. Arley, H. ten Kate, and M. V. Savarina, A note on climatological normals, *Tech. Note 84*, World Meteorol. Organ., Geneva, Switzerland, 1967. (Also available as *Rep. WMO-No. 208.TP.198*.)

Kelly, D. L., J. T. Schaefer, R. P. McNulty, C. A. Doswell III, and R. F. Abbey, Jr., An augmented tornado climatology, *Mon. Weather Rev., 106*, 1172–1183, 1978.

Linehan, U. J., Tornado deaths in the United States, *Tech. Pap. 30*, 48 pp., Natl. Weather Serv., Silver Spring, Md., 1957.

McNulty, R. P., D. L. Kelly, and J. T. Schaefer, Frequency of tornado occurrence, in *Preprints, 11th Conference on Severe Local Storms*, pp. 222–226, American Meteorological Society, Boston, Mass., 1979.

Novlan, D. J., and W. M. Gray, Hurricane-spawned tornadoes, *Mon. Weather Rev., 102*, 476–488, 1974.

Schaefer, J. T., D. L. Kelly, and R. F. Abbey, Tornado track characteristics and hazard probabilities, in *Wind Engineering*, edited by J. E. Cermak, pp. 95–110, Pergammon, New York, 1980.

Schaefer, J. T., D. L. Kelly, and R. F. Abbey, Development and application of a minimum assumption tornado hazard probability model, *J. Clim. Appl. Meteorol., 25*, 1934–1945, 1986.

Snider, C. R., Michigan tornadoes 1834–1975, internal report, 37 pp., Natl. Weather Serv. Office, Ann Arbor, Mich., 1975.

Tukey, J. W., *Exploratory Data Analysis*, 535 pp., Addison-Wesley, Reading, Mass., 1977.

A 110-Year Perspective of Significant Tornadoes

THOMAS P. GRAZULIS

St. Johnsbury, Vermont 05819

1. EVOLUTION OF THE PROJECT

A tornado documentation project began in 1980 as an effort to resolve differences between the two independently designed tornado data bases used by the Nuclear Regulatory Commission (NRC). One was established for the NRC at the University of Chicago (UC) by T. T. Fujita for the years 1916–1985. In addition, the NRC assisted the National Severe Storms Forecast Center (NSSFC) in adding F scale ratings to the existing NSSFC data base in Kansas City, from 1950 to the present. The final task of this effort (hereinafter called the project) was to locate and list all significant tornadoes from 1880 through 1989. Significant is defined here as all tornadoes doing confirmable F-2 or greater damage or causing a death.

The inherent subjectivity of the Fujita scale gave rise to thousands of F scale rating differences, as the two data bases were compiled independently by teams of students. For the period 1950–1970 (the years before the introduction and use of the Fujita scale by *Fujita* [1971]), nearly half of all tornadoes were assigned a different rating. In a few drastic cases the ratings differed by as much as four F scale rating points, as UC graduate students worked from *Storm Data* (in Chicago) and undergraduate NSSFC students worked from both *Storm Data* and newspaper microfilm (in their home states during the summer of 1975). (*Storm Data* is published by the U.S. Department of Commerce, National Climatic Data Center, in Asheville, North Carolina.)

The discrepancies were found originally by *McDonald and Abbey* [1979]. Research by this author, aided by computer-assisted comparisons at Texas Tech, revealed the extent and nature of the differences. For instance, both data sets initially had 40 violent (F-4–F-5) tornadoes in Oklahoma, but only seven were in common. There were about 2000 events that differed by two or more F scale rating points.

The NRC phase of the project was completed under the guidance of R. Hadlock at Battelle, Pacific Northwest Laboratories, and resulted in the publication of *Grazulis* [1984]. The two data sets were made compatible for violent events (F-4–F-5) and brought to within one F scale rating point for all others. A best possible F-4–F-5 data set was crucial to NRC assessment of 10^{-7} maximum wind speed probabilities. *Schaefer et al.* [1986] noted that 4% (about 1300 tornadoes) of the NSSFC data base had a rating change. These were based on the recommendations of the author, in this first phase of the project. A similar number of changes were made at UC. No other changes were made after 1983.

There remained, however, about 5000 events differing by one F scale rating, many of them across the important dividing point between weak (F-0–F-1) and strong (F-2–F-3) events. These differences, and all F-2–F-5 events since 1880, were explored in the second phase of the project, funded by the National Science Foundation (NSF). It culminated in the work of *Grazulis* [1991], which describes the 12,209 tornadoes found to be "significant" from 1880 to 1989.

The entire effort required about 500 days of full-time on-the-road library work, spread over nine years. The search for information required the use of about 35,000 reels of newspaper microfilm. Most of this microfilm is archived in central state historical libraries, except in Texas, New York, and Massachusetts, where visits to local libraries and newspaper offices were required.

2. PRE-1915

The original purpose for attempting to expand the useful tornado data base from 1916 back to 1880 was to provide the NRC with a long violent tornado data set. Since there are only 10–20 F-4/F-5 tornadoes per year, it was assumed that the relatively few documented tornadoes for that period would include the violent events. In addition, it was thought that virtually all killer tornadoes could be located and that such a data set would provide at least some insight into long-term trends. The project documented 2432 significant tornadoes (including 1060 killer tornadoes) for the years 1880–1915. These paths are plotted in Figure 1. Tornado and severe weather occurrence dates were found in the *Monthly Weather Review*, *Monthly Climate and Crop Report* (the

Fig. 1. A subjective comparison of 1880–1915 tornado paths with 1916–1985 risk from *Grazulis* [1991]. The tornado paths are for 2432 significant tornadoes, 1880–1915. Isopleths are drawn for a 120-mph (53.7 m s^{-1}) maximum wind speed, at 10^{-4} probability, as derived from *Fujita* [1987]. Areas with relatively low risk for 1916–1985 and high activity for 1880–1915 include the upper Missouri River Valley from Omaha to southwest Minnesota and eastern South Dakota, the Ozark Plateau of southeast Missouri and northwest Arkansas, two relatively small areas of central Kansas and along the Kansas-Nebraska border, and northwest Illinois.

precursor of *Climatological Data*), the *Annual Report of the Chief of the U.S. Weather Bureau*, 19th-century reports such as those of *Finley* [1884, 1889], the author's files, and much personal correspondence. For every known or suspected tornado in this period, an attempt was made to seek out local newspaper descriptions. No previously unlisted tornado was added unless a sufficiently detailed funnel description could be located in a local newspaper.

Attempting to use this rather sparse data set for quantitative estimates should be done with great caution. However, some useful qualitative observations can be made. Figure 1 also contains the outline of the 10^{-4} maximum wind speed probability for 140 mi hr^{-1} (mph) (62.6 m s^{-1}) from *Fujita* [1987]. This isopleth is the midpoint of the nominal F-2 windspeed range and encloses the major tornado risk areas for the period 1916–1985. A rough comparison of the 1916–1985 risk can now be made with the distribution of the worst pre-1916 activity. The purpose of this comparison is to locate areas of high pre-1916 activity that were not active in more recent periods. Figure 1 identifies five such areas. This adds a cautionary note to the use of current risk analysis maps. The long-term risk from strong and violent tornadoes for some areas of the United States may be underestimated by *Fujita* [1987], as well as by other recent risk analysis models.

This pre-1916 period has, unfortunately, additional variables and uncertainties that may render the data little more than a curiosity. For instance, the greater tornado concentration in Georgia may be influenced by the distribution of the post-Reconstruction Army Signal Corps in Georgia. This 1880–1915 period was largely before the development of poured concrete foundations, thus adding more uncertainty to Fujita scale F-3 or greater rating standards. In addition, preservation of local newspapers in the south, especially Mississippi, was very poor until about 1910. The role of racism in the documentation of deaths also may be a more important variable than was originally assumed.

3. 1916–1949

For this data period the UC data base, derived according to *Fujita* [1987], was the starting place. A local newspaper description of every event was sought out. The high success rate in locating these descriptions (over 90%) may be due to the possibility that officially listed events were those tornadoes which were widely reported in newspapers. As with 1880–1915, tedious searching and the inadvertent location of new events (while searching for information about known tornadoes) expanded the data base for this period. *Fujita* [1987] lists 3465 F-2 or greater tornadoes for this period. *Grazulis* [1991] lists 4028 such events for the 1916–1949 period. About 200 of those tornadoes are previously unlisted events. Over 2000 previously unlisted tornadoes were located, the majority of which were F-0/F-1 events.

4. 1950–1989

The project's philosophy in reviewing official NSSFC ratings for this period was somewhat different for the 1950–1970 period than for 1971–1989. Virtually all official F scale ratings prior to 1970 (pre-Fujita scale) were based on old newspapers and *Storm Data* descriptions. The ratings for that period were made without benefit of on-site surveys or published studies, such as that of *Minor et al.* [1977], which brought engineering concerns the forefront.

No official NSSFC ratings were accepted outright for the 1950–1970 period. The original student rating sheets were obtained from NSSFC and studied. These contained some notes from newspaper searches. More credence was placed in the official rating from 1971 to 1976, but there was still the obvious lack of concern for engineering principals. Local newspaper descriptions were sought out for about 7000 events from 1950 to 1976.

For the years 1971 to 1976 the philosophy of National Weather Service (NWS) personnel, or the state climatologists, apparently was to default to F-2 on borderline F-1–F-2 damage. This was the case for tornadoes with odd types of damage that were not described in the official Fujita scale descriptions. Most of the largely overrated 1974 events (of which the project lowered 113 ratings) stand in bold contrast to the tornadoes of April 3–4, 1974, where Fujita's personal ratings were quite the opposite. His F scale ratings for this so-called superoutbreak are among the most conservative in the entire data base. (Both data sets used Fujita's personal ratings for that outbreak.) Many of his F-1 ratings for that day would probably be given an F-2 in most states today. It is Fujita's standards for April 3, 1974, the insights from *Minor et al.* [1977], and discussions with other meteorologists and engineers that guided the creation of the project's strict standards.

In the 1976–1989 period, engineering considerations slowly filtered into the F scale rating process, and a wholesale review of all ratings was not necessary. Problems did arise here, as they will in the future, with the difficult-to-rate, short-path, borderline F-1/F-2 events that hit nonengineered structures.

5. Opinions

The following statements are opinions, developed during the NRC work and the author's previous historical research, and were present at the start of the NSF phase of the project. They might be considered by the reader as biases affecting the final results of the project.

1. Large numbers of F-2-ranked tornadoes in the NSSFC and UC data bases (especially for the years 1950–1976) are overrated. Hundreds of tornadoes, having only partially unroofed homes, collapsed old barns, and overturned mobile homes were incorrectly assigned an F-2 rating.

2. Some F scale ratings after 1976 have state-to-state inconsistencies and some lack of appreciation for engineering principles.

3. The entire research and rating effort should be done by a single person, to maximize the concept of consistency. This applies not only to the project but also to current documentation at NSSFC.

4. The year 1880 is the best starting point for the documentation, as it was *Finley's* [1884] first year of documenting over 100 tornadoes.

5. The turn-of-the-century distribution pattern of farms provided a better tornado intensity detection grid than farms do today. In 1900, there were tens of thousands of 160-acre (0.6 km^2) homesteads, each with a barn and house. Today, most have been replaced by 2000-acre (8.1 km^2) agribusiness or expanded family farms, sometimes with no house and an absentee operator. Thus the distribution of tornadoes during the first half of the century may show less population bias than that during the later half.

6. Urbanization tends to increase both the number and intensity of significant tornadoes and the number of reported weak tornadoes. Short-path significant tornadoes that would hit nothing in rural areas can hit buildings in urbanized areas and be given a higher F scale rating. Urbanization increases the number of buildings hit by longer-track significant tornadoes. The potential for longer-track tornadoes to reveal their F-3 or greater winds increases as more targets are placed in their path. Weak or poorly formed shear zone tornadoes that might have gone undocumented in rural areas will produce urban damage and be recorded.

7. The farther back in time the work extended, the greater the likelihood of F scale rating errors. This is due to great differences in building practices and construction methods in times past. A stress of consistency will result, however, in data that is useful in making qualitative judgments of risks and trends.

6. Fujita Scale Standards

It was apparent, very early in the project, that the Fujita scale standards are inadequate for creating a consistent data set. The current standards, as listed in each issue of *Storm Data*, focus on house damage, but with less than adequate detail. There are no specifics on how to rate different types of house construction. There is no mention of what percentage of a roof needed to be "removed" for an F-2 rating (not a trivial matter when newspapers may refer to "unroofed homes" that only lost shingles). However, less than half of all F-2 tornadoes are rated as such on the basis of house damage. There is no standard that clarifies just what degree of damage constitutes a "destroyed" barn or a "destroyed" mobile home, so an expanded set of rules of thumb have been developed to enhance consistency.

Rather than try to expand the list of the F-2 damage descriptions, it is useful to expand and clarify the types of damage that are not considered to be F-2 damage and must be assigned an F-1 rating. The project became largely an effort to eliminate the F-0/F-1 events. A list of these guidelines is presented below.

The destruction of the following items is considered F-1, unless an engineering site survey is completed: (1) boat docks, marinas, and boat houses; (2) sheds, even if called "barns"; (3) barns with two or more walls remaining intact; (4) mobile homes which still have a recognizable shape; (5) single or isolated trees; (6) chimneys; (7) sheet metal commercial buildings; (8) homes or other buildings swept away while in highway transit; (9) attached garages and carports; (10) small chicken or brooder houses; (11) private single-engine airplanes; (12) athletic field grandstands; (13) telephone poles; (14) Quonset-type storage buildings; (15) drive-in movie screens; (16) windmills and oil derricks; (17) homes destroyed by falling trees; and (18) conveyor belts or irrigation systems.

The following conditions are also considered as F-1: (1) homes with over 20% of roof remaining; (2) homes "cleanly" unroofed (as if not nailed down); (3) cattle lifted or rolled along the ground; (4) boats carried into trees; (5) objects driven into wood; (6) collapsed barns; (7) blown-down unreinforced concrete block walls; (8) unroofed tourist cabins; (9) a house rotated on, or shifted off a foundation, but not unroofed; (10) unroofed fast-food restaurants; and (11) barns blown over but not swept away.

When this list was applied to NSSFC data (1950–1989) some 2567 (out of 8273) F-2–F-5 ratings were found to belong in the F-1 category, a larger-than-expected number.

7. NSSFC-Project Differences

Figure 2 graphs the comparison of NSSFC data and project data for F-2–F-5 tornadoes for 1950–1989. Figure 3 displays the differences in total events for each year. The NSSFC curve (top) in Figure 2 shows a steady, rather dramatic decline in F-2–F-5 tornadoes beginning in 1974, after steadily high numbers since 1953. The project data (bottom) shows an irregular series of peaks and valleys from 1950 through 1984. It is concluded here that the NSSFC decline is a false one, having nothing whatsoever to do with climate or meteorology. It is the product of early-year inexperience, lack of engineering considerations in the F scale rating of tornadoes, and default philosophy in odd and borderline situations. The short-term decline in both data sets, during the late 1980s, is apparently valid. (That decline apparently was reversed in 1990.)

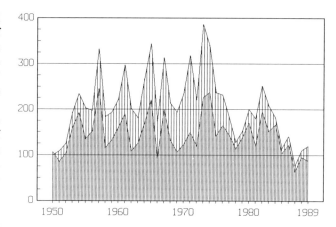

Fig. 2. A comparison of the number of NSSFC F-2–F-5 tornadoes (top hatched area) with project F-2–F-5 tornadoes (bottom hatched area), 1950–1989.

8. Killer Tornadoes and Deaths

Figures 4, 5, and 6 show that killer tornadoes in most categories peaked in the 1920s. Figure 7 shows a gradual decline in the height of the peak killer tornado years since 1909. This trend was interrupted by 1974, which was heavily influenced by the remarkable outbreak on April 3 of that year. Figures 8 and 9 show the trend in total deaths to be similar to that in killer tornadoes. The last four decades cover the period of NWS watches and warnings. In Figure 4 the trend for all killer tornadoes shows no decline in these last four columns. Figures 5 and 6, for tornadoes killing at least 10 and 50 people respectively, show a definite decline. In Figure 6, both single events, during 1960–1979, were in Mississippi, and both (February 21, 1971: 58 killed, 160 path miles (257 km); March 3, 1966: 57 killed, 70 path miles (113 km)) were probably tornado families with fewer than 50 deaths for any single member. The combination of NSSFC watches and warnings (since the early 1950s) and the dis-

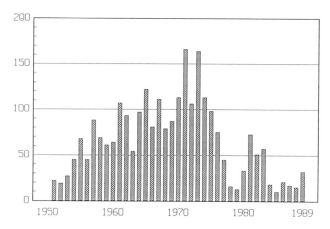

Fig. 3. The annual number of differences in F-2–F-5 tornadoes between NSSFC and project records, 1950–1989.

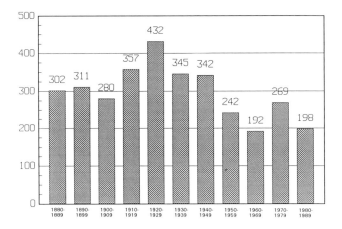

Fig. 4. Tornadoes with one or more deaths by decade.

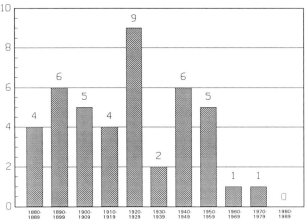

Fig. 6. Tornadoes with 50 or more deaths by decade.

semination of information through schools and the media apparently have done a remarkable job of reducing the number of multiple-death killer tornadoes, despite an increasing population. The totals deaths by decade (Figure 8) is rather stable, except for the peak in the 1920s, from 1880 to 1949. Maintaining the current downward trend through another decade may be a difficult challenge for the NSSFC. Figure 9 shows the annual death tolls. The total has exceeded 300 deaths in a single year 16 times since 1880, but only once since the NSSFC was established in Kansas City in 1954. The 1980s were the first decade in over a century to have no single year with 150 or more deaths.

The decline in multiple-death killer tornadoes since the 1920s may reflect both a decline in rural population following the Great Depression and an increase in awareness levels brought about by the spread of radio, television, broader public education, spotter groups, and finally the watch/warning system. One can also speculate that the decline in Figure 8 begins in 1940s and coincides with the formation of organized spotter groups after World War II. There is a chance, of course, that the violence of tornadoes themselves peaked in the 1920s and that an undetected decline in intensity has resulted in a decline in all death categories. Another possible meteorological reason is a shift of the most violent activity away from vulnerable populated areas. While neither of these reasons seem likely, there is a shred of evidence that the intensity of major outbreaks peaked in the 1916–1949 period. The project uncovered 33 tornado outbreaks during 1880–1989, with 20 or more significant tornadoes each. The count for each period is as follows: 1880–1915, six outbreaks; 1916–1949, 16 outbreaks; and 1950–1989, 11 outbreaks. The 1916–1949 period included seven of the 10 worst outbreaks. This 1916–1949 peak may be a result of the author's misjudgment of building construction (with a corresponding increase in F scale ratings) and/or a more favorable distribution of homes and barns in rural areas. This is a highly subjective and speculative field.

As noted by *Doswell and Burgess* [1988], the problem of proper analysis of tornado families is unresolved. If all of the pre-1950 long-track tornadoes listed by the project as single

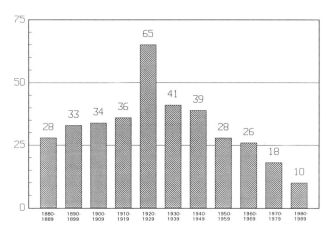

Fig. 5. Tornadoes with 10 or more deaths by decade.

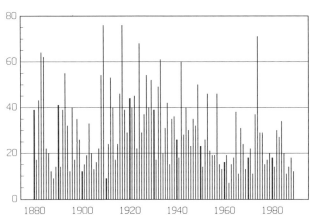

Fig. 7. Annual killer tornadoes, 1880–1989. The start of each decade is marked by a heavy line.

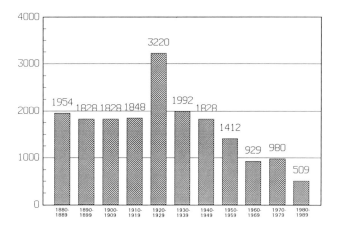

Fig. 8. Total tornado deaths by decade.

events were broken properly into families, the number of killer tornadoes in the three listed categories (Figures 4, 5, and 6) would probably increase or stay the same. This would make the decline in the past 80 years even more pronounced. Most killer tornadoes produced deaths in small concentrated areas, such as residential areas or clusters or rural homes. For example, the April 24, 1908, tornado (with 143 deaths in Louisiana and Mississippi) actually may be a family of four killer tornadoes, with 24, 64, 33, and 22 deaths each. A similar breakdown of post-1950 events would have little impact, except for the two events in Mississippi listed above. Some of the increase in the number of significant tornadoes, as shown in Figures 10 and 11, may be due to better analysis of tornado families.

9. A Tale of Two Cities

If long-term, small-scale maxima and minima exist, they are effectively masked by an endless variety of population variations and human activities. However, there are a few areas that combine tornado minima and population maxima

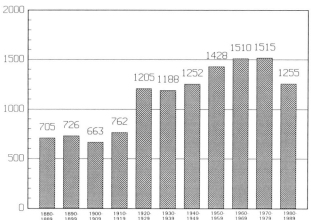

Fig. 10. Significant tornadoes by decade.

(or vice versa) that are interesting. The most intriguing tornado minimum is that over Fort Worth, Tarrant County, Texas (and possibly some counties to the west), when compared to neighboring Dallas County. The count of all NSSFC tornadoes does not indicate a tornado minimum at Fort Worth. Tarrant County (home of the National Weather Service Southern Region Office) leads Dallas County in all tornadoes, 55–53. Prior to 1976, tornadoes were overrated in Tarrant County, distorting the NSSFC statistics for F-2–F-5 tornadoes. However, project statistics show Dallas County with a 23 to seven lead in significant tornadoes. For killer tornadoes, Dallas leads eight to one and is ahead in deaths 34 to one. Counties to the north and northeast, with a total population equal to Tarrant County, outnumber it about 50 to one in killer tornadoes. The city of Fort Worth, sitting in what is often called "tornado alley," has had a sizable population for over a century without a killer tornado. As seen in Figure 12, the one death was in a rural area southeast of the city. Since 1976, when NSSFC and project data are in

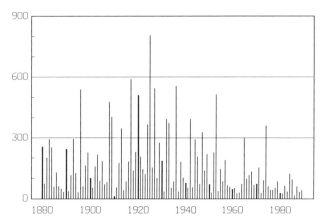

Fig. 9. Annual tornado deaths, 1880–1989. The start of each decade is marked by a heavy line.

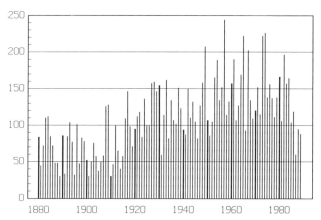

Fig. 11. Annual significant tornadoes, 1880–1989, project count. The start of each decade is marked by a heavy line.

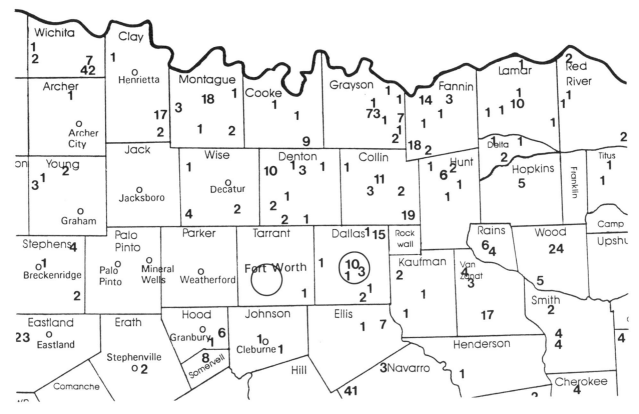

Fig. 12. A map of the deaths from each killer tornado in north-central Texas, 1880–1989. Some larger towns, west of Fort Worth, have their names listed.

agreement, the significant total is Dallas six, Tarrant zero. There is something peculiar here. Either it is a classic example of blind chance sparing a town, or there is a permanent (semipermanent?) minimum for strong and violent tornadoes west of Dallas. No speculation is made concerning the meteorology behind this difference, if indeed there is an identifiable reason. With an atmosphere in chaotic, turbulent motion, something such as the principle of strange attractors [*Gleick*, 1987] may lead to long-lasting maxima and minima without a directly identifiable cause.

10. RECOMMENDATIONS

The NSSFC should develop a comprehensive set of guidelines for the use of the Fujita scale. This, fortunately, seems to be underway. Preliminary work by *Bunting and Marshall* [1991] on such a guide uses some of the basic philosophy outlined in section 6. A similar list was supplied to them from our unpublished report to NSF in 1987. Of equal importance is that the NSSFC decide whether it has a damage-based data base or an estimated wind speed based data base. A clear statement of policy, backed by insistence on consistent procedures, should be made by NSSFC. Some states assign F-2 ratings to nondamaging (or F-1 damaging) tornadoes, on the basis of spotter/chaser intuitive estimates of rotational speed. For instance, Colorado routinely applies F-2 ratings to tornadoes that did no damage, if they were visually impressive. Kansas does not, despite having as many opportunities as Colorado to do so. This makes it difficult to compare the two states and distorts risk analysis maps. Perhaps two F scale ratings should be archived, one for damage and one for estimated wind speed, however cumbersome.

This matter and others were raised by *Doswell and Burgess* [1988] in their examination of several unresolved issues in tornado climatology. Some of the concerns discussed in that paper were major obstacles to the completion of the project. One of the conclusions reached by this author was at odds with a conclusion reached by Doswell and Burgess. They suggest that there may be an advantage to having many people involved in the ratings effort, assuming that errors and biases would be more likely to cancel out for the aggregate data base. This would produce national averages that are less prone to individual bias. However, national averages may be little more than a curiosity and of no practical use in regional risk assessment. The author maintains that biased or not, consistency is the only means by which progress is possible. Thus this author also recommends that all ratings be reviewed at NSSFC by a single individual with the experience and authority to override any ratings made in an NWS forecast office.

11. A Final Note

The large number of significant NSSFC tornadoes found by this project to be less than significant does, at first glance, seem excessive. Since the UC data tape has only 300 fewer F-2–F-5 tornadoes than that of the NSSFC (1950–1979), this must mean that the UC data are almost equally overrated. That is, indeed, the conclusion reached by this author. From 1950–1970, *Storm Data* used words like "destroyed" and "unroofed" without regard to their eventual use in quantitative judgments. The Fujita-trained students at UC had no choice but to take *Storm Data* at its word. Both data bases have interesting and convoluted histories. The author used every means practical to maintain consistency. While hundreds of the NSSFC ratings are obviously overrated, it would be difficult to make a powerful objective defense for most of the 2567 rating changes. It would also be equally hard to make a strong case for restoring any particular event to its original F-2 rating. There is a huge "gray area" here. The objective from the beginning was not the unachievable goal of "correctness." It was the more achievable goal of "consistency," which would reveal trends and uncover possibilities for future research.

References

Bunting, W., and T. Marshall, A resource guide for conducting damage surveys, paper presented at Warning Coordination Meteorologist Short Course, Natl. Oceanic and Atmos. Admin./Natl. Weather Serv., Norman, Okla., Feb. 26–28, 1991.

Doswell, C. A., III, and D. W. Burgess, Some issues of United States tornado climatology, *Mon. Weather Rev.*, *116*, 495–501, 1988.

Finley, J. P., Report on the character of six hundred tornadoes, *Prof. Pap. VII*, Signal Serv., Washington, D. C., 1884.

Finley, J. P., State tornado charts, *Am. Meteorol. J.*, *5*, 446–476, 501–507, 545–551, 1889.

Fujita, T. T., Proposed characterization of tornadoes and hurricanes by area and intensity, *Res. Pap. 91*, Satell. Mesometeorol. Res. Proj., Univ. of Chicago, Chicago, Ill., 1971.

Fujita, T. T., U.S. Tornadoes, part 1, 70 Year Statistics, report, 122 pp., Satell. and Mesometeorol. Res. Proj., Univ. of Chicago, Chicago, Ill., 1987.

Gleick, J., *Chaos, Making a New Science*, 354 pp., Penguin, Middlesex, England, 1987.

Grazulis, T. P., Violent tornado climatography, 1880–1982, *Rep. NUREG/CR-3670*, 165 pp., U.S. Nucl. Regul. Comm., Washington, D. C., 1984.

Grazulis, T. P., *Significant Tornadoes, 1880–1989*, vol. I and II, 970 pp., Tornado Project, St. Johnsbury, Vt., 1991.

McDonald, J. R., and R. F. Abbey, Jr., Comparison of the NSSFC and DAPPL tornado data tapes, in *Preprints, Eleventh Conference on Severe Local Storms*, American Meteorological Society, Boston, Mass., 1979.

Minor, J. E., J. R. McDonald, and K. C. Mehta, The tornado: an engineering-oriented perspective, *Tech. Memo. ERL NSSL-82*, 192 pp., Natl. Oceanic and Atmos. Admin., Boulder, Colo., 1977.

Schaefer, J. T., D. L. Kelly, and R. F. Abbey, A minimum assumption tornado hazard probability model, *J. Clim. Appl. Meteorol.*, *25*, 1934–1945, 1986.

Discussion

K. CRAWFORD, SESSION CHAIR

Oklahoma Climatological Survey

PAPER H1

Presenter, T. P. Grazulis, private consultant [*Grazulis et al.*, this volume, advances in tornado climatology, hazards, and risk assessment since Tornado Symposium II]

(C. Anderson, North Carolina State University) At one time, I lived in Wisconsin and used to consider tornado climatology there. I noticed some very strange reporting, also. There were a number of counties along the Lake Michigan shore that showed no reported tornadoes from 1915 to 1950. Did you find something like this as well?

(Grazulis) That's a curious thing, because if you look at risk analysis in Wisconsin, there is a clear minimum along the eastern side of the state, except for the last 15 years. There are major shifts in the climatology, filling up these apparent minima; there are other examples I can't think of specifically at the moment. I don't think it's reporting; it appears that there really were no tornadoes in that area of Wisconsin for that time. We have to be cautious with short data bases. Given a map of tornado occurrence, is that a prediction? I have yet to see a map that I feel comfortable with as a prediction for the next 20 years.

(L. Twisdale, Applied Research Associates) I'd like to make one comment on saying that risk assessments are 10-million-year projections. That's an incorrect interpretation of probability of exceedance; what you're trying to do is estimate the probability of wind speed exceedance over the design life of a plant, which might be anywhere from 1 to 30 years. It's like throwing darts at a board; you're considering a population of rare events over an area and trying to estimate the chances of a certain target being hit. There's a lot of randomness and spatial variability in that. The projection is not whether or not a tornado is going to occur in the next 10 million years, it's the chances of a structure being hit.

The Tornado: Its Structure, Dynamics, Prediction, and Hazards.
Geophysical Monograph 79
This paper is not subject to U.S. copyright. Published in 1993 by the American Geophysical Union.

(Grazulis) Yes, I understand that, and everyone should appreciate that distinction. There really is no effort to project 10 million years. However, it effectively presents the absurdity of the problem better. There isn't time for me to try to explain it. That wind speed probability may be off in some places by as much as 80–100 mph from what I believe to be the true meteorological wind speed, because it's being handled purely statistically.

(Twisdale) We recognize that the wind speed estimates are not inherently correct, and that's why we put adjustments on those probabilities.

(Grazulis) You need enormous analysis and caution on this. The details of how to estimate the true wind speed are really out of my league. I simply look at distributions and what has happened.

(C. Doswell, National Severe Storms Laboratory) It's fascinating to work with this data base, and I think you and I both agree that it's a mess. You've talked about Ted Fujita's monumental effort to account for a variety of factors in adjusting the data, and offered some interesting suggestions of your own. Do you have any plans to create an adjusted data base that's consistent with your view of the true hazard distribution?

(Grazulis) If three or four very small adjustments could be made to the Kansas City data base, you could begin to create a data base that would really work. The problem is that Allan Pearson created that data base without realizing what the problems would be 20 and 30 years later. It's a fossil of the early ideas. There are several things I'd change. Number one, I think there need to be two F-scale estimates; one for the actual damage done, and the other is the subjective adjustment for the intensity. The second adjustment: did it hit a building? Another: was it or was it not spawned by a supercell? We've discussed the famous Sunray, Texas, event and it was given an F-4 rating even though it never hit anything. It's very hard to compare Colorado and Kansas statistics, because in Colorado, the rating is very subjective, while in Kansas, the rating is strictly based on damage. The data base needs adjustment.

(Allan Pearson, private consultant) Having gone through the North Dakota data base, if it weren't for outdoor privies, no one would know what events occurred there. [laughter] There is absolutely nothing in those publications; it's the best we could do. I've told you this before, but I don't think there's any way to go back and adjust those past events. How people use the data is the problem.

(Grazulis) I'm not saying there's any way to go back and fix it up. The very sophisticated climatological data adjustment models built by Ted [Fujita] and Joe [Schaefer] and Jim [MacDonald] are like building Rolls-Royces to carry manure by the teaspoonful; my work has been confined to composting [lauughter]. You just can't go back and fix it; you can make it more useful, but you can't improve it very much and you've got to be very careful how you use the data.

Paper H2

Presenter, J. Dessens, Université Paul Sabatier [*Desseus and Snow*, this volume, A comparative description of tornadoes in France and the United States]

(T. Grazulis, private consultant) Did you only plot the F1 through F5s?

(Dessens) Yes, only F2 through F5s.

(Grazulis) But there were some F0s and F1s?

(Dessens) Yes, certainly, we have many more F0s. We have perhaps 100 per year, but F0s cause little or no damage.

(Grazulis) The tornado distribution in France appears to be similar to New York and Pennsylvania.

(Dessens) Yes, I agree.

(R. Petersen, Texas Tech University) Are there any attempts in France or adjacent countries to warn people for these tornadoes, or do they come as a complete surprise to people?

(Dessens) No attempt is made to make tornado warnings in France. For hail, yes, but not for tornadoes. After a tornado the Meteorological Offices study the situation, but only afterward.

Paper H3

Presenter, Z. Xu, Chinese Academy of Meteorological Sciences [*Xu et al.*, this volume, Tornadoes of China]

(R. Petersen, Texas Tech University) China has many typhoons. It there any evidence that any of these tornadoes have occurred with typhoons coming on land?

(Xu) Yes, often tornadoes occur in front of typhoons. This is discussed in my paper.

(Petersen) Are most of the tornadoes you showed associated with typhoons, or just many of them?

(Xu) From July to September, typhoons often occur and tornadoes often are related to those typhoons. However, tornadoes also occur with convergence lines, squall lines, and in the warm sector of extratropical cyclones.

(D. Burgess, National Severe Storms Laboratory) The radar echoes you showed seemed to be a supercell thunderstorm. Was there large hail associated with that storm?

(Xu) When a tornado occurs, the associated hailstorms often are very strong. More often, the storm is a hailstorm without a tornado. But if a tornado occurs, usually the hail event is significant.

(A. Court, California State University, Northridge) In this country and in western Europe, the tornado has been understood to be a rotating storm for only about 150 years. How old is the understanding of the tornado in China? Is that knowledge relatively old, or has it been effectively introduced by Western meteorologists as a separate kind of storm?

(Xu) In our ancient times, we have recorded the events for a long time, but only reported the situation for damage, and not very accurately.

(Court) It was recognized that it turned?

(Xu) Yes, it was known to be rotating.

Paper H4

Presenter, M. Leduc, (M. J. Newark, Tornado incidence in Canadian cities, not in this volume)

(T. Grazulis, private consultant) It's so satisfying to have someone agree with me. [laughter] Did you do any breakdown or distributions by F scales?

(Leduc) I know there is, but I don't have the figures.

(Grazulis) Well, Edmonton comes to mind. Are there any other examples out there?

(Leduc) For example, in Ontario, we've had six F3 or greater outbreaks since 1979. In southern Ontario, we've been talking about one F3 or greater event every 4 years, but the statistics the last 10 years now indicate it might be more like every 1 or 2 years. Certainly, the numbers are going up as we learn more about it.

(J. Golden, National Oceanic and Atmospheric Administration, Office of the Chief Scientist) There's a host of intriguing things in your talk. Would you say a little more about that frequency maximum map. I gather that the Alberta maximum is fairly new, but what about that "intermountain" maximum? I know that area is pretty mountainous.

(Leduc) I'm afraid I can't help you very much on that one. This isn't my paper, after all.

(Golden) Governments tend to react to big events, sometimes overreacting. I understand your government is contemplating the deployment of Doppler radars to protect your biggest cities. In fact, I understand that you've recently Dopplerized the C-band radar in Edmonton. What are your plans in this regard?

(Leduc) Well, of course, the plans are subject to budget constraints. In fact, there just was a fairly significant budget cut in March. There was a plan last year that we would have four new Doppler radars by 1996, with eventually the network of 20 radars being Dopplerized. The Edmonton Doppler will be up and running this summer, and we have the Doppler radar north of Toronto. The radar at McGill is, I believe, about to be Dopplerized.

(D. Burgess, National Severe Storms Laboratory) When I was in Kananaskis, I remember seeing a videotape of a very impressive tornado, but on the audio, there was radio communication suggesting an amateur radio network, storm spotters, and a successful warning for that storm. Could you comment on your warning system? Was that just an exception, or was it organized in many areas?

(Leduc) Well, I'm the severe weather meteorologist at the Ontario Weather Centre, so our program is excellent. [laughter]. Speaking for Ontario especially, in the past 2 or 3 years, we've been organizing the ham radio people into spotter networks. We had a terrific success on August 28, 1990. We were fortunate to have a spotter see the first funnel develop and had tornado warnings out for over an hour before that storm did the worst of its damage.

(J. Anderson, Prairie Weather Centre) I think I can help out with the tornado he's referring to: that was the Saskatoon tornado. The warning went out just as it was reaching the rope stage. We don't really have a well-established ham radio weather network. I went chasing two nights ago here in Oklahoma, and I was really impressed by that network. I think maybe we can try and encourage some of our people to do it for us. The year before last was the first time we got some of our younger meteorologists to go out and chase storms; we haven't seen any tornadoes yet. Right now we rely on a pre-established network of watchers, usually from government and provincial agencies, and we can phone them. It's actually most unhelpful in most cases; we typically have about three people in a 50 km square area and they report looking out the window and seeing it's dark and stormy outside.

PAPER H6

Presenter, G. R. Miller, National Weather Service Forecast Office [*Miller*, this volume, Oregon tornadoes: More fact than fiction]

(J. Golden, National Oceanic and Atmospheric Administration, Office of the Chief Scientist) You were joking that it took days to confirm that the Vancouver event was a tornado . . .

(Miller) No, I wasn't! [laughter] It took them 2 days before they officially came out and said it was a tornado.

(Golden) Where I live, in the Washington, D. C., metro area, we had an outbreak of "something" this year, during the winter. At the time, the damage reports indicated downburst damage, although people reported possible funnels and touchdowns. It was at least 2 days before the staff at the forecast office got out and did their surveys and confirmed at least two of them were, indeed, tornadoes. Small ones, and in fact one of them touched down closer to my house than I've ever experienced after 6 years of living here in Oklahoma . . .

(Chair) Joe, do you have a point? [laughter]

(Golden) There are some tools you can use. The people at SELS [Severe Local Storms] have done a remarkable job; Jack Hales has been very successful with forecasting tornadoes on the west coast, particularly California. I'd also suggest that rapid scan satellite data will be of significant help to you.

PAPER H7

Presenter, R. L. Livingston, National Weather Service, Kansas City [*Schaefer et al.*, this volume, The stability of climatological tornado data]

(D. Burgess, National Severe Storms Laboratory) I can't help but comment, since you didn't, about the trend. You talked about the recent decreasing trend in tornadoes, but you didn't show last year, which I believe had the greatest number of tornadoes in the data base.

(Livingston) That's probably true. I think there was a number of years in the time trend that showed spikes. I don't think that's unusual at all. I think you need to look at a longer time period before you infer the trend. Years like 1973 showed such spikes and I believe that's a normal characteristic.

(C. Anderson, North Carolina State University) Do you have any explanation for the Missouri minimum? It appears over and over gain.

(Livingston) We've been scratching our heads for a number of years over that. Tom [Grazulis] talked about it this morning; he's looked at it and can't come up with anything. The Ozark mountains may divert the winds. Population density is low, but the population in other parts of Missouri is low and yet they don't show that pronounced a minimum. It's a real puzzle.

Paper H8

Presenter, T. P. Grazulis, private consultant [*Grazulis*, this volume, A 110-year perspective of significant tornadoes]

(S. Vasiloff, National Severe Storms Laboratory) I'd like to offer the prospect of terrain normalization. In photogrammetry, the maximum windspeeds in tornadoes are tens of meters above the surface. If a tornado moves through a valley and hits a house on a cliff, the "observation point" moves from a flat surface into a higher wind speed regime in the tornado. Can you comment on this?

(Grazulis) Yes, I'd like to avoid the idea! [laughter] Topography doesn't alter the frequency of tornadoes, but wind speed measurements certainly are influenced by topography. It's another complication, definitely.

Aerial Survey and Photography of Tornado and Microburst Damage

T. T. FUJITA

The University of Chicago, Chicago, Illinois 60637

B. E. SMITH

National Severe Storms Forecast Center, Kansas City, Missouri 64106

1. INTRODUCTION

The tornado as defined in the *Glossary of Meteorology* [*Huschke*, 1959] is "a violent rotating column of air, pendant from a cumulonimbus cloud, and nearly always observable as a funnel cloud or tuba." In reality, however, no funnel cloud can be confirmed in blinding rain or during a dark night. Furthermore, a well-defined funnel on the ground does not always leave behind a continuous damage swath produced by a single vortex traveling on the ground.

In explaining the break in a vortex swath, the terms skipping and lifting were used frequently, implying that a tornado funnel intensifies or weakens within a very short distance. During posttornado interviews we often hear "the tornado leveled my neighbor's house but it skipped over my house." In the wake of the Palm Sunday tornadoes of April 11, 1965, Fujita and his associates conducted their coordinated aerial photography over the vast areas of the northern U.S. Midwest, becoming suspicious that a tornado does not skip or lift within short distances, but rather its wind structure is very complicated.

The objective of the aerial survey/photography in 1965 by the Fujita group at the University of Chicago was to determine multiscale airflows in and around tornado funnels and to identify nontornadic damaging winds induced by severe thunderstorms. During the 27 years since then, over 300 damage swaths have been flown and mapped photogrammetrically (Figure 1). A total number of 30,000 aerial photographs were taken from low-flying aircraft, mostly Cessna.

2. DETERMINATION OF MULTISCALE AIRFLOWS OF TORNADOES

Although the news media had taken numerous aerial photos of structures, the first aerial photos of well-defined circular marks, left behind by the North Platte Valley tornado of June 27, 1955, were reported by *Van Tassel* [1955]. He assumed that the grey circles on plowed fields (Figure 2) were produced by a single object caught in the tornado funnel, computing 216 m/s rotational speed of the object. Similar ground marks were photographed in the wake of the Shelby, Iowa, tornado of May 5, 1964, and reported by *Prosser* [1964]. These marks gave him an impression that an enormous vacuum cleaner had swept the ground clean of vegetation, loose soil, and other movable objects.

A major advance in the interpretation of the circular/cycloidal marks was made by taking zoom photos from 200 m above ground level (AGL) and visiting the sites on the paths of the Palm Sunday tornadoes of April 11, 1965, and the Barrington, Illinois, tornado of April 21, 1967. As evidenced in Figure 3, a circular mark was neither a scratch mark nor a band of cleaned-up bare ground. Instead, it was a band of debris deposit consisting of short pieces of corn crops, dry leaves, chicken feathers, etc. The maximum height of the deposit was less than 5 to 10 cm. In explaining the mechanism of the debris band, *Fujita* [1971] proposed the concept of a suction vortex in the tornado (Figure 4).

The diameter of a suction vortex is at least 1 order of magnitude smaller than that of the parent tornado. By virtue of its spinning motion and small diameter, the vortex gathers up near-ground debris toward its rotation axis, but it fails to pick up the debris on the ground at the center of rotation, leaving behind a narrow band of debris deposit along the path of the vortex center.

Because the shape of the cycloidal mark is a simple function of the velocity ratio, rotational velocity V divided by the translational velocity U, *Fujita et al.* [1970] generated the shapes of the ground mark by changing the velocity ratio from 1 to 10 (Figure 5). No loop will form when the velocity ratio is 1.00, but a suction vortex stays momentarily at one spot, creating a stepping spot (Figure 6). As the velocity

---

The Tornado: Its Structure, Dynamics, Prediction, and Hazards.
Geophysical Monograph 79
Copyright 1993 by the American Geophysical Union.

Fig. 1. Tracks of tornadoes surveyed by the Fujita group during the 27-year period 1965–1991. The first aerial photography was conducted immediately after the Palm Sunday tornadoes of April 11, 1965.

ratio increases, the size of the loop increases (Figure 7), reaching a near-circular loop when the ratio approaches 10 (Figure 8).

The maximum horizontal wind speed inside an orbiting suction vortex is the sum of U, V, and S, the spinning velocity of the suction vortex. A strong suction vortex in a residential area could induce a one- to two-house-wide swath in which houses could be wiped off their foundations (Figure 9). On the contrary, several "lucky" houses located between intersecting paths of multiple suction vortices could be left untouched (Figure 10). These damage patterns cannot be explained by the so-called skipping phenomenon of a tornado. Threatened by such a tornado, one should not open windows because there is no way of guessing the direction of oncoming suction vortices. At this point, the evidence of aerial surveys did alter one of the traditional tornado safety rules.

A large number of aerial photos showed the existence of cycloidal marks in the swaths of many large-core tornadoes.

Fig. 2. An aerial photo of the circular ground mark assumed to be the scratch mark by a single object caught in the tornado funnel. From *Van Tassel* [1955].

Fig. 3. (Top) Aerial photo of the cycloidal mark of the Barrington, Illinois, tornado of April 21, 1967. (Bottom) The cycloidal mark photographed on the ground. Both photos by Ted Fujita.

Their frequencies far exceeded our initial expectation. Nevertheless, pictures of tornadoes showing suction vortex funnels had been very rare until the Jumbo Tornado Outbreak of April 3–4, 1974. Since then, a large number of multiple-vortex (suction vortex) pictures (Figure 11) have become available from various parts of the United States. These pictures, along with cycloidal marks, were analyzed by *Fujita et al.* [1976], *Agee et al.* [1975, 1977], and many others.

3. EVIDENCE OF SUCTION VORTICES

Wind effects of suction vortices on the ground can be photographed from a low-flying aircraft. Their appearances vary with the scattering angle of the sunlight, appearing either relatively darker or lighter as an aircraft circles around the target. As has been well known, tracks of orbiting suction vortices appear as a group of cycloidal curves (Figure 12). On the other hand, a stationary suction vortex leaves behind a pattern of high winds indicating the existence of either a small (Figure 13) or a large (Figure 14) eye at the location where the vortex center had existed momentarily.

We also observed the path of an isolated vortex mark suggesting a single-loop motion of the suction vortex (Figure 15). An interesting vortex signature is the path of twin vortices which traveled side by side while rotating slowly around their common center (Figure 16). Another remarkable aerial photo shows a curved path with five intensification spots along the centerline (Figure 17). The picture also shows that the initial vortex disappeared, being taken over by the new vortex which flattened the corn crop along its path. This picture evidences the rapidly changing nature of an orbiting vortex which could cause unexpected damage.

Apparently, the smaller the vortex, the stronger the vertical winds relative to the horizontal winds around a small vortex. Figure 18 shows the corn crop pushed over by the Hobart, Indiana, tornado of June 30, 1977. A telephoto view of the strong shear zone reveals the existence of several tiny vortices, 1 m to 2 m in the core diameter (Figure 19), in which several corn plants were pulled off the ground. Convergence inside the core of an axisymmetric vortex is approximated by

$$\text{Conv} = u/D$$

where u denotes the inflow velocity and D denotes the core diameter. When convergence is 2 s^{-1}, a 2 m/s vertical wind is expected at 1 m AGL, and a 4 m/s vertical wind is expected at 2 m AGL. Vertical winds of these magnitudes would be able to pull loosened young plants out of the ground.

One of the best examples of a small tornado with dominantly vertical winds just above the ground is seen in a video sequence of the Minneapolis tornado of July 18, 1986, which was taken from a low-flying helicopter. *Fujita and Stiegler* [1986] pointed out that a tree in the field caught by a small tornado funnel which was 3 m in diameter on the ground tilted for about 1 s before it was blown down when the funnel moved away from the tree (Figure 20).

A small-core tornado less than 10 m in diameter, east of Denver, Colorado, on June 30, 1987, was investigated by *Wakimoto and Wilson* [1989] using both aerial and ground photographs. In spite of the herringbone pattern of damage due to the storm motion, their photographs suggest the existence of an appreciable inflow into the small core, implying a strong rising motion inside the small core.

The library of the tornado data collected by the Fujita group during the past 30 years indicates that tornadoes in general are more complicated than earlier conceived. It is often very difficult to distinguish a suction vortex from its parent tornado (Figure 21). Furthermore, their appearance and structure keep changing very rapidly within a matter of seconds (Figure 22).

During the 1976 Symposium on Tornadoes at Texas Tech University, it was noticed that there was a basic disagreement between meteorologists and structural engineers on the mechanism of structural damage by tornado winds. Most engineers, at that time, approximated tornado winds as

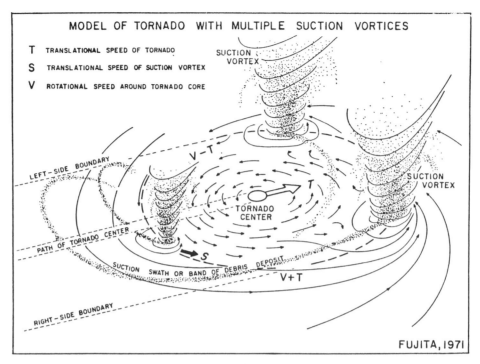

Fig. 4. A model of a tornado with three suction vortices orbiting around the core of the parent tornado. From *Fujita* [1971].

straight-line winds. This assumption is valid when vertical winds are negligibly smaller than horizontal winds, such as in the case of hurricanes and downbursts. Since the near-ground convergence is approximated by the inflow velocity divided by the vortex diameter, the straight-line wind assumption becomes invalid for most small vortices such as small tornado, suction vortex, dust devil, etc. A structure in such a small but intense vortex could be rapidly torn apart vertically under high-speed vertical winds just above the ground (Figure 23).

4. MICROBURST, INDUCER OF NONTORNADIC DAMAGING WINDS

Fujita's aerial survey and photography of the Jumbo Outbreak Tornadoes of April 3–4, 1974, played an important role in developing his concept of the downburst. After the tornadoes, when Fujita was circling over an area of reported tornado damage, he found a diverging pattern of uprooted trees (Figure 24); thereupon he reached the conclusion that the damage was caused by a strong downdraft as it impacted on the tree-covered ground.

While investigating the Eastern Airlines Flight 66 accident (landing) on June 24, 1975, at John F. Kennedy Airport in New York, *Fujita* [1976] attempted to apply his downburst concept in explaining the strong tailwind and downwind shears encountered simultaneously by the accident aircraft. Horace R. Byers, Fujita's mentor professor, was the first person who supported his downburst concept, agreeing to write a joint paper [*Fujita and Byers*, 1977]. Shortly thereafter, Fernando Caracena applied the downburst concept to the probable cause of the Continental Airlines Flight 426 accident (takeoff) on August 7, 1975, at the Stapleton, Denver Airport. The results of the joint research on three aircraft accidents is found in the work of *Fujita and Caracena* [1977].

Fujita had great confidence in the downburst concept, backed by numerous aerial photos of the starburst damage; nevertheless, a large number of nonbelievers expressed their controversial views, summarized by *West* [1979]. Most meteorologists who expressed strong opposition probably did not have the opportunity to fly over areas of tornado damage. Since then, Fujita has trained his group and initiated extensive downburst-hunting flights.

The training and flights have been very successful, establishing the existence of multiscale airflows [*Fujita and Wakimoto*, 1981]. By the end of the 1970s, solid evidence of downburst winds and their horizontal scales was established, on the basis of aerial photos and the National Center for Atmospheric Research's Doppler radars operated during the Northern Illinois Meteorological Research on Downburst (NIMROD), the landmark experiment which ended most controversies that prevailed at that time. At the termination of the experiment, the downburst concept was subdivided into microburst (<4-km horizontal size) and macroburst (>4 km).

An aerial photo of a large microburst (Figure 25) shows an

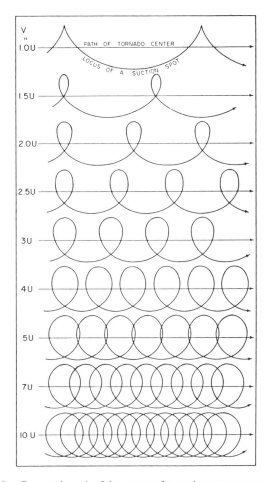

Fig. 5. Geometric path of the center of a suction vortex computed by changing the velocity ratio from 1 to 10. From *Fujita et al.* [1970].

Fig. 7. Cycloidal ground marks left behind by suction vortices with velocity ratio between 2 and 3. Photo by Ted Fujita after the West Lafayette, Indiana, tornado of March 20, 1976.

extensive area of diverging winds which blew down numerous corn plants. Frequently, a small microburst touches down with a sharp boundary of the wind speed increase from less than 25 m/s to 40 m/s within a 5- to 10-m distance (Figure 26). The estimated divergence at the boundary should reach 1.5 to 3.0 s^{-1}, suggesting that the parent downdraft descended to the treetop height without weakening significantly. A swath of high winds which deflected off of the sloping roof of a farm building in Indiana (Figure 27) also suggests that a downdraft descended to the rooftop height. An extremely small microburst may be only 30 to 50 m wide and 100 to 300 m long, with an appearance of a rush of diverging jet (Figure 28). The parent downward current

Fig. 6. Stepping spots (West Lafayette, Indiana, tornado of March 20, 1976) where orbiting suction vortices pause momentarily when the velocity ratio is 1.0. Refer to Figure 5. Photo by Ted Fujita.

Fig. 8. Near-circular cycloidal marks of the Goessel, Kansas, tornado (F5) of March 13, 1960, as the tornado was traveling at 20 m/s. Aerial photo by Duane Stiegler.

Fig. 9. (Top) An arc of a suction vortex track left in the residential section by the Wichita Falls, Texas, tornado of April 10, 1979. (Bottom) An enlargement of the boxed area. Photo by Ted Fujita. This damage is similar to the Lubbock, Texas, tornado case reported by *Fujita* [1970].

Fig. 10. One-, two-, and three-house groups left untouched by the Wichita Falls, Texas, tornado of April 10, 1979, because they were located between a number of intersecting tracks of suction vortices. Photo by Ted Fujita.

Fig. 11. Six suction vortices inside the Wichita Falls, Texas, tornado of April 10, 1979. Three vortices at the far side are those forming in the inflow region of the tornado airflow. Copyrighted photo by Floyd Styles.

Fig. 12. Typical cycloidal ground marks of the suction vortices orbiting around the core of a traveling tornado. Magnet, Nebraska, tornado of May 6, 1975. From *Fujita* [1981].

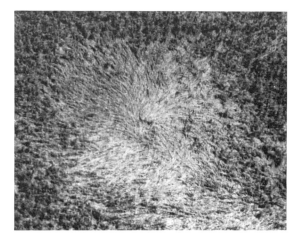

Fig. 13. Small eye of a stationary suction vortex inside the Mattoon Lake, Illinois, tornado of August 21, 1977, made visible by the pattern of blown down corn crops. From *Fujita* [1981].

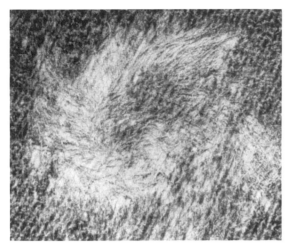

Fig. 14. Large eye of a stationary suction vortex inside the Bloomer, Wisconsin, tornado of July 30, 1977. From *Fujita* [1978].

Fig. 15. A single-loop motion of a suction vortex in the Bloomer, Wisconsin, tornado of July 30, 1977. Photo by Greg Forbes.

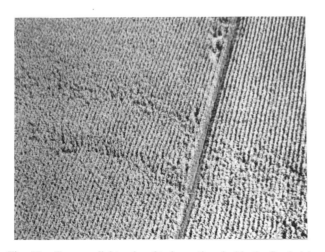

Fig. 16. Two parallel tracks of twin vortices inside the Sandwich, Illinois, tornado of June 30, 1977. Photo by Ted Fujita.

Fig. 17. Paths of two suction vortices which were joined smoothly as the first vortex weakened and was taken over by the second vortex. Bright dots along the centerline of the first vortex denote successive intensifications. Photo by Ted Fujita after the Bloomer, Wisconsin, tornado of July 30, 1977.

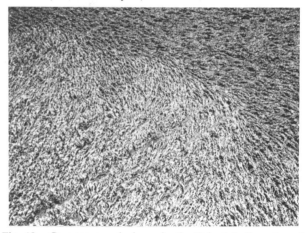

Fig. 18. Corn crops pushed over by the Hobart, Indiana, tornado of June 30, 1977, which moved from right (west) to left (east) across the picture. Photo by Ted Fujita.

Fig. 19. Telephoto view of the strong shear zone in Figure 18, showing a tiny suction vortex which pulled several corn crops off the ground. Photo by Ted Fujita.

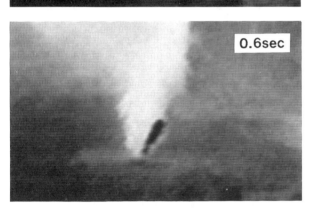
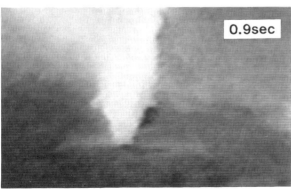

Fig. 20. A tree trying to stand up in the small funnel of the Minneapolis, Minnesota, tornado of July 18, 1986. Four selected frames of the video taken from the helicopter of KARE TV. Courtesy of Paul Douglas, Chief Meteorologist, News 11.

Fig. 21. Suction vortices and their parent tornado amalgamated into a complex system of vortices.

Fig. 22. Two frames of the movie in Figure 21. These frames are 2 s apart, showing the rapid change of the vortex system. Figures 21 and 22 were enlarged from the movie of the first Lomira, Wisconsin, tornado of April 21, 1974. Courtesy of Larry Floeter.

Fig. 23. At 2:05 P.M. PST on February 19, 1980, a small funnel cloud moved over the Fresno, California, Airport. Although the funnel was not on the ground, the roof of an airport building was blown upward and broken into pieces, suggesting the existence of strong vertical winds beneath the funnel cloud. The damage path extended from the airport into the residential areas of Fresno, California. Courtesy of Peter Stommel.

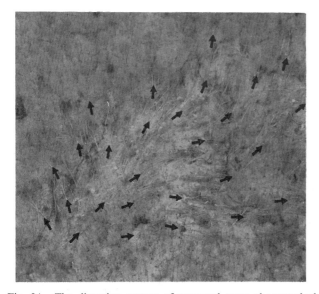

Fig. 24. The diverging pattern of uprooted trees photographed near Beckley, West Virginia, where tornado damage by one of the Jumbo Outbreak Tornadoes of April 3, 1974, had been reported. We believe that this starburst damage was located at the root of the downdraft which induced a microburst. Photo by Ted Fujita.

Fig. 25. The Danville, Illinois, microburst of September 30, 1977, which blew down corn plants in a large area. From *Fujita* [1978].

Fig. 26. A small but intense microburst found near the Cornell, Wisconsin, tornado of July 30, 1977. The 100-m-wide and 130-m-long damage area was located on Brunet Island in the Chippewa River. Photo by Ted Fujita.

Fig. 27. A microburst airflow deflected by a tin roof during the downburst storm of September 30, 1977. Photo by Ted Fujita in Kingman, Indiana. From *Fujita* [1978].

Fig. 28. A forest in northern Wisconsin blown down by a rush of diverging winds embedded inside the large downburst of July 4, 1977. Photo by Ted Fujita.

Fig. 29. A photograph of a microburst outflow with a ring vortex along the spreading edge. Photo taken on July 1, 1978, from near Wichita, Kansas. Courtesy of Mike Smith of Weather Data, Inc.

Fig. 30. A trace of mesocyclone winds with large radii of curvature photographed near the touchdown location of the Sadorus, Illinois, tornado of March 20, 1976, investigated by *Fujita et al.* [1976]. Are such winds the precursor of a tornado touchdown? Photo by Ted Fujita.

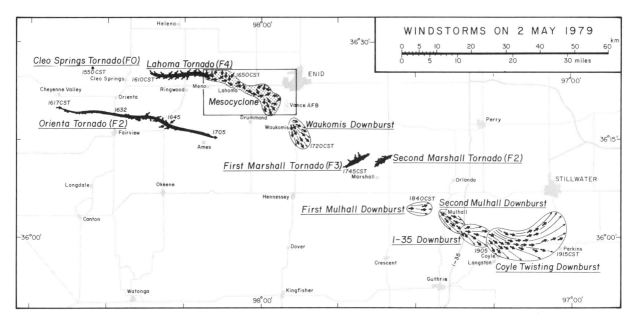

Fig. 31. The Coyle, Oklahoma, twisting downburst of May 2, 1979, extending from Mullhall to Coyle to the south of Stillwater, Oklahoma. The center of the mesocyclone, which did not spawn a tornado, moved along the north edge of the twisting downburst.

should have reached very close to the top of the forest before diverging violently.

Microbursts in progress have been photographed by many, but the most dramatic sequence of photographs was taken by Mike Smith on July 1, 1978 (Figure 29), and reported by *Fujita* [1985].

5. MESOCYCLONE AND TWISTING DOWNBURST WINDS ON THE GROUND

During the overflights of an area of wind damage, we often found traces of twisting downburst winds with large radii of curvature. These winds are often located in the areas of either pretouchdown or postliftoff of tornadoes, suggesting that they are probably induced by the parent mesocyclones of tornadoes (Figure 30).

One of the most significant twisting downbursts was the Coyle, Oklahoma, twisting downburst of May 2, 1979. Its area was approximately 10 km wide and 30 km long (Figure 31), indicating that the surface winds were induced by a traveling mesocyclone which did not spawn a tornado. So far, we have not researched the mesocyclone winds on the ground. However, an intensive effort in search of large radius of curvature winds on the ground will be important for a better understanding of the velocity data from the future NEXRAD covering the United States.

6. TORNADO-MICROBURST INTERACTION

Prior to the NIMROD experiment in 1978, *Fujita* [1978] documented a number of microbursts in the proximity of tornado tracks flown over for the purpose of aerial survey, photography, and mapping. Documented tornadoes with nearby microbursts were the Canton, Illinois, tornadoes of July 23, 1975; the Earlville, Illinois, tornadoes of June 30, 1977; and the Mattoon Lake, Illinois, tornado of August 21, 1977. Thereafter, *Forbes and Wakimoto* [1983] investigated the Springfield, Illinois, area tornadoes of August 6, 1977, revealing that 7 out of 18 tornadoes mapped from the air were located on the left (cyclonic shear) side of microbursts.

The Windsor Locks tornado of October 3, 1979, surveyed by Roger Wakimoto, Duane Stiegler, and Pete McGurk of the Fujita group was associated with eight microbursts, all located on the right-hand side of the 30-km-long, F4 tornado which moved from south to north across the Massachusetts-Connecticut state line (Figure 32). The Teton-Yellowstone tornado of July 21, 1987, surveyed by Brian Smith, Jim Partacz, and Bradley Churchill was analyzed in detail by *Fujita* [1989], revealing that there were 72 microbursts located mostly on the right-hand side of the 40-km-long path of the F4 tornado.

The interaction between a tornado and a nearby microburst was first evidenced while taking aerial photos of the path of the Rainsville, Indiana, tornado of April 3, 1974. Figure 33 shows that the course of the tornado deviated by 30° toward the east as microburst winds from the northwest blew toward the tornado (Figure 33). Since then, no action pictures of a tornado interacting with a microburst were obtained until April 12, 1991, when KAKE TV obtained video scenes from Lincoln, Kansas (Figure 34). Another interaction was confirmed by the aerial photos by Duane Stiegler. In this case, the Hesston, Kansas, tornado of March 13, 1990, deviated its track, being pushed by a

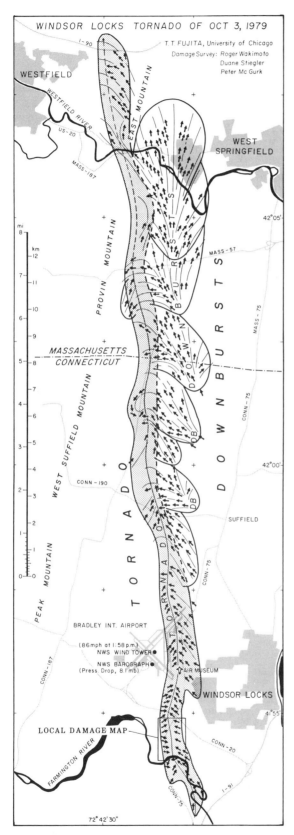

Fig. 32. The Windsor Locks tornado of October 3, 1979, which traveled from south to north. Seven microbursts were mapped on the right-hand side of the tornado track.

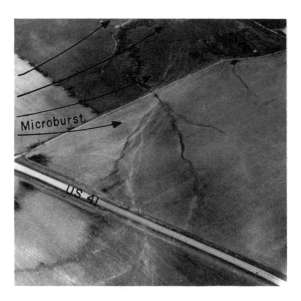

Fig. 33. Swirl marks of the Rainsville, Indiana, tornado of April 3, 1974. Undisturbed ground marks extend until crossing U.S. Highway 41. Thereafter, a microburst airflow from the left (northwest) of the picture pushed the track toward the right. From *Fujita* [1978].

Fig. 34. Two video scenes of the April 12, 1991, tornado and a nearby microburst near Lincoln, Nebraska. Courtesy of Mike Phelps. (Top) Formative stage of a tornado and a microburst on the right side. (Bottom) The tornado after touchdown was pushed toward the left by microburst winds.

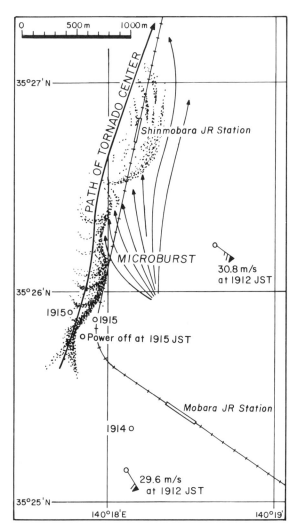

Fig. 35. The Mobara, Japan, tornado of December 11, 1990. The tornado was pushed toward the left by a microburst which touched down on the right-hand side of the tornado track. The interactive process is similar to that shown in Figure 34.

microburst located on the right side of the tornado track. Such interactions will alter or contaminate the velocity pattern of tornadoes.

In analyzing the Mobara, Japan, tornado of December 11, 1990, it was found that the tornado was pushed off the straight-line track by an F1 microburst on the right-hand side of the tornado (Figure 35). This tornado was rated by the Japan Meteorological Agency as the worst tornado in 50 years. A research Doppler radar of the Meteorological Research Institute of Japan indicated that the direction of the positive-negative velocity couplet rotated 45° while the large tornado was being pushed off the track by a microburst. It is obvious that the existence of a nearby microburst does alter the Doppler-velocity field, making the NEXRAD data interpretation difficult for tornado warnings.

7. Conclusions

An organized effort of fact-finding aerial survey, photography, and mapping of selected U.S. tornadoes by the Fujita group gave rise to the identification of the multiscale surface winds associated with tornadoes. These wind systems are the mesocyclone, tornado, and suction vortex which are blended into a complicated system of vortices.

Also identified and clarified were downbursts which are subclassified into microbursts and macrobursts based on their horizontal dimensions. It has been recognized that the microburst is the inducer of the intense wind shear which endangers aircraft during takeoff and landing operations. It is likely that the microburst would not have been identified and confirmed in the 1970s had there been no aerial photos of strange starburst damage found in the vicinity of tornado tracks.

The renewed aerial survey and photography of the damage in storm-affected areas in future years will be important in evaluating the NEXRAD data, velocity in particular, in relation to the estimated surface winds in different scales.

Acknowledgments. Long-term, continuous support of a specific subject is very difficult to obtain. Fortunately, the aerial survey of tornadoes and microbursts has been supported for 27 years, since 1965, without interruption by five agencies as partial objectives of their grants/contracts to T.T.F. as principal investigator at the University of Chicago. These objectives are "interpretation of meteorological satellite data," "design of satellite sensors for severe-storms detection," "understanding of tornado winds for nuclear safety," and "ground-truth and detection of microburst for air safety." The principal investigator wishes to express his sincere appreciation to those agencies which rendered their long-term support in achieving the research presented in this paper. Specific agencies which contributed in part to the ground-truth aerial surveys are as follows: National Science Foundation under grants ATM 78-01074, 79-21260, 81-09828, and 85-16705 (1978–1990); Nuclear Regulatory Commission under contracts NRC-04-74-239 and 04-82-004 (1974–1987); Office of Naval Research under Storm Data contract N00014-86-K-0374 (1986–1989); NOAA/NESDIS under grants 04-4-158-1, NA80-AAD0001, NA85-AADRA064, and NA90-AADRA511 (1973–1991); and National Aeronautics and Space Administration under grant NGR 14-001-008 (1962–1990). This paper is funded by grant NA90AADRA511 from the National Oceanic and Atmospheric Administration.

References

Agee, E. M., C. R. Church, C. M. Morris, and J. T. Snow, Some synoptic aspects and dynamic features of vortices associated with the tornado outbreak of 3 April 1974, *Mon. Weather Rev.*, *103*, 318–333, 1975.

Agee, E. M., J. T. Snow, F. S. Nickerson, P. R. Clare, C. R. Church, and L. A. Schaal, An observational study of the West Lafayette, Indiana, tornado of 20 March 1976, *Mon. Weather Rev.*, *105*, 893–907, 1977.

Forbes, G. S., and R. M. Wakimoto, A concentrated outbreak of tornadoes, downbursts, and implications regarding vortex classification, *Mon. Weather Rev.*, *111*, 220–235, 1983.

Fujita, T. T., The Lubbock tornadoes: A study of suction spots, *Weatherwise*, *23*, 160–173, 1970.

Fujita, T. T., Proposed mechanism of suction spots accompanied by tornadoes, in *Preprints, Seventh Conference on Severe Local*

Storms, Kansas City, pp. 208–213, American Meteorological Society, Boston, Mass., 1971.

Fujita, T. T., Jumbo tornado outbreak of 3 April 1974, *Weatherwise*, 27, 116–126, 1974.

Fujita, T. T., Spearhead echo and downburst near the approach end of a John F. Kennedy airport runway, New York City, *SMRP Res. Pap. 137*, 51 pp., Univ. of Chicago, Chicago, Ill., 1976.

Fujita, T. T., Manual of downburst identification for Project NIMROD, *SMRP Res. Pap. 156*, 104 pp., Univ. of Chicago, Chicago, Ill., 1978.

Fujita, T. T., Tornadoes and downbursts in the context of generalized planetary scales, *J. Atmos. Sci.*, 38, 1511–1534, 1981.

Fujita, T. T., The downburst-microburst and macroburst, *SMRP Res. Pap. 210*, 122 pp., Univ. of Chicago, Chicago, Ill., 1985.

Fujita, T. T., The Teton-Yellowstone tornado of 21 July 1987, *Mon. Weather Rev.*, 117, 1913–1940, 1989.

Fujita, T. T., and H. R. Byers, Spearhead echo and downburst in the crash of an airliner, *Mon. Weather Rev.*, 105, 129–146, 1977.

Fujita, T. T., and F. Caracena, An analysis of three weather-related aircraft accidents, *Bull. Am. Meteorol. Soc.*, 58, 1164–1181, 1977.

Fujita, T. T., and D. J. Stiegler, Tornado of Minneapolis, Minnesota on July 18, 1986, *NOAA Storm Data*, 28, 10–13, 1986.

Fujita, T. T., and R. M. Wakimoto, Five scales of airflow associated with a series of downbursts on 16 July 1980, *Mon. Weather Rev.*, 109, 1438–1456, 1981.

Fujita, T. T., D. L. Bradbury, and C. F. Van Thullenar, Palm Sunday tornadoes of April 11, 1965, *Mon. Weather Rev.*, 98, 29–69, 1970.

Fujita, T. T., G. S. Forbes, and T. A. Umenhofer, Close-up view of 20 March 1976 tornadoes: Sinking cloud tops to suction vortices, *Weatherwise*, 29, 116–131, 1976.

Huschke, R. E. (Ed.), *Glossary of Meteorology*, p. 585, American Meteorological Society, Boston, Mass., 1959.

Prosser, N. E., Aerial photographs of a tornado path in Nebraska, May 5, 1964, *Mon. Weather Rev.*, 92, 593–598, 1964.

Van Tassel, E. L., The North Platte Valley tornado outbreak of June 27, 1955, *Mon. Weather Rev.*, 83, 255–264, 1955.

Wakimoto, R. M., and J. W. Wilson, Non-supercell tornadoes, *Mon. Weather Rev.*, 117, 1113–1140, 1989.

West, S., Are downbursts just a lot of hot air?, *Sci. News*, 115, 170–171, 1979.

Lessons Learned From Analyzing Tornado Damage

TIMOTHY P. MARSHALL

Haag Engineering Company, Dallas, Texas 75381

1. INTRODUCTION

In the past two decades, much has been learned by studying building damage in the wake of tornadoes. Still, there are problems at the basic level in deciding whether the damage was caused by tornadoes or straight-line winds. Past studies of wind damage have frequently revealed that objects were transported along straight paths in tornadoes and followed curved trajectories in straight-line winds. The flight characteristics of objects and the variability of the terrain are only two of the many factors which cause this uncertainty. Thus damage investigators must exercise caution when using a single point of damage to determine the type of wind field.

Assigning F scale numbers to structures based on the degree of damage is a subjective visual procedure. However, when trying to derive the intensity of the winds, it is important to consider how well the buildings are constructed and to recognize weak links or flaws within such structures. Large variabilities in the strength of wood-framed buildings will yield an F scale number with no greater confidence than plus or minus one F scale.

Studies by *Minor et al.* [1977a] and *Minor and Mehta* [1979] have dispelled some of the myths associated with tornadoes and the damage they cause. Cooperative efforts between engineers and meteorologists have continued to yield a better understanding of tornado/structure interaction. Still, the process of disseminating this knowledge is slow and confusing. Many of the popular beliefs about tornadoes conflict with what we know today. We still read about opening windows as a tornado approaches, and yet people are told to board up their windows when a hurricane threatens. Tornado safety rules tell us to stay away from auditoriums, but in hurricane situations, officials still place people in them. This confusion stems from a perception that wind damage from tornadoes is somehow different than wind damage from hurricanes. Some of the lessons learned in analyzing tornado damage are the subject of this paper.

The Tornado: Its Structure, Dynamics, Prediction, and Hazards.
Geophysical Monograph 79
Copyright 1993 by the American Geophysical Union.

2. WIND IS WIND

Aerodynamic forces are induced as air flows over and around buildings. As a result, the greatest outward (or uplift) wind pressures occur around windward walls, roof corners, eaves, and ridges. The damage due to wind typically involves the removal of wall cladding and roof coverings at these locations. Damage surveys by *McDonald and Marshall* [1983] after tornadoes and *Savage* [1984] after hurricanes have revealed the same types of building response regardless of the phenomenon creating the wind.

Mehta et al. [1975] and *Abernathy* [1976] determined that large-span structures, such as auditoriums and gymnasiums, are quite vulnerable in high winds owing to their large surface areas which induce large loads. Such buildings have been just as susceptible to wind damage in hurricanes as in tornadoes. The general consensus now is that people should avoid shelter in auditoriums and gymnasiums during any type of windstorm.

3. BUILDINGS DO NOT EXPLODE

It was once thought that the low pressure within tornadoes caused buildings to explode. This theory was based on the erroneous assumption that a building somehow remains structurally intact after passing the radius of maximum winds on the periphery of the tornado. Furthermore, the theory assumes that the building remained sealed such that the barometric pressure inside the building can become significantly greater than outside.

Studies of tornado damage presented by *Mehta* [1976] and *Minor* [1976] indicated that building damage initiates from wind pressure breaching the building, not from low barometric pressure. The wind typically enters the building through broken windows or doors. Evidence of mud, insulation, glass shards, and wood missiles inside buildings that remain partially intact indicate wind had entered the buildings. Openings on the windward side of a building actually increase the internal wind pressures, resulting in additional uplift on the roof (Figure 1). Thus persons are no longer advised to open their windows in advance of a tornado. Another reason is that flying debris will likely break the

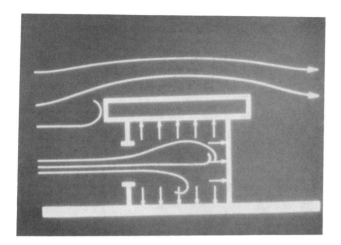

Fig. 1. This simplified schematic shows the airflow (long arrows) extending over and into a building with a windward wall opening. The pressure on the building's interior actually increases, resulting in additional uplift or outward forces (small arrows). Thus opening windows when a tornado approaches could actually be detrimental to the structure. Figure courtesy of the Institute for Disaster Research at Texas Tech University.

windows anyway; thus people should use any advance warning time to seek appropriate shelter rather than opening windows.

4. OBJECTS TWIST ACCORDING TO THEIR OWN PROPERTIES

The fact that a tree, house, or object is twisted during a tornado does not indicate that the varying direction of the wind caused the damage. Although the primary wind flow in a tornado at the ground is rotational, the rotating wind field extends over a diameter much larger than the dimension of most objects. The width of an average house is much smaller than the diameter of an average tornado. Thus at any given instant, a building in the tornado path would receive winds that are approximately unidirectional. Tornado damage studies by *Minor et al.* [1977b] and *Minor* [1982] have indicated that twisted buildings are usually the result of variations in the strength of foundation anchorage, not the rotating winds. Often the bathroom plumbing provides the greatest anchorage of a house to the foundation, and the house will pivot around this point. *McDonald* [1971] concluded that a twisted house was more likely the result of different resistances in foundation anchorage rather than the spiraling winds (Figure 2). Such damage has been known to occur even in straight-line winds from severe thunderstorms.

In the northern hemisphere the greatest wind velocities typically occur on the right sides of cyclonically rotating tornadoes as the effects of translation are added to the rotation. Using computer simulations, *Metcalf* [1978] has shown that fast translating, weak tornadoes can leave straight-line damage paths. *Marshall* [1985a] found such straight-line damage trajectories in the debris left behind in the Mesquite, Texas, tornado.

In addition, the variations of building strength, orientation, number and type of openings, roof type, degree of shielding, and impact by neighboring objects are known to influence trajectories in the damage path. Thus it is important to study a large area of damage with several points of reference before drawing definitive conclusions about whether a tornado caused the damage. Furthermore, these ideas should be kept in mind when assigning an F scale rating.

5. TYPICAL WOOD FRAME BUILDING FAILURES

Damage investigations by *Marshall* [1985b] and *Liu et al.* [1989] have identified certain building connections that have failed in windstorms, particularly at the following locations: (1) wall/foundation, (2) wall stud/bottom plate, (3) roof joist/top plate, and (4) rafter/top plate. Uplift forces are often not considered when the connections are utilized. Each member of a structure should be thought of as a link in a chain; the weakest link usually initiates failure.

Inadequate wall/foundation anchorage has meant failure of large portions of structures. Conventional wood structures, especially rural dwellings, sometimes have little to no anchorage to their foundations. *Minor* [1981] has shown examples of unanchored buildings that have moved laterally off their foundations in strong winds. See Figure 3. Properly installed anchor bolts in slabs would help secure walls to the foundation and provide greater resistance against lateral movement.

Wooden wall framing usually is straight nailed to the bottom plates. As a result, laterally applied forces distributed over the height of the wall cause rotation of this

Fig. 2. This premanufactured home was rotated off its concrete block foundation during the Hereford, Texas, tornado. The house actually pivoted about bathroom plumbing which was the only significant point of floor-to-ground anchorage. The rotation or "twist" of this object was best explained by the variations inherent within the structure rather than rotating wind currents. Note the lack of damage to roof shingles and felt paper underlayment along walls indicative of relatively low wind velocities. Photograph courtesy of the Institute for Disaster Research at Texas Tech University.

connection, and the nail ends pull out. A stronger connection would be to install straps and braces to put nails or bolts in shear. Significant resistance to racking failure can be achieved by installing plywood sheets in wall corners of wood-framed structures.

Wooden roof joists/top plate and rafter/top plate connections are usually toe nailed. Such connections typically fail in tension, causing large sections of the roof to become displaced (Figure 4). Properly installed straps or braces are needed to place fasteners in shear, not tension, in order to provide greater resistance against uplift (Figure 5). As wind velocities increase with height above the ground, roof systems usually experience strong wind uplift pressures. *Conner et al.* [1987] showed various illustrations in using straps and tie-downs in securing roofs to perimeter walls. Pull tests conducted by *Canfield et al.* [1991] have shown a dramatic increase in the strength of the rafter-top plate connection when metal rafter ties were used instead of simple toe nailing.

The best time to install connections to resist uplift forces is during construction. Proper placement of anchors, braces, and connections is essential to anchor floors to foundations, walls to floors, and roofs to walls. Increasing the wind uplift resistance of a building after it is constructed is more expensive, more difficult, and often less effective. *Sherwood* [1972] presents several construction details showing how to install anchors and braces to resist wind forces.

6. METAL BUILDING PROBLEMS

Studies of metal building performance after windstorms have revealed several weak links within such structures. When failure of a weak link leads to breach of the building containment, the damage to the building increases signifi-

Fig. 3. Lateral displacement of this unanchored house during the Grand Island, Nebraska, tornado caused the wooden floor structure to lose support and fall into the basement. Lack of damage to surrounding trees and roof shingles indicated house movement occurred at wind velocities of probably less than 100 mph (161 km/h).

Fig. 4. The roof on this house was not anchored very well and was uplifted and displaced during peripheral winds in the Mesquite, Texas, tornado on December 13, 1984. Wind velocities of less than 100 mph (161 km/h) can cause such residential damage.

cantly. An extensive study by *Mehta et al.* [1971] on metal building performance after the Lubbock, Texas, tornado found that inward buckling of overhead doors frequently led to loss of roof and wall corner cladding. Openings in the windward side of a metal building resulted in increased interior wind pressures, especially when there were no openings on the remaining building faces. Similarly, open bays "catch" the wind, causing increased wind pressures on cladding. Similar failure initiation points in metal structures have been documented in hurricane damage by *Perry et al.* [1989] and *Ellifritt* [1984].

7. UNREINFORCED MASONRY PROBLEMS

The absence of steel reinforcement in concrete block masonry makes such a system vulnerable to lateral wind loads. Even load-bearing block masonry walls have often collapsed owing to the lack of steel reinforcement and cell grouting. Examples of such failures have been shown by *Sparks et al.* [1989] and *Hogan and Karwoski* [1990] in both tornado and hurricane damage. Bond beams atop masonry walls have performed poorly in past windstorms. *McDonald and Marshall* [1983] had attributed the failure of an entire roof system to a failed bond beam system on a masonry structure. Failure initiated as the mortar joint below the bond beam failed in tension.

Absence of brick ties to secure masonry veneers has led to wall failures at relatively low wind velocities (Figure 6). Even when brick ties have been properly installed, corrosion and fatigue over time can reduce the performance of such ties.

8. ROOF SYSTEM PROBLEMS

A number of damage studies have been conducted involving the removal of various roof coverings. A common failure

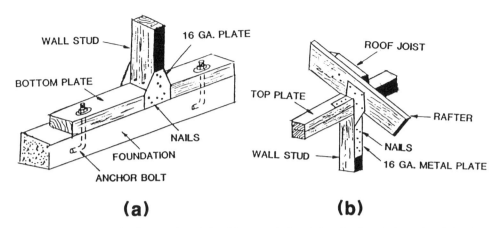

Fig. 5. Connection details to resist wind uplift: (*a*) wall-foundation detail and (*b*) roof-wall detail.

initiation point on roof systems occurs where the roof membranes are attached to edges and corners. *McDonald and Smith* [1990] attributed several roof system failures to the lifting and peeling of metal edge flashings. Uplift of the roof edges allows the wind to penetrate underneath the roof membrane, resulting in pressure rise beneath the membrane and removal of the roof covering.

Another failure initiation point has been traced to the lack of attachment between insulation board and the roof deck. In instances involving the application of hot bitumen on a metal roof deck, it is important that insulation board be installed immediately after application of the bitumen in order to develop a significant bond. A number of roof failures have been documented by the author where the applied asphalt had cooled prior to the installation of the insulation board. In other instances, an insufficient amount of bitumen had been applied to the metal deck. This led to virtually no bond between the insulation board and the deck.

Failure of roof systems has occurred when an adhesive was used to secure the roof membrane to the insulation board. As the membrane was uplifted, failure initiated in the layer of insulation board just beneath the location where the adhesive was absorbed. In this construction the insulation board was inadvertently used to resist some of the wind uplift forces. Similar roof system failures have been documented by *Kramer* [1985].

9. Flying Debris Problems

The presence of flying debris or missiles in the wind field can greatly increase the damage to structures by creating openings in the building where the wind can enter. *McDonald* [1976] has documented a wide range of missiles found in the damage paths of tornadoes. Among the most common missiles are wooden boards, sheet metal, and roof gravel. Missile drop tests conducted by *Thompson* [1973] showed that small missiles could penetrate walls with single sheathing at speeds as little as 32 mph (51 km/h).

10. Summary

It is important to consider how well buildings are constructed when interpreting the damage after a windstorm. Variations in damage along the path are not always explained by a change in wind velocity and could be attributed to weak links in building construction. Engineering studies over the past two decades have yielded increasing evidence that the extent and type of anchors, braces, and connections used in buildings correlate well with how buildings behave in the wind. A more accurate F scale rating can be obtained by considering the structural strengths and weaknesses of damaged buildings.

Fig. 6. Damage to a two-story brick veneer wall in the Mesquite, Texas, tornado. There were no brick ties to anchor the wall to the wooden frame. The wall toppled on the leeward or "suction" side.

References

Abernathy, J. J., Protection of people and essential facilities, in *Proceedings of the Symposium on Tornadoes*, pp. 407–418, Texas Tech University, Lubbock, 1976.

Canfield, L., S. Niu, and H. Liu, Uplift resistance of various rafter-wall connections, *For. Prod. J.*, 41(7/8), 27–34, 1991.

Conner, H., et al., Roof connections in houses: Key to wind resistance, *J. Struct. Div. Am. Soc. Civ. Eng.*, *113*, 2459–2473, 1987.

Ellifritt, D. S., Performance of metal buildings in Houston-Galveston area, in *Hurricane Alicia One Year Later*, pp. 117–123, American Society of Civil Engineers, New York, 1984.

Hogan, M., and A. Karwoski, Masonry performance in the coastal zone, in *Hurricane Hugo One Year Later*, pp. 195–206, American Society of Civil Engineers, New York, 1990.

Kramer, C., Wind effects on roofs and roof coverings, in *Proceedings of the Fifth U.S. National Conference on Wind Engineering*, pp. 17–34, 1985.

Liu, H., H. Saffir, and P. Sparks, Wind damage to wood frame houses: Problems, solutions, and research needs, *J. Aerosp. Transp. Div. Am. Soc. Civ. Eng.*, *2*, 57–69, 1989.

Marshall, T. P., Damage analysis of the Mesquite, Texas tornado, in *Fifth Conference on Wind Engineering*, session 4B, pp. 11–18, American Society of Civil Engineers, New York, 1985a.

Marshall, T. P., Performance of structures during Hurricane Alicia and the Altus tornado, in *Proceedings of the Department of Energy Natural Phenomena Hazards Mitigation Conference*, pp. 140–150, Department of Energy, Washington, D. C., 1985b.

McDonald, J. R., The Hereford tornado: April 19, 1971, report, 15 pp., Tex. Tech Univ., Lubbock, 1971.

McDonald, J. R., Tornado generated missiles and their effects, in *Proceedings of the Symposium on Tornadoes*, pp. 331–348, Texas Tech University, Lubbock, 1976.

McDonald, J. R., and T. P. Marshall, Damage survey of the tornadoes near Altus, Oklahoma, on May 11, 1982, *Publ. 68D*, Tex. Tech Univ., Lubbock, 1983.

McDonald, J. R., and T. L. Smith, Performance of roofing systems in Hurricane Hugo, report, 42 pp., Inst. for Disaster Res., Tex. Tech Univ., Lubbock, 1990.

Mehta, K. C., Windspeed estimates: Engineering analyses, in *Proceedings of the Symposium on Tornadoes*, pp. 89–103, Texas Tech University, Lubbock, 1976.

Mehta, K. C., J. R. McDonald, J. E. Minor, and A. J. Sanger, *Response of Structural Systems to the Lubbock Storm*, Storm Res. Rep. 3, 427 pp., Texas Tech University, Lubbock, 1971.

Mehta, K. C., J. E. Minor, J. R. McDonald, B. R. Manning, J. J. Abernathy, and U. Koehler, *Engineering Aspects of the Tornadoes of April 3–4, 1974*, 110 pp., National Academy of Sciences, Washington, D. C., 1975.

Metcalf, D. R., Simulated tornado damage windfields and damage patterns, M.S. thesis, 67 pp., Tex. Tech Univ., Lubbock, 1978.

Minor, J. E., Applications of tornado technology in professional practice, in *Proceedings of the Symposium on Tornadoes*, pp. 375–392, Texas Tech University, Lubbock, 1976.

Minor, J. E., Effects of wind on buildings, in *The Thunderstorm in Human Affairs*, pp. 89–109, University of Oklahoma Press, Norman, 1981.

Minor, J. E., Advancements in the perceptions of tornado effects (1960–1980), in *Preprints of the Twelfth Conference on Severe Local Storms*, pp. 280–288, American Meteorological Society, Boston, Mass., 1982.

Minor, J. E., and K. C. Mehta, Wind damage observations and implications, *J. Struct. Div. Am. Soc. Civ. Eng.*, *105*, 2279–2291, 1979.

Minor, J. E., K. C. Mehta, and J. R. McDonald, 1977a: The tornado: An engineering oriented perspective, *NOAA Tech. Memo.*, ERL-NSSL-82, 196 pp., 1977a.

Minor, J. E., K. C. Mehta, and J. R. McDonald, Engineering oriented examinations of the tornado phenomenon, in *Preprints of the Tenth Conference on Severe Local Storms*, pp. 438–445, American Meteorological Society, Boston, Mass., 1977b.

Perry, D. C., J. R. McDonald, and H. Saffir, Strategies for mitigating damage to metal building systems, *J. Aerosp. Transp. Div. Am. Soc. Civ. Eng.*, *2*, 71–87, 1989.

Savage, R. P., *Hurricane Alicia: Galveston and Houston, Texas, August 17 and 18, 1983*, 158 pp., Committee on Natural Disasters, National Academy Press, Washington, D. C., 1984.

Sherwood, G., Wood structures can resist hurricanes, in *Civil Engineering*, pp. 91–94, American Society of Civil Engineers, New York, 1972.

Sparks, P. R., H. Liu, and H. Saffir, Wind damage to masonry buildings, *J. Aerosp. Transp. Div. Am. Soc. Civ. Eng.*, *2*, 186–198, 1989.

Thompson, R. G., The response of residential wall construction concepts to missile impact, report, 59 pp., Dep. of Civ. Eng., Tex. Tech Univ., Lubbock, 1973.

Survey of a Violent Tornado in Far Southwestern Texas: The Bakersfield Valley Storm of June 1, 1990

GARY R. WOODALL AND GEORGE N. MATHEWS

National Oceanic and Atmospheric Administration/National Weather Service, Lubbock, Texas 79401

1. INTRODUCTION

During the late afternoon of June 1, 1990, a violent tornado struck northern and eastern Pecos County, Texas. This tornado killed two people, injured 21, and caused over $5 million in damage. The tornado passed over the unincorporated ranching community of Bakersfield Valley, so the tornado will be referred to as the Bakersfield Valley (BV) tornado.

On June 4 and 5, 1990, personnel from the National Weather Service office in Midland, Texas (MAF), and the National Weather Service forecast office in Lubbock, Texas (LBB), performed a ground and aerial survey of the tornado damage. The survey revealed a damage track 22 miles (~35 km) long and an average of 0.7 mile (~1.1 km) wide. There were several isolated occurrences of F4 damage on the Fujita damage scale [*Fujita*, 1981]. Results of the survey suggested that this was probably the largest and strongest tornado to strike western Texas since the Lubbock tornado of 1970.

Section 2 of this work reviews the synoptic and mesoscale environment of the BV tornado. Section 3 outlines the life history of the storm as viewed by satellite and radar. Section 4 describes the results of our damage survey in detail, and section 5 presents some discussion and concluding remarks.

2. SYNOPTIC AND MESOSCALE ENVIRONMENT

At 1200 UTC (all times are given in UTC) on June 1, 1990, a number of parameters conducive to severe convection were assembling over West Texas [*Miller*, 1972; *National Severe Storms Forecast Center*, 1991]. The 1200 composite analysis (Figure 1) illustrates these parameters. A 20 m s^{-1} southerly low-level jet was centered over the region. Dew points at 850 mbar were greater than 15°C over all but the Texas Panhandle. A 700-mbar dry punch was moving toward

The Tornado: Its Structure, Dynamics, Prediction, and Hazards.
Geophysical Monograph 79
This paper is not subject to U.S. copyright. Published in 1993 by the American Geophysical Union.

the area from the southwest. A major disturbance at 500 mbar was located over Idaho, with several shortwaves rotating around the large 500-mbar low. An area of difluence was present over western Texas at 200 mbar, with the left front quadrant of a 200-mbar jet streak [*Uccellini*, 1990] moving toward the area.

Area soundings taken at 1200 (Figure 2) also suggested severe weather potential [*Doswell*, 1982]. The sounding at MAF had a low-level moist layer with a layer of drier air at midlevels (above 750 mbar). Although the winds in the lowest 3 km of the MAF sounding were less than 15 m s^{-1}, the winds veered more than 90° in the lowest 3 km. The sounding from El Paso (ELP) indicated dry air and westerly to southwesterly winds at all levels.

During the midmorning, a mesoscale convective system (MCS) developed near Childress, in the southeastern panhandle. The MCS propagated slowly southeastward, leaving much of northwest Texas in cool, stable outflow. Meanwhile, the dryline became very well defined in southwest Texas, near and to the south of MAF, as low-level moisture was mixed out west of the dryline. See Figure 3 for a surface analysis at 2100, shortly after the time thunderstorms broke out southwest of MAF.

Had the MCS's cool outflow not affected the region, much of West Texas would likely have seen significant severe weather. As it was, though, the main severe weather threat area was shifted to the south of MAF, where strong surface heating and the pronounced dryline acted to focus the development of strong convection.

3. MORPHOLOGY OF THE STORM

The first low-level radar echoes appeared to the southwest of MAF around 1900. By 2015 (Figure 4), a supercell had developed in southwestern Crane County, with a tornado reported 8 miles (~13 km) west of Crane at about this time. New cells had developed south of the supercell, between McCamey and Grandfalls. The cell which went on to become the BV supercell was one of these cells and developed on the Pecos River west of McCamey.

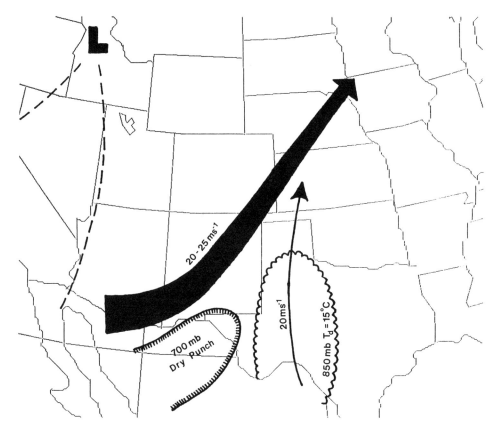

Fig. 1. Upper air composite analysis for 1200 UTC on June 1, 1990. Thin arrow marks 850-mbar jet. Scalloped line denotes 850-mbar dew point temperatures of 15°C or greater. Ticked line marks 700-mbar dry intrusion. Broad arrow shows 500-mbar jet. Broad L denotes 500-mbar low center, and dashed lines denote 500-mbar shortwaves.

By 2130 (Figure 5) the Crane County supercell had lost some of its radar structure. The low-level reflectivity gradient relaxed, and the curvature earlier seen on the echo's southern flank was no longer visible. The BV storm acquired radar characteristics associated with a supercell, namely, a tight low-level reflectivity gradient and a broad concavity on the echo's south side. The cells were moving from 270° at about 8 m s^{-1}. A hook echo was evident at the southwest flank of the BV supercell, near the town of Girvin.

The BV supercell had taken on the appearance of a high-precipitation (HP) supercell [*Moller et al.*, 1990] by 2230 (Figure 6). The echo had become large (70 km × 60 km along its long and short axes). The radar echo had taken on a kidney bean–shaped appearance, with a broad hook echo and a tight reflectivity gradient on the southeastern flank of the cell. Unfortunately, radar data at higher-elevation angles were not available. The BV tornado had touched down 10 min prior to this radar observation. As the tornado moved toward Iraan, eyewitnesses noted that the tornado was at least partially embedded in precipitation. The storm was now propagating to the right of the mean wind, moving still at 8 m s^{-1} but from about 300°.

Another supercell developed to the west-southwest of the BV supercell by 2330 (Figure 7). The BV tornado lifted at about the time this radar observation was made. The BV supercell had lost much of its structure by this time and was propagating roughly parallel to the mean wind. The storm produced large hail after this time, but no further tornadoes were reported.

4. DAMAGE SURVEY RESULTS

Personnel at the National Weather Service (NWS) forecast office LBB were notified of the casualties in Pecos County late in the evening of June 1, 1990. Because both of the fatalities and most of the injuries were in vehicles and because little additional information was available to the NWS, it was concluded that the BV tornado was probably not a large, violent tornado. Thus the NWS concluded that delaying a damage survey until Monday, June 4, 1990, would be acceptable.

The damage survey team was shocked at what it discovered when it arrived in Pecos County on June 4, 1990. A map of the tornado track, with documented damage and estimated F scale damage intensities, is shown in Figure 8. The tornado track was 22 miles (~35 km) in length, with an

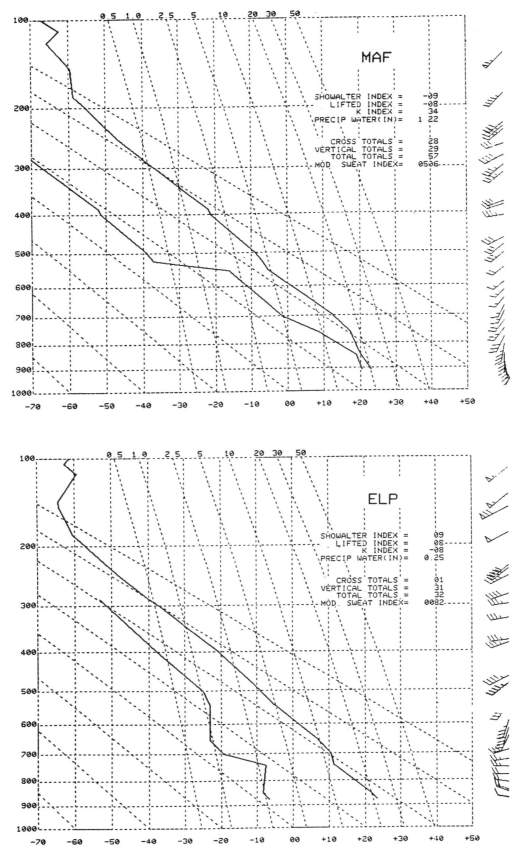

Fig. 2. Area soundings for 1200 UTC on June 1, 1990. Soundings plotted in pseudoadiabatic diagram format. MAF, Midland, Texas; ELP, El Paso, Texas.

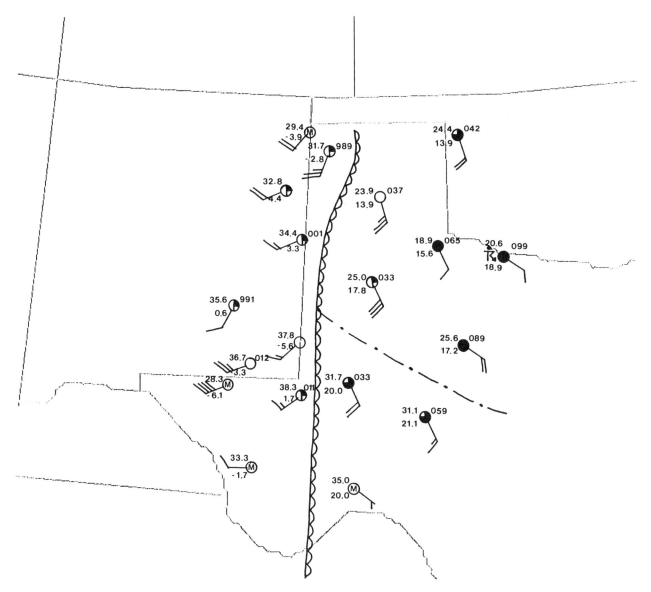

Fig. 3. Surface analysis at 2100 UTC on June 1, 1990. Dry bulb and dew point temperatures in degrees centigrade. Sea level pressures given in tenths of millibars with leading 9 or 10 omitted. Full wind barbs indicate 5 m s^{-1}; half barbs denote 2.5 m s^{-1}. Dryline and outflow boundary plotted in standard notation. Asterisk indicates location of BV tornado.

average width of 0.7 mile (~1.1 km). The maximum path width was 1.3 miles (~2 km). There were several isolated instances of F4 damage.

Bunting and Marshall [1991] and *Minor et al.* [1977] note that numerous variables exist when attempting to assign a rating to damaged structures. The design and construction of a building are perhaps the two biggest factors contributing to overall building strength. Orientation of the building to the wind and variations in the strength of construction materials also affect the building's ability to withstand strong wind. When all of these factors are combined, it becomes apparent that a variation of ±1 F rating can be expected when assigning damage ratings. As with all damage surveys, this fact should be kept in mind during the following paragraphs.

The times indicated in the following summary were acquired as follows. Reliable eyewitness times were obtained at the Co-Op building, the McKenzie residence, the Kimbrough residence, and the locations of the vehicle fatalities. The radar-derived cell motion of 8 m s^{-1} was used to estimate the remaining times along the damage path.

The tornado first touched down at 2220, approximately 9 miles (~14.5 km) south of McCamey. The first 2 miles (~3 km) of the tornado track were marked by intermittent F0 damage, inflicted mainly on telephone poles and lightweight structures.

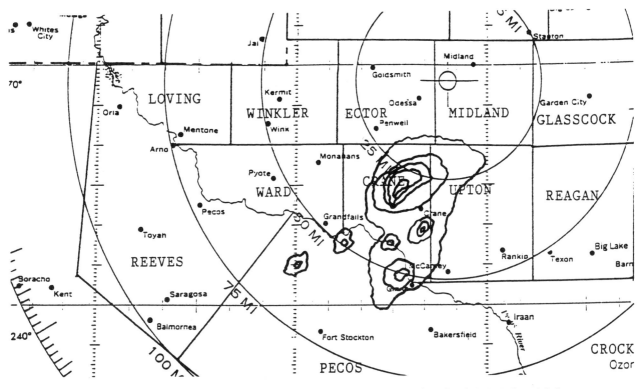

Fig. 4. Midland, Texas, radar overlay taken at 2030 UTC on June 1, 1990. D/VIP levels 2, 3, 4, 5, and 6 shown.

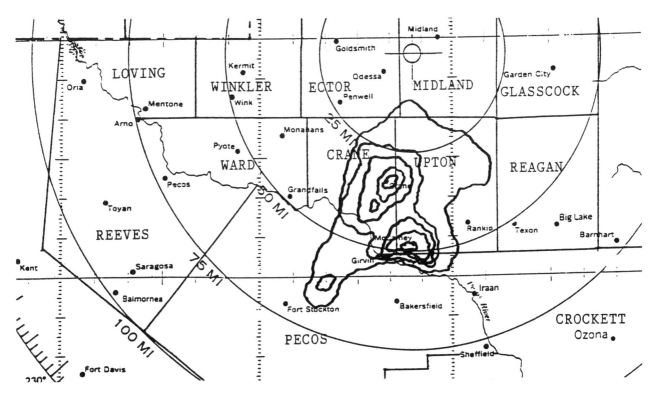

Fig. 5. As in Figure 4, but for 2130 UTC.

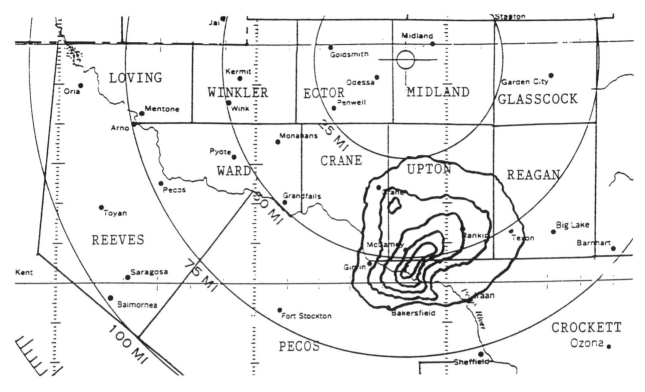

Fig. 6. As in Figure 4, but for 2230 UTC.

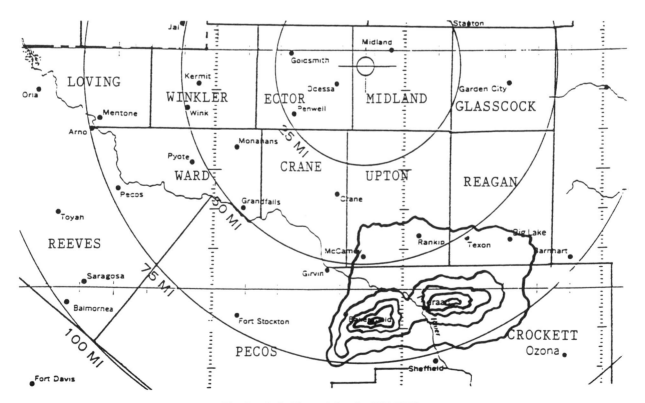

Fig. 7. As in Figure 4, but for 2330 UTC.

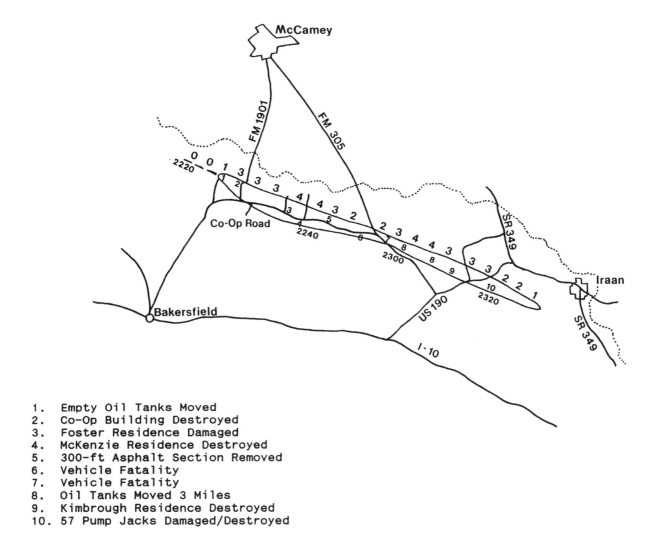

Fig. 8. Damage track of the BV tornado. Numbers inside track indicate documented damage locations. Numbers above track denote estimated F scale rating, and numbers below track indicate approximate time (CDT).

At 2225 the tornado began a period of dramatic increase in size and intensity. Two miles west of Farm to Market Road (FM) 1901, a pair of empty 500-barrel oil tanks were moved by the tornado. One tank was moved about 50 feet (~15 m) to the northeast, while the second tank was carried nearly 0.5 mile (~0.8 km) to the north. Roughly 0.25 mile (~0.4 km) north of the tanks' starting point, several ranch outbuildings were destroyed.

Between 2225 and 2230 the tornado struck a farm cooperative building (Figure 9). The building was of corrugated metal and I-beam construction. The building was heavily damaged, and there was evidence of vegetation scouring in this area. This location marked the starting point of a 0.5 × 8 mile (0.8 × 14 km) swath in which the vegetation (grass, mesquite trees, etc.) was completely removed by the tornado's wind and debris. Because the cooperative building was of fairly lightweight construction, we rated the damage as marginal F3.

After leaving the farm cooperative building, the tornado moved along a secondary highway called Co-Op Road. Between 2230 and 2240, 4 miles (~6.5 km) southeast of the cooperative building, the tornado struck a residence owned by the Foster family (Figure 10). The Foster home was a 5-year-old brick building with an adjacent barn and workshop constructed of corrugated metal. The home and the workshop were heavily damaged, but the barn suffered only minor damage. Extensive vegetation scouring was evident in this area. Damage to the Foster home was rated as F3.

Fig. 9. Farm Co-Op building damage (F3); view to west.

Fig. 11. McKenzie residence damage (F4); view to east.

The Fosters were home when the tornado struck. They had no warning of the tornado, however, until it appeared to their west (electric power had already been lost). The Fosters quickly took shelter into the interior portion of the building and escaped with only minor injuries.

A few moments later, the tornado struck the McKenzie residence, located 1 mile (~1.6 km) southeast of the Foster residence (Figure 11). The McKenzie home was a 3-year-old, two-story, well-engineered home of solid adobe brick construction. The home was destroyed by the tornado. The survey team rated this damage as F4. All that was left was the staircase and some cabinets at the northeast corner of the building. Papers from the McKenzie home were found in Big Lake, 55 miles (~88 km) east-northeast of Bakersfield Valley. The damage track at this time was over 1 mile (~1.6 km) in width.

The McKenzies were also at home when the tornado struck. Mrs. McKenzie, who taught science in high school, immediately recognized the large tornado when it appeared to their west. Like the Fosters, the McKenzies quickly moved into the interior part of the ground floor of their home and, amazingly, escaped with only minor injuries.

Almost 2 miles (~3 km) east of the McKenzie residence, the tornado removed a 300-foot (~90 m) section of asphalt from Co-Op Road (Figure 12). The removal of asphalt by a tornado is a function not only of the tornado's strength, but also of the age of the asphalt and the quality of road engineering. While asphalt removal is not unique, the removal of a large area of roadway is suggestive of a violent tornado (A. R. Moller, personal communication, 1990).

Fig. 10. Foster residence damage (F3); view to northeast.

Fig. 12. Section of asphalt removed from Co-Op Road (F4); view to east.

Fig. 13. Pickup truck in which fatality occurred, near Co-Op Road and FM 305 (F2).

Extensive vegetation scouring continued into this area. Suction marks, suggesting a multiple-vortex structure, were also evident. The damage to the road and vegetation, which was inflicted at approximately 2245, rated as F4.

After the asphalt removal, the BV tornado showed some signs of weakening. However, the tornado became a killer as it approached the intersection of Co-Op Road and FM 305. At roughly 2250 a man driving a pickup truck on Co-Op Road was caught by the tornado. The truck was blown 150 feet (~45 m) off the road, and the driver was killed (Figure 13). Several minutes later, a family traveling along FM 305 apparently drove into the tornado. The tornado had possibly become embedded in precipitation by this time. The family's pickup truck was blown 500 feet (~150 m) off the road by the tornado, and the father was killed. Damage in this area was estimated as only F2.

The tornado crossed FM 305 at about 2300. The tornado seemed to increase in intensity after crossing FM 305. The tornado entered an oil field, hit a tank battery, broke three of the 500-barrel oil tanks loose from the battery, and rolled them 3 miles (~5 km) across a field. Consultation with local authorities revealed that the tanks were full when they were moved and that they weighed approximately 70 tons each. Two of the tanks were pushed onto a hillside, while one tank was pushed over the hill. Weak suction marks were evident near the tank battery. The survey team rated this damage as marginal F4.

From 2315 to 2330 the BV tornado moved through the Yates Oil Field and damaged or destroyed 57 oil pumps. Numerous other pieces of oil field equipment were damaged. The tornado also struck a residence owned by the Kimbrough family (Figure 14). The Kimbrough house was picked up, carried roughly 30 feet (~9 m), and dropped by the

Fig. 14. Kimbrough residence damage (F3); view to north.

tornado. The building was of fairly lightweight construction and was not well engineered, so the damage in this area was rated as only F2–F3. The Kimbroughs were not home when the tornado struck. A laborer in their workshop escaped injury by hiding under a drill press when the tornado approached.

The damage to the oil field was the last documented damage caused by the BV tornado. The BV tornado lifted at 2330, 5 miles (~8 km) west-southwest of Iraan.

Eyewitnesses near the farm cooperative building and at the Foster/McKenzie residences described the tornado as a turbulent, debris-filled cylinder extending up to the cloud base. The tornado possibly appeared similar to the Wichita Falls, Texas, tornado of April 10, 1979 [*U.S. Department of Commerce*, 1980]. By the time the tornado reached US 190 west of Iraan, eyewitnesses noted that the vortex had become embedded in precipitation and that it was difficult to discern. Unfortunately, no pictures or video were taken of the tornado.

5. Discussion and Conclusions

On June 1, 1990, a violent multiple-vortex tornado struck northern and eastern Pecos County, in far southwestern Texas. This tornado, named the Bakersfield Valley (BV) tornado, had a 22-mile (~35 km) path length and an average path width of 0.7 mile (~1.1 km). There was evidence of multiple-vortex structure in the damage path. Several instances of F4 damage were noted. Such occurrences are rare; during the preceding 40-year period, the Saragosa tornado of 1987 was the only violent tornado documented in this region [*National Severe Storms Forecast Center*, 1990].

Damage was observed and documented in the BV tornado which was not seen in the Saragosa tornado. The removal of asphalt from Co-Op Road and the extensive scouring of vegetation are examples of damage which was documented in the BV tornado but not in the Saragosa tornado. On the basis of this additional damage, we suspect that the BV tornado was at times stronger than the Saragosa tornado. However, insufficient conclusive evidence existed to warrant an F5 rating for BV.

Spotter operations were hampered by the remote location and rugged geography in the vicinity of BV. However, emergency management operations in and around Iraan and McCamey were handled exceptionally well. People who found themselves in or near the tornado's path generally employed proper protective measures. These measures, repeatedly stressed by the NWS's public preparedness programs, undoubtedly contributed to the low fatality toll.

The NWS must continue to stress tornado safety procedures, especially tornado safety in vehicles. Admittedly, vehicle safety will probably remain a major problem in tornado safety, even with the addition of National Oceanic and Atmospheric Administration weather radios in some vehicles. People outside of the main tornado threat areas must continue to be reminded that they are indeed vulnerable to the type of destruction which occurred in Bakersfield Valley on June 1, 1990.

Acknowledgments. John Wright of NWS Midland, Texas, led the damage survey of June 4–5, 1990, and provided the radar data. The Pecos County Sheriff's Department gave us the initial tour of the damage area. The Texas Department of Public Safety provided an aerial tour of the damage track. Alan Moller of NWS Fort Worth, Texas, and Tim Marshall of Haag Engineering in Dallas, Texas, assisted with the F scale estimates. A special thanks to the employees of Bakersfield Co-Op, the Foster family, the McKenzie family, the Kimbrough family, the Iraan News, the Fort Stockton Pioneer, and the Odessa American for their eyewitness and news accounts of the tornado. Melody Woodall drafted the figures.

References

Bunting, W., and T. Marshall, A resource guide for conducting damage surveys, in *WCM Manual*, National Weather Service, Southern Region, Fort Worth, Tex., 1991.

Doswell, C. A., III, *The Operational Meteorology of Convective Weather*, vol. I, *Operational Mesoanalysis*, NOAA Tech. Memo. NWS NSSFC-5, 160 pp., National Severe Storms Forecast Center, Kansas City, Mo., 1982.

Fujita, T. T., Tornadoes and downbursts in the context of generalized planetary scales, *J. Atmos. Sci.*, *38*, 1511–1532, 1981.

Miller, R. C., Notes on analysis and severe storms forecasting procedures of the Air Force Global Weather Central, Tech. Rep. 200, revised, Air Weather Service, U.S. Air Force, Washington, D. C., 1972.

Minor, J. E., J. R. McDonald, and K. C. Mehta, The tornado: An engineering perspective, NOAA Tech. Memo. ERL NSSL-82, 196 pp., Natl. Severe Storms Lab., Norman, Okla., 1977.

Moller, A. R., C. A. Doswell III, and R. Przybylinski, High-precipitation supercells: A conceptual model and documentation, in *Preprints, 16th Conference on Severe Local Storms*, pp. 52–57, American Meteorological Society, Boston, Mass., 1990.

National Severe Storms Forecast Center, Tornado breakdown by counties: 1950–1989, report, 108 pp., Kansas City, Mo., 1990.

National Severe Storms Forecast Center, Storm scale processes: What to look for when forecasting severe storms, report, 40 pp., Kansas City, Mo., 1991.

Uccellini, L. W., The relationship between jet streaks and severe convective storm systems, in *Preprints, 16th Conference on Severe Local Storms*, pp. 121–130, American Meteorological Society, Boston, Mass., 1990.

U.S. Department of Commerce, Red River Valley tornadoes of April 10, 1979, *Natur. Disaster Surv. Rep. 80-1*, 60 pp., Natl. Oceanic and Atmos. Admin., Rockville, Md., 1980.

An Observational Study of the Mobara Tornado

H. NIINO, O. SUZUKI, T. FUJITANI, H. NIRASAWA, H. OHNO, I. TAKAYABU, AND N. KINOSHITA

Meteorological Research Institute, Tsukuba 305, Japan

T. MUROTA AND N. YAMAGUCHI

Building Research Institute, Tsukuba 305, Japan

1. INTRODUCTION

According to statistics from 1961 to 1982 [*Mitsuta*, 1983], an average of 18 tornadoes per year occur in Japan. This number is much less than 771, which is the 30-year annual mean in the United States [*Fergusson et al.*, 1989]. Since the United States is about 25 times as wide as Japan, this makes the number density about 1.7 times greater in the United States than in Japan.

Tornadoes in Japan are generally believed to be weaker than those in the United States. In fact, there have been no tornadoes ranked F4 or more since 1950 [*Fujita*, 1971; *Mitsuta*, 1983]. However, the relative frequencies of F0, F1, F2, and F3 tornadoes in Japan and the United States are quite similar. One possible explanation for the absence of F4 tornadoes in Japan may be as follows: While most of the violent tornadoes in the United States are known to be spawned by supercell storms [*Browning*, 1964], this type of storm has rarely been observed in Japan. This may be mainly due to the difference in the abundance of moisture. In the United States the atmosphere is usually drier than in Japan, which is surrounded by oceans. Consequently, cloud bases are generally higher over the United States (except for some eastern states), and storms can have stronger density contrast between the ambient air and the downdraft-induced cold air. This density contrast together with southeasterly flow at low levels is important for producing a strong convergence on the right flank of the storms. Various studies on the supercell storms [e.g., *Browning*, 1964] have also revealed that the existence of a dry southwesterly flow in the middle troposphere is essential to maintain their typical structure. Thus it seems of interest to examine the differences of tornadic storms in Japan and the United States. Since the number of tornadoes per year is about 40 times smaller in Japan than in the United States, not many detailed studies of tornadoes and tornadic storms have been made so far. One exception is the study by *Fujita et al.* [1972], who made a damage survey and mesoscale analysis of the Omiya tornado on July 7, 1971. In order to examine differences of tornadic storms in Japan and the United States, it is necessary to increase our knowledge about tornadic storms in Japan.

On the evening of December 11, 1990, one of the severest tornadoes after the World War II era hit Mobara City, 55 km southeast of Tokyo and 75 km south of the Meteorological Research Institute (MRI) in Tsukuba (see Figure 1). One person died, 73 people were injured. 82 houses were totally destroyed, 161 houses were severely damaged, and 1504 houses were lightly damaged. On December 12, scientists from MRI and the Building Research Institute (BRI) started to make detailed damage surveys and to interview the occupants.

During the evening of December 11, the single Doppler radar at MRI was operating and detected a mesocyclone. As will be shown in the following, this mesocyclone produced the Mobara tornado. Therefore it will hereinafter be called tornado cyclone (TC) [*Brooks*, 1949]. A considerable amount of meteorological data was collected from the Japan Meteorological Agency, local governments, schools, fire departments, and elsewhere to analyze the mesoscale features which led to the tornadogenesis.

In the following the characteristics of the tornado derived from the damage survey and the results of the mesoscale analysis will be described. This is the first study in Japan in which a TC detected by a Doppler radar was analyzed together with surface meteorological data.

The Tornado: Its Structure, Dynamics, Prediction, and Hazards.
Geophysical Monograph 79
Copyright 1993 by the American Geophysical Union.

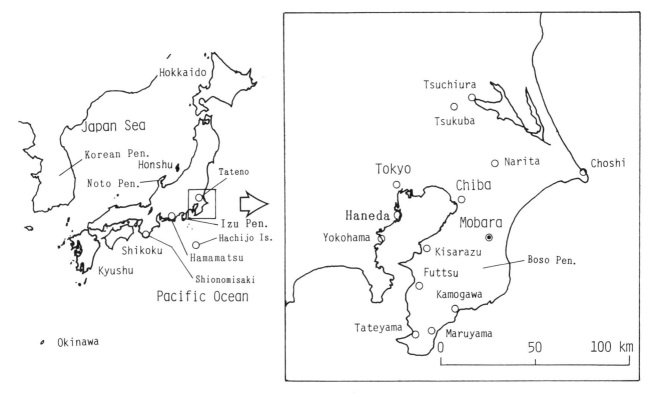

Fig. 1. Maps of Japan and the Kanto District.

2. Characteristics of the Tornado

2.1. Direct Observation

The tornado started at about 1913 JST (see section 2.6) when it was already dark. There were frequent lightning flashes, and the tornado itself seemed to emit a blue light occasionally. The funnel cloud illuminated by lightning was recorded on video by Kazumi Sasaki from a distance of 1 km. The video showed that the tornado was rotating anti-clockwise and moved approximately northward.

2.2. Damage Characteristics

Figure 2 shows the distribution of damaged houses based on the data provided by the Mobara City government office. The first damage to a house occurred 500 m south of the intersection of Route 128 and Route 409. The width of the damage path increased as the tornado moved northward. The damage path extended to the area around the Japan Railways (JR) Shin-Mobara Station and further to the north. The length of the damage path was about 6.5 km, and its average width was about 500 m (the maximum width was 1200 m).

Most of the severe damage to houses was concentrated in the Takashi area near the corner where the JR Sotobou Line changes direction from northwest to north northeast. Figure 3 shows the damage in the Takashi area as photographed from a helicopter on the morning of December 12.

Careful examination of Figure 2 shows that the damage distribution has an eastward protuberance of about 500 m near the Mobara City Gymnasium north of the Takashi area (see Figure 4). It is also noteworthy that several houses in the east part of the protuberance were completely destroyed or severely damaged. This point will be discussed later.

2.3. Meteorological Records

Meteorological records near the tornado path were taken at several schools and at other observation points where there was no means to accurately register the times. The times described in this subsection are mostly based on those recorded on barographs and anemographs and will involve some uncertainty.

Wind speed and direction. A maximum wind speed of 30.8 m/s was recorded at 1910 JST by an anemometer at a height of 10 m on the farm of the Mobara Agricultural High School (point D in Figure 2). The wind direction was SE before 1850, then changed to NE, and was SE just before the tornado cut the electric power supply. The anemometer at the Mobara Transforming Station of the Tokyo Electric Power Company (point F) recorded a maximum speed of 29.5 m/s at about 1920 JST. The variation of the wind direction at point F was quite similar to that at point D. The anemometer of the Environmental Division of the Chiba Prefecture (point B) showed that the wind direction just before the interruption of electrical power supply was NE. The wind directions at these three points are consistent with the anticlockwise rotation of the tornado observed in the video.

Pressure. A sudden pressure drop of about 9 hPa was recorded at about 1900 JST at the Chosei High School (point C), about 400 m east of the tornado path. This pressure drop is considered to be due to both the tornado and the TC. At point D, 900 m east of the path, a pressure drop of 2 hPa was recorded at about 1900 JST, and at the Fujimi Junior High School (point E), about 2000 m west of the path, a drop of 2.5 hPa was recorded. These pressure drops are considered to be due to the TC.

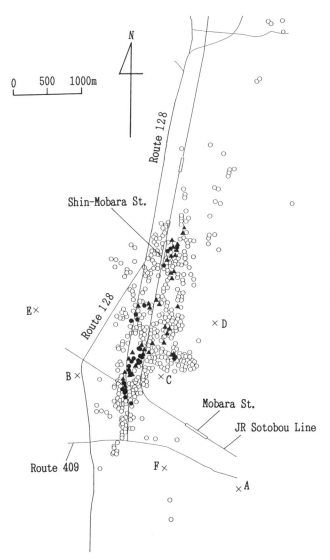

Fig. 2. Distribution of the damaged houses and the points where meteorological data were recorded. The open circles show lightly damaged houses; the solid triangles, severely damaged houses; and the solid circles, completely destroyed houses.

Fig. 3. Damage in the Takashi area (courtesy of Kyodo News Service).

2.4. Tornado Path

Figure 4 shows the spatial distribution of directions of the strongest winds derived from the damage survey. The wind directions were estimated from the directions in which road signs, poles, trees, and farm crops were leaning or had fallen. The wind directions on the east side of the dashed lines are S~SE and those on the west side W~NE. Thus the wind direction changes markedly across the dashed lines, which are essentially convergence lines. These lines are located on the northwest side of the severely damaged area in Figure 2. These results suggest that the center of the tornado vortex moved along the dashed lines and the tornado had a cyclonic rotation, also in accordance with the video images and the records of wind directions.

The tornado moved directly north northeastward from 500 m south of the intersection of Route 128 and Route 409, where the first damage was noted, to the south of the Mobara City Gymnasium. The tornado path, as determined from the distributions of wind directions and damage, became obscure near the gymnasium. The path became evident again near Route 128 to the north of the gymnasium and extended north northeastward. The tornado crossed the JR Sotobou

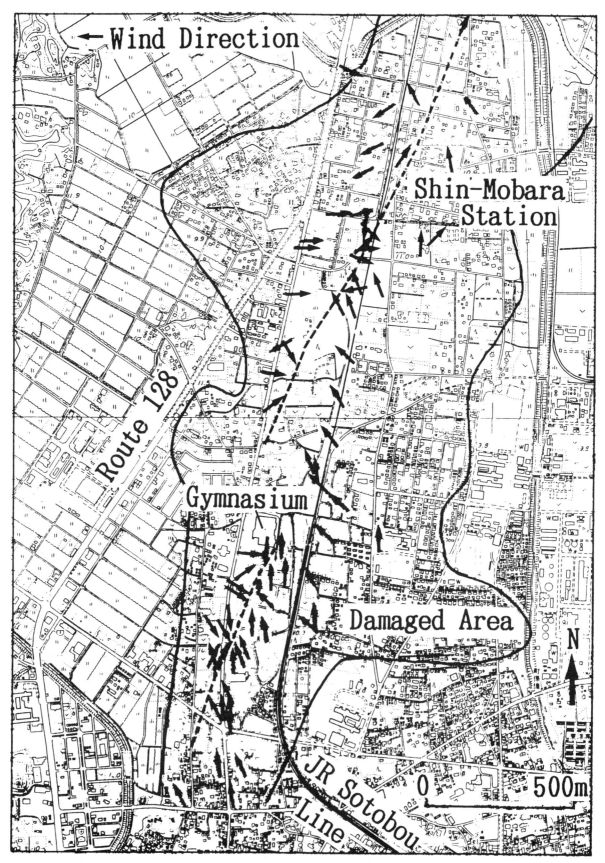

Fig. 4. Distribution of wind directions obtained from the damage survey. The dashed lines show convergence lines. The damaged area is enclosed by solid lines.

Line just south of the Shin-Mobara Station and continued on to the north.

The tornado path seemed to jump near the gymnasium where the eastward protuberance of the damage distribution is seen. Two possible reasons for this may be considered: First, the gymnasium, dimensions of which are 77 m in width, 94 m in length, and 21 m in height, could have blocked the inflow near the ground to the tornado vortex. This might have resulted in the generation of two vortices on the east and northwest sides of the gymnasium. Second, either an intensification of the mesocyclone circulation or an increase of surface roughness may have led to the structural change of the tornado vortex. Laboratory experiments show that in either case a transition among a thin laminar vortex, a thick turbulent vortex, and multiple vortices can occur [e.g., Church et al., 1979; Leslie, 1977]. Further investigation concerning the cause of this structural change is needed.

2.5. Maximum Wind Speed

Maximum wind speeds were estimated from damage to structure. The implicit assumptions here are that the structures were damaged by pressure forces due to winds which were steady and uniform in the vertical direction. Since most of the structures used for the estimation of the wind speed were no taller than 3 m, the estimated wind speeds are likely to have been affected by local conditions such as the proximity of other buildings.

The locations of the structures together with the estimated wind direction are indicated by solid arrows in Figure 5, and the estimated wind speeds are given near the arrows. The largest wind speed near the ground was determined from the twisting of a road sign at point H and is 78 m/s; hence the Fujita scale of the Mobara tornado was at least F3.

2.6. Translational Velocity

One of the unique features of the present study is that the time data recorded by a single computer at the control center of a building security company was used to determine the translational velocity of the tornado. The computer records times and places at which alarm signals due to damage to houses or interruption of power are reported. The accuracy of the time data is 1 s.

The earliest alarm signal, which signified an interruption of electrical power supply, came from the Hayano area, 1 km south of the intersection of Route 128 and Route 409, 2 s before 1913 JST (see Figure 6). This time was adopted as the time of the tornadogenesis in the present study.

Four signals which signified direct damage to buildings came from the Takashi area at 1, 7, 10, and 37 s after 1915 JST. The distances between the four points in the Takashi area and the point in the Hayano area together with the time differences gave estimates for the translational speed of 13.4, 14.8, 16.5, and 17.8 m/s, the average being 15.6 m/s.

A further estimate of the translational speed was made from the time lag between the stoppage of power and the

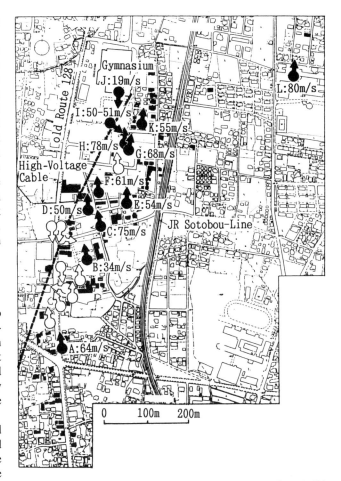

Fig. 5. Distribution of the wind speeds and wind directions (solid arrows) estimated from damage to structures. The open arrows are data from Figure 4. The buildings and houses which were completely destroyed or severely damaged are shaded.

onset of damage to a house near the Shin-Mobara Station. According to the Tokyo Electric Power Company, the stoppage occurred between 1915 and 1916 JST, when the high-voltage cable was cut. The time lag was estimated to be 90–100 s from an account provided by an occupant of the house and a repetition of her movements during the event. The distance of 1600 m between the high-voltage cable and her house gave the translational velocity of 16.2–18.0 m/s, which agrees reasonably well with the previous estimation. Thus one may conclude that the tornado moved north-northeastward at the speed of 16 m/s.

The translational velocity of the tornado is closely related to the movement of the TC which spawned the tornado. As will be described in section 3, the translational velocity of the TC obtained by the single Doppler radar at MRI between 1909 and 1917 JST was 14.9 m/s, which again agrees well with that of the tornado.

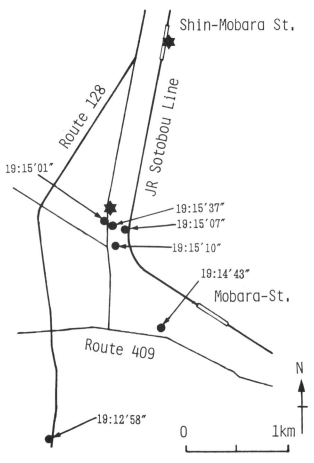

Fig. 6. Locations of buildings and houses where the damage or stoppage of electricity was monitored by a computer of a building security company. The southernmost star shows the location where the high-voltage cable was cut, and the northernmost star shows the location of the house where the eyewitness account was given by its occupant.

3. MESOSCALE ANALYSIS

3.1. Synoptic Conditions

Figure 7 shows the surface weather map about 1 hour prior to the tornado in Mobara City. It is seen that the tornado occurred in the warm sector of a developing extratropical cyclone (cf. Figure 1). At 0900 JST a cyclone with central pressure 1004 hPa was located east of the Korean Peninsula (see Figure 1). It moved eastward while continuing to develop and reached the Noto Peninsula by 1500 JST. At 1500 JST a new cyclone was generated on the cold front extending southwestward from the original cyclone. It moved eastward along the southern coast of the island of Honshu. The new cyclone reached 20 km north of Tokyo at 2100 JST.

3.2. Parent Storms

During the evening of December 11, two isolated strong cells moved northeastward off the southern coast of Honshu according to the Haneda radar (Figure 8). The eastern cell, hereinafter referred to as cell A, became detectable by the Tokyo radar at 1322 JST, when it was 80 km south of Hamamatsu. It lasted for more than 6 hours and maintained an isolated structure throughout its lifetime. After 1730, when it reached the Boso Peninsula, a vault-shaped echo region was clearly identified on both the Haneda and Narita radars (see Figure 1), and a mesocyclone circulation was found from an analysis of the surface wind and pressure data. Thus cell A showed many interesting features that are characteristic of a supercell, which is very rare in Japan. It produced at least one tornado in Kamogawa and heavy winds in Maruyama and Choshi. The analysis of the phenomena associated with cell A is reported elsewhere [*Niino et al.*, 1991].

The western cell, hereinafter referred to as cell B, became detectable by the Tokyo radar by 1500 JST, when it was 70 km south of Hamamatsu. It crossed the Izu Peninsula at around 1700 JST. It merged with a cell to the north and exhibited a fairly complex structure after it reached the Boso Peninsula at around 1815 JST. Cell B produced heavy winds at Futtsu and Kimitsu from 1830 to 1850 JST and the tornado in Mobara City from 1913 to 1920 JST.

3.3. Tornado Cyclone

The single Doppler radar at MRI was operating after 1840 JST. Its resolution is 2 km in both the vertical and crossbeam

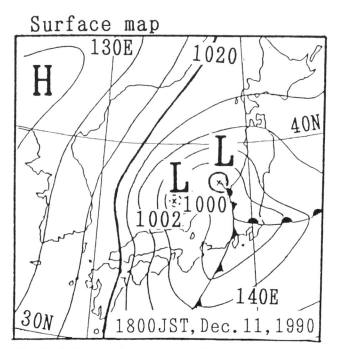

Fig. 7. Surface weather map at 1800 JST, December 11, 1990.

directions and 0.5 km in the beamwise direction. A region of high vorticity which corresponds to the TC was tracked from 1840 to 1924 at 1 km above ground level (AGL). The core diameters of the tornado cyclone, as defined by the distance between wind maxima, are shown in Figure 9 by the bars. The center of each bar indicates the center of the TC at successive times. The number to the right of the center gives the time in JST, and the number to the left gives the maximum vorticity in units of 10^{-3} s^{-1}.

The black regions show the damage paths due to heavy winds in Futtsu and Kimitsu and due to the tornado in Mobara. These areas of damage occurred when the TC passed close to those regions. When the tornado or the damage due to heavy winds occurred, the diameter of the TC was relatively small. The open circles show the locations of barographs. Each barograph recorded a sudden pressure drop of about 2 hPa at about the time when the TC passed near the barographs.

Figure 10 shows the surface wind flow and the temperature distribution derived from surface meteorological data at 1900 JST. In addition to the wind data at 1900 JST shown in the figure, wind data generated by the time-space conversion technique [*Fujita*, 1963] from 1850 to 1910 JST were used to draw the flow pattern, where the storm system was assumed to move from 236.5° at a speed of 16 m/s.

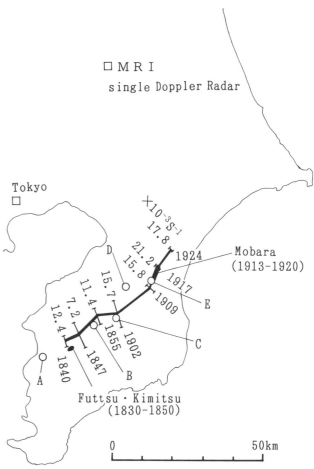

Fig. 9. The movement of the TC accompanied by the Mobara tornado as revealed by the MRI single Doppler radar. The thick line shows the path of the center of the TC.

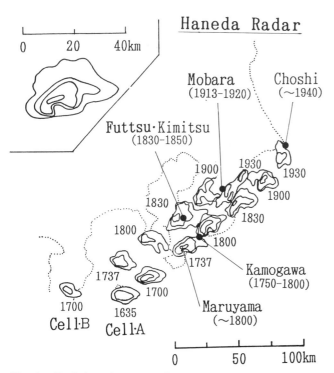

Fig. 8. Evolution of two tornadic storms as seen by the Haneda radar. The contour lines are drawn for precipitation intensities of 4, 16, and 64 mm/h. The radar echo of cell A at 1737 JST is enlarged and shown in the upper left corner, where the contour lines are drawn for 1, 4, 16, and 64 mm/h. The times and locations of damage due to tornadoes and heavy winds are also shown.

The temperature contour lines (dashed lines) were drawn on the basis of data from the Automated Meteorological Data Acquisition System (AMeDAS), whose stations are distributed for each 20-km by 20-km square on average. The contour interval is 1°, and the numbers on the contour lines are temperatures in degrees Celsius. The data show a fairly steep temperature gradient over the southern part of the Boso Peninsula.

The wind field at 1900 JST clearly shows a convergence line (indicated by the double line) where the warm southerly wind meets the cold northerly wind. The southwest end of the convergence line nearly coincides with the location of the TC at 1902 in Figure 9.

Figure 11 shows the areal distribution of precipitation intensity derived from the Tokyo radar divided into 2.5-km by 2.5-km grid spaces. The position of the convergence line in Figure 10 coincides with the southeast edge of the region where the precipitation rate is greater than 16 mm/h. This suggests that the convergence line was produced by precipitation-induced cold moist air spreading southeastward

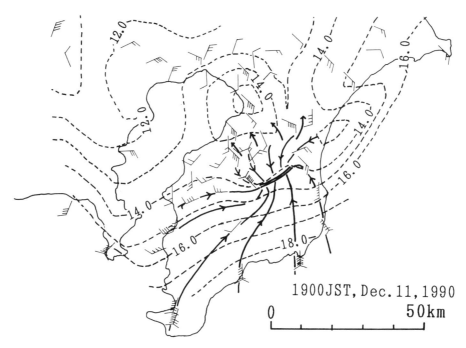

Fig. 10. The wind flows and the temperature distribution near the surface at 1900 JST, December 11, 1990. The thick solid lines with arrows show the wind flows. The surface winds at certain AMeDAS stations and at several stations of local governments at 1900 are shown. One barb corresponds to 1 m/s.

against relatively warm southerly wind. It is also interesting to note that there is a region of relatively low intensity precipitation at the southwest end of the convergence line, which corresponds to the location of the TC.

If we compare the characteristic features of this storm with the model for tornadic thunderstorms by *Lemon and Doswell* [1979], the main difference is that no evidence of a rear flank downdraft is found in Figure 10. Several reasons

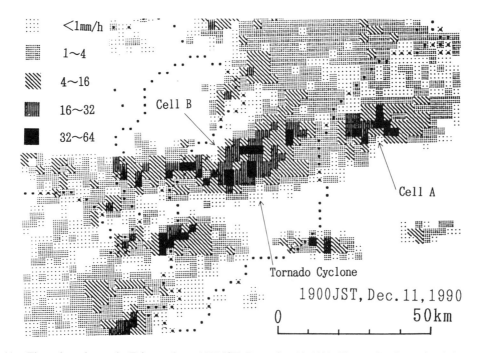

Fig. 11. The radar echo on the Tokyo radar at 1900 JST, December 11, 1990. The weak echo region is located 7.5 km (three grid squares) north of the tip of the arrow labeled "tornado cyclone."

for this can be considered: First, cell B merged with a northern cell before it reached the Boso Peninsula and exhibited a complex structure (see Figure 11). Thus it may not have been a typical supercell and does not need to have shown a structure similar to Lemon and Doswell's model, although a TC had been detected for more than 40 min. Second, given that the atmosphere over Japan is generally more humid than that over the United States, there was no dry southwesterly flow in the middle troposphere, and a significant downdraft and gust front could not develop. This may then have resulted in a different structure for the tornadic storms. To arrive at any more definite conclusions, further research on tornadic storms in Japan will be required.

4. Summary and Conclusions

The characteristics of the Mobara tornado on December 11, 1990, were derived from a damage survey; the mesoscale features which led to the tornadogenesis were analyzed from the surface meteorological data, a single Doppler radar, and three conventional radars.

The cyclonically rotating tornado moved north-northeastward at the speed of 16 m/s and produced a damage path 6.5 km in length and 1.2 km in maximum width. The maximum wind speed near the ground was estimated to be at least 78 m/s. A sudden widening of the damage path and a discontinuity in the tornado path were found near the Mobara City Gymnasium; these developments seem to suggest that the tornado vortex experienced some structural change when the vortex approached the gymnasium.

The single Doppler radar at MRI detected the TC at 1 km AGL at least 33 min before the tornado formed and was able to track it for 44 min. In the cities of Futtsu and Kimitsu, damage to several houses was produced by heavy winds when the TC passed nearby. The degree of coincidence between the location and damage path of the Mobara tornado and the location and path of the TC was excellent. Sudden pressure drops were observed along the path of the TC when it passed near barographs. Analysis of the surface wind fields showed that the TC was located at the southwest end of a convergence line which formed between the warm southerly wind and the cold northwesterly air flowing out of the region of intense precipitation. A region of weak radar reflectivity also seemed to correspond to the location of the TC.

This is the first study in Japan in which a TC was detected by a single Doppler radar and at the same time the flow field around the TC was analyzed on the basis of surface meteorological data. In the United States it is generally believed that there are two kinds of tornadogenesis: one is associated with supercell storms, and the other is associated with shear lines [*Wakimoto and Wilson*, 1989]. Since moisture content over Japan and over the United States are generally quite different, it is of interest to examine what percentage of tornadoes are generated by supercells or by shear lines in Japan. It is also of interest to examine how the structure of a supercell storm in Japan, if a unique structure exists, compares with the classical model for those in the United States. We hope to answer these questions in the near future when we have accumulated several more case studies like the present one.

Acknowledgments. The authors are grateful to the Prefectural Office of Chiba, the City Office of Mobara, the Japan Meteorological Agency, the Mobara Agricultural High School, the Chosei High School, the Fujimi Junior High School, the Tokyo Electric Power Company, and the information media for providing the various kinds of data concerning the tornado event. The authors are also grateful to H. Nagamatsu (BRI) for cooperation in the damage survey, to K. Sasaki for providing a copy of his video tape, and to anonymous referees for useful comments. The present investigation was made as part of a cooperative project between MRI and BRI entitled "An Investigation on the Characteristics of the Air Flows Associated With the Tornado in Mobara City."

References

Brooks, E. M., The tornado cyclone, *Weatherwise*, 2, 32–33, 1949.

Browning, K. A., Air flow and precipitation trajectories within severe local storms which travel to the right of winds, *J. Atmos. Sci.*, 21, 634–639, 1964.

Church, C. R., J. T. Snow, G. L. Baker, and E. M. Agee, Characteristics of tornado-like vortices as a function of swirl ratio: A laboratory investigation, *J. Atmos. Sci.*, 36, 1755–1776, 1979.

Fergusson, E. W., F. P. Ostby, and P. W. Leftwich, Jr., The tornado season of 1986, *Mon. Weather Rev.*, 117, 221–230, 1989.

Fujita, T. T., *Analytical Mesometeorology: A Review*, Meteorol. Monogr. 27, pp. 77–125, American Meteorological Society, Boston, Mass., 1963.

Fujita, T. T., Proposed characterization of tornadoes and hurricanes by area and intensity, *SMRP Res. Pap. 91*, 42 pp., Dep. of Geophys. Sci., Univ. of Chicago, Chicago, Ill., 1971.

Fujita, T. T., K. Watanabe, K. Tsuchiya, and M. Shimada, Typhoon-associated tornadoes in Japan and new evidence of suction vortices in a tornado near Tokyo, *J. Meteorol. Soc. Jpn.*, 50, 431–453, 1972.

Lemon, L. R., and C. A. Doswell III, Severe thunderstorm evolution and mesocyclone structure as related to tornado genesis, *Mon. Weather Rev.*, 107, 1184–1197, 1979.

Leslie, F. M., Surface roughness effects on suction vortex formation: A laboratory simulation, *J. Atmos. Sci.*, 34, 1022–1027, 1977.

Mitsuta, Y. (Ed.), Studies on wind disasters caused by tatsumaki (tornadoes and waterspouts) and severe local storms in Japan (in Japanese with English abstract), final report, Spec. Res. Proj. for Nat. Disaster, 124 pp., Minist. of Educ., Tokyo, 1983.

Niino, H., Y. Ogura, I. Takayabu, and H. Nirasawa, A mesoscale analysis of the tornadoes and heavy winds in the Chiba Prefecture on December 11, 1990 (in Japanese), in *Proceedings of the 1991 Spring Meeting of the Japan Meteorological Society*, Japan Meteorological Society, Tokyo, 1991.

Wakimoto, R. G., and J. W. Wilson, Non-supercell tornadoes, *Mon. Weather Rev.*, 117, 1113–1140, 1989.

Discussion

GREG FORBES, SESSION CHAIR

Pennsylvania State University

PAPER I1

Presenter, Ted Fujita, University of Chicago [*Fujita and Smith*, this volume, Aerial survey and photography of tornado and microburst damage]

(Dave Keller, National Severe Storms Laboratory) All of your damage slides looked like they were taken under sunny and calm conditions. I assume that your surveys were conducted on the day after the tornado. What is the range of weather conditions on the day after the event, and how do the conditions affect your damage surveys?

(Fujita) We like to fly on the next morning, but getting an aircraft that soon is difficult. To survey the Plainfield [Illinois] tornado [August 28, 1990] damage, we took off at 1 P.M. A few days ago, we did a survey of the damage from the March 27th [1991] Lamont [Illinois] tornado, which was on the ground for 16 miles. The next day was clear, but the turbulence was terrible. We had to learn how to hold a camera steady while the aircraft was shaking. Quite often, the weather the next day is very good, but windy and turbulent. Maybe you have to withstand 1.5 g's all the time. When we came to Oklahoma to survey the Lahoma and Orienta tornadoes [of May 2, 1979], we had to wait in Oklahoma City for 2 days because of dust being blown from [New] Mexico. We couldn't fly because visibility was less than 1 mile. So dust may be a problem the next day in Oklahoma. But east of Oklahoma, the next day is best. It's not cloudy, but very turbulent.

PAPER I2

Presenter, Tim Marshall, Haag Engineering Co. [*Marshall*, this volume, Lessons learned from analyzing tornado damage]

(Don Burgess, National Severe Storms Laboratory) One concern that I have about engineering analysis is that very often the estimate is for the minimum wind speed required

The Tornado: Its Structure, Dynamics, Prediction, and Hazards.
Geophysical Monograph 79
Copyright 1993 by the American Geophysical Union.

for doing the damage. Thus we may be underestimating the actual wind speed.

(Marshall) That is true. We may also make a totally erroneous F scale rating on the basis of the location of the building [in the tornado path].

(Joe Golden, National Oceanic and Atmospheric Administration, Office of the Chief Scientist) Why do you think that it has taken so long to get some of the simple and inexpensive procedures, that you described and that have been around for 20 years, into common construction practice? Is it because building codes do not require them? Is the insurance industry to blame?

(Marshall) It is all of the above. The insurance industry does not promote this because they offer replacement cost coverage. This does not encourage better construction. Building codes are great, but cities and municipalities are hurting financially, so they do not have the staff to enforce them. So, I am probably going to have to preach the same message for the next 20 years. This is symptomatic of the public's conceptions of tornadoes and hurricanes. They first learned about tornadoes exploding buildings and opening windows, and we spend the next 20 years trying to correct that. But people are very reluctant to give up the old beliefs.

(J. T. Lee, Cooperative Institute for Mesoscale Meteorological Studies) John Weaver wrote a paper [on anchoring roofs to walls] some 12 years ago [available from the National Severe Storms Laboratory]. The original instruction on windows was to open the windows on the opposite side to the direction from which the tornado was approaching, because if the windows [on the windward side] blew in, the wind would blow through the building and the structure would be saved [by reducing the pressure force acting to blow out the leeward wall]. Somehow, a misconception arose that the windows on the windward side, not on the leeward side, were the ones supposed to be opened.

(Marshall) I think that it was just an oversimplification. People just said, "Well, heck, in order to make my house stay there, I will open all the windows."

Paper I4

Presenter, George Marshall, National Weather Service, Lubbock, Texas (for Gary Woodall of same affiliation) [*Woodall and Mathews*, this volume, Survey of a violent tornado in far southwestern Texas: The Bakersfield Valley storm of June 1, 1990]

(Joe Golden, National Oceanic and Atmospheric Administration, Office of the Chief Scientist) After the Saragosa [southwest Texas] tornado [of May 22, 1987], people noted that violent tornadoes are extremely rare in this part of the country. Do you think that it was just coincidence that another violent tornado occurred just 4 years later, or have we undersampled tornadoes in this part of Texas? Also, could you elaborate on the poison gas that you referred to?

(Marshall) I don't doubt that we miss some small tornadoes, but I hope that we haven't missed any of this magnitude. It's just coincidence that two violent tornadoes occurred so close together in time. Concerning the poisonous gas, on the one side of the ridge, where the 57 pump jacks were knocked down, the well was not under pressure. On the other side of the ridge, that oil was under pressure and had hydrogen sulfide, which is very potent. I don't know if the gas would have travelled to Iran. The tornado decayed about 4 miles southwest of Iran, but part of the storm passed over Iran. People at Iran told me that only two whiffs of that gas and you are dead. So the tornado could have been much more disastrous.

Paper I5

Presenter, H. Niino, Meteorological Research Institute, Tsukuba, Japan [*Niino et al.*, this volume, An Observational study of the Mobara tornado]

(Chris Church, Miami University of Ohio) Do you have any explanation for why the structure of the tornado changed?

(Niino) Part of the reason why the structure changed may be that the tornado was passing the last building. But the vortex structure may have changed anyway because the path width was increasing continuously. I don't have a definite answer.

(Church) I would like to suggest that there was an increase in swirl at that particular point. Upstream the tornado had a narrow core. The increase in swirl caused a vortex breakdown to descend very close to the surface, producing a maximum in intensity ...

(Niino) That is one possibility.

(Church) ... and that might have been triggered by a change in surface roughness. A transition from a very rough surface with a lot of housing to one that is smoother seems to be a dangerous place.

(Don Burgess, National Severe Storms Laboratory) Was the antenna tilted to measure the vertical extent of the high-vorticity region?

(Niino) Unfortunately, Mobara City is about 75 km south of our institute, and the beam width there is about 2 km. The results that I presented were for 0 to 2 km. Since the tornado occurred last December, we haven't carefully looked at all the data yet. But at some time the circulation extended to at least 2 to 4 km. We are planning to make a detailed investigation into this question.

Damage Mitigation and Occupant Safety

JAMES R. MCDONALD

Institute for Disaster Research, Texas Tech University, Lubbock, Texas 79409

INTRODUCTION

Each year tornadoes kill and injure a significant number of people and, at the same time, cause millions of dollars in property damage. Although timely watches and warnings issued by the National Severe Storms Forecast Center and local National Weather Service office have resulted in significant reductions in deaths and injuries in the last 10 years [*National Oceanic and Atmospheric Administration*, 1991] the need for readily available shelters remains an important issue. The NEXRAD Doppler radars that are scheduled for installation across the country beginning in 1993 have the potential for increasing the lead time for tornado warnings, which should allow people to take precautions and seek shelter when tornadoes threaten (*Golden*, 1989). However, early warnings are of little value if there are no places available for protection against the storms.

Property damage from severe windstorms, including tornadoes, continues to escalate each year. The magnitude of the losses has reached a level where insurance alone is no longer able to cover the private losses. The costs to local, state, and federal governments are having detrimental effects on the economics of the affected regions. The potential for damage in the future is even more ominous.

The objective of this paper is to review the state of the art for providing occupant protection and for mitigating tornado damage. Both the state of knowledge and the extent of implementation of this knowledge into practice will be examined. Future research needs are also indicated.

OCCUPANT PROTECTION

The term "occupant protection" applies in this context to anyone needing protection when severe weather threatens. The first shelters against tornadoes in the United States were root cellars found on almost every farm in the midwest. These underground dugouts served as cool damp places for the storage of foodstuffs, as well as tornado shelters. As a large segment of the farm population moved to the city, cellars were constructed in backyards but were used mostly for shelter.

When shelters that are specifically designed to withstand tornado-induced forces are not available, people must rely on those areas in a building that are least likely to be damaged, where the chances of survival are relatively large. Such areas are designated "protective areas." When designated by a knowledgeable structural engineer, a protective area is the best available space for occupant protection in a particular building.

Tornado Shelters

Two forms of shelters are described herein: an in-residence shelter and a community shelter. The in-residence shelter is an aboveground closet, bathroom, or hallway that has been strengthened to resist the wind forces and missile impacts produced by very intense tornadoes. The community shelter can be freestanding or incorporated in a building.

In-residence shelters. The concept of an in-residence shelter was developed when engineers at Texas Tech University frequently observed that a small interior room or closet nearly always remained standing, even though the rest of the house was completely destroyed [*Kiesling et al.*, 1977]. Engineering attention to the room could assure that the room would remain intact against virtually all tornadoes.

Resisting the wind-induced forces is relatively simple. Special consideration had to be given to the impact of flying debris. The design criteria for in-residence shelters is based on the wind speeds associated with a tornado whose intensity corresponds to an F-4 Fujita scale tornado (F-5 being the most intense) [*Defense Civil Preparedness Agency*, 1975]. Tornado statistics from 1916 to 1979 suggest that at least 99% of all recorded tornadoes are less than or equal to F-4 intensity. The criteria translates into a tornado with a 260 mi/hr (mph) (1 mph = 1.6 km/hr) wind speed. The atmospheric pressure induced forces can be negated by providing sufficient venting area in the wall or roof of the shelter. Adequate venting area allows the pressure to equalize as the tornado passes over the shelter.

The Tornado: Its Structure, Dynamics, Prediction, and Hazards.
Geophysical Monograph 79
Copyright 1993 by the American Geophysical Union.

Fig. 1a

Fig. 1b

Fig. 1. Two design concepts for in-residence shelters: (*a*) concrete masonry construction, recommended for new construction, and (*b*) plywood construction on walls and ceiling, recommended for existing construction.

Missile impact resistance presented a more complex problem. A wide range of missiles, including roof gravel, insulation, sheet metal, pieces of timber, pipes, automobiles, and storage tanks, have been observed in tornado damage paths. Ordinary residential walls are capable of resisting roof gravel, insulation, and sheet metal. Walls of heavily reinforced masonry or concrete are required to resist the pipe, automobile, and larger missiles. Furthermore, damage investigations through typical residential areas revealed that the most common missiles were pieces of timber from damaged or destroyed roof structures. For these reasons the missile criteria for shelter design is a 2 × 4 timber plank weighing 15 lb (1 lb = 0.45 kg) impacting on end at 100 mph. On the basis of extensive tests at Texas Tech University [*McDonald and Bailey*, 1985; *McDonald and Kiesling*, 1988], two shelter designs have emerged: one for installation in an existing home and one more suitable for new construction [*Institute for Disaster Research*, 1989]. (Plans for construction of in-residence shelter can be requested from the Institute for Disaster Research, Texas Tech University, Lubbock, Texas 79409-1023.) Figure 1 illustrates the two concepts.

The shelter for an existing home has four layers of $\frac{3}{4}$ inch (1 inch = 2.54 cm) plywood attached to the ceiling and walls from inside the room. These four layers of plywood are capable of stopping the design missile. The advantage of this concept is that all construction related to the shelter can be done from inside the room or closet.

With construction of a new house, 8-inch concrete masonry, grouted and reinforced with one $\frac{3}{8}$-inch-diameter (number 3) reinforcing bar in each vertical cell, is used for the walls. A 4-inch reinforced concrete slab is normally used for the roof of the shelter, although the four layers of $\frac{3}{4}$-inch plywood would be acceptable, if properly anchored.

Common features of the two shelter concepts include a sliding door capable of resisting the missile impact and vent pipes to allow equalization of pressure. The shelter must be securely anchored to a concrete floor slab. The in-residence shelter is not recommended with a pier and beam foundation unless a special foundation is designed for the shelter itself.

Statistics are not available on how many in-residence shelters have been built. The Institute for Disaster Research at Texas Tech University distributes more than 100 sets of plans each year. There are many advantages to in-residence shelter, the main one being convenient access from any location in the house without having to go outdoors. The space can be utilized for other purposes, such as a closet or bathroom. Costs of an in-residence shelter range from $1500 to $3000.

Several commercial ventures to build and market proprietary shelters have been attempted. Aboveground commercial concepts that have been evaluated at Texas Tech include a conically shaped concrete structure, a reinforced bed frame (shelter bed), and a plate steel structure. Figure 2 shows the concrete shelter and the shelter bed. Neither of these has been economically successful.

Fig.2a

Fig. 2. Two examples of proprietary tornado shelter concepts: (a) conical reinforced-concrete shelter, where door as shown did not meet tornado missile impact criteria, and (b) shelter bed consisting of a heavy steel box that is bolted to the concrete foundation. Neither concept has achieved financial success.

Fig. 2b

Fig. 3. Freestanding aboveground community shelter constructed of reinforced masonry walls and a roof with 4-inch concrete slab supported on closely spaced open web steel joists.

Community shelters. Many situations arise where there simply are no safe places for people to seek protection from a severe windstorm. Manufactured home (mobile home) communities are the first that come to mind. Because of the uncertainty of the effectiveness of ground anchors and the large area to weight ratio of manufactured homes, occupants are advised to evacuate when severe thunderstorm or tornado warnings are issued. Underground or above ground shelters with a capacity of up to 200 people can be built to the same criteria stated for the in-residence shelters. Allowing approximately 6 square feet (1 ft^2 = 0.096 m^2) per person, a shelter 30 feet × 40 feet (9.1 m × 12.2 m) in plan will hold approximately 200 people.

The reinforced and grouted masonry walls of an aboveground shelter are capable of resisting impact of the 2 × 4 timber missile, as well as the 260-mph wind-induced forces. Venting is provided to equalize the atmospheric pressure change forces. A concept illustrated in Figure 3 uses open web steel joists to support a 4-inch-thick concrete roof slab. By supporting the slab on the joists, it does not have to be designed as a long-span structural element. Slab weight and thickness offset uplift forces and missile impacts. Careful attention must be paid to the structural integrity of the building, so that no weakness exists in the load paths. This type of shelter has been constructed in manufactured home communities, at schools where protective areas in the school building are not adequate to hold the students and staff, and at manufacturing plants where processing hazards preclude the use of the building itself for protection of people. A large number of people working in a light-frame steel building may require construction of a community shelter to provide for their safety.

Community-type shelters should be readily accessible. Although somewhat more expensive, two smaller shelters placed in close proximity to the potential occupants are better than one large shelter located some distance away. Community shelters are often incorporated into larger buildings. One architectural engineering firm in Lubbock, Texas, has successfully incorporated community shelters in a number of school construction projects [*Harris and Mehta*, 1991].

Protective Areas

The identification of protective areas in existing buildings grew from experience gained in wind damage investigations. By combining the experience of many damage studies, performances of certain classes of buildings are reasonably well understood. Factors that contribute to adequate performance of structures in windstorms are (1) degree of engineering attention to the design, (2) construction material, (3) reserve strength, (4) ductility, and (5) redundancy.

The engineer identifying protective areas first gains an understanding of the building construction and the structural system. Possible failure modes are identified. With the failure modes in mind, areas of relative safety for building occupants are identified on building floor plans. Consideration is given to accessibility, as well as safety from hazards that may exist in the building. In buildings without basements, protective areas are designated in the interior of buildings with at least two walls between the protective areas and the outside. Small rooms are preferred, because the partition walls are less likely to collapse and are more likely to support the roof elements, should they collapse.

Protective areas can be designated in schools, nursing

homes, office buildings, shopping malls, and other public places. An elementary school teacher in Huntsville, Alabama (1989), moved her class from a gymnasium which had a large, open span to a designated protective area and probably saved the lives of her 13 students. The gym roof collapsed to the floor under the tornadic forces. In the designated protective areas the students received only a few minor scratches, even though the school building was almost totally destroyed.

Efforts should be put forth to get protective areas identified in all public buildings. Floor plans and signs are needed to quickly direct building occupants to the protective areas when severe weather threatens.

DAMAGE MITIGATION

Twenty or more years of research and experience in dealing with tornado damage has not resulted in a significant reduction of tornado damage. In fact, because of escalating building costs and urban growth, the damage continues to increase. This section addresses the question; What is required to mitigate tornado damage?

Design of Structures

Knowledgeable architects and engineers are needed, who will insist on construction practices that will lead to wind-resistant construction. Wind design in the past has received very little attention in the engineering or architectural classroom, because the problem is not perceived as being serious. More emphasis on wind loading is needed in undergraduate and graduate courses. Continuing education courses for practicing engineers and architects should be offered.

In the last 10 years, more and more companies are coming to realize that tornado-resistant design is necessary to protect facilities that are critical to the company's operation. NCR Corporation designed its corporate computer facility in Dayton, Ohio, to be tornado-resistant. The American Airlines computer facility in Tulsa, Oklahoma, and Indiana Bell in Indianapolis, Indiana, along with other high-value facilities, have been designed to resist the effects of tornadoes. The Department of Energy is engaged in a natural phenomena hazards project to evaluate and upgrade all of their facilities. Those facilities, judged to be moderate or high hazard, relative to injury of people or damage to the environment, consider tornadoes in the design criteria, if tornadoes are a viable threat at the site.

The process of tornado-resistant design involves classifying the facility with regard to importance or hazard threat, performing a site-specific hazard probability assessment, and selecting a design wind speed and design procedure (for example, allowable stress design or strength design) that are consistent with established performance goals. Connections and anchorage details must assure continuity in the load paths. Finally, consideration must be given to impact resistance of tornado-generated missiles. This process is used by the Department of Energy (DOE) for design and evaluation of its facilities [Kennedy et al., 1990].

Peer reviews of tornado-resistant designs are recommended for those facilities that have high potential impact on the environment (for example, nuclear power plants, certain critical DOE facilities). The designs must then be executed using good construction practices that also involve strict quality control.

Evaluation of Existing Structures

Years of wind damage documentation have produced a body of knowledge of building performance which allows engineers to evaluate the potential performance of existing buildings in windstorms [Menta et al., 1981]. Two levels of assessment are possible: one qualitative (level I) and the other quantitative (level II). The level I procedure is based on a walk-down of the facility. Depending on the structural system, components, and cladding, potential failure modes are postulated, on the basis of performance of similar systems in windstorms. The level I procedure is ideal for identifying protective areas in buildings, for developing emergency plans, or as a preliminary assessment prior to conducting a level II procedure. With the level II procedure, failure modes are postulated, and by performing engineering calculations, the wind speeds to produce the postulated damage can be calculated. If a tornado hazard assessment has been performed for the site, the probabilistic risks of the damage can be determined. Knowing the risks is important, if decisions on extensive retrofit must be made when a facility does not meet acceptable criteria.

Strengthening and retrofit of existing structures have not been used to any great extent to improve wind resistivity. Repair of wind-damaged structures seldom incorporate improvements in the designs. The buildings are simply rebuilt exactly as they were before the storm. Retrofit projects for earthquake resistance in California are becoming popular. Efforts are needed in the future to encourage improved building designs following major windstorm events. The problem is particularly difficult in the case of tornadoes. Once hit by a tornado, people tend to think their chances of being struck again are extremely small.

Reluctance to Invest in Tornado Protection

The major reason progress has been slow in tornado damage mitigation is the reluctance of owners to invest in tornado protection. From a layman's view the risks of tornado damage seem very low. Given the low risks, there are very few significant financial incentives, because insurance or federal aid minimizes financial losses. Building code requirements do not include consideration of tornadoes in design. Furthermore, there is a general perception that added tornado protection is very expensive and disruptive. Insurance rates for extended coverage do not reward owners for investing in more than the minimum wind resistivity in buildings required by building codes.

Another factor that contributes to reluctance of an owner to design a new building or retrofit an older one is the excessive conservatism that sometimes creeps into the design process. Points of conservatism can be heaped one upon the other, especially in evaluating the resistance of a structure. Conservative values of material properties, conservative load factors (or factors of safety), and conservative limits on structural response all tend to add costs to a project. Reluctance to utilize inelastic response for wind design also adds to costs.

ADDITIONAL RESEARCH NEEDS

Convincing the professional community, as well as the owners of the built environment, to construct more wind-resistant new buildings and to strengthen and retrofit existing ones is a societal problem. Solution to this problem is long-term and, in fact, will never be complete.

There are several areas where additional technical research, from an engineering perspective, is needed. Highly resistant, but economical, roofing systems are needed. Current roof compliance tests do not correctly represent the true effects of the wind on roofs.

Designing a building to resist tornado-induced loads tends to preclude the use of windows in exterior walls. Flying debris tends to break the windows, allowing wind and water to circulate through the building. Research is needed to develop missile-resistant shutters or grills that could be closed when tornadoes threaten. The devices need to be capable of stopping the 2 × 4 timber plank missile. Heavier missiles, comparable to a 3-inch-diameter steel pipe, are not as common as the timber plank and do not fly as readily in the wind.

CONCLUSIONS

Twenty years of tornado research have produced much knowledge than can be used to design new buildings and retrofit existing ones to be more resistant to tornado-induced loads. Use of this knowledge would provide better protection of occupants and would mitigate property damage.

While there are still a number of technical questions to be answered, the major problem is in implementation of the knowledge into practice. Until this is done on a broad scale, the damage from tornadoes each year will continue to be very high.

REFERENCES

Defense Civil Preparedness Agency, *Interim guidelines for building occupant protection from tornadoes and extreme winds*, Rep. *TR-83A*, U.S. Dep. of Def., Washington, D. C., 1975.

Golden, J. H., The coming deluge of wind data for wind engineers, in *Proceedings, 6th U.S. National Conference on Wind Engineering*, Houston, Tex., 1989.

Harris, H. W., and K. C. Mehta, Design of tornado protective areas in schools, paper presented at Ninta, Structures Congress 1991, Am. Soc. Civ. Eng., Indianapolis, Indiana, April 29 to May 1, 1991.

Institute for Disaster Research, Inresidence shelter for protection from extreme winds, pamphlet, Tex. Tech Univ., Lubbock, 1989.

Kennedy, R. P., S. A. Short, J. R. McDonald, M. W. McCann, Jr., R. C. Murray, and J. R. Hill, Design and evaluation guidelines for department of energy facilities subjected to natural phenomena hazards, *Rep. UCRL 15910*, U.S. Dep. of Energy, Washington, D. C., 1990.

Kiesling, E. W., K. C. Mehta, and J. E. Minor, Protection of property and occupants in windstorms, *Rep. 27D*, Inst. for Disaster Res., Tex. Tech Univ., Lubbock, 1977.

McDonald, J. R., and J. R. Bailey, Impact resistance of masonry walls to tornado-generated missiles, paper presented at Third North American Masonry Conference, The Masonry Soc. and Univ. of Tex., Arlington, Tex., June 3–5, 1985.

McDonald, J. R., and E. W. Kiesling, Impact resistance of wood and wood products subjected to simulated tornado missiles, paper presented at International Conference on Timber Engineering, Wash. State Univ., Seattle, Wash., Sept. 19–22, 1988.

Mehta, K. C., J. R. McDonald, and D. A. Smith, Procedures for predicting wind damage to buildings, *J. Struct. Div. Am. Soc. Eng.*, *107* (ST11), 2089–2096, 1981.

National Oceanic and Atmospheric Administration, *Climatological Data, National Summary*, Environmental Data Service, Washington, D. C., 1991.

Tornado Fatalities in Ohio, 1950–1989

THOMAS W. SCHMIDLIN

Department of Geography, Water Resources Research Institute, Kent State University, Kent, Ohio 44242

1. INTRODUCTION

Ohio is at the eastern edge of the "tornado alley" of the United States, yet the risk of killer tornadoes is greater in Ohio than some states within the core of the tornado alley [*Abbey*, 1976]. This is due to the higher population density in Ohio and the relatively high frequency of violent tornadoes in portions of Ohio [*Allen*, 1981; *Schaefer et al.*, 1986]. These factors, along with recent Ohio tornado disasters, spurred me to intensify research into minimizing the risk of death from tornadoes. *White and Haas* [1975, p. 280] suggested that research on the tornado hazard should emphasize social ramifications of the hazard. *Riebsame et al.* [1986] called on the atmospheric sciences community to be more effective in reducing hazard vulnerability, in part by encouraging historical hazards analysis. This statewide examination of tornado mortality patterns over four decades addresses these issues in tornado hazards, incorporating the literature and methods of atmospheric sciences, natural hazards, and epidemiology.

The goal of community tornado preparedness programs and tornado watches and warnings issued by the National Weather Service is the reduction of personal injury and death from tornadoes. An axiom of epidemiology is that adverse health effects of natural disasters do not occur randomly within a population but occur in a somewhat predictable pattern clustered in time, in space, or in certain groups of persons [*Binder and Sanderson*, 1987]. Knowledge of the attributes of persons killed by tornadoes and the circumstances of the deaths may be useful to evaluate preparedness programs and tornado-warning methods [*Ferguson et al.*, 1987; *Sanderson*, 1989]. This knowledge may identify risk groups of persons who should be targeted in tornado preparedness and warning and who need special attention during tornado disasters.

The goals of this research were to (1) establish a data base of the persons killed by tornadoes in Ohio during the period 1950–1989, (2) summarize their age, sex, race, cause of death, and location when the storm struck, (3) identify risk groups in the population, and (4) make recommendations for improving tornado preparedness and warning in Ohio. This paper reports on the results of the research.

2. REVIEW OF LITERATURE

Previous research on mortality due to tornadoes focused on specific storms [*Mandelbaum et al.*, 1966; *Beelman*, 1967; *Abbey and Fujita*, 1981; *Glass et al.*, 1980; *Centers for Disease Control (CDC)*, 1985, 1986, 1988, 1991; *Topp and Sauve*, 1988; *Carter et al.*, 1989]. While useful, the studies that describe mortality from only one tornado event suffer from studying only a small impacted population in a relatively small geographic region at one point in time. The attributes of persons killed in particular storms may reflect unique circumstances of the community or impacts on a special population. Conclusions concerning risk factors are difficult to draw from a review of these studies because the reported summary statistics on deaths were generally brief, incomplete, and inconsistent among studies [*Sanderson*, 1989]. A study of tornado fatalities on a statewide basis, sampling many disasters that occurred in various circumstances over several decades, has not been published. The National Severe Storms Forecast Center (NSSFC) began keeping data on the personal attributes of tornado fatalities in 1985 [*Ferguson et al.*, 1987]. The CDC has examined tornado mortality of specific events but maintains no national archive of tornado fatalities (L. M. Sanderson, personal communication, 1988).

White and Haas [1975, p. 276] observed that tornado fatality rates were decreasing in the United States but geographic differences in the death rates were not explained by differences in tornado occurrence. They suggested that regional differences in the death rates could be caused by differences in tornado severity, urbanization, building construction, preparedness, hospital facilities, warning systems, and the distinctive behavioral characteristics of individuals. *Sims and Baumann* [1972] concluded that cultural differences between the northern and the southern regions of the United States were partly responsible for the higher death rate from tornadoes in the south. Their work suffered from a small and narrowly stratified sample [*Davies-Jones et al.*, 1973], but it is reasonable to expect that cultural differences within society will affect risk of death from tornadoes and other hazards. Casualty rates in disasters have been shown to differ between the sexes, among age groups, and among race or ethnic groups. Females have been shown to have higher death rates than males in some tornado disasters [*Glass et al.*, 1980; *Abbey and Fujita*, 1981; *Ferguson et al.*, 1987] but lower than males in other tornadoes [*Beelman*, 1967; *Topp and Sauve*, 1988; *Carter et al.*, 1989]. This variety of results relating sex and tornado mortality reinforces the earlier contention that results from individual tornado events cannot be used to represent general patterns of risk and mortality.

The young and old suffer higher rates of disaster casualties than other age groups [*Friedsam*, 1962]. While few studies of tornado mortality have identified children as a high-risk group, young children are especially vulnerable to death from head trauma, a major cause of serious injury in tornadoes. The elderly have been shown to be at greater risk of injury or death in weather disasters [*Moore*, 1958; *Abbey and Fujita*, 1981; *CDC*, 1985; *Sanderson*, 1989; *Carter et al.*, 1989]. This may be due to their high incidence of mobility-limiting chronic disease, lower likelihood of receiving warning due to isolated living arrangements or disability, and reluctance to evacuate [*Friedsam*, 1962]. Risk-taking behavior and responses during a disaster are a function of personality [*White and Haas*, 1975, p. 101], so it is reasonable to expect that in addition to age and sex, culture and ethnicity may be factors in tornado mortality. Several studies have shown that ethnicity affects access and response to warnings and mortality in weather disasters [*Moore*, 1958; *Perry et al.*, 1982; *Aguirre*, 1988].

The location of an individual at the time the tornado strikes can affect the extent of injuries and risk of death. The percentage of tornado fatalities occurring in a house has been variously reported to range from 14% [*CDC*, 1991] to 56% [*CDC*, 1986] and in motor vehicles from 10% [*CDC*, 1986] to 60% [*Glass et al.*, 1980]. *Edwards* [1985] reported that only 5% of the U.S. population lived in mobile homes, yet the proportion of tornado fatalities that occurred in mobile homes nationwide was 46% in 1984. The cause of death from tornadoes is most often head and chest trauma [*Mandelbaum et al.*, 1966; *Beelman*, 1967; *Glass et al.*, 1980; *Carter et al.*, 1989]. These injuries result from impact with objects that were made airborne by the wind, commonly called missiles or projectiles, from the person becoming airborne and striking the ground or a fixed object, or from crushing injuries in collapsed buildings or vehicles. Additional knowledge of the causes of death in tornadoes may lead to improved safety and shelter procedures.

It is clear from the literature that personal attributes, such as age, sex, race or ethnicity, access to warnings, location when the tornado strikes, and others, may affect tornado mortality rates. Numerous inconsistencies appear in the literature because the previously published studies focused on individual tornado events where local circumstances may have affected mortality patterns. Different methods of investigation and reporting of results among the published literature preclude simply combining those studies into a conclusive regional or national summary of tornado mortality risk factors. This research addresses that dilemma by examining mortality patterns in many killer tornadoes that occurred in Ohio over a 40-year period.

3. Methods

A list of the dates, locations, and strengths of all tornadoes in Ohio during the period 1950–1989 was provided by NSSFC. This list included the number of fatalities caused by each tornado. A total of 172 fatalities were indicated in this NSSFC list. Newspapers from the days after the storm were examined initially to obtain the name, age, and sex of victims. This information was found to be unreliable. Therefore county coroner records and death certificates on file at county health departments and the Ohio Department of Health were examined. The personal attributes of age, sex, race, location when the storm struck, and cause of death were obtained from the death certificates, when possible, for each tornado fatality. Deaths that were not directly related to a tornado or that resulted from injuries after the storm, such as heart attacks, traffic accidents, cleanup injuries, or fires, were not considered as tornado fatalities in this study.

This research procedure revealed several errors in the NSSFC count of fatalities and tornadoes. When the adjustments were tallied, the number of persons killed by tornadoes in Ohio during the period 1950–1989 was determined to be 155, not the 172 reported in the NSSFC listing. All subsequent figures quoted in this paper utilize the adjusted fatality counts. These 155 deaths were caused by 33 tornadoes on 16 days.

4. Results

4.1. Characteristics of Killer Tornadoes

Figure 1 shows the paths of the 33 killer tornadoes. The unglaciated Appalachian plateau of southeastern Ohio is hilly and wooded with a low population density compared to the rest of the state. The reported tornado frequency is lowest in this portion of Ohio [*Schmidlin*, 1988], and relatively few tornadoes caused fatalities in southeastern Ohio. The median length of the tornado paths from the first damage point to the last damage point was 40 km, much longer than the typical Ohio tornado path of about 2 km [*Schmidlin*, 1988]. Most (80%) tornado fatalities were caused by violent tornadoes (F4–F5), yet these storms comprised only 3% of reported Ohio tornadoes [*Schmidlin*, 1988]. The greatest number of people killed by a single tornado was 32 by NSSFC tornado 3 on April 3, 1974. The deadliest day was April 11, 1965, when 55 were killed by seven tornadoes. This represented 35% of total fatalities in the 40-year period. The median number of deaths per killer tornado was two. The

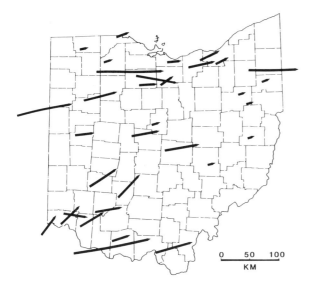

Fig. 1. Paths of Ohio tornadoes that caused fatalities during 1950–1989. Paths are shown from first touchdown to last lift-off as provided by NSSFC, although the paths were often not continuous on the ground. Some long-path tornadoes may actually represent more than one tornado.

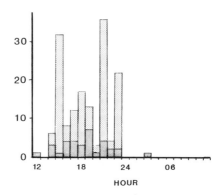

Fig. 2. Number of fatal tornadoes (heavy stippling) and tornado fatalities (light stippling) by hour (EST) of tornado touchdown.

median number of deaths per day when tornado fatalities occurred was 3.5.

Tornado fatalities occurred in 14 (35%) of the 40 years from 1950 to 1989. Nearly 93% of deaths and 78% of fatal tornadoes occurred during April, May, and June. April had 45% of the fatal tornadoes and 65% of the deaths. There were no tornado fatalities during January, February, July, October, or December. During the late autumn and winter, only 4% of the strong or violent (F2–F5) tornadoes caused fatalities, compared to 20% during other months. This is in contrast to previous studies for Indiana [*Keyser et al.*, 1977] and the entire United States [*Galway and Pearson*, 1981] that showed a high proportion of cold season tornadoes caused fatalities. The pattern of hourly occurrences of fatal tornadoes (Figure 2) is similar to the general diurnal cycle of tornado occurrence in Ohio [*Schmidlin*, 1988]. Nearly all (98.7%) of the fatalities resulted from tornadoes that touched down between 1400 and 2400 EST. Sixty-three percent of the fatalities occurred with tornadoes that struck during the dark hours between sunset and sunrise. Risk of death from tornadoes may be related to the diurnal cycle of society, including sleep, work, school, transportation, and attention to radio or television [*CDC*, 1991].

4.2. *Age of Victims*

The age-specific mortality rate [*Friedman*, 1987, p. 11] was calculated to determine whether some age groups were at higher risk of tornado death than others. This was calculated as

(average annual fatalities per group/group population) $\times 10^7$

Results are given in Figure 3. The very young and the elderly die at a higher rate than teenagers and middle-aged adults. The death rate for children under 5 years old was twice the rate for children 5 to 9 years old and 3 times the death rate for children 10 to 19 years old. The death rate for children under 1 year old (not shown) was about 10, higher than any other age group except those over 80 years old. Forty percent of the fatalities under 5 years old died along with a sibling, and at least 55% were with a parent or care-giver who also died in the storm. There is no evidence that any of those under 5 years old were alone at the time the storm struck. Age-specific mortality rates fell to 1.7 for ages 10 to 19 years, the lowest of any age group, then rose and leveled out at 3.0 to 3.5 for ages 20 to 59 years. Mortality rates doubled to over 6.0 for the ages 60 to 79 years. Persons aged 65 years and older comprised 9.8% of Ohio's population but accounted for 23.9% of tornado fatalities. These high tornado mortality rates among the elderly in Ohio are in agreement with previous studies of disaster mortality. A portion of these excess deaths may be due to the generally poorer health of the aged which can affect their reaction to warnings and ability to survive injuries.

The age-specific mortality rate for persons aged 25 to 34 years was 3.7. This was 50% higher than the mortality rate in the 20- to 24-year age group and about 10% higher than

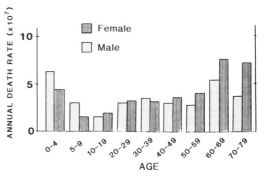

Fig. 3. Ohio tornado mortality rate by age group and sex.

mortality rates in the 35- to 44- and 45- to 54-year age groups. Since persons aged 25 to 34 years are more likely to have small children, their higher mortality rate may result from putting themselves at risk to gather and shelter their children. Of the 21 fatalities in the 25- to 34-year age group, eight (38%) died along with one of their children. It is not known how many other persons in this age group died while successfully protecting a child's life. The sacrifice of parents' lives in protecting their children was evident in the Saragosa tornado [*CDC*, 1988]. This aspect should be investigated in future tornado disasters.

4.3. *Sex of Victims*

Mortality rates were nearly the same for males and females in the age groups of 10 to 19 years, 20 to 29 years, and 30 to 39 years (Figure 3). At ages less than 10 years, males had a mortality rate 57% higher than females. At ages of 40 years and older, females generally had a progressively higher mortality rate than males. The higher mortality rates among young male children than among female children was not expected from reading the disaster literature. The sex-based difference in mortality rates among young children may be caused by a lower ability of boys to survive trauma, less efficient evasive behavior among boys than girls in disasters, more aggressive protective measures taken by adults toward girls than toward boys, or other factors. Further research is needed to identify the causes of higher tornado mortality rates among boys than girls. The higher mortality rates for females than males over age 50 was expected. The general reason given for the higher mortality rates among older females has been the brittle nature of bones in postmenopausal women [*Glass et al.*, 1980] resulting in more severe traumatic injuries.

4.4. *Race of Victims*

Most of the Ohio tornado victims were white. Only 3.2% were black, well below the state's 9.1% proportion of blacks, weighted over the 1960, 1970, and 1980 census data. All of the five black victims were killed on April 3, 1974, and four were employees, students, or visitors at Central State University, a predominantly black institution struck by a tornado. The low proportion of blacks among Ohio tornado fatalities may be explained by the generally segregated population patterns in the state. The 1980 Ohio population was over 99.0% white in 22 counties comprising 25% of the state's area. Among the rural regions of Ohio, only 0.9% of the population was black in 1980, while 27.3% of the central-city urban population was black. Therefore it was unlikely that large numbers of blacks would have been tornado victims in Ohio unless a violent tornado struck a black urban neighborhood or predominantly black institution, such as Central State University.

4.5. *Locations of Victims When Struck*

The general location of the victim at the time the tornado struck was determined for 152 (98%) of the Ohio fatalities

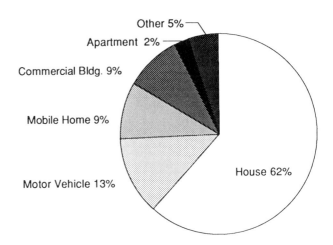

Fig. 4. Location of Ohio tornado fatalities.

(Figure 4). Most (62%) of the victims were in a house (not including mobile homes) when the tornado struck. Information on the precise location of victims within structures was not available for most victims in this study but should be collected in future disasters. Twelve of the fatalities in vehicles were in cars, five died on a commercial bus, and one fatality was noted in both a van and a small truck. The relative risk of persons in vehicles cannot be determined without knowing the percentage of the population that was in a vehicle in the storm's path when the tornado struck. Of the Ohio tornado fatalities that occurred at home, 13% occurred in mobile homes. This was 9.2% of all Ohio tornado fatalities, a relatively low rate of death in mobile homes compared to results from other studies. However, the proportion of housing units in Ohio that are mobile homes is only 2.5%, so the proportion of Ohio fatalities that occurred in mobile homes affirms the well-known risk of those locations. The relatively low percentage (9.2%) of Ohio tornado fatalities that occurred in mobile homes compared to percentages reported from some southern states may result from the relatively low number of Ohio housing units that are mobile homes. Only 4.6% of the victims were struck at their place of employment, away from the home, perhaps because most fatal tornadoes struck after traditional working hours. The greatest number of fatalities in one location was five. *Logue et al.* [1981] suggested that higher mortality of women in tornado disasters may be due to women being in more vulnerable locations when the storm struck, but there was no statistical difference in these Ohio data between male and female fatalities in their location when the storm struck.

4.6. *Cause of Death*

Head injuries were the primary cause of death for 49% of the Ohio victims (Figure 5). Crushing chest injuries and trauma and multiple injuries, each likely to include some head and chest injuries, were also common causes of death. Most victims of asphyxia suffocated under the debris of

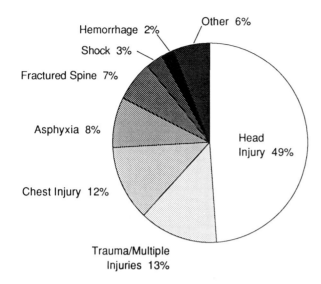

Fig. 5. Primary cause of death from Ohio tornadoes.

collapsed buildings, but two young brothers drowned after being blown from their parents' car into a ditch. One person was electrocuted while extricating herself from a destroyed house.

5. CONCLUSIONS AND RECOMMENDATIONS

This examination of 155 tornado fatalities that occurred in Ohio during a 40-year period is the first large-scale summary of the personal attributes of tornado victims. Newspapers were found to be an unreliable source of data on tornado fatalities, so death certificates were used.

Tornado fatalities have occurred over a wide area of Ohio, in rural and urban settings, but were least common in the hilly southeast. Most fatalities occurred during April, May, and June, and nearly all occurred between the hours of 1500 and 2400 EST. The age groups at highest risk of tornado death were the very young (<5 years), especially males, and the elderly (>65), especially females. There is some indication that parents of small children die at higher rates than expected, perhaps because of placing themselves in danger while attempting to protect their children. Most fatalities occurred in houses and motor vehicles. Mobile homes were the location for 9.2% of Ohio fatalities. This is much less than in most other studies, but only 2.5% of Ohio housing units are mobile homes, so this still represents a higher death rate in mobile homes than other forms of housing. Head injuries accounted for nearly half of the fatalities, with trauma, chest injuries, and asphyxia accounting for significant portions of the tornado deaths.

The following recommendations are made to reduce mortality from tornadoes in Ohio. They arise from the results of this research and other referenced literature.

1. Continued efforts toward providing efficient warning through an unmistakable and audible community siren should be made. The National Oceanic and Atmospheric Administration (NOAA) Weather Radio should continue to be a priority for the National Weather Service and disaster planners.

2. Laws requiring underground community shelters in mobile home communities should be enacted and enforced, since tie-down anchors are ineffective in winds over 22 m/s [Glass et al., 1980].

3. The unsafe nature of motor vehicles in a tornado should be stressed more in our mobile society.

4. Tornado safety procedures for public transportation should be formulated and practiced, especially for school buses. Students should not board school buses during a tornado warning, and students already boarded should evacuate and enter a sturdy building.

5. The vulnerable nature of small children should be stressed to parents and care-givers. Family tornado drills should be practiced to familiarize children with procedures and locations of safe shelter. The differences between fire and tornado safety procedures in the home should be stressed to children.

6. More attention should be given among the elderly to their education in tornado hazards, access to warnings, and availability of shelter. This will become increasingly important as the proportion of elderly persons grows in our society.

7. While tornado fatalities did not occur during the study period in an Ohio school, daycare center, nursing home, office building, or factory, special attention should be given to warning and sheltering persons in those facilities. Each facility should have a designated "storm watcher." This person would be trained by the National Weather Service as a storm spotter, monitor threatening weather through visual observation and NOAA Weather Radio, and issue a tornado warning for the building if a tornado is observed. Regular tornado drills should be held, regardless of the disruption caused by drills in these facilities.

We are now extending this data base of Ohio tornado fatalities back through the nineteenth century to examine changes in tornado mortality patterns as Ohio evolved from a frontier state into the industrial era. Additional research is planned to compare the historical tornado fatality patterns in Ohio to those in other regions of the United States where the population, culture, and tornado history differ from those of Ohio.

Since historical research on tornado fatalities is limited to data on death certificates, it is important that additional relevant data on modern victims be gathered soon after each fatal storm. Such data should include income, education, awareness of storm warnings, exact location when struck, responsibility for care of family members, and previous serious health problems.

The disparity in mortality rates between young boys and young girls and between men and women over 50 years old needs further research. The efficiency of various electronic media in conveying warnings needs attention. More research on the effects of culture, education, and income may allow focused targeting of populations for tornado preparedness and warning. A determination of the percentage of the affected local population in houses, vehicles, mobile homes, or offices during a tornado would allow mortality rates to be determined for those locations. Finally, the continued efforts of meteorologists and disaster planners to improve tornado warnings is an important means of reducing tornado mortality.

Acknowledgments. Appreciation is extended to the National Severe Storms Forecast Center, Herman Butler of the Ohio Department of Health, the staff of the Ohio Historical Society, county coroner offices, county health departments, Daniel Rooney, and Jeanne Appelhans Schmidlin. A portion of this research was funded with a Summer Research Appointment from the Research Council of Kent State University. Travel to Tornado Symposium III was supported by the College of Arts and Sciences at Kent State University.

REFERENCES

Abbey, R. F., Jr., Risk probabilities associated with tornado windspeeds, in *Proceedings of the Symposium on Tornadoes: Assessment of Knowledge and Implications for Man*, edited by Richard E. Peterson, pp. 177–236, Texas Tech University, Lubbock, Tex., 1976.

Abbey, R. F., Jr., and T. T. Fujita, Tornadoes as represented by the tornado outbreak of 3–4 April 1974, in *Thunderstorms: A Social, Scientific, and Technological Documentary*, vol. 1, *The Thunderstorm in Human Affairs*, edited by E. Kessler, pp. 47–88, National Oceanic and Atmospheric Administration, Washington, D. C., 1981.

Aguirre, B. E., The lack of warnings before the Saragosa Tornado, *Int. J. Mass Emerg. Disasters*, 6, 65–74, 1988.

Allen, B. S., Regionalization of tornado hazard probability, M.S. thesis, 170 pp., Tex. Tech Univ., Lubbock, 1981.

Beelman, F. C., Disaster planning: Report of tornado casualties in Topeka, *J. Kansas Med. Soc.*, 68, 153–161, 1967.

Binder, S., and L. M. Sanderson, The role of the epidemiologist in natural disasters, *Ann. Emerg. Med.*, 16, 1081–1084, 1987.

Carter, A. O., M. E. Millson, and D. E. Allen, Epidemiologic study of deaths and injuries due to tornadoes, *Am. J. Epidemiol.*, 130, 1209–1218, 1989.

Centers for Disease Control, Tornado disaster—North Carolina, South Carolina, March 28, 1984, *Morbidity Mortality Weekly Rep.*, 34, 205–206, 211–213, 1985.

Centers for Disease Control, Tornado disaster—Pennsylvania, *Morbidity Mortality Weekly Rep.*, 35, 233–235, 1986.

Centers for Disease Control, Tornado disaster—Texas, *Morbidity and Mortality Weekly Rep.*, 37, 454–456, 461, 1988.

Centers for Disease Control, Tornado disaster—Illinois, 1990, *Morbidity Mortality Weekly Rep.*, 40, 33–36, 1991.

Davies-Jones, R., J. Golden, and J. Schaefer, Psychological response to tornadoes: Letter, *Science*, 180, 544, 1973.

Edwards, C. M., Statewide tornado drills—Have we gone far enough?, in *Preprints, 14th Conference on Severe Local Storms*, pp. J5–J6, American Meteorological Society, Boston, Mass., 1985.

Ferguson, E. W., F. P. Ostby, and P. W. Leftwich, Jr., Annual summary: The tornado season of 1985, *Mon. Weather Rev.*, 115, 1437–1448, 1987.

Friedman, G. D., *Primer of Epidemiology*, 305 pp., McGraw-Hill, New York, 1987.

Friedsam, H. J., Older persons in disaster, in *Man and Society in Disaster*, edited by G. W. Baker and D. W. Chapman, pp. 151–182, Basic Books, New York, 1962.

Galway, J. G., and A. Pearson, Winter tornado outbreaks, *Mon. Weather Rev.*, 109, 1072–1080, 1981.

Glass, R. I., R. B. Craven, D. J. Bregman, B. J. Stoll, N. Horowitz, P. Kerndt, and J. Winkle, Injuries from the Wichita Falls tornado: Implications for prevention, *Science*, 207, 734–738, 1980.

Keyser, D. A., E. M. Agee, and C. R. Church, The modern climatology of Indiana tornadoes, *Proc. Indiana Acad. Sci.*, 86, 380–390, 1977.

Logue, J. N., M. E. Melick, and H. Hansen, Research issues and directions in the epidemiology of health effects of disasters, *Epidemiol. Rev.*, 3, 140–162, 1981.

Mandelbaum, I., D. Nahrwold, and D. W. Buyer, Management of tornado casualties, *J. Trauma*, 6, 353–361, 1966.

Moore, H. E., *Tornadoes Over Texas: A Study of Waco and San Angelo in Disaster*, 334 pp., University of Texas Press, Austin, 1958.

Perry, R. W., M. K. Lindell, and M. R. Greene, Crisis communications: Ethnic differentials in interpreting and acting on disaster warnings, *Social Behav. Personality*, 10, 97–104, 1982.

Riebsame, W. E., H. F. Diaz, T. Moses, and M. Price, The social burden of weather and climate hazards, *Bull. Am. Meteorol. Soc.*, 67, 1378–1388, 1986.

Sanderson, L. M., Tornadoes, in *The Public Health Consequences of Natural Disasters 1989*, pp. 39–49, Centers for Disease Control, U.S. Department of Health and Human Services, Atlanta, Ga., 1989.

Schaefer, J. T., D. L. Kelly, and R. F. Abbey, A minimum assumption tornado-hazard probability model, *J. Clim. Appl. Meteorol.*, 25, 1934–1945, 1986.

Schmidlin, T. W., Ohio tornado climatology, 1950–85, in *Preprints, 15th Conference on Severe Local Storms*, pp. 523–524, American Meteorological Society, Boston, Mass., 1988.

Sims, J. H., and D. D. Baumann, The tornado threat: Coping styles of the north and south, *Science*, 176, 1386–1392, 1972.

Topp, F. M., and A. G. Sauve, *The Edmonton Tornado, 31 July 1987: The Police Perspective*, 220 pp., Edmonton Police Department, Edmonton, Canada, 1988.

White, G. F., and J. E. Haas, *Assessment of Research on Natural Hazards*, 485 pp., MIT Press, Cambridge, Mass., 1975.

Calculation of Wind Speeds Required to Damage or Destroy Buildings

HENRY LIU

Department of Civil Engineering, University of Missouri-Columbia, Columbia, Missouri 65211

1. INTRODUCTION

Determination of wind speeds required to damage or destroy a building is important not only for the improvement of building design and construction but also for the estimation of wind speeds in tornadoes and other damaging storms. For instance, since 1973 the U.S. National Weather Service has been using the well-known Fujita scale (F scale) to estimate the maximum wind speeds of tornadoes [*Fujita*, 1981]. The F scale classifies tornadoes into 13 numbers, F-0 through F-12. The wind speed (maximum gust speed) associated with each F number is given in Table 1. Note that F-6 through F-12 are for wind speeds between 319 mi/hr (mph) and the sonic velocity (approximately 760 mph; 1 mph = 1.6 km/kr). However, since no tornadoes have been classified to exceed F-5, the F-6 through F-12 categories have no practical meaning [*Fujita*, 1981].

While underestimating the maximum wind speeds in tornadoes and other severe storms can lead to unsafe design of buildings, overestimating such wind speeds leads to wasteful constructions. For instance, prior to 1970, nuclear power plants in the United States were designed to resist tornado wind speeds higher than 600 mph. Such designs have been excessively costly to the public and have contributed to the diminution of the cost effectiveness of nuclear power. Another harmful effect caused by exaggerated wind speeds of tornadoes is the hesitance of the public to accept tornado-resistant design and construction. The general public still believes that tornado wind speeds are so high that any attempts to improve building performance in tornadoes are futile. This attitude has even permeated into building codes. Most building codes in the United States contain an escape clause to the effect that buildings are not designed to resist tornadoes. Such a clause releases the builder, the architect, and the engineer from any liability for any buildings they have designed or constructed, in the event that the building fails in a tornado or a reported tornado.

Because of the foregoing reasons, it is essential that the maximum wind speeds of tornadoes and other severe storms be neither overestimated nor underestimated. This points to the importance of realistic determination of wind speeds of severe storms.

The state of the art in determining wind speeds of tornadoes from damaged structures is described by *Mehta* [1976] and *Minor et al.* [1977]. These publications indicate that ordinary engineering estimates of tornado wind speed are based on the analysis of simple structures such as traffic signs, or engineered buildings, structures whose behaviors are more predictable and hence given higher "credence levels." These studies have found that the wind speeds for high F scale tornadoes (F-3 and above) given in Table 1 appear to be grossly overstated. Additional study is needed to improve the accuracy of engineering estimates of wind speed from damaged buildings, not only the engineered buildings but also the nonengineered and marginally engineered buildings which constitute the bulk of buildings damaged in wind storms.

The purpose of this paper is to explore how to predict as accurately as possible the minimum wind speed required to damage or destroy a building, and to make such predictions more a science than an art. Improved accuracy of such wind speed predictions will benefit not only the engineer who must design and construct safe buildings and structures at reasonable costs, but also the meteorologist who must report to his (her) organization and to the public the estimated maximum wind speed associated with each damaging storm. Understanding the basic characteristics of tornadoes and other wind storms will also be enhanced by improved accuracy in such wind speed estimates.

TABLE 1. F Scale Classification According to Wind Speed

	F Number						
	F-0	F-1	F-2	F-3	F-4	F-5	F-6 to F-12
Maximum gust speed, mph*	40–72	73–112	113–157	158–206	207–260	261–318	319 to sonic velocity

Note: 1 mph = 1.6 km/hr.

2. Wind Loading and Building Response

Accurate determination of the wind speed from building damages observed, and the converse task of predicting building damages from known wind speeds, require a good knowledge of wind engineering, structural engineering, and construction practice. Some knowledge of meteorology is also helpful in accomplishing this goal.

Wind-engineering knowledge is needed for the determination of the wind loads (pressure and forces) on buildings, and a knowledge of structural engineering and construction practice is essential to predicting the building response: how buildings react to known wind loads and how buildings get damaged or fail in high winds. Both the loading and the response must be accurately determined before one can accurately determine the extent of damage associated with a given wind speed.

2.1. Wind Loading

Wind creates pressure on the surface of buildings, and the pressure generates forces (loads). Different parts of a building encounter different wind pressures and loads. For instance, when wind direction is perpendicular to a building, the windward wall experiences an external pressure that is higher than the ambient atmospheric pressure at the ground level, whereas the leeward and the side walls experience an external pressure lower than ambient. Thus the relative pressure (i.e., wind pressure on the building minus the ambient pressure) on the windward wall is positive, while it is negative on leeward and side walls. The relative pressure on ordinary roofs is negative except on the windward part of a steep roof. Figure 1 depicts the pressure distribution over a typical rectangular building with a flat roof. Note that arrows pointing toward the building represent positive pressure (simply "pressure"), whereas arrows pointing away from the building represent "suction" (negative pressure).

The pressure discussed in the previous paragraph is that generated on the external surfaces of buildings. It is called the "external pressure." Wind also generates a pressure inside buildings termed the "internal pressure." The wind loads on building cladding (i.e., walls, roofs, and ceilings) depend on the difference between the external and the internal pressures acting on opposite sides of the cladding. Generally, the net wind load, F, on any part of cladding is

$$F = A(P_e - P_i) \quad (1)$$

where A is the cladding area, P_e is the average external pressure on the part of cladding for which F is calculated, and P_i is the internal pressure. When F is positive, the net wind load F is directed toward the building, and when F is negative, the wind load acts away from the building.

It is important to realize that the external pressure P_e depends on building geometry, not on cladded openings. In contrast, the internal pressure P_i depends strongly on cladding openings. For instance, with a window or door open on the windward wall, the internal pressure rises (becomes positive). However, when cladding openings are on leeward and side walls, the internal pressure drops (becomes negative). Figure 2 shows the variation of internal pressure with cladding openings. Very high positive internal pressure can be generated by opening a door or a window on the windward wall in high winds. Such high internal pressure often contributes to the failure of lightweight roofs.

In general, the wind pressure P on a building, for both internal or external presssures and for both pressure and suction, is proportional to the square of the wind velocity (speed) V in the following manner:

$$P = C_p \rho V^2/2 \quad (2)$$

where ρ is the density of the air and the proportionality constant C_p is called the "dimensionless pressure" or "pressure coefficient." Substituting (2) into (1) yields

$$F = A(C_{pe} - C_{pi})\rho V^2/2 \quad (3)$$

where C_{pe} and C_{pi} are the external or internal pressure coefficients, respectively.

Equation (3) shows that accurate determination of wind

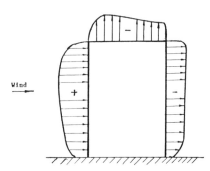

Fig. 1. Wind-generated pressure around a block-type building.

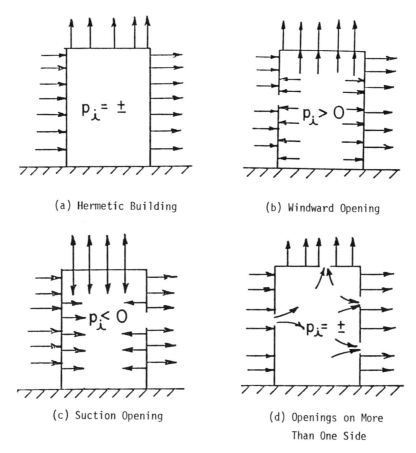

Fig. 2. Variation of equilibrium internal pressure with opening. (Note: Wind is from left to right. Arrows show directions of pressure. Arrows outside the building indicate external pressure, whereas arrows inside indicate internal pressure.)

loads on buildings requires an accurate knowledge of the external and internal pressure coefficients, C_{pe} and C_{pi}, respectively. Values of C_{pe} are normally determined from wind tunnel tests and are given in handbooks and standards such as those of the *American Society of Civil Engineers (ASCE)* [1988]. Values of C_{pi} for a given building, on the other hand, can be calculated in a simple manner if the cladding opening conditions are known, as discussed by *Liu* [1991] and *Liu and Darkow* [1989].

Accurate determination of the wind loads on a building in high wind requires the use of (3) on various parts of the building, using different values of C_{pe} for different building parts. The value of C_{pi}, in contrast, is uniform throughout the interiors of a building except in cases where the building has tightly sealed rooms, which give rise to different internal pressures in different rooms.

2.2. Building Response

Different types of buildings respond differently to high winds. Some buildings are wind-sensitive; others are wind-resistant. For instance, manufactured housing (mobile homes), ordinary wood frame houses, ordinary buildings with sheet metal roofs, and ordinary buildings that use unreinforced masonry walls are known to be particularly sensitive to wind. They are often damaged or destroyed by winds at less than 100 mph (gust speed at roof height). In contrast, structurally engineered buildings (i.e., buildings that have been designed through proper structural analysis), such as reinforced concrete buildings or steel-framed buildings (even for skyscrapers) are seldom destroyed or severely damaged by wind.

The wind resistance of buildings depends not only on the type of buildings but also on construction details. A good knowledge of various ways buildings are constructed is essential to the determination of building responses to high winds. Because building codes govern the construction practice of buildings, the kind of building code used in a particular city or county has a profound effect on the wind resistance of buildings. For instance, some wood frame houses built in rural counties in the nation where no building codes exist may use nails to anchor walls to foundations, a totally inadequate practice. Such houses can be blown off

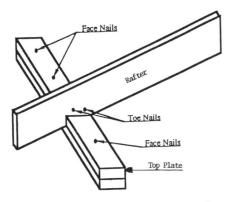

Fig. 3. Ordinary rafter-to-top-plate joint. (Note that two toenails are used on one side of the rafter, and one toenail on the opposite side not shown).

foundations and be destroyed in an 80-mph wind. (All wind speeds referred to in this section and hereafter, unless otherwise specified, are the gust speeds at the mean roof height of a house.) No building codes would allow such dangerous construction.

On the other hand, houses in southern Florida built according to the South Florida Building Code [*Saffir*, 1987] and houses in the coastal regions of North Carolina built according to the 1986 North Carolina Uniform Residential Building Code [*Sparks*, 1987] (two of the best codes in providing wind-resistant houses) should be able to resist high winds of the order of 140 mph without substantial damage. This is due to the stringent requirements of these codes with respect to roof tie-down and other construction details. In most cities in the United States governed by "model codes" such as the Uniform Building Codes, the Standard Code (Southern Code), and the Basic Code (BOCA), a wood frame house will lose its roof in 80- to 120-mph winds because of the toenailed rafter-to-top-plate connection, and the roof failure often causes walls to fail as well. Note that toenailing is a common practice in roof connection sanctioned by most building codes. Each toenailed joint contains three toenails which are nails driven at a 45° angle to lumber surface near the bottom edge of the lumber (see Figure 3). As analyzed by *Conner et al.* [1987] and *Canfield et al.* [1991], the use of rafter ties instead of toenails greatly increases the wind resistance of wood frame roofs.

The foregoing discussion shows that the wind speeds that damage or destroy a building depend to a large extent on construction deals which in turn depend on building codes. Trying to determine wind speed from building damages without consideration of the particular building code used in the damage area, or construction details of individual damaged houses, as normally done in F scale classification of tornadoes, can result in grossly inaccurate wind speed estimates.

3. Determining Wind Speed Required to Damage Buildings

3.1. Weak Links and Failure Modes

Wind damage investigations conducted by various engineering groups have established that most wind-sensitive buildings have particular weak points (the "weak links") that are responsible for the failure of such buildings in high winds. For instance, it is known that the most common failure of wood frame houses is roof failure, or the collapse of the entire house triggered by roof failure. Once the roof of a house is lifted off by winds, the walls of the house are no longer tied together on the top and they may be blown over by the wind, resulting in total building failure. Roof failure of such houses, on the other hand, is often caused by the failure of connections, especially the connections between the roof and the walls (the rafter-to-top-plate joint; see Figure 3). This particular joint normally consists of three toenails; it cannot resist the large uplift forces generated by high winds.

The first step toward determining the wind speed that has damaged a given building is to conduct a detailed wind damage investigation to determine how the building failed (the failure mode). Normally, the failure mode is related to one of the weakest links of the building. Once the weakest link responsible for building failure is identified, an engineering analysis can be conducted to calculate the corresponding minimum wind speed that can cause the damage. The weak links of wood frame houses, masonry buildings, and metal buildings are discussed in detail by *Liu et al.* [1989], *Sparks et al.* [1989], and *Perry et al.* [1989], respectively.

3.2. Equations for Calculating Wind Speed From Damage

Different failure modes require different equations for determining the failure wind speed. The equations for only one mode of failure, roof blow-off due to roof-to-wall-joint failure, will be described here. More equations for other modes of failures are given by *Liu* [1990].

For wind perpendicular to the ridge of a gable roof house (Figure 4), the vertical pullout force on the windward joints per linear length of the roof along the eave is

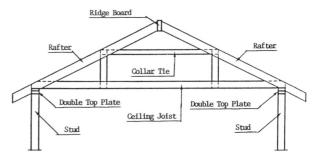

Fig. 4. Typical construction of the wood frame of a gable roof.

$$F_{1v} = Bq \cos \theta \left[\left(1 - \frac{B_0^2}{B^2}\right) \frac{(C_{p1} - C_{p2})G}{4 \cos^2 \theta} - \left(1 + \frac{B_0}{B}\right) G C_{p1} \right.$$
$$\left. + \frac{B_0}{B} G C_{ps1} + \frac{B_0^2}{4B^2} G(C_{ps1} - C_{ps2}) + G_i C_{pi} \right]$$
$$- B \cos \theta \left(w + w_0 \frac{B_0}{B}\right) \quad (4)$$

where B is the length of the rafter or top chord of the truss measured from the ridge to the wall; B_0 is the length of the overhanging part of the rafter or top chord; q is the stagnation pressure $0.5\rho V^2$ where ρ is the air density and V is the fastest-mile wind speed at the mean roof height; θ is the roof slope (assuming a gable roof); G is the gust response factor for external pressure; G_i is the gust response factor for internal pressure; C_{p1} is the average external pressure coefficient on the windward part of the roof; C_{p2} is the average external pressure coefficient on the leeward part of the roof; C_{ps1} and C_{ps2} are the mean pressure coefficients for the windward soffit and leeward soffit, respectively; C_{pi} is the internal pressure coefficient; w is the weight per unit area of the roof in the region between the walls; and w_0 is the weight per unit area of the roof overhangs. Likewise, the pullout force on the leeward joints per linear length of the roof along the eave is

$$F_{2v} = Bq \cos \theta \left[\left(\frac{B_0^2}{B^2} - 1\right) \frac{(C_{p1} - C_{p2})G}{4 \cos^2 \theta} \right.$$
$$- \left(1 + \frac{B_0}{B}\right) G C_{p2} - \frac{B_0^2}{4B^2} (C_{ps1} - C_{ps2})G$$
$$\left. + \frac{B_0}{B} G C_{ps2} + G_i C_{pi} \right] - B \cos \theta \left(w + w_0 \frac{B_0}{B}\right)$$
$$(5)$$

For a flat roof without overhang, (4) and (5) reduce to

$$F_v = F_{1v} = F_{2v} = B[(G_i C_{pi} - G C_{pe})q - w] \quad (6)$$

where $C_{pe} = C_{p1} = C_{p2}$ and B is half of the roof length in the wind direction. From (6), F_v is positive when w is less than $(G_i C_{pi} - G C_{pe})q$ and vice versa. This means the roof pulls the joints upward only when the wind-generated uplift on the roof is greater than the weight of the roof.

For a flat roof with overhangs, (4) and (5) yield

$$F_{1v} - F_{2v} = GqB \left(\frac{B_0^2}{B^2} + \frac{B_0}{B}\right)(C_{ps1} - C_{ps2}) \quad (7)$$

Because $(C_{ps1} - C_{ps2})$ is positive, (7) shows that $(F_{1v} - F_{2v})$ is positive (or F_{1v} is greater than F_{2v}) for flat roofs with overhang, and $F_{1v} = F_{2v}$ when no overhang exists.

For the horizontal component of force on joints, the following equations may be used: For the windward joints,

$$F_{1H} = khq(C_{p1w} - C_{pi}) \quad (8)$$

For the leeward joints,

$$F_{2H} = khq(C_{pi} - C_{p2w}) \quad (9)$$

In the above equations, C_{p1w} and C_{p2w} are the external pressure coefficients for the windward and leeward walls, respectively; h is the wall height; and k is the fraction of the out-of-plane force distributed to the top of the wall or the roof-to-wall joints. Note that k equals 0.5 if we assume the horizontal wind load on each wall to be equally shared by the supports on the top and the bottom of the walls. The horizontal forces F_{1H} and F_{2H} are in the wind direction when positive and in the opposite direction of wind when negative. From (8) and (9) the horizontal forces on windward and leeward joints, F_{1H} and F_{2H}, depend to a large extent on the internal pressure of the building: larger internal pressure causes a larger force on the leeward joints, F_{2H}, and a smaller force on the windward joints, F_{1H}.

Once the vertical forces and horizontal forces on the joints are determined from the foregoing equations, the resultant force is

$$F = (F_H^2 + F_v^2)^{1/2} \quad (10)$$

and the direction of this force is

$$\phi = \tan^{-1} F_H/F_v \quad (11)$$

where ϕ is the angle measured from the vertical. The quantities F, F_H, and F_v in (10) and (11) can be for either windward or leeward joints.

Once the force per length, F, is found from (10), the force on each joint, be it the three-toenail joint or any other joint, can be found from

$$F_j = F/n \quad (12)$$

where n is the number of joints on each side of the roof per linear width of the roof along the eave or ridge. The force F_j can be compared to the strength of the joint in order to determine whether the joint is adequate to resist the wind load. Joint strength can be found from published independent sources such as *Canfield* [1989].

3.3. Procedure and Example

Equations (4) through (12) can be used for calculating the force on each roof-to-wall joint once the wind speed V is known. The converse problem of determining the wind speed V from observed roof damage can be solved by using the same equations in the same order through a trial-and-error approach. The approach assumes various wind speeds to determine the various corresponding forces on the joint. By comparing the calculated joint forces with the anticipated or tested pullout strength of the joint, the minimum wind speed that can cause roof failure can be ascertained. This is illustrated in the following example.

Suppose a wood frame house has lost its gable roof in a

wind storm. An examination of the damage shows that the whole roof was lifted off the walls with the latter left standing. Failure was caused by toenails being pulled out of the top plate, a common mode of failure. From the damage pattern, debris scatter, and bending of nearby vegetation, it was determined that the damaging wind was from southwest to northeast, which is in the direction prependicular to the ridge/eave of the building. Thus the equations derived in the previous section hold.

From damage investigation it was found that $B_0 = 2.0$ feet (1 foot = 0.3 m), $B = 12$ feet, $L = 20.8$ feet, $\theta = 30°$, h (wall height) = 15 feet, $w = 5$ psf (1 psf = 48 Pa), and $w_0 = 4$ psf. From *ASCE* [1988], $C_{p1} = -0.2$, $C_{p2} = -0.7$, $C_{p1w} = 0.9$, and $C_{p2w} = -0.5$ for this building. The pressure coefficients under the roof overhangs are approximately $C_{ps1} = 0.9$ and $C_{ps2} = -0.5$. Because a large windward window failed during the storm, high internal pressure is anticipated, and hence we choose $C_{pi} = +0.7$. The gust factors G and G_i are assumed to be unity because we are interested in the gust speed rather than the fastest-mile speed used in ordinary engineering designs.

Using the aforementioned quantities, (4), (5), (8), and (9) yield, respectively,

$$F_{1v} = 13.04q - 58.9 = 0.0143V^2 - 58.9 \quad (13)$$

$$F_{2v} = 13.11q - 58.9 = 0.0144V^2 - 58.9 \quad (14)$$

$$F_{1H} = 1.5q = 0.00165V^2 \quad (15)$$

$$F_{2H} = 9.0q = 0.0099V^2 \quad (16)$$

Let us first try the value of $V = 100$ feet/s (70 mph). Equations (13)–(16) yield $F_{1v} = 84.1$ lb, $F_{2v} = 85.1$ lb, $F_{1H} = 16.5$ lb, and $F_{2H} = 99$ lb (1 lb = 0.45 kg). From (10) the resultant forces on the windward and leeward joints are $F_1 = 85.7$ lb and $F_2 = 130.5$ lb, respectively. This shows that the critical joint for this case is the leeward joint.

Suppose the roof frame consists of wood trusses spaced 24 inches (61 cm) apart; the number of joints per linear width of the roof is $n = 0.5$. Then, from (12) the uplift on each leeward joint is $F_j = 130.5/0.5 = 261$ lb. Suppose each joint consists of three 8d toenails as required by the local building code. From tests reported by *Cranfield* [1989], the rafter-to-toe-plate joint using 3–8d toenails has average pullout resistance strength of 264 lb, which is only slightly higher than that calculated from the 100-feet/s wind. Using the same procedure, it can be shown that the average failure force of 264 lb is reached when the gust speed at the roof level of this house reaches 100.5 feet/s.

The foregoing example, plus numerous other examples studied by the writer, indicate that it takes no more than a 120-mph wind, often much less speed, to severely damage or destroy an ordinary wood frame house, a building with a sheet metal roof, or a building with nonreinforced masonry walls, constructed according to existing building codes. This raises serious doubt about the high wind speeds given for F-3 and higher-scale tornadoes on the basis of damages to ordinary (nonengineered and marginally engineered) buildings. If most such buildings can be destroyed by a 120-mph wind, then F-3 and higher-number tornadoes cannot be justified on the basis of wind damage to ordinary houses, no matter how severe the damage appears to be. Furthermore, *Golden* [1976, p. 39] conducted a detailed analysis of tornadoes and concluded: "Even allowing for errors and conservative nature of the tracers used in photogrammetric studies of tornadoes, I should conclude that maximum wind speeds in tornadoes are no more than 110 m/s (240 to 250 mph) in the lowest 10–20 m above ground." All these confirm a growing consensus in the wind-engineering profession and by many meteorologists, including T. T. Fujita, that the wind speeds associated with F-3 and higher-scale tornadoes appear overrated.

4. Concluding Remarks

The foregoing procedure for determining wind speed must be based on a detailed wind damage investigation to provide the basic data needed such as the geometry and the type of building; the mode of failure; the construction details of the failed parts; the conditions of windows, doors, and other exterior openings immediately before the occurrence of the wind damage; and the wind direction at the building site. In addition, if the strength of the parts or the joints causing the failure is uncertain, tests must be performed to ascertain such strength. The damage investigation should be conducted professionally and scientifically in a manner recommended by *Liu* [1990]. For instance, the direction of the wind that has damaged a building should not be determined simply by debris scatter. The damage pattern, rather than the debris scatter pattern, is often a more reliable indication of the damaging-wind direction. Whether a window or door was opened or damaged before or after a building was damaged cannot be determined without careful consideration of evidence and without interviewing the surviving occupants of the building about what happened just before the building failed.

Even after taking all the precautions and cares required of a well-conducted engineering investigation of wind damage to buildings and after performing all the engineering calculations illustrated here, it should be realized that the wind speed calculated is still an approximation which can easily be in error by more than 20%, mainly because of the uncertainties in some of the data collected and because of the simplifying assumptions used. The margin of error can be reduced by calculating the wind speeds from several damaged neighboring buildings and other structures, and then taking the average of the speeds found.

Furthermore, one should realize that the wind speed calculated from a failed structure or structural component is the minimum, rather than the actual, wind speed. Any wind speed higher than that calculated could have caused the structure to fail. On the other hand, the wind speed calculated from a structure that has survived a wind storm without major damage gives the maximum wind speed; any wind speed lower than that would have produced the same result. Therefore determination of wind speed from failure analysis requires an analysis of both failed and unfailed neighboring structures or parts of structures. If the failure wind speeds for a failed and an unfailed structure (or part) exposed to the same wind are V_1 and V_2, respectively, the actual wind speed must be between these two values.

Engineering calculation of wind speed based on structural damage and analysis, such as described here, is inexact and crude. This is so because of possible uncertainties involved in the determination of wind direction and opening conditions at the time of damage, and uncertainties with regard to materials' properties and construction quality. In spite of that, such calculation still provides the most accurate and reliable wind speed estimate of any damaging storm in situations where no direct measurements of wind speed are taken or available and where photogrammetric analyses of wind speed are absent. Using such estimates in the case of tornadoes can provide the meteorologist with a tool to recalibrate the wind speeds in the F scale classifications, a task that the meteorological profession should undertake with the help of wind engineers.

More research is needed to improve the method reported and illustrated herein, and application of this method to wind damage investigations is recommended for the future. Misuse of this method can be avoided or kept to a minimum by taking proper precautions discussed herein and in more detail by *Liu* [1990].

Acknowledgment. This study resulted from a research project entitled "Response of Wood-Frame Houses to High Winds," supported by the Natural and Man-Made Hazard Mitigation Program, National Science Foundation, under grant 88-08425.

References

American Society of Civil Engineers, Minimum design loads for buildings and other structures, *Stand. 7*, 94 pp., New York, 1988.

Canfield, L. R., Ultimate strength of various rafter ties, report for M.S. degree, 48 pp., Dep. of Civ. Eng., Univ. of Missouri, Columbia, 1989.

Canfield, L. R., S. H. Niu, and H. Liu, Uplift resistance of various rafter-wall connections, *For. Prod. J., 41*(7/8), 27–34, 1991.

Conner, H. W., D. S. Gromala, and D. W. Burgess, Roof connections in houses: Key to wind resistance, *J. Struct. Eng., 113*(12), 2459–2473, 1987.

Fujita, T. T., Tornadoes and downbursts in the context of generalized planetary scales, *J. Atmos. Sci., 38*(8), 1511–1534, 1981.

Golden, J. H., An assessment of windspeeds in tornadoes, in *Proceedings of the Symposium on Tornadoes: Assessment of Knowledge and Implications for Man*, pp. 5–42, Texas Tech University, Lubbock, 1976.

Liu, H., How to conduct scientific investigation of wind damage to non-engineered buildings (2nd draft), report prepared for the Task Committee on Wind Damage Investigation, 65 pp., Am. Soc. of Civ. Eng., New York, 1990.

Liu, H., *Wind Engineering—A Handbook for Structural Engineers*, 209 pp., Prentice Hall, Englewood Cliffs, N. J., 1991.

Liu, H., and G. L. Darkow, Wind effect on measured atmospheric pressure, *J. Atmos. Oceanic Technol., 6*(1), 5–12, 1989.

Liu, H., H. S. Saffir, and P. R. Sparks, Wind damage to wood-frame houses: Problems, solutions and research needs, *J. Aerosp. Eng., 2*(2), 57–70, 1989.

Mehta, K. C., Windspeed estimates: Engineering analysis, in *Proceedings of the Symposium on Tornadoes: Assessment of Knowledge and Implications for Men*, pp. 89–103, Texas Tech University, Lubbock, 1976.

Minor, J. E., J. R. McDonald, and K. C. Mehta, The tornado: An engineering-oriented perspective, *Tech. Memo. ERL NSSL-82*, 196 pp., Natl. Severe Storms Lab., Natl. Oceanic and Atmos. Admin., Norman, Okla., 1977.

Perry, D. C., J. R. McDonald, and H. S. Saffir, Strategies for mitigating damage to metal building systems, *J. Aerosp. Eng., 2*(2), 71–87, 1989.

Saffir, H. S., State of Florida and South Florida building codes to prevent hurricane damage, in *Proceedings of the WERC/NSF Symposium on High Winds and Building Codes*, pp. 115–128, Engineering Extension, University of Missouri, Columbia, 1987.

Sparks, P. R., The North Carolina residential building code and its wind load requirements, in *Proceedings of the WERC/NSF Symposium on High Winds and Building Codes*, pp. 107–114, Engineering Extension, University of Missouri, Columbia, 1987.

Sparks, P. R., H. Liu, and H. S. Saffir, Wind damage to masonry buildings, *J. Aerosp. Eng., 2*(4), 186–199, 1989.

Risk Factors for Death or Injury in Tornadoes: An Epidemiologic Approach

SUE ANNE BRENNER AND ERIC K. NOJI

Centers for Disease Control, Atlanta, Georgia 30333

INTRODUCTION

On August 28, 1990, between 3:15 and 3:45 P.M., a tornado beat a path of destruction through the Will County, Illinois, towns of Plainfield, Crest Hill, and Joliet [*National Weather Service*, 1991]. The parent severe thunderstorm formed on the Illinois-Wisconsin border and moved southeastward across northeast Illinois. Over its 4-hour lifetime it produced several other less damaging tornadoes, as well as large hail and strong straight line winds. The Plainfield tornado was rated as an F-5 tornado, with a path length of 16.5 mi (0.62 mi = 1 km) and path width of 700 yd (1.09 yd = 1 m). No tornado warning was issued by the National Weather Service. It was the worst tornado in Illinois in more than 20 years and one of the most violent in U.S. history. The tornado severed electrical power to 65,000 homes and businesses, cut off phone service to 10,000 residences, and caused more than $200 million worth of damage. When the tornado hit, few people were in a protected area such as the basement or inner area of a house. As a result of the storm's impact, 302 people were injured, including 80 persons who were hospitalized and survived and 28 who died.

METHODS

To prevent such casualties during future tornadoes, the Centers for Disease Control conducted an investigation to assess the risk factors for injury or death. Researchers abstracted 350 emergency room and inpatient medical records from eight hospitals. We obtained further information for this report from the American Red Cross, coroners, and newspapers.

An impact-related injury or death was defined as an injury or death caused by the direct mechanical effects of the tornado. Postimpact injuries were defined as injuries occurring within 48 hours of the tornado that would not have occurred in the absence of the tornado (such as injuries sustained by walking through the debris, during cleanup, or as a result of the loss of electrical power).

The Tornado: Its Structure, Dynamics, Prediction, and Hazards.
Geophysical Monograph 79
This paper is not subject to U.S. copyright. Published in 1993 by the American Geophysical Union.

RESULTS

The majority of persons whom we interviewed reported having had little or no time to reach protective areas.

Of persons injured during the impact phase, 301 sought medical attention at hospitals within 48 hours of the disaster, 193 (64%) were treated in an emergency department and released, 80 (27%) were admitted to a hospital and survived, five (2%) were hospitalized and later died, and 23 (8%) were killed instantly.

Although most impact-related deaths occurred instantaneously, four impact-related injuries resulted in death 2–8 weeks after the tornado, including the death of one man who died 8 weeks later because of complications of a chest contusion suffered during the impact phase. At least 44 injuries and one death were due to postimpact events. The postimpact death involved a man who put a gas-powered generator into his garage after electricity was lost. The automatic garage door was open due to loss of power; however, when power was restored, the garage door closed, resulting in the man's asphyxiation from carbon monoxide produced by the generator.

Of the 28 persons who died from impact-related injuries, eight were younger than 20 years of age (range: 1 month to 69 years; mean: 34 years); 14 were male. Nine people died in a large apartment complex, eight in vehicles, five in schools, four in houses, and two outside. We could not determine whether most of those who died in the apartment complex were inside or outside when the tornado hit, since several of them were reported to have been blown from the second floor. Three people died at one high school, where at least 10 students avoided death or severe injury by crouching against the only hallway wall that did not collapse.

Most victims were treated at one of eight local hospitals, with several of the more severely injured transferred to tertiary-care facilities in Chicago. Of those injured in the tornado, 221 patients were treated in one local emergency room, with the next highest total being 38 at a second hospital. This unequal distribution of patients occurred because many victims reached hospitals by their own devices, effectively bypassing local ambulance services. Fortunately, since the disaster occurred during the first hospital's change of shift, twice as many health care personnel were available to care for the onslaught of patients. In addition, numerous physicians and nurses volunteered their services.

Discussion

Previous studies of tornadoes have shown that persons older than 60 years of age are more likely to be injured than are people younger then 20, presumably as a result of preexistent medical illnesses, decreased mobility, decreased ability to comprehend and respond rapidly to tornado warnings, and greater susceptibility to injury from comparable amounts of mechanical energy [*Glass et al.*, 1980]. In the Plainfield disaster the relatively high proportion of deaths and injuries among persons younger than 20 years of age (37.9%) compared with those older than 65 years (9.7%) may reflect the population affected (primarily a suburban, family-oriented community, with a median age of 27.5 years), the time of day (3:15 P.M., when homemakers and young children are at home), and the lack of warning, making it almost equally likely that those with and without disabilities would be in a dangerous area.

The injuries responsible for most of the deaths in this tornado included head injuries, multiple fractures, and arterial lacerations. In one study [*Glass et al.*, 1980] the most common primary diagnoses noted in the hospital records of seriously injured patients included trauma to the head, extremities, and thorax and severe lacerations and abrasions. As in our study, they were unable to determine whether injuries were due to high-velocity projectiles or to the collapse of structures. In the study noted, *Glass* [1980, p. 736] states, "most patients who received major abrasions and lacerations had not covered themselves with blankets, pillows, or mattresses"; in our study the victims did not have sufficient time to take such preventive measures.

The results of epidemiologic studies of tornadoes have shown that people who attempt to drive their cars away from their homes to escape a tornado have an increased risk of death and severe injury, as do occupants of mobile homes [*Glass et al.*, 1980]. There are anecdotal accounts of people surviving a tornado's impact by leaving their cars for ditches or overpasses. Unfortunately, no systematic data analysis has been undertaken to determine whether one should stay inside a vehicle or leave the vehicle for a ditch or flat ground if no building is nearby for sheltering. A current National Weather Service recommendation suggests leaving one's car for a ditch if caught in the path of a tornado. No recommendations exist, however, for occupants of vehicles in areas lacking ditches or other protection.

The degree of building failure is highly correlated with the risk of death or injury [*Glass et al.*, 1980; *Duclos and Ing*, 1989]. Buildings may fail in several ways: translation (lateral movement of the entire structure), racking (lateral collapse of the structure), overturning, material failure, and connection separation [*Defense Civil Preparedness Agency*, 1979]. Apartments often sustain heavy damage, probably because of their long-span roofs and other structural factors. Because of the difficulty in locating apartment occupants who have lost their homes, previous disaster researchers have not compared the risk of being in an apartment with that of being in a single-family home. *Glass et al.* [1980] found that people inside brick homes suffered serious injuries less frequently than those inside wood frame homes. On the other hand, *Duclos and Ing* [1989] found that people were more likely to be injured if they were occupants of a house with walls constructed of wood (see also *Glass et al.* [1980]). In these two studies, information on housing materials was obtained from the building occupants themselves; in neither study were other sources of information about housing structure used (for example, from architects, wind engineers, local land use offices, tax assessors, etc.).

The findings in this report are consistent with the results of previous investigations. These include the following observations: (1) automobiles are particularly lethal; (2) most deaths occur during the impact phase of a tornado disaster; (3) tornadoes can result in large economic losses; and (4) timely warnings and appropriate protective actions may significantly reduce the number of deaths and serious injuries.

We make the following recommendations to minimize tornado-related morbidity: (1) institution of adequate early warning systems in regions at risk of tornado disasters (improved methods of detecting tornadoes, such as "next generation radar," in all highly tornado-prone areas), (2) construction of tornado-resistant shelters or structurally sound areas within buildings, (3) development of disaster preparedness and response plans, (4) identification of occupant behaviors that maximize the possibility of survival when a tornado strikes, and (5) institution of public education programs that teach appropriate protective actions during tornadoes. We are currently studying the contribution of specific housing characteristics and occupant behaviors to the risk of death or injury during a tornado disaster.

Acknowledgments. The authors express gratitude to the following organizations for their assistance in carrying out this study: the American Red Cross, the Illinois Department of Public Health, the Will County Health Department, and the Illinois Emergency Services Disaster Agency. The staffs of the following hospitals are also thanked: St. Joseph Medical Center, Silver Cross Hospital, Joliet; Loyola University Medical Center, May; Christ Hospital and Medical Center, Oak Lawn; Copley Memorial Hospital, Aurora; Edward Hospital, Naperville; Palos Community Hospital, Palos Heights; and Morris Hospital, Morris.

References

Defense Civil Prepardness Agency, Wind-resistant design concepts for residences: Guidelines for homeowners and builders, *Rep. TR-83*, p. 17, U.S. Gov. Print. Office, Washington, D. C., 1979.

Duclos, P. J., and R. T. Ing, Injuries and risk factors for injuries from the May 29, 1982 tornado, Marion, Illinois, *Int. J. Epidemiol.*, *18*, 213–219, 1989.

Glass, R. I., R. B. Craven, D. J. Bregman, B. J. Stoll, N. Horowitz, P. Kerndt, and J. Winkle, Injuries from the Wichita Falls tornado, *Science*, *207*, 734–738, 1980.

National Weather Service, The Plainfield/Crest Hill tornado, report, Natl. Oceanic and Atmos. Admin., U.S. Dep. of Commer., Silver Spring, Md., May 1991.

Design for Occupant Protection in Schools

HAROLD W. HARRIS

BGR Architects/Engineers, Lubbock, Texas 79411

KISHOR C. MEHTA AND JAMES R. MCDONALD

Texas Tech University, Lubbock, Texas 79409

INTRODUCTION

The authors of this paper live in Lubbock, Texas. Lubbock is situated on the High Plains of west Texas at the southwest end of the so-called "tornado alley" which stretches from Texas through Oklahoma and Kansas into the midwest. Schools with severe storm protection capabilities, as illustrated in the examples, were designed by BGR Architects/Engineers (A/E).

People everywhere, and particularly residents of "tornado alley," are very concerned about protection from tornadoes. School administrators often ask for information about the location of relatively safe areas in their existing buildings or about the possibility of providing protective areas in new construction. In the past they have often been told that there was no economical way to construct this protection.

We now know that economical protection for occupants from severe storms can be incorporated into buildings during the design process. There have been many buildings, such as nuclear power plants, atomic weapons assembly facilities, and disaster control centers, which have been constructed with nearly 100% tornado protection. This protection has been achieved at a very high cost. Although this cost is acceptable for these facilities because of their high potential for causing a widespread disaster, it cannot be justified for other buildings. However, there is a great need for protection for the occupants of schools, public buildings, nursing homes, and residences. A fairly high degree of protection can be achieved in designated shelter areas within these buildings for a cost that is not prohibitive.

We already have enough knowledge about design of buildings for forces that tornadoes impose on structures. Research has established a reasonable forecast of the recurrence interval for tornadic wind speeds. This makes it possible for us to make rational choices of design parameters for buildings with differing occupancies and economic restraints.

The authors of this paper have been involved in the design and construction of at least seven schools and various other buildings with occupant protective areas. These buildings have been constructed over a 20-year period.

RATIONALE FOR SCHOOL BOARDS

We have had the design guidelines that would allow us to build protection for the occupants of buildings for more than 10 years. However, there have been relatively few public buildings constructed with substantial protection for occupants from severe storms and tornadoes. This protection can only be afforded when owners become willing to build it into their buildings. In most cases this would not be a problem if the potential owners were aware that the building could and would meet the following criteria.

Cost of construction must be acceptable. Costs can be kept relatively low if construction for occupant protection is properly planned. The protective areas would be built of the same materials as the remainder of the building. A proliferation of differing materials and building trades almost always increases the construction costs. Protective areas should be built using techniques well known in the building industry. Builders become accustomed to certain details and construction types and can build more economically using them. Adequate strength can usually be achieved without going to more exotic materials and methods of construction. The protective area should be no larger than necessary to provide quick and easy access for all the occupants of the building. Costs can be reduced by using as much inherent protection in the building as is available. Speed of construction is always a large factor in costs; therefore the protective area should be designed so that its construction can proceed at the same rate as the rest of the building.

Fig. 1. High School building in Xenia, Ohio, destroyed by tornado on April 4, 1974.

It is also important that the design of the protective area does not interfere with the normal function of the building. It might be desirable from a protection standpoint to provide heavy doors and massive screens in corridors, but if these items present a major inconvenience or hazard to daily use, it is likely that they will provide a major objection to the construction of the protective area.

School administrators and the public may fear that a protective area will look like a concrete bunker. In fact, protective areas that have been built are constructed in such a way that they can barely be distinguished from the rest of the building.

For economy, protective areas should be formed by strengthening spaces that have normal functions to provide.

Fig. 2. Undamaged interior hallway of the high school in Xenia, Ohio.

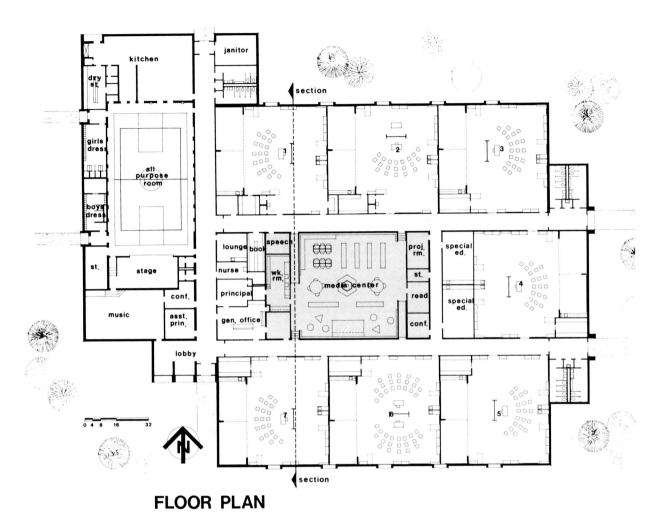

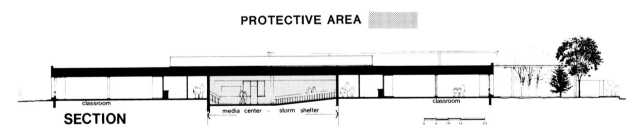

Fig. 3. Hereford Elementary School where media center located in the center of the building was strengthened for occupant protection.

If strengthening of protective areas is kept as an alternate bid in construction documents, the owner is assured of the actual cost of the protection.

BACKGROUND

Personnel of the Institute for Disaster Research at Texas Tech University, including the authors, have conducted on-site windstorm damage documentation at more than 60 locations around the country since 1970. More than half of these windstorms were tornadoes. These documentation efforts have been published as papers in technical journals or reports [*Mehta et al.*, 1971, 1981; *Minor and Mehta*, 1979; *Fujita and McDonald*, 1978; *McDonald and Marshall*, 1982]. These documentation experiences, in aggregate, have indi-

Fig. 4. Architectural rendering of junior/senior high school in Sudan, Texas.

cated that custom-engineered frames of buildings do not collapse, even in the most intense tornadoes. In addition, interiors of the buildings have provided good protection to occupants even when no consideration of tornado-resistant design was given to construction of the building. As an example, a high school in Xenia, Ohio, was virtually destroyed in the tornado of April 4, 1974 [Mehta et al., 1976], as indicated in Figure 1. The first-floor interior hallway of that school, as shown in Figure 2, sustained no damage and could provide good occupant protection. These observations suggest that occupant protective areas should be located in the interior as far as possible.

Fujita and Pearson [1973] developed the FPP scale rating to categorize intensity and size of tornadoes. *Tecson et al.* [1979] categorized every recorded tornado for the period between 1916 and 1978. This categorization of intensity of tornadoes shows that most of the tornadoes are of intensity of F-4 or less. It would be reasonable to design occupant protective areas that would resist tornado intensity of F-4 and could possibly resist even more intense tornadoes with some damage.

Design Considerations

The buildings that are described in this paper were all designed using the "Interim guidelines for building occupant protection" published in 1975 [Mehta et al., 1975]. Wind pressures, rapid internal pressure changes, and wind-borne missile criteria are specified in these guidelines. Recommendations include a design wind speed of 260 mi/hr (mph) (1 mph = 1.6 km/hr), an internal design pressure of 205 psf (1 psf = 47.9 Pa) acting outward unless venting is provided, and a design missile consisting of a 2 inch × 4 inch × 12 ft (1 inch = 2.54 cm) long piece of lumber moving parallel to its long axis at 100 mph. These design force criteria are based on categorization of tornadoes and data obtained from field research in the aftermath of a large number of tornadoes. It is believed that they exceed the forces encountered in 98% of the recorded tornadoes.

Experience gained from the construction of each building has been used to refine subsequent designs. The cost of construction of these protective areas has decreased as refinements are introduced.

It is important that tornado protective areas be easily accessible. The accuracy of weather warning continues to improve; however, the time available to move to a protective area after a warning has been issued varies from seconds to a few minutes. There are occasions when a tornado strikes inhabited areas before a warning can be issued. In cases of very short warning times, it is extremely important to have an easily accessible protective area. Schools often conduct tornado drills for the school population. In one of the schools described in this paper (Littlefield Elementary), all the students can be moved to the protective area in 1–2 min.

Protective areas are usually constructed on the same floor as the occupancy they serve. Moving several hundred students up or down stairs is time consuming. Ease of access often dictates a protective area near the center of the occupied space, fed directly by corridors. If heavily populated spaces, such as school classrooms or media centers, are constructed as protective areas, many occupants may already be sheltered when a warning is issued. This could be even more significant in a case where no warning is received before a tornado strikes. Obviously, from a protection standpoint the ideal case would be for the entire building to

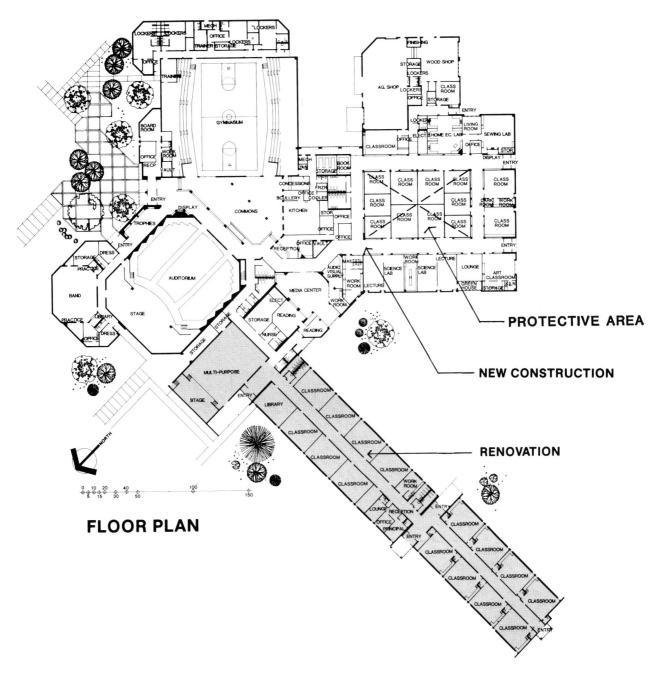

Fig. 5. Protective area in junior/senior high school in Sudan, Texas, is surrounded by corridors on all sides.

be strengthened, but this is seldom economically feasible. Difficult or time-consuming movements may discourage the use of sufficient drills, and in the case of an actual storm may cause a delay in getting occupants to safety.

The most important function of the protective area is to provide dependable protection from tornadoes. This does not necessarily mean absolute protection, but it should be much better protection than would be available to its occupants anywhere else in the building. It should prevent injury to nearly all of those it shelters, except from those rare storms whose intensity exceeds the design criteria. A properly designed protective area can provide substantial safety from most severe tornadoes.

One damage mitigation strategy for the protective area is to locate it as far from the exterior walls as possible. Walls of conventional construction provide some shielding from dam-

Fig. 6. Exterior view of the primary school in Littlefield, Texas.

aging forces. Physical distance from the exterior allows dissipation of missile energy before the missiles strike the walls of the protective area. Some of the protective areas shown in this paper utilize a method of connecting the conventional roof deck in areas outside the protective area to the roof of the protective area with a very strong hinge-type connection. We believe that forcing the debris of the collapsed structure to remain in contact with the walls of the protective area will provide additional bracing and protection from missiles. If the collapsing roof of corridors is forced to drape over shelter walls, it can be effective in preventing the entry of missiles into the protective area through doors.

The most obvious strength requirement of the protective area is to prevent its lateral collapse due to horizontal wind forces. Two force-resisting systems have been employed by the authors in the design of protective areas. Early designs utilized heavy steel frames to support the walls and roof. This was a conventional approach to the problem, but the large structural elements sometimes projected into corridors and interfered with the movement of people. Later designs employed a series of room-sized boxes with heavily reinforced walls and roofs. Calculations indicate that great strength can be attained in this way.

An important problem in tornado-resistant design is to provide adequate resistance to uplift of the roof structure. Roofs of conventional construction are designed to mostly resist downward gravity-induced loads. With high wind velocities the net upward loads become great, and the low strength of conventional roof systems to resist these upward loads makes this a critical condition. Resistance to uplift can be provided by adequately bracing the bottom portion of the roof beams and joists, by providing extra stiff members, and by providing extra weight in the roof by providing thick roof decks of normal-weight concrete. The heavy roof decks provide extra missile protection from above. It is also essential to provide positive connection between the roof member and the wall structure to resist high localized uplift pressures.

Another very important function of the protective area is to provide protection from airborne missiles. Much research has been conducted at Texas Tech University, by one of this paper's authors and by others, on the resistance of various wall types to missile penetration [McDonald and Bailey, 1985]. Only two wall types were deemed safe and practical to resist airborne missile penetration. A 6-inch-thick moderately reinforced concrete wall provides resistance to the design missile, but in the area where the example buildings were built, this wall was considered to be more expensive than masonry construction. All shielding walls in the protective areas designed by BGR A/E were constructed of 8-inch thick concrete masonry with all cells filled with normal-weight concrete and each cell reinforced.

A possible source of injury to occupants of a shelter is entry of missiles through its door openings (we do not put windows in protective areas). There are several methods of providing protection from this hazard. We could provide massive doors for each room of the protective area. This

Fig. 7. Floor plan of primary school in Littlefield, Texas, where strengthening of protective area cost less than 4% of the total building cost.

would provide protection but would be fairly costly. Another more important objection to this method of protection may be the difficulty in managing these doors during normal operation of the facility. An alternate method of protecting openings would be to construct screening walls to block direct paths of missiles. The screen walls could be constructed with strength equal to the walls of the protective area. They would offer the advantage of eliminating the problem of operating the heavy doors but would provide disruption to the normal movement of people. This method also will require the widening of corridors, resulting in a larger building and increased cost. These two methods should provide adequate protection, and each should be considered. The buildings designed by BGR A/E used conventional doors without screening walls. This method was economical and avoided the disadvantages of restricted movement and difficult operation of the doors. It resulted in some added exposure to injury for the occupants. We feel

Fig. 8. Classroom in primary school of Littlefield, Texas, is strengthened for occupant protection.

that the technique of causing the roof of the collapsed structure to drape over the walls and doors of the protective area will provide a degree of protection, as will the placement of the protective area away from the exterior of the building. An evaluation of the possible solutions to this and other design problems must balance the degree of protection against the possibility that a shelter of any kind might not be built if economic or convenience criteria are not met.

Examples of Schools Incorporating Occupant Protective Areas

Each of the example schools was designed by BGR A/E with the senior author as structural engineer. In each case the protective area was presented to the school board as a refuge from severe storms, rather than as a total protection from tornadoes. There has been no active marketing for this type of construction, but there have been many inquiries from school districts because of their knowledge that neighboring districts have constructed protective areas. The schools presented here are typical of the schools that have been constructed.

Hereford Elementary School in Hereford, Texas, was constructed in 1974. This school has a maximum capacity of 600. The media center indicated in the floor plan (see Figure 3) was strengthened to provide storm protection. This protective area is located near the center of the building and has corridors approaching it from all four sides. Wind loads are resisted by heavy steel bents spanning the media center. The bents support a 6-inch-thick concrete roof deck and are enveloped with heavily reinforced masonry walls. The shelter contains 4300 square feet (1 ft^2 = 0.093 m^2) of space.

The junior/senior high school in Sudan, Texas (see architectural rendering in Figure 4), was constructed in 1981–1982. This school has a capacity of 400 students and serves as a focus for community activity with its large auditorium and competitive gymnasium. Occasionally, twice the student population assembles in the school for a community activity. Ten classrooms in the middle of the building (shown in floor plan in Figure 5) were strengthened to provide a protective area of 7500 square feet. Each classroom has an 8-inch-thick concrete-filled masonry wall on all four sides and has a heavy concrete roof supported on steel joists. These "boxes" are all linked together. The protective area is surrounded by corridors that extend to all parts of the building. The conventional roof decks over the adjoining corridors are attached to the protective area in such a way that a collapse of the surrounding structure will leave the corridor roof draped over the walls of the protective area.

The primary school of Littlefield Independent School District was completed in August 1990 at a total cost of $2,133,900 and has a capacity of 550 students (see photograph of exterior in Figure 6). Five adjoining classrooms and a resource room in the interior (see floor plan in Figure 7) were strengthened to provide more than 5500 square feet of protective space. The cost of strengthening the protective area was determined by an additive alternate at the time of bidding to be $79,000. This only added 3.8% to the total cost of the building. The protective area is located away from the exterior of the building and has easy access from all parts of the school. The doors are of conventional classroom construction without screens or massive storm doors. The adjoining conventional construction is designed to drape over the protective area in the event of a collapse of the remainder of the building. As illustrated in Figure 8, the protective area does not look any different than an ordinary classroom.

SUMMARY

It is possible to provide a high level of protection from tornadoes for occupants of schools and by extension for many other types of buildings. The design should provide for accessible, usable, and affordable shelter space. An additional cost of less than 4% seems to be attainable for this type of construction.

REFERENCES

Fujita, T. T., and J. R. McDonald, Tornado damage at the Grand Gulf, Mississippi nuclear power plant site: Aerial and ground surveys, *Rep. NUREG/CR-0383*, U.S. Nucl. Regul. Comm., Washington, D. C., 1978.

Fujita, T. T., and A. D. Pearson, Results of FPP classification of 1971 and 1972 tornadoes, *Preprints, Eighth Conference on Severe Local Storms*, pp. 142–145, American Meteorological Society, Boston, Mass., 1973.

McDonald, J. R., and J. R. Bailey, Impact resistance of masonry walls to tornado-generated missiles, in *Proceedings, at Third North American Masonry Conference*, University of Texas, Arlington, Tex., 1985.

McDonald, J. R., and T. P. Marshall, Damage surveys of tornadoes near Altus, Oklahoma on May 11, 1982, report, Inst. for Disaster Res., Tex. Tech Univ., Lubbock, 1982.

Mehta, K. C., J. R. McDonald, J. E. Minor, and A. J. Sanger, Response of structural systems to the Lubbock storm, *Storm Res. Rep. 03*, 428 pp., Tex. Tech Univ., Lubbock, 1971.

Mehta, K. C., J. E. Minor, J. R. McDonald, and D. B. Ward, Interim guidelines for building occupant protection from tornadoes and extreme winds, *Rep 83-A*, 24 pp., Def. Civ. Prep. Agency, Baltimore, Md., 1975.

Mehta, K. C., J. E. Minor, and J. R. McDonald, Wind speed analysis of April 3–4, 1974 tornadoes, *J. Struct. Div.*, 102(ST9), 1709–1724, 1976.

Mehta, K. C., J. R. McDonald, R. D. Marshall, J. J. Abernetny, and D. Boggs, *The Kalamazoo Tornado of May 13, 1980*, 54 pp., National Academy Press, Washington, D. C., 1981.

Minor, J. E., and K. C. Mehta, Wind damage observations and implications, *J. Struct. Div.*, 105(ST11), 2279–2291, 1979.

Tecson, J. J., T. T. Fujita, and R. F. Abbey, Jr., Statistics of U.S. tornadoes based on the DAPPLE tornado tape, in *Preprints, 11th Conference on Severe Local Storms*, pp. 227–234, American Meteorological Society, Boston, Mass., 1979.

Discussion

TIM MARSHALL, SESSION CHAIR

Haag Engineering Company

PAPER J1

Presenter, Jim McDonald, Texas Tech University [*McDonald*, this volume, Damage mitigation and occupant safety]

(Arnold Court, California State University) Can you tell me how to park my car to mitigate tornado damage?

(McDonald) I don't subscribe to the theory that all tornado winds come from the southwest, so putting your car on the northeast side of a structure wouldn't help. We know that winds in a tornado can blow from any direction.

(Court) Should I park against a fence or against a building?

(McDonald) The building is probably better than the fence, but I'm not going to say that's going to help you a whole lot. If you were at home, I would recommend that you put your car in your garage. This is because we have seen many cases where the automobile parked at the curb is blown into the house and collapses a wall or a chimney.

(Rudy Engleman, private citizen) Would you comment on the recent construction practice of using foam material or wafer board for siding, rather than plywood?

(McDonald) If a weaker material is being used as opposed to plywood, which normally gives additional shear resistance to a building, I would not think that such construction would be good practice. If you're not making other provisions for shear resistance, then you're creating a problem for yourself.

(Chair) There is general complacency in building practice out there. We see more wafer board going up rather than plywood. We see a tendency toward "economic construction" and complacency. Look at the graphs Tom Grazulis showed this morning about the number of 50 or more fatality tornadoes. There has been a large reduction and people are no longer as interested in tornado safety. Don't blame the contractor. He does what people are willing to pay for. It goes back to the education process.

PAPER J2

Presenter, Tom Schmidlin, Kent State University [*Schmidlin*, this volume, Tornado fatalities in Ohio, 1950–1989]

(Mat Biddle, Oklahoma University School of Geography) I wanted to ask if you considered ethnicity instead of race, or if not, why not. I would guess that ethnicity would be important, especially going back to Grazulis' study. What is an F2 barn? If your parents were German Mennonite and you lived in Fulton County, you've got an F4 barn, not an F2 barn. These types of things could also be applied to housing and fatalities.

(Schmidlin) I'm sure you're right. Unfortunately, the death certificates do not give that information. All you can go by is the last name, and of course, that gets too confusing; so you cannot separate ethnicity from a historical standpoint. That kind of information would need to be gathered in a quick response survey, perhaps within a week of the disaster.

(Larry Twisdale, Applied Research Associates) You certainly derived many statistics from your data. Did you do any significance tests? Looking at your chart on age of victims, the only thing that looked significant to me, given the small sample of your data, was the older age group.

(Schmidlin) You make a good point. One thing that lent credibility to the age conclusions is that they were consistent over time, not just from one event. Also, remember this is not a sample; this is the population. That was the death rate for that 40-year period.

(Twisdale) Yes, but it is a sample of larger nationwide population. I'm not sure you can make some of the conclusions you have made. I think you have overworked the data and border on speculation.

(Schmidlin) The only thing I can say is that the data support the results. Things weren't different in different areas of the state, and they didn't vary from time to time.

Paper J3

Presenter, Henry Liu, University of Missouri [*Liu*, this volume, Calculation of wind speeds required to damage or destroy buildings]

(Joe Golden, National Oceanic and Atmospheric Administration, Office of the Chief Scientist) I want you to deliver something you promised in your title, Henry. What are the highest wind speeds so far in your investigations from damage analysis of tornadoes?

(Liu) We just started the program last year and investigated three storms in Missouri, only one labeled a tornado. I found the highest wind speed to be approximately 100 mph. In the case of the Columbia, Missouri, tornado in November, I was shocked, and so was my colleague, Dr. Darkow, a senior meteorologist, that it was classified as an F3 tornado. That places the wind speed much, much higher than it appeared to be from our investigation.

(Arnold Court, California State University) Over what time period and how big a volume are you talking about when you mention wind speed failure? Are you talking gusts or average wind speed?

(Liu) I'm talking about gust speed and at roof level.

(Court) Over a square foot?

(Liu) No. I mean the fastest speed at roof height. The engineer uses the mean roof height wind speed to get his pressure coefficient and so forth.

(Court) This assumes the same wind speed over the entire length of the roof?

(Liu) Yes, it is the same speed. In most of the storms I have investigated there is fairly widespread damage. The width of the tornado is wide compared to the size of the roof.

Paper J4

Presenter, Sue Brenner, Centers for Disease Control [*Brenner and Noji*, this volume, Risk factors for death or injury in tornadoes: An epidemiologic approach]

(Tom Schmidlin, Kent State University) Did you look at any interaction between age of house and age of people in the house?

(Brenner) No, I did not look at that. We also did not look at interaction of socioeconomic factors or other things that could have factored into the building construction of the house. We just used tax assessor data, which I realize is not particularly accurate, but it was a good source of a large amount of data for our statistical analysis.

Tornado Forecasting: A Review

CHARLES A. DOSWELL III

National Severe Storms Laboratory, Norman, Oklahoma, 73069

STEVEN J. WEISS AND ROBERT H. JOHNS

National Severe Storms Forecast Center, Kansas City, Missouri 64106

1. INTRODUCTION

Present-day operational tornado forecasting can be thought of in two parts: anticipation of tornadic potential in the storm environment and recognition of tornadic storms once they develop. The former is a forecasting issue, while the latter is associated with warnings (or so-called nowcasting). This paper focuses on the forecasting aspect of tornadoes by dealing primarily with the relationship between the tornadic storm and its environment (Recognition and detection issues are treated by *Burgess et al.* [this volume]). We begin with a short history of tornado forecasting and related research in section 2; in section 3 we provide an overview of current tornado forecasting procedures within the Severe Local Storms (SELS) Unit at the National Severe Storms Forecast Center (NSSFC). In section 4 we give a short summary of 35 years of SELS tornado and severe thunderstorm forecast verification. In section 5 we describe our current understanding of the connection between tornadoes and their environment. We conclude in section 6 with our thoughts about the future of tornado forecasting.

2. SHORT HISTORY OF TORNADO FORECASTING

Our historical review necessarily must be brief; interested readers can consult *Schaefer* [1986] for additional details about the history of severe weather forecasting in general; another review by *House* [1963] is somewhat dated but provides excellent background material. Although tornado forecasting has its roots in the nineteenth century, stemming mostly from the work of J. P. Finley (see *Galway* [1985] for more on Finley), it was not until the early 1950s that serious tornado forecasting began. Before then, the use of the word "tornado" in public forecasts was prohibited, largely be-

cause of the perception that tornado forecasts would cause public panic. It is clear that modern tornado forecasters owe a great deal to the pioneering efforts of *Fawbush and Miller* [1952, 1954], two Air Force officers who had some early tornado forecasting successes at Tinker Air Force Base in Oklahoma in the late 1940s. On the civilian side, work was proceeding [e.g., *Showalter and Fulks*, 1943; *Lloyd*, 1942], but until 1952 the civilian weather service (then called the Weather Bureau) still was reluctant to use the word "tornado" in any forecast. The successes of Fawbush and Miller clearly paved the way for a civilian tornado forecasting program.

The first civilian tornado forecasts began with the formation of a specialized unit as part of the Weather Bureau Analysis (WBAN) Center in Washington, D. C., during March of 1952 (see *Galway* [1973, 1989] for more details). This unit became the Severe Local Storms (SELS) Center in early 1953 and moved to Kansas City, Missouri, in August 1954, eventually forming part of the National Severe Storms Forecast Center (NSSFC) in 1966.

When SELS first came into being, the relationship between the synoptic scale environment and the tornado was not well understood; forecasting was essentially empirical. Various forecasters and researchers observed that certain meteorological elements, detectable within the large-scale data networks (surface and aloft), tended to be present in many tornado events [e.g., *Fawbush et al.*, 1951; *Beebe and Bates*, 1955]. These "features" included static instability, significant extratropical cyclones, abundant low-level moisture, jet streams, surface convergence boundaries, and so forth.

Early on, however, it became clear that no single set of such features was present with each and every tornado event; rather, particular collections of elements were associated with particular groups of cases. In effect, pattern recognition became the basis for forecasting. This approach

The Tornado: Its Structure, Dynamics, Prediction, and Hazards.
Geophysical Monograph 79
This paper is not subject to U.S. copyright. Published in 1993 by the American Geophysical Union.

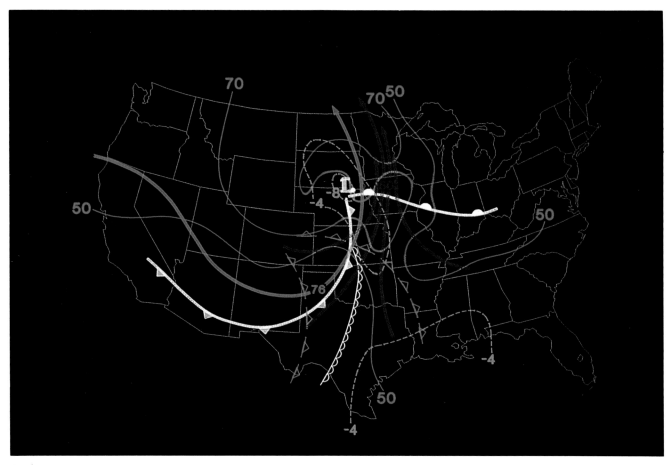

Plate 1. An example of a composite chart; in this case it is a composite prognosis depicting 12-hour forecast positions of surface and upper air features at 0000 UTC April 27, 1991, based on initial data from 1200 UTC April 26, 1991. Solid symbols are conventional surface frontal features, green lines denote 50 and 70% mean (surface to 500 mbar) relative humidity, red streamlines are at 850 mbar with a 50-knot (25 m s^{-1}) maximum indicated, blue stream line is the 500 mbar jet stream axis, with a 76-knot (38 m s^{-1}) maximum indicated, blue dashed frontal symbols depict 500 mbar thermal trough axes, and orange lines as isopleths of the forecast lifted index.

DOSWELL ET AL. 559

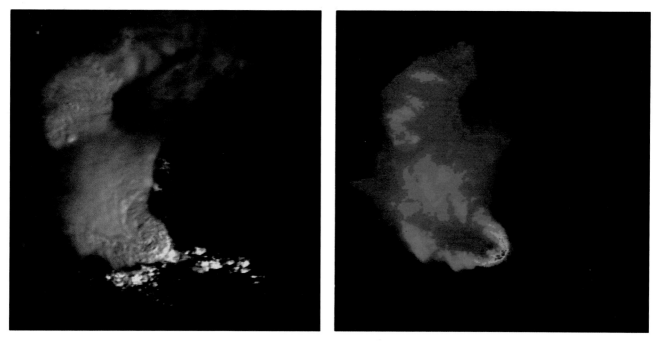

Plate 2. Example of an "enhanced V" signature from a satellite image [from *Setvák and Doswell*, 1991], showing the visible appearance (left) and the enhanced thermal infrared appearance (right), the latter of which emphasizes the signature.

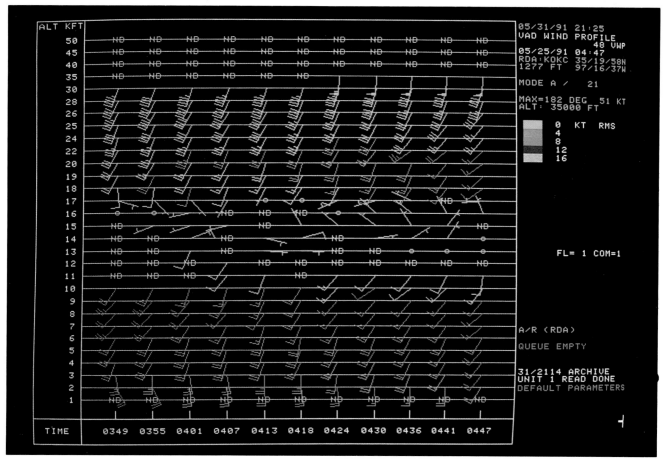

Plate 3. An example of a time-height cross section of wind from a steerable Doppler radar using the velocity-azimuth display algorithm.

SEVERE THUNDERSTORM AND TORNADO PARAMETER WORKSHEET					
AREA	ADVISORY NUMBER	DATE		ADVISORY VALID	
LA-MS-AL	86	21 FEB 71		21 / 18 z to 22 / 06 z	
PARAMETER	1200Z ANAL		0000Z PROG		REMARKS/VERIFICATION
	VALUE	RATING	VALUE	RATING	
SWEAT	500-550	S+	500-666	S	MAX PROGGED OVER CNTRL & NRN MS
TOTALS	58	S+	58-60	S	
LIFTED INDEX	-5	S+	-7	S	
PVA	30°	M+	40°	S	
500 MB HT FALLS	-200M	S	-200M	S	
500 MB JET	95K	S-	90K	S	
850 MB MOISTURE	11°	M+	13	S	
850 TEMP RIDGE	W OF MOIST RIDGE	S	W OF MOIST RIDGE	S	
LO-LEVEL JET	45-55K	S	50K	S	
700 MB DRY INTRUSION	—	S	—	S	
700 MB NO-CHANGE TEMP	WINDS CROSS ≤ 20°	W+	WINDS CROSS 20-40°	M	0000Z DATA ACTUALLY STRONG S
WINDS VEER WITH HEIGHT	YES	—	YES	—	
WINDS INCREASE WITH HEIGHT	YES	—	YES	—	
INTERSECTING UP AND LO JETS	YES	—	YES	—	
SFC DEW POINT	62°	M+	66°	S	
SFC PRESSURE THREAT AREA	1008	M+	1002	S	
FALLING PRESSURE	YES	—	YES	—	
INCREASING SFC TEMP	YES	—	YES	—	
INCREASING DEW POINT	YES	—	YES	—	
THICKNESS RIDGE	YES	—	YES	—	
THICKNESS NO-CHANGE	YES	—	YES	—	
MESO OR SYNOP PATTERN	FAVORABLE		FAVORABLE	—	
REMARKS	MARKED DIFLUENCE OVER THREAT AREA AT 1200Z AND PROGGED FOR 0000Z. NUMEROUS TORNADOES OCCURRED FROM N.E. LA INTO MS AFTN AND EVNG.				

Fig. 1. Example of a checklist for severe weather.

was applied to synoptic maps and vertical soundings to develop what *Schaefer* [1986] calls a "forecast rote." The essential reference on this forecasting approach is by *Miller* [1972], who describes map types commonly associated with major severe weather outbreaks. The notion of indicating the location and orientation of the various features of interest on a single map, the so-called composite chart (Plate 1), is the cornerstone of the forecast rote. It is noteworthy that the composite chart, as employed in severe weather forecasting, specifically attempts to establish the interaction between features aloft and at the surface. Thus it is a product with a long history of addressing what *Mass* [1991] considers a common deficiency in synoptic analysis, namely, the failure to depict three-dimensional relationships among features at different levels. Another form of this method is the checklist (also described by *Miller* [1972]), as shown in Figure 1.

Unfortunately, there was little information in this approach that allowed forecasters to make a direct connection between what they saw in their analyzed weather data and the storms responsible for producing the tornadoes. On the scale of a weather map, a tornado is a microscopic dot. There is no information on a weather map distinguishing a tornadic from a nontornadic storm, except insofar as there is some association between the storm and its environment.

What weather map typing gives one is an association; it does little to explain cause and effect. What was the basis for making the operational distinction between tornadic and nontornadic situations? For forecasting purposes, the connection between the storm and its environment has been and largely remains by means of a synoptic climatology, generally derived quite subjectively.

When the techniques of the 1950s were being developed, there was little comprehension of the structure and evolution of tornadic storms and what relationship existed between the tornadic storm and the tornado. Weather radars were a brand-new technology, and no scientific basis existed to use a radar for understanding tornadic storms, much less detecting them when present. Real-time radar in a national tornado forecasting unit was a distant dream, as were real-time satellite views of storms and computer-based analysis and forecasting.

The radar observations of the 1950s created a great deal of interest in learning more about tornadic storm structure and evolution. Research surface mesonetworks had been established in the early 1950s to pursue some ideas about tornadoes [e.g., *Tepper*, 1950, 1959], but radar rapidly became the primary means for observing tornadic storms. Thus, while forecasters concentrated on empirical methods based on synoptic-scale surface and upper air data, the focus of tornado-related research dealt with storm-scale processes as revealed by radar. This schism between research and operational goals grew with time; by the early 1960s an institutionalized fission of the tornado research and forecasting communities was created with the establishment of the National Severe Storms Laboratory, created from the Weather Bureau's National Severe Storms Project [see *National Severe Storms Project Staff*, 1963] in 1964.

The research produced a picture of the tornadic storm as a "supercell" (as detailed by *Browning and Fujita* [1965]), a type of convective storm that differed significantly from other, nontornadic storms in its radar structure and evolution. Although it became apparent that not all supercells produced tornadoes and that not all tornadoes came from supercells [see *Doswell and Burgess*, this volume], supercells clearly were prolific tornado producers in comparison to other convective storms.

In these pioneering studies it also was found that supercells favored certain environments, although the reason for this association remained somewhat unclear. In spite of these gains, the knowledge of tornadic storms developed during this research was not readily incorporated in the operational environment (some of the research results have had an impact in some operational detection and warning programs, although even there, progress has been slow); research and operations seemed unable to communicate effectively. Because they concentrated on different scales and data streams, most forecasters and researchers no longer spoke the same language.

By the mid-1970s, numerical cloud modeling had become capable of fully three-dimensional, time-dependent storm simulations. We think the flowering of these models has signaled the beginning of the end to the barrier separating basic storm-scale research from operational forecasting, although this was not widely recognized at the time. We believe this because the cloud models can be used to explore how the characteristic features of a simulated storm depend on the larger environment in which it develops. Subsequent cloud model-based research indeed has been quite successful in developing the storm-environment connection for the first time [e.g., *Weisman and Klemp*, 1982, 1984].

Another critical source of insight into convective storms has been research Doppler radar observations. While Browning's work [e.g., *Browning*, 1964] made innovative use of reflectivity information to infer storm flow, the detailed velocity field information has confirmed the basic supercell storm structures deduced from non-Doppler radar studies [e.g., *Brandes*, 1977] and has been quite important in validating concepts developed from numerical cloud models [e.g., *Weisman and Klemp*, 984].

A third important research development of relevance has been the deployment of "storm chase" teams: groups of meteorologists attempting to observe tornadoes and tornadic storms firsthand. This has produced an unprecedented number of detailed visual observations, including many storms of the nontornadic variety. For the first time, scientists have been able to relate events (tornadic and nontornadic) observed directly in the field to structures seen in large-scale weather maps. It should be obvious that tornado forecasting is an essential part of a storm chase; thus storm chasers have become contributors to forecasting research [e.g., *Weaver and Doesken*, 1991; *Davies and Johns*, 1991; *Brady and Szoke*, 1988].

In spite of the proliferation of new technologies in the workplace and the burgeoning research developments, the decades following the 1950s have not seen much change in operational tornado forecasting techniques. Rather than supporting qualitative changes in the way tornado forecasting is done, new observing and analysis tools have been used to increase the precision and timeliness of the forecasting approaches primarily developed in the 1950s. The new observations most often have been used to identify new ways to detect severe storms (e.g., the satellite-observed "enhanced V" signature noted by *McCann* [1983]; see Plate 2) as well as to enhance recognition of previously known elements (e.g., intersecting thunderstorm-generated outflow boundaries, as described by *Purdom* [1982]). These new observations have improved severe storm detection and recognition, but they have not been very useful in forecasting. Their value in forecasting is compromised by the fact that most studies of their use have focused on cases where storm events actually happened; cases where the features (e.g., outflow boundaries) were present but nothing happened have been studied only occasionally [e.g., *Stensrud and Maddox*, 1988].

Moreover, computers have been used mostly to speed and enhance subjective analysis techniques developed decades earlier (mostly by automated data plotting) rather than to create new techniques. In effect, for operational forecasting, the computer often has been asked to duplicate electronically what used to be done manually.

In recent years the research and operational sides of tornado forecasting have begun to collaborate once again. For example, recent research into streamwise vorticity [*Davies-Jones*, 1984] is being applied directly in assessing tornado potential [e.g., *Johns et al.*, 1990; *Davies-Jones et al.*, 1990; *Davies and Johns*, this volume] operationally.

3. CURRENT SELS TORNADO FORECASTING PROCEDURES

Present-day SELS tornado forecasting comprises three steps: the "Second Day Severe Thunderstorm Outlook" (hereinafter referred to as the DY2 AC), the "First Day Convective Outlook" (hereafter referred to as the DY1 AC), and severe thunderstorm/tornado watches. This suite of SELS products has evolved over time. The products are partially described by *Weiss et al.* [1980], but for a full description the interested reader should consult the National Weather Service Operations Manual, chapter C-40.

The Convective Outlooks (or ACs) are regularly issued (and updated) general forecasts of severe thunderstorm potential for relatively large areas. (For official purposes, a severe thunderstorm is defined as one which produces one or more of the following: hail $\geq 3/4$ inch (2 cm) in diameter, measured winds ≥ 50 knots (25 m s^{-1}), "damaging" winds (involving some subjective judgment of effects required to meet the threshold), a tornado. Heavy rain, large quantities of subthreshold hail, funnel clouds, frequent lightning, etc. are not considered to meet the official criteria (see discussion by *Doswell* [1985].) Watches, on the other hand, are issued only as needed (in the judgment of the SELS lead forecaster) and are more specific in terms of timing, location, and expected types of severe weather. The basic premise is that as the time of the event approaches, it is possible to refine the forecasts of severe thunderstorm type, timing, and location. While this premise seems logical, it is not necessarily valid; the relevant scales decrease as the event develops, first shrinking from synoptic scale to mesoscale and then on to the convective storm scale. However, the data available to the forecaster do not undergo an increase commensurate with this scale decrease. It is not uncommon for forecasting to become more difficult as the time of the event approaches (as discussed by *Doswell et al.* [1986]). Generally, it is during the watch phase that SELS attempts to distinguish between tornadic and nontornadic storms.

3.1. *Forecasting Procedures: Convective Outlooks*

Operational SELS tornado forecasting employs three general approaches: synoptic pattern recognition, meteorological parameter assessment (checklists), and climatology. These are the tools that developed historically as noted in section 1. Specialized, synoptic pattern-specific, or geographically localized forecasting techniques (for some examples, see *Doswell* [1980], *Hales* [1985], *Johns* [1984], *Weiss* [1985], *Hirt* [1985], and *Weiss* [1987]) also contribute to the ACs.

As more is learned about the physical processes resulting

in tornadoes and/or severe thunderstorms, parameters considered operationally relevant have been changing to reflect that new understanding. Thus, for example, vorticity advection, emphasized by *Miller* [1972] and questioned by *Maddox and Doswell* [1982], is giving way to helicity-related parameters, as discussed by *Davies-Jones et al.* [1990]. Continuing efforts to refine the climatological information about tornadoes [*Kelly et al.*, 1979] and nontornadic severe thunderstorms [*Kelly et al.*, 1985] are aimed in part at improving operational forecasting; recall that climatology, modified by knowledge of the synoptic pattern, is a traditional basis for distinguishing tornadic from nontornadic situations.

For the long lead times of the ACs (up to 52 hours in the case of the DY2 AC), the primary input to these products is the numerical weather prediction model guidance from the National Meteorological Center (NMC), of which NSSFC (and hence SELS) is a part. With diminishing lead times, and especially with regard to the watches, diagnostic evaluation of surface and upper air data becomes dominant over model prognosis. In conjunction with the analysis of surface and sounding data, the remotely sensed data (such as satellite, radar, and lightning ground strike location) that have become available in ever-growing amounts are increasingly important. These data, especially satellite imagery, are useful in assessing the numerical model initial conditions [e.g., *Hales*, 1979] and in data voids (most oceanic regions and some sparsely populated land areas, as well as when conventional data are missing or contaminated with convection).

SELS continues to employ many parameters designed to summarize information contained in the data; such parameters often are called "indices" (see for example, *Miller* [1972], *Galway* [1956], and *Showalter* [1953] for some of the myriad thermodynamic indices measuring static instability). Recently, a more comprehensive parameter than the traditional indices for static instability is coming into use in SELS: the potential buoyant energy (PBE, also called the convective available potential energy, or CAPE; [see *Moncrief and Miller*, 1976] is the "positive" area on a sounding associated with the buoyant part of a lifted parcel ascent between the level of free convection (LFC) and the equilibrium level [see *Doswell et al.*, 1982]).

Nearly all the parameters (past and present) used on composite charts and/or checklists can be shown to be associated with (1) synoptic and mesoscale upward motion, (2) sufficient moisture and lapse rate for a parcel to be positively buoyant, and (3) vertical wind shear structure. Therefore the scientific connection between the parameters used and the physical processes can be made, even if forecasters have not always recognized that connection.

A key notion employed in tornado forecasting is that of "limiting factors." Once a preliminary general threat area has been defined, it is refined by considering what factors make it unlikely that some parts of the original threat area actually will experience severe weather. Obviously, anything precluding thunderstorms will preclude tornadoes. (As noted by *Doswell and Burgess* [this volume], some atmo-

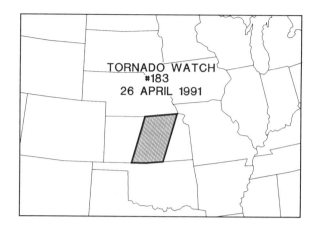

Fig. 2. Plot of tornado watch 183 on April 26, 1991 (compare Figure 3).

spheric vortices are not associated with deep, moist convection. These are not considered to be tornadoes. In tornado forecasting, vertical wind shear structure is becoming the key factor in distinguishing tornadic from nontornadic events, so this becomes a critical limiting factor in delineating tornado threat areas.

In a few cases (typically fewer than 10 days per year), tornado "outbreaks" are forecast in the ACs. Such forecasts began in the mid-1970s, following the April 3–4, 1974, outbreak. A separate public version of the AC is issued in such situations. Although it is impossible to be completely general regarding tornado outbreak conditions that might result in such an AC, they typically are associated with what we call "synoptically evident" patterns. Tornadoes may be mentioned in the outlooks when moderate or greater thermal instability is likely to be coupled with favorable vertical wind shear structures [*Davies-Jones et al.*, 1990; *Johns et al.*, 1990; *Leftwich*, 1990].

3.2. *Forecasting Procedures: Watches*

The foremost SELS public forecast products are the tornado and severe thunderstorm watches (an example of which is shown in Figure 2). These forecasts usually take the form of quadrilaterals covering on the order of 20,000 miles (about 52,000 km^2) and are valid for time periods of several hours. Some statistical information about watches in 1990 is shown in Table 1.

In order to convey information about SELS forecaster interpretations to the forecasting community (not to the public), SELS Mesoscale Discussions (MDs) are issued as needed. The MDs provide a narrative of probable weather developments and possible forthcoming watch issuance; MDs began in 1986. Another information-conveying, nonpublic product is the watch-related Status Report, begun in the 1950s, which has several aims: to keep the field offices informed about severe weather conditions in and near an issued watch area, to clear those parts of the watch where the severe weather threat has ended, and to provide information about additional, follow-on watches.

In general, SELS forecasters must deduce the character of the subsynoptic scale processes relevant to severe weather watches from limited operational data: hourly surface observations, satellite images, and radar displays. These current sources provide the highest operationally available space and time resolution for the task at hand. Subjective surface analyses locate and track features believed relevant [see *Miller*, 1972; *Doswell*, 1982] to tornado and severe thunderstorm forecasting. These features then are related to the satellite imagery using the advanced interactive computer system called NSSFC's VAS Data Utilization Center (VDUC [see *Browning*, 1991]). Recently, lightning ground strike data have become available [*Mosher and Lewis*, 1990] and are used in SELS for defining, locating, and monitoring convection, as a supplement to satellite and radar data.

In the past it was common to differentiate tornadic from nontornadic situations using parameters related to the strength of the winds aloft [e.g., *Miller*, 1972]. This created a seasonal bias to the frequency of issuing tornado watches, with increased likelihood of a tornado watch in the winter and spring (and to a lesser extent in the fall) when strongly baroclinic disturbances are present in severe weather situations. During the summer, with weaker synoptic scale disturbances, the tendency was toward severe thunderstorm (instead of tornado) watches. Although this method matches climatology reasonably well [see *Kelly et al.*, 1979], there was no understanding of the processes relevant to tornadogenesis being employed, largely because that understanding was not available. It was not easy to justify the choice in any given situation, except by experience and climatology.

It is only within the last several years that this situation has begun to change. At present, recognition of the supercell environment is becoming the cornerstone of SELS tornadic versus nontornadic decision-making. This philosophy also reflects a lack of knowledge about nonsupercell tornadoes at present, at least compared to supercell tornadoes; as noted by *Doswell and Burgess* [this volume], the study of the nonsupercell tornado has only just begun.

On the other hand, the high degree of association between supercells and tornadoes has made it possible to identify situations that are likely to produce tornado outbreaks. Outbreak-related tornadoes usually are produced by supercells, and the tornadoes that form from supercells are

TABLE 1. SELS Watches in 1990

	Value
Number of tornado watches	249
Number of severe thunderstorm watches	496
Average watch duration	4.86 hours
Average watch area	76,519 km^2 = 29,544 (statute mi)^2
Median lead time (begin time minus issue time)	31 min

```
BULLETIN - IMMEDIATE BROADCAST REQUESTED
TORNADO WATCH NUMBER 183
NATIONAL WEATHER SERVICE KANSAS CITY MO
1210 PM CDT FRI APR 26 1991

A..THE NATIONAL SEVERE STORMS FORECAST CENTER HAS ISSUED A TORNADO
WATCH FOR

   PARTS OF CENTRAL AND EASTERN KANSAS

EFFECTIVE THIS FRIDAY AFTERNOON AND EVENING UNTIL 800 PM CDT.

THIS IS A PARTICULARLY DANGEROUS SITUATION WITH THE POSSIBILITY OF
VERY DAMAGING TORNADOES. ALSO..LARGE HAIL...DANGEROUS LIGHTNING AND
DAMAGING THUNDERSTORM WINDS CAN BE EXPECTED.

THE TORNADO WATCH AREA IS ALONG AND 65 STATUTE MILES EAST AND WEST
OF A LINE FROM 45 MILES EAST SOUTHEAST OF MEDICINE LODGE KANSAS TO
45 MILES NORTHEAST OF CONCORDIA KANSAS.

REMEMBER...A TORNADO WATCH MEANS CONDITIONS ARE FAVORABLE FOR
TORNADOES AND SEVERE THUNDERSTORMS IN AND CLOSE TO THE WATCH AREA.
PERSONS IN THESE AREAS SHOULD BE ON THE LOOKOUT FOR THREATENING
WEATHER CONDITIONS AND LISTEN FOR LATER STATEMENTS AND POSSIBLE
WARNINGS.

B..OTHER WATCH INFORMATION..THIS TORNADO WATCH REPLACES SEVERE
THUNDERSTORM WATCH NUMBER 181. WATCH NUMBER 181 WILL NOT BE IN
EFFECT AFTER 100 PM CDT.

C...TORNADOES AND A FEW SVR TSTMS WITH HAIL SFC AND ALF TO 3 IN.
EXTRM TURBC AND SFC WND GUSTS TO 75 KNOTS. A FEW CBS WITH MAX TOPS
TO 600. MEAN WIND VECTOR 23040.

D...LN TCU DVLPG FM SW OF CNK TO BTWN RSL AND SLN ATTM. EXPCT RAPID
TSTM DVLPMT WITHIN NEXT HR ALG DRYLN/CDFNT WITH SUPERCELLS AND
TORNADO DVLPMT LKLY.

E...OTR TSTMS...CONT WW NR 182. WW LKLY TO BE RQRD WITHIN NEXT HR
OR TWO OVR PTNS WRN AND CNTRL OK. WW LKLY TO BE RQRD LATER THIS
AFTN OVR PTNS ERN KS AND WRN MO.

...JOHNS
```

Fig. 3. Tornado watch message 183, issued on April 26, 1991, using the "enhanced" wording. Parts A and B are transmitted to the public and so are in plain text, whereas parts C–E are not made public and are written in contractions to save characters.

capable of the highest damage potential associated with any tornado. These cases, not surprisingly, are exemplified best by "classical" severe weather patterns, that is, those that we have called synoptically evident. (Note that a situation we label as synoptically "evident" should not be automatically equated with an "easy" forecast. No real-world forecast situation is ever easy, except in retrospect!) Thus SELS has had the option (since the early 1980s) to issue tornado watches that have "enhanced wording" to highlight the tornado threat (see Figure 3). Let us define a tornado day as a day with one or more tornadoes and a "big" tornado day as a day with two or more violent (F4-F5) tornadoes. The interanannual variation in tornado days is rather small, averaging about 175 days per year (see Figure 4), whereas big tornado days fluctuate considerably from year to year (Figure 5), with an average frequency of about 5 such days per year. Enhanced wording is used when tornado outbreaks are expected, and outbreaks typically have two or more violent tornadoes, meeting our criterion for a big tornado day. Thus the enhanced wording in tornado watches is not commonly employed.

The role of vertical wind shear-related parameters in tornado forecasting has made the development and operational availability of additional sources for vertical wind structure quite critical. Therefore the demonstration network of vertical wind profilers [*Gage and Balsley*, 1978] presently being implemented [*National Weather Service*,

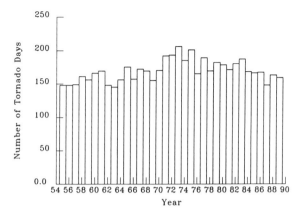

Fig. 4. Plot of yearly totals of tornado days (i.e., days with one or more reported tornadoes).

1987] is of great interest to operational forecasters. Also, the wind profiling capability [see *Rabin and Zrnic'*, 1980] of the WSR-80 (NEXRAD) Doppler radars in the prestorm, clear air environment may be extremely valuable.

4. 35 Years of Tornado Forecast Verification

Since the essential aspect of tornado forecasting is to distinguish between tornadic and nontornadic situations, this will be the primary issue discussed here. There are numerous other aspects of tornado and severe thunderstorm watch verification that we cannot dwell on here; some will be presented in a future publication. Our data consist of the final SELS log of severe weather reports and records of tornado and severe thunderstorm watches, covering the 35-year period 1955–1989, inclusive.

It should be noted that a tornado watch/event is, in a sense, also a severe thunderstorm watch/event. That is, a tornado-producing storm is, by definition, a severe thunderstorm. Therefore a tornado in a severe thunderstorm watch verifies that watch, but a nontornadic event does not verify

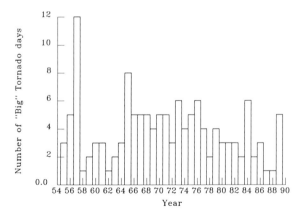

Fig. 5. Plot of yearly totals of "big" tornado days (i.e., those days with at least two or more violent (F4-F5) tornadoes).

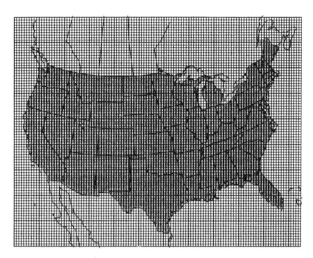

Fig. 6. Illustrating the manually digitized radar (MDR) grid with those blocks considered in the verification shaded gray.

a tornado watch. However, for the purposes of this paper we have not accounted for this fact in the verification; that work is not yet complete, but it will be reported upon in the aforementioned future publication.

Another important aspect of tornado forecast verification is that tornado watches are area forecasts that typically cover several tens of thousands of square kilometers, whereas tornadoes affect only a few square kilometers even in major events. This disparity in coverage means that successful tornado watches (i.e., those with tornadoes in them) are mostly "false alarms" in the sense that the vast majority of the forecast area is unaffected. The original watch verification schemes considered them as area forecasts, so that a single tornado effectively verified the entire area of the watch. Recently, as described by *Weiss et al.* [1980], watch verification has been changed such that a single report verifies only a portion of the total watch area/time. However, this new scheme still does not incorporate information about areas outside the watch.

The verification scheme used herein (first described by *Doswell et al.* [1990b]) is based on the so-called manually digitized radar (MDR) grid shown in Figure 6. Each MDR box is roughly 40 km (25 nautical miles) on a side. (Since the grid is defined on a polar stereographic map projection, the grid boxes vary in size across the map by as much as about 10%. The nominal size applies only at 60°N latitude, where the map scale factor is unity.) Every watch has been broken down into MDR grid boxes, using the convention that if the centroid of an MDR box is within the watch, that MDR box is considered to be within the watch. The valid time of the watch is broken down by hours; if a watch begins on the hour or within the first 29 min of the hour, the watch is considered valid for the whole hour, whereas watches beginning 30 min or more after the hour apply to the next whole hour. The MDR box hour is the basic unit of the verification, and it naturally gives a somewhat "grainy" picture. We have

TABLE 2. Contingency Table for Severe Thunderstorm and Tornado Watch Box-Hours, as Described in the Text

	Observed			
	Tornado	Severe Thunderstorm	Nothing	Total
Forecast				
tornado	n_{11}	n_{12}	n_{13}	$n_{1.}$
severe thunderstorm	n_{21}	n_{22}	n_{23}	$n_{2.}$
nothing	n_{31}	n_{32}	n_{33}	$n_{3.}$
Total	$n_{.1}$	$n_{.2}$	$n_{.3}$	$n_{..}$

tested the effect of increasing the resolution both in space and time and found that for our verification purposes, it is detectable but does not affect the overall patterns. There are 4533 MDR boxes over the United States (boxes over water are not counted), and each nonleap year has 39,709,080 MDR box hours.

For verification purposes, a severe weather report is considered to verify an entire MDR box hour if it occurs anytime within that hour. Reports other than the first in that MDR box hour are ignored unless they are of a different type (the two types of reports are "tornado" and "nontornadic severe thunderstorm"). If one or more tornadoes occur within a given box hour, it is counted as a "tornado hit" irrespective of any concurrent nontornadic severe thunderstorm reports. If one or more nontornadic severe events occur within a given box hour, it is counted as a "severe thunderstorm hit."

Our basic tool for verification is the 3 × 3 contingency table shown in Table 2, derived via the above process for every MDR box. From the information contained within this basic table, a wide variety of summary measures, histograms, maps, etc. can be constructed, of which we obviously have room for only a small fraction. Although a single number cannot express all of the content implied in Table 2, we shall use the Heidke skill score as a summary measure of skill (see *Doswell et al.* [1990a, b] for details). As indicated in Table 3 (and Figure 10), skill scores have increased by nearly an order of magnitude over the 35 years. If we compare the first decade (1955–1964) with the last decade (1980–1989) of our record (Figure 7), it can be seen that the primary "tornado alley" skill maximum has persisted, but additional centers of relative skill have developed in North Carolina, New York state, Montana, and Idaho. It is at present difficult to know how to interpret these results; however, spatial distribution of verification scores clearly is influenced by the distribution of severe weather reports (compare Figure 7 and Figure 8).

The reporting of nontornadic severe weather has increased markedly with time (Figure 9), whereas tornado reporting has remained more nearly constant. While the skill has climbed more or less steadily during the 35 years under consideration (Figure 10), how much of this skill is attributable to enhanced reporting? To attempt to account for this "inflation," we did the following. If one does not distinguish for the moment between tornadoes and nontornadic severe thunderstorms, then the contingency table (Table 2) reduces to the 2 × 2 table shown in Table 4.

We take the number of severe reports in 1955 as the standard, denoted by $(x + y)_{55}$. (As is the case in accounting for currency inflation, this does not imply that there is anything special about 1955. It simply represents a base, or reference state. We could just as easily have adjusted toward a 1989 standard, with no material difference in our conclusions.) Subtract this from the number of severe reports in the ith year, $(x + y)_i$, to obtain the difference d_i. This difference is an estimate of the number of events for the ith year that would have gone unreported in 1955, so these are all put into the z box (refer to Table 4) in the contingency table. We assume that the ratio of the x box to the y box remains the same, and we redo the skill score verification on the revised table. Of course, some of the d_i might actually belong in the w box, but since w is typically much larger than the other

TABLE 3. Actual 3 × 3 Contingency Tables for Years 1955 and 1989 Showing the Number of Watch-Box Hours as Described in the Text

	Observed			
	Tornado	Severe Thunderstorm	Nothing	Total
		1955		
Forecast				
tornado	85	62	39,696	39,843
severe thunderstorm	40	56	78,580	78,676
nothing	420	604	39,589,357	39,590,561
Total	545	722	39,707,813	39,709,080
		1989		
tornado	237	1372	61,036	62,645
severe thunderstorm	102	1965	96,325	98,392
nothing	393	4255	39,543,395	39,548,043
Total	732	7592	39,700,756	39,709,080

Heidke skill score is 0.00317 for 1955 data and 0.0348 for 1989 data.

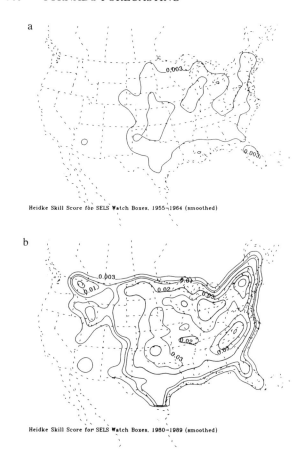

Fig. 7. Maps of smoothed Heidke skill score for (a) 1955–1964 and (b) 1980–1989.

Table 3, the occurrence of tornadoes in severe thunderstorm watches has increased with time, but only by about a factor of 2. This relatively small increase appears in part to be a consequence of the relatively modest rate of increase in tornado reporting overall. However, the number of correctly forecast tornado events has increased by about a factor of 3, suggesting some increase in the skill of discrimination. The occurrence of tornadoes without any watch of either type actually decreased during the period, although such total misses still constitute the majority of observed tornadic events. Since most tornadoes are weak, and because the probability of detection for tornadoes in watches is lowest for weak tornadoes [*Leftwich and Anthony*, 1991], most of these missed events are weak tornadoes and probably are not associated with supercells. *Galway* [1975] has observed that the majority of tornado deaths occur in watches, which

entries in the table, ignoring this creates only a negligible error. While we temporarily have lumped tornadoes and nontornadic severe thunderstorm events together for this purpose, the Heidke skill scores for the 2 × 2 and 3 × 3 versions of the table are not markedly different [see *Doswell et al.*, 1990b]. The results of this procedure show that overall, when report "inflation" has been accounted for, the skill of severe weather watches has improved by about 50% rather than the order of magnitude increase without correcting for this inflation.

Turning to the specific issue of tornado versus severe thunderstorm forecasts, the asymmetry of the forecasts is important. That is, as long as one or more tornadoes occur in a tornado watch, there is no problem with having many reports of nontornadic severe thunderstorm events in that tornado watch (element n_{12} in Table 2). On the other hand, having tornadoes occur in severe thunderstorm watches is an undesirable event (element n_{21} in Table 2). Of course, there is an easy way to prevent this from ever occurring: always issue only tornado watches. Doing so would represent no attempt to discriminate between tornadic and nontornadic severe thunderstorms; this clearly is not desirable, nor does it reflect what has been done. As can be seen in

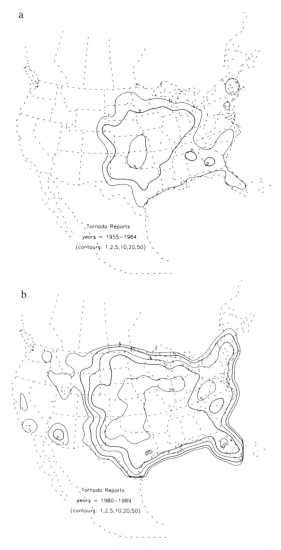

Fig. 8. Maps of smoothed tornado occurrence for (a) 1955–1964 and (b) 1980–1989.

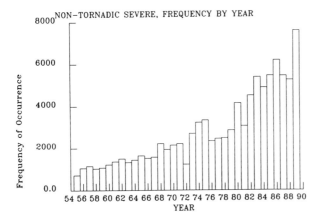

Fig. 9. Plot of yearly totals of nontornadic severe weather.

remains true to this day, suggesting that watches capture most of the significant tornadoes.

5. CURRENT UNDERSTANDING OF TORNADIC STORMS

Our understanding of tornadic storms and how they interact with their environment is in an exciting state of transition. As *Doswell and Burgess* [this volume] indicate, we have come to realize that tornadoes can occur in many ways and are not limited to supercell events. Not all tornado reports represent the same meteorological phenomenon, and we are just beginning to understand how nonsupercell tornadoes might arise [e.g., *Wakimoto and Wilson*, 1989].

Even for supercell events, however, mesoscale variability in atmospheric structure can be crucial in estimating the chances for tornadoes (see *Burgess and Curran* [1985] for a case study example). Mesoscale details, often slipping more or less unnoticed through the present-day observing system, can be the difference between correct and incorrect forecasts. The sorts of events used by *Miller* [1972] to exemplify tornadic environments are not as difficult to predict, because

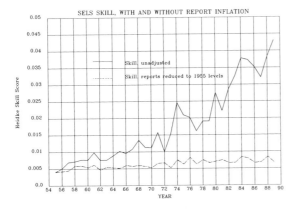

Fig. 10. Plot of yearly values in Heidke skill score, both unadjusted and adjusted (as described in the text) for "inflation" of nontornadic severe weather reports.

TABLE 4. Table 1 Contingency Table With No Distinction Between Tornadoes and Severe Thunderstoms

	Observed		
	Severe	Nothing	Total
Forecast			
severe	x	y	$x + y$
nothing	z	w	$z + w$
Total	$x + z$	$y + w$	$x + y + z + w$

Since no distinction is made, all forecasts and observations are combined in the generic "severe weather" category.

they are characterized by large-scale environments readily identified as being favorable for supercells with today's data. Such synoptically evident events are not frequent during the tornado season, however (see also *Maddox and Doswell* [1982]).

On some tornado days (typically those with large CAPE but weak to marginal shear), while severe storms are likely, the apparent tornado potential is not high. A notable recent example is August 28, 1990, on which the violent Plainfield, Illinois, event occurred. In such environments (many of which produce no significant tornadoes) some mesoscale process creates a local environment such that isolated storms become tornadic in ideal conditions, but those conditions are uncharacteristic of a large area. In contrast, major outbreak days (such as the Kansas-Oklahoma outbreak of April 26, 1991, that produced the Andover, Kansas, killer tornado, among others) have widespread supercell-favorable environments and many supercells.

Said another way, more than 95% of the tornado days per year are not synoptically "evident"; most tornado events involve mesoscale processes that are difficult to anticipate at present [see *Doswell*, 1987; *Rockwood and Maddox*, 1988]. The forecaster must formulate a correct prior assessment of tornado potential, then carefully monitor the observations and watch for crucial developments that might escape notice even when they occur.

Recent research has suggested that the vertical wind shear structure is the most crucial element in supercells. Thus the source of rotation in supercell-related tornadoes seems to be the vertical wind shear in the environment. Static instability does not seem particularly useful in distinguishing supercell from nonsupercell events, since supercells occur within a broad range of instabilities [*Johns et al.*, 1990, this volume], although it may be important in determining other aspects of severe thunderstorm potential.

To the extent that supercells are responsible for most strong and violent tornadoes, the current task of tornado forecasting hinges on predicting environments favorable for supercells. This begins with the requirement that deep, moist convection be possible; in turn, this depends on having sufficient moisture, instability, and lift such that potentially buoyant parcels reach their LFCs. A large group of the empirical forecasting "rules of thumb" have physical explanations rooted in the required presence of deep, moist convection. Beyond this, for supercells it appears that the development of a supercell's deep, persistent mesocyclone takes longer than the 20- 40-min lifetime of an ordinary convective cell. Long lifetimes for convective events are the result of propagation, with new convective cells developing in preferred locations relative to existing cells. Preferential convective development is now known to be related to the vertical wind shear [see *Weisman and Klemp*, 1982, 1984]. It is also important that the combination of vertical wind shear and storm motion produces enough storm-relative helicity [e.g., *Davies-Jones et al.*, 1990] to allow the mesocyclone to reach down to the surface [*Brooks et al.*, this volume]. Despite some recent progress, these ideas have yet to be validated in forecasting practice, in part owing to a lack of mesoscale detail in the observations.

Although nonsupercell events have not yet received the research attention given to supercells, it appears that at least some of them may be related to mesoscale processes associated with terrain features [e.g., *Brady and Szoke*, 1988]. To whatever extent nonsupercell tornadoes are associated with topography, their prediction may be correspondingly straightforward [see *Doswell*, 1980; *Weaver and Doesken*, 1991]. As of this writing, it is not known what fraction of all tornadoes fall into this category of being terrain related.

6. FUTURE DIRECTIONS

We have indicated that tornado forecasting is presently in a state of rapid change. We have emphasized how important scientific understanding has become in the tornado prediction problem. It is the new technologies that will make new science possible, but breakthroughs will not arise simply by switching on new systems. Systematic research will be needed to achieve technology's promise, and a commitment to transferring new understanding to operations is required. As new ideas are developed, their forecast value must be verified rigorously and the ideas must be modified on the basis of the results of the verification.

6.1. Technological Tools

There are several new observing technologies to which the operational weather services in the United States already are committed. The WSR-88D radar network is the keystone for the future health of tornado warnings, but what about tornado forecasting? With the current emphasis on the importance of vertical wind shear in supercell storms, the clear air wind profiling capability of the WSR-88Ds (e.g., see Plate 3) becomes potentially valuable. Since this capability is in general limited to low levels, it nicely complements the vertical wind profilers, which do not provide low-level winds. An opportunity for increasing our wind observations is in the use of automated instruments on commercial aircraft, which gather high-resolution data during ascent and descent (in effect creating wind and temperature soundings near major air terminals). Altogether, these enhancements to observations of the mesoscale flow variations should improve our capacity to diagnose and anticipate the wind shear structure.

Widespread deployment of automated surface observations should enhance the resolution of our surface data. The new technologies that make automated surface observations feasible have made it realistic to propose the operation of what we now view as research-density mesonetworks over much of the country as we enter the new century. In the past, processes observed with high-resolution networks [see *Fujita*, 1963] were not resolved in operational networks. With the proliferation of such networks in operations, it should be possible to put into practice the concepts derived from the research network observations of the past, as well as to do new research on a wider range of situations than the research networks, with their limited area and time of operations, could sample.

As of this writing, we are on the verge of having a nationwide network of lightning ground strike detectors and a space-based lightning mapper [*Turman and Tettelbach*, 1980] that will allow observations of intracloud and intercloud lightning as well as ground strikes. The ultimate value of such information in tornado forecasting remains unknown. If such data are to be of value in forecasting, we must integrate the lightning data with the rest of the observations in determining how lightning ground strike information relates to storm severity [see *MacGorman*, this volume].

Multispectral satellite observations of an unprecedented scope are promised for the near future, also. We believe the real value in such observations is not in trying to emulate direct measurement (e.g., rawinsonde) data but in using the new data in ways that are consistent with their character (e.g., layer-averaged variables [see *Fuelberg and Olson*, 1991]). The most complete and useful observations will involve the union of all observing technologies (e.g., Earth- and space-based remote sensing technologies such as wind and thermodynamic profilers and Doppler radars, as well as various direct measurements such as rawinsondes and aircraft measurements), a task easier said than done.

Another technological tool is the operational meteorological workstation. Although new observing technologies are about to unleash a torrent of new data on the operational work environment, the same basic technologies also give us the capacity to absorb and integrate it. If a future operational workstation is to have a positive impact on forecasting, it must meet two requirements. First, the workstation hardware must have sufficient data processing resources to deal effectively in real time with the data transfer rates associated with the new observations. Second, the workstation software must make this torrent of data available in operationally useful ways.

The last technological tool for the future we shall discuss is the numerical prediction model. There can be no doubt that numerical prediction models will continue to assume ever greater roles in the tornado forecast problem. Clearly, the tornado is many orders of magnitude smaller than the environmental processes that give it birth, so scale interac-

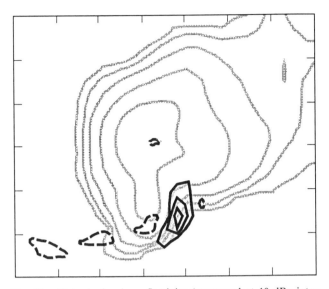

Fig. 11. Output of radar reflectivity (contoured at 10 dBz intervals) and vorticity (contoured at 0.0025 s^{-1} interval, with the zero contour suppressed) from a numerical cloud model; tic marks are 5.45 km apart. The model began with horizontally homogeneous initial conditions, using a forecast thermodynamic sounding and hodograph on May 26, 1991. The forecast input is valid for northwestern Oklahoma, and a supercell tornadic storm developed that afternoon near Woodward, Oklahoma.

tion is a crucial question. However, it will be many years (if ever) before a single numerical prediction model will encompass all these scales explicitly and simultaneously.

The real impact of mesoscale numerical prediction models in operations remains to be seen, but experimental real-time predictions with mesoscale models are underway [see *Warner and Seaman*, 1990]. Even numerical cloud models may have a role in operational forecasting (see Figure 11). See *Brooks et al.* [1992] for an extended discussion on this topic. The capacity for such modeling can only increase, so such models could come to be viewed as essential diagnostic tools [see *Keyser and Uccellini*, 1987] of the forecaster. The forecaster of the future must be much more than a passive recipient of model "guidance" [see *Snellman*, 1977]; the forecaster should be using numerical models in a way analogous to using pencils and paper weather charts. The computer should not be used to do what can be done with a pencil on paper, however. Rather, the numerical models should form the basis for operationally useful exploration of the intricate, nonlinear relationships associated with weather-making processes.

6.2. *Productive Areas for Research*

We believe that tornado forecasting eventually will outgrow its purely associative, empirical roots. The new observing systems mean that forecasters and researchers will be sharing the same data sets in the near future and that the application of meteorological science to tornado forecasting probably will be a more natural process than it is now. Such an outcome depends on (1) successful research aimed at increasing understanding of storm-environment relationships, (2) a meaningful education and training program for operational forecasters, and (3) a collective will to overcome a decades-long schism between research and operations.

An issue of considerable concern is the forecasting of the detailed thermodynamic structure of the storm environment, be it tornadic or not. Present models and observations give only a synoptic scale picture that often is inadequate to resolve the important mesoscale details. Of course, on some days it is possible to do an adequate job with the subjective and objective forecasting tools at hand. The processes by which moisture, momentum, and static stability change are well known in principle, but those processes are not necessarily well handled even in research-level modeling, much less in operational practice. Some of this stems from lack of resolution, both in our models and in our observations; some of the inadequacies arise from deficiencies in the physical parameterizations used in numerical models. The parameterization of convection is a crucial issue in the large-scale forecast evolution of static stability; release the convection too soon and the model never develops realistic convective instability, whereas if the parameterized convection is "turned on" too late, the forecast instability can reach unrealistically high levels.

If we continue to depend on physical parameterizations in our numerical models, as we almost certainly shall, it is obvious that improved parameterizations are needed. We believe that research aimed at improving physical understanding is preferable to improving parameterizations in the long run; parameterization is, in effect, a concession made in ignorance, even when it is a necessary concession. Of course, improved physical understanding can have a positive impact on parameterization schemes, as well.

Sensible and latent heat fluxes can have an enormous impact on static stability and moisture availability. The operational model "boundary layer physics" parameterizations leave much to be desired from the viewpoint of a tornado forecaster, especially in data-void areas like the Gulf of Mexico [see *Lewis et al.*, 1989].

In addition to the thermodynamic fields, it is clear that with the apparent importance of vertical wind structure for supercells, wind forecasting is a key problem in numerical prediction. Owing to their poor resolution using present operational data, mesoscale structures in the wind field are created by, rather than included initially in, the model. Our experience suggests that current operational weather prediction models are rather unsuccessful in predicting those important details reliably. The models have their successes, but there does not seem to be any consistency in them; the predicted details are not reliable, at least for the needs of a tornado forecaster.

As the new, more detailed observations accumulate, we may be able to develop new understanding of these mesoscale processes, heretofore inadequately sampled. This understanding may lead to improvements in mesoscale numerical prediction that will be essential to the tornado forecast problem. However, an additional complication to forecasting details of the wind structure is the impact from convection. *Maddox* [1983], *Ninomiya* [1971], and others have shown that persistent deep convection can alter the surrounding environment. This means that new convection developing near preexisting storms will encounter a different wind and thermodynamic environment than the convection which preceded it.

Given all of the new technological and associated scientific developments that are likely to take place in the next decade, it is quite plausible to be optimistic about the future of tornado forecasting. There can be no doubt that considerable progress will be made during the next 10 years, and we look forward to those developments.

Acknowledgments. The authors wish to thank Robert A. Maddox (NSSL) and the anonymous reviewers for their helpful comments on previous versions of this paper. Thanks are also extended to Dave Keller (NSSL) for his efforts in the verification study and Joan Kimpel (Cooperative Institute for Mesoscale Meteorological Studies) for her help in drafting, as well as Louis Wicker (University of Illinois and National Center for Supercomputing Applications) and Harold Brooks (National Research Council) for their help with Figure 11.

REFERENCES

Beebe, R. G., and F. C. Bates, A mechanism for assisting in the release of convective instability, *Mon. Weather Rev.*, *83*, 1–10, 1955.

Brady, R. H., and E. Szoke, The landspout—A common type of Northeast Colorado tornado, in *Preprints, 15th Conference on Severe Local Storms*, pp. 312–315, American Meteorological Society, Boston, Mass., 1988.

Brandes, E. A., Flow in severe thunderstorms observed by dual-Doppler radar, *Mon. Weather Rev.*, *105*, 113–120, 1977.

Brooks, H. E., C. A. Doswell III, and R. P. Davies-Jones, Environmental helicity and the maintenance and evolution of low-level mesocyclones, this volume.

Brooks, H. E., C. A. Doswell III, and R. A. Maddox, On the use of mesoscale and cloud-scale models in operational forecasting, *Weather Forecasting*, *7*, 120–132, 1992.

Browning, K. A., Airflow and precipitation trajectories within severe local storms which travel to the right of the winds, *J. Atmos. Sci.*, *22*, 664–668, 1964.

Browning, K. A., and T. Fujita, A family outbreak of severe local storms—A comprehensive study of the storms in Oklahoma on 26 May 1963, Part 1, *U.S. Air Force Spec. Rep. 32*, AFCRL-65-695(1), 346 pp., L. G. Hanscom Field, Bedford, Mass., (Available as *NTIS AD 623787* from Natl. Tech. Inf. Serv., Springfield, Va.)

Browning, P. A., The VDUC interactive computer system at the National Severe Storms Forecast Center, in *Preprints, 7th International Conference on Interactive Information and Processing Systems for Meteorology, Oceanography, and Hydrology*, pp. 204–207, American Meteorological Society, Boston, Mass., 1991.

Burgess, D. W., and E. B. Curran, The relationship of storm type to environment in Oklahoma on 26 April 1984, in *Preprints, 14th Conference on Severe Local Storms*, pp. 208–211, American Meteorological Society, Boston, Mass., 1985.

Burgess, D. W., R. J. Donaldson, Jr., and P. R. Desrochers, Tornado detection and warning by radar, this volume.

Davies, J. M., and R. H. Johns, Some wind and instability parameters associated with strong and violent tornadoes, 1, Wind shear and helicity, this volume.

Davies-Jones, R. P., Streamwise vorticity: The origin of updraft rotation in supercell storms, *J. Atmos. Sci.*, *41*, 2991–3006, 1984.

Davies-Jones, R. P., D. W. Burgess, and M. Foster, Test of helicity as a tornado forecast parameter, in *Preprints, 16th Conference on Severe Local Storms*, pp. 588–592, American Meteorological Society, Boston, Mass., 1990.

Doswell, C. A., III, Synoptic scale environments associated with High Plains severe thunderstorms, *Bull. Am. Meteorol. Soc.*, *60*, 1388–1400, 1980.

Doswell, C. A., III, The operational meteorology of convective weather, vol. I, Operational mesoanalysis, *NOAA Tech. Memo. NWS NSSFC-5*, 172 pp., Natl. Severe Storms Forecast Center, Kansas City, Mo., 1982.

Doswell, C. A., III, The operational meteorology of convective weather, vol. II, Storm scale analysis, *NOAA Tech. Memo. ERL ESG-15*, 240 pp., Natl. Severe Storms Forecaster Center, Kansas City, Mo., 1982.

Doswell, C. A., III, The distinction between large-scale and mesoscale contributions to severe convection: A case study example, *Weather Forecasting*, *2*, 3–16, 1987.

Doswell, C. A., III, and D. W. Burgess, Tornadoes and tornadic storms: A review of conceptual models, this volume.

Doswell, C. A., III, J. T. Schaefer, D. W. McCann, T. W. Schlatter, and H. B. Wobus, Thermodynamic analysis procedures at the National Severe Storms Forecast Center, in *Preprints, 9th Conference Weather Forecasting and Analysis*, pp. 304–309, American Meteorological Society, Boston, Mass., 1982.

Doswell, C. A., III, R. A. Maddox, and C. F. Chappell, Fundamental considerations in forecasting for field experiments, in *Preprints, 11th Conference on Weather Forecasting and Analysis*, pp. 353–358, American Meteorological Society, Boston, Mass., 1986.

Doswell, C. A., III, R. Davies-Jones, and D. L. Keller, On summary measures of skill in rare event forecasting based on contingency tables, *Weather Forecasting*, *5*, 576–585, 1990a.

Doswell, C. A., III, D. L. Keller, and S. J. Weiss, An analysis of the temporal and spatial variation of tornado and severe thunderstorm watch verification, in *Preprints, 16th Conference on Severe Local Storms*, pp. 294–299, American Meteorological Society, Boston, Mass., 1990b.

Fawbush, E. J., and R. C. Miller, A mean sounding representative of the tornadic airmass environment, *Bull. Am. Meteorol. Soc.*, *33*, 303–307, 1952.

Fawbush, E. J., and R. C. Miller, The types of air masses in which North American tornadoes form, *Bull. Am. Meteorol. Soc.*, *35*, 154–165, 1954.

Fawbush, E. J., R. C. Miller, and L. G. Starrett, An empirical method of forecasting tornado development, *Bull. Am. Meteorol. Soc.*, *32*, 1–9, 1951.

Fuelberg, H. E., and S. R. Olson, An assessment of VAS-derived retrievals and parameters using in thunderstorm forecasting, *Mon. Weather Rev.*, *119*, 795–814, 1991.

Fujita, T., Analytical mesometeorology: A review, *Meteorol. Monogr.*, *5(27)*, 77–125, 1963.

Gage, K. S., and B. B. Balsley, Doppler radar probing of the clear atmosphere, *Bull. Am. Meteorol. Soc.*, *58*, 1074–1093, 1978.

Galway, J. G., The lifted index as a predictor of latent instability, *Bull. Am. Meteorol. Soc.*, *37*, 528–529, 1956.

Galway, J. G., Severe local storm forecasting in the United States, SELS workshop notes, 13 pp., Natl. Severe Storms Forecast Center, Kansas City, Mo., 1973.

Galway, J. G., Relationship of tornado deaths to severe weather watch areas, *Mon. Weather Rev.*, *103*, 737–741, 1975.

Galway, J. G., The evolution of severe thunderstorm criteria within the Weather Service, *Weather Forecasting*, *4*, 585–592, 1989.

Galway, J. G., J. P. Finley: The first severe storms forecaster, *Bull. Am. Meteorol. Soc.*, *66*, 1389–1395, 1985.

Hales, J. E., Jr., A subjective assessment of model initial conditions using satellite imagery, *Bull. Am. Meteorol. Soc.*, *60*, 206–211, 1979.

Hales, J. E., Jr., Synoptic features associated with Los Angeles tornado occurrences, *Bull. Am. Meteor. Soc.*, *66*, 657–662, 1985.

Hirt, W. D., Forecasting severe weather in North Dakota, in *Preprints, 14th Conference on Severe Local Storms*, pp. 328–331, American Meteorological Society, Boston, Mass., 1985.

House, D. C., Forecasting tornadoes and severe thunderstorms, *Meteorol. Monogr.*, *5(27)*, 141–155, 1963.

Johns, R. H., A synoptic climatology of northwest flow severe weather outbreaks, II, Meteorological parameters and synoptic patterns, *Mon. Weather Rev.*, *112*, 449–464, 1984.

Johns, R. H., J. M. Davies, and P. W. Leftwich, An examination of the relationship of 0–2 km AGL "positive" wind shear to potential buoyant energy in strong and violent tornado situations, in *Preprints, 16th Conference on Severe Local Storms*, pp. 593–598, American Meteorological Society, Boston, Mass., 1990.

Johns, R. H., J. M. Davies, and P. W. Leftwich, Some wind and instability parameters associated with strong and violent tornadoes, 2, Variations in the combinations of wind and instability parameters, this volume.

Kelly, D. L., J. T. Schaefer, R. P. McNulty, C. A. Doswell III, and R. F. Abbey, Jr., An augmented tornado climatology, *Mon. Weather Rev.*, *106*, 1172–1183, 1979.

Kelly, D. L., J. T. Schaefer, and C. A. Doswell III, The climatology of non-tornadic severe thunderstorm events in the United States, *Mon. Weather Rev.*, *113*, 1997–2014, 1985.

Keyser, D., and L. W. Uccellini, Regional models: Emerging research tools for synoptic meteorologists, *Bull. Am. Meteorol. Soc.*, *68*, 306–320, 1987.

Leftwich, P. W., Jr., On the use of helicity in operational assessment of severe local storm potential, in *Preprints, 16th Conference on Severe Local Storms*, pp. 306–310, American Meteorological Society, Boston, Mass., 1990.

Leftwich, P. W., Jr., and R. W. Anthony, Verification of Severe Local Storms Forecast Issued by the National Severe Storms Forecast Center: 1990, *NOAA Tech. Memo. NWS NSSFC-31*, 9 pp., 1991. (Available at Natl. Severe Storms Forecast Center, Kansas City, Mo.)

Lewis, J. M., C. M. Hayden, R. T. Merrill, and J. M. Schneider, GUFMEX: A study of return flow in the Gulf of Mexico, *Bull. Am. Meteorol. Soc.*, *70*, 24–29, 1989.

Lloyd, J. R., The development and trajectories of tornadoes, *Mon. Weather Rev.*, *70*, 65–75, 1942.

MacGorman, D. R., Lightning in tornadic storms: A review, this volume.

Maddox, R. A., Large-scale meteorological conditions associated with midlatitude, mesoscale convective complexes, *Mon. Weather Rev.*, *111*, 1475–1493, 1983.

Maddox, R. A., and C. A. Doswell III, An examination of jet stream configurations, 500 mb vorticity advection and low level thermal advection patterns during extended periods of intense convection, *Mon. Weather Rev.*, *110*, 184–197, 1982.

Mass, C. F., Synoptic frontal analysis: Time for a reassessment?, *Bull. Am. Meteorol. Soc.*, *72*, 348–363, 1991.

McCann, D. W., The enhanced-V: A satellite observable severe storm signature, *Mon. Weather Rev.*, *111*, 887–894, 1983.

Miller, R. C., Notes on the analysis and severe-storm forecasting procedures of the Air Force Global Weather Central, *Tech. Rep. 200* (revision), 190 pp., Air Weather Serv., Scott Air Force Base, Ill., 1972.

Moncrieff, M. W., and M. J. Miller, The dynamics and simulation of tropical cumulonimbus and squall lines, *Q. J. R. Meteorol. Soc.*, *102*, 373–394, 1976.

Mosher, F. R., and J. S. Lewis, Use of lightning location data in severe storm forecasting, in *Preprints, Conference on Atmosphere Electricity*, pp. 692–697, American Meteorological Society, Boston, Mass., 1990.

National Severe Storms Project Staff, Environmental and thunderstorm structure as shown by National Severe Storms Project observations in spring 1960–61, *Mon. Weather Rev.*, *91*, 271–292, 1963.

National Weather Service, 1987: Assessment plan for the uses of Wind Profiler technology in NWS operations, internal planning document, 70 pp., Office of Meteorol., Silver Springs, Md., 1987.

Ninomiya, K., Mesoscale modification of synoptic situations from thunderstorm development as revealed by ATS III and aerological data, *J. Appl. Meteorol.*, *10*, 1103–1121, 1971.

Purdom, J. F. W., Subjective interpretation of geostationary satellite data for nowcasting, in *Nowcasting*, edited by K. Browning, pp. 149–166, Academic, San Diego, Calif., 1982.

Rabin, R. M., and D. S. Zrnic', Subsynoptic-scale vertical wind revealed by dual Doppler-radar and VAD analysis, *J. Atmos. Sci.*, *37*, 644–654, 1980.

Rockwood, A. A., and R. A. Maddox, Mesoscale and synoptic scale interactions leading to intense convection: The case of 7 June 1982, *Weather Forecasting*, *3*, 51–68, 1988.

Schaefer, J. T., Severe thunderstorm forecasting: A historical perspective, *Weather Forecasting*, *1*, 164–189, 1986.

Setvák, M., and C. A. Doswell III, The AVHRR channel 3 cloud top reflectivity of convective storms, *Mon. Weather Rev.*, *119*, 841–847, 1991.

Showalter, A. K., A stability index for thunderstorm forecasting, *Bull. Am. Meteorol. Soc.*, *6*, 250–252, 1953.

Showalter, A. K., and J. R. Fulks, Preliminary report on tornadoes, 162 pp., U.S. Weather Bur., Washington, D. C., 1943.

Snellman, L. W., Operational forecasting using automated guidance, *Bull. Am. Meteorol. Soc.*, *58*, 1036–1044, 1977.

Stensrud, D. J., and R. A. Maddox, Opposing mesoscale circulations: A case study, *Weather Forecasting*, *3*, 189–204, 1988.

Tepper, M., A proposed mechanism of squall lines: The pressure jump line, *J. Meteorol.*, *7*, 21–29, 1950.

Tepper, M., Mesometeorology—The link between macroscale atmospheric motions and local weather, *Bull. Am. Meteorol. Soc.*, *40*, 56–72, 1959.

Turman, B. N., and R. J. Tettelbach, Synoptic-scale satellite lightning observation in conjunction with tornadoes, *Mon. Weather Rev.*, *108*, 1878–1882, 1980.

Wakimoto, R. M., and J. W. Wilson, Non-supercell tornadoes, *Mon. Weather Rev.*, *117*, 1113–1140, 1989.

Warner, T. T., and N. L. Seaman, A real-time, mesoscale numerical weather-prediction system used for research, teaching, and public service at the Pennsylvania State University, *Bull. Am. Meteorol. Soc.*, *71*, 792–805, 1990.

Weaver, J. F., and N. J. Doesken, High Plains severe weather—Ten years after, *Weather Forecasting*, *6*, 411–414, 1991.

Weisman, M. L., and J. B. Klemp, The dependence of numerically simulated convective storms on vertical wind shear and buoyancy, *Mon. Weather Rev.*, *110*, 504–520, 1982.

Weisman, M. L., and J. B. Klemp, The structure and classification of numerically simulated convective storms in directionally varying wind shears, *Mon. Weather Rev.*, *112*, 2479–2498, 1984.

Weiss, S. J., On the operational forecasting of tornadoes associated with tropical cyclones, in *Preprints, 14th Conference on Severe Local Storms*, pp. 293–296, American Meteorological Society, Boston, Mass., 1985.

Weiss, S. J., On the relationship between NGM mean relative humidity and the occurrence of severe local storms, in *Preprints, 15th Conference on Severe Local Storms*, pp. J111–J114, American Meteorological Society, Boston, Mass., 1987.

Weiss, S. J., D. L. Kelly, and J. T. Schaefer, New objective verification techniques at the National Severe Storms Forecast Center, in *Preprints, 8th Conference on Weather Forecasting and Analysis*, pp. 412–419, American Meteorological Society, Boston, Mass., 1980.

Some Wind and Instability Parameters Associated With Strong and Violent Tornadoes
1. Wind Shear and Helicity

JONATHAN M. DAVIES

Pratt, Kansas 67124

ROBERT H. JOHNS

National Severe Storms Forecast Center, Kansas City, Missouri 64106

1. INTRODUCTION

Although the vertical wind profile through the troposphere has been recognized to be important in tornado development since the beginning of tornado forecasting efforts in the 1950s, only recently have researchers begun to investigate more detailed characteristics of wind profiles contributing to low-level mesocyclone formation and tornado production in supercell thunderstorms. *Davies-Jones et al.* [1990] provide an overview of recent work completed in this area. From modeling results and storm observations, it appears that both (1) the wind profile in the low levels (i.e., the storm inflow layer) and (2) the strength of the wind field and shear extending through a deeper layer of the troposphere (i.e., through middle levels) are important to supercell-induced tornado development.

Regarding the low-level wind profile, mean shear [*Rasmussen and Wilhelmson*, 1983; *Davies*, 1989] is one parameter that has been used for assessing the veering of winds with height and turning of the hodograph associated with supercell thunderstorms. More recently, storm-relative helicity [*Lilly*, 1986; *Davies-Jones et al.*, 1990], which addresses the importance of the wind fields viewed from a storm's frame of reference, has become recognized as an effective parameter for measuring rotational potential in the low-level wind field. Helicity in the storm inflow layer is associated with streamwise vorticity [*Davies-Jones*, 1984] which, when "ingested" and tilted into the storm updraft, induces rotation. Also, numerical simulations by *Brooks and Wilhelmson* [1990], *Brooks et al.* [1992], and *McCaul* [1990]

demonstrate the development of a mesolow in the inflow region of tornadic supercells, initiated by the dynamical interaction of the updraft with shear-induced vertical pressure forces [*Rotunno and Klemp*, 1982] that are largely a result of low-level curvature shear and helicity. This feature serves to intensify the vertical velocity of the growing updraft, as well as to strengthen inflow into the storm, an important requirement noted by *Lazarus and Droegemeier* [1990].

The strength of the wind fields and shear through a layer deeper than the low levels is also significant in supercell development. While low-level curvature shear is an important factor for producing the vertical pressure gradient mentioned in the prior paragraph, *Weisman and Klemp* [1986] note that it is also necessary for the vertical wind shear to extend through middle levels of the troposphere to sustain a rotating updraft via shear-induced pressure forces. One measure of this deeper ambient shear is the U parameter in the denominator of the Bulk Richardson Number [*Weisman and Klemp*, 1982, 1986]. *Doswell* [1991] also notes that middle-level winds of sufficient strength are a necessary component of hodographs that can support supercell development. A similar requirement is mentioned by *McCaul* [1991], who notes that 700 mbar wind speeds show good correlation with tropical cyclone tornado outbreaks. Strong middle-level winds may move precipitation downwind out of the upper portion of the updraft (or in some cases tilt the updraft so that precipitation falls out downwind), thereby eliminating a potential impediment to the development of a strong and sustained updraft.

On the basis of the above discussion, the following wind parameters were chosen for examination in this paper: (1) low-level mean shear (0–2 km/0–3 km/0–4 km above ground

The Tornado: Its Structure, Dynamics, Prediction, and Hazards.
Geophysical Monograph 79
Copyright 1993 by the American Geophysical Union.

level (AGL)), (2) storm-relative low-level helicity (0–2 km/0–3 km/0–4 km AGL), (3) Bulk Richardson Number shear (U) (boundary layer through 6 km AGL), and (4) mean wind speed in the middle levels (3–6 km AGL). *Johns et al.* [1990] (henceforth referred to as JDL) assembled a large and comprehensive data set of strong and violent mesocyclone-induced tornadoes (242 cases) for the purpose of examining the associated mean shear and buoyancy values. The same data set is utilized in this study to examine the aforementioned wind parameters, with suggestions for possible application to operational forecasting.

2. Case Selection Methodology

As described by JDL, Storm Data was examined systematically for the 10-year period April 1980 through March 1990. All nonlocalized tornado outbreaks that involved six or more tornadoes were selected, with the added requirement that two or more of the tornadoes be F2 or greater in intensity and separated by at least 60 nautical miles (111 km). Also, any F3 or greater intensity tornado accompanied by an uncontaminated proximity sounding was selected. These criteria were used in an attempt to eliminate most tornadoes of nonsupercell origin [*Wakimoto and Wilson*, 1989]. In large outbreaks, more than one case was selected if the constituent tornadoes were separated by ≥200 nautical miles (370 km) in distance or ≥10 hours in time.

3. Determination of Parameter Values

As discussed in detail by JDL, parameters computed from rawinsonde observations were either used directly or interpolated, depending on tornado case time and distance relative to soundings. In general, if a tornado case occurred within 3 hours and 75 statute miles (121 km) of an uncontaminated sounding in the warm sector, parameters computed from the observation were used directly. If the tornado case was more removed in distance, yet within 3 hours of sounding time, an interpolation was performed between parameter values from two or more soundings, depending on the availability and location of adjacent warm sector rawinsonde observations. If the tornado case was more removed in time, an interpolation of sounding parameter values in time and distance was performed by observing the evolution of wind features between sounding times, provided that wind fields were well defined on the synoptic scale. Cases that did not meet these criteria or that involved significant amounts of missing data were not accepted into the data set.

The limitations of the coarse rawinsonde network are well known. In an attempt to incorporate localized data, surface wind observations were also examined for each case. If the wind direction from a surface observation nearby in the warm sector exhibited significant backing (i.e., ≥20°) when compared to the surface wind of the sounding(s) associated with the case, the wind from the surface observation was blended into the sounding wind profile(s) for computing parameter values such as helicity and mean low-level shear.

In spite of this effort to incorporate local data, the authors recognize that mesoscale variations not detected by the sounding network or surface observations can have major impact on thunderstorm behavior. Therefore parameter values obtained in this study can be viewed as only roughly representative of the actual prestorm wind environments for each case. By limiting the study to cases from nonlocalized tornado outbreaks, the probability is increased that rawinsonde network observations sampled at least some aspects of the wind environments supporting the observed tornadic supercell thunderstorms.

4. Mean Shear Results

From *Rasmussen and Wilhelmson* [1983], mean shear (S) is essentially

$$S = \text{hodograph length (m s}^{-1}\text{)/depth of layer (m)}$$

"Positive" mean shear [*Davies*, 1989], as amended by JDL, modifies this definition by setting the shear magnitude to zero for those hodograph segments where the ground-relative winds back significantly with height. Unlike storm-relative helicity, mean shear does not consider storm motion and is computationally very sensitive to the fine-scale wind structure. Recognizing these limitations, positive mean shear can, in many cases, provide an estimated assessment of rotational potential when a storm motion is not available.

Average positive mean shear magnitudes ($\times 10^{-3}$ s^{-1}) were computed for the 242 cases in the JDL data set using three atmospheric depths (AGL) for comparison:

0- 2-km average positive shear	13.6
0- 3-km average positive shear	10.7
0- 4-km average positive shear	9.1

The fact that the 0- 2-km average positive shear is larger than both the 0- 3-km and 0- 4-km averages suggests that in most cases a majority of the low-level wind shear is below 2 km AGL, similar to results from *Davies* [1989] using a smaller data set.

Grouping cases by tornado intensity (205 strong and 37 violent tornado cases), average positive mean shear values were

	Strong(F2/F3)	Violent(F4/F5)
0- 2-km average positive shear	13.4	14.7
0- 3-km average positive shear	10.5	11.7
0- 4-km average positive shear	9.0	10.0

These values suggest that, on an average, violent tornadoes occur with higher values of low-level positive mean wind shear than do tornadoes of lesser intensity.

5. Estimation of Storm Motion for Storm-Relative Helicity Computations

In order to calculate storm-relative helicity a storm motion is required. Storm motions originally were not obtained for the JDL data set. For this study the authors have examined a subset of cases for which radar imagery was available at the National Severe Storms Forecast Center. This subset of 31 cases contains tornadoes that occurred during the period 1988–1990. Two of the cases from 1990 were not in the original JDL data set. Mean environmental winds (0- 6-km AGL [after *Weisman and Klemp*, 1986]) were computed from the sounding data for these cases, and comparisons were made with the storm motions obtained from the radar data. Though the storm motion sample is small, the cases appear to represent all seasons and most areas of the country east of the Rocky Mountains. It should be noted that the storm motion deviations here are not exactly comparable to *Maddox* [1976], *Darkow and McCann* [1977], and *Bluestein and Parks* [1983], who computed the mean wind through much deeper layers that generally encompassed the troposphere.

5.1. *Observed Storm Motions From the Data Subset*

The directional deviations within the 31 case subset vary from 2° left to 42° right of the 0- 6-km mean wind. Speed deviations vary from 58 to 158% of the 0- 6-km mean wind speed. It is notable that five of the cases moved faster than the mean wind speed; four of these were associated with bow echo structures [see *Doswell et al.*, 1990]. From these results it is apparent that for any given sample of tornado-producing supercell storms there can be a wide range of deviant motions when compared to the 0- 6-km mean environmental wind. This fact illustrates that without a better understanding of the complex problem of storm motion it will be difficult to implement a predictive cell movement procedure that will be successful in all cases.

Despite this problem a closer examination of the storm motion subset reveals some useful general information. Mean deviant storm motion for the 31 cases is 20° right of the 0- 6-km mean wind at 89% of the mean wind speed (or 20R89, as deviations will henceforth be annotated). Roughly half of the subset (15 cases) exhibits a directional deviation within 5° of this mean. This suggests that when considering supercell development, a motion deviation assumption of 20R85 or 20R90 would be a reasonably accurate forecast motion estimate roughly 50% of the time.

Some storm motion differences are apparent when grouping cases by both season and geographical locale. From the data subset, tornadic storms in the Great Plains during March through June (eight cases) display a mean motion of 27R80, while tornadic storms east of the Mississippi during November through February (11 cases) exhibit a mean motion of 18R97. The Great Plains storms have a directional deviation spread ranging from 13° to 37° right of the mean wind (with only one of eight cases less than 20° right) and a speed deviation spread ranging from 61 to 122% of the mean wind (only one case exceeded 100% of the mean wind speed). In contrast, the cool season storms east of the Mississippi have a smaller directional deviation spread, ranging from 12° to 24° right of the mean wind (six of 11 cases moved less than 20° right) and a speed deviation spread ranging from 77 to 122% (three of 11 cases moved at 100% or more of the mean wind speed). This difference suggests that the strength of the wind fields (which varies by season and locale) may play a significant part in storm motion.

5.2. *Storm Motion Related to Wind Field Strength*

In the 31 case subset, extreme right-moving storms (i.e., >25° right) are found to be associated with significantly weaker wind fields (0- 6-km mean wind speeds averaging 32 knots (16.5 m s^{-1})) than the other cases (0- 6-km mean wind speeds averaging 46 knots (23.7 m s^{-1})). This tendency for tornadic storms in relatively weak wind fields (i.e., 0- 6-km mean wind magnitudes <30–35 knots (15.5–18.0 m s^{-1})) to deviate increasingly right of the mean wind also can be seen in a study by *Darkow and McCann* [1977]. If their Figures 3a and 1a are compared, it becomes apparent that their extreme right-moving storms occurred in significantly weaker wind fields when compared to their data set as a whole.

The weakest 0- 6-km mean wind speeds found in the 31 case subset were near 20 knots (10.3 m s^{-1}), suggesting that mean wind speeds of lesser magnitude may be too weak to support strong or violent tornadoes in supercell thunderstorms. For purposes of this discussion, relatively "weak" wind fields that can support significant supercell-induced tornadoes will be defined as those with 0- 6-km mean wind speeds ≤30 knots (15.5 m s^{-1}) and ≥20 knots (10.3 m s^{1}).

5.3. *Suggested Storm Motion Assumptions*

On the basis of the above discussion, some rough first guess assumptions can be suggested regarding storm movement for right-moving cells in a given wind regime that is potentially tornadic. For environments characterized by relatively strong wind fields (i.e., 0- 6-km mean wind speeds >30 kt) a reasonable first guess movement would be a motion such as 20R85, similar to the 31-case subset mean deviant motion of 20R89. The storms in the subset that are not extreme right movers yield a mean deviant motion of 16R86, also close to 20R85. When weaker wind fields are involved (i.e., 0- 6-km mean wind speeds ≤30 knots (15.5 m s^{1})), a motion closer to 30R75, as used by *Maddox* [1976], is probably more appropriate. This is suggested by the fact that the storms in the 31-case subset that were extreme right movers traveling at less than the mean wind speed yielded a mean deviant motion of 34R76, close to a 30R75 assumption. It should be strongly emphasized that these estimates are only general reference points and do not address many other factors that affect storm motion, such as the orientation of boundaries and low-level forcing.

For this study, these rough assumptions will provide starting points, in lieu of actual observed storm motions, for examination of storm-relative helicity utilizing the JDL data set.

6. STORM-RELATIVE HELICITY RESULTS

For this study, the authors have computed integrated total storm-relative helicity (H) for specific layers. The computation used is that of *Davies-Jones et al.* [1990]:

$H = -2 \times$ the signed area swept out by the storm-relative wind vector while ascending through a layer

Units are $m^2 \, s^{-2}$ (identical to $J \, kg^{-1}$). Helicity is useful in measuring rotational potential because it has a sound physical link to overall updraft rotation and is sensitive to storm motion while remaining computationally insensitive to the fine-scale wind structure.

First, observed helicity values obtained from the 31-case storm motion subset will be examined. Then, utilizing the storm motion refinement assumptions from section 5, helicity values will be examined using the entire 242-case JDL data set.

6.1. *Observed Helicity From the Storm Motion Subset*

Helicity magnitudes were computed for the observed storm motions of the 31 subset cases and the same atmospheric layers (AGL) used previously in computing mean shear. Mean helicity magnitudes ($m^2 \, s^{-2}$) for the three layers were as follows.

0- 2-km mean observed helicity	392
0- 3-km mean observed helicity	416
0- 4-km mean observed helicity	428

Note that most of the low-level helicity resides below 2 km AGL, similar to the mean shear results in section 4. An average of 94% of the 0- 3-km helicity is located within the 0- 2-km layer, and an average of 92% of the 0- 4-km helicity is located within the 0- 2-km layer.

The 31 cases (21 strong and 10 violent tornado cases) were also examined from the standpoint of helicity and tornado intensity:

	Strong(F2/F3)	Violent(F4/F5)
0- 2-km mean observed helicity	359	460
0- 3-km mean observed helicity	369	519
0- 4-km mean observed helicity	378	539

Results suggest that violent tornadoes, on an average, occur with larger helicity magnitudes than do strong tornadoes. This agrees with the general helicity guidelines presented by *Davies-Jones et al.* [1990] for strong and violent tornado development.

6.2. *Helicity Using the JDL Data Set and an Assumed Storm Motion*

For computing helicity, storm motions were estimated by applying a 20R85 assumption when 0- 6-km mean wind speeds were >30 knots (15 m s^{-1}), and a 30R75 assumption when 0- 6-km mean wind speeds were ≤30 knots (15 m s^{-1}) (see section 5). Helicity magnitudes were computed for the same three lower tropospheric layers examined using the 31 case storm motion subset, yielding the following mean values for the 242 cases.

0- 2-km mean assumed helicity	332
0- 3-km mean assumed helicity	356
0- 4-km mean assumed helicity	375

As with the 31-case subset, it is notable that so much of the low-level helicity is below 2 km AGL.

Grouping cases by tornado intensity (205 strong and 37 violent tornado cases), helicity values were computed using the same motion assumptions:

	Strong(F2/F3)	Violent(F4/F5)
0- 2-km mean assumed helicity	317	415
0- 3-km mean assumed helicity	339	452
0- 4-km mean assumed helicity	357	478

As with observed helicity from the 31 case subset, the larger data set yields similar results, showing violent tornadoes occur, on an average, in environments with larger helicity values than do strong tornadoes. However, notable exceptions certainly do occur, as will be seen in section 10.

For the 29 overlapping cases between the storm motion subset and the JDL data set, over half the mean helicity values computed using the motion assumptions were within 15% of the observed values, and all but eight cases were within 20% of the observed value. It therefore appears that the 20R85/30R75 assumptions based on general strength of the wind fields are useful starting points for generating helicity output.

Using this approach, Figure 1 illustrates a helicity forecast derived from the National Meteorological Center's Limited Fine Mesh (LFM) model, blending winds aloft forecast data with surface winds from Model Output Statistics (MOS) forecast guidance. For each output station the 0- 6-km mean wind direction and speed are printed along with helicity values computed using storm motion deviations of 20R85 and 30R75 relative to the 0- 6-km mean wind. This would allow a forecaster to see how the helicity changes with increasing storm deviation to the right at different locations plotted on the same map. On the basis of the strength of the 0- 6-km mean wind or other factors the forecaster then could select for analysis the deviation deemed most appropriate for the situation. This output format would also help the forecaster assess whether the risk is widespread or localized. If helicity values are large regardless of deviation, the potential for low-level rotation is more extensive than if helicity values become significant only with extreme right deviation. This or a similar format could be used to generate helicity output maps from sources such as rawinsonde data, profiler and WSR-88D wind data, and numerical model forecasts.

7. BULK RICHARDSON NUMBER SHEAR (U) RESULTS

As discussed in section 1, the Bulk Richardson Number (BRN) shear, U [*Weisman and Klemp*, 1982, 1986], is a parameter that measures the ambient environmental shear that is important for supercell development:

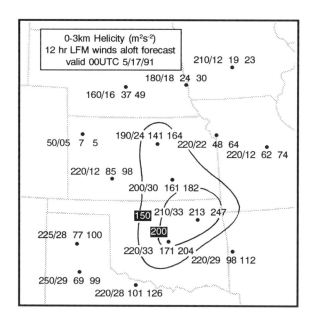

Fig. 1. Example of 12-hour forecast of 0- 3-km AGL integrated storm-relative helicity, derived from LFM forecast winds aloft data and surface data from MOS guidance valid at 0000 UTC May 17, 1991, for sites at approximate dot locations. Three computations are indicated for each site: the first number/group is the 0- 6-km AGL mean wind direction and speed in degrees and knots; the second and third numbers are helicity values derived using storm motions of 20R85 and 30R75 (see text for explanation of annotation), respectively. The 30R75 helicity (third number) is analyzed for values ≥ 150 m^2 s^{-2}. In this particular case the LFM guidance verified reasonably well, and damaging supercell-induced tornadoes occurred near Wichita and Tulsa.

U = the straight vector difference between the 0- 6-km AGL mean wind and the boundary layer wind (i.e., 0- 500-m AGL mean wind)

Units are m s^{-1}. It is important to note that although the BRN correlates reasonably well with observed storm type (e.g., supercell versus multicell), it is a poor predictor of storm rotation in low levels because it does not account for some aspects of the wind profile, such as low-level curvature shear [*Lazarus and Droegemeier*, 1990].

Figure 2 shows the distribution of U for the JDL data set. The majority of cases (70%) are associated with U values greater than 18, with one case exceeding 40. None of the cases has a U value less than 12. This suggests that U values much less than 12 m s^{-1} may not be able to support supercell-induced tornadoes.

8. MIDDLE-LEVEL WIND SPEED RESULTS

Doswell [1991] specifies middle-level winds as those in roughly the 700- to 500-mbar layer. For consistency when dealing with surface elevations ranging from the high plains to ocean coastal areas, this study considers the mean measured wind speed in the 3- 6-km AGL layer as representative of middle levels.

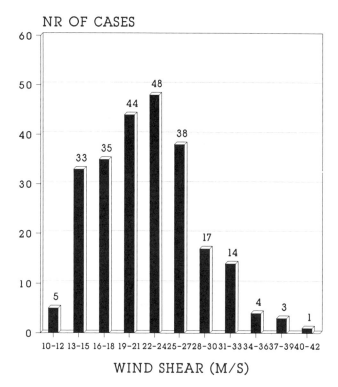

Fig. 2. Distribution of cases from JDL data set, grouped according to U magnitude (Bulk Richardson Number shear) (in units of meters per second).

The average mean 3- 6-km wind speed for the JDL data set is 49 knots (24 m s^{-1}). The average mean 3- 6-km wind speed for relatively weak wind cases (0- 6-km mean wind speed ≤ 30 knots (15 m s^{-1})) is 34 knots (17.5 m s^{-1}). The lowest mean 3- 6-km wind speed encountered in the data set is 20 knots (11 m s^{-1}), which agrees with Doswell's observation that middle-level winds should generally be 20 knots (11 m s^{-1}) or greater regarding hodographs that indicate potential for supercell development.

9. CASE STUDIES

Horizontal fields generated from sounding data for parameters selected for this study will be examined briefly here for two cases from the 31-case storm motion subset, one involving strong wind fields and one involving less intense wind fields. These serve as examples of how wind parameters might come together diagnostically to indicate potential for tornadic supercell development. However, the forecaster needs to remember that atmospheric wind profiles are changing constantly. The reader is referred to *Doswell* [1991] for a discussion concerning the problems of forecasting changes to the hodograph/wind profile.

9.1. Raleigh, North Carolina, November 28, 1988

Davies [1989] examined this case from the perspective of positive mean shear, contrasting the shear in several different low-level layers. For comparison, the same case will be examined here using the wind parameters selected for this study.

Using 0000 UTC sounding data (6 hours prior to the Raleigh tornado), helicity fields were generated using a deviant storm motion assumption of 20R85 (the 0- 6-km wind fields were quite strong). These fields were computed for the 0- 2-km, 0- 3-km, and 0- 4-km AGL layers (only the 0- 2-km and 0- 4-km fields are shown; see Figures 3a and 3b). While the 0- 2-km field appears more focused than patterns from deeper layers, all three analyses place a maximum over the Greensboro-Raleigh area, as did similar analyses for positive mean shear by Davies [1989]. Tornadoes later occurred with storms in North Carolina and southern Virginia where helicity values were very large (600 $m^2 s^{-2}$ or greater) but not with storms that occurred earlier in Georgia. The lack of tornadoes further south where helicity magnitudes were less but still significant (values around 300 $m^2 s^{-2}$) was probably due to the fact that instability (not shown) was weak or limited mainly to coastal areas, whereas weak to moderate instability extended inland further north across North Carolina. Part 2 of this paper [Johns et al., this volume] examines combinations of instability and helicity associated with strong and violent tornadoes.

Helicity fields were also computed using the observed motion of the Raleigh storm (24R86). The results are very similar to the fields derived using an assumed motion. This can be seen by comparing the 0- 2-km fields in Figures 3a and 4 (the 0- 3-km and 0- 4-km observed helicity fields are not shown). In this case the use of an assumed storm motion of 20R85 relative to the 0- 6-km mean wind worked quite well in offering a useful diagnostic depiction of the helicity.

Bulk Richardson Number shear U (not shown) was strong with a maximum of more than 28 m s^{-1} over central North Carolina. Middle-level winds (not shown) were also strong, ranging from 40 to more than 80 knots (20–40 m s^{-1}) over the eastern United States. Considering just the wind criteria in this case (apart from other factors such as instability), all parameters are more than adequate for supercell development, most over a large area. It is the helicity pattern that is most helpful in narrowing the threat area, particularly the 0- 2-km field. This is typical of tornado cases involving strong wind fields.

9.2. Central Kansas/Southeast Wyoming, May 24, 1990

Figure 5 shows the 0- 2-km assumed helicity field (the 0- 3-km and 0- 4-km fields are not shown) from 0000 UTC May 25, 1990, sounding data, when significant supercell-induced tornadoes were occurring over central Kansas and southeast Wyoming. The wind fields are weaker than in the Raleigh case (0- 6-km mean wind ≤30 knots (15 m s^{-1}) over a majority of the area), so a 30R75 motion assumption was

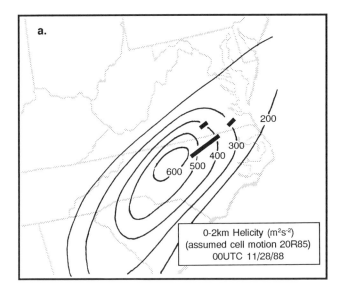

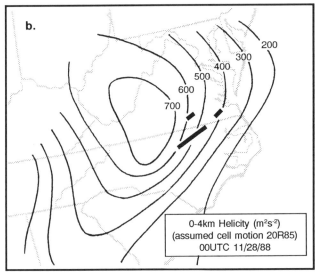

Fig. 3. Horizontal distribution of storm-relative helicity subjectively analyzed from 0000 UTC November 28, 1988, sounding data, using an assumed storm motion of 20R85. Helicity is integrated through (a) 0–2 km AGL and (b) 0–4 km AGL. Units of contours are $m^2 s^{-2}$. Heavy lines represent tracks of significant tornadoes which occurred 6–8 hours after 0000 UTC.

applied. Note that even with the coarse resolution of the rawinsonde network, two general areas of maximum helicity are evident, each roughly corresponding to the locations of tornado occurrences. The assumed helicity depictions compare well with analyses generated using observed storm motions (the observed helicity field for 0–2 km is shown in Figure 6).

An analysis of BRN shear U (Figure 7) indicates two maximum areas/ridges of strong shear through middle levels, one over Kansas and another extending from western Colorado into the northern high plains. Middle-level wind speeds

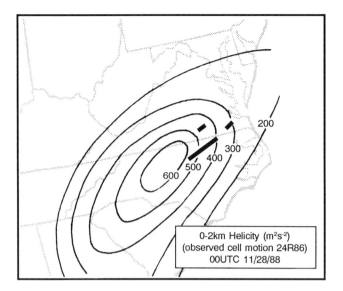

Fig. 4. Horizontal distribution of storm-relative helicity as in Figure 3a, except observed motion of Raleigh tornadic storm (24R86) is used. Tornado tracks are as in Figure 3.

(3- 6-km AGL averages) in Figure 8 reveal two streams of maximum winds, each correlating well with the U parameter shear maxima in Figure 7. These wind maxima are moving downstream across the areas of tornado occurrence, providing shear through middle levels and additional support for supercell thunderstorms.

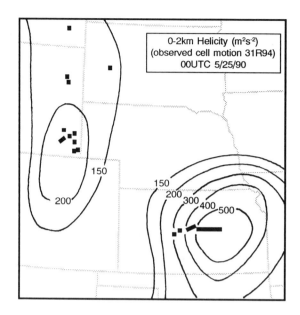

Fig. 6. Horizontal distribution of storm-relative helicity as in Figure 5, except observed motion of tornadic storm in central Kansas (31R94) is used. Tornado occurrences are as in Figure 5.

10. Extremes of Hodographs Associated With Supercell-Induced Tornadoes

Examination of the large JDL data set exposed the authors to a wide range of wind profiles that were associated with strong and violent tornadoes. While many hodographs were

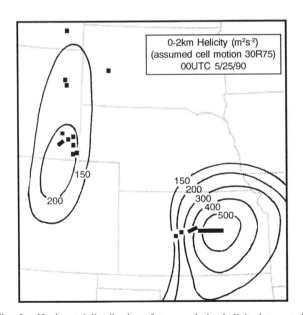

Fig. 5. Horizontal distribution of storm-relative helicity integrated through 0–2 km AGL, subjectively analyzed from 0000 UTC May 25, 1990, rawinsonde network data using an assumed storm motion of 30R75. Units of contours are $m^2 s^{-2}$. Square dots and heavy lines represent approximate locations of tornadoes occurring within 2–3 hours of 0000 UTC.

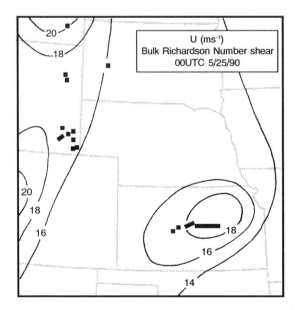

Fig. 7. Horizontal distribution of U (Bulk Richardson Number shear) subjectively analyzed from 0000 UTC May 25, 1990, rawinsonde network data. Units of contours are meters per second. Tornado occurrences are as in Figure 5.

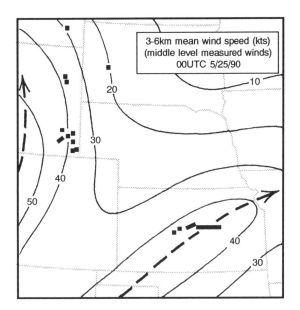

Fig. 8. Horizontal distribution of mean 3- 6-km AGL wind speed subjectively analyzed from 0000 UTC May 25, 1990, rawinsonde network data. Units of contours are knots. Heavy dashed lines are axes of maximum wind speed. Tornado occurrences are as in Figure 5.

similar in scope and shape, the extremes of hodographs encountered are significant and are worth discussing briefly.

Figure 9 shows two low-level hodographs from the JDL data set that are quite different from each other in size and scope. Both hodographs were associated with tornadoes of F4 intensity. While not obvious in a ground-relative frame, the directional turning for both hodographs is nearly the same when shifted to a storm-relative frame (roughly 100° for Figure 9a, and 90° for Figure 9b). The most notable contrast between the two cases is in the magnitudes of helicity (745 versus 140 $m^2 s^{-2}$, based on observed storm motions) and buoyancy (convective available potential energy, CAPE) (300 versus 4500 J kg^{-1}). In the BNA example the extreme wind shear and helicity appear to have been responsible for producing updraft rotation, in spite of meager instability. At the opposite end of the spectrum the extreme instability in the OMA example may have somehow played a role in enhancing the rotational potential of a wind environment exhibiting weak helicity yet significant directional turning in low levels. While undetected local changes favoring rotation [see *Doswell*, 1991] also could have been involved in the latter case, the fact remains that these widely differing hodographs were associated with violent tornadoes. These two cases illustrate that combinations of helicity and instability are an important consideration in assessing the potential for supercell-induced tornadoes (see part 2 of this paper).

11. DISCUSSION

The wind parameters examined in this study appear to relate well to supercell-induced tornado development. Significant or optimum parameter values as suggested in the literature from theoretical deduction, numerical modeling, or storm observations appear to be in reasonable agreement with the results from the large data set assembled by the authors. These wind parameters address the contributions of wind fields to several aspects of supercell and tornado development (e.g., updraft rotation, shear-induced vertical pressure gradient, and storm inflow). However, there are many other parameters apart from the wind profile that are important for development of supercells and supercell-induced tornadoes. These include thermodynamic factors such as atmospheric instability and the potential buoyant energy available to an updraft (discussed in part 2 of this paper), as well as the humidity of middle-level air entrained into downdrafts. A forecaster also must address issues such as dynamic forcing and capping inversions to determine whether thunderstorms will develop in the first place. Because of these factors one cannot produce a forecast concerning supercell-induced tornadoes from examination of wind parameters alone. However, it does appear that sufficient critical or "optimum" values of the wind parameters must be present for thunderstorms to develop into supercells that produce tornadoes.

While statistical correlations of the selected wind parameters to tornado occurrence and likelihood are beyond the scope of this study, some subjective comments can be offered regarding their relative usefulness in forecasting.

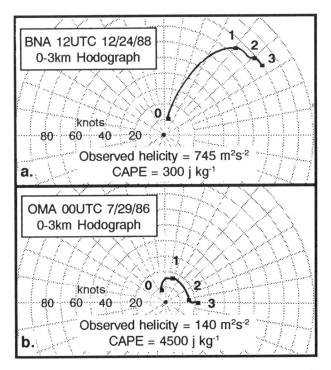

Fig. 9. Ground-relative low-level hodographs (0–3 km AGL) for (a) Nashville, 1200 UTC December 24, 1988, and (b) Omaha 0000 UTC July 29, 1986. Each ring increment represents 10 knots (5 m s^{-1}).

Helicity is probably the most crucial wind parameter for indicating tornadic supercell potential. The numerical simulations of *Brooks and Wilhelmson* [1990] tend to reinforce this impression. Their results show that storms with similar BRN values but differing shear curvature profiles in the low levels often develop differing circulations. Their storm simulations with low-level curvature shear (significant low-level helicity) developed low-level mesocyclones typical of observed tornadic supercells, while simulations involving rectilinear low-level shear (weak or negligible low-level helicity) did not produce low-level mesocyclones. Both types of simulations had similar values of U (vertical "straight-vector difference" shear through middle levels), and both produced supercell storms with identifiable middle-level mesocyclones. Although *Brooks et al.* [this volume] did encounter some numerically modelled storms in high-helicity environments that failed to produce low-level mesocyclones, it nevertheless appears that low-level curvature shear and helicity are crucial factors for the development of mesocyclones at low levels.

Positive mean shear correlates well with helicity in most cases involving moderate to strong wind fields and is a useful alternate calculation in such situations. As the positive mean shear becomes greater, the inference is that a larger range of storm motions can support tornadic supercell development. However, when wind fields are relatively weak (yet adequate to support supercell development), helicity is the preferable parameter because it can provide a useful depiction of the enhancement of rotational potential due to a storm's local motion.

Because deeper vertical shear (through middle levels) is an important ingredient for supporting supercells, the BRN shear and the strength of the middle-level winds also have importance in a forecast setting. Although most cases exhibiting significant low-level helicity also will be associated with significant wind fields through a deeper layer (due to the baroclinity and dynamics associated with weather systems), it is possible that significant low-level helicity can be present without deeper shear through middle levels when weaker weather disturbances are involved. Hence it is important to examine the middle-level wind fields in addition to the low-level helicity.

Returning to storm-relative helicity as a crucial parameter, key questions are (1) how to estimate in advance of storm development a reasonable storm motion for the more right-moving storms that may occur and (2) what layer is most appropriate for computing low-level helicity.

Regarding storm motion estimation, section 5 and recent operational experience suggest that applying one exclusive "preset" storm deviation in all situations is inappropriate for estimating helicity operationally. From examination of the 31-case storm motion subset the idea that supercells in weaker wind environments tend to deviate more from the mean wind has some validity. Therefore it would seem appropriate for a forecaster to be able to "phase in" larger deviations as wind fields become weaker for purposes of computing helicity. The 20R85/30R75 scheme suggested in sections 5 and 6 and in Figure 1 is one simple way of addressing this.

The depth of the layer that is used to compute low-level helicity depends largely on what constitutes the primary inflow layer for a given thunderstorm. This is because the streamwise vorticity [*Davies-Jones*, 1984] associated with helicity in the inflow layer is tilted into an updraft to produce rotation. The definition of inflow layer is probably related to the level of free convection (LFC) in a thunderstorm's near environment, among other factors. Using a mobile sounding system, *Bluestein et al.* [1989] found the lapse rate within the updraft and wall cloud of a tornadic thunderstorm to be "wet adiabatic" above the LFC, which was measured between 1.5 and 2 km AGL. Because the LFC varies considerably between "wet" environments and somewhat "drier" environments (e.g., along the dry line), the depth of the relevant inflow layer for computing helicity probably also varies from case to case. This suggests using the layer below the LFC for computation of helicity, which may be a useful subject for future study.

Operationally, different studies have used different layers for computing helicity. *Davies-Jones et al.* [1990] use 0-3-km AGL for the general inflow layer, while *Woodall* [1990] computes helicity for a deeper layer (0- 4-km AGL). From this study, the vertical distribution of helicity (section 6) and the case studies (section 9) suggest that useful results can often be obtained using the 0- 2-km AGL layer. On the basis of the examination of data set and recent operational experience the following general comments are offered.

1. Helicity derived from the 0- 2-km layer often tends to focus the horizontal pattern into a more useful area diagnostically (see Figure 3), while helicity patterns from deeper layers tend to spread out. This is particularly true in situations involving strong wind fields. In weak wind fields, helicity patterns derived from different layers (particularly 0–2 km and 0–3 km) often are nearly identical.

2. In high plains environments that tend to be "drier," resulting in higher LFCs, the 0- 3-km or 0- 4-km layer works better in capturing the inflow layer and relevant areas of helicity.

3. In hurricane/tropical cyclone tornado environments (which are associated with low LFCs), the largest low-level shear and helicity is often in the bottom 1–1.5 km [*McCaul*, 1991], suggesting that a relatively shallow layer be used for helicity computation.

The merits of one layer versus another for computing helicity are not dealt with here; the important point is that the inflow layer is not fixed from situation to situation. This suggests that operational helicity programs might be improved somewhat by including output information about the vertical distribution of helicity, as well as offering computation options involving different layers.

As noted earlier, the potential for tornadic supercell development does not depend on wind parameters alone. Part 2 of this paper [*Johns et al.*, this volume] will examine combinations of instability and shear/helicity that are associated with strong and violent supercell-induced tornadoes.

Acknowledgments. The authors wish to thank Grant Bean, Dave Higginbotham, and other NSSFC computer staff in helping to assemble the large data set used for this study. Ken Howard, NSSL, ERL; Mike Ryba, WSO DDC; and Joe Schaefer, SSD, Central Region NWS, are also acknowledged for their help with this project.

REFERENCES

Bluestein, H. B., and C. R. Parks, A synoptic and photographic climatology of low-precipitation severe thunderstorms in the southern plains, *Mon. Weather Rev.*, *111*, 2034–2046, 1983.

Bluestein, H. B., E. W. McCaul, Jr., G. P. Byrd, G. R. Woodall, G. Martin, S. Keighton, and L. C. Showell, Mobile sounding observations of a tornadic thunderstorm near the dryline: The Gruver, Texas storm complex of 25 May, 1987, *Mon. Weather Rev.*, *117*, 244–250, 1989.

Brooks, H. E., and R. B. Wilhelmson, The effects of low-level hodograph curvature on supercell structure, in *Preprints, 16th Conference on Severe Local Storms*, pp. 34–39, American Meteorological Society, Boston, Mass., 1990.

Brooks, H. E., C. A. Doswell III, and R. P. Davies-Jones, Environmental helicity and the maintenance and evolution of low-level mesocyclones, this volume.

Darkow, G. L., and D. W. McCann, Relative environmental winds for 121 tornado bearing storms, in *Preprints, 11th Conference on Severe Local Storms*, pp. 413–417, American Meteorological Society, Boston, Mass., 1977.

Davies, J. M., On the use of shear magnitudes and hodographs in tornado forecasting, in *Preprints, 12th Conference on Weather Forecasting and Analysis*, pp. 219–224, American Meteorological Society, Boston, Mass., 1989.

Davies-Jones, R. P., Streamwise vorticity: The origin of updraft rotation in supercell storms, *J. Atmos. Sci.*, *41*, 2991–3006, 1984.

Davies-Jones, R. P., D. W. Burgess, and M. Foster, Test of helicity as a tornado forecast parameter, in *Preprints, 16th Conference on Severe Local Storms*, pp. 588–592, American Meteorological Society, Boston, Mass., 1990.

Doswell, C. A., III, A review for forecasters on the application of hodographs to forecasting severe thunderstorms, *Natl. Weather Dig.*, *16*, 2–16, 1991.

Doswell, C. A., III, A. R. Moller, and R. W. Przybylinski, A unified set of conceptual models for variations on the supercell theme, in *Preprints, 16th Conference on Severe Local Storms*, pp. 40–45, American Meteorological Society, Boston, Mass., 1990.

Johns, R. H., J. M. Davies, and P. W. Leftwich, An examination of the relationship of 0–2 km AGL "positive" wind shear to potential bouyant energy in strong and violent tornado situations, in *Preprints, 16th Conference on Severe Local Storms*, pp. 593–598, American Meteorological Society, Boston, Mass., 1990.

Johns, R. H., J. M. Davies, and P. W. Leftwich, Some wind and instability parameters associated with strong and violent tornadoes, 2, Variations in the combinations of wind and instability parameters, this volume.

Lazurus, S. M., and K. K. Droegemeier, The influence of helicity on the stability and morphology of numerically simulated storms, in *Preprints, 16th Conference on Severe Local Storms*, pp. 269–274, American Meteorological Society, Boston, Mass., 1990.

Lilly, D. K., The structure, energetics, and propagation of rotating convective storms, II, Helicity and storm stabilization, *J. Atmos. Sci.*, *43*, 126–140, 1986.

Maddox, R. A., An evaluation of tornado proximity wind and stability data, *Mon. Weather Rev.*, *104*, 133–142, 1976.

McCaul, E. W., Jr., Simulations of convective storms in hurricane environments, in *Preprints, 16th Conference on Severe Local Storms*, pp. 334–339, American Meteorological Society, Boston, Mass., 1990.

McCaul, E. W., Jr., Buoyancy and shear characteristics of hurricane-tornado environments, *Mon. Weather Rev.*, *119*, 1954–1978, 1991.

Rasmussen, E. N., and R. B. Wilhelmson, Relationships between storm characteristics and 1200 GMT hodographs, low-level shear, and stability, in *Preprints, 13th Conference on Severe Local Storms*, pp. J5–J8, American Meteorological Society, Boston, Mass., 1983.

Rotunno, R., and J. B. Klemp, The influence of the shear-induced vertical pressure gradient on thunderstorm motion, *Mon. Weather Rev.*, *110*, 136–151, 1982.

Wakimoto, R. M., and J. W. Wilson, Non-supercell tornadoes, *Mon. Weather Rev.*, *117*, 1113–1140, 1989.

Weisman, M. L., and J. B. Klemp, The dependence of numerically simulated convective storms on vertical wind shear and buoyancy, *Mon. Weather Rev.*, *110*, 504–520, 1982.

Weisman, M. L., and J. B. Klemp, Characteristics of isolated convective storms, in *Mesoscale Meteorology and Forecasting*, edited by P. S. Ray, pp. 331–358, American Meteorological Society, Boston, Mass., 1986.

Woodall, G. R., Qualitative forecasting of tornadic activity using storm-relative environmental helicity, in *Preprints, 16th Conference on Severe Local Storms*, pp. 311–315, American Meteorological Society, Boston, Mass., 1990.

Some Wind and Instability Parameters Associated With Strong and Violent Tornadoes
2. Variations in the Combinations of Wind and Instability Parameters

ROBERT H. JOHNS, JONATHAN M. DAVIES,[1] AND PRESTON W. LEFTWICH

National Weather Service, NOAA, National Severe Storms Forecast Center, Kansas City, Missouri 64106

1. INTRODUCTION

Meteorologists have long known that both potential buoyant energy and the strength and vertical profile of the tropospheric wind fields are important in the process of tornado development [e.g., *Miller*, 1972]. Further, operational experience suggests that the combinations of wind parameters and instability parameters vary considerably from one tornado situation to another. Numerical models suggest that the type of storm that develops in a given situation (e.g., an isolated supercell) is related to both the vertical wind profile and the potential buoyant energy of the air in the updraft entrainment layer [*Weisman and Klemp*, 1982, 1986]. Observational studies by *Rasmussen and Wilhelmson* [1983] and *Leftwich and Wu* [1988] have examined the wind shear/potential buoyant energy relationship in association with tornado development. These studies involved limited data sets and were concerned with the wind shear in a relatively deep layer of the troposphere (0- 4-km above ground level (AGL)). Recently, interest has focused on the nature of the wind fields in shallower layers of the lower troposphere, layers that more nearly correspond to the updraft entrainment region [*Davies*, 1989; *Davies-Jones et al.*, 1990]. Utilizing a large comprehensive data set, *Johns et al.* [1990] (hereafter referred to as JDL) examined the relationship between 0- 2-km AGL "positive" wind shear (PWS) and convective available potential energy (CAPE) [*Moncrieff and Green*, 1972] in association with strong and violent tornado development. In this paper (part 2) the authors review the initial work of JDL and examine additional wind and potential buoyant energy parameter relationships associated with the data set compiled by JDL.

Positive wind shear (PWS), helicity, and deep tropospheric mean wind shear (U) [see *Weisman and Klemp*, 1982] are the three wind parameters with which the relationship with instability is examined. The wind parameters are defined and the method of calculation is discussed in part 1 [*Davies and Johns*, this volume]. The primary instability parameter examined in this study is CAPE. Surface parcel lifted index (SPLI) values based on lifted surface parcel temperatures [*Hales and Doswell*, 1982; *Bothwell*, 1988] are also examined since the SPLI is widely utilized operationally.

2. DETERMINATION OF BUOYANT ENERGY PARAMETER VALUES

2.1. Calculation of CAPE

Values of potential buoyant energy for soundings utilized in this study have been determined by an algorithm described by *Doswell et al.* [1982]. The values are essentially equivalent to CAPE with the lifted layer being the lowest 100 mbar AGL. To make representative estimates of CAPE in the vicinity of events considered for inclusion in the data set, these guidelines were followed:

1. If a tornado case occurred within $1\frac{1}{2}$ hours of the sounding time, within 40 nautical miles (74 km) of the sounding site, and in the same air mass as the sounding site, the CAPE value from the sounding was used directly.

2. For tornado cases not satisfying the conditions in guideline 1, the thermodynamic profiles of the surrounding soundings were examined. Interpolation of a CAPE value for both the time and location of the tornado event from the CAPE values computed for the regularly scheduled soundings was attempted only if (1) the horizontal temperature gradients at the standard levels above the boundary layer appeared to be relatively weak and (2) surface data and the temperature and moisture stratification in the boundary layer of the soundings indicated that any instability maximum (center or axis) associated with the case was adequately sampled by at least two nearby stations.

---

[1] Also at Davies, Incorporated, Pratt, Kansas 67124.

The Tornado: Its Structure, Dynamics, Prediction, and Hazards.
Geophysical Monograph 79
Copyright 1993 by the American Geophysical Union.

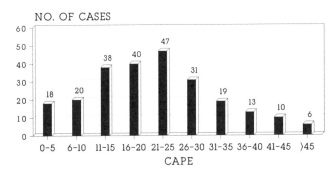

Fig. 1. Distribution of the convective available potential energy (CAPE) values associated with the 242 strong and violent tornado cases in the data set. Number above each bar is the number of cases in that particular range of CAPE values.

3. For most tornado cases not satisfying guideline 1 or 2, proximity soundings were constructed by using proximity surface data representative of the inflow air and interpolating (for time and/or location) from the surrounding soundings both the boundary layer thermodynamic profiles and the standard level data above the boundary layer. The CAPE values for these tornado cases were derived from the constructed soundings. Soundings were constructed for three fourths (181 cases) of the cases utilized in the data set.

4. In a few instances, missing data or difficulties in interpolation resulted in potential cases being excluded from the data set.

The procedure for determining the CAPE for a case differs from that utilized to determine the mean shear and helicity values (see part 1). One of the reasons for the difference is that in determining CAPE, it was often necessary to consider the temperature profile through the depth of the entire troposphere. Also, the thermodynamic patterns (particularly for moisture) can be quite complicated in the boundary layer. These complications required that proximity soundings be constructed in most cases in order to arrive at a representative CAPE value.

2.2. Surface Parcel Lifted Index (SPLI) Calculations

An estimate of the instability may be obtained by computing SPLI values. For all of the tornado cases in which CAPE was successfully computed, SPLI values were also computed. The following method was employed to obtain SPLI values.

1. Surface temperature, dew point, and pressure values near the time and place of the event and representative of the air mass from which storm inflow was occurring were obtained.

2. A 500-mbar temperature for the time and over the place of event occurrence was obtained by interpolation from radiosonde constant level data.

3. By using a skew-T diagram the surface values for each event were combined to produce a measurement of the pseudo-adiabat which the surface parcel would follow if lifted to saturation. The difference between the temperature at which this adiabat crosses 500 mbar and the environmental 500-mbar temperature is the SPLI value.

3. RESULTS

3.1. Potential Buoyant Energy Distribution

Figure 1 illustrates that the strong and violent tornadoes are associated with an extremely wide range of CAPE values. Values in the JDL data set range from 200 to 5300 J kg^{-1} with about two thirds (64%) of the cases exhibiting values between 1000 and 3000 J kg^{-1}. The data set displays seasonal and geographical differences in the values of CAPE associated with strong and violent tornado development. The cold season (November 1 to March 31) cases occur mostly in the eastern portions of the southern Plains and Gulf coastal region (see JDL) and exhibit CAPE values that are mostly weak to moderate (Figure 2). Ninety-five percent

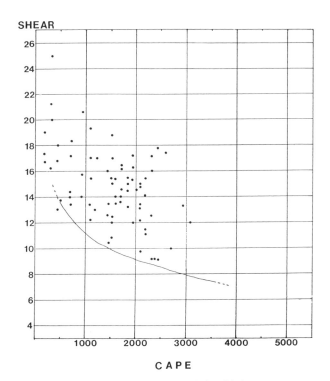

Fig. 2. Scatter diagram showing the relationship between convective available potential energy (CAPE) in joules per kilogram and 0-2-km AGL positive wind shear ($\times 10^{-3}$ s^{-1}) for the 75 cold season (November 1 to March 31) tornado cases. Solid curved line is a suggested lower limit of combined CAPE/low-level shear value that would support the development of strong or violent mesocyclone-induced tornadoes [after *Johns et al.*, 1990].

of the cold season cases are associated with CAPE values of less than 2500 J kg^{-1}.

Cases during the warm season (May 15 to August 31) generally occur farther west and north than the cold season cases (see JDL) and exhibit a wide range of CAPE values, from a weak 500 J kg^{-1} to a strong 5300 J kg^{-1} (Figure 3). The two cases exhibiting CAPE values of less than 1000 J kg^{-1} were associated with tropical cyclones. The low CAPE values in these cases agree with the findings of *McCaul* [1991]. The two cases associated with warm season derechos [*Johns and Hirt*, 1987] exhibit characteristically high CAPE values. Despite the wide range of values, note that over two thirds (68%) of the warm season cases are associated with CAPE values of 2500 J kg^{-1} or greater. This agrees with the findings of *Rasmussen and Wilhelmson* [1983].

Surface parcel lifted index (SPLI) values also suggest that strong and violent tornadoes are associated with a wide range of instability. Values in the JDL data set range from -1 to -14 (Figure 4). However, a large majority of the cases (72%) are associated with values from -5 to -10.

3.2. Bulk Richardson Numbers

Weisman and Klemp [1982, 1986] have shown that the type of storm that develops in a given environment is at least partially related to a Bulk Richardson Number (BRN) defined as

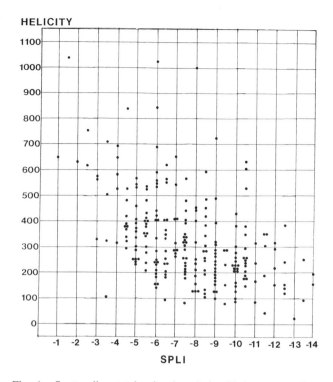

Fig. 4. Scatter diagram showing the relationship between surface parcel lifted index (SPLI) in degrees Celsius and 0- 2-km AGL helicity in m^2 s^{-2} utilizing the 20R85/30R75 storm motion assumption method for all 242 cases in the data set.

$$\mathrm{BRN} = \frac{B}{\frac{1}{2}U^2},$$

where B is the buoyant energy (CAPE) for a lifted parcel in the storm's environment, and U is a measure of the vertical wind shear through a relatively deep layer (0- 6-km AGL). Results from numerical modeling experiments and a limited number of storm observations have suggested that growth of supercells is confined to values of BRN between 10 and 40 [*Weisman and Klemp*, 1986].

Figure 5 illustrates the range of BRN values associated with the strong and violent tornadoes in the JDL data set. Almost one half (47%) of the BRN values are less than 8. The number of cases diminishes as the BRN values become larger, with most cases associated with values of less than 40 (94%). These results agree well with the findings of *Riley and Colquhoun* [1990], who also examined a large number of cases.

The question arises as to why there are so many cases with very low BRN values. The prevailing theory has been that in such an environment the very strong shear would inhibit the growth of the updraft, thus not allowing for a deep rotating convective column to develop and be sustained. One potential explanation for this apparent contradiction may lie with the nature of the BRN. As pointed out by *Weisman and Klemp* [1982] and *Lazarus and Droegemeier* [1990], the

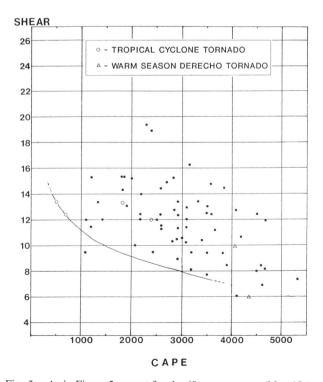

Fig. 3. As in Figure 2 except for the 69 warm season (May 15 to August 31) tornado cases. The open circles represent tornado cases associated with tropical cyclones. The open triangles represent tornado cases associated with warm season derechos [after *Johns et al.*, 1990].

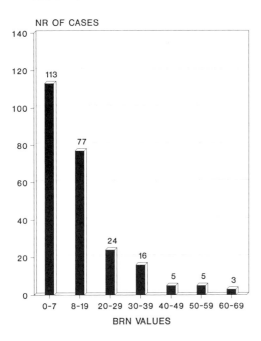

Fig. 5. Distribution of Bulk Richardson numbers [*Weisman and Klemp*, 1982] for all 242 strong and violent tornado cases.

BRN is a "bulk" measure of the ambient shear and does not account for detailed aspects of the wind profile, particularly low-level curvature. Modeling experiments [*Weisman and Klemp*, 1984; *Brooks and Wilhelmson*, 1990; *McCaul*, 1990] indicate that low-level curvature shear and storm-relative helicity interacting with the deeper tropospheric shear can enhance strongly the intensity of an updraft. McCaul's simulations of storms in the low-buoyancy environments of tropical cyclones indicate that shear-induced pressure forces [*Rotunno and Klemp*, 1982], related at least in part to the low-level curvature shear, can be up to 3 times as important as buoyancy in controlling updraft strength. Therefore it appears likely that in many situations where the BRN is a very low value, the low-level curvature shear plays a crucial role in helping to sustain a rotating convective column.

Although the above simulations suggest how a supercell might be maintained in a low BRN environment, there is still the question of initial development, or how an updraft can sustain itself in its earliest stages. Operational forecasting experience suggests a possible explanation for this question. The authors have observed cases where tornado-bearing supercells are associated with (or evolve from) complex convective structures, sometimes on a scale much larger than the supercell itself. Others have observed this type of storm structure (or evolution) also [e.g., *Nolen*, 1959; *Burgess and Curran*, 1985; *July*, 1990; *Moller et al.*, 1990; *Przybylinski et al.*, 1990]. Further, *Doswell et al.* [1990] have proposed that the "classic" isolated supercell is just one storm type in a broad range of storm types associated with supercell circulations. Examination of the radar imagery (low-elevation reflectivity data) associated with the 31 strong and violent tornado cases in the data subset described in part 1 supports the hypothesis of Doswell et al. Only 10 cases (32%) appear to be associated with "classic" isolated supercells, while the large majority of cases (68%) appear to be associated with a variety of multicellular systems including lines, spiral bands, clusters, and bow echoes. These results suggest that in strong and violent tornado situations, supercell circulations associated with complex multicellular convective structures are quite common.

Recall that the BRN is related to the type of storm structure that develops in a given situation [*Weisman and Klemp*, 1986]. The 10–40 range of BRN values that Weisman and Klemp have associated with supercell development applies to isolated convection (i.e., the "classic" isolated supercell). While an isolated supercell is not likely to develop in an environment with a very low BRN, the larger multicellular convective structures described in the previous paragraph can be initiated and sustained in such an environment by synoptic scale forcing (e.g., a squall line ahead of a cold front). Once developed, irregularities and differential movements in the storm outflow patterns [see *Doswell et al.*, 1990, Figure 4] may allow some updrafts to encounter enhanced inflow and develop rotation. Once rotation is established, the effect of shear-induced pressure forces (discussed earlier in this section) can help to sustain this rotation.

3.3. *CAPE and 0- 2-km AGL Positive Wind Shear Relationship*

While the BRN calculations involve the wind shear through a deep layer (0- 6-km AGL), recent interest has focused on the importance of both the shear and nature of the hodograph in the updraft inflow layer [e.g., *Davies*, 1989; *Davies-Jones et al.*, 1990; JDL]. JDL calculated 0- 2-km AGL positive wind shear for the purpose of indirectly estimating the rotational potential of the environmental wind field (i.e., the strength of the 0- 2-km AGL positive shear is related to the range of storm motions that would support supercell development).

To investigate the relationship of the low-level environmental shear to potential buoyant energy in strong and violent tornado situations, JDL constructed scatter diagrams. From Figures 2, 3, and 6 it is evident that there is a pattern to the combinations of 0- 2-km AGL PWS and CAPE associated with strong and violent tornadoes (Figure 6) and that there are seasonal variations in this pattern (Figures 2 and 3). For example, the cold season cases (Figure 2), which occur primarily from the eastern portion of the southern Plains into the Gulf coastal region, are generally associated with the combination of strong 0- 2-km AGL PWS values and weak to moderate CAPE values. Therefore the data points for cold season cases are concentrated in the upper left portion of the scatter diagram.

When considering all cases (Figure 6), the pattern of data

points suggests that progressively stronger 0- 2-km AGL PWS values are associated with strong and violent tornado development as the CAPE decreases. In a few cases where the PWS was very strong, greater than 16×10^{-3} s^{-1}, CAPE values as low as 200–300 J kg^{-1} were observed. At the other extreme, in situations where the CAPE was greater than 4000 J kg^{-1}, strong and violent tornadoes have been associated with PWS values as low as 6×10^{-3} s^{-1}. Despite the wide range of values there appears to be an optimum range of 0- 2-km AGL PWS values associated with strong and violent tornado development for any particular CAPE value. Further, JDL have shown that for violent tornadoes (F4 and F5 intensity) shear values are generally in the higher two thirds of any given optimum range.

3.4. CAPE and Helicity Relationship

Helicity has become an important parameter for evaluating the rotational potential of air in the storm inflow layer [e.g., *Davies-Jones et al.*, 1990]. To calculate storm-relative helicity a storm motion vector is needed. On the basis of the actual storm motions and the associated environmental conditions from a 31-case data subset, a method for assuming the storm motions from precursor conditions has been proposed (see part 1). This method assumes a storm motion of 20R85 (see part 1 for notation) for cases where the 0- 6-km AGL mean wind speeds are greater than 30 knots (15 m s^{-1}) and a motion of 30R75 for cases where the 0- 6-km AGL mean wind speeds are of lesser intensity. Although this

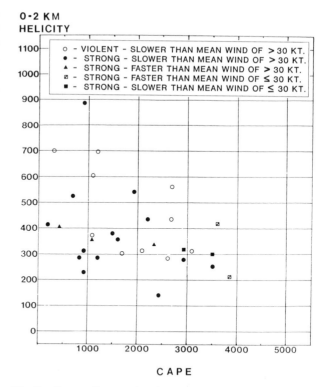

Fig. 7. Scatter diagram showing the combinations of convective available potential energy (CAPE) in joules per kilogram and 0- 2-km AGL helicity in m^2 s^{-2} utilizing observed storm motions for a 31-case data subset. Open circles represent violent tornadoes (F4-F5). All squares (solid and hatched) represent cases where the 0- 6-km AGL mean wind speed is equal to or less than 30 knots (15 m s^{-1}). Triangles and hatched squares represent cases in which the associated supercell moves faster than the 0- 6-km AGL mean wind speed.

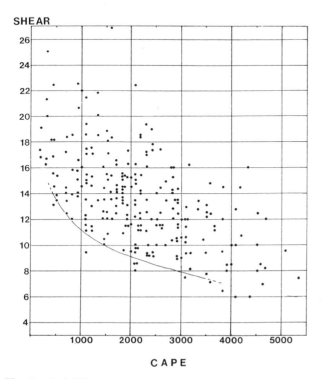

Fig. 6. As in Figure 2 except for all 242 cases in the data set [after *Johns et al.*, 1990].

method for assuming storm motions appears to be rather "crude," it does represent a refinement over applying a single assumed deviation, such as 30R75 [*Maddox*, 1976].

On the basis of the findings in part 1, the authors have chosen to construct and compare scatter diagrams displaying combinations of CAPE and helicity using the following criteria: (1) 0- 2-km AGL helicity using the observed storm motions for the 31-case data subset, (2) 0- 3-km AGL helicity using the observed storm motions for the 31-case data subset, (3) 0- 2-km AGL helicity using the 20R85/30R75 storm motion assumption method for the 242-case JDL data set, and (4) 0- 3-km AGL helicity using the 20R85/30R75 storm motion assumption method for the 242-case JDL data set.

Although the data sample is not large, the scatter diagram (Figure 7) depicting the combinations of 0- 2-km AGL helicity using observed storm motion and CAPE for the 31 subset cases displays a pattern that is qualitatively similar to the pattern for combinations of 0- 2-km AGL PWS and CAPE for the 242 JDL cases (Figure 6). There is a tendency for cases exhibiting strong values of low-level helicity to be

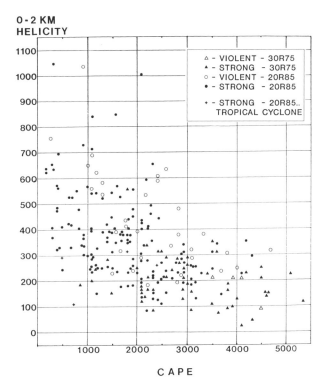

Fig. 8. Scatter diagram showing the combinations of convective available potential energy (CAPE) in joules per kilogram and 0- 2-km AGL helicity in $m^2\,s^{-2}$ utilizing the 20R85/30R75 storm motion assumptions for the 242 cases in the JDL data set. All triangles (open and solid) represent cases in which the assumed storm motion is 30R75, while the assumed storm motion for the remainder of the cases is 20R85. The open circles and open triangles represent violent tornadoes (F4-F5). The crosses represent cases associated with tropical cyclones.

associated with environments characterized by weak buoyancy, and vice versa. A scatter diagram (not shown) depicting 0- 3-km AGL helicity for the 31 subset cases is very similar to Figure 7.

In Figure 7, note that those cases in the subset which are associated with the weakest 0- 6-km AGL mean winds (equal to or less than 30 knots (15 m s^{-1})) and depicted as solid and hatched squares) are also associated with very high values of CAPE. This implies that supercell-induced strong and violent tornadoes occurring in a weak wind environment usually are associated with very strong to extreme instability.

When the 20R85/30R75 storm motion assumptions for computing helicity are applied to all 242 cases in the JDL data set, the resultant scatter diagram depicting combinations of 0- 2-km AGL helicity and CAPE (Figure 8) has a strong similarity to the one for 0- 2-km AGL PWS and CAPE (Figure 6). A progressively higher range of values for both helicity and PWS is associated with strong and violent tornado development as the CAPE decreases. However, the rate of change appears to be more gradual in the case of helicity and CAPE.

Since a number of researchers have used 0- 3-km AGL as the approximate inflow layer, a scatter diagram similar to Figure 8 has been prepared (not shown) with the only difference being that helicity has been computed using the 0- 3-km AGL layer. The scatter diagram patterns for both assumed inflow layers (0- 2-km and 0- 3-km AGL) are similar.

The four cases associated with tropical cyclones (indicated by crosses in Figure 8) tend to have relatively low values of both CAPE and helicity, resulting in the points residing in the lower left portion of the scatter diagram (one case in particular is far below the other points). The fact that tropical cyclone tornado cases normally are associated with low CAPE values is well documented [*McCaul*, 1991]. However, the reason for the accompanying low helicity values in Figure 8 is unclear. It may be related to the vertical distribution of helicity (helicity density) and the depth of the effective inflow layer [*McCaul*, 1991] (also see part 1). Also, there is a possibility that the motions of tornadic storms in tropical cyclone situations are estimated poorly by the storm motion assumptions utilized to compute helicity.

3.5. SPLI and Helicity

Figure 4 is a scatter diagram depicting the combinations of SPLI and 0- 2-km AGL helicity utilizing the 20R85/30R75 storm motion assumptions for all 242 cases in the JDL data set. The pattern is very similar to that for when CAPE is used as the instability parameter (Figure 8). The slightly greater scattering of points in Figure 4 is probably a result of the SPLI being a less precise method of estimating instability. Nevertheless, Figure 4 appears to have value for the operational forecaster who is computing SPLI values on an hourly basis and does not have access to the finer-incremented model forecasts of CAPE.

4. DISCUSSION

From the JDL data set it has been determined that strong and violent tornadoes are associated with an extremely wide range of potential buoyant energy (CAPE) values. Most of the cases with low values of CAPE occur during the cooler months of the year and almost always are associated with dynamic weather systems and strong tropospheric wind fields (and wind shear). A few others with low CAPE values are associated with tropical cyclones. Both of these environments result in very low Bulk Richardson Number (BRN) values. As a result, when all strong and violent tornadoes are considered, almost half are associated with a BRN of less than 8. Until recently, the general perception has been that the BRN associated with supercell development should fit into a certain range, generally between 10 and 40. However, the results of this study (and the one by *Riley and Colquhoun* [1990]) suggest that while the BRN has utility in determining when the initiation of isolated supercells is possible, it is not a good discriminator for supercell development in general. This is particularly so for strong shear/low buoyancy situations (BRN from 0 to 8), which need to be recognized more widely as having potential for supercell-induced tornadoes.

The importance of low-level positive shear (PWS) and helicity in strong and violent tornado development has been discussed in part 1. When the combinations of these wind parameters with CAPE are considered, a definite pattern emerges. In cases where the values of CAPE are very high, the associated PWS values are usually relatively low. Generally higher values of PWS are associated with strong and violent tornado development as the CAPE decreases. At very low CAPE values, cases are associated with relatively high values of PWS. For any particular CAPE value there appears to be a range of PWS values that is optimum for strong and violent tornado development.

The relationship between CAPE and helicity in strong and violent tornado development is similar to that for CAPE and PWS. That is, generally higher values of helicity are associated with strong and violent tornadoes as the CAPE decreases. If it is assumed that most tornadoes in the data set are associated with supercells, then it can be stated that for any particular CAPE value there appears to be a range of helicity values that is optimum for inducing strong rotation and low-level mesocyclones in supercell storms (as suggested by *Lazarus and Droegemeier* [1990]).

The fact that there is considerable scatter on the PWS/CAPE and helicity/CAPE diagrams emphasizes that there are other factors involved in supercell-induced tornado development. One of these factors is the downdraft circulations affecting the supercell. It generally is accepted that if a supercell is to produce a mesocyclone-associated tornado, the development of a sufficiently strong rear flank downdraft (RFD) is necessary. The strength of the rear flank downdraft is dependent on conditions, particularly relative humidity, in its middle level source region.

Another factor also involves downdraft outflows, but usually on a larger scale. The intense downdraft and outflow that develops with some bow echoes affects storm motion, accelerating the storm structure. In some cases this results in the bow echo "experiencing" increased helicity, which may induce a supercell that is associated with the bow echo structure. Recall that in the 31-case data subset there were four cases associated with bow echoes that moved faster than the mean wind speed (from part 1). Note that the 20R85/30R75 storm motion assumptions used in constructing the helicity/CAPE diagram do not take this type of storm motion into account.

These two examples involve some effects of storm downdrafts. There are probably many more factors involved in supercell-induced tornado development (some of which are mentioned in part 1). This suggests that much additional work needs to be done toward understanding the processes that initiate and support storm rotation and mesocyclone-induced tornado development.

Acknowledgments. The authors especially appreciate the many hours of work volunteered by Grant Bean, Dave Higginbotham, and others of the NSSFC computer staff in assembling the massive data set for this study. The authors also wish to thank Bill Henry, NWSTC; Ken Howard, NSSL, ERL; Mike Ryba, WSO DDC; and Joe Schaefer, SSD, Central Region, NWS, for their help with this project and Deborah Haynes for helping with manuscript preparations.

REFERENCES

Bothwell, P. D., Forecasting convection with the AFOS data analysis programs (ADAP-version 2.0), *NOAA Tech. Memo. NWS SR-122*, 91 pp., Natl. Weather Serv. S. Reg., Fort Worth, Tex., 1988.

Brooks, H. E., and R. B. Wilhelmson, The effects of low-level hodograph curvature on supercell structure, in *Preprints, 16th Conference on Severe Local Storms*, pp. 34–39, American Meteorological Society, Boston, Mass., 1990.

Burgess, D. W., and E. B. Curran, The relationship of storm type to environment in Oklahoma on 26 April 1984, in *Preprints, 14th Conference on Severe Local Storms*, pp. 208–211, American Meteorological Society, Boston, Mass., 1985.

Davies, J. M., On the use of shear magnitudes and hodographs in tornado forecasting, in *Preprints, 12th Conference on Weather Forecasting and Analysis*, pp. 219–224, American Meteorological Society, Boston, Mass., 1989.

Davies, J. M., and R. H. Johns, Some wind and instability parameters associated with strong and violent tornadoes, 1, Wind shear and helicity, this volume.

Davies-Jones, R. P., D. W. Burgess, and M. Foster, Test of helicity as a tornado forecast parameter, in *Preprints, 16th Conference on Severe Local Storms*, pp. 588–592, American Meteorological Society, Boston, Mass., 1990.

Doswell, C. A., III, J. T. Schaefer, D. W. McCann, T. W. Schlatter, and H. B. Wobus, Thermodynamic analysis procedures at the National Severe Storms Forecast Center, in *Preprints, 9th Conference on Weather Forecasting and Analysis*, pp. 304–309, American Meteorological Society, Boston, Mass., 1982.

Doswell, C. A., III, A. R. Moller, and R. W. Przybylinski, A unified set of conceptual models for variations on the supercell theme, in *Preprints, 16th Conference on Severe Local Storms*, pp. 40–45, American Meteorological Society, Boston, Mass., 1990.

Hales, J. E., Jr., and C. A. Doswell III, High resolution diagnosis of instability using hourly surface lifted parcel temperatures, in *Preprints, 12th Conference on Severe Local Storms*, pp. 172–175, American Meteorological Society, Boston, Mass., 1982.

Johns, R. H., and W. D. Hirt, Derechos: Widespread convectively induced windstorms, *Weather Forecasting*, 2, 32–49, 1987.

Johns, R. H., J. M. Davies, and P. W. Leftwich, An examination of the relationship of 0-2 km AGL "positive" wind shear to potential buoyant energy in strong and violent tornado situations, in *Preprints, 16th Conference on Severe Local Storms*, pp. 593–598, American Meteorological Society, Boston, Mass., 1990.

July, M. J., Forcing factors in the violent tornado outbreak of May 5, 1989: A study in scale interaction, in *Preprints, 16th Conference on Severe Local Storms*, pp. 72–77, American Meteorological Society, Boston, Mass., 1990.

Lazarus, S. M., and K. K. Droegemeier, The influence of helicity on the stability and morphology of numerically simulated storms, in *Preprints, 16th Conference on Severe Local Storms*, pp. 269–274, American Meteorological Society, Boston, Mass., 1990.

Leftwich, P. W., Jr., and Wu X., An operational index of the potential for violent tornado development, in *Preprints, 15th Conference on Severe Local Storms*, pp. 472–475, American Meteorological Society, Boston, Mass., 1988.

Maddox, R. A., An evaluation of tornado proximity wind and stability data, *Mon. Weather Rev.*, 104, 133–142, 1976.

McCaul, E. W., Jr., Simulations of convective storms in hurricane environments, in *Preprints, 16th Conference on Severe Local*

Storms, pp. 334–339, American Meteorological Society, Boston, Mass., 1990.

McCaul, E. W., Jr., Buoyancy and shear characteristics of hurricane-tornado environments, *Mon. Weather Rev.*, *119*, 1954–1978, 1991.

Miller, R. C., Notes on analysis and severe storms forecasting procedures of the Air Force Global Weather Central, *Tech. Rep. 200*, (rev.), 1972.

Moller, A. R., C. A. Doswell III, and R. Przybylinski, High-precipitation supercells: A perceptual model and documentation, in *Preprints, 16th Conference on Severe Local Storms*, pp. 52–57, American Meteorological Society, Boston, Mass., 1990.

Moncrieff, M. W., and J. S. A. Green, The propagation and transfer properties of steady convective overturning in shear, *Q. J. R. Meteorol. Soc.*, *98*, 336–352, 1972.

Nolen, R. H., A radar pattern associated with tornadoes, *Bull. Am. Meteor. Soc.*, *40*, 277–279, 1959.

Przybylinski, R. W., S. Runnels, P. Spoden, and S. Summy, The Allendale, Illinois tornado—January 7, 1989—One type of an HP supercell, in *Preprints, 16th Conference on Severe Local Storms*, pp. 331–357, American Meteorological Society, Boston, Mass., 1990.

Rasmussen, E. N., and R. B. Wilhelmson, Relationships between storm characteristics and 1200 GMT hodographs, low level shear, and stability, in *Preprints, 13th Conference on Severe Local Storms*, pp. J5–J8, American Meteorological Society, Boston, Mass., 1983.

Riley, P. A., and J. R. Colquhoun, Thermodynamic and wind related variables in the environment of United States tornadoes and their relationship to tornado intensity, in *Preprints, 16th Conference on Severe Local Storms*, pp. 599–602, American Meteorological Society, Boston, Mass., 1990.

Rotunno, R., and J. B. Klemp, The influence of the shear-induced pressure gradient on thunderstorm motion, *Mon. Weather Rev.*, *110*, 136–151, 1982.

Weisman, M. L., and J. B. Klemp, The dependence of numerically simulated convective storms on vertical wind shear and buoyancy, *Mon. Weather Rev.*, *110*, 504–520, 1982.

Weisman, M. L., and J. B. Klemp, The structure and classification of numerically simulated convective storms in directionally varying wind shears, *Mon. Weather Rev.*, *112*, 2479–2498, 1984.

Weisman, M. L., and J. B. Klemp, Characteristics of isolated convective storms, in *Mesoscale Meteorology and Forecasting*, edited by P. S. Ray, pp. 331–358, American Meteorological Society, Boston, Mass., 1986.

Diurnal Low-Level Wind Oscillation and Storm-Relative Helicity

ROBERT A. MADDOX

Environmental Research Laboratories, NOAA, National Severe Storms Laboratory, Norman, Oklahoma 73069

1. INTRODUCTION

Thunderstorms characterized by significant mesocyclones occur within local environments that exhibit strong, storm-relative low-level winds which veer substantially with height [e.g., *Weisman and Klemp*, 1984; *Burgess and Curran*, 1985; *Davies-Jones et al.*, 1990]. Such an environment, i.e., one with high storm-relative helicity [*Lilly*, 1983, 1986; *Davies-Jones*, 1984; *Brooks et al.*, this volume; *Davies-Jones*, 1991], is favorable for the occurrence of intense tornadoes when and if deep moist convection occurs. All atmospheric processes that act to modify the character of the low-level, vertical wind profile can lead to enhanced or diminished storm-relative helicity and thereby affect the structure and character of thunderstorms, should they occur. Since changes in storm motion also affect relative helicity, it is assumed for simplicity that changes in the lower troposphere winds have little affect upon the movement of thunderstorms.

Thunderstorms that produce intense tornadoes of F3 rating or higher [*Fujita*, 1971] tend to occur during middle to late afternoon. For example, F3 or greater tornadoes occur most frequently in the central United States between 1700 and 1900 local standard time [*Fujita*, 1987]. The geographical locations of F3 and greater tornadoes that have occurred between 1200 and 1800 central standard time (CST) are shown in Figure 1. During the afternoon, winds near the surface are strong and often backed relative to geostrophic. If the winds between 2 and 3 km above the ground are strong and highly veered relative to the surface wind during the late afternoon, the environment may be characterized by high storm-relative helicity.

However, not all major supercell thunderstorms and tornadoes occur during the afternoon (refer to Figure 2). As illustrated in Figure 2, many fewer intense tornadoes occur between midnight and sunrise than occur during the afternoon, and the geographical preference is clearly shifted to the south central states. The reasons for the geographical

The Tornado: Its Structure, Dynamics, Prediction, and Hazards.
Geophysical Monograph 79
This paper is not subject to U.S. copyright. Published in 1993 by the American Geophysical Union.

shift are not clear. However, it is known that all diurnal wind circulations (e.g., sea breeze, nocturnal jet, mountain-valley breeze) are expected to be most pronounced at about 30° latitude north or south, where the inertial period is nearly equal to the forcing period [*Rotunno*, 1983]. It is likely that the well-known diurnal oscillation of the low-level jet [e.g., *Blackadar*, 1957; *Wexler*, 1961; *Bonner*, 1968] can, under some conditions, act to increase storm-relative helicity after dark and help support the nighttime occurrence of intense tornadoes.

This possibility is explored first by considering typical afternoon conditions that favor supercell thunderstorms. The vertical wind profiles for the afternoon setting then are compared with profiles modified in ways that typify the development of the nocturnal low-level jet. It is shown that storm-relative helicity can be enhanced during the evening and nighttime hours because of the development of the nocturnal jet. The relevance of the diurnal cycle for understanding observed storm behavior is discussed.

2. TYPICAL SUPERCELL ENVIRONMENTS

The early empirical models of severe thunderstorm environments [e.g., *Fawbush and Miller*, 1954; *Newton*, 1963] highlighted the importance of the low-level jetstream for the occurrence of intense storms. The role of veering winds with height also was noted and related to differential advections of temperature and moisture, which serve to destabilize the atmosphere and support the development of deep convective storms [*Beebe and Bates*, 1955]. Studies of proximity soundings taken within several tens of kilometers of tornadic thunderstorms [*Beebe*, 1958; *Darkow and Fowler*, 1971; *Maddox*, 1976] clearly showed the presence of strong and highly veered low-level winds (refer to Figure 3). More recent work [*Lilly*, 1983, 1986; *Davies-Jones*, 1984] established the importance of storm-relative helicity for the development of supercell storms. *McCaul* [1991] has shown that proximity hodographs for tornadoes occurring within hurricanes also are characterized by strong veering and helicity (see Figure 4).

Storm-relative helicity is proportional to the area of the

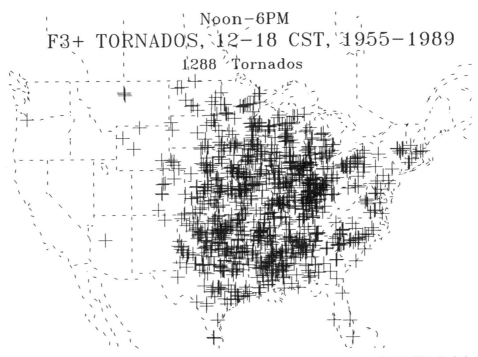

Fig. 1. Locations of occurrence of F3 and greater intensity tornadoes between 1200 and 1800 CST. Period of record is 1955–1989.

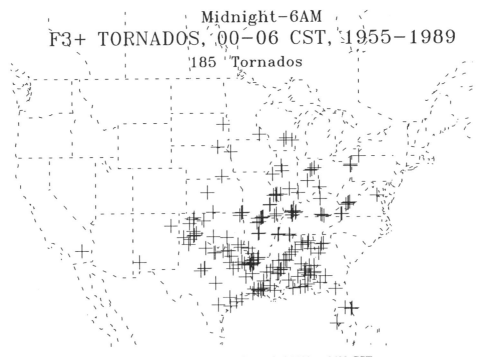

Fig. 2. Same as Figure 1 but for period 0000 to 0600 CST.

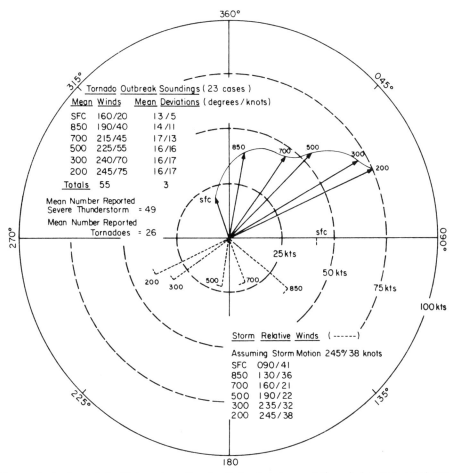

Fig. 3. Composite hodograph for 23 tornado outbreaks. Estimated storm-relative wind vectors are shown below the hodograph [from *Maddox*, 1976].

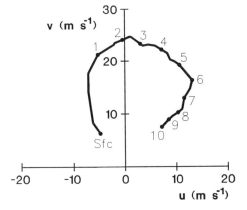

Fig. 4. Mean hodograph for hurricane tornado proximity soundings. See *McCaul* [1991] for details. The height (in kilometers) above surface is indicated.

low-level wind hodograph that lies between the head of a storm motion vector and the environmental wind hodograph. This area is computed from the surface to some arbitrarily determined level chosen to capture the inflow layer for convective storms, typically 3 km [*Davies-Jones et al.*, 1990; *Doswell*, 1991]. The storm-relative wind hodograph for a composite of 28 supercell environments [*Brown*, 1990] is contrasted with three similar hodographs for long-track, damaging derechos [*Johns and Hirt*, 1987] in Figure 5. The supercell environment exhibits large helicity, while the derecho straight-line wind storm environment exhibits almost none.

Thus the typical environments of tornadic storms possess substantial storm-relative helicity. A pronounced low-level jet also typifies these environments [*Newton*, 1963; *Miller*, 1967] and can contribute to the helicity, especially when it is strongly veered relative to the surface wind. The develop-

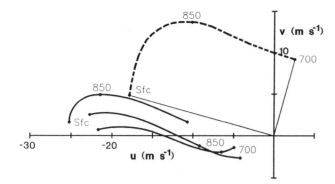

Fig. 5. Composite hodograph for 28 supercell environments (dashed) presented in the storm-relative reference frame [from *Brown*, 1990]. Also shown are system-relative hodographs for three long-lived derecho events (solid lines; refer to *Johns et al.* [1990]).

ment of the low-level jet in severe thunderstorm situations characterized by strong large-scale baroclinicity and highly ageostrophic winds has been discussed by *Uccellini and Johnson* [1979]. In these situations, intense isallobaric accelerations of the low-level winds and the veering of the wind with height beneath the exit region of the upper level jet can result in a high-helicity environment.

During late spring and early summer, when large-scale storm systems are not intense, late afternoon conditions still lead frequently to environments characterized by high helicity. An example of this type of situation is illustrated in Figure 6. A weak subsynoptic surface low often develops at the northern extent of the dry line (Figure 6a) with severe storms and tornadoes occurring to the north and east of this feature [*Tegtmeier*, 1974]. The local isallobaric field acts to increase the mean low-level winds as the southerly pressure gradient increases and to back the winds (Figure 6b). These changes often occur in the presence of substantial low-level warm temperature advection and its associated veering of the geostrophic wind with height (Figure 6c). The net result can be a hodograph characterized by high storm-relative helicity over a distinctly mesoscale region during the middle to late afternoon.

The morning and afternoon hodographs from the Binger tornado (F4) day in central Oklahoma illustrate well this type of diurnal evolution of the hodograph (Figure 7). As *Doswell* [1987] discussed, these types of mesoscale storm episodes are naturally more difficult to anticipate than are severe thunderstorm and tornado outbreaks that usually occur within highly baroclinic, synoptic weather systems. Indeed, *Anthony* [1990] has shown that the forecasters at the National Severe Storms Forecast Center have considerably higher verification scores for tornado outbreak days.

3. THE NOCTURNAL JET AND STORM-RELATIVE HELICITY

It is observed by storm chasers, whose activities tend to be confined to the open spaces of the Plains states, that poorly organized storms occasionally change markedly around sunset, with supercell events developing during the evening to early nighttime hours (C. A. Doswell and R. P. Davies-Jones, personal communication, 1992). However, evolution from supercell storms into multicell storms or mesoscale systems (e.g., derechos [*Johns and Hirt*, 1987]) at or after dusk is more typically observed. These contrasting

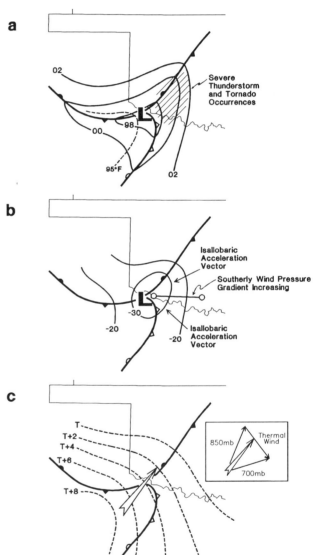

Fig. 6. (a) Typical late spring synoptic pattern associated with supercell storms in the southern Plains. Fronts, isobars, an isotherm, and the severe weather threat area are shown. Approximate time of map is 1500 CST [after *Tegtmeier*, 1974]. (b) Typical three hourly pressure fall field associated with this pattern [after *Tegtmeier*, 1974]. (c) Typical mean isotherms for the 1.5- 3.0-km layer associated with this pattern. Broad arrow indicates mean wind vector in the layer, while inset shows the thermal wind vector and associated veering through the layer due to geostrophic warm advection.

observations lead to an obvious question: what role, if any, does the diurnal wind cycle play in determining the local, storm-relative helicity and thereby storm organization?

The typical scenario for low-level jet development (in the absence of any evolving large-scale baroclinic forcing) begins as radiational cooling decouples the surface contact layer from the well-mixed afternoon boundary layer. This results in an inertial oscillation, because of the large imbalance between frictional and Coriolis forces, that accelerates and veers the winds though a layer that may extend up to 3 km above the surface. The flow becomes supergeostrophic and the low-level jet reaches its maximum intensity in the early morning hours [*Blackadar and Buajitti*, 1957; *Hoxit*, 1975]. *Uccellini* [1980] has shown that many of the cases used to study the diurnal cycle of the low-level jet under "quiescent" conditions also have been characterized by an evolving baroclinic environment that influenced the development of the jet. This complicates interpretation of the cooling/decoupling aspects of the diurnal circulation features. However, he found that even in the face of complicating baroclinic processes, the maximum wind speeds in low-level jet streaks usually occurred during the early morning hours.

Although the broad character and climatological aspects of the low-level jet are well known [*Bonner*, 1968; *Reiter*, 1969], particularly under synoptically stagnant summer conditions, the detailed temporal and spatial structures of the jet have not been well observed. *Hoecker* [1963] and *Bonner* [1966] used serial observations from a special two-dimensional pibal line operated during spring of 1961 to show the complicated structures and evolution of the southern Plains low-level jet for several case events. *Maddox* [1985] suggested that the mesoscale structures and timing of the diurnal jet may be influenced strongly by local radiative conditions, particularly the effects of mean relative humidity

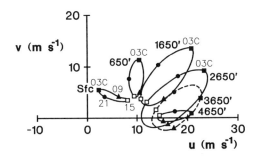

Fig. 8. Average diurnal variation from the mean wind at Oklahoma City, Oklahoma, and Wichita, Kansas, for 29 quiescent summer days [after *Blackadar and Buajitti*, 1957]. The winds were averaged relative to the 700-mbar wind, which first was rotated to be from the west in all cases.

and cloud cover, leading to very complex and dramatically varying evolutions of the low-level jet and the hodograph on a case-by-case basis; again, see the case analyses by *Hoecker* [1963] and *Bonner* [1966]. The special observations reported by *Stensrud et al.* [1990] also appear to have captured examples of such local variability.

The work of *Blackadar and Buajitti* [1957] can be used to illustrate that the diurnal wind cycle leads also to a cycle in storm relative helicity. This is apparent from examination of Figure 8, where the amplitude of the average diurnal cycle is largest from 1000 to 3000 feet (300–900 m) above the surface. This diurnal cycle can lead to a rapid increase in storm-relative helicity after 1500 local time that maximizes in the early morning, if it is assumed that middle tropospheric "steering" winds, and thus the general movement of thunderstorms, remain relatively unaffected by the diurnal cycle. It must be remembered that these changes are "average" summertime features and that their relative importance and magnitude varies markedly from case to case, as shown by *Uccellini* [1980].

For example, consider the four hypothetical scenarios shown in Figure 9. In each of these four cases an initial afternoon storm-relative hodograph is modified by the same representative diurnal changes (as indicated by Figure 8) to produce a 0300 local time hodograph. The initial profiles range from a straight-line hodograph (i.e., one which is characterized by no turning with height of the wind shear vectors) to highly curved ones. In the first case (Figure 9a), changes in the wind profile after dark lead to little change in storm-relative helicity and a trend toward a straight-line hodograph below 750 m. However, the other three cases (Figures 9b–9d) lead to substantial increases in storm-relative helicity. Thus the meteorological situation that develops during the afternoon and the resulting local wind profile relate directly to how the environment will evolve, in a storm-relative helicity sense, as radiational heating decreases rapidly at the end of the day. It appears that in most situations the component of the diurnal wind cycle driven by frictional decoupling alone acts to increase low-level helicity.

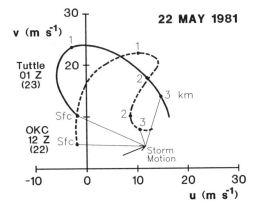

Fig. 7. Morning and evening hodographs associated with the Binger, Oklahoma, tornadic supercell of May 22, 1981. The morning sounding is from Oklahoma City, while the afternoon sounding was taken at Tuttle, Oklahoma (approximately 15 miles southwest of Oklahoma City) (hodographs provided by D. Burgess).

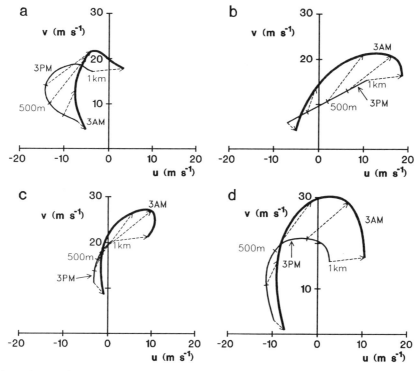

Fig. 9. Hodographs showing hypothetical changes in storm-relative low-level winds between 1500 and 0300 local time. The diurnal changes imposed are similar to those indicated in Figure 8 and are the same in each panel.

Although we have precious few observations of the evolution of the vertical wind profile with high time and space resolution, the capabilities of radar wind profilers and the new National Weather Service (NWS) Doppler radars (WSR-88D) bring promise that, at least in the time domain, we will soon be able to observe the diurnal wind cycle with more detail than ever in the past. *Stensrud et al.* [1990] have noted that the radar wind profilers deployed in the central United States do not sample the winds well from the surface to 500–750 m. This means that supplemental systems or modified operational procedures will be required to capture important details of low-level jet structures and evolution when the nose of the jet occurs very near the surface.

The clear air wind data shown in Figure 10 were gathered by the first NWS WSR-88D radar operating at Twin Lakes, Oklahoma, this past spring. The resulting hodographs (Figure 11) show that storm-relative helicity in the environment over central Oklahoma essentially doubled between sunset and midnight. There were no thunderstorms in the state on this particular night, so the possible thunderstorm-related

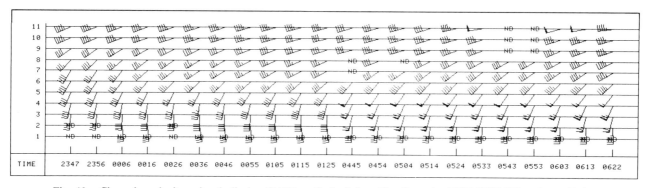

Fig. 10. Clear air, velocity azimuth display (VAD) vertical wind profiles from the NWS WSR-88D radar at Twin Lakes, Oklahoma (approximately 25 miles east-southeast of Oklahoma City). Period extends from 2347 UTC (1747 CST) on March 25 to 0622 UTC (0022 CST) on March 26, 1991. Winds are in meters per second; full barb indicates 5 m s^{-1}, and pennant indicates 25 m s^{-1}. Note the time break at 0125 UTC.

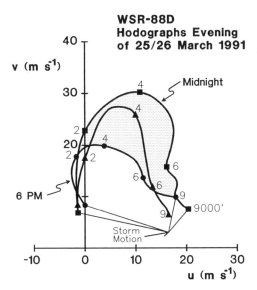

Fig. 11. Three hodographs derived from the velocity azimuth display (VAD) winds show a marked increase in storm-relative helicity between 1800 and 0000 CST. Middle tropospheric winds were used to estimate a likely storm velocity (no convection actually occurred). Profile times are (circles) 0006 UT, (triangles) 0306 UT, and (squares) 0603 UT. Stippled area shows increase in helicity from 1800 to midnight.

effects of these changes in the hodograph as the low-level jet developed remain a matter of speculation.

4. Discussion

This brief presentation illustrates that at least in certain situations the development of nocturnal low-level jet streaks can have dramatic effects upon the storm-relative helicity of the environment within which storms may be occurring. These changes can, depending upon the specific meteorological details of any particular situation, act either to increase or to diminish the likelihood of highly organized, supercell thunderstorms. Because of the crude time and space resolution of current observing systems it is not possible at this time to quantify these effects; indeed, because of observational limitations we do not yet have fine-resolution documentation of the behavior and structures of the low-level jet over a wide variety of geographical and meteorological settings. However, this paper illustrates that the well-known diurnal cycle of the low-level jet may be influencing the severe storm environment in many and more complex ways than have been considered in the past.

Naturally, the effects discussed in this paper occur in concert with a wide range of other atmospheric processes that also act to modify the low-level winds. These include sloping terrain, spatial variations in surface character, spatial and temporal variations in cloudiness, synoptically evolving pressure and height fields, etc. Further, the diurnal wind cycle is usually out of phase with the cycle in convective instability. This means that the updrafts of storms which occur during the night may be decoupled from the near-surface layer, if it has stabilized. The greatest storm-relative helicity often resides in this layer, further complicating the understanding of nighttime thunderstorm behavior. It is not surprising that detailed prediction of the character and evolution of individual thunderstorms remains elusive.

Acknowledgments. Discussions with David Stensrud and Harold Brooks helped the author to improve this presentation. Suggestions provided in reviews by Doug Lilly and Lou Uccellini have helped the author clarify the central thesis of the paper. David Keller provided the maps of tornado occurrences. The figures were carefully prepared by Joan Kimpel. The hodographs from the Binger tornado day were provided by Don Burgess. The WSR-88D wind observations were made available by Ron Alberty of the NEXRAD Operational Support Facility in Norman. The author thanks all of these persons for their help and support.

References

Anthony, R., Trends in severe local storm watch forecasting performance at the National Severe Storms Forecast Center, in *Preprints 16th Conference on Severe Local Storms*, pp. 281–286, American Meteorological Society, Boston, Mass., 1990.

Beebe, R. G., Tornado proximity soundings, *Bull. Am. Meteorol. Soc.*, *39*, 195–201, 1958.

Beebe, R. G., and F. C. Bates, A mechanism for assisting in the release of convective instability, *Mon. Weather Rev.*, *83*, 1–10, 1955.

Blackadar, A. K., Boundary layer wind maxima and their significance for the growth of nocturnal inversions, *Bull. Am. Meteorol. Soc.*, *38*, 283–290, 1957.

Blackadar, A. K., and K. Buajitti, Theoretical studies of diurnal wind structure variations in the planetary boundary layer, in *Studies of Wind Structure in the Lower Atmosphere*, edited by J. E. Miller, pp. 45–79, Dep. Meteorol. and Ocean., New York Univ., 1957.

Bonner, W. D., Case study of thunderstorm activity in relation to the low-level jet, *Mon. Weather Rev.*, *94*, 167–178, 1966.

Bonner, W. D., Climatology of the low-level jet, *Mon. Weather Rev.*, *96*, 833–850, 1968.

Brooks, H. E., C. A. Doswell III, and R. P. Davies-Jones, Environmental helicity and the maintenance evolution of low-level mesocyclones, this volume.

Brown, R. A., Characteristics of supercell hodographs, in *Preprints, 16th Conference on Severe Local Storms*, pp. 30–33, American Meteorological Society, Boston, Mass., 1990.

Burgess, D. W., and E. B. Curran, The relationship of storm type to environment in Oklahoma on 26 April 1984, in *Preprints, 14th Conference on Severe Local Storms*, pp. 208–211, American Meteorological Society, Boston, Mass., 1985.

Darkow, G. L., and M. G. Fowler, Tornado proximity sounding wind analysis, in *Preprints, 7th Conference on Severe Local Storms*, pp. 148–151, American Meteorological Society, Boston, Mass., 1971.

Davies-Jones, R. P., Streamwise vorticity: The origin of updraft rotation in supercell storms, *J. Atmos. Sci.*, *41*, 2991–3006, 1984.

Davies-Jones, R. P., and H. Brooks, Mesocyclogenesis from a theoretical perspective, this volume.

Davies-Jones, R. P., D. W. Burgess, and M. Foster, Test of helicity as a tornado forecast parameter, in *Preprints, 16th Conference on Severe Local Storms*, pp. 588–592, American Meteorological Society, Boston, Mass., 1990.

Doswell, C. A., III, The distinction between large-scale and mesoscale contributions to severe convection: A case study example, *Weather Forecasting*, *2*, 3–16, 1987.

Doswell, C. A., III, A review for forecasters on the application of hodographs to forecasting severe thunderstorms, *Natl. Weather Dig.*, *16*, 2–16, 1991.

Fawbush, E. J., and R. C. Miller, The types of air masses in which North American tornadoes form, *Bull. Am. Meteorol. Soc.*, *35*, 154–165, 1954.

Fujita, T. T., Proposed characterization of tornadoes and hurricanes by area and intensity, *SMRP Res. Pap. 91*, 42 pp., Univ. of Chicago, Chicago, Ill., 1971.

Fujita, T. T., U.S. tornadoes. Part One: 70-year statistics, *SMRP Res. Pap. 218*, 122 pp., Univ. of Chicago, Chicago, Ill., 1987.

Hoecker, W. H., Three southerly low-level jet streams delineated by the Weather Bureau special pibal network of 1961, *Mon. Weather Rev.*, *9*, 573–582, 1963.

Hoxit, L. R., Diurnal variations in planetary boundary-layer winds over land, *Boundary Layer Meteorol.*, *8*, 21–38, 1975.

Johns, R. H., and W. D. Hirt, Derechos: Widespread convectively induced windstorms, *Weather Forecasting*, *2*, 32–49, 1987.

Johns, R. H., K. W. Howard, and R. A. Maddox, 1990: Conditions associated with long-lived DERECHOS—An examination of the large-scale environment, in *Preprints, 16th Conference on Severe Local Storms*, pp. 408–412, American Meteorological Society, Boston, Mass., 1990.

Lilly, D. K., Dynamics of rotating thunderstorms, in *Mesoscale Meteorology—Theories, Observations and Models*, edited by D. K. Lilly and T. Gal-Chen, pp. 531–543, D. Reidel, Hingham, Mass., 1983.

Lilly, D. K., The structure, energetics and propagation of rotating convective storms, II, Helicity and storm stabilization, *J. Atmos. Sci.*, *43*, 126–140, 1986.

Maddox, R. A., An evaluation of tornado proximity wind and stability data, *Mon. Weather Rev.*, *104*, 133–142, 1976.

Maddox, R. A., The relation of diurnal, low-level wind variations to summertime severe thunderstorms, in *Preprints, 14th Conference on Severe Local Storms*, pp. 202–207, American Meteorological Society, Boston, Mass., 1985.

McCaul, E. W., Jr., Bouyancy and shear characteristics of hurricane-tornado environments, *Mon. Weather Rev.*, *119*, 1954–1978, 1991.

Miller, R. C., Notes on analysis and severe-storm forecasting procedures of the Military Weather Warning Center, *Tech. Rep. 200*, Air Weather Serv., 94 pp., Scott Air Force Base, Ill., 1967.

Newton, C. W., Dynamics of severe convective storms, *Meteorol. Monogr.*, *5*(27), 33–58, 1963.

Reiter, E. R., Tropopause circulation and jetstreams, in *World Survey of Climatology*, vol. 4, *Climate of the Free Atmosphere*, edited by D. F. Rex, pp. 85–193, Elsevier, Amsterdam, 1969.

Rotunno, R., On the linear theory of the land and sea breeze, *J. Atmos. Sci.*, *40*, 1999–2009, 1983.

Stensrud, D. J., M. H. Jain, K. W. Howard, and R. A. Maddox, Operational systems for observing the lower atmosphere: Importance of data sampling and analysis procedures, *J. Atmos. Oceanic Tech.*, *7*, 930–937, 1990.

Tegtmeier, S. A., The role of the surface, sub-synoptic low pressure system in severe weather forecasting, M.S. thesis, 66 pp., School of Meteorol., Univ. of Okla., Norman, 1974.

Uccellini, L. W., On the role of upper tropospheric jet streaks and leeside cyclogenesis in the development of low-level jets in the Great Plains, *Mon. Weather Rev.*, *108*, 1689–1696, 1980.

Uccellini, L. W., and D. R. Johnson, The coupling of upper and lower tropospheric jet streaks and implications for the development of severe convective storms, *Mon. Weather Rev.*, *107*, 682–703, 1979.

Weisman, M. L., and J. B. Klemp, The structure and classification of numerically simulated convective storms in directionally varying wind shears, *Mon. Weather Rev.*, *112*, 2479–2498, 1984.

Wexler, H., A boundary layer interpretation of the low-level jet, *Tellus*, *13*, 368–378, 1961.

Tornadoes: A Broadcaster's Perspective

Tom Konvicka

Lanford Companies, KALB-TV, Alexandria, Louisiana 71309

1. Introduction

Tornadoes are quite common in the United States. In 1990 the Severe Local Storms Unit (SELS) of the National Severe Storms Forecast Center (NSSFC) tallied 1132 tornadoes; this is the highest total on record (SELS smooth log, internal report, 1990). This number represents only tornadoes that were properly documented. Thus it is likely the actual number of tornadoes in the United States for 1990 is somewhat (perhaps even considerably) higher.

Research, operational, academic, and various "private" entities each contribute to addressing hazards associated with tornadoes. The purpose of this paper is to present some tornado-related concerns from the author's viewpoint as a broadcast meteorologist. Section 2 discusses the essential role a broadcast meteorologist must play in educating the public about tornadoes. Section 3 covers the function of the broadcast meteorologist before, during, and after a tornadic event. Section 4 analyzes some issues regarding interaction between broadcast meteorologists and their peers in the operational sector in the context of severe local storms. Section 5 highlights two effects the ongoing technological boom could cause in the broadcasting arena.

2. Role of Broadcast Meteorologists in Public Education

Meteorologists, be they in the operational, research, academic, or broadcasting areas, can play an important role in helping to increase public knowledge and awareness of tornadoes. The broadcaster's function, however, may be especially vital since in most areas the local broadcast meteorologist is the most visible representative of the profession. In smaller markets (where no National Weather Service (NWS) office exists and only one or two television stations are present) the broadcast meteorologist may well be the sole example of his/her profession. The importance of proper public education is evident when tornado-related fatality statistics are analyzed. It is generally accepted that a rather dramatic drop in the tornado death toll has occurred during the past 20 years and that proper education of the public is at least partially responsible. For these reasons, broadcast meteorologists must be aggressive leaders in the attempt to educate the public about tornadoes (and weather in general).

One primary vehicle for accomplishing the goal of a successful campaign on tornado education is the school visit. This visit can include preschool through college age groups. In the school setting, many eager and impressionable minds await the meteorologist's program. The agenda begins with an explanation of how thunderstorms develop. Then, a video potpourri of tornadoes in action stimulates audience interest. A demonstration model such as a "tornado in a bottle" enlightens students by showing them a simple physical analogue which behaves, albeit crudely, like a natural tornado. The presentation concludes by answering questions, sharing experiences, and reviewing safety rules.

A second avenue the broadcast meteorologist must explore to accomplish his/her goal of public education is a visit to local civic organizations. It is common for these meetings to include local decision makers. Some of these people may even be involved with handling weather-related disasters. A comprehensive program is appropriate. As in the school presentation, the tornado video is the most popular component of the agenda.

Another favorite aspect of the meteorologist's visit is the question and answer session. Many tornado-related anecdotes are told, and it is up to the meteorologist to find quickly a plausible explanation for these experiences. Yet, following the meeting, more serious discussion may ensue, and the broadcaster may learn that some members of the civic organization are involved with civil defense, the 911 program, or the NWS. This example of "networking" with people enables the broadcast meteorologist to be more in touch with those involved with the meteorological profession as well as those exterior to the meteorological community who must also deal with tornado hazards.

A third aspect of public education is the production of written materials. This could include almost anything rang-

The Tornado: Its Structure, Dynamics, Prediction, and Hazards.
Geophysical Monograph 79
Copyright 1993 by the American Geophysical Union.

ing from a one-page pamphlet on tornado safety rules to a full-length book describing several tornado-related topics in detail. The task of writing some of these items can be time-consuming but is quite beneficial to both the public and the meteorologist. The pamphlet on tornado safety rules can be located in the home where it is easily accessible during a tornado warning. The full-length book can provide hours of interesting and enjoyable reading. By authoring such material the broadcast meteorologist enlists the power of the written word to help carry out tornado education. Hence television meteorologists need not be limited to the usual verbal method of conveying information. In addition to employing the written word as a communication method, the television meteorologist is increasing his/her knowledge of tornadoes by researching the topic. An important end result is to solidify credibility with viewers.

3. TORNADIC EVENTS:
THE BROADCAST METEOROLOGIST AT WORK

Besides fostering public education, another critical responsibility the broadcast meteorologist must accept is proper coverage of tornadic events. For the discussion presented in this section, the process a broadcast meteorologist engages in when covering tornadic episodes can be divided into three phases: anticipation, recognition, and postevent. The anticipation phase may be defined as the time when dynamic/thermodynamic conditions suggest potential for tornado development. As such, then, the anticipation phase is a forecasting stage. The recognition phase begins when echoes appear on the radar display and must be interpreted as either severe or nonsevere. The postevent phase occurs after tornadoes have dissipated or have moved out of the viewing area.

3.1. *The Anticipation Phase*

Successful performance during tornadic episodes begins with proper anticipation of the event. It is essential that both the operational personnel and the weather staff at the local television station correctly diagnose tornadic potential present in the atmosphere. This is one process whose value will not be rendered obsolete by ongoing technological advances. *Foster* [1990] comments that even when the NEXt Generation of Weather RADars (NEXRAD) becomes fully operational, successful anticipation of an event will allow the full potential of the NEXRAD system to be realized during warning situations.

As one might expect, much of the information used by the broadcast meteorologist during the anticipation phase originates at the NSSFC. The first inkling of a tornado threat is often discussed in either the First Day Severe Thunderstorm Outlook or the Second Day Convective Outlook. These outlooks are designed to assist NWS field offices with the task of forecasting convective weather. These products are transmitted from the NSSFC and are easily accessible to all broadcast meteorologists. By evaluating these products and by interpreting computer model output for themselves, broadcast meteorologists can, like NWS forecasters, become aware of the salient features which may lead to an outbreak of tornadoes. Some broadcasters are tempted to consider only the available graphical guidance. Although this graphical product easily delineates the severe local storm threat area, it does not provide insight into the conditions operating to bring about potential for tornadoes.

As time passes and it becomes possible for SELS forecasters at NSSFC to refine some of the details of the evolving situation, a Mesoscale Discussion is usually disseminated. This product, like the earlier convective outlooks, is easily received via alphanumeric data services. The Mesoscale Discussion is recommended reading for the serious broadcast meteorologist because it gives details, based on the interpretation of SELS forecasters, concerning the evolution of factors which may be responsible for producing tornadoes in the near future. At this point the broadcast meteorologist should already have performed (1) sounding analysis and interpretation and (2) subsynoptic analyses of available surface data. A helpful resource for the author on the topics of sounding analysis and surface subsynoptic analysis is given by *Doswell* [1982]. By performing these two fundamental tasks, broadcast meteorologists exercise their skills as practitioners of meteorology. Also, the telecaster's ability to understand the suite of valuable convective products from SELS will be augmented. Thus the individual feels that he/she is becoming more mature and competent as a professional, and his/her level of on-air confidence rises. Viewers will notice the difference in knowledge and confidence.

As parameters continue to mesh, a tornado watch is issued by SELS. During a tornado watch it is important to convey to viewers the ideas contained in the following statements.

1. Continue normal activities but be aware of nearby thunderstorms and the weather changes they bring.
2. It is common to have little or no warning of a tornado. Be prepared to move quickly to a place of safety if you perceive danger is imminent.

The concepts contained in these two statements properly inform the public on the situation without creating unnecessary alarm.

At this point the qualified [*American Meteorological Society*, 1991a] broadcast meteorologist, who is also well trained in the field of severe storms meteorology, must analyze critically the reasoning given by SELS forecasters as the basis for issuing the watch. It is vital that broadcast meteorologists be able to think for themselves. Viewers are best served when independent thoughtful synthesis and unique knowledge of local meteorological effects provided by the local broadcast meteorologist are combined with expertise from SELS forecasters. Even if future government plans to transfer watch responsibility from SELS to NWS field offices [*Friday*, 1988] become reality, the broadcaster will still need to be capable of following the anticipation phase independently.

3.2. The Recognition Phase

For the discussion presented in this section, assume that thunderstorms have developed and must be monitored for indications of tornadic activity. Again, it is desirable for the broadcast meteorologist to follow storm motion and evolution independently, but only if he/she is qualified and has the appropriate equipment (a conventional or a Doppler weather radar) to use.

Of fundamental importance during the recognition phase is the association of rotation in a storm with certain radar signatures. The pendant and hook echo [*Stout and Huff*, 1953], the line-echo wave pattern, or LEWP [*Nolen*, 1959], the bow echo [*Fujita*, 1978], and the derecho [*Johns and Hirt*, 1987] should be familiar to the broadcaster. In addition, the review by *Lemon* [1980] is highly recommended to the serious meteorologist who desires to understand methods of recognizing radar signatures associated with severe storms.

One important facet of the recognition phase from the broadcasting viewpoint concerns how to relay information to the viewer. The choice involves using a "crawl" statement (a message moved across the bottom of the television screen) or a live, unscheduled interruption of programming. A "crawl" offers the advantage of allowing the viewer to continue to watch the program without interruption. However, in rapidly changing situations with multiple warnings, the "crawl" is not an efficient method of communication. The live on-air appearance is preferred for tornado warnings. In this instance, the implication to viewers is that the weather situation is becoming increasingly dangerous and that it is appropriate for them to react by being more conscious of weather conditions. Most viewers will not react to a severe weather threat unless they believe there is danger or someone they trust tells them about it in a personal manner. The on-air appearance offers this personal touch. The maximum amount of information about the weather situation is included during an on-air update. It is crucial for the public to receive the maximum amount of information possible, in the most personal manner, from someone who has followed the recognition phase (and the anticipation phase) independently. The viewer's reaction to the information given may mean the difference between life and death.

3.3. The Postevent Phase

Once tornadic thunderstorms have moved through the area and the Watch Status Report from SELS has lifted the watch, the chore of locating and assessing damage begins. The broadcast meteorologist can contribute valuable service during the postevent phase. In working directly with news reporters (who have little or no meteorological training) the meteorologist can speculate on the forces which caused the visible damage and can provide expertise on the phenomenon deemed responsible. Also, it is best to have the meteorologist critically review a reporter's package for accuracy before it airs on a newscast. In this way, the viewer has a better chance of getting correct information. In larger television markets, where weather personnel are more numerous, a broadcast meteorologist may actually be the one to survey and report on how the viewing area was affected by the tornadoes. Subsequently, the broadcaster should work with NWS personnel to assure that the event is properly documented and takes its place on the climatological record.

The motivated broadcast meteorologist may consider writing a case study review of the event. The case study could be presented, for example, at a conference on severe local storms or at a weather analysis and forecasting conference. The published version of the paper would appear in the respective conference preprint volume. As he/she writes about the episode, the process of fusing observations with concepts and reading the literature allows the individual to become more knowledgeable about severe local storms. Results of the case study can be applied to future severe storm episodes with the outcome being an enhanced performance level during the anticipation and recognition phases.

4. BROADCAST METEOROLOGISTS AND NWS

It is appropriate to discuss the sometimes thorny question of interplay between the NWS and broadcast meteorologists in the context of severe local storms. Each entity plays an essential role in ensuring that timely warning of impending severe weather is received; these respective functions have already been clearly defined [*American Meteorological Society*, 1991b]. It is precisely for these reasons that the broadcast meteorologist should seek to develop a positive relationship with the local NWS office and should not feel obliged to criticize the NWS as an inept competitor for the public's attention.

It has already been stated that the roles of the broadcast meteorologist and the NWS are to remain distinct during a severe local storm threat. It should be pointed out, however, that there are circumstances which are not so easily resolved. Specifically, the author has been faced with the following instances: (1) a person calls the television station and reports sighting a tornado, (2) a member of a spotter's network calls and reports sighting a tornado, (3) there is no NWS warning in a situation that appears to deserve one, and (4) NWS issues a warning that appears unwarranted.

Several factors enter into the decision on whether or not a tornado sighting by a caller is legitimate. First, the sighting must be reported in a competent manner. One method of assessing competence of the report is to ask the caller (1) Did you actually see the tornado? and (2) How do you know it was a tornado and not something else? Generally, these two questions provide enough information to decide whether or not to pursue the matter any further. If the description seems to hold some credibility, a fresh look at the radar is necessary. This is another reason why it is recommended that the qualified broadcast meteorologist follow the recognition stage independently. If he/she has no skill in weather radar interpretation, or no radar to interpret, then ignorance becomes the basis for the decision and the viewer's best interest is not well served. While questioning the caller and looking at weather radar, the meteorologist should wait for a second report of a tornado. The concurrent testimony of two dependable witnesses is considered to be sufficient for placing a call to the NWS office with warning responsibility and reporting the information to them. An on-air bulletin is delivered after the report has been relayed to the NWS.

It is not uncommon for broadcast weather departments to have an array of severe storm spotters. These individuals, when properly trained, are of great value because they are able to see detail which can be missed by weather radar and satellite photographs. Proper training includes a working knowledge of the material covered in the National Oceanic and Atmospheric Administration series of spotter tapes and pamphlets. A report of a tornado from a member of the spotter's network can be considered reliable enough to immediately relay to the NWS. In fact, the broadcast meteorologist should encourage these individuals to place calls to the NWS first and the television station second.

Another point of interaction between the NWS and the broadcast meteorologist concerns those times when no NWS warning exists in a situation that appears, at least to the broadcaster, to merit one. It is best to be direct with viewers. Therefore the author endorses a live on-air bulletin that would convey ideas included in the following statement.

> KALB-TV 5 Doppler weather radar is monitoring a thunderstorm with the POTENTIAL to produce a tornado. This storm is located at *X POSITION* and is expected to affect *CITIES AND TOWNS* during the next *NUMBER* of minutes. If you are in the expected path of this storm please be keenly aware of sudden changes in the wind. If you feel danger is imminent, then move quickly to your basement, an interior hallway, or closet. Do not wait for us to broadcast an official warning from the National Weather Service. We will continue to follow this situation closely and will inform you on any new developments.

In effect, this approach represents a step between the watch and warning [*Friday*, 1988]. Also, this type of bulletin makes viewers aware of a potential threat but allows the broadcaster to stop short of issuing his/her own warning.

A fourth concern occurs when it appears the NWS has issued a warning that is not necessary. Again, honesty with the viewer is best. They should be informed that the warning exists but that questions concerning the validity of the warning are present. A typical example is a F0–F1 tornado spawned by a pulse thunderstorm. In this circumstance, the author believes proper procedure is to broadcast the warning but communicate that this type of tornado is usually weak and short-lived. Thus a distinction is made between the F0–F1 tornado and the F3–F5 variety. It is an opportune time to remind viewers that all tornadoes deserve respect but some command more concern than others. In this way, the broadcast meteorologist fulfills his/her role as a partner in the public/private relationship by broadcasting the tornado warning. In addition, the broadcaster maintains his/her integrity with viewers by being honest and keeps his/her credibility intact by not falling victim to the "media hype" temptation.

5. Two Possible Future Trends in Broadcast Weather

Continuing technological advances will bring profound yet beneficial changes to the entire meteorological profession. The NEXRAD system, wind profilers, a series of new weather satellites, and the continued growth of the computer industry constitute the majority of the technological revolution. The broadcasting sector must directly address the challenges presented by the present wave of technology. One way to accomplish this would be for a policy on hiring only qualified meteorologists to staff the broadcast weather center. A similar plan has already been implemented by the NWS in an attempt to develop a more professional work force [*Friday*, 1988]. It is both logical and necessary for the television industry to follow the NWS lead in this matter. The hazards associated with tornadoes demand maximum use of all available data in order to warn the public accurately. Those broadcasters not acquainted with all the new technology (and its proper uses) will not be able to deliver the quality of information that will be expected by viewers. The competitive nature of the broadcasting business will dictate the employment of qualified meteorologists to deliver weather information, especially during inclement conditions.

In the future, in weather conscious broadcasting markets across the United States, the need for individuals with specialized knowledge of severe local storms will increase. In this setting, the viewer is assured that the broadcaster delivering critical information is not only a meteorologist but a specialist in the field of severe local storms. Perhaps even a "severe storms specialist" certification program could be designed by input from the National Severe Storms Laboratory, the NSSFC, and broadcast meteorologists. It is evident that with the increase in technology comes more responsibility, not less, for the broadcast meteorologist.

6. Conclusion

This paper has addressed several facets of tornadoes from the perspective of a broadcast meteorologist. The role of the broadcast meteorologist in fostering public education will remain vital for some time to come. Telecasters must not forget that their highly visible position can enable them to hold considerable sway over the opinion viewers have concerning the tornado threat. A crucial element in whether or not the best information possible is being conveyed to the public during a tornado threat is the individual's independent performance. Clearly, proper coordination with the NWS reduces the chance for misunderstanding during warning situations, assures that public/private sector roles remain distinct and, when appropriate, permits the broadcast meteorologist valuable input to the warning/no warning decision. Also, documentation of tornadic events can be increased with better teamwork between the NWS and media. With the present boost of technology in its infancy, it is hoped that television stations will keep pace with the NWS by raising standards for employment in the broadcast weather department.

The author's experience indicates that there is a gap separating the broadcasting, operational, and research communities. However, the author believes technological improvements will help result in a fusion of these splintered groups by increasing their interdependence. Motivated and qualified broadcast meteorologists should work with peers in the operational, research, and academic sectors to assure that future growth of the meteorological profession is harmonious.

Acknowledgments. The author wishes to thank cognizant editor Charles A. Doswell III and two anonymous reviewers for their patience and for the professional manner in which they accomplished their duties. This paper would not have been published without their dedicated effort.

References

American Meteorological Society, What is a meteorologist?, *Bull. Am. Meteorol. Soc.*, 72, 61, 1991a.

American Meteorological Society, Policy statement on the Weather Service/private sector roles, *Bull. Am. Meteorol. Soc.*, 72, 393–397, 1991b.

Doswell, C. A., III, The operational meteorology of convective weather, vol. I, Operational mesoanalysis, *NOAA Tech. Memo., NWS NSSFC-5*, 158 pp., Natl. Severe Storms Forecast Center, Kansas City, Mo., 1982.

Foster, M. P., NEXRAD operational issues: Meteorological considerations in configuring the radar, in *Preprints, 16th Conference on Severe Local Storms*, pp. 189–192, American Meteorological Society, Boston, Mass., 1990.

Friday, E. W., Jr., The National Weather Service Severe Storms Program: Year 2000, in *Preprints, 15th Conference on Severe Local Storms*, pp. J1–J8, American Meteorological Society, Boston, Mass., 1988.

Fujita, T. T., Manual of downburst identification for Project NIMROD, *SMRP Res. Pap. 156*, 42 pp., Univ. of Chicago, Chicago, Ill., 1978.

Johns, R. H., and W. D. Hirt, Derechoes: Widespread convectively induced windstorms, *Weather Forecasting*, 2, 32–49, 1987.

Lemon, L. R., Severe thunderstorm radar identification techniques and warning criteria, *NOAA Tech. Memo., NWS NSSFC-3*, 60 pp., Natl. Severe Storms Forecast Center, Kansas City, Mo., 1980. (Available as *NTIS PB81-23809* from Natl. Tech. Inf. Serv., Springfield, Va.)

Nolen, R. H., A radar pattern associated with tornadoes, *Bull. Am. Meteorol. Soc.*, 40, 277–279, 1959.

Stout, G. E., and F. A. Huff, Radar records an Illinois tornado, *Bull. Am. Meteorol. Soc.*, 34, 281–284, 1953.

The "Short Fuse" Composite: An Operational Analysis Technique for Tornado Forecasting

JIM JOHNSON

Weather Service Office, Dodge City, Kansas 67801

1. INTRODUCTION

The concept of composite analysis for use in prediction of severe storms is far from new and probably reached maturity in the late 1960s to early 1970s through the work of *Miller* [1972]. A real value of the composite analysis is that it leads the meteorologist through a careful analysis of the initial atmospheric conditions [*Doswell*, 1982]. The "short fuse" composite is an attempt to extend the composite method to hourly surface data sets with the explicit purpose of making a tornado forecast.

In forecasting the tornado environment it is clear that no "mandatory" or "magic" numbers will suffice. Rather, it is through the interpretation and understanding of the analyzed fields that one comes to recognize those mesoscale features which are most closely associated with an incipient tornado.

This technique is designed for use by an operational forecaster, who must function under the constraints of data sets that are often incomplete due to observation and communication failures, as well as computer run-stream priorities. It is possible that with the tremendous volume of new data sets becoming available to the forecaster, the saturation point cannot be far away. Therefore any method which allows a large set of atmospheric information to be digested simultaneously, especially on the mesoscale where the data density only compounds the problem, is worthy of consideration.

The idea of the "short fuse" composite grew from an attempt to refine the use of the Automation of Field Operations and Services (AFOS) Data Analysis Program (ADAP) to forecast only tornadoes. *Bothwell* [1988] has provided the forecaster with an excellent forecast tool which takes advantage of the density of surface observations, thus giving much finer resolution in a shorter time frame (hourly). He outlines a systematic approach to analysis of the 15 ADAP products that he presents in the form of a decision tree for forecasting convection. The "short fuse" composite uses only five of the ADAP products in a fast graphical approach to forecast only tornadoes. It is meant to be used with, not to replace, the ADAP decision tree.

2. THE "SHORT FUSE" COMPOSITE

The process begins with a typical *Miller* [1972] analysis of synoptic scale features at 0000 UTC and 1200 UTC, to which it may be helpful to add an analysis of storm-relative helicity. *Davies-Jones* [1983, 1984], *Davies-Jones et al.* [1990], and *Woodall* [1990] have shown the value of helicity as a forecast tool. Areas identified as favorable for significant rotating convection by this composite analysis are tracked hourly with the "short fuse" composite technique.

By using the analyzed parameters from ADAP, the following statement was seen to define a relatively small geographic area that contained a large proportion of the tornado occurrences.

> That area where the surface moisture flux convergence 1.5 g kg^{-1} h^{-1} isopleth overlaps the downwind side of the axis of potential temperature advection within the plateau (axis) of highest instability and where the cap inversion is less than 2°C.

Perhaps the most significant portion of the threat area so defined is with the overlapping of the downwind side of the axis of potential temperature (θ) advection by the 1.5 g kg^{-1} h^{-1} surface moisture flux convergence isopleth. It is this region where one can anticipate the greatest low-level directional shear or the greatest positive subcloud layer shear [*Davies*, 1989] to coincide with the strongest inferred upward motion and sufficient moisture for significant convection. Removal of the convective cap and inclusion of the most thermodynamically unstable air may add more to the chance of significant convection than it does to the likelihood of tornadoes. The implication is that storms developing in or moving into this environment would have the best chance of producing tornadoes.

The five charts for the composite and the parameters analyzed are as follows. (The AFOS graphics selected for

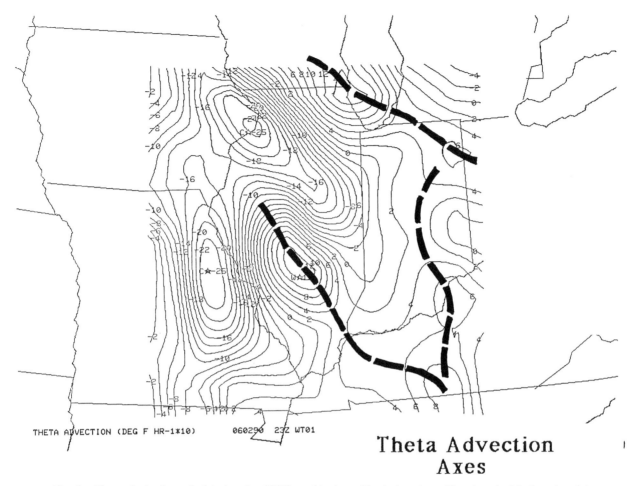

Fig. 1. The analysis of a typical θ advection (STA) graphic chart. The dashes show ridges (axes) of θ advection. It is also helpful to mark the centroid of maximum θ advection.

the composite will be identified according to the last three characters of the nine character group with the first six (NMCGPH) assumed.) (1) Surface cap inversion (SSC): select the 2°C isopleth, (2) Surface moisture flux convergence (SMC): select the 1.5 g kg^{-1} h^{-1} isopleth, (3) θ advection (STA) (Figure 1): analyze the axes of maximum θ advection values greater than zero and, using the surface wind vector streamlines (SSW), locate the downwind side of the axes, (4) Surface lifted index to 500 mbar (SSL) (Figure 2): select two or more isopleths that identify the "plateaus" (axes) of highest instability (greatest negative values). Finally, the parameters are transferred onto the composite map and the threat area is drawn in.

3. Results of the Case Studies

Since ADAP is a locally run computer applications program and is not archived (except occasionally on the local level), finding the data sets collected during actual tornado events proved to be a challenge. Twenty-seven cases were found with sufficient ADAP data to permit analysis. Several of them provided an opportunity to use the technique in real-time forecasting.

This technique is empirical; the derived fields of ADAP products were compared to a number of actual tornado events for which the data were available, and an isopleth which contained the majority of the events was selected.

The results were as follows.

Cap inversion. Of 66 tornadoes (27 cases), all but two tornadoes were first reported where the cap was weaker than 2°C during the preceding hour.

Surface moisture flux convergence. Of 66 tornadoes (27 cases), all but four tornadoes were first reported where the surface moisture flux convergence was 1.5 g kg^{-1} h^{-1} or greater during the preceding hour.

Downwind side of the θ advection ridge. Of 66 tornadoes (27 cases), all but two tornadoes were first reported on or just downwind of the axes of maximum advection of potential temperature during the preceding hour.

"Plateau" (axis) of greatest instability. Of 66 tornadoes (27 cases), all but four tornadoes were first reported upon the

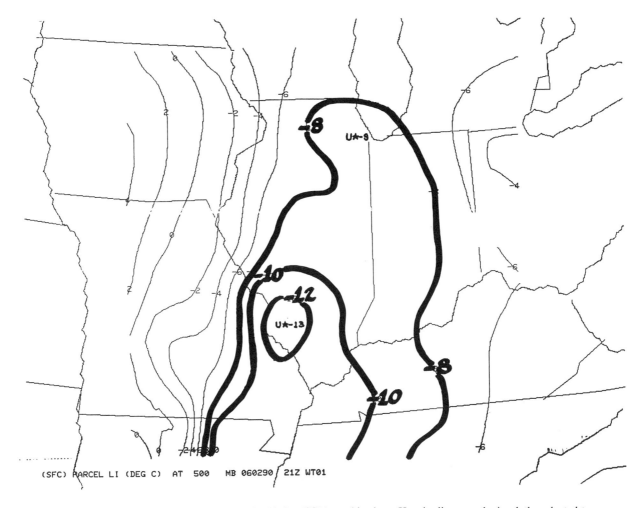

Fig. 2. The analysis of a typical surface lifted index (SSL) graphic chart. Heavier lines are the isopleths selected to define the "plateau" of greatest instability.

plateau of highest instability, with all but 14 first reported inside the isopleth of the highest instability. Again, this was from the hour immediately preceding the event.

4. Three Specific Cases

An example of a very small threat area was found November 28, 1988, at 0600 UTC (Figure 3), the day of the Raleigh-Durham tornado. Moisture was pooling along the east side of the mountains as south-southeast wind advected the θ axis into the moisture pool. Instability was not as extreme as in other cases. This storm likely would not have been forecast by the threat area on a purely subjective usage of the composite, since the threat area did not materialize until 0600 UTC, the time of the first tornado report. However, the chart-to-chart change from the preceding several hours did show the developing and nearly stationary moisture pool and the θ ridge moving northwestward into it.

In a second case, a relatively large threat area was indicated at 2200 UTC May 5, 1989 (Figure 4). Two F4 tornadoes occurred during the hour (2200–2300 UTC). It is not known if the F4 occurring at 0001 UTC would have been forecast or not since no ADAP data were available past 2200 UTC.

Composite charts from 2100 UTC and 2300 UTC on the day of the major tornado outbreak of June 2–3, 1990, are given in Figures 5 and 6. The initial tornadoes were reported on the downwind side of the strong θ advection and on the downwind side of the surface moisture flux convergence gradient [*Waldstreicher*, 1989] during the hour following the 2100 UTC composite threat area. Two hours later, the threat area had become elongated by southerly surface winds as moisture flux convergence continued to increase. During the next 2 hours, 10 tornadoes were reported. The threat area identified all but three of them, and the three misses were within a few kilometers of the threat area.

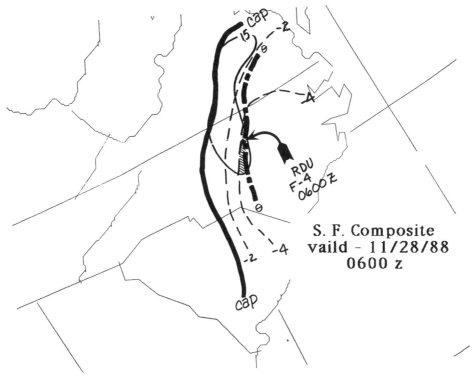

Fig. 3. A "threat area" is hatched in, while the arrow indicates the reported tornado (with time and F scale strength). Light dashes are the plateau of instability. The heavy solid line is the 2°C cap. The heavy broken line is the θ axis and the light solid line is the 1.5 g kg^{-1} h^{-1} surface moisture flux convergence isopleth. The "threat area" materialized at 0600 UTC, the time of the first tornado report.

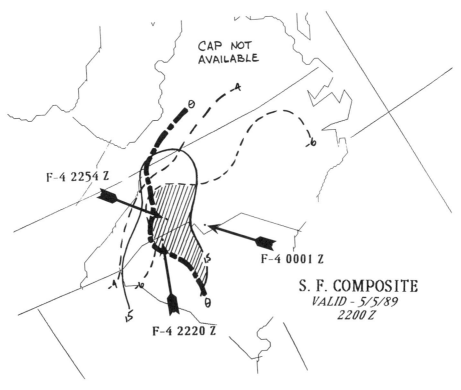

Fig. 4. Another outbreak in the Carolinas, this time with a large "threat area". The F4 tornado that occurred at 0001 UTC on May 6, 1989, was just outside the 2-hour-old "threat area." Had the cap strength (SSC) been available, perhaps the area would have been smaller.

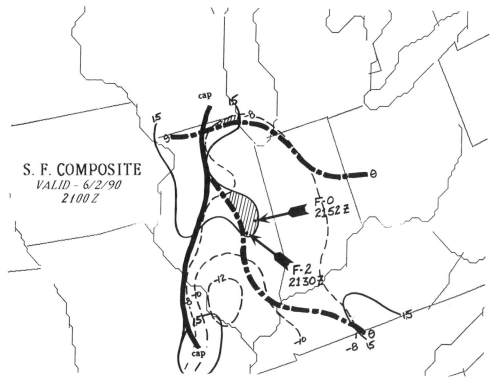

Fig. 5. Two "threat areas" are analyzed, with the northernmost being on the edge of the data fields where the objective analysis may be in some doubt. For this reason, the tornado that occurred just north of Chicago and inside this small "threat area" was not included in the verification.

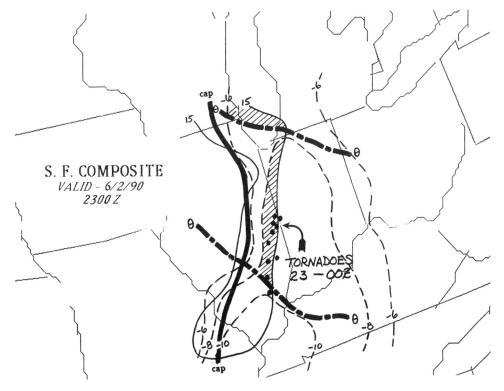

Fig. 6. A strong axis of θ advection was over southern Illinois at 2300 UTC (see Figure 1). The initial tornadoes from 2300 UTC began almost directly atop the strongest part of the θ axis. Analysis is as in Figure 3.

TABLE 1. Results of Simple Verification

	1 Hour After Composite Time	2 Hours After Composite Time
Total cases	27	9
Total tornadoes	66	26
Total within threat areas	56	17
Probability of detection	85%	65%

5. Verification of the Results

Only simple verification was attempted on the composite forecasts from the 27 cases studied. To relate the size of the threat area versus the number of tornadoes within the area is difficult since the threat areas were usually mesosynoptic to subsynoptic in scale and the tornado itself much smaller. For this reason, the only computation performed was probability of detection (POD). The POD is simply the number of correctly forecast events divided by the total number of events. The POD was computed using the number of tornadoes that occurred both in and out of the threat area for the hour immediately following the composite time and again for the hour beginning one hour after the composite time (Table 1).

As a control, two of the 27 cases had no severe thunderstorms and no tornadoes. A further six cases had severe thunderstorms but no tornadoes. The "short fuse" composite correctly indicated no tornadic activity for all of these null cases. In fact, every time a threat area developed, there was at least one tornado in it.

6. Summary

A POD of 85% for the initial hour after the composite time is quite good, even if one considers that ADAP requires 15 min to produce the analysis. Additional cases are being collected and analyzed to add more credence to the composite scheme presented here, but it bodes well for the technique that in all cases where the threat area did materialize, at least one tornado did occur within the threat area. Further, the technique often identified the initial convection that subsequently became tornadic and in several cases identified very small threat areas (of the order of 2–4 counties) in which a lone tornado was observed.

The technique presented has been in use at the weather service office in Dodge City, Kansas, for nearly six months as of this writing, and the results have continued to be encouraging. Spring 1991 was a particularly busy severe weather season, and the technique was at least partly responsible for the timely issuance of several verified tornado warnings. Most importantly, the warnings were issued before the event with a much greater degree of confidence, and use of the technique allowed probable tornado situations to be separated from probable severe thunderstorm events.

Acknowledgments. The author wishes to thank the following: For assistance in collection of data, Ron Przybylinski, Jeff Waldstreicher, and Dick Livingston; for encouragement and enthusiasm, Steve Letro, Ed Berry, and Joe Schaefer; for trying the technique in operations, the staff at WSO Dodge City, Kansas, and Don Baker at WSFO Lubbock, Texas; for piquing my interest in severe weather, Bob Miller, Bob Maddox, and Chuck Doswell; and finally, for introducing me to the wonders of dynamic meteorology, Doug Sinton and Pete Lester.

References

Bothwell, P. D., Forecasting convection with the AFOS data analysis program (ADAP-version 2.0), *NOAA Tech. Memo. NWS SR-122*, 91 pp., Sci. Serv. Div., Natl. Weather Serv. S. Reg., Fort Worth, Tex., 1988.

Davies, J. M., On the use of shear magnitudes and hodographs in tornado forecasting, in *Preprints, 12th Conference on Weather Forecasting and Analysis*, pp. 219–224, American Meteorological Society, Boston, Mass., 1989.

Davies-Jones, R., The onset of rotation in thunderstorms, in *Preprints, 13th Conference on Severe Local Storms*, pp. 215–218, American Meteorological Society, Boston, Mass., 1983.

Davies-Jones, R., Streamwise vorticity: The origin of updraft rotation in supercell storms, *J. Atmos. Sci.*, *41*, 2991–3006, 1984.

Davies-Jones, R., D. W. Burgess, and M. Foster, Test of helicity as a tornado forecast parameter, in *Preprints, 16th Conference on Severe Local Storms*, pp. 588–592, American Meteorological Society, Boston, Mass., 1990.

Doswell, C. A., III, The operational meteorology of convective weather, vol. I, Operational mesoanalysis, *NOAA Tech. Memo. NWS NSSFC-5*, Natl. Severe Storms Forecast Center, Kansas City, Mo., 1982.

Miller, R. C., Notes on analysis and severe-storm forecasting procedures of the Air Force Global Weather Central, *Tech. Rep. 200* (revision), 190 pp., Air Weather Serv., Scott Air Force Base, Ill., 1972.

Waldstreicher, J. S., A guide to utilizing moisture flux convergence as a predictor of convection, *Natl. Weather Dig.*, *14*, 20–35, 1989.

Woodall, G. R., Qualitative forecasting of tornadic activity using storm-relative environmental helicity, *NOAA Tech. Memo. NWS SR-127*, Sci. Serv. Div., Natl. Weather Serv. S. Reg., Fort Worth, Tex., 1990.

The Plainfield, Illinois, Tornado of August 28, 1990: The Evolution of Synoptic and Mesoscale Environments

WILLIAM KOROTKY

National Severe Storms Forecast Center, Kansas City, Missouri 64106

RON W. PRZYBYLINSKI

National Weather Service Forecast Office, St. Louis, Missouri 63304

JOHN A. HART

National Weather Service Forecast Office, Charleston, West Virginia 25311

1. INTRODUCTION

During the afternoon of August 28, 1990, a devastating tornado ripped through the southwestern suburbs of Chicago. The F5 intensity tornado was responsible for 29 deaths and over 300 injuries and caused property damage exceeding 160 million dollars (with much of the destruction occurring between Plainfield and Crest Hill, Illinois). The tornadic thunderstorm produced a nearly continuous path of severe weather for more than 4 hours as it tracked across northern Illinois into central Indiana.

The environment over northern Illinois exhibited limited tornadic potential at 1200 UTC August 28. A combination of large-scale features, including an approaching midlevel trough and a southward moving cold front (both forecast to interact with an extremely unstable airmass as strong mid and upper level winds moved over the region), indicated increasing potential for severe thunderstorm development across northern Illinois during the afternoon. Although the overall setting (i.e., strong instability and relatively weak "bulk" shear through the lowest 6 km) appeared to favor nonsupercell storms (with large hail and damaging wind as the primary severe weather threat), poststorm analysis of conventional radar images indicated supercell storm signatures along much of the path traversed by the tornadic storm. (Supercell storms are used here as a general term to include all convective storms that develop persistent mesocyclones. Storms that do not develop this dynamical structure are referred to as nonsupercell storms.)

This paper will investigate the synoptic and mesoscale conditions associated with the Plainfield tornadic event. Thermodynamic and vertical wind shear characteristics of the evolving tornadic environment will be examined with the aid of the SHARP (Skew T/Hodograph Analysis and Research Program) Workstation [*Hart and Korotky*, 1991], a software package that allows interactive analysis of upper air sounding data. This paper will focus on those aspects of the changing storm environment that appeared to highlight the growing tornadic potential, with a special emphasis placed on low-level storm-relative characteristics (e.g., low-level storm relative flow, magnitude and orientation of the low-level shear, vorticity, and helicity characteristics of the storm inflow layer). Forecast implications will be discussed in an effort to increase the state of readiness for similar rare event occurrences. Finally, the study will underscore the important role of interactive software (e.g., the SHARP Workstation) for diagnosing enhanced convective potential in a changing environment.

2. DATA

The SHARP Workstation is a sounding analysis program that generates Skew T plots and hodographs from observed upper air sounding data. The software also produces a comprehensive inventory of sounding-derived parameters and allows all aspects (thermodynamic and kinematic) of the sounding to be modified in an interactive manner. SHARP was used in poststorm analysis to examine upper air sound-

The Tornado: Its Structure, Dynamics, Prediction, and Hazards.
Geophysical Monograph 79
This paper is not subject to U.S. copyright. Published in 1993 by the American Geophysical Union.

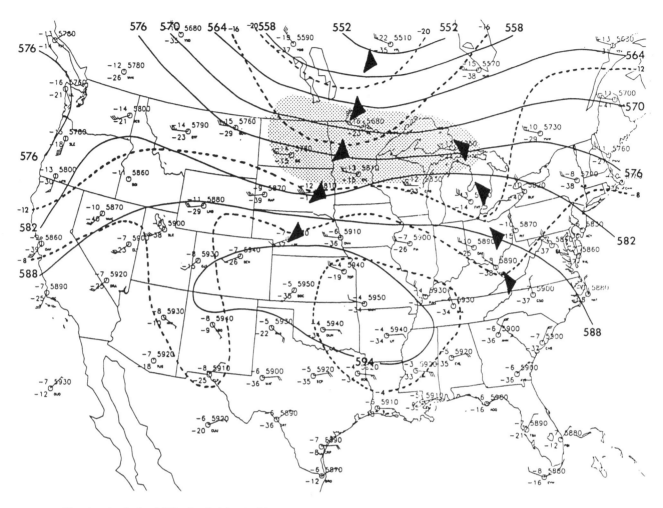

Fig. 1. Analysis of 500-mbar heights (solid contours every 60 m) and temperature (dashed contours every 4°C) for 1200 UTC August 28, 1990. Station reports show wind velocity (1/2 barb = 2.5 m s^{-1}, barb = 5 m s^{-1}, triangular barb = 26 m s^{-1}), temperature, and dew point (degrees Celsius). Axes of cold troughs indicated with solid triangles. Shaded area represents wind speeds equal to or greater than 50 knots (25 m s^{-1}).

ings for stations across the Great Lakes region at 1200 UTC August 28. Plainfield, Illinois (PLF), proximity soundings were created for 1200 UTC and 2000 UTC by modifying the 1200 UTC Peoria, Illinois (PIA), Skew T and hodograph with data representing the estimated thermodynamic and vertical wind profiles over PLF at 1200 UTC and 2000 UTC. Data for the PLF proximity soundings were obtained by linearly interpolating in space and time between Green Bay, Wisconsin (GRB) and PIA, using 1200 UTC and 0000 UTC sounding data. Surface values of temperature, dew point, and wind for the PLF proximity soundings were estimated from surface observations. SHARP also was used to produce an extensive summary of thermodynamic variables, convective indices, and wind-shear-related parameters for selected atmospheric layers (for both the 1200 UTC and 2000 UTC PLF proximity soundings).

Several derived fields incorporating hourly surface observations and the 1200 UTC August 28, 1990, regional upper air analyses were produced by ADAP (Automation of Field Operations and Services (AFOS) Data Analysis Program [*Bothwell*, 1988]). The output was used in conjunction with radar, satellite, and surface analyses to examine the changing thermal, moisture, stability, and wind patterns at the surface prior to the tornadic event.

3. METEOROLOGICAL SETTING

The 500-mbar pattern at 1200 UTC August 28 was characterized by a low-amplitude ridge extending from California to the Carolinas (Figure 1). The northern extent of the ridge had been eroded by a series of short-wave troughs moving eastward across western and central Canada, with another in this progression of short waves approaching the western Great Lakes at 1200 UTC. A band of strong westerly winds associated with the approaching trough stretched from Montana to Minnesota, with a 60-knot (31 m s^{-1}) jet core nosing

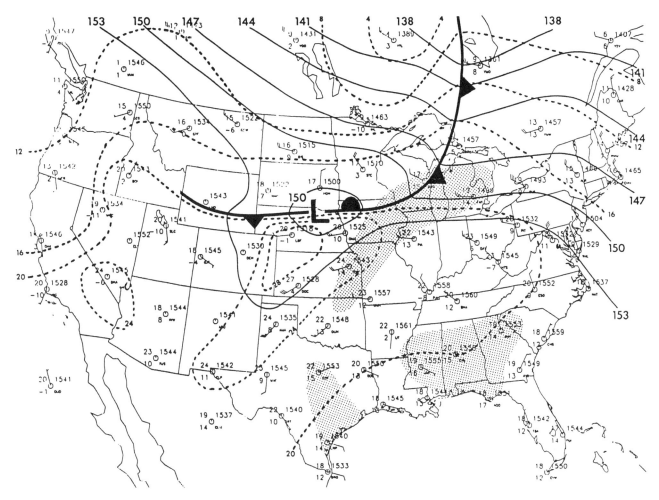

Fig. 2. Analysis of 850 mbar heights (solid contours every 30 m) and temperature (dashed contours every 4°C) for 1200 UTC August 28, 1990. Station reports show wind velocity (1/2 barb = 2.5 m s^{-1}, barb = 5 m s^{-1}), temperature, and dew point (degrees Celsius). Shaded area represents dew points equal to or greater than 14°C.

into Minnesota from the Dakotas. A difluent pattern was present over the western Great Lakes region at 250 mbar (not shown), with strong winds around 80-knots (41 m s^{-1}) over northern Illinois. The approaching trough caused a northwesterly component to develop in the upper flow over the Great Lakes region and allowed a surface cold front to move southeastward across northern Illinois during the afternoon of August 28. For the 12 hours ending at 1200 UTC, height falls and cold advection associated with the changing upper flow pattern caused 500-mbar temperatures to decrease from −4°C to −7°C at PIA and from −8°C to −12°C at GRB.

The 850-mbar chart for 1200 UTC August 28 revealed a significant thermal gradient across the upper Mississippi Valley and western Great Lakes, associated with a cold front moving through the region (Figure 2). Northwesterly winds at GRB and a decrease in temperature/dew point during the previous 12-hour period indicate that the front had passed GRB just prior to 1200 UTC. An axis of higher dew points was noted from the lower Great Lakes region to the southern Plains, with highest dew points in the vicinity of the front near GRB.

Soundings for stations across the Great Lakes region were characterized by very steep lapse rates (especially from ~800 to ~600 mbar) and considerable convective available potential energy (CAPE). A thermal cap near 850 mbar (evident in the 1200 UTC August 28 PLF sounding, Figure 3) was strong enough to prevent large-scale overturning of the atmosphere by widespread but ordinary (nonsevere) convection, thereby allowing heat and moisture to increase steadily across northern Illinois prior to the arrival of the cold front. However, the area of negative buoyancy below the level of free convection (LFC) was not impenetrable; the projected effects of frontal forcing and diurnal heating appeared strong enough to allow isolated deep convection during the afternoon. The 1200 UTC PLF sounding indicated a "best" lifted index (lifting a parcel containing the highest wet bulb potential temperature within the lowest 150 mbar; i.e., from approximately 1200 feet (365 m) above ground level (AGL))

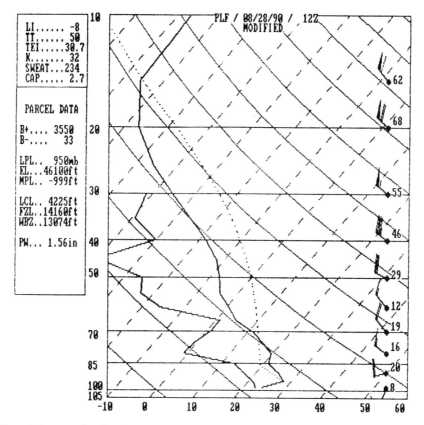

Fig. 3. PLF proximity sounding for 1200 UTC August 28, 1990, as displayed by the SHARP Workstation. CAPE is represented by B+ and depicted as the area between the parcel ascent curve (dotted contour) and the temperature curve (from the level of free convection to the equilibrium level (EL)). See *Hart and Korotky* [1991] for a thorough discussion of display characteristics, terms, and units associated with the Skew T presentation.

of −8 at 500 mbar with CAPE evaluated near 3550 J kg^{-1}. It is noteworthy that the sounding indicated even greater instability above 500 mbar, with a best lifted index of −12 at 300 mbar and −10 at 200 mbar. Thus the 1200 UTC PLF sounding indicated extreme instability, even without the aid of diurnal heating. With projected afternoon temperatures in the mid 90s (degrees Fahrenheit) (mid 30s degrees Celsius) and projected dewpoints in the mid 70s (degrees Fahrenheit) (mid 20s degrees Celsius) the thermodynamic structure of the 1200 UTC PLF sounding displayed a potential for uncommonly large instability across northern Illinois before the forecast arrival of the cold front (especially for a surface-based parcel). Kinematic analysis of the PLF sounding indicated considerable speed shear in the vertical wind profile between 600 and 500 mbar, but the strongest winds were near and above 400 mbar (approximately 7 km). Moderate speed and curvature shear was noted within the first kilometer, but the wind shear pattern was rather weak above this layer to around 600 mbar (Figures 3 and 4). A bulk measure of the shear through 6 km (19 m^2 s^{-2}, used in the calculation of the Bulk Richardson Number (BRN) [*Weisman and Klemp*, 1982]) was rather weak relative to the large instability, yielding a large BRN, as we shall discuss below.

4. ANALYSIS

Visible satellite imagery (Figure 5) and surface analysis (Figure 6) revealed a complex pattern of features across northern Illinois at 1900 UTC. The most important features included a cold front across northern Illinois (Figure 6), a gradual wind shift line south of the cold front (confirmed with time sections of surface data, not shown), and a moisture convergence gradient across north central Illinois (Figure 7). Surface dew points ranged from the mid 70s (degrees Fahrenheit) (mid 20s degrees Celsius) to around 80°F (27°C) across much of north central Illinois. However, dew points were around 70°F (20°C) even north of the cold front, with the true moisture gradient still in Wisconsin. More subtle features revealed by satellite imagery included several converging lines of cumulus across north central Illinois (Figure 5), but these features were difficult to resolve in the surface analysis. The ADAP lifted index pattern (Figure 7) depicted extreme instability across the lower Great Lakes region, with surface-based lifted indices between −12 and −15 across much of Wisconsin, Illinois, Indiana, and Michigan. Operational experience suggests that instability of this magnitude rarely occurs over such a large

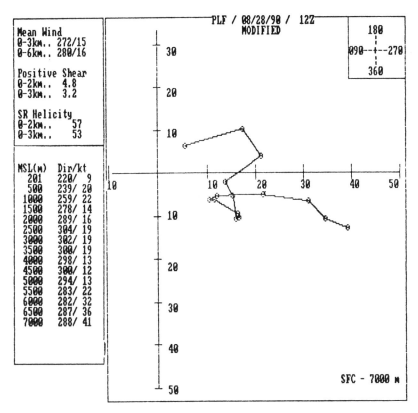

Fig. 4. PLF proximity hodograph for 1200 UTC August 28, 1990, as displayed by the SHARP Workstation. Units along axes in knots. Data points on hodograph are every 500 m. See *Hart and Korotky* [1991] for a thorough discussion of display characteristics, terms, and units associated with the hodograph presentation.

geographical area! The incipient Plainfield storm intensified rapidly in the vicinity of the cold front near Rockford, Illinois (RFD) between 1800 and 1900 UTC. One or two small F1 tornadoes were reported just west of RFD at 1842 UTC, with damaging wind and large hail reported at RFD between 1900 and 1945 UTC [*National Oceanic and Atmospheric Administration*, 1991].

The PLF proximity sounding for 2000 UTC (Figure 8) reflected an estimated surface temperature (92°F) (33°C) and dew point (78°F) (25°C) consistent with the low-level air most likely entrained into the storm as it approached Plainfield. The sounding revealed that important thermodynamic changes had occurred across northern Illinois between 1200 UTC and 2000 UTC, with CAPE (now representing the surface parcel) increasing from under 4000 J kg^{-1} (1200 UTC) to almost 7000 J kg^{-1} (2000 UTC). It should be noted that we could have determined CAPE for a parcel representing the mean thermal and moisture properties of the lowest 100 mbar of the sounding (e.g., the lifted index used by the Severe Local Storms Unit of the National Severe Storms Forecast Center), in which case CAPE would have been around 4200 J kg^{-1} at 2000 UTC. However, we felt that the air being lifted into the updraft by a fast-moving gust front was probably related to properties much closer to the surface. This is supported by the near equivalence between measured cloud bases (3000 feet (920 m) AGL) and the lifted condensation level (LCL, 3100 feet (945 m) AGL) calculated by lifting a surface parcel associated with the 2000 UTC PLF sounding.

The PLF proximity hodograph for 2000 UTC (Figure 9) reflected the estimated vertical wind profile, representative surface wind (300°/5 knots (2 m s^{-1})) and observed storm motion (from radar observations, 308°/32 knots (16 m s^{-1})) near Plainfield just prior to the tornadic phase of the storm. In addition to producing thermodynamic parameters, SHARP was used to calculate the storm-relative, vertically integrated helicity (hereafter referred to as helicity) and the low-level storm-relative flow structure of a storm moving through the estimated PLF environment with the observed Plainfield storm motion. A computed helicity of 165 m^2 s^{-2} through the lowest kilometer (AGL) of the PLF sounding was within the 0- to 3-km threshold for weak mesocyclone development suggested by *Davies-Jones et al.* [1990] but less than the 0- to 3-km threshold established by *Lazarus and Droegemeier* [1990]. However, the storm-relative inflow (resulting largely from storm motion) was quite significant (28 knots (14 m s^{-1})) through this layer.

By 2100 UTC the Plainfield tornado had just lifted, but the parent thunderstorm was still producing damaging wind and large hail as it continued across northeastern Illinois. Sur-

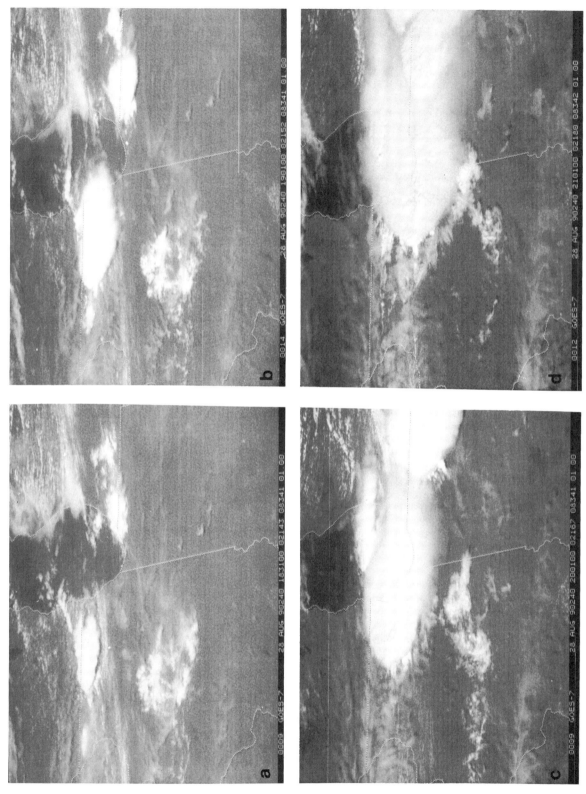

Fig. 5. Visible GOES satellite imagery showing the development and movement of the Plainfield tornadic thunderstorm across northern Illinois at (a) 1831 UTC, (b) 1901 UTC, (c) 2001 UTC, and (d) 2101 UTC on the afternoon of August 28, 1990.

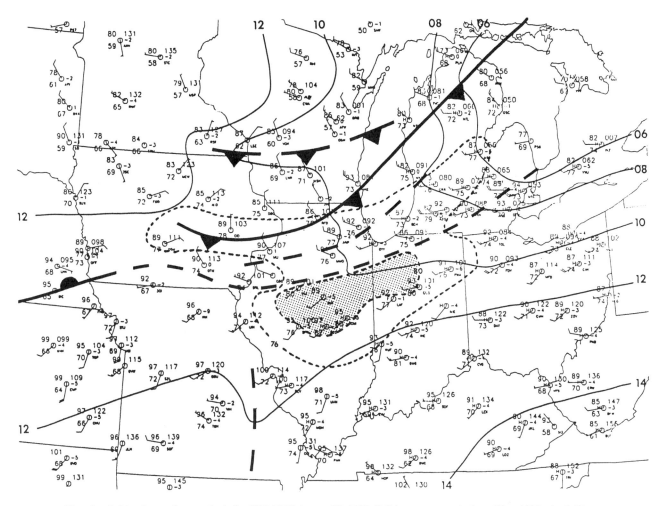

Fig. 6. Subjective surface analysis for 1900 UTC August 28, 1990. Solid contours are isobars (12 = 1012 mbar). Fronts indicated with the usual symbols. Large dashed contours are boundaries. Station reports show wind velocity (1/2 barb = 2.5 m s^{-1}, full barb = 5 m s^{-1}), temperature, and dew point (degrees Fahrenheit). Smaller dashed contours represent 76°F and 80°F isodrosotherms. Shaded area represents dew points equal to or greater than 80°F.

face analysis (Figure 10) and PPI radar film (not shown) used in conjunction with a sequence of radar tracings (Figure 11) indicated a convectively generated mesohigh with rapidly expanding outflow along its leading edge. The storm-produced mesohigh and gust front are further supported by satellite imagery (Figure 5) and by the surface moisture flux convergence/divergence pattern at 2100 UTC (Figure 12). Time sections of surface observations (not shown) indicated a gradual northwesterly shift in the surface winds ahead of the fast-moving gust front (Figure 11). This suggests that the ground-relative surface wind field of the mesoscale environment was not enhancing convergence and inflow. Rather, the movement of the storm and its attendant outflow created significant low-level convergence and inflow despite a northerly component to the surface winds just ahead of the gust front!

Analysis of conventional radar images (Figure 11) [*National Oceanic and Atmospheric Administration*, 1991] indicated reflectivity features commonly associated with supercell storms between 1900 UTC and 2100 UTC (e.g., right rear appendage with associated high reflectivity, inflow notch characterized by a strong low-level reflectivity gradient on the updraft flank of the storm, pronounced weak echo region with a periodic bounded weak echo region, etc.). Although the tornadic storm appeared to have interacted with several converging lines of cumulus (noted earlier) as it approached Plainfield, the operational surface observational network is not sufficiently dense to resolve the details.

5. Discussion

Observations and numerical simulations [*Weisman and Klemp*, 1982] indicate that the dynamic organization of storms is strongly influenced by instability and by the vertical structure of the horizontal wind. Weak vertical wind shear supports limited convective scale organization and

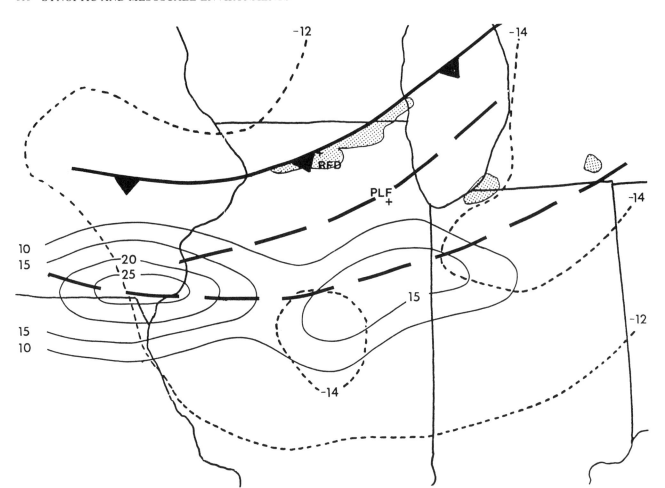

Fig. 7. Automation of Field Operations and Services Data Analysis Program (ADAP) surface moisture flux convergence (solid contours depict $g\ kg^{-1}\ h^{-1} \times 10$), ADAP lifted index to 500 mbar (dashed contours in degrees Celsius), and surface cold front (with usual symbols) for 1900 UTC August 28, 1990. Large dashed contours represent surface boundaries. Shaded area indicates active convection.

generally leads to loosely organized nonsupercell storms. However, stronger shear favors progressively better organization (well-organized nonsupercells and supercell storms) by maximizing boundary layer convergence and by promoting dynamical processes that result in storm rotation.

The BRN [*Weisman and Klemp*, 1982] has been used to measure the relative importance of instability and vertical wind shear for a particular environment and generally is associated closely with observed storm type (nonsupercell or supercell storms) for modeled values of CAPE between 1000 and 3500 J kg^{-1}. For this range of CAPE, numerical simulations and a limited number of storm observations suggest a tendency for supercell storms to form when the BRN is less than 45. On the other hand, BRN values exceeding 45 are generally associated with nonsupercell convection.

The Plainfield environment exhibited substantial instability at 1200 UTC (i.e., CAPE around 3550 J kg^{-1}, Figure 3), but a bulk measure of the shear through 6 km (i.e., the vector difference between the 0- to 6-km AGL density-weighted mean wind and the 0- to 500-m density-weighted mean wind) was weak (19 m^2 s^{-2} [*Weisman and Klemp*, 1982]) relative to the large CAPE. The resulting high BRN for PLF (189) suggested a tendency for nonsupercell storms and indicated a severe weather threat primarily involving damaging straight-line wind and large hail. While the BRN apparently failed to forecast the potential for supercell growth within this environment, recall that Weisman and Klemp did not model environments where CAPE exceeded 3500 J kg^{-1}. In fact, high-CAPE environments have yet to be modeled successfully (H. E. Brooks, personal communication, 1991). By 2000 UTC the PLF sounding indicated CAPE approaching 7000 J kg^{-1} (Figure 8), with an associated BRN of 234.

Recently, much interest has focused on the vertical wind shear structure of the low-level storm environment, specifically on the region associated with entrainment into the

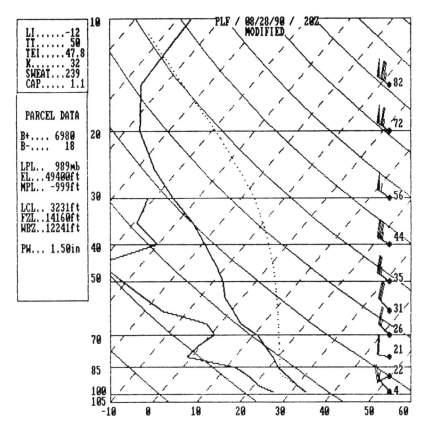

Fig. 8. Same as Figure 3 for 2000 UTC August 28, 1990.

updraft [*Brooks and Wilhelmson*, 1990; *Davies*, 1989; *Davies-Jones et al.*, 1990; *Johns et al.*, this volume; *Lazarus and Droegemeier*, 1990]. Shear characteristics within this inflow region appear to play a major role in determining both the potential for a storm to develop supercell characteristics (i.e., persistent mesocyclone) and the potential for a particular supercell to develop a tornadic low-level circulation. Numerical simulations [*Brooks and Wilhelmson*, 1990] indicate that midlevel rotation does not guarantee a tornadic low-level mesocyclone. *Lazarus and Droegemeier* [1990] found that sufficient helicity in the low-level environment of simulated storms results in strong storm rotation but suggested that rotation and storm longevity also depend on sufficient low-level storm inflow (i.e., greater than 19 knots (10 m s^{-1}) [*Davies-Jones*, 1984]. The helicity and inflow structure of the estimated Plainfield storm environment will be discussed in greater detail; however, it can be noted that the flow into the developing tornadic storm (resulting largely from storm motion characteristics) was rather significant (around 30 knots (15 m s^{-1})) within the lowest kilometer (AGL).

5.1. *Characteristics of the Storm-Relative Environment*

For a given storm motion, streamwise vorticity refers to the component of horizontal vorticity parallel to the storm inflow and plays an important role in determining the potential for the updraft to rotate [*Davies-Jones*, 1984]. Helicity is related to streamwise vorticity [*Doswell*, 1991] and measures the rotation potential that can be realized by a storm moving through the vertically sheared environment (i.e., streamwise vorticity associated with the inflow layer can be drawn into and correlated with the updraft core to produce a persistent mesocyclone). Since the storm reference frame is physically important, the actual movement of a storm can exert a significant influence on these characteristics. Consequently, storms that "experience" helical low-level environments favor the persistent rotating updrafts characteristic of supercell storms and represent the greatest tornadic threat.

Helicity values between 150 [*Davies-Jones et al.*, 1990] and 250 m^2 s^{-2} [*Lazarus and Droegemeier*, 1990] within the 0- 3-km layer appear to mark a transition region between nonsupercell and supercell storms (given that adequate CAPE and forcing are also present). However, *Davies and Johns* [this volume] found that much of the 0- to 3-km helicity associated with an extensive set of modified tornado proximity soundings occurred within the lowest 2 km. We suggest that the relevant layer for determining the significance of inflow properties should be the layer associated with entrainment into the updraft, which is probably related to the LFC for a given storm environment [*Bluestein et al.*, 1989]. Accord-

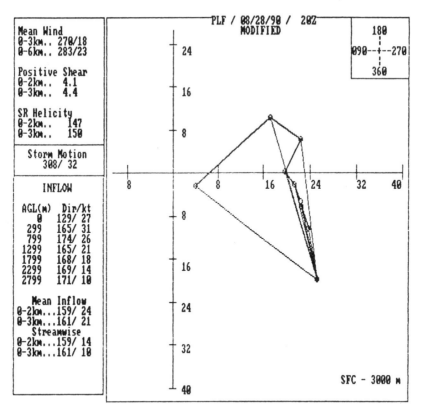

Fig. 9. Same as Figure 4 for 2000 UTC, August 28, 1990.

ingly, estimated inflow characteristics for the layer beneath the LFC were determined from the 2000 UTC PLF sounding.

For this case, the movement of the developing storm appeared to exert the strongest influence on the low-level storm environment. While the topic of storm motion is beyond the scope of this paper, some general comments can be made. The Plainfield storm appeared to deviate from the mean steering flow. However, the amount of this deviation (both direction and magnitude) depends on how "steering flow" is defined. Steering flow has been represented in the research community and operationally by the mean flow through the approximate cloud-bearing layer but also by the 0- to 6-km density-weighted mean wind. Since the observed storm motion (308°/32 knots (16 m s^{-1})) was significantly faster than the estimated 0- to 6-km mean wind over PLF at 2000 UTC (283°/23 knots (12 m s^{-1})), it might seem reasonable to suggest that the actual vertical wind profile in the environment of the tornadic storm was stronger than the interpolated vertical wind profile for 2000 UTC. However, another possibility is that the 0- 6-km layer was too shallow to represent a true steering flow for the Plainfield storm. A deep layer mean wind (approximating the depth of the cloud layer) over PLF at 2000 UTC (288°/45 knots (23 m s^{-1})) was much closer to the reported storm motion, especially after application of an adjustment factor [*Leftwich*, 1990] representing observed mean deviant supercell storm motions (i.e., storm motion at 75% of the mean wind speed and 30° to the right of the mean wind direction, 318°/34 knots (17 m s^{-1})).

For the Plainfield case it appears that the 0- 6-km mean wind was unduly affected by the lower troposphere, where wind fields were relatively light. Given the extraordinary height attained by the Plainfield storm (over 65,000 ft, or 19.8 km) and the presence of stronger winds above 400 mbar, it seems plausible that a deeper (cloud layer) mean wind would better represent the steering flow for this environment. In any case, storm motion can be influenced by a number of additional factors, including the dynamical structure (and evolution) of the storm and external features (e.g., boundaries, etc.) with which the storm may interact [*Doswell et al.*, 1990]. Since the relative importance of factors affecting storm motion for a particular environment are still unclear, we suggest that the important issue (from an operational point of view) is not the deviation of a storm from some perceived steering flow (aside, perhaps from the association of supercell characteristics with apparent deviant storm motion). Rather, we believe it is more important operationally to recognize the impact that an observed storm motion can have on the evolution of possible storm type. Since storm inflow characteristics for a given environment can be maximized within a certain range of possible storm motions (as they were with the Plainfield storm), it is important to recognize the limits of this range. Given a reasonably esti-

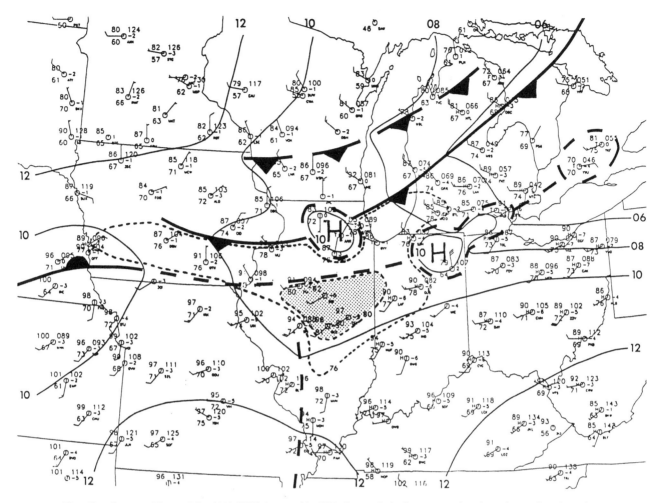

Fig. 10. Same as Figure 6 for 2100 UTC August 28, 1990. Large dashed contours also show the outflow boundary (northeastern Illinois) associated with the Plainfield storm.

mated vertical wind profile, the interactive features of the SHARP Workstation can be used to establish a range of storm motions representing the greatest threat for supercell development. If the observed storm motion associated with developing convection falls within this "higher threat" range of motions, and additional analyses support a growing severe weather threat, the storm environment may be changing to (or may have changed to) a state favoring supercell growth.

The 2000 UTC August 28 PLF sounding indicated marginal helicity (165 m^2 s^{-2}) from the surface to the LFC (approximately 0–1 km). However, we are comparing 0- to 1-km helicity values with 0- to 3-km rotation thresholds. Since we are measuring these "marginal" values over a relatively shallow inflow layer, it is possible that they represent more than marginal potential. *Lazarus and Droegemeier* [1990] suggested that storm rotation and longevity depend not only on helicity but also on the specific combination of storm inflow and shear vorticity producing the helicity. Accordingly, the mean inflow (around 15 m s^{-1}) within the first kilometer appears to be highly significant.

Of course, the helicity and shear values presented here represent only an approximation, using data from widely spaced upper air sites. The potential for supercell development within the environment "experienced" by the moving Plainfield storm may have been enhanced considerably by storm scale processes involving intense convection and/or by the interaction of intense convection with external features (e.g., boundaries, etc.). Still, it is noteworthy that computed helicity and inflow through the layer representing the updraft entrainment region of the Plainfield storm were within a range compatible with storm rotation.

5.2. CAPE and Storm-Relative Helicity

Although strong mesocyclones often are associated with large values of helicity, there appears to be a relationship between helicity and CAPE that contributes to the formation of strongly rotating updrafts for any particular environment [*Lazarus and Droegemeier*, 1990]. As a result, intense rotating updrafts can form with relatively weak instability if

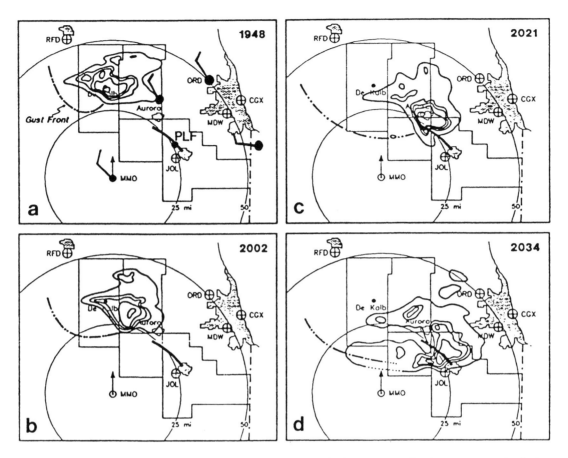

Fig. 11. Tracings of the reflectivity patterns from the Marseilles, Illinois (MMO) PPI film for (a) 1948 UTC, (b) 2002 UTC, (c) 2021 UTC, and (d) 2034 UTC, August 28, 1990. All panels represent VIP levels 2–6 at an elevation angle of 0.5° [after *National Oceanic and Atmospheric Administration*, 1990]. Bold contour indicates tornado track. Stations in Figure 11a show wind velocity (1/2 barb = 2.5 m s^{-1}, full barb = 5 m s^{-1}). Gust front associated with the tornadic Plainfield storm is depicted by contour alternating between dashes and two dots.

high values of helicity are supported by strong inflow. On the other hand, marginal helicity still appears to support storm rotation if CAPE and inflow characteristics are sufficiently large. *Johns et al.* [this volume] examined tornado proximity soundings for an extensive data set involving strong and violent tornado occurrences to observe the relationship between low-level helicity and CAPE. The distribution of data presented in that study indicates that progressively stronger (weaker) 0- to 2-km helicity values are associated with strong and violent tornado development as the CAPE decreases (increases). Further, for any particular value of CAPE there appears to be a range of helicity values compatible with the initiation of strong and violent tornadoes. Accordingly, the *Johns et al.* [this volume] data support a possible relationship between CAPE and helicity. The pretornadic environment in the vicinity of PLF exhibited marginal helicity through the storm inflow layer (165 m^2 s^{-2}), but the thermodynamic contribution of instability was extraordinary (possibly near 7000 J kg^{-1}). On the basis of these findings, it appears that high values of CAPE and strong low-level, storm inflow may have been sufficient to foster supercell development within the marginally sheared Plainfield environment.

5.3. The Supercell Environment

Tornadic supercells can develop in a wide variety of environments, some of which do not exhibit "clear-cut" supercell potential (i.e., highly baroclinic systems with associated strong wind shear; see *Doswell et al.* [1990] for a discussion of supercell environments). Of course, the determination of low-level inflow characteristics is more reliable when significant baroclinicity contributes strong vertical wind shear through a deep atmospheric layer (because strong synoptic scale features can often be observed by operational surface and upper air networks). However, tornadic supercells also can develop when the synoptic scale indicates weak or marginal shear, especially if the environment is characterized by anomalously high CAPE. Under such conditions, dynamical processes associated with in-

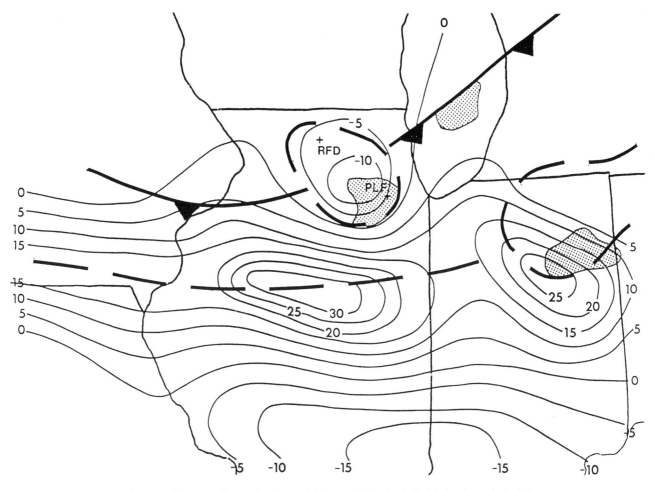

Fig. 12. Same as Figure 7 for 2100 UTC August 28, 1990. Lifted index is not included.

tense convection (e.g., storm-scale rotation, vertical pressure gradient accelerations, etc.) and external features (e.g., surface pressure troughs, thermal/moisture boundaries, wind shift/convergence lines, fronts, etc.) with which a developing storm may interact can greatly enhance tornadic potential on the storm scale (see *Doswell et al.* [1990] for a discussion of internal and external storm processes; also see *Maddox et al.* [1980] for a discussion involving thunderstorm/thermal boundary interactions), where critical information is often missing. Consequently, mesoscale environments that can support supercell characteristics will continue to escape detection (especially when significant low-level shear is localized).

5.4. Summary of the Plainfield Environment

The prestorm environment near Plainfield suggested limited tornadic potential at 1200 UTC on August 28. The large-scale pattern indicated a potential for severe thunderstorm development, but a "bulk" measure of the 0- 6-km bulk was rather weak relative to the large thermodynamic instability. Consequently, the "measurable" environment favored nonsupercell storms, and the main threat appeared to be from large hail and damaging straight-line wind. However, the environment changed significantly after 1200 UTC, with CAPE increasing to almost 7000 J kg^{-1} prior to the tornadic event. The vertical wind shear also increased during this period, although the amount of increase remains uncertain. Of course, VAD (velocity azimuth display) wind profiles and wind profilers would have provided critical prestorm information regarding important changes in the vertical wind profile across the region affected by the Plainfield storm.

Satellite imagery, surface observations, and ADAP fields revealed a complex pattern of surface features across northern Illinois prior to the tornado. The intensifying tornadic storm appeared to interact with several converging lines of cumulus as it approached Plainfield, but the observing network was not sufficiently dense to resolve any specific cell/boundary interactions. The most significant "measurable" influence on the evolution of the low-level storm environment was related to characteristics of the observed storm motion (and associated gust front), which resulted in strong low-level inflow and significant low-level vorticity (despite a northerly component to the surface wind field of the mesoscale environment ahead of the gust front). Although 0- to 3-km helicity was only marginally compatible with storm rotation prior to the tornado, it appears that high values of CAPE and strong inflow characteristics within the lowest kilometer may have been sufficient to promote supercell development. Finally, the tornadic potential of the developing storm may have been enhanced by storm scale processes involving the internal properties associated with intense convection and/or by the interaction of intense convection with external environmental features.

6. Forecast Implications

The central theme of this paper emphasizes the importance of storm-relative inflow characteristics, especially for large-scale environments not clearly indicating tornadic potential. Since properties of the inflow are largely determined by the movement of a storm through the vertically sheared environment, it is important to evaluate the range of storm motions that will maximize low-level inflow characteristics for a given environment and to recognize the impact of a given storm motion on the evolution of possible storm type for that environment.

Supercell storms can develop in a wide variety of environments (some of which are not obviously supercell situations). As a result, tornadic supercell development may still be possible when large-scale conditions indicate weak or marginal shear, especially if the environment is characterized by anomalously high CAPE. Since the processes contributing to the development of tornadic supercells in high-CAPE environments are not well understood, it would seem reasonable to view such environments with an extra measure of caution, especially if cell/boundary interactions are possible.

Finally, if the large-scale pattern indicates a potential for severe thunderstorm development in a high-CAPE environment, the mesoscale and storm scale possibilities become very important. Although storm scale processes and interactions are not resolved by conventional observing networks, forecasters must recognize the potential within a particular environment for these storm scale processes to occur. If the large-scale environment shows less than favorable conditions for the development of tornadic supercells, an awareness of storm scale potential can increase the state of readiness for important radar signatures, spotter reports, etc.

Acknowledgments. The authors wish to thank Bob Johns and Steve Weiss, NSSFC, for carefully reviewing this manuscript and offering many insightful suggestions. Thanks also to Charles A. Doswell III for guidance; to Jeff Waldstreicher, NWS Eastern Region Scientific Services Division, for reviewing our presentation; and to Brian Smith, NSSFC, for providing satellite and radar pictures of the event. Additional thanks to the editors for patience and to Deborah Haynes for initial manuscript preparations.

References

Bluestein, H. B., E. W. McCaul, Jr., G. P. Byrd, G. R. Woodall, G. Martin, S. Keighton, and L. C. Showell, Mobile sounding observations of a thunderstorm near the dryline: The Gruver, Texas storm complex of 25 May 1987, *Mon. Weather Rev.*, *117*, 244–250, 1989.

Bothwell, P. D., Forecasting convection with the AFOS data analysis program (ADAP-version 2.0), *NOAA Tech. Memo. NWS SR-122*, 91 pp., Sci. Serv. Div., Natl. Weather Serv., S. Reg., Fort Worth, Tex., 1988.

Brooks, H. E., and R. B. Wilhelmson, The effects of low-level hodograph curvature on supercell structure, in *Preprints, 16th Conference on Severe Local Storms*, pp. 34–39, American Meteorological Society, Boston, Mass., 1990.

Davies, J. M., On the use of shear magnitudes and hodographs in tornado forecasting, in *Preprints, 12th Conference on Weather Forecasting and Analysis*, pp. 219–224, American Meteorological Society, Boston, Mass., 1989.

Davies, J. M., and R. H. Johns, Some wind and instability parameters associated with strong and violent tornadoes, 1, Wind shear and helicity, this volume.

Davies-Jones, R. P., Streamwise vorticity: The origin of updraft rotation in supercell storms, *J. Atmos. Sci.*, *41*, 2991–3006, 1984.

Davies-Jones, R. P., and D. W. Burgess, and M. Foster, Test of helicity as a tornado forecast parameter, in *Preprints, 16th Conference on Severe Local Storms*, pp. 588–592, American Meteorological Society, Boston, Mass., 1990.

Doswell, C. A., III, A review for forecasters on the application of hodographs to forecasting severe thunderstorms, *Natl. Weather Dig.*, *16*, 2–16, 1991.

Doswell, C. A., III, A. R. Moller, and R. W. Przybylinski, A unified set of conceptual models for variations on the supercell theme, in *Preprints, 16th Conference on Severe Local Storms*, pp. 40–45, American Meteorological Society, Boston, Mass., 1990.

Hart, J. A., and W. D. Korotky, The SHARP Workstation version 1.5, A Skew-t/Hodograph Analysis and Research Program for the IBM and compatible PC, 58 pp., *NOAA Eastern Reg. Comput. Programs*, Natl. Weather Serv., Bohemia, NY, N. Y., 1991.

Johns, R. H., J. M. Davies, and P. W. Leftwich, Some wind and instability parameters associated with strong and violent tornadoes, 2, Variations in the combinations of wind and instability parameters, this volume.

Lazarus, S. M., and K. K. Droegemeier, The influence of helicity on the stability and morphology of numerically simulated storms, in *Preprints, 16th Conference on Severe Local Storms*, pp. 269–274, American Meteorological Society, Boston, Mass., 1990.

Leftwich, P. W., Jr., On the use of helicity in operational assessment of severe local storm potential, in *Preprints, 16th Conference on Severe Local Storms*, pp. 306–310, American Meteorological Society, Boston, Mass., 1990.

Maddox, R. A., L. R. Hoxit, and C. F. Chappell, A study of tornadic thunderstorm interactions with thermal boundaries, *Mon. Weather Rev.*, *109*, 171–180, 1980.

National Oceanic and Atmospheric Administration, The Plainfield/Crest Hill Tornado, Northern Illinois, August 28, 1990, *Nat. Disaster Surv. Rep.*, 119 pp., Natl. Weather Serv., Silver Spring, Md., 1991.

Weisman, M. L., and J. B. Klemp, The dependence of numerically simulated convective storms on vertical wind shear and buoyancy, *Mon. Weather Rev.*, *110*, 504–520, 1982.

Characteristics of East Central Florida Tornado Environments

BARTLETT C. HAGEMEYER AND GARY K. SCHMOCKER

National Weather Service Office, Melbourne, Florida 32935

1. INTRODUCTION

The diagnosis of the dynamic and thermodynamic structure of the environment in which tornadic thunderstorms develop, as well as potential mechanisms to initiate intense convection, are crucial to severe storm forecasting. Over the years many researchers such as *Beebe* [1958], *Darkow* [1969], and *Taylor and Darkow* [1982] have used mean upper air sounding data to investigate the structure of the atmosphere in proximity to tornadic thunderstorms. Recent studies indicate renewed interest in the use of tornado proximity soundings to relate thermodynamic and dynamic variables to tornado intensity [*Riley and Colquhoun*, 1990; *Johns et al.*, 1990] and the examination of the structural characteristics and evolution of different types of tornado proximity soundings [*Schaefer and Livingston*, 1988, 1990].

Characteristically, these studies combine data collected from a large area of the country and contain little data from peninsular Florida. *Byers and Rodebush* [1948] recognized the uniqueness of the peninsular Florida environment and the need for investigation of dynamic mechanisms that result in a United States maximum of thunderstorms in central Florida. Recently, *Golden and Sabones* [1991], using Doppler radar and mesonet data, investigated two tornadic waterspouts near Cape Canaveral in east central Florida. However, a specific, systematic study of the environment of central Florida tornadoes has not been done despite their significance and relatively frequent occurrence [*Kelly et al.*, 1978]. Past work has been primarily confined to case studies of tornadoes in south Florida in the wet season. *Gerrish* [1967] produced a mean tornado sounding for Miami for the wet season, and case studies of wet season tornadoes in south Florida were investigated by *Hiser* [1968], *Gerrish* [1969], *Golden* [1971], and *Holle and Maier* [1980].

These researchers have documented that low-level convergence boundaries, particularly intersecting outflow boundaries, are a triggering mechanism for tornadic thunderstorms in the wet season. *Wakimoto and Wilson* [1989] have proposed a model to explain nonsupercell tornado development in which low-level boundaries play a crucial role in tornadogenesis. This theory may have important applications in central Florida.

The tracking of boundaries with Doppler radar, satellite, and mesonet data aids in the short-term prediction of potentially tornadic thunderstorms, but few boundaries, or boundary intersections, actually result in tornadic development [*Holle and Maier*, 1980]! More information concerning the characteristics of the overlying dynamic and thermodynamic structure in these wet season situations is needed if forecasters are to have much success in assessing the tornado threat.

Over all of central Florida most of the strong and violent tornadoes (F2–F5 [*Fujita*, 1981]) occur in the dry season, and a majority of these occur in the morning [*Schmocker et al.*, 1990]. A more complete understanding of the environment of these tornadoes, which are responsible for most of the deaths and injuries in central Florida, is also needed.

The current situation is one where tornado forecast techniques and conceptual models developed from mean upper air data over the Great Plains and Midwest are applied in central Florida with limited success. This investigation consists of the determination of the mean atmospheric structure of two significant central Florida tornado environments. The results presented should lead to improved severe weather forecasts for central Florida.

2. METHODOLOGY OF CASE SELECTION

This study is unique in that the area of investigation was restricted to the 10 county warning area (CWA) of future National Weather Service (NWS) weather forecast office (WFO) Melbourne in east central Florida. The location of the Melbourne CWA and upper air stations used in this study are shown in Figure 1. On the basis of a climatological investigation of this area by *Schmocker et al.* [1990], tornado characteristics were divided into two seasons: dry season (November through April) and wet season (May through October). The hourly distribution of central Florida tornadoes by season (Figure 2) clearly shows a diurnal afternoon

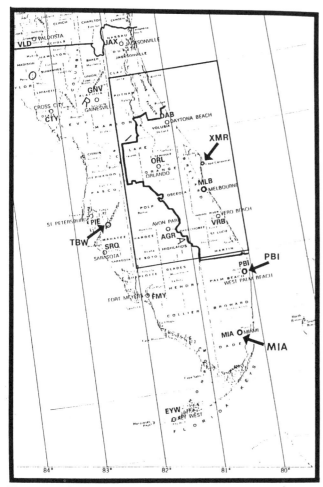

Fig. 1. Peninsular Florida. The location of the future WFO Melbourne 10 county warming area is enclosed by a bold line in east central Florida. Upper air stations are indicated by bold arrows: Cape Canaveral (XMR), Miami (MIA, moved to PBI in 1977), Tampa Bay (TBW), and West Palm Beach (PBI).

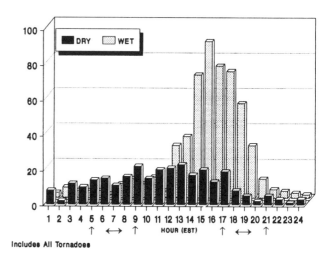

Fig. 2. Hourly distribution of dry season and wet season tornadoes within 125 nautical miles of future WFO Melbourne. The time intervals of this studies cases (1200 UTC (0700 EST) and 0000 UTC (1900 EST) ±2 hours) are shown by the arrows along the x axes.

maximum in the wet season while the dry season is characterized by fewer tornadoes but a more even hourly distribution with significant morning activity.

The distributions of F2–F5 tornadoes and those that have caused injury and death specifically in future WFO Melbourne's CWA are shown in Figure 3. After peaking in March and April, strong tornado activity drops off sharply in May, as the influence of vigorous mid-latitude disturbances diminishes and the transition from dry to wet season takes place. The increase in August and September is due to tornadoes associated with tropical cyclones. Climatological analyses indicate that an attack on the problem by seasonal and diurnal divisions is necessary.

Upper air data were available for 0000 and 1200 UTC for TBW, PBI, and MIA (note that the MIA site was deactivated and moved 100 km north to PBI in 1977). Only 1200 UTC soundings were available for XMR since 1980. With these data limitations in mind, tornado candidate cases were considered by looking at first spatial, then temporal, selection criteria. Printouts of all tornadoes reported in Florida supplied by the Verification Section, National Severe Storms Forecast Center (NSSFC), were reviewed for cases of tornado touchdowns in the Melbourne CWA (Figure 1) that occurred within ±2 hours of standard observation times (0000 UTC (1900 EST) and 1200 UTC (0700 EST)) from 1980 to 1988. This resulted in 16 0000 UTC cases and seven 1200 UTC cases and reflects the fact that fewer tornadoes occur at 1200 UTC than at 0000 UTC (see Figure 2). Dry season morning tornadoes often are stronger than wet season tornadoes, so five additional 1200 UTC cases were added by searching the tornado records back to 1975 to get a larger sample. Twenty-eight candidate cases were thus identified (16 0000 UTC and 12 1200 UTC) and soundings obtained. Six soundings were removed because of poor data quality and/or contamination by deep convective. Four tropical cyclone cases were also removed from this data set. This selection criterion yielded nine 1200 UTC dry season cases and nine 0000 UTC wet season cases.

The upper air station nearest to a reported tornado was designated the proximity sounding for each season. Since 0000 UTC wet season proximity data were not available for Cape Canaveral, but 1200 UTC data were, a 1200 UTC XMR wet season precedent sounding (12 hours prior to tornado touchdown, ±2 hours) was examined to provide useful information on tornado precursor conditions. Soundings were then pressure averaged at 50-mbar intervals to 200 mbar. Mean soundings and diagnostic parameters for the seasonal atmospheres were computed, and vertical profiles

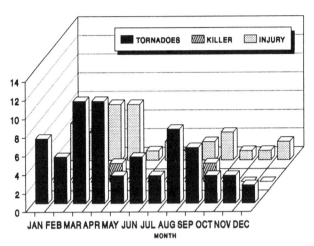

Fig. 3. Monthly distribution of strong and violent (F2–F5) tornadoes, and killer and injury tornadoes reported within future WFO Melbourne's CWA.

of temperature (T), dew point (T_d), and wind components (U and V) were constructed for dry and wet season tornado proximity and wet season tornado precedent, atmospheres. These seasonal profiles were then compared to seasonal mean atmospheres.

3. Results

3.1. Case Tornado Characteristics

Of the nine dry season cases, five had F2 tornadoes, two had F1 tornadoes, and two had F0 tornadoes. Two cases had multiple tornadoes. Of the nine wet season cases, one had multiple tornadoes, four had F1 tornadoes, and five had F0 tornadoes. The mean time of occurrence was 1220 UTC (0720 EST) and 2240 UTC (1740 EST) for the dry and wet season cases, respectively.

3.2. Mean Soundings and Diagnostic Parameters

Skew-T, Log-P plots and hodographs of the dry and wet season mean tornado proximity soundings are shown in Figure 4. A summary of diagnostic parameters derived from the mean soundings are shown in Table 1.

The general thermodynamic structure of the mean dry season sounding (Figure 4) has both similarities and differences when compared with the classic Midwestern tornadic environment [*Doswell*, 1982]. Both environments display a pronounced deep, dry layer in the middle and upper levels overlaying a moist layer. However, the Florida dry season profile exhibits a deeper moist layer than is the case with the "classic" profile. The most notable difference is the lack of any mean capping inversion or steep lapse rate overlying a well-mixed moist layer. An inspection of all individual soundings in the data set revealed the presence of a few minor inversion layers, but none of the low-level moist layers were capped by an inversion.

Dry season stability indices are less than for typical Midwestern tornado cases [*Miller*, 1972]. This is probably because midtropospheric temperatures are generally warmer over Florida; also, because these are morning tornado cases, there is not much contribution to destabilization from diurnal heating.

The dry season hodograph shows strong shear in the lower levels and winds veering with height. This is similar to the mean F1 and F2 tornado hodographs documented by *Riley and Colquhoun* [1990]. A comparison of the wind profile with *Miller*'s [1972] key tornado forecast parameters reveals that the low, middle, and upper level jets of the mean dry season soundings are all in the strong category.

The wet season mean proximity sounding and hodograph (Figure 4) are unlike any of the classic tornado environments [*Miller*, 1972] and represent a regional hybrid. There is a general similarity to the mean dry season sounding in that a distinct dry layer overlies the moist layer; thus it is most unlike *Newton*'s [1980] type C and *Miller*'s [1972] type II for the Gulf Coast and southeastern regions which are typified by high moisture through the troposphere.

The wet season proximity hodograph is quite different from the dry season. It exhibits very weak shear in the lower levels, and winds are nearly unidirectional from the west. Tornadic thunderstorms forming in this type of an environment would likely be of the nonsupercell variety [*Wakimoto and Wilson*, 1989].

3.3. Relative Atmospheric Profiles

To determine, what, if anything, is unusual about the dry and wet season proximity and wet season precedent tornado atmospheres profiles of potential temperature θ, wet-bulb potential temperature θw, U, and V were computed and compared to seasonal means [*U.S. Department of Defense*, 1983].

3.3.1. *Dry season.* Comparisons of mean dry season profiles to adjusted seasonal means (average for all months with cases, January through April) are shown in Figures 5a–5d. Potential temperature values are significantly higher than mean values below 800 mbar and above 400 mbar but are only slightly above mean between about 700 and 450 mbar indicating less static stability below 700 mbar when compared to normal. More notable is θ_w greatly exceeding normal values below 650 mbar (Figure 5b). The mean atmosphere has a degree of convective instability with a θ_w minimum around 800 mbar, but the depth of the moist layer in the mean tornado atmosphere is about twice as deep, and convective instability is much greater.

The high U values of the mean tornado profile, reaching a maximum of 45 m s^{-1} at 200 mbar, are not significantly higher than seasonal means (Figure 5c). Indeed, there are only

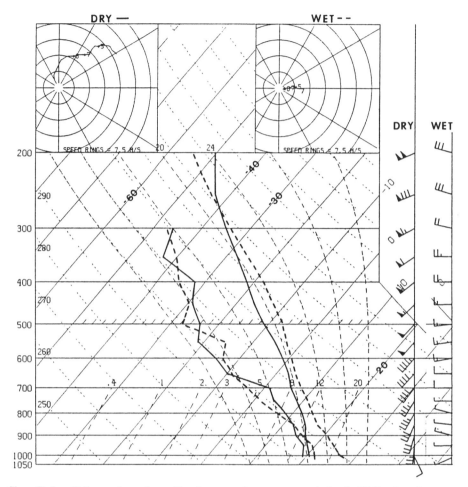

Fig. 4. Skew-T, Log-P thermodynamic profiles for mean dry season proximity (solid lines) and mean wet season proximity (dashed lines) tornado environments. Hodographs and vertical wind profiles (half wind barb 2.5 m s^{-1}, a full wind barb 5 m s^{-1}, and a pendant 25 m s^{-1} are also shown.

minor differences (<5 m s^{-1}) from the surface to 250 mbar. This is not the case with the southerly wind (V) components (Figure 5d), where mean seasonal V is nearly zero and is greatly exceeded by the mean tornado environment at all levels. Very strong shear of V is found in the lowest 100 mbar, and there is an indication of a midlevel southwest jet between 600 and 400 mbar and an upper jet at 200 mbar.

3.3.2. *Wet season.* Comparisons of the mean wet season precedent and proximity profiles to adjusted seasonal means (average for months with cases, May through July) are shown as Figures 6a–6d. Except for differences at the surface due to diurnal heating the potential temperature values of the precedent and proximity atmospheres are very close to the mean seasonal values (Figure 6a) indicating there is little difference in the vertical temperature structure between a nontornado and tornado day in the wet season. Comparisons of θ_w profiles (Figure 6b) show the depths of the moist layers are nearly the same with θ_w minima around 650 mbar in all three atmospheres. Proximity wet-bulb potential temperature values are 2° higher at the surface and the same at 650 mbar when compared to the tornado precedent sounding taken 12 hours earlier. This indicates a much greater degree of convective instability and illustrates how diurnal heating can nearly double CAPE (see Table 1) between 1200 UTC (904 J/kg) and 0000 UTC (1683 J/kg).

The most outstanding feature of the wet season tornado kinematic environment was found to be the existence of significantly increasing shear, and westerly winds greatly exceeding seasonal means, in the mid and upper troposphere (Figure 6c). The trend and magnitude of the V component is generally very close to seasonal means (Figure 6d).

All nine wet season cases had westerly winds in the mid and upper troposphere, and eight cases were westerly from the surface to 200 mbar. This dominance of westerly flow cases has several causes. *Hagemeyer* [1991] found that lower tropospheric flow is westerly over central Florida into June and that persistent easterly flow does not appear until well into the wet season. Most cases presented here oc-

TABLE 1. Mean Diagnostic Parameters

	Dry Prx	Wet Pre	Wet Prx
Freezing level (m AGL)	4104	4351	4477
Wet-bulb zero (m AGL)	3360	3410	3419
Showalter index (°C)	−0.8	−1.1	1.5
Lifted index (°C)	−1.8	−3.5	−3.4
Totals index (°C)	48	48	43
Cross totals index (°C)	23	22	19
Vertical totals index (°C)	25	26	24
K Index (°C)	33	31	26
SWEAT Index	325	194	168
Precipitable water (cm)	3.8	4.2	3.9
LCL height (m)	386	596	882
LCL mixing ratio (g/kg)	13.9	15.8	16.1
LFC height (m)	2584	2109	1484
CCL height (m)	631	1187	1236
EL temperature (°C)	−26.2	−51.1	−55.0
EL level height (m)	8150	11727	12346
Convective temperature (°C)	24.1	30.6	31.2
CAPE (J/kg)	164	904	1683
Bulk Richardson Number	1.5	166	576
Mean 0- 6-km wind (m/s)	215/19	260/06	267/04
Assumed storm motion	245/14	290/04	297/03
Absolute helicity (m^2/s^2)	250	10	06
Mean 3–10 km wind (m/s)	230/28	268/12	271/09
Mean shear 850–200 mbar (m/s)	34	14	13

Parameters were computed from mean dry and wet season proximity and from mean wet season precedent, tornado soundings. AGL is above ground level.

curred early in the wet season when westerly disturbances are more likely compared to late in the wet season when easterly flow dominates. Additionally, undisturbed easterly flow early in the wet season tends to be drier and to have a shallower moist layer than later in the wet season [*Hagemeyer*, 1991] and is thus less likely to produce strong thunderstorms and tornadoes.

There is also a bias toward west flow cases on the east coast in the ±2 hours from 0000 UTC selection criterion used in this study that can be explained by reviewing a study of spatial patterns of south Florida convection, without regard to severity, done by *Blanchard and Lopez* [1985]. They identified three basic patterns of convection over south Florida during the summer. Type I exhibits weak southeast flow with early development of convection along the East Coast Sea Breeze convergence zone (ECSB) which moves inland and merges with the West Coast Sea Breeze convergence zone (WCSB) west of the central peninsula. Type II exhibits stronger, dry, east flow with early passage of the ECSB with limited convection which is quickly advected to the west coast to merge with the WCSB and move out into Gulf. Type III exhibits southwest to northwest flow over central Florida with the westerly flow advecting the WCSB inland while the ECSB remains anchored along the east coast by ambient flow, causing stronger circulations and convergence. Outflow boundaries from the WCSB convection in the central peninsula interact with the ECSB and other outflow boundaries to enhance convection before dissipation during the evening. Type III days exhibit greater echo coverage, and dissipation takes place much later in the day, whereas on type I and II days, convective activity develops earlier and moves through east central Florida before the 0000 UTC (±2 hours) selection criteria.

The wet season tornado cases presented here are clearly westerly flow type III cases, but it is important to note that the wind speeds between 500 and 200 mbar on the mean wet season tornado precedent sounding are twice as high as for the mean type III day sounding for MIA and PBI produced by *Blanchard and Lopez* [1985].

Stronger upper level winds and shear appear to be an important factor in significant wet season tornado cases. The strong, damaging, tornado that struck Miami on June 17, 1959, developed in the evening in westerly flow, and *Hiser* [1968] found that the most outstanding synoptic weather feature at the time of the tornado was a 47 knot (24 m s^{-1}) westerly wind speed maximum at 12 km.

Although the wet season data sample is small (five F0 and four F1 cases) there does appear to be a positive relationship between higher middle and upper tropospheric wind speeds and shear and tornado strength. The maximum upper level winds associated with F1 tornadoes were not lower than 20 m s^{-1}, while values as low as 10 m s^{-1} were associated with F0 tornadoes. More research is planned on this aspect of the wet season environment.

Tornadoes and waterspouts also occur in easterly flow, but no east flow wet season cases are included in this study, because, as stated earlier, they would tend to occur in east central Florida well before 2 hours prior to 0000 UTC. Research is planned on these east flow events in the future.

4. Concluding Remarks

We have examined dry and wet season environmental profiles relative to seasonal means. However, to put this information to good use operationally, forecasters must diagnose the whole picture in detail and understand the interactions and physical processes involved to the degree that technology allows. Key factors that effect whether thunderstorms on one day may be tornadic or not depend on the characteristics of the overlying airmass and on the probable existence, and strength, of low-level triggering boundaries. While low-level boundaries are necessary for convective initiation, the overlying atmosphere plays an important role in the development of tornadic thunderstorms. NEXRAD/WSR-88D with its ability to detect, track, and possibly quantify the strength of low-level boundaries, as well as diagnose the dynamic environment at much higher spatial and temporal resolutions than the current upper air network, offers the promise of improving short-term forecasting of tornadic thunderstorms in east central Florida.

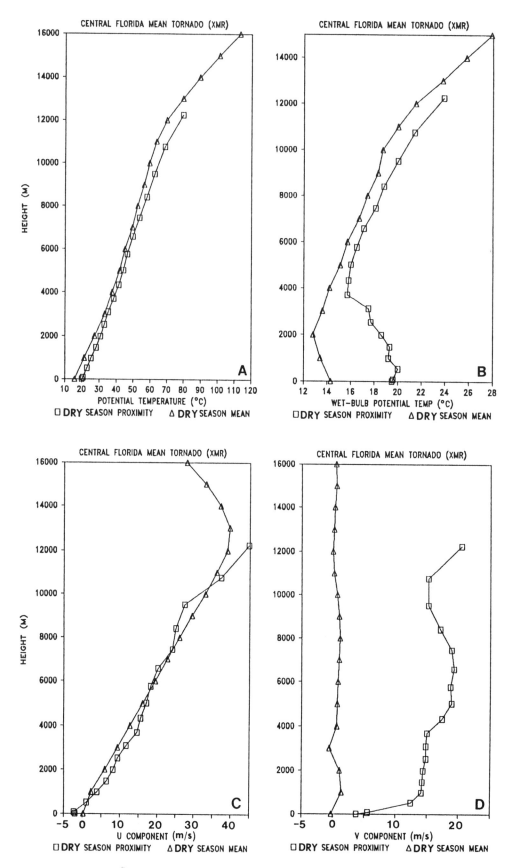

Fig. 5. Mean vertical profiles of (a) θ, (b) θ_w, (c) U, and (d) V for mean dry season tornado proximity cases and the mean dry season environment.

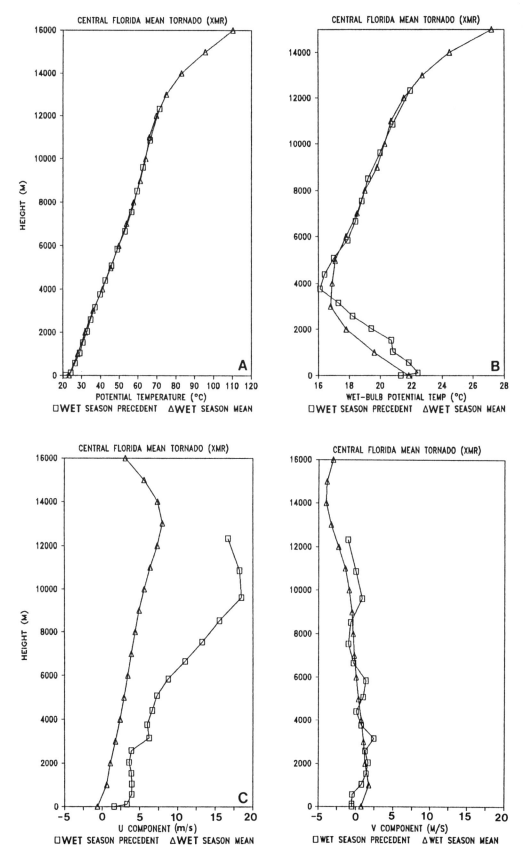

Fig. 6. Mean profiles of (a) θ, (b) θ_w, (c) U, and (d) V for mean wet season tornado precedent and proximity cases and the mean wet season environment.

Acknowledgments. Cape Canaveral data were provided by Hal Herring, Computer Sciences Raytheon Corporation. Daniel Smith provided Tampa Bay, Miami, and Palm Beach data. Barry Schwartz provided Skew-T, Log-P plots of mean soundings and computations of diagnostic parameters. Preston W. Leftwich, Jr., provided calculations of helicity. Jo Ann Carney and Karen Hileman assisted in data tabulation and figure preparation. Thanks to Joe Golden for his advice and encouragement to continue the study. Paul Hebert and Mike Sabones provided helpful reviews of the paper. Special thanks to Ron Holle, Irv Watson, and Raul Lopez for providing an extensive collection of papers relating to Florida convection for our use. Financial support for this paper was provided by the National Weather Service Southern Region, Harry Hassel, Regional Director.

References

Beebe, R. G., Tornado proximity soundings, *Bull. Am. Meteorol. Soc.*, *39*, 195–201, 1958.

Blanchard, D. O., and R. E. Lopez, Spatial patterns of convection in south Florida, *Mon. Weather Rev.*, *113*, 1282–1299, 1985.

Byers, H. R., and H. R. Rodebush, Causes of thunderstorms of the Florida Peninsula, *J. Meteorol.*, *5*, 275–280, 1948.

Darkow, G. L., An analysis of over 60 tornado proximity soundings, in *Preprints, 6th Conference on Severe Local Storms*, pp. 218–221, American Meteorological Society, Boston, Mass., 1969.

Doswell, C. A., III, The operational meteorology of convective weather, vol. 1, Operational mesoanalysis, *NOAA Tech. Memo. NWS NSSFC-5*, NaH. Severe Storms Forecast Center, Kansas City, Mo., 1982.

Fujita, T., Tornadoes and downbursts in the context of generalized planetary scales, *J. Atmos. Sci.*, *38*, 1511–1534, 1981.

Gerrish, H. P., Tornadoes and waterspouts in the south Florida area, paper presented at Army Conference on Tropical Meteorology, U.S. Army, Coral Gables, Fla., June 8–9, 1967.

Gerrish, H. P., Intersecting fine lines and a south Florida tornado, in *Preprints, 6th Conference on Severe Local Storms*, pp. 188–191, American Meteorological Society, Boston, Mass., 1969.

Golden, J. H., Waterspouts and tornadoes over south Florida, *Mon. Weather Rev.*, *99*, 146–154, 1971.

Golden, J. H., and M. E. Sabones, Tornadic waterspout formation near intersecting boundaries, in *Preprints, 25th Conference on Radar Meteorology*, pp. 178–181, American Meteorological Society, Boston, Mass., 1991.

Hagemeyer, B. C., A lower-tropospheric thermodynamic climatology for March through September: Some implications for thunderstorm forecasting, *Weather Forecasting*, *6*, 254–270, 1991.

Hiser, H. W., Radar and synoptic analysis of the Miami tornado of 17 June 1959, *J. Appl. Meteorol.*, *7*, 892–900, 1968.

Holle, R. L., and M. W. Maier, Tornado formation from downdraft interaction in the FACE mesonetwork, *Mon. Weather Rev.*, *108*, 1010–1028, 1980.

Johns, R. H., J. M. Davies, and P. W. Leftwich, An examination of the relationship of 0-2 km AGL "positive" wind shear to potential buoyant energy in strong and violent tornado situations, in *Preprints, 16th Conference on Severe Local Storms*, pp. 593–598, American Meteorological Society, Boston, Mass., 1990.

Kelly, D. L., J. T. Schaefer, R. P. McNulty, C. A. Doswell III, and R. F. Abbey, Jr., An augmented tornado climatology, *Mon. Weather Rev.*, *106*, 1172–1183, 1978.

Miller, R. C., Notes on analysis and severe storm forecasting procedures of the Air Force Global Weather Central, *Tech. Rep. 200* (revision), 190 pp., Air Weather Serv., Scott Air Force Base, Ill., 1972.

Newton, C. W., Overview on convective storm systems, in *Proceedings of CIMMS Symposium*, edited by Y. K. Sasaki et al., pp. 3–107, University of Oklahoma Press, Norman, 1980.

Riley, P. A., and J. R. Colquhoun, Thermodynamic and wind related variables in the environment of United States tornadoes and their relationship to tornado intensity, in *Preprints, 16th Conference on Severe Local Storms*, pp. 599–602, American Meteorological Society, Boston, Mass., 1990.

Schaefer, J. T., and R. L. Livingston, Structural characteristics of tornado proximity soundings, in *Preprints, 15th Conference on Severe Local Storms*, pp. 537–540, American Meteorological Society, Boston, Mass., 1988.

Schaefer, J. T., and R. L. Livingston, The evolution of tornado proximity soundings, in *Preprints, 16th Conference on Severe Local Storms*, pp. 96–101, American Meteorological Society, Boston, Mass., 1990.

Schmocker, G. K., D. W. Sharp, and B. C. Hagemeyer, Three initial climatological studies for WFO Melbourne, Florida: A first step in the preparation for future operations, *NOAA Tech. Memo. NWS SR-132*, 52 pp., Natl. Weather Serv., Fort Worth, Tex., 1990.

Taylor, G. E., and G. L. Darkow, Atmospheric structure prior to tornadoes derived from proximity and precedent upper air soundings, *U.S. Nucl. Regul. Comm. Rep.*, *NUREG/CR-2359*, 95 pp., 1982.

U.S. Department of Defense, Cape Canaveral, Florida Range Reference Atmosphere 0–70 km altitude, 203 pp., Secretariat, Range Commanders Council, White Sands Missile Range, N. Mex., 1983.

Wakimoto, R. M., and J. W. Wilson, Non-supercell tornadoes, *Mon. Weather Rev.*, *117*, 1113–1140, 1989.

Discussion

D. MCCARTHY, SESSION CHAIR

National Weather Service Forecast Office

PAPER L1

Presenter, C. A. Doswell III, National Severe Storms Laboratory [*Doswell*, this volume, Tornado forecasting: A review]

(H. Volkman, private consultant) Who's leading the effort to get SELS [Severe Local Storms; see session H, paper H6] out of the tornado and severe thunderstorm watches? We need to know in case there's some influence we can have to divert this change?

(Doswell) How honest do I want to be? [laughter] The effort is essentially one of politics and economics, the origins of which have little to do with the science of meteorology or the practical issues of forecasting the weather. Basically, it's a question of where we are to invest our limited resources. If you want to have an influence, write letters to your congressman.

(W. Bonner, COMET) I'd like to make it clear that there is no intent to destroy or do away with SELS. The intent is to take advantage of the Doppler radars and so forth to blur the distinction between watches and warnings. There should be an element of prediction with warnings, i.e., those with lead time. You often end up trying to tell simple stories, perhaps too simply in this case; the lead time with a warning overlaps with the forecast aspects of a watch. The National Severe Storms Forecast Center [NSSFC] in this story becomes much more of a mesoscale guidance center. The question is, do products go directly from NSSFC to the public, or do they go out like NMC [National Meteorological Center] products as guidance to the local offices, who then issue whatever statements are required? My own feeling was that this is an issue you could argue philosophically . . .

(Doswell) And I have.

(Bonner) Yes, you and I have discussed it. I have been beaten up by SELS lead forecasters on this, trying to represent Weather Service policy. As the modernization and restructuring demonstration takes place, this will be the test period. The issue probably will be resolved gradually.

(Doswell). My response to Bill's [Bonner's] comments is that the actual resolution of such questions often has little to do with the outcome of a test; they tend to be dominated by politics and economics. Further, I don't believe you can extrapolate the results of a test at a single location around the entire country. If you watch the Weather Channel, you see flash flood and winter storm watches issued by the local offices, the result of which I find to be a national disgrace. They're a patchwork quilt, where the forecast for a weather event walks up to a state boundary, stops, and then picks up again somewhere else. This is the classic warning and coordination problem. The severe thunderstorm and tornado watches, along with the hurricane watches, are issued from a central location. Whether they're right or not (and as we've seen, that's an interesting and complex issue), at least we're not sending conflicting signals to the public. I think there's a good argument in favor of the watch responsibility staying in the hands of SELS.

PAPER L2

Presenter, J. M. Davies, private consultant [*Davies and Johns*, this volume, Some wind and instability parameters associated with strong and violent tornadoes, 1, Helicity and mean shear magnitudes]

(R. Davies-Jones, National Severe Storms Laboratory) Helicity seems to work well in the wintertime but not quite so well in the summertime. You get tornadoes in the summertime with high CAPE [convective available potential energy] and low helicity. There may be a sampling problem, but such events with low helicity do seem to occur. Have you found this?

(Davies) Yes, I've found that, so far. That's been my general experience. We have to get better with our observation networks for nowcasting purposes.

(Davies-Jones) Maybe the storms are different from the wintertime storms.

(Davies) That's a very good point. I think that there are storms which occur with relatively low helicity. The

The Tornado: Its Structure, Dynamics, Prediction, and Hazards.
Geophysical Monograph 79
This paper is not subject to U.S. copyright. Published in 1993 by the American Geophysical Union.

hodographs are still in a nice, curved shape, but it's a small curve, and you have strong buoyancy. I'd like to have some of the modelers look at that situation and see what you get with low helicity (say, 150 m^2 s^{-2}) and high CAPE (say, 5000 J kg^{-1}). That's the opposite of what you get in the cool season, with very low CAPE and strong helicity.

(Davies-Jones) One more comment: the layer over which you compute helicity doesn't really matter very much, as long as it's deep enough.

(Davies) I was wondering if there was much importance to the vertical distribution of the helicity (or the shear).

(Davies-Jones) The lower it is, the better it is.

Paper L3

Presenter, R. H. Johns, National Severe Storms Forecast Center [*Johns et al.*, this volume, Some wind and instability parameters associated with strong and violent tornadoes, 2, Variations in the combinations of wind and instability parameters]

(J. Jarboe, National Weather Service Training Center) Since helicity is so dependent on the storm motion vector, I was wondering if you could integrate your data with the observations of Ted Fujita. He's flown down many tornadic storm paths, and it would seem you could get an accurate storm motion from his data.

(Johns) Part of the problem with that is not many storms in this data set involve long-track tornadoes, and so most of them have not been surveyed directly. Big outbreaks are surveyed, of course, but I'm not certain if we could ever get much more path information for tornadoes that way. What we can do, now that we in SELS are getting radar data, is to collect the radar data and track the cells directly. We eventually should be able to build up to a good data set that way.

Open Discussion

C. DOSWELL, SESSION CHAIR

National Severe Storms Laboratory

PRESENTATION 1

Presenter, P. Desrochers (Air Force Geophysics Laboratory): Determination of which mesocyclones will be tornadic, using single Doppler radar observations of a new parameter: Excess rotational kinetic energy (ERKE)

(C. Anderson, North Carolina State University) Don Burgess showed yesterday that as range from the radar increases, the radar sees higher and higher into the storm. You seem to be saying that mesocyclones high in the storm will determine whether or not that storm will produce a tornado. In the Huntsville [Alabama] tornado [November 15, 1989] the storm produced a mesocyclone for 2 hours before the Huntsville tornado; it produced funnel clouds and hail, but it certainly didn't produce a tornado within 20 minutes. I don't believe your predictions can be as accurate as you have shown, especially with respect to false alarms.

(Desrochers) We're keying in here on mesocyclones at low levels...

(Anderson) How are you going to see at low levels at ranges over 100 km?

(Desrochers) With violent tornadoes, we see that low-level intensification is reflected at mid-levels, also.

(Anderson) I don't see that intensification at mid-levels necessarily means you're getting intensification at low levels.

(Desrochers) No, that's not what we're saying. It's not at mid-levels that it's important, but at low levels. Ed Brandes showed a good figure yesterday of a tornadic storm where the mesocyclone at low levels intensified up to about 4 km. The low-level intensification is reflected at moderate heights within the storm and that's what we're detecting, and why we get the lead times we do at these ranges.

(Anderson) I'm interested in your results and I'd like to see a lot more data. You've got maybe 20 cases; I'd like to see 100 cases.

(Desrochers) Yes, we would, too. That's what we're working on.

(D. Burgess, National Severe Storms Laboratory) I have a concern about the algorithms at their current level of development. There are things that we might not expect and might be out there when we collect a lot of data. One of these is cyclonically rotating downdrafts. We haven't really considered that in the past in algorithm development. They have been observed, a few in Oklahoma, many more in Colorado. A human observer, as he looks to identify the updraft region in a storm, isn't going to be confused by a cyclonic downdraft, but an automated algorithm, without developing some code to approximate what a human does, may not recognize that. There are potentially a lot of such problems, and we need to remember that in this stage of algorithm development.

(P. Joe, Atmospheric Environment Service of Canada) I don't have a problem with ERKE; I think it points out that a change in the algorithm is needed. I do echo Charlie Anderson's comment. Thirty mesocyclones out of 8000 tornadoes per year is a pretty limited data set, especially when it comes from Oklahoma, where it's highly biased.

(Desrochers) I would agree. What can you say about a small data set? When you have consistency, though, you can get a good feeling for it. We haven't been let down yet, and we're hoping it will continue. Maybe we'll be surprised.

PRESENTATION 2

Presenter, M. Weisman (National Center for Atmospheric Research): Some cautionary comments on the use of helicity to assess supercell potential, and the use of the time changes in storm-relative helicity

(R. Davies-Jones, National Severe Storms Laboratory) I agree with Morris that helicity is intended as a nowcast parameter. If you have a lot of helicity to begin with, using the initial storm motion, the potential is high for a severe mesocyclone. Most mesocyclones don't produce a tornado for over an hour. Once the mesocyclone gets cranking, the more the storm deviates in its motion. Storm rotation and motion are interlocked, as is well known.

(Weisman) I think my concern with this is with a few of the formal NEXRAD products, where the only indication of helicity for nowcasting is a number in a corner of the display, as opposed to consideration of the processes that go into the production of severe weather. I'm advocating displaying a hodograph, not just a number.

(Davies-Jones) At the Weather Service in Norman, now, we have an interactive program where the forecaster can update the storm motion and see a revised estimate of helicity. We plan to be able to change the hodograph winds as well, so the hodograph can be amended with new wind information, such as profiler winds, VAD [velocity azimuth display] winds, new surface wind observations, and model forecast winds.

(Chair) I agree with Morris, in the sense that we should not expect a single number calculated from the sounding to give us anything but a poor representation of the total information content of the data. I certainly hope that helicity will not become the next Lifted Index, which everybody uses mindlessly. [applause]

(J. Davies, private consultant) I think helicity can be useful in forecasting as well as nowcasting. Gary Woodall has been circulating a program with the National Weather Service that calculates helicity density using LFM [limited-area fine-mesh model] winds, that has shown some utility. Of course, the LFM is questionable in a lot of cases. I've adapted his program to my system at home and I've found some usefulness for the helicity forecasts based on the model winds and some rough estimates of storm motion. If helicity values using most storm motions are high, this suggests a stronger probability of tornadoes. You're trying to anticipate the helicity as the situation changes with time.

(Weisman) I agree and I want to stress again that it is the change in helicity over the life of the storm that is most important. It's when the storm changes its motion that you get really concerned.

(M. Leduc, Atmospheric Environment Service of Canada) I agree very much with the previous comment. It's a matter of continuous testing, using hourly data, pireps, or whatever. You're sort of pretuned to watch for certain things to happen on radar; when it happens, then you're ready to jump. You're going to be testing all the time for every different area and cell. With the King Doppler radar, we'll have an interactive program to look at the hodograph and a variety of shear parameters.

(S. Goodman, NASA) The Huntsville, Alabama, tornado case [November 15, 1989] pointed out that there was a lot of conflicting or imperfect information getting to the forecasters who had to put out the warning. Someone asked Chuck Doswell on the first day "If 50% of all mesocyclones put out tornadoes, do you put out tornado warnings with detected mesocyclones?" It appears that the public responds more to tornado than to severe thunderstorm warnings, so if you have indications from the time rate of change in the data that tornadoes are more likely, then that would be a useful nowcasting tool.

Presentation 3

Presenters, R. Wilhelmson and L. Wicker (National Center for Supercomputing Applications and University of Illinois): Modelling of long-lived, low-level mesocyclonic and tornadic vortices in supercells

(C. Anderson, North Carolina State University) I wanted to ask Bill McCaul about this earlier. This pressure-perturbed updraft goes against what we learned earlier in numerical modelling, where the vertical perturbation pressure gradient opposes the buoyancy and reduces the vertical velocity. I'm not sure how this comes out of the models.

(E. W. McCaul, Universities Space Research Association) The pressure perturbations would tend to counteract the buoyancy if the environment had no shear. In a sheared environment, air parcels can be accelerated upwards by this perturbation pressure if the patterns of buoyancy and shear are related in the vertical in just the right way. It's most noticeable when there's not that much environmental buoyancy available. Lou Wicker has been using CAPEs [convective available potential energy] of around 2900 J kg^{-1}; I've done a series of simulations with CAPEs of around 800 J kg^{-1}. I also see dramatic differences in storm behavior, depending on the vertical distributions of the buoyancy and shear. You can get steady updrafts as weak as 6 m s^{-1} and as strong as 25 m s^{-1}, depending on how shear and buoyancy are related in the vertical.

(Anderson) Is that set of conditions rare, or is it common? How does that fit in the spectrum of mesocyclones?

(McCaul) Convection in shear is extremely common. Mesocyclones only form when there's pretty strong shear available. Vertical distribution variations need a lot more study. I think there are a lot of cases where buoyancy is weak, but it's located in the lower part of the troposphere where the vertical shears are strongest. That can promote strong storms in situations where it might not be expected.

(Chair) I'm delighted to see this result, since it suggests that CAPE as a number is not so important as the vertical distribution of buoyancy. It pounds another nail in the particular coffin I'm trying to build. [laughter]

(M. Leduc, Atmospheric Environment Service of Canada) I've seen that, specifically, with instability. Last summer, we saw three squall lines with Lifted Indices of 0 or -1, but virtually dry adiabatic below about 750 mbar and lots of shear. They all three produced long-lived squall lines, and we kept thinking that this shouldn't happen with this little instability. We keep seeing this happen.

(H. Brooks, National Severe Storms Laboratory). Charlie, I think I have an answer to your question. We've been looking at the accelerations on parcels that end up as the maximum updraft, and also the maximum buoyancy parcel (since the maximum buoyancy ends up displaced from the maximum updraft), in a sheared environment. One of the things we find is that for the maximum updraft parcels, at low levels (below 4 km), the majority of the acceleration is due to the vertical pressure gradient. The parcels that go through the maximum buoyancy location, on the other hand, encounter an adverse pressure gradient over most of their trajectory. Helicity in this case is a measure of the amount of curvature in the hodograph. For 21 simulations using the same thermodynamic profile, considering the maximum updraft in each simulation, in those storms with more curvature in the hodograph, the stronger the updraft. It's not just the shear, because we see different behavior with storms that have the same amount of linear shear but different curvature. In the Beltrami model, the amplification due to the circular hodograph over a kilometer depth is about 2.2 times the square root of your initial updraft perturbation, from whatever means. In a sheared environment, the vertical pressure gradient acceleration can contribute 50% or more of the total updraft.

(R. Rotunno, National Center for Atmospheric Research) Charlie, this is what I was emphasizing in my review paper. In a sheared environment, there are lifting pressure gradients on the storm flanks. When observed, it is found that the buoyancy is actually negative at cloud base. This requires low pressure somewhere at mid-levels to lift the air to its level of free convection. In some situations, this lifting pressure gradient can be quite substantial.

(Anderson) Everyone is jumping on me for making that comment. [laughter] If you have an outbreak of tornadoes with, say, five tornadic ells, that outbreak may be accompanied by 20 or more nontornadic cells. If you want to make that argument about the hodograph, why aren't these others being affected by the same conditions?

(Rotunno) I don't know, but they can't all make it.

(D. Burgess, National Severe Storms Laboratory) Charlie, I've thought about that. My answer is that the atmosphere is not homogeneous. In the model the atmosphere is homogeneous, and perhaps every storm initialized that way would behave the same way. But out there where the outbreak is taking place, that atmosphere is not homogeneous, and that's got to be the answer to your question.

(Chair) Furthermore, I'd argue that when we see big outbreaks, most of the cells are supercells and most do produce tornadoes, at least in the big ones. "Most" does not mean "all," of course.

Presentation 4

Presenter, A. Siemon (State University of New York at Albany): Unusual aspects of the Plainfield, Illinois, tornadic storm [August 28, 1990], including its cloud-to-ground lightning characteristics

(J. Jarboe, National Weather Service Training Center) I've done some helicity calculations for this storm. You said you had a very low shear of +4. Is that the environmental shear or the storm-relative shear? Since the storm was moving rapidly from the northwest, the number I found was more on the order of 170. You also said the storm wasn't capable of being warned for. Had you looked at the storm-relative flow, it would have indicated that the hook should be on the leading edge.

(Siemon) How commonly is such a feature observed [directed to the audience]? You guys have looked at a lot of hook echoes. How often have you seen major tornadogenesis along the leading edge of a storm? I have an eyewitness who, 3 minutes before the tornado, was in sunshine.

(D. Burgess, National Severe Storms Laboratory) It's very common in storms moving from the northwest.

(Siemon) It is? That's very interesting.

(K. Brewster, University of Oklahoma) I think some of the things that occurred in this storm show how long-lived storms react to the inflow of moisture. This was moving from the northwest with good moisture inflow from the southwest.

(Siemon) The environmental wind flow was west northwest at low levels.

(Brewster) But relative to the storm, the flow at low levels was from the southeast, since it was moving so quickly from the northwest. Some modeling results don't take into account the interaction with nearby storms, so the near-storm environment may be quite different.